CALCIUM PROTEIN SIGNALING

ADVANCES IN EXPERIMENTAL MEDICINE AND BIOLOGY

Recent Volumes in this Series

Volume 249
MINERAL ABSORPTION IN THE MONOGASTRIC GI TRACT:
Chemical, Nutritional, and Physiological Aspects
Edited by Frederick R. Dintzis and Joseph A. Laszlo

Volume 250
PROGRESS IN POLYAMINE RESEARCH:
Novel Biochemical, Pharmacological, and Clinical Aspects
Edited by Vincenzo Zappia and Anthony E. Pegg

Volume 251
IMMUNOBIOLOGY OF PROTEINS AND PEPTIDES V: Vaccines—Mechanisms, Design, and Applications
Edited by M. Zouhair Atassi

Volume 252
DRUGS, SYSTEMIC DISEASES, AND THE KIDNEY
Edited by Alberto Amerio, Pasquale Coratelli, Vito M. Campese, and Shaul G. Massry

Volume 253A
PURINE AND PYRIMIDINE METABOLISM IN MAN VI, Part A:
Clinical and Molecular Biology
Edited by Kiyonobu Mikanagi, Kusuki Nishioka, and William N. Kelley

Volume 253B
PURINE AND PYRIMIDINE METABOLISM IN MAN VI, Part B:
Basic Research and Experimental Biology
Edited by Kiyonobu Mikanagi, Kusuki Nishioka, and William N. Kelley

Volume 254
MECHANISMS OF LYMPHOCYTE ACTIVATION AND IMMUNE REGULATION II
Edited by Sudhir Gupta and William E. Paul

Volume 255
CALCIUM PROTEIN SIGNALING
Edited by H. Hidaka

Volume 256
ENDOTOXIN
Edited by Herman Friedman, T. W. Klein, Masayasu Nakano, and Alois Nowotny

---

CALCIUM PROTEIN SIGNALING

Edited by

H. Hidaka

Nagoya University
Nagoya, Japan

Associate Editors

E. Carafoli

Swiss Federal Institute of Technology
Zurich, Switzerland

A. R. Means

Baylor College of Medicine
Houston, Texas

and

T. Tanaka

Mie University School of Medicine
Mie, Japan

PLENUM PRESS • NEW YORK AND LONDON

Library of Congress Cataloging in Publication Data

International Symposium on Calcium-Binding Proteins in Health and Disease (6th: 1988: Nagoya-shi, Japan)
Calcium protein signaling / edited by H. Hidaka.
p. cm.—(Advances in experimental medicine and biology; v. 255)
"Proceedings of the sixth International Symposium on Calcium-Binding Proteins in Health and Disease, held July 24–28, 1988, in Nagoya, Japan"—T.p. verso.
Bibliography: p.
Includes index.
ISBN 978-1-4684-5681-3 ISBN 978-1-4684-5679-0 (eBook)
DOI 10.1007/ 978-1-4684-5679-0
1. Calcium-binding proteins—Congresses. 2. Calcium—Physiological effect—Congresses. 3. Cellular control mechanisms—Congresses. I. Hidaka, Hiroyoshi, 1938- . II. Title. III. Series.
QP552.C24I57 1988 89-16115
574.87′5—dc20 CIP

Proceedings of the Sixth International Symposium on
Calcium-Binding Proteins in Health and Disease,
held July 24–28, 1988, in Nagoya, Japan

Softcover reprint of the hardcover 1st edition 1989
A Division of Plenum Publishing Corporation
233 Spring Street, New York, N.Y. 10013

PREFACE

The Sixth International Symposium on Calcium-Binding Proteins in Health and Disease was held in Nagoya, Japan, July 24-28, 1988. Four hundred and seventy-two persons participated in this symposium. This large attendance is proof indeed of the growing interest in this field of research. Previous meetings were held in Jablonna, Poland in 1973; Ithaca, New York in 1977, Madison, Wisconsin in 1980; Trieste, Italy in 1983 and in Asilomar, California in 1986. The scientific program of this 1988 meeting included forty-one oral presentations given by invited specialists, ten round table presentations and one hundred and eighty-seven poster presentations. The program was set-up by the Organizing Committee and the members were Hidaka, H. (Japan), Forsen, S. (Sweden), Klee, C. B. (U. S. A.), Means, A. R. (U. S. A.), Norman, A. W. (U. S. A.) and Sykes, B. D. (Canada). The members of the Advisory Committee, Carafoli, E. (Switzerland), Gergeley, J. (U. S. A.), Kretsinger, R. H. (U. S. A.), MacLennan, D. H. (Canada), Siegel, F. L. (U. S. A.), Vanaman, T. C. (U. S. A.), Wasserman, R. H. (U. S. A.) and Williams, R. J. P. (England) provided important suggestions and advice to the Committee. This volume includes contents of the talks given by the invited speakers and some, but not all of the free communications. The contributions are grouped according to subject and based on the classification made by the editor. Much focus was placed on the molecular biology of clamodulin, calbindin, calcium pump ATPase of plasma membrane, phospholamban, parvalbumin, calcineurin, protein kinase C, myosin light chain kinase, calmodulin-dependent cyclic nucleotide phosphodiesterase and the vitamin D receptor. Voltage-sensitive calcium channels, calcium-activated neutral protease, gelsolin and caldesmon also were extensively discussed. Factors related to cell biology and the pharmacology of calmodulin, myosin light chain kinase, Ca^{2+}/Calmodulin-dependent protein kinase II and inositol phosphate metabolism also received much attention. It is fully expected that calcium-binding proteins in health and disease will be a field of scientific research which will continue to hold great interest for biomedical science investigators.

Hiroyoshi Hidaka
Nagoya Japan

ACKNOWLEDGMENTS

The generous financial support by the following is thankfully acknowledged:

Organizations:

- Asahi Chemical Industry Co., Ltd.
- Bayer Yakuhin, Ltd.
- Chugai Pharmaceutical Co., Ltd.
- Eisai Co., Ltd.
- Farmitalia Carlo Erba K.K.
- Fujisawa Pharmaceutical Co., Ltd.
- Hokuriku Seiyaku Co., Ltd.
- Kyowa Hakko Kogyo Co., Ltd.
- Mitsui Pharmaceuticals, Inc.
- Nippon Organon K.K.
- Otsuka Pharmaceutical Co., Ltd.
- Sankyo Co., Ltd.
- Sanwa Kagaku Kenkyusho Co., Ltd.
- Smith Kline Beckman Japan Ltd.
- Taisho Pharmaceutical Co., Ltd.
- Takeda Chemical Industries, Ltd.
- Tanabe Seiyaku Co., Ltd.
- Tobishi Pharmaceutical Co., Ltd.
- Toyo Jozo Co., Ltd.
- Yamonouchi Pharmaceutical Co., Ltd.

Individual:

- Takashi Tanimura
- Toshio Moriki

With gratitude, the final cooperation of the members of Department of Pharmacology, Nagoya University School of Medicine and Department of Molecular and Cellular Pharmacology, Mie University School of Medicine is also acknowledged.

CONTENTS

Molecular Regulation of Calcium Channels

Molecular Mechanisms Involved in Calcium Protein Signaling

Genes of Calcium-Binding Proteins

Molecular Pathophysiology of Calcium Metabolism

MOLECULAR PHARMACOLOGY OF PROTEIN KINASE C

Hiroyoshi Hidaka and Masatoshi Hagiwara

Department of Pharmacology
Nagoya University School of Medicine
Showa-ku, Nagoya 466
Japan

INTRODUCTION

When Castagna et al.[1] demonstrated that the potent tumor promoter 12-O-tetradecanoylphorbol-13-acetate (TPA) could stimulate protein kinase C activity, protein kinase C caught the attention of many scientists and has resulted in a vast and ever-growing literature on the role of protein kinase C in numerous biologic processes induced by TPA.[2] Indeed, the hypothesis that protein kinase C is the one of the binding protein of the phorbol esters was attractive for cancer researchers and helped rationalize the broad range of biological activities that has been observed previously for the phorbol esters. However, could we attribute all the biological effects induced by phorbol esters to the activation of the function of protein kinase C. The gel overlay experiment using [^{3}H]12-O-tetradecanoylphorbol-13-acetate, for example, did not give a single band of protein kinase C, suggesting that we must have more pharmacological tools for elucidating the function of protein kinase C. Therfore we have started to synthesize a specific inhibitor of protein kinase C. Since synthesizing H-7 in 1986,[4] we have succeeded in utilizing the inhibitor as an affinity ligand and could purified protein kinase C reproducibly. By using this enzyme preparation, we have determined that structure of isozyme of protein kinase C and obtained monoclonal antibodies against three isozymes. We established enzyme immunoassay method and determined tissue distribution of these three isozymes.

PHARMACOLOGY OF PROTEIN KINASE C INHIBITORS

Since the initial identification of protein kinase C, a number of natural or synthesized modulators of protein kinase C have been reported (Table 1). In order to use a phramacological probe to elucidate the physiological role of protein kinase C. Protein kinase C inhibitors must meet definite criteria: (a) direct binding to protein kinase; (b) strict specificity for the protein kinase; and (c) cell membrane permeability. Calmodulin antagonists, polypeptides, polyamines, local anesthetics, doxorubicin (adriamycin) and other lipids are lipophylic and interact with phospholipids, resulting in an indirect inhibition of protein kinase C. Moreover, these inhibitors are relatively nonselective in their action on phospholipid/Ca^{2+}- and calmodulin/Ca^{2+}-dependent enzymes. Staurosporin showed very potent inhibition of protein kinase C, but it seems to be difficult to penetrate the cell membrane.[14] Although it was reported that sphingosine specifically ihibited protein kinase C, recently Jefferson & Schulman exhibited that sphingosine showed calmodulin antagonistic effect.[17] Aminoacridines potently inhibited protein kinase C, but the inhibitory mechanism is very complex and the number of aminoacridine-binding sites on protein kinase C is more than two.[16]

Table 1. Protein Kinase C Inhibitors

Inhibitors	Binding Site	References
Isoquinolinesulfonamides		
H-7	C3	3
H-9	C3	3
Calmodulin Antagonists		
Phenothiazines	-	4
W-7	-	5
Calmidazolium	-	6
Polypeptides		
Cytotoxins	-	7
Polymixin B	-	8
Neurotoxins	-	8
Polyamines		
Spermine	-	9
1, 12-Diaminododecane	-	9
Local Anesthetics		
Dibucaine	-	10
Antibiotics		
Adriamycin	-	4
Neomycin	-	11
Others		
Palmitoylcarnitine	-	4
Alkyllysophosphlipid	-	12
Gangliosides	-	13
Sphingosine	C1	14
Staurosporin	C3	15
Quercetin	-	16
Aminoacridines	C1 & C3	17

On the contrary to above compounds, one of isoquinolinesulfonamides, 1-(5-isoquinolinesulfonyl)-2-methylpiperazine (H-7) exhibits a relatively selective inhibition of protein kinase C. Kinetic analysis by double reciprocal plots revealed that the inhibition of protein kinase C by H-7 was competitive with respect to ATP and noncompetitive with respect to the phosphate acceptor. The K_i value of this compound against protein kinase C is 6.0 μM. Radiolabelled isoquinolinesulfonamide derivatives are incorporated into the cells[3] and fluoroscopical investigations revealed that hydroxy H-7 penetrated the cell membrane and was distributed mainly in the cytoplasm.[19] Evidence for their binding to the ATP binding site of a protein kinase with a binding ratio of 1:1 was obtained using the gel-permeation binding assay.[20, 21] Moreover, only the isoquinolinesulfonamide, among protein kinase C inhibitors, can be used as an affinity ligand for purification of protein kinase C.

AFFINITY PURIFICATION AND CLONING OF PROTEIN KINASE C

In order to demonstrate the specific binding of H-7 to protein kinase C, we developed a simple procedure for the purification of protein kinase C using affinity chromatography with an affinity ligand as the analogue of H-7. The adsorbent was synthesized by attaching the aminoresidue of N-(2-aminoethyl)-5-isoquinolinesulfonamide to cyanogen bromide activated sepharose.[22] This derivative is also a potent inhibitor of protein kinase C as H-7. A 750 fold purification was achieved by three step procedures, with DEAE, butyl-sepharose and affinity chromatography. The resultant preparation was essentially homogenous, as indicated by polyacrylamide gel electrophoresis under conditions of denaturation with sodium dodecyl sulfate. Using this affinity chromatography method, 15 μg of protein kinase C was obtained from 3 rabbit brains. The purified enzyme which gave a single band on SDS-polyacrylamide gel electrophoresis, was digested with lysylendopeptidase. The peptides were separated by reverse phase HPLC and four peptide sequences we determined by gas-phase sequencer. These sets of oligonucleotide mixtures containing predicted sequences of protein kinase were made with a DNA synthesizer and used as hybridization probes. A rabbit brain cDNA library was screened using the synthetic oligonucleotide mixtures and four types of positive clones (α, β_1, β_2, γ) were identified.[23, 24]

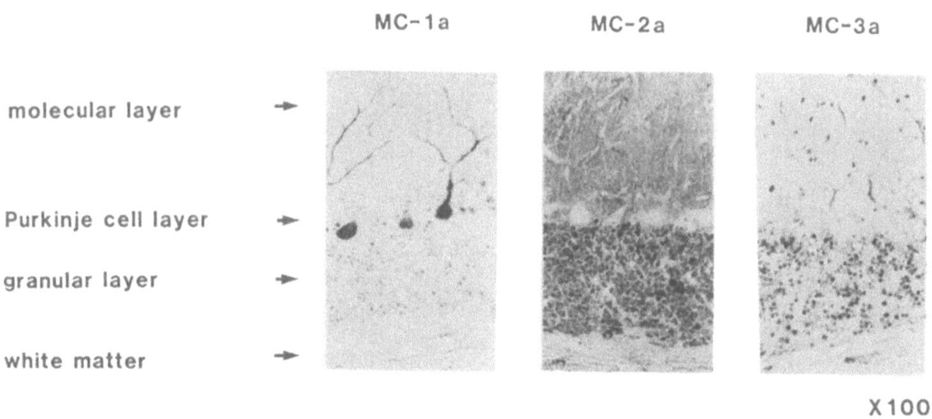

Fig. 1. Immunomicroscopic localization of protein kinase C reacting MC-1a, MC-2a and MC-3a in rabbit cerebellum

IMMUNOCYTOCHEMICAL AND IMMUNOENZYMOMETRIC STUDIES OF PROTEIN KINASE C

On the other way, monoclonal antibodies were prepared by injecting the purified protein kinase C into mice. Monoclonal antibodies raised against rabbit brain protein kinase C were classified into at least three types, MC-1a,

MC-2a, and MC-3a, which selectively interact with hydroxyapatite column chromatographically resolved isozymes type I, II, and III of protein kinase C, respectively, determined on the basis of immunoblotting experiments.[24] Immunoprecipitating analysis revealed that MC-1a, -2a, and -3a interacted with protein kinase C expressed in the brain and MC-2a and -3a, but MC-1a immunoprecipitated the protein kinase from rabbit spleen.Immuno-cytochemical localization of the protein kinase isozymes in the rabbit cerebellum and spleen was studied using these monoclonal antibodies (Fig. 1). Staining with MC-1a was restricted to Purkinje cells of the rabbit cerebellum, and there was no reaction product in the Purkinje cells with either MC-2a or MC-3a. Cells in the granular layer of the cerebellum were labeled with MC-2a, and oligodendroglia in the white matter of the cerebellum was stained with MC-3a. The rabbit spleen contained no product reacting with MC-1a, but MC-2a and MC-3a reacted with lymphocytes in the mantle zone and the periarterial lymphatic sheath of the white pulp, respectively. Cell type-dependent expression of protein kinase C in the rabbit cerebellum suggest that types I, II and III protein kinase C are products of α–, β–, γ–, type genes, respectively.

To correlate the presence of protein kinase C isozymes with specialized cellular function or human disease, we attempted to determine the quantity of the three types of isozymes, in various tissues. Since quantification of protein kinase C isozymes by immunoblotting is not so reliable, we set up a two-site enzyme immunoassay (EIA) method for detecting type I, II and III of protein kinase C. The soluble protein kinase C isozymes were selectively immobilized by type-specific monoclonal antibodies, MC-1a, -2a and -3a which bind to the regulatory domain (N terminal side) of protein kinase C. The amount of each isozyme was then determined using a horseradish peroxidase-conjugated polyclonal antibody (PC-40) raised against the C terminal peptide of protein kinase C (Fig. 2). By adding increasing concentrations of the antigen, the range of the assay proved to be 0.51-51 nM, 0.081-8.1 nM, 0.31-31 nM for type I, II, and III, respectively. This sandwich method was used to determine the level of protein kinase C isozymes in rabbit tissues (Table 2). Type I was mainly present in the cerebrum and cerebellum, the highest amount of type II

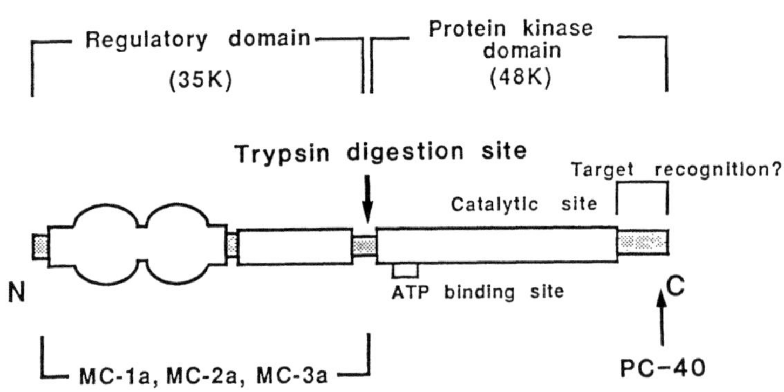

Fig. 2A. Epitope of protein kinase C antibody
Immunoblot analysis of the trypsin-digested protein kinase C revealed that all the monoclonal antibodies bound to the regulatory domain (33 or 35 kDa) of protein kinase C. On the contrary PC-40, anti-C terminal peptide polyclonal antibody, recognized the catalytic domain (48 kDa) of the enzyme (Fig. 2).

isozyme was present in blood platelets (1007.0 ng/ml protein) while type III was the predominant isozyme in the lung. This two site immunoenzymometric assay is expected to be useful for examining the differential role of protein kinase C isozymes.

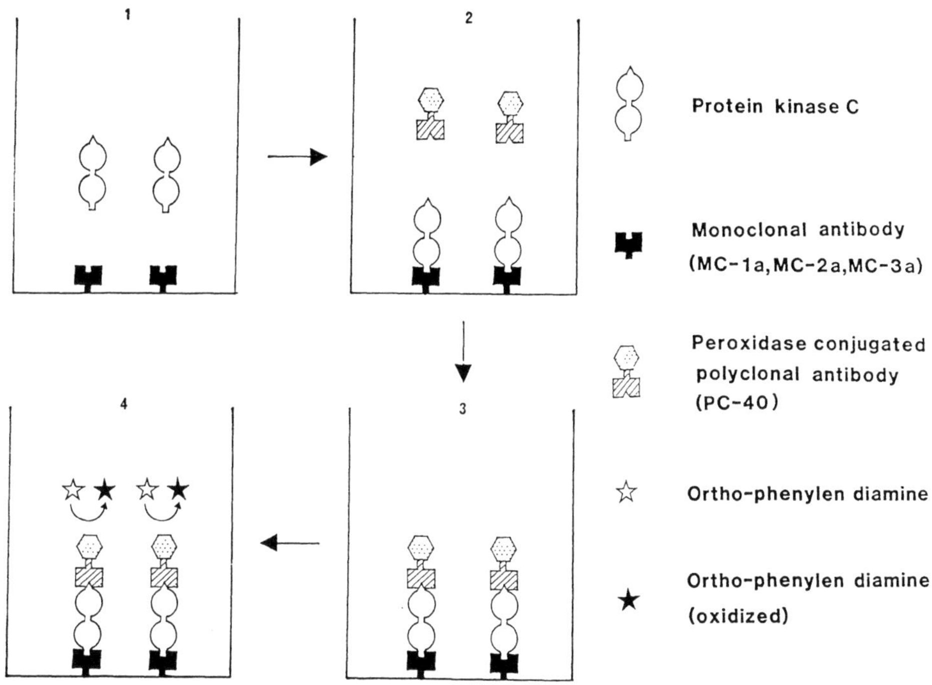

Fig. 2B. Procedures of two-site enzyme immunoassay of protein kinase C isozymes.
(1) Five micrograms of monoclonal antibodies in 0.1 ml of buffer A; 0.1 M sodium phosphate (pH 7.4, 0.1% NaN_3) were placed in wells of microtiter plates. After overnight incubation at 4°C, excess monoclonal antibody solutions were removed and the wells were washed three times with 0.35 ml of PBS. Each of the monoclonal antibody-coated microtiter plate wells were incubated with 0.35 ml of blocking buffer (5% BSA in buffer A) overnight at 4°C. (2) After three washings with PBS, 0.1 ml of sample solutions containing various amounts of purified protein kinase C isozymes were added. The wells were incubated for 2h at 4°C and washed three times with PBS. (3) PC-40 conjugated with horseradish-peroxidase (100 ng in 0.1 ml), 10 mM sodium phosphate buffer (pH 7.4) containing 0.15 M NaCl and 1% BSA was added to each well and the preparations incubated 2 h at 4°C. (4) Each well was washed as described and horseradish-peroxidase activity bound to the well was assayed. The enzyme reaction was initiated by the addition of 0.1 ml of 0.1 M citratephosphate buffer (pH 5.1) containing 3 mM H_2O_2 and 1 mg/ml *o*-phenylenediamine. After 10 min incubation at room temperature, the enzyme reaction was halted by adding 0.1 ml of 2N H_2SO_4. The amounts of oxidation of *o*-phenylene-diamine were measured by adsorption at a wavelength of 492 nm.

Table 2. Distribution of cytosolic protein kinase C isozymes in rabbit tissues

Tissues	Protein kinase C content (ng/mg protein)		
	Type I	Type II	Type III
Cerebrum	698.1 ± 25.5	918.0 ± 53.9	739.6 ± 87.1
Cerebellum	256.7 ± 13.2	82.9 ± 18.1	139.0 ± 17.2
Lung	15.7 ± 3.9	19.1 ± 4.1	70.4 ± 12.3
Heart	16.5 ± 4.1	1.9 ± 0.6	20.1 ± 4.2
Liver	6.1 ± 1.5	2.3 ± 1.5	0.4 ± 0.5
Kidney	10.5 ± 2.1	1.0 ± 0.7	36.7 ± 7.3
Spleen	10.5 ± 1.3	13.3 ± 31.2	137.7 ± 13.4
Testis	7.4 ± 3.3	1.6 ± 0.8	26.1 ± 7.7
Skeletal Muscle	5.6 ± 1.1	0.6 ± 0.4	4.5 ± 1.7
Platelet	23.6 ± 4.2	1007.0 ± 145.6	159.4 ± 56.0

Contents of protein kinase C isozymes in the various tissues were determined by enzyme immunoassay. The soluble protein fraction was extracted in 50 mM sodium phosphate buffer (pH 7.4) containing 1% BSA, 0.1 M NaCl, 20 mM EDTA, 1 mM $MgCl_2$ and 0.001 % leupeptin, then 0.1 ml of the extract (100-500 μg protein) was used for the enzyme immunoassay. Results are expressed as ng of rabbit brain protein kinase C isozyme equivalents/mg protein. Each value represents the mean and standard error of duplicate determinations of four preparations.

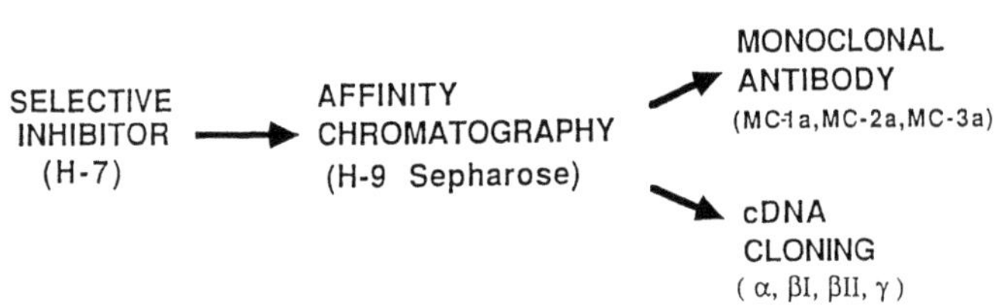

Fig. 3. Development of molecular pharmacology of protein kinase C.

CONCLUSION

Since we initially synthesize H-7 (1984), we have developed the investigation of protein kinase C (see Fig. 3). The fruitful results showed that our first choice to use isoquinolinesulfonamides as pharmacological probes was correct and that the strategy presented in Fig. 3 can be taken for the investigation of other protein kinases. Recently, we discovered the MLC-kinase inhibitors[26, 27, 28] casein kinase Inhibitor, and calmodulin kinase II inhibitor (unpublished). The remarkable progress in investigation of these protein kinases should be expected.

REFERENCES

1. M. Castagna, Y. Takai, K. Kaibuchi, K. Sano, U. Kikkawa, and Y. Nishizuka, J. Biol. Chem. 257: 7847 (1982).
2. Y. Nishizuka, Science: 225: 1365 (1984)
3. H. Hidaka, M. Inagaki, S. Kawamoto, and Y. Sasaki, Biochemistry 23: 5036 (1984).
4. M. B. C. Wise, and J. F. Kuo, Biochem. Pharmacol. 32: 1259 (1983).
5. T. Tanaka, T. Ohmura, T. Yamakado, and H. Hidaka, Mol. Pharmacol. 22: 408 (1982).
6. G. J. Mazzei, R. C. Schatzman, R. S. Turner, W. R. Volger, and J. F. Kuo, Biochem. Pharmacol. 33: 125 (1984).
7. J. F. Kuo, R. L. Raynor, G. J. Mazzei, R. C. Schatzman, R. S. Turner, and W. R. Kem, FEBS Lett. 153: 183 (1983).
8. G. J. Mazzei, N. Katoh, and J. F. Kuo, Biochem. Biophys. Res. Commun. 109: 1129 (1982).
9. D. F. Qi, R. C. Schatzman, G. J. Mazzei, R. S. Turner, T. L. Raynor, S. Liano, and J. F. Kuo, Biochem. J. 213: 281 (1983).
10. T. Mori, Y. Takai, R. Minakuchi, B. Yu, and Y. Nishizuka, J. Biol. Chem. 255: 8378 (1980).
11. M. Hagiwara, M. Inagaki, K. Kanamaru, H. Ohta, and H. Hidaka, J. Pharmacol. Exp. Ther. 244: 355 (1988).
12. D. M. Helfman, K. C. Darnes, J. M. Kinkade, Jr., W. R. Volger, M. Shoji, and J. F. Kuo, Cancer Res. 43: 2955 (1983).
13. J. Y. H., Kim, J. R. Goldenring, R. J. DeLorenzo, and R. K. Yu, J. Neurosci. Res. 15: 159 (1986).
14. Y. A. Hannun, C. R. Loomis, A. H. Merrill, Jr., and R. M. Bell, J. Biol. Chem. 261: 12604 (1986).
15. T. Tamaoki, H. Nomoto, I. Takahashi, Y. Kato, M. Morimoto, and F. Tomita, Biochem. Biophys. Res. Commun. 135: 397 (1986).
16. A. K. Srivastava, Biochem. Biophys. Res. Commun. 131: 1 (1985).
17. Y. A. Hannun, and R. M. Bell, J. Biol. Chem. 11: 5124 (1988).
18. A. B. Jefferson, and H. Schulman, J. Biol. Chem. 263: 15241 (1988).
19. M. Hagiwara, M. Inagaki, J. Takahashi, T. Yoshida, and H. Hidaka, Pharmacology 36: 365 (1988).
20. M. Hagiwara, M., Inagaki, and H. Hidaka, Mol. Pharmacol. 31: 523 (1987).
21. H. Ohta, T. Tanaka, and H. Hidaka, Biochem. Pharmacol. 37: 2704 (1988).
22. M. Inagaki, M. Watanabe, and H. Hidaka, J. Biol. Chem. 260: 2922 (1985).
23. S. Ohno, H. kawasaki, S. Imajoh, K. Suzuki, M. Inagaki, H. Yokokura, T. Sakoh, and H. Hidaka, Nature 325: 161 (1987).
24. S. Ohno, H. Kawasaki, Y. Konno, M. Inagaki, H. Hidaka, and K. Suzuki, Biochemistry 27: 2083 (1988).
25. H. Hidaka, T. Tanaka, K. Onoda, M. Hagiwara, M. Watanabe, H. Ohta, Y. Ito, M. Tsurudome, and T. Yoshida, J. Biol. Chem. 263: 4523 (1988).
26. M. Saitoh, T. Ishikawa, S. Matsushima, M. Naka, and H. Hidaka, J. Biol. Chem. 262: 7796 (1987).
27. M. Hagiwara, M. Inagaki, M. Watanabe, M. Ito, K. Onoda, T. Tanaka, M. Hagiwara, and H. Hidaka, Mol. Pharmacol. 32: 7 (1987).
28. M. Hagiwara, M. Mamiya, and H. Hidaka, J. Biol. Chem. (in press).

PROTEIN KINASE C IN CELL GROWTH AND DIFFERENTIATION

J.F. Kuo, M. Shoji*, Z. Kiss, P.R. Girard, E. Deli, K. Oishi, and W.R. Vogler*

Departments of Pharmacology and *Medicine
(Hematology/Oncology)
Emory University School of Medicine
Atlanta, Georgia 30322, U.S.A.

It has been shown that human promyelocytic leukemia cell line HL60[1] and acute myeloblastic leukemia cell line KG-1[2] can be induced to terminally differentiate to macrophage-like cells by 12-0-tetradecanoylphorbol-13-acetate (TPA). Because TPA is a potent activator of protein kinase C (PKC) and the enzyme is the only known high affinity receptor for the tumor promoter[3], it is likely that the TPA effects are PKC-dependent. TPA, however, is ineffective in inducing differentiation of a HL60 subline HL60BII[4] and a KG-1 subline KG-1a[5]. The molecular mechanisms underlying the responsiveness or resistance to the differentiating effect of TPA remain unclear. The density and affinity of cell surface receptors for phorbol ester appear to be uninvolved, because there are no significant differences in the binding properties of phorbol ester between KG-1 and KG-1a cells[6] or between HL60 and HL60BII cells[4]. The post-receptor events that occur after binding of TPA to the plasma membrane PKC, which might be distinct for the sublines, perhaps are critical in determining the sensitivity of cells to the differentiating effect of TPA. In this paper we summarize our recent studies on the PKC system in these and other cell lines as well as in developing nervous tissues, aiming to elucidate a role of PKC in cell growth and differentiation.

Although the number of phorbol ester binding sites on the cell surface of intact KG-1 and KG-1a cells have been shown to be similar[5], we found[7] that the PKC activity in the soluble and particulate fractions of KG-1 cells was about 3 times higher than that in the corresponding fractions from KG-1a cells (Fig. 1). TPA caused a PKC translocation (within 30 min) to the particulate fraction, with a maximal effect seen at 3 and 30 nM for KG-1 and KG-1a cells, respectively (Fig. 1). The particulate enzyme decreased to the control levels at 200 nM TPA for both cell lines. The soluble PKC, on the other hand, progressively decreased as a function of TPA concentration in both cell lines.

Translocation and down-regulation of PKC was further investigated immunocytochemically (Fig. 2). TPA (30 nM) caused

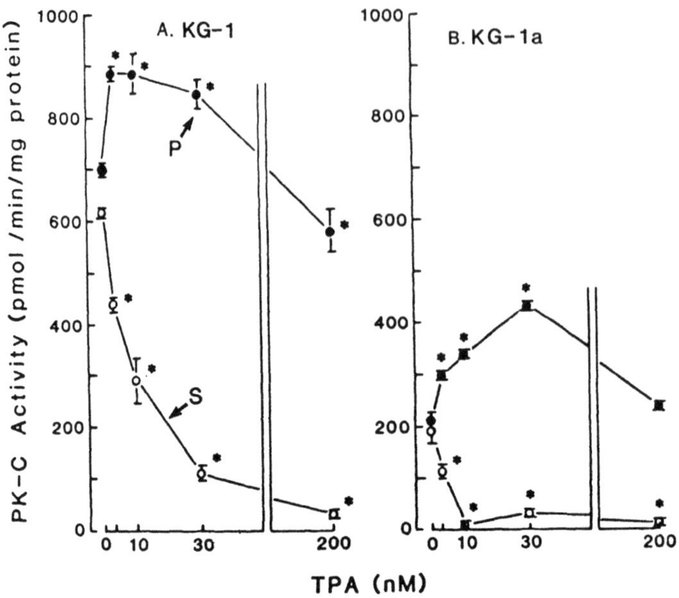

Fig. 1. Concentration-dependent effect of TPA on PKC activity in the soluble (S) and particulate (P) fractions of KG-1 and KG-1a cells. The cells were treated for 30 min with varying concentrations of TPA, as indicated. *Significantly different from the controls, i.e. no TPA treatment ($p < 0.01$). Taken from Ref. 7.

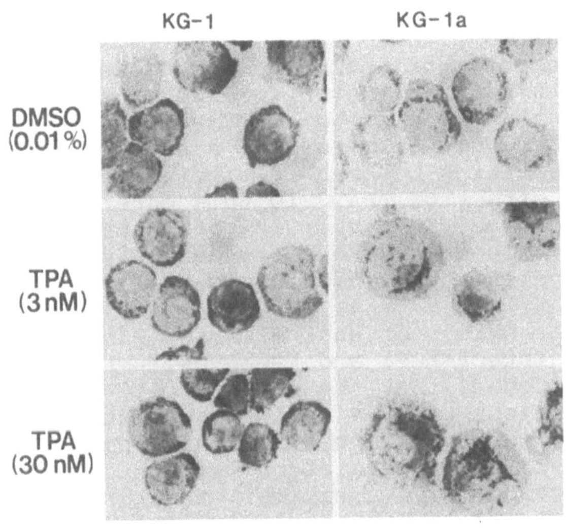

Fig. 2. Immunocytochemical localization of PKC in KG-1 and KG-1a cells. The cells were treated without (0.01% DMSO, solvent control) or with TPA (3 or 30 nM) for 30 min. PKC was visualized with immunoperoxidase method. Taken from Ref. 7.

a translocation of the enzyme to the plasma membrane in 20-30% of KG-1 cells in 30 min. The effect, however, appeared to be less obvious at a lower concentration (3 nM) of TPA. The remaining population appeared to be a mixture of cells in which PKC was already down-regulated or the response to TPA had yet to occur. PKC translocation to the plasma membrane was also observed in KG-1 cells treated with higher concentrations (up to 200 nM) of TPA (data not shown). In contrast, PKC was found to localize in the perinuclear region (probably the Golgi apparatus and endoplasmic reticulum) in almost all cells of TPA-resistant KG-1a exposed to low concentrations (3 or 30 nM) of TPA (Fig. 2). This pattern of immunostaining was essentially the same as that of the control cells, although the TPA-treated cells exhibited a decreased cytoplasmic staining compared to the control cells, probably reflecting the down-regulation and/or translocation of cytoplasmic PKC (Fig. 2). KG-1a cells, however, exhibited some plasma membrane translocation of PKC when incubated with high concentrations (100-200 nM) of TPA (data not shown). Therefore, it seems that KG-1a cells were relatively resistant to TPA with respect to PKC translocation to the plasma membrane, and this relative unresponsiveness was partially overcome by higher TPA concentrations. We reported earlier that TPA caused a plasma membrane translocation of PKC in TPA-responsive HL60 cells[8,9], but a perinuclear redistribution of the enzyme was noted for cell lines resistant to the differentiating effect of TPA including K562, CHO and E7SKS[9] and 3T3-L1[10]. It appeared that an initial inward redistribution of PKC to perinuclear region (as opposed to an outward movement to plasma membrane as in KG-1 and HL60 cells) might be a common initial response characteristic of cell lines unresponsive to this particular effect of TPA.

Taking advantage of the fact that KG-1 cells become adherent when the cells have committed to differentiate, we investigated PKC localization in the adherent and non-adherent subpopulations of KG-1 cells after the cells were incubated with 10 nM TPA for 30 min. We found that the enzyme was localized in the plasma membrane of the adherent cells, whereas it was localized primarily in the perinuclear region of the non-adherent cells (Fig. 3). It should be noted that adherent cells were smaller and more rigid than the non-adherent counterpart. In comparison, the immunostaining was found mainly in the cytoplasm of control cells treated with 0.01 % dimethylsulfoxide (DMSO, the solvent for TPA) for 0 or 30 min or treated with TPA (10 nM) for 0 min (Fig. 3). The cells incubated with preimmune serum showed only a slight staining (Fig. 3), indicating a specificity of the PKC antibodies.

TPA is metabolized only very slowly in cells and its effects are hence long lasting. In contrast, the second messenger diacylglycerol is rapidly metabolized by diacylglycerol kinase, yielding only transient effects. An inhibitor of diacylglycerol kinase, R59022, has been shown to potentiate the effect of OAG on platelets[11]. In order to maximize the action of OAG on protein phosphorylation and to better compare it with that of TPA, R59022 (25 μM) was used in combination with OAG in the studies; R59022, by itself, was without effect. We found[12] that the patterns of protein phosphorylation in the TPA-responsive KG-1 cells and the TPA-resistant KG-1a cells under the basal conditions (i.e. in the absence of PKC activators) were remarkably similar (Fig. 4). In

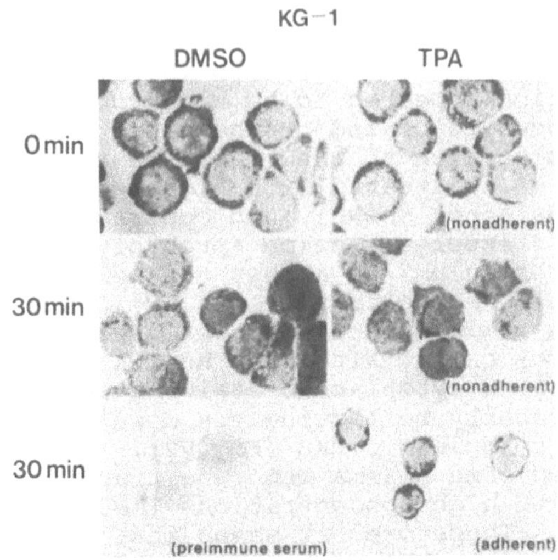

Fig. 3. Immunocytochemical localization of PKC in the adherent and non-adherent subpopulations of KG-1 cells. The cells were incubated without (0.01% DMSO) or with TPA (10 nM) for 0 or 30 min. Taken from Ref. 7.

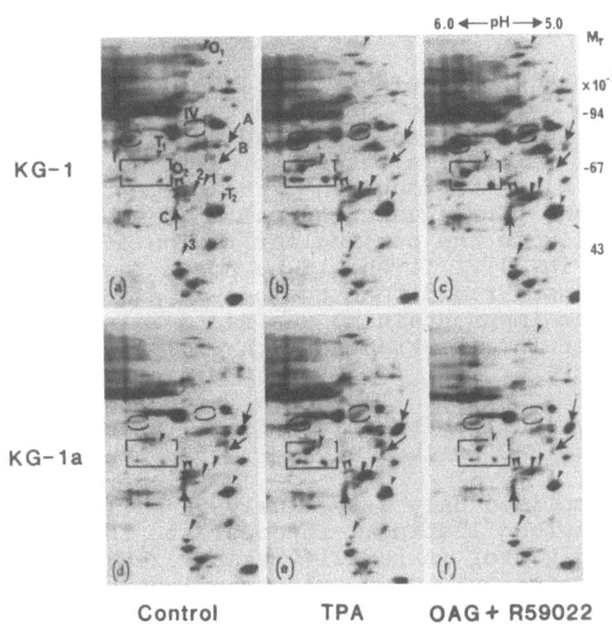

Fig. 4. Comparative effects of TPA and OAG on protein phosphorylation in KG-1 and KG-1a cells. The cells were labeled with ^{32}Pi for a total of 7 h and 45 min with or without 100 nM TPA or OAG (20 μg/ml) + 25 μM R59022 present during the final 45 min of the labeling time. Taken from Ref. 12.

KG-1 cells, TPA (Fig. 4b) or OAG (in the presence of R59022) (Fig. 4c) stimulated phosphorylation of a similar but non-identical set of proteins. Phosphorylation of protein A (pI = 5.1; M_r = 76,000) and proteins in blocks I (pI = 5.45 to 5.65; M_r = 60,000 to 64,000), III (pI = 5.6; M_r = 80,000), and IV (pI = 5.2; M_r = 84,000) were similarly stimulated by these agents. In KG-1a cells, both TPA (Fig. 4e) and OAG (Fig. 4f) stimulated phosphorylation of proteins in blocks I, III and IV, although the extent of their stimulation was lower compared to that seen for KG-1 cells. Of interest, phosphorylation of protein A (and probably that of proteins B and C), either basal or stimulated, was much higher in KG-1a cells than in KG-1 cells. 8-Bromo cAMP (1 mM), dibutyryl cAMP (1 mM) or prostaglandin E_2 (0.1 μM) plus isobutyryl methylxanthine (50 μM), agents that directly or indirectly activate cAMP-dependent protein kinase and do not induce differentiation, on the other hand, had no effect on the protein phosphorylation in KG-1 and KG-1a cells (data not shown).

As in the case of TPA and OAG shown above in Fig. 4, other PKC activators (mezerein and bryostatin) had similar effects on protein phosphorylation in KG-1 and KG-1a cells[12]. For the reason of brevity, only their effects on phosphorylation of protein A and block I proteins are compared (Table 1). It is clear that PKC activators stimulated phosphorylation of block I proteins to a greater extent in KG-1 cells than that in KG-1a cells, whereas the reverse was true for protein A. It is attractive to speculate that higher phosphorylation of protein A accompanied by a lower phosphorylation of block I proteins in KG-1a cells might be specifically related to resistance of the cells to the differentiating effect of TPA. Mezerein and bryostatin, like TPA, were capable of inducing differentiation of KG-1 cells but OAG (plus R59022) was unable to do so (Table 2)[12]. Kreutter et al.[13] also reported that, in HL60 cells, both TPA and OAG stimulated phosphorylation of many common proteins, but repeated addition of OAG to the culture (attempting to maintain effective cellular level of OAG) failed to mimic TPA in inducing the cell maturation. It is of interest that Merrill et al.[14] showed that repeated addition of dioctanoylglycyrol, another synthetic permeable diacylglycerol, caused HL60 cell differention, although the effect was less pronounced than that of TPA and the results were quite variable. It appeared that PKC activation can be dissociated from TPA-induced differentiation of HL60 and KG-1 cells, perhaps reflecting the unique pleiotropic actions of the phorbol ester.

The biochemical changes triggering cell differentiation are still largely unclear. PKC has been shown to associated with or translocated to the nucleus[9,10,15-17], and topoisomerase II[18], polymerase II[19], lamin B[20] and nuclear protein 1b[21] were substrates for PKC. An increased synthesis of phosphatidylcholine[22] and sphingomyelin[23] and a decreased synthesis of phosphatidylserine[21] have been shown to be characteristic of differentiating HL60 cells. It appeared that phosphorylation of nuclear proteins, in addition to that of plasma membrane proteins[24] by a TPA-stimulated mechanism is required for differentiation to occur. As shown previously, TPA (10-200 nM) induced a rapid (1-30 min) PKC translocation from cytoplasm to plasma membrane in HL60 cells[8,9]. In the experiments described below[17], we examined the temporal distribution of PKC in HL60 cells during the entire course of

Table 1. Comparative effects of various PKC activators on the phosphorylation of representative proteins in KG-1 and KG-1a cells[a].

PKC activator	^{32}P-Labeling of proteins (-fold increase)			
	A		Block I	
	KG-1	KG-1a	KG-1	KG-1a
None (control)	1.0 (23)[b]	1.0 (66)	1.0 (172)	1.0 (136)
TPA (2 nM)	1.9	2.9	1.9	1.0
TPA (20 nM)	3.1	4.4	2.8	1.4
TPA (100 nM)	3.2	4.2	2.9	1.9
TPA (500 nM)	3.4	4.2	3.1	2.0
Mezerein (20 nM)	3.4	4.4	2.0	1.1
Mezerein (500 nM)	4.1	5.0	3.4	1.7
Bryostatin (20 nM)	2.6	3.6	2.1	1.0
Bryostatin (500 nM)	3.1	4.8	2.3	1.2
OAG (20 μg/ml) + R59022 (25 μM)	1.9	2.4	3.5	1.4

[a]Data presented are means of two determinations of ^{32}P activity in excised spots from the gels, as shown in Fig. 4. All experiments were performed at the same time, and thus the values are directly comparable. Taken from Ref. 12.
[b]Numbers in parentheses are the actual radioactivity (cpm) measured.

Table 2. Comparative effects of various PKC activators on differentiation (adherence) of KG-1 and KG-1a cells[a].

PKC activator	Adherent cells (% of total)[b]	
	KG-1	KG-1a
None (control)	5 ± 2	< 1
TPA (20 nM)	72 ± 5	< 1
Mezerein (20 nM)	66 ± 3	< 1
Bryostatin (20 nM)	54 ± 7	< 1
OAG (20 μg/ml) + R59022 (25 μM)	ND[c]	ND[c]

[a]The cells were treated for 48 h with various PKC activators as indicated. Taken from Ref. 12.
[b]Mean ± SE of the percentage of cells which became adherent in 3 to 5 dishes for each treatment.
[c]ND, not determined due to 60-70% cell death in 48 h.

differentiation induced by a very low concentration (1 nM) of TPA, hoping to slow the TPA-mediated processes so that they can be examined more closely and accurately. We observed that at day 0 of the treatment (i.e. before or no treatment) with 1 nM TPA, the PKC immunostaining was largely cytoplasmic, but it became more concentrated around nuclear membrane and perinuclear region at day 1 (Fig. 5). PKC was then became associated with the nuclear membrane and, to a lesser extent, in plasma membrane at day 2, and was heavily localized in both plasma and nuclear membranes at day 3. By day 5, most of the immunoreactivity was detected in cytoplasm and nuclear membrane although some staining was also evident in plasma membrane. Most of the cells (about 70%) became adherent to substratum (characteristic of differentiating cells) during days 2-5, and immunocytochemical micrographs of PKC in these adherent subpopulation of the cells are shown (Fig. 5). Patterns of temporal changes in PKC redistribution during the course of differentiation, similar to those shown above in Fig. 5, were also observed in a separate series of experiments using immunofluorescence technique to visualize PKC (Fig. 6). We noted repeatedly that a 30 min-treatment of HL60 cells with 100 nM TPA (followed by washing of cells) was sufficient to induce both PKC translocation to plasma membrane and, within 48 h, cell attachment. At a very low concentration of 1 nM, similar effects on both parameters were observed only if the cells were treated for at least 4 h. It appears that the length of TPA treatment required for the induction of cell differentiation strictly correlated with the time needed for PKC translocation to plasma membrane, suggesting a relationship between the two events. We and others noted that changes in metabolism of phospholipids[21-23], following the early partitioning of PKC to plasma membrane in response to TPA. Modifications of membrane lipids, and thereby the membrane structures and properties, are likely to be a precondition for cell adhesion. Nuclear localization of PKC in HL60 cells was associated with enhanced phosphorylation of nuclear proteins of 31, 35, 40 and 48 kDa[17]. The identities of these phosphorproteins and their functional significance in gene regulation related to cell differentiation, however, are unclear at present.

It is well documented that differentiation of HL60 cells to macrophages/monocytes-like cells is induced by TPA (1), while acquisition of granulocytic properties can be achieved by DMSO[25], retinoic acid[26] or cAMP derivatives[27]. Recent evidence supports a two-step model for the induction of terminal maturation. It appears that the early events, which are associated with growth arrest and commitment to differentiate, are not differentiation lineage-specific[28]. The choice of a specific differentiation pathway is regulated by late events. The existence of mechanisms which are not specific to differentiation pathways would suggest a certain degree of interactions between the two sets of differentiation inducers. Indeed, DMSO and retinoic acid have been shown to potentiate the effect of TPA on monocytic differentiation of HL60 cells[29]. We investigated the possible effect of cAMP on TPA-induced HL60 cell differentiation[24]. It was found that 0.6 mM dibutyryl cAMP, or 100 nM prostaglandin E_2 (a cAMP-elevating agent), greatly potentiated the effect of 1 nM TPA, which was accompanied by an enhanced translocation of PKC to perinuclear region and an accelerated decrease in phosphorylation of a plasma membrane protein (protein Ia; pI = 5.6, M_r = 66,000).

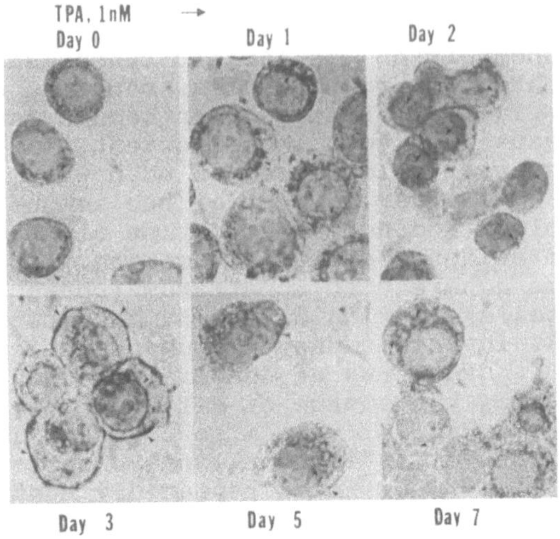

Fig. 5. Immuocytochemical evidence showing temporal redistribution of PKC during HL60 cell differentiation. The cells were treated with 1 nM TPA for up to 7 days. The cells at days 0 and 1 were in supension (undifferentiated), at days 2, 3 and 5 were adherent to substratum (differentiating), and at day 7 became detached and were in suspension (differentiated). Visualization of PKC was performed by immuniperoxidase technique. Taken from Ref. 17.

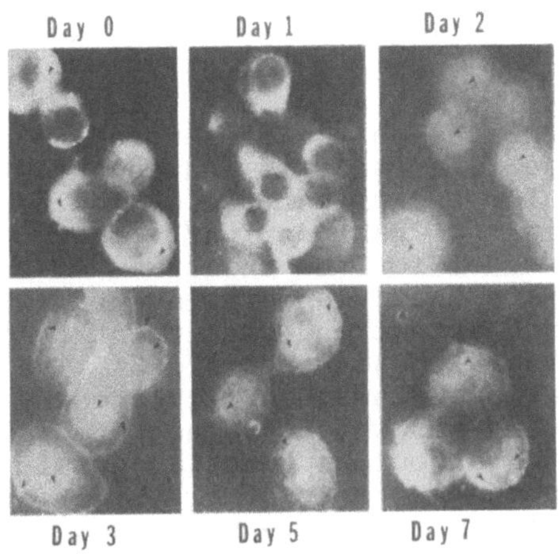

Fig. 6. Immunocytochemical evidence showing temporal redistribution of PKC during HL60 cell differentiation. The cells were treated as in Fig. 5, but PKC was visialized using the immunofluorescence technique. Taken from Ref. 17.

Table 3. Comparative effects of PKC inhibitors on TPA-induced down-regulation of soluble PKC in HL60 and K562 cells[a].

Cell line and treatment	Soluble PKC activity (% control)
HL60 Cells	
Experiment 1:	
None (control)	100 ± 3
TPA (30 nM)	35 ± 4
ET-18-OCH_3 (20 μM)	98 ± 2
TPA (30 nM) + ET-18-OCH_3 (20 μM)	74 ± 5[b]
Experiment 2:	
None (control)	100 ± 4
TPA (10 nM)	24 ± 6
BM 41.440 (20 μM)	95 ± 4
TPA (10 nM) + BM 41.440 (20 μM)	66 ± 5[b]
K562 cells	
None (control)	100 ± 5
TPA (30 nM)	19 ± 4
BM 41.440 (20 μM)	99 ± 5
TPA (30 nM) + BM 41.440 (20 μM)	21 ± 3

[a]The cells were incubated with or without 20 μM ET-18-OCH_3 or BM 41.440 for 1 h followed by incubating with TPA (10 or 30 nM) for 10 min.
[b]Significantly different from the values seen in the presence of TPA alone ($p < 0.01$; $n = 3$).

Table 4. Inhibition of TPA-induced differentation (adherence) of leukemia cells by PKC inhibitors[a].

Addition	Adherenct cells (% of total)	
	KG-1	HL60
Experiment 1:		
None (control)		< 1
TPA (20 nM)		76 ± 8
ET-18-OCH_3 (75 μM)		< 1
TPA (20 nM) + ET-18-OCH_3 (75 μM)		16 ± 11[b]
Experiment 2:		
None (control)	< 1	< 2
TPA (10 nM)	26 ± 1	45 ± 2
BM 41.440 (5 μM)	< 1	< 3
TPA (10 nM) + BM 41.440 (5μM)	19 ± 1[b]	31 ± 2[b]

[a]In Experiment 1, the cells were preincubated with ET-18-OCH_3 for 10 min prior to incubation with TPA for additional 50 min. The cells were then washed twice with the fresh medium and cultured for 16 h. In Experiment 2, the cells were preincubated with BM 41.440 for 1 h, washed once with the fresh medium and cultured in the presence of TPA for 2 days. The cell viability was 84-98% for both experiments.
[b]Significantly lower than TPA alone ($p<0.05;n = 3$).

Table 5. Comparative effects of some agents on PKC and Na,K-ATPase activity[a].

Agent	IC_{50} or K_i (μM)	
	PKC	Na,K-ATPase
Ouabain	NE[b]	0.9
Polymyxin B	2	NE[b]
ET-18-OCH_3	12	18
BM 41.440	9	12
Sphingosine	17	9
Lysophosphatidylcholine	60[c]	22

[a]The reactions for both enzymes were started with the addition of ATP.
[b]NE, no effect up to 100 μM.
[c]Stimulatory at low concentrations (< 25 μM)[33].

The presence and distribution of PKC in developing cerebellar cortex of postnatal rat and in cultures of rat sympathetic ganglia was also examined in order to further elucidate involvements of the enzyme system in cell growth and differentiation[30]. In the cerebellar cortex of 2-, 4-, and 6-day old rats, PKC immunostaining was observed in the areas of early-forming presynaptic terminals and growth cones. No staining was evident in the cortical proliferative zone. Beginning at 10 days postnatal, nuclear staining, not apparent at earlier stages, was prominent in Purkinje cells In neuronal cultures of dissociated rat sympathetic ganglia, PKC was immunolocalized in cell bodies and bundles of neuronal processes. Immunoreactivity was particularly striking in growth cones of extending neurites and in axonal varicosities. These results suggest a role for PKC in neuronal growth following cell proliferation and in synaptic function. The appearance of nuclear staining in later developmental stages suggests that the enzyme may be involved in the promotion and maintenance of the differentiated state of neurons.

Another approach to the problem of TPA-induced cell differentiation is the use of PKC inhibitors, such as the ether lipid analogs ET-18-OCH_3 (1-0-octanoyl-2-0-methyl-glycero-3-phosphocholine) and BM 41.440 (1-S-hexadecyl-2-methoxymethyl-glycero-3-phosphocholine). Both lipids are antileukemic agents exhibiting a cytocidal activity toward HL60 cells but not K562 cells (for a review, see Ref. 31). We noted that TPA increased phosphorylation of many proteins in HL60 cells including block I proteins and that ET-18-OCH_3 counteracted the TPA effect[32]. TPA, on the other hand, was unable to stimulate phosphorylation of block I proteins in K562 cells and ET-18-OCH_3 was without effect[32]. It should be noted that K562 cells, unlike HL60 cells, are unresponsive to the differentiating effect of TPA. We also observed that ET-18-OCH_3 or BM 41.440 blocked the TPA-induced down-regulation of soluble PKC in HL60 cells (Table 3). Although TPA also caused a marked depletion of soluble PKC in K562 cells, this TPA effect was not overcome by BM 41.440 (Table 3). Perhaps it was not unexpected that the two PKC inhibitors blocked the TPA effect in inducing differentiation of HL60 or KG-1 cells (Table 4)[32]. Merrill et al.[14] reported that another PKC inhibitor, sphingosine, could also inhibit the TPA-induced HL60 cell maturation.

More recently we found that the three PKC inhibitors mentioned above as well as lysophosphatidylcholine, an activator and inhibitor of PKC[33], were equally potent inhibitors of Na,K-ATPase purified from brain and other tissues (Table 5). Because the ATPase is also a major membrane enzyme critical in the maintenance and regulation of cellular activities, its inhibition could have a profound biological consequence. The exact molecular mechanisms underlying the inhibition of TPA-induced leukemia cell differentiation by ET-18-OCH_3, for example, should await further clarification and should be interpreted with cautions.

Acknowledgments: We thank Ms.R. Brooks for skillful preparation of the manuscript. The original work from the authors' laboratories was supported by USPHS Grants CA-29850, CA-36777, NS-17608 and HL-15696.

References

1. E. Huberman, and M.F. Callahan, Proc. Natl. Acad. Sci. U.S.A. 76: 293-297 (1979).
2. H.P. Koeffler, M. Bar-Eli, and M.C. Territo, Cancer Res. 41: 919-926 (1981).
3. Y.Nishizuka, Nature (London) 308: 693-698 (1984).
4. J.B. Weinberg, M.A. Misukonis, and B.J. Goodwin, Cancer Res. 44: 976-980 (1984).
5. H.P. Koeffler, P. Billing, A.J. Lusis, R. Sparkes, and D.W. Golde, Blood, 56: 265-273 (1980).
6. R.I. Lehrer, L.E. Cohen, and H.P. Koeffler, Cancer Res., 43: 3563-3566 (1983).
7. M. Shoji, P.R. Girard, P.A. Charp, H.P. Koeffler, W.R. Vogler, and J.F. Kuo, Cancer Res. 47: 6363-6370 (1987).
8. M. Shoji, P.R. Girard, G.J. Mazzei, W.R. Vogler, and J.F. Kuo, Biochem. Biophys. Res. Commun. 135: 1144-1149 (1986).
9. P.R. Girard, V.L. Stevens, P.J. Blackshear, A.L. Merrill, J.G. Wood, and J.F. Kuo, Cancer Res. 47: 2892-2898 (1987).
10. D.L. Halsey, P.R. Girard, J.F. Kuo, and P.J. Blackshear, J. Biol. Chem., 262: 2234-2243 (1987).
11. B.C. Courcelles, P. Roevens, and H. Van Belle, J. Biol. Chem., 260: 15762-15770 (1985).
12. Z. Kiss, E. Deli, M. Shoji, H.P. Koeffler, G.R. Pettit, W.R. Vogler, and J.F. Kuo, Cancer Res., 47: 1302-1307 (1987).
13. D. Kreutter, A.B. Caldwell, and M.J. Morin, J. Biol. Chem., 260: 5979-5984 (1985).
14. A.H. Merrill, A.M. Sereni, V.L. Stevens, Y.A. Hannun, and R.M. Bell, J. Biol. Chem., 261: 12610-12615 (1986).
15. S. Capitani, P.R. Girard, G.J. Mazzei, Kou, J.F., Berezney, R., and F.A. Manzoli, Biochem. Biophys. Res. Commun., 142: 367-375 (1987).
16. J.G. Wood, P.R. Girard, G.J. Mazzei, and J.F. Kuo, J. Neurosci., 6: 2571-2577 (1986).
17. Z. Kiss, E. Deli, and J.F. Kuo, FEBS Lett., 231: 41-46 (1988).

18. N. Sahyoun, N. Wolf, J. Besterman, T.S. Hsieh, M. Sander, H. Levine, K.J. Chang, and P. Cuatrecasas, Proc. Natl. Acad. Sci. U.S.A., 83: 1603-1607 (1986).
19. L.F. Chung, R.H. Cooper, P. Yau, E.M. Bradbury, and R.Y. Chuang, Biochem. Biophys. Res. Comun., 145: 1376-1383 (1987).
20. A.P. Fields, G.R. Pettit, and W.S. May, J. Biol. Chem., 263: 8253-8260 (1988).
21. Z. Kiss, E. Deli, and J.F. Kuo, Biochem. J., 248: 649-656 (1987).
22. R.A. Cooper, A.K. Braunwald, and A.L. Kuo, Proc. Natl. Acad. Sci. U.S.A., 79: 2865-2869 (1982).
23. Z. Kiss, E. Deli, and J.F. Kuo, Arch. Biochem. Biophys., 265: 000-000 (1988), in press.
24. E. Deli, Z. Kiss, and J.F. Kuo, FEBS Lett., 231: 407-412 (1988).
25. S.J. Collins, F.W. Ruscetti, R.B. Gallagher, and R.C. Gallo, Proc. Natl. Acad. Sci. U.S.A., 75: 2458-2462 (1978).
26. T.R. Breitman, S.E. Selonick, and S.J. Collins, Proc. Natl. Acad. Sci. U.S.A., 77: 2936-2940 (1980).
27. T.J. Chaplinski, and J.E. Niedel, J. Clin. Invest. 70: 953-964 (1982).
28. A. Yen, M. Forbes, G. DeGala, and J. Fishbaugh, Cancer Res., 47: 129-134 (1982).
29. E. Fibach, T. Peled, A. Treves, A. Kornberg, and E. Rachmilewitz, Leukemia Res., 6: 781-790 (1982).
30. P.R. Girard, J.G. Wood, J.E. Freschi, and J.F. Kuo, Develop. Biol., 126: 98-107 (1988).
31. W.E. Berdel, R. Andreesen, and P.G. Munder, Synthetic alkyl-phospholipids analogs: A new class of antitumor agents, in: "Phospholipids and Cellular Regulation", vol. 2, J.F. Kuo, ed., CRC Press, Boca Raton, pp. 41-74 (1985).
32. Z. Kiss, E. Deli, W.R. Vogler, and J.F. Kuo, Biochem. Biophys. Res. Commun., 142: 661-666 (1987).
33. K. Oishi, R.L. Raynor, P.A. Charp, and J.F. Kuo, J. Biol. Chem., 263: 6865-6871 (1988).

ROLES OF PROTEIN KINASE C ISOZYMES IN CELLULAR REGULATION

Kuo-Ping Huang, Freesia L. Huang, Hiroki Nakabayashi, and Yasuyoshi Yoshida

Section on Metabolic Regulation, Endocrinology and Reproduction Research Branch, National Institute of Child Health and Human Development, National Institutes of Health, Bethesda, Maryland 20892

INTRODUCTION

The importance of protein kinase C (PKC) in the regulation of cellular function was fully recognized after the discovery of diacylglycerol (DG) as an activator of PKC and that the tumor-promoting phorbol esters elicit their pleiotropic responses through binding to PKC as a phorbol ester receptor (for review see 1, 2). Both DG and phorbol ester stimulate PKC activity by increasing the affinity of the enzyme for Ca^{2+} and phospholipid so that the enzyme becomes active under the physiological conditions. This hypothetical scheme appears to provide a satisfactory explanation for the activation of PKC inside the cells.

Recently, several groups have discovered the molecular diversity of PKC at both the protein and gene levels (3-12). There are at least seven cloned PKC cDNAs and three PKC isozymes have been identified from several sources. The type I, II, and III PKCs separated by hydroxylapatite column chromatography have been shown to be the products of γ, β, and α genes, respectively (13,14). The purified type II PKC contains the products of βI and βII cDNA derived from alternative splicing of a single mRNA (15). These three PKC isozymes account for the majority of Ca^{2+}/phospholipid/DG-dependent protein kinase activity found in the mammalian tissues. In addition to the above-mentioned four cDNAs, α, βI, βII, and γ, three other homologous members of the PKC gene family, δ, ϵ, and ξ, have been identified from rat brain gene library using a mixture of α, βII, and γ cDNAs as probes under low stringency conditions (12). These new members of rat brain cDNA as well as a ϵ cDNA-related rabbit subspecies (16) have been shown to code for protein kinases dependent on phospholipid and DG but independent of Ca^{2+}. The native protein kinases corresponding to these new subspecies have not been identified.

In spite of the general consensus that PKC is involved in the regulation of numerous cellular functions, detailed information regarding the role of each PKC isozyme is not yet available. In this article we will discuss possible functional roles of the three major PKC isozymes. Our studies encompass the biochemical and immunological characterizations of these enzymes in several tissues and cultured cells.

RESULTS

Preparation of PKC isozymes and specific antibodies against each isozyme

Three major PKC isozymes have been purified to near homogeneity by a procedure including ion-exchange, affinity, molecular sieve, and hydroxylapatite column chromatographies (3,17). The type I, II, and III PKCs were eluted from the hydroxylapatite column at KPO_4 concentrations of 0.07, 0.08, and 0.15 M, respectively (Fig. 1). The Ca^{2+}/phosphatidylserine (PS)/DG-dependent kinase activity peaks correspond with the [^{3}H]phorbol 12,13-dibutyrate (PDBu) binding. The purified kinases all have the same molecular weight of 82,000 and undergo autophosphorylation in the presence of Ca^{2+}, PS, and DG. Specific antibody against each PKC isozyme was purified from a preparation of polyclonal antibody raised against a mixture of all three isozymes by immunoabsorption. The purified type I, II, and III monospecific antibodies recognize mainly their respective immunogens (Fig. 2) and the expression products in COS cells transfected by γ, βI and βII, and ∝ cDNAs, respectively. These results indicate that these antibodies recognize the epitopes of these enzymes at the variable regions so that distinction among them is possible.

1

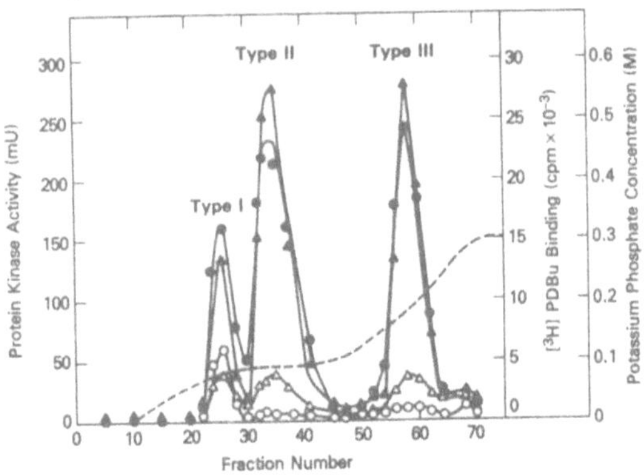

2

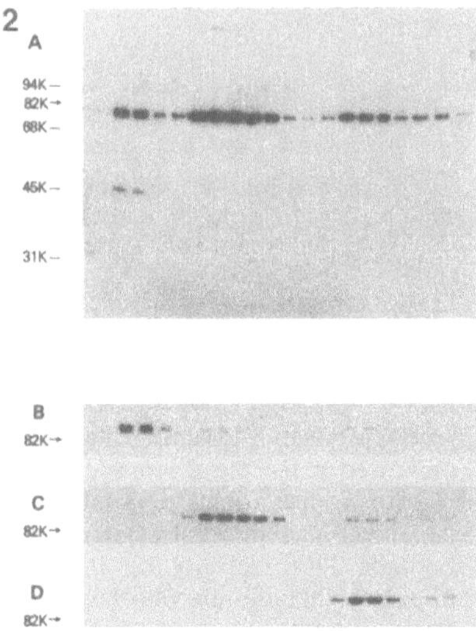

Fig. 1. (left) Separation of three PKC isozymes by hydroxylapatite column chromatography. Fractions were collected for the measurement of PKC activity with Ca^{2+}, PS, and DG (●) or with EGTA (0). [^{3}H]PDBu binding was measured with Ca^{2+} and PS (▲) or with PS and EGTA (Δ). The conductivity (- - -) was measured with an on-line conductivity meter.

Fig. 2. (right) Immunoblot analysis of PKC isozymes isolated from hydroxylapatite column by polyclonal and isozyme-specific antibodies. Fractions as shown in Figure 1 were analyzed by using goat anti-rat brain PKC polyclonal (panel A), type I- (panel B), type II- (panel C), or type III- (panel D) specific antibodies. The immunoreactive bands were detected with ^{125}I-protein A.

Distinct expression of PKC isozymes

The levels of PKC isozymes were quantified by immunoblot analysis with [^{125}I]protein A (18). Among the various rat tissues tested the brain contained higher levels of all three PKC isozymes than the peripheral tissues. In particular, the type I PKC was found mainly in the CNS and especially enriched in hippocampus, amygdala, cerebral cortex, and cerebellum. In cerebellum, the type I PKC is present in the Purkinje cells, the type II PKC in granule cells, and the type III enzyme in both the Purkinje cells and granule cells (Fig. 3). Though both the type I and III PKC are found in the Purkinje cells, these two enzymes appear to have different subcellular localization; the type I PKC is in both the cell body and dendrites, whereas the type III enzyme is mainly in the cell body. The coexistence of multiple PKC isozymes in a cell type has also been detected in the pyramidal cells of hippocampus and cerebral cortex and granule cells of dentate gyrus, all of which contain three PKC isozymes. Several cell lines derived from the peripheral tissues, such as EL-4 mouse thymoma, HL-60 human promyelocytic leukemia, 2H3 rat basophilic leukemia, and PC-12 pheochromocytoma cells, were found to contain both type II and III PKCs, whereas 3T3-L1 fibroblast and COS cells contained only the type III PKC. It appears that the PKC isozymes have distinct cellular distribution and, frequently, multiple subspecies of PKC are present in a cell type.

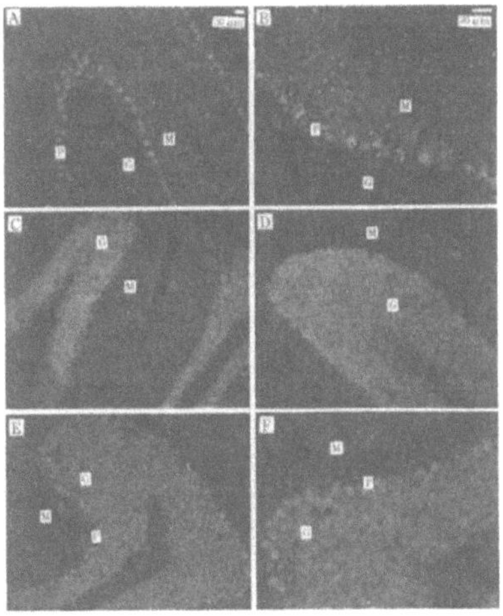

Fig. 3. Indirect immunofluorescent staining of rat cerebellum. Frozen cerebellum sections were stained with monospecific antibodies against type I (A and B), type II (C and D), and type III (E and F) PKC. The symbols are: P, Purkinje cells; M, molecular layer; and G, granular layer. The scale bar is 30 μm.

The expressions of PKC isozymes in the CNS and peripheral tissues are developmentally controlled. In rat brain, both type II and III enzymes were found to increase progressively from 3 days before birth up to 2-3 weeks of age and remained constant thereafter. The expression of type I PKC, however, displayed a different developmental pattern; it was very low within one week, and an abrupt increase was observed between 2 and 3 weeks of age (Fig. 4). Both thymus and spleen contain relatively high level of the type II PKC and the expression of this isozyme in these two tissues are different. In thymus, the type II enzyme was found to be maximal shortly after birth; whereas in spleen the same enzyme was very low within 2 weeks of age, and a significantly increase was observed between 2 and 3 weeks (18).

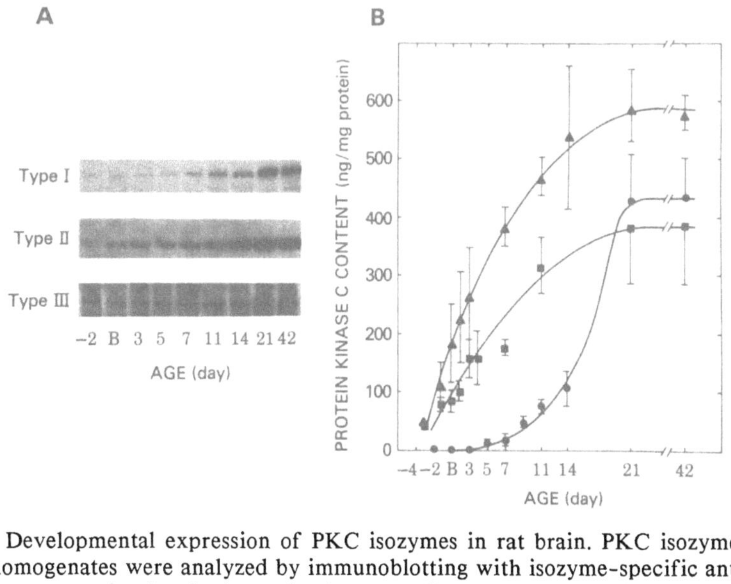

Fig. 4. Developmental expression of PKC isozymes in rat brain. PKC isozymes in the rat brain homogenates were analyzed by immunoblotting with isozyme-specific antibodies (A). The immunoreactive bands shown in the autoradiograms correspond to the 82-kDa native PKC. Also shown are the quantitative estimation of the type I (●), II (▲), and III (■) PKC (B).

Distribution of PKC isozymes in cerebral cortex of monkey brain involved in the visual information processing

PKC has been implicated in the control of neurotransmitter release and neuroplasticity (19-22). Regional distribution of PKC isozymes along the occipitotemporal cortex of visual information processing pathway in monkey brain were analyzed by immunoblotting with isozyme-specific antibodies (Fig. 5). The type I enzyme was found to have increasing concentration rostral along the cerebral cortex of occipital and temporal lobes, and peaked at perirhinal, entorhinal, and temporal pole areas. Unlike the type I enzyme, both the type II and III PKC were more evenly distributed in the cortex along the visual information processing pathway. Levels of type I PKC present in the cortical regions of perirhinal, entorhinal, and temporal pole were almost ten times more than that in the striate and prestriate of primary visual cortex. Neurobehavioral studies have demonstrated that the perirhinal and entorhinal cortices of the temporal lobe participate more than the striate and prestriate cortices of the occipital lobe in the storage of visual representation (23), we predict that the type I PKC may be involved in the mnemonic function.

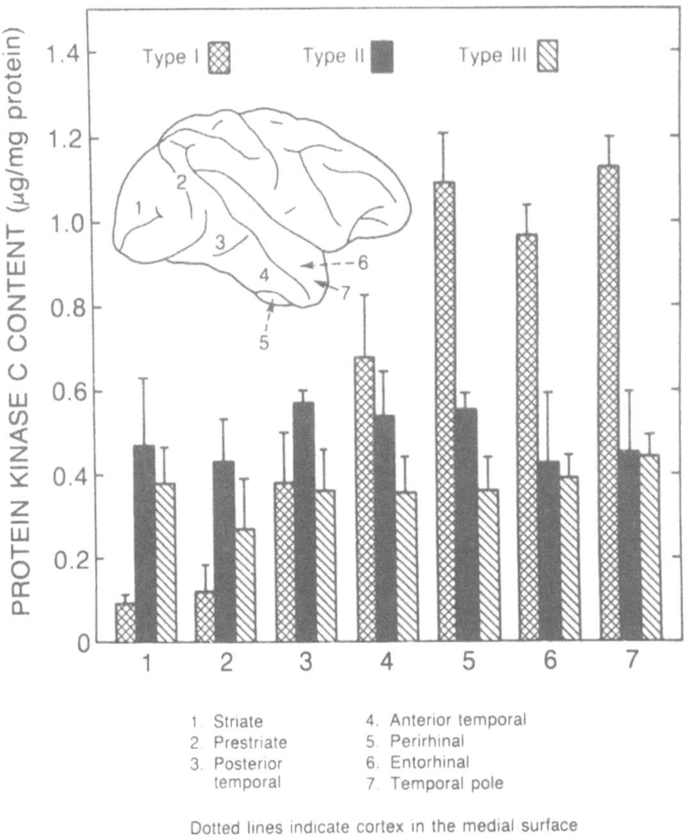

Fig. 5. Relative contents of type I, II, and III PKC in monkey cerebral cortex of visual information processing pathway. PKC isozymes in the homogenates of the various regions were quantified by immunoblotting with ^{125}I-protein A.

Differential down-regulation of PKC isozymes in RBL 2H3 cells

RBL 2H3 cells can be stimulated to secret histamine by aggregation of the IgE receptors either directly with oligomers of IgE that cross link receptors or indirectly with antigen when cells are primed with the appropriate antigen-specific IgE (24). The secretory response of 2H3 cells appears to require an increase in the intracellular level of Ca^{2+} and activation of PKC as demonstrated by the synergism between the Ca^{2+} ionophore and PMA, although neither alone appears to provide a sufficient signal for secretion (25). Incubation of 2H3 cells with PMA also results in a translocation of PKC from the cytosol to the particulate fraction and subsequent degradation of the kinase. The 2H3 cells contained both the type II and III PKC as determined by the isozyme-specific antibodies (Fig. 6). In unstimulated cells the type II enzyme was present mainly in the cytosol, whereas the type III enzyme was found both in the cytosol and particulate fractions. Addition of 20 nM of PMA resulted in a rapid loss of the type II

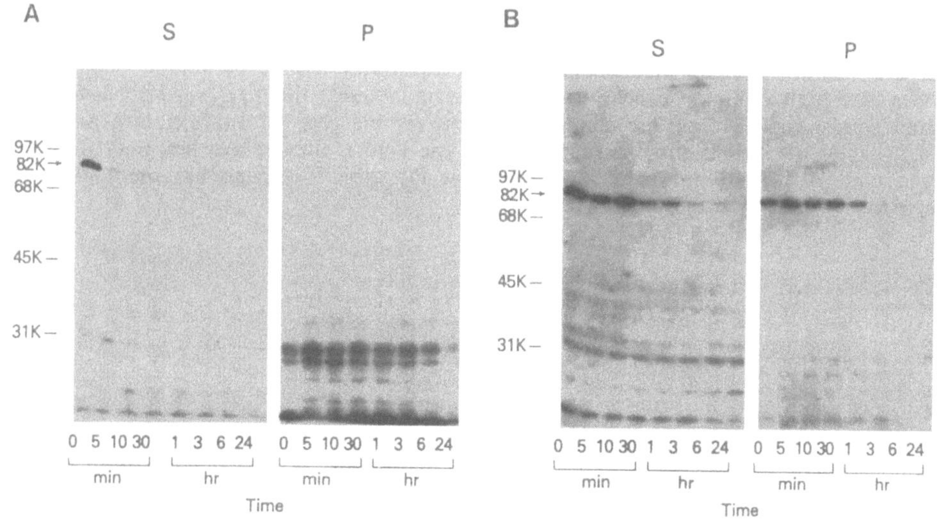

Fig. 6. PMA-mediated down-regulation of PKC isozymes in RBL 2H3 cells. RBL 2H3 cells were exposed to 20 nM of PMA for the time period indicated. Cytosolic (S) and particulate (P) fractions (60 μg protein each), prepared from 8 x 10^6 cells, were analyzed by immunoblot with type II- (panel A) and type III- (panel B) specific antibodies.

PKC from the cytosol. After 5 min of incubation with PMA, over 90% of the cytosolic type II enzyme was lost and presumably degraded. In comparison, the PMA-mediated degradation of the type III enzyme was much slower than that of the type II enzyme. Significant degradation of the type III enzyme was observed only after 10-30 min of incubation with PMA. These results indicate that the type III PKC is more resistant to PMA-induced degradation than the type II enzyme in 2H3 cells.

Biochemical characterization of PKC isozymes

Protein kinase and phorbol ester binding activities of the three types of PKC were compared using the detergent-phospholipid mixed micelles assay (26). All three isozymes appear to be alike in their phorbol ester-binding activities as indicated by their comparable K_d (between 15-20 nM) for phorbol 12,13-dibutyrate and requirement for Ca^{2+} and phospholipid. However, differences with respect to the effector-mediated stimulation of protein kinase activity were detected among these enzymes. The type I enzyme could be stimulated by cardiolipin to a greater extent than that of the type II or III enzyme. In the presence of cardiolipin, the concentrations of dioleoylglycerol and PDBu required for half-maximal activation, $A_{1/2}$, of the type I enzyme were nearly an order of magnitude lower than those for the type II and III enzymes (27). In the presence of PS, differences in the $A_{1/2}$ of DG and PDBu for the three isozymes were less significant than those measured in the presence of cardiolipin. Nevertheless, the $A_{1/2}$ of these two activators for the type I enzyme were lower than those for the type II and III enzymes. At high levels of PS (greater than 15 mol %), binding of PDBu to the type I enzyme evoked a corresponding stimulation of the kinase activity, whereas binding of this phorbol ester to the type II and III PKC produced a lesser degree of kinase stimulation (Fig.7), indicating that binding of additional PDBu molecule is required for maximal activation of the latter two isozymes.

The three PKC isozymes appeared to have similar substrate specificity when the phosphorylation sites of histone H1 and myelin basic protein were analyzed by two-dimensional peptide mapping analysis. Phosphorylation of histone IIIS and myelin basic protein by all these enzymes were dependent on Ca^{2+}, PS, and DG, whereas those of poly(lysine, serine) (PLS) and poly(arginine, serine) were independent of the activators. Detailed analysis of the kinetic parameters of these enzymes with different protein substrates revealed some differences among them (Table I). The K_m values of histone IIIS for the type I and II enzymes are approximately one-half of that for the type III enzyme. The K_m of myelin basic protein for the type II enzyme is almost twice as much as those for the type I and III enzymes, and the K_m of PLS for the type I enzyme is one-half of those for the type II and III enzymes. The K_m values of protamine for all three isozymes are comparable. Among the four substrates tested, both type I and II enzymes have highest V_{max} with myelin basic protein, whereas the V_{max} with histone IIIS and myelin basic protein are almost equivalent for the type III enzyme. In general, the K_m values of various protein substrates for the type I enzyme are less than those for the type II and III enzymes, indicating that the type I enzyme has the highest affinity for all these substrates.

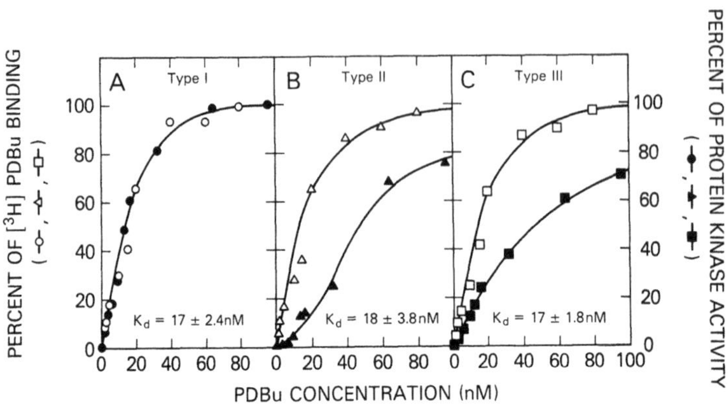

Fig. 7. Correlation between PDBu binding and kinase activity for PKC isozymes. Binding of [^{3}H]PDBu (open symbols) and measurement of the kinase activity (filled symbols) were carried out in the presence of 15 mol% PS. Binding of PDBu in the presence of 100 nM [^{3}H]PDBu and the kinase activity in the presence of 800 nM PDBu were taken as 100%.

TABLE I. Kinetic parameters of PKC isozymes with different protein substrates. PKC activity was measured in the presence of 30 mM Tris-Cl buffer, pH 7.5, 6 mM magnesium acetate, 0.12 mM [γ-^{32}P]ATP, 0.25 mM EGTA, 0.4 mM $CaCl_2$, 0.04% Nonidet P 40, 15 mol% PS, 4 mol% DG, 0.01-1.0 mg/ml histone IIIS, myelin basic protein (MBP), or protamine, and the enzymes. The kinase activity with PLS was measured under the same condition containing 2 mM EGTA without $CaCl_2$, PS, and DG.

Protein substrate	Type I		Type II		Type III	
	V_{max}	K_m	V_{max}	K_m	V_{max}	K_m
	units/mg	µg/ml	units/mg	µg/ml	units/mg	µg/ml
Histone IIIS	2,301±126	139.9±20	5,238±330	176±27	10,852±1,247	346 ± 42
MBP	4,229±149	75.1±6.6	9,908±880	167±31	10,239±329	58.9±5.2
PLS	650±22	24.4±2.2	2,321±75	62.5±8.5	2,720±235	61.1±9.2
Protamine	464±22	31.6±2.4	1,931±54.5	48.3±2.5	1,919±51	38.5±2.2

DISCUSSION

PKC is a ubiquitous enzyme present in the various animal tissues. Among the seven cloned PKC genes identified so far, four of them, α, βI, βII, and γ, appear to code for the majority of the Ca^{2+}, PS, and DG-dependent PKC, type I, II, and III, identified in the mammalian tissues. The type I PKC, encoded by the γ gene, has been identified mainly in the CNS, whereas the type II and III PKC, encoded, respectively, by βI and βII, and α genes are present in both CNS and peripheral tissues. Because of the high concentrations of type I PKC in the neurons of hippocampus, amygdala, cerebral cortex, and cerebellum, regions important for learning and memory, we investigated the possibility that this enzyme might be involved in the mnemonic function. This hypothesis was tested by measuring the PKC isozyme levels in the cortex regions of the monkey brain important for visual information processing. It appears that the brain regions important for visual information processing and storage contain higher concentration of type I PKC than those of the primary visual cortex. The type II and III PKC are, however, uniformly distributed in the cortex regions of this pathway. It was also found that during brain development the type I PKC increased parallel with the progress of synaptogenesis. In addition, the type I PKC was found to associated in a large extent with the particulate fraction in CNS, presumably in the neuronal dendrites. These circumstantial evidences seem to support the role of the type I PKC in the functioning of memory.

The three PKC isozymes are not only heterogeneously localized in different tissues and cells, they are also unevenly distributed in the various subcellular locales (28). This is most evident from the immunocytochemical staining of rat cerebellum Purkinje cells where we found the type I PKC in both the cell body and dendrites and the type III PKC mainly in the cell body. It is likely that these two enzymes may function differently in the Purkinje cells where the type I PKC transduces signal at both the cell body and postsynaptic sites at the dendrites and the type III PKC mainly at the cell body. In RBL 2H3 cells the type II and III PKC also appear to have different subcellular localization; the type II enzyme is predominantly cytosolic and the type III enzyme is associated with both cytosolic and particulate fractions. Upon treatment of 2H3 cells with PMA the type II PKC is preferentially degraded, whereas the type III enzyme is degraded at a slower rate. The physiological implication of differential down-regulation of PKC isozymes in these cells is unknown; however, it is tempting to speculate that the activation and eventual degradation of the type II PKC may be involved in the ligand-induced secretory response of these cells. This supposition is based on the observation that several type II PKC-deficient 2H3 cell variants seem to be defective in the antigen-mediated secretory response.

In spite of the similarities among the three PKC isozymes with respects to their requirement for Ca^{2+} and binding affinity for PDBu, subtle differences are detectable among these enzymes. We notice that the type I PKC exhibits a higher kinase activity in the presence of cardiolipin without DG or PDBu than the other two isozymes. In addition, in the presence of cardiolipin, the concentration of DG or PDBu required for half-maximal activation of the type I PKC is less than that for the type II and III PKC. Similarly, in the presence of PS the type I PKC is also more sensitive to activation by these two activators. These findings point to the possibility that each of the three PKC isozymes exist in their unique cellular environment and possess different biochemical characteristics may respond differently to the various modulators. This hypothesis may provide a basis for the different responses observed upon stimulation of PKC by different agonists (29-33).

REFERENCES

1. Nishizuka, Y. (1986) Science 233, 305-312
2. Berridge, M.J. (1987) Annu. Rev. Biochem. 56, 159-193
3. Huang, K.-P., Nakabayashi, H., and Huang, F.L. (1986) Proc. Natl. Acad. Sci. U.S.A. 83, 8535-8539
4. Parker, P.J., Coussens, L., Totty, N., Rhee, L., Young, S., Chen, E., Stabel, S., and Ullrich, A. (1986) Science 233, 853-859
5. Coussens, L., Parker, P.J., Rhee, L., Yang-Feng, T.L., Chen, E., Waterfield, M.D., Francke, U., and Ullrich, A. (1986) Science 233, 859-866
6. Ono, Y., Kurokawa, T., Kawahara, U., Nishimura, O., Marumoto, R., Igarashi, K., Sugino, Y., Kikkawa, U., Ogita, K., and Nishizuka, Y. (1986) FEBS Lett. 203, 111-115
7. Ono, Y., Kurokawa, T., Fujii, T., Kawahara, K., Kikkawa, U., Ogita, K., Sugino, Y., and Nishizuka, Y. (1986) FEBS Lett. 206, 347-352
8. Knopf, J.L., Lee, M.-H., Sultzman, L.A., Kriz, R.W., Loomis, C.R., Hewick, R.M., and Bell, R.M. (1986) Cell 46, 491-502
9. Housey, G.M., O'Brian, C.A., Johnson, M.D., Kirschmeier, P., and Weinstein, I.B. (1987) Proc. Natl. Acad. Sci. U.S.A. 84, 1065-1069
10. Ohno, S., Kawasaki, H., Imajoh, T., Suzuki, K., Inagaki, M., Yokokura, H., Sakoh, T., and Hidaka, H. (1987) Nature 325, 161-166
11. Kubo, K., Ohno, S., and Suzuki, K. (1987) FEBS Lett 223, 138-142
12. Ono, Y., Fujii, T., Ogita, K., Kikkawa, U., Igarashi, K., and Nishizuka, Y. (1988) J. Biol. Chem. 263, 6927-6932
13. Huang, F. L., Yoshida, Y., Nakabayashi, H., Knopf, J.L., Young,S.W. III, and Huang,

K,-P. (1987) Biochem. Biophys. Res. Commun. 149, 946-952
14. Kikkawa, U., Ono, Y., Ogita, K., Fujii, T., Asaoka, Y., Sekiguichi, K., Kosaka, Y., Igarashi, K., and Nishizuka, Y. (1987) FEBS Lett. 217, 227-231
15. Ono, Y., Kikkawa, U., Ogita, K., Fujii, T., Kurokawa, T., Asaoka, Y., Sekiguchi, K., Ase, K., Igarashi, K., and Nishizuka, Y. (1987) Science 236, 1116-1120
16. Ohno, S., Akita, Y., Konno, Y., Imajoh, S., and Suzuki, K. (1988) Cell 53, 731-741
17. Huang, F. L., Yoshida, Y., Nakabayashi, H., and Huang, K.-P. (1987) J. Biol. Chem. 262, 15714-15720
18. Yoshida, Y., Huang, F. L., Nakabayashi, H., and Huang, K.-P. (1988) J. Biol. Chem. 263, 9868-9873
19. Malenka, R.C., Madison, D.V., and Nicoll, R.A. (1986) Nature 321, 175-177
20. Malenka, R.C., Ayoub, G.S., and Nicoll, R.A. (1987) Brain Res. 403,198-203
21. Routtenberg, A. (1985) Behav. Neural Biol. 44, 186-200
22. Akers, R.F., Lovinger, P., Colley, P., Linden, D., and Routtenberg, A. (1986) Science 231, 587-589
23. Mishikin, M. (1982) Phil. Trans. R. Soc. London Ser.B. 298, 85-95
24. Beaven, M.A., Rogers, J., Moore, J.P., Hesketh, T.R., Smith, G.A., and Metcalfe, J.C. (1984) J. Biol. Chem. 259, 7129-7136
25. Beaven, M.A., Guthrie, D.F., Moore, J.P., Smith, G.A., Hesketh, T.R., and Metcalfe, J.C. (1988) J. Cell Biol. 105, 1129-1136
26. Hannun, Y., Loomis, C.R., and Bell, R.M. (1985) J. Biol. Chem. 260, 10039-10043
27. Huang, K.-P., Huang, F.L., Nakabayashi, H., and Yoshida, Y. (1988) J. Biol.Chem. in press
28. Huang, F.L., Yoshida, Y., Nakabayashi, H., Young, W.S.III., and Huang, K.-P. (1988) J. Neurosci. in press
29. Yamamoto, S., Gotoh, H., Aizu, E., and Kato, R. (1985) J. Biol. Chem. 260,14230-14234
30. Kreutter, D., Caldwell, A.B., and Morin, M.J. (1985) J. Biol. Chem. 260, 5979-5984
31. Kiss, Z., and Luo, Y. (1986) FEBS Lett. 198, 203-207
32. Ramsdell, J.S., Pettit, G.R., and Tashjian, A.H.,Jr. (1986) J. Biol. Chem. 261, 17073-17080
33. Kolesnick, R.N., and Paley, A.E. (1987) J. Biol. Chem. 262, 9204-9210

ACTIVATION OF PROTEIN KINASE C BY SHORT CHAIN PHOSPHOLIPID MICELLES

Jeffrey M. Walker and Julianne J. Sando

Department of Pharmacology
Program in Biophysics, and Cancer Center
University of Virginia
Charlottesville, Virginia, U.S.A. 22908

I. INTRODUCTION

Protein Kinase C (PKC) plays an important role in the regulation of cellular responses to external stimuli (reviewed in 1). Inactive cytosolic kinase translocates to the membrane in response to the phospholipase C-mediated release of diacylglycerol (DAG) and mobilization of Ca^{2+}. Treatment of cells with phorbol esters, a potent class of tumor promoters, also causes a translocation of kinase (2,3). Once active, PKC phosphorylates a large number of both membrane and cytosolic proteins (1). It is unclear whether these phosphorylation events occur at the membrane, or whether the kinase is released from the membrane subsequent to activation.

Ca^{2+} was initially thought to act through regulation of enzyme-membrane binding. *In vitro*, the activation of PKC requires the presence of micromolar concentrations of Ca^{2+} and phospholipid (4). The addition of DAG or phorbol ester, which bind to the same site on the enzyme (5), reduces the Ca^{2+} required to more physiologically relevant levels (6,7). Because the enzyme also showed a preference for negatively charged phospholipids, it was suggested by several groups that Ca^{2+} might mediate enzyme-lipid binding through charge interactions. Hannun, Bell and coworkers refined this model and suggested that a direct phosphatidylserine (PS) - Ca^{2+} - enzyme bridge is formed during the activation process (8-10).

While membrane binding appears to be a prerequisite for kinase activation, it is not sufficient. Hannun et al. demonstrated a Ca^{2+}-dependent binding of PKC to PS/Triton mixed micelles, but found that the addition of DAG was required to observe substrate phosphorylation (8). Wolf et al. (11) and Bazzi and Nelsestuen (12) examined enzyme binding to erythrocyte membranes and PS/phosphatidylcholine (PC) vesicles, respectively. Both groups observed PKC-membrane binding at Ca^{2+} concentrations at least one order of magnitude below those concentrations required for kinase activity. Together, these results suggest a multi-step activation process involving at least two Ca^{2+}-dependent steps.

II. PKC-LIPID INTERACTIONS

Several types of interactions are possible between proteins and lipid bilayers including 1) specific protein-lipid interactions, 2) hydrophobic interactions, and 3) interactions between the protein and bilayer/water interface. To assess the relative importance of these interactions in the activation of PKC, we have examined kinase activity in the presence of short

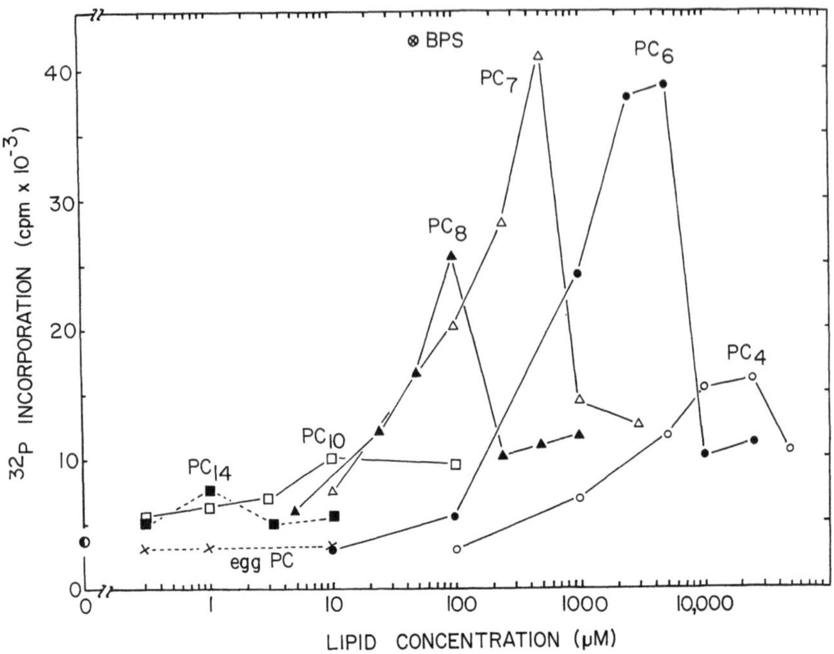

Fig. 1. Dependence of PKC activation on acyl chain length and lipid concentration in the PC series. Kinase activity was assayed with histone as substrate. From Walker and Sando, Ref. 14.

chain phospholipids.[1] Short chain phospholipids can provide either monomeric or micellar phospholipid, depending on the concentration used. We have found that short chain PSs and short chain PCs are both able to support full kinase activity, but only when the lipid is present as a micelle, and not as a monomer (14,15). As shown in Fig. 1 and Table 1, a close correlation exists between the lipid concentration required for maximum kinase activity and the critical micelle concentration of the activating species. This result, and the lack of specificity for either head group composition or acyl chain length argues for the absence of a specific phospholipid binding site on PKC.

The striking contrast in the ability of short and long chain PCs to support kinase activity indicates that the micelle structure possesses certain unique properties that are important for kinase activation. Micelles formed from short chain PC are dynamic, loosely packed structures which may permit relatively unhindered access of protein to the interfacial and hydrophobic regions. (In this respect, short chain PSs behave more like short chain PCs than like bovine PS.) Stable enzyme association with biological membranes, which contain more tightly packed long chain phospholipids, may require perturbation of local lipid structure. Ca^{2+} has the ability to cause such perturbations in vesicles containing long chain PS (16). A model illustrating the difference in behavior of long and short chain lipid systems is shown in Fig. 2. It is likely that additional properties of the interfacial region (e.g. charge density, or lipid

---

[1]Short chain phospholipids have been used extensively in the study of phospholipase A_2 for similar reasons. For review see Ref. 13.

TABLE 1. COMPARISON OF PEAK PKC ACTIVATING CONCENTRATIONS AND CMC OF SHORT AND LONG CHAIN PHOSPHOLIPIDS

Lipid	Lipid concentration for maximum PKC activity[a] Isozymes HA2	HA3	cmc
PC_6	1 - 5 mM	2.5 mM	8.6 mM[b]
PC_7	0.5 mM	0.5 mM	0.8 mM[b]
PC_8	0.1 mM	0.1 mM	0.2 mM[b]
PC_{10}	Inactive	Inactive	5 uM[c]
PC_{14}	Inactive	Inactive	-
PC_{16}	-	-	0.47 nM[c]
egg PC	Inactive	Inactive	pM[c]
PS_6	0.5 mM	1.5 mM	4 mM, 20 mM[b,d]
PS_8	75 uM	0.1 mM, 1 mM[e]	0.15 mM, 0.7 mM[b,d]
Bovine PS	50 uM	50 uM	pM[c]

[a] Assayed under the conditions described (14,15), and including the presence of 5.2 mol % dioctanoylglycerol.
[b] Determined under the conditions of the kinase assay, including the presence of Ca^{2+}. The conditions and method are described elsewhere (14).
[c] Estimates for the pure lipid (see Ref. 30).
[d] Biphasic aggregation curves were observed in the presence of Ca^{2+}. Values are given for both transitions (see Ref.15).
[e] A complex activation profile was observed (Ref. 15).

orientation) contribute to the stability of the enzyme phospholipid association since other micelle-forming agents such as detergents do not support kinase activity.

It is unclear which activation step is regulated by DAG. Although high concentrations of DAG disrupt bilayer structure (17), very low concentrations have been demonstrated to bind specifically to PKC (5). Wolf et al. (11) reported that phorbol esters reduce the Ca^{2+} requirement for binding of enzyme to erythrocyte membranes and stabilize the bound enzyme against subsequent chelator-induced disassociation. Bazzi and Nelsestuen (12) found that DAG had little if any effect on the Ca^{2+} dependency of either the enzyme association or disassociation events.

Phorbol esters and DAGs do not bind to or activate PKC in the absence of phospholipid, suggesting a primary requirement for a hydrophobic environment to expose the enzyme site to these activators. We have observed, however, that the specificity for a particular DAG (15), as well as the level of DAG-independent activity (Fig. 3), is determined in part by the type of phospholipid used in the kinase assay. In the PC_6 system, for example, both dioctanoyl and dioleoylglycerol enhance kinase activity, with dioleoylglycerol being the more effective activator. In contrast, dioleoyl glycerol is totally ineffective as an activator in the presence of PS_6, while dioctanoylglycerol is effective. Both DAGs are equally effective in the presence of bovine PS. Phorbol esters are less effective than DAGs in the short chain phospholipid systems. The discrepancy between the results of Wolf et al. (11) and Bazzi and Nelsestuen (12) may have been a result, at least in part, of these phospholipid-dependent differences. The varying effectiveness of the DAGs under different conditions may suggest that the structure of the DAG binding site and the local phospholipid environment are

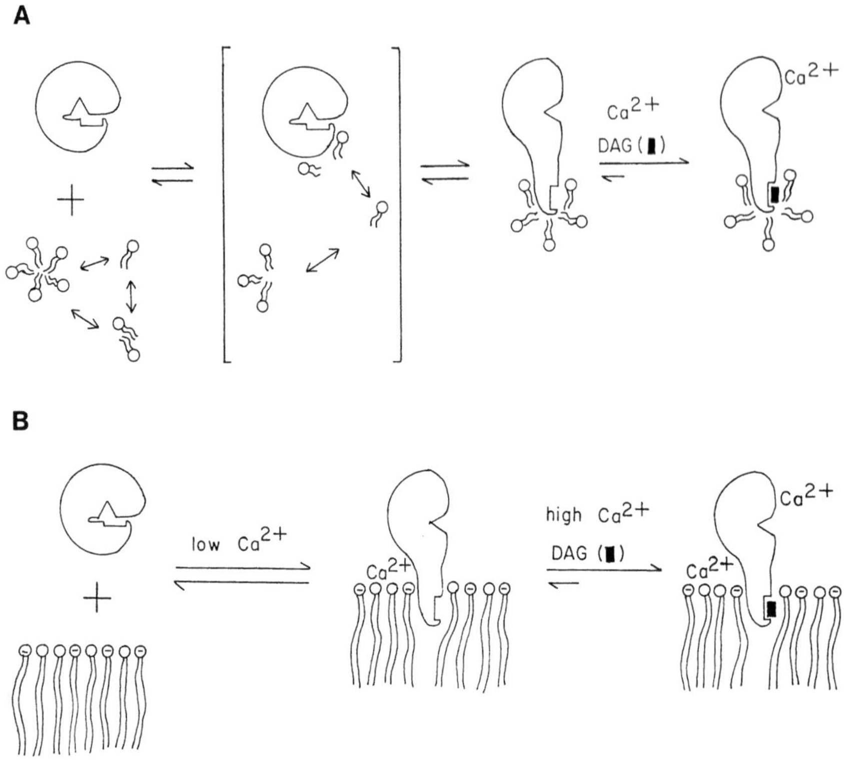

Fig. 2. A model for the activation of PKC in (A) short and (B) long chain phospholipid systems.

each influenced by the other. Alternatively, the effect of bound DAG on kinase activity may be determined in part by the environment of the enzyme. Direct phorbol ester binding assays are currently under way to discriminate between these possibilities.

III. PKC-Ca^{2+} INTERACTIONS

The activation of PKC by short chain PCs illustrates that, at least in certain model systems, Ca^{2+}-PS interactions are not essential for membrane binding or for subsequent activation steps. We have observed, however, that there is a shift from Ca^{2+}-independent to Ca^{2+}-dependent PKC activity as the structural complexity of the lipid system increases (i.e. with increasing acyl chain length and head group charge). It is possible that in the more tightly packed phospholipid structures, Ca^{2+} facilitates enzyme insertion, either directly through an enzyme-Ca^{2+}-PS bridge (8-10), or indirectly by a Ca^{2+}-induced perturbation of the local lipid structure. Ca^{2+} does not significantly influence the lipid structure of PC micelles or bilayers (14,16), a fact that may account for the inability of PC bilayers to support kinase binding (12) or activity (4,14).

Attempts to further define the process of PKC activation have recently been complicated by the discovery of at least seven different isozymes (reviewed in Ref. 18). Three of the most abundant of these isozymes have been isolated and shown to exhibit quantitative differences in their cofactor requirements (18-21). An elevated Ca^{2+}- and phospholipid-independent activity is especially prominent in one of these isozymes

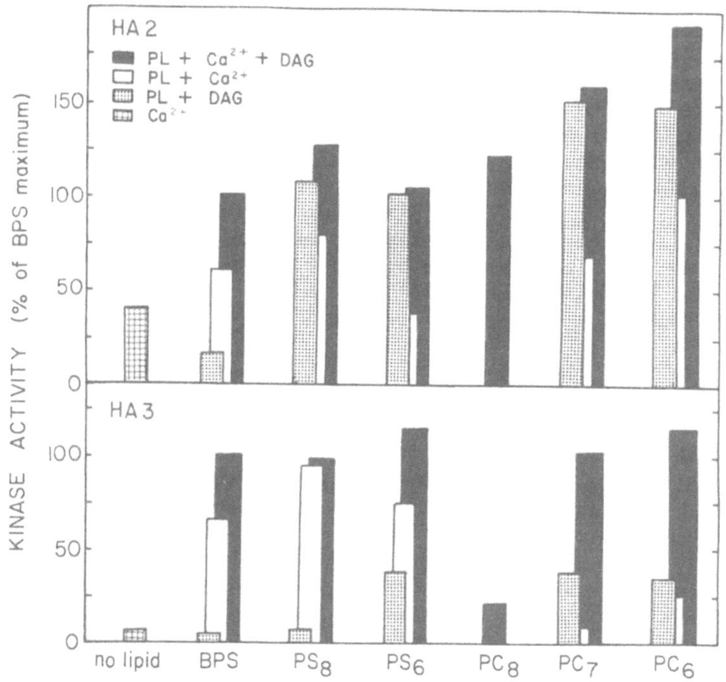

Fig. 3. The effect of phospholipid type on Ca^{2+} and DAG requirements for the activation of two different PKC isozymes. Activity was assayed under optimum conditions for each phospholipid as described (15).

(Type II). The majority of previous studies on PKC activation have been done with enzyme isolated from brain tissue, which is known to contain all three of these isozymes. The purification procedure employed by each laboratory will affect the types and relative amounts of these isozymes in the final enzyme preparation. To eliminate this problem, as well as to investigate possible reasons for the differences in cofactor requirements, we have separated two of these isozymes (HA2 and HA3) and compared their behavior in different lipid systems.[2] A number of differences in activation by short chain phospholipids and DAG were observed (e.g. Fig. 3 and Ref. 15). The most prominent differences, however, were in Ca^{2+} requirements. These differences have important implications with regard to activation mechanisms.

The substantial Ca^{2+}-independent activity of the HA2 isozyme in the presence of short chain phospholipids (Fig. 3) suggests that Ca^{2+} may be required only for membrane binding, and even then only in long chain phospholipid systems. Once associated with lipid, the enzyme assumes its activated state in the absence of Ca^{2+}. In contrast, Ca^{2+} significantly increases HA3 isozyme activity even in the presence of the phospholipid

---

[2] We believe that these preparations correspond to PKC types II and III (see Ref. 18 for nomenclature). The purification procedure and rational for this isozyme assignment are described elsewhere (15).

structures most permissive to protein insertion. This suggests that the HA3 isozyme requires a second Ca^{2+}-dependent step for activation. Direct enzyme-membrane/micelle binding studies will be required to determine whether the HA3 isozyme also requires Ca^{2+} for the initial membrane association step.

The reason for the second Ca^{2+} requirement is unknown. PKC does not require calmodulin for activation (22). Since the Ca^{2+} requirement for HA3 is not influenced by the type of phospholipid present (15), it is unlikely that this requirement involves Ca^{2+}-phospholipid binding. The second Ca^{2+} requirement may reflect binding of Ca^{2+} directly to PKC, binding of Ca^{2+} to substrate, or a Ca^{2+}-regulated interaction between the enzyme and substrate. Although the sequences of the protein kinase C isozymes contain no clearly identifiable Ca^{2+} binding domain (18), the possibility of Ca^{2+} binding sites on the enzyme remains to be tested directly.

IV. LOSS OF PKC ACTIVITY AT HIGH PHOSPHOLIPID CONCENTRATIONS

An interesting and unexpected feature of kinase activation by short chain phospholipids is the inability to maintain kinase activity at high lipid concentrations (Fig 1). A similar activity drop was observed when HA2 was assayed in the presence of short chain PS.[3] There are several possible explanations for this phenomena. Bazzi and Nelsestuen have recently proposed that the Ca^{2+} and phospholipid requirements for PKC activation are determined by the ability of these cofactors to aggregate the kinase substrates (23,24). A need for aggregated substrate on the surface of the micelle and a dilution of these aggregates with increasing lipid concentration could account for a drop in kinase activity. As Fig. 4 illustrates, however, an identical lipid-dependent activation profile is obtained for PKC autophosphorylation in the absence of exogenous substrates.

A requirement for PKC dimerization on the micelle surface or within the micelle would also result in a sharp drop in kinase activity at high micelle/PKC ratios. Indirect evidence from other laboratories suggests that PKC exists as a monomer when bound to PS/Triton micelles (8), and when autophosphorylation occurs in PS/Triton micelles (25) or in bovine PS alone (26). There are alternative interpretations for these results, however, and we are currently using the short chain phospholipid system to re-examine the possibility that PKC functions as a dimer (or higher order aggregate). Finally, the unexpected drop in kinase activity may result from a change in micelle structure at higher lipid concentrations. While Tausk et al. (27) have observed changes in micelle size at higher concentrations for PC_8 and PC_7, no such changes were observed for PC_6. We cannot, however, rule out significant structural changes when enzyme and substrate are also present.

V. SUMMARY

PKC (80 kDa) can be cleaved by limited proteolysis into distinct catalytic (50 kDa) and regulatory (32-35 kDa) fragments. After cleavage, the catalytic fragment is active in the absence of Ca^{2+}, phospholipid, or DAG while the regulatory fragment is found associated with phospholipid and continues to bind phorbol esters in a Ca^{2+}- and PS-dependent manner (28,29). In the holoenzyme, the association of the regulatory domain with the membrane may be important to release the catalytic domain from inhibition by the regulatory domain (Fig. 2). We have presented evidence indicating that effective membrane binding occurs through interaction with the hydrophobic

[3]Activation of HA3 by short chain PS was more complex, with slightly higher lipid concentrations required for maximum activity, and no clear drop in activity within the lipid concentration range tested. Possible explanations and implications of this unusual case are discussed elsewhere (15).

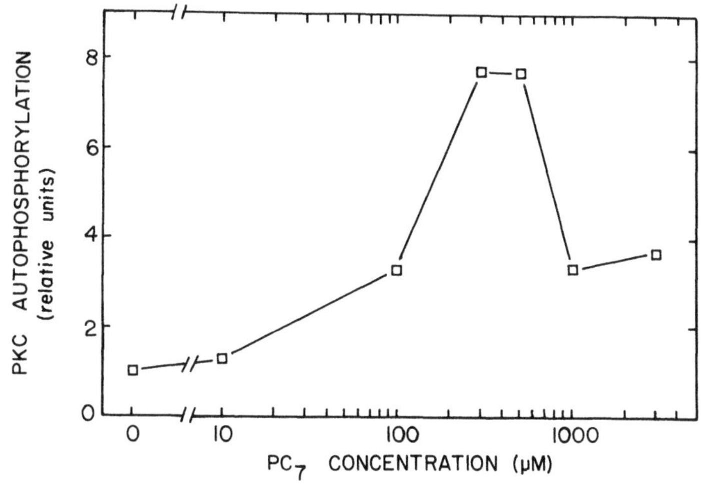

Fig. 4. PC_7 dependence of PKC autophosphorylation. ^{32}P-phosphate incorporation was determined as described in Ref. 15 and expressed relative to incorporation in the absence of phospholipid.

and/or interfacial regions of the bilayer, and does not result from binding to individual phospholipids. In vivo and in vitro studies suggest that the binding event is carefully regulated. An important function of Ca^{2+} may be to modify the local structure of the membrane, and thus affect the ability of PKC to associate with it. For at least one of the isozymes, however, Ca^{2+} may also play an additional role at a site distant from the membrane, suggesting the possibility that the isozymes may be differentially regulated.

ACKNOWLEDGEMENTS

This work was supported by NIH grants GM31184, T32-GM07267, and ACS grant FRA-306.

REFERENCES

1. Nishizuka, Y. (1986) Science 233, 305-312.
2. Kraft, A.S., Anderson, W.B., Cooper, H.L., and Sando, J.J., (1982) J. Biol. Chem. 257, 13193-13196.
3. Kraft, A.S. and Anderson, W.B. (1983) Nature 301, 621-623.
4. Takai, Y., Kishimoto, A., Iwasa, Y., Kawahara, Y., Mori, T., Nishizuka, Y., Tamura, A., and Fujii, T. (1979) J. Biochem. 86, 575-578.
5. Sharkey, N.A., Leach, K.L. and Blumberg, P.M. (1984) Proc. Natl. Acad. Sci. U.S.A. 81, 607-610.
6 Kishimoto, A., Takai, Y., Mori, T., Kikkawa, U., and Nishizuka, Y. (1980) J. Biol. Chem. 255, 2273-2276.
7. Castagna, M., Takai, Y., Kaibuchi, K. Sano, K., Kikkawa, U. and Nishizuka, Y. (1982) J. Biol. Chem. 275, 7847-7851.
8. Hannun, Y.A., Loomis, C.R., and Bell, R.M. (1986) J. Biol. Chem. 260, 10039-10043.
9. Hannun, Y.A., Loomis, C.R., and Bell, R.M. (1986) J. Biol. Chem. 261, 7184-7190.
10. Ganong, B.R., Loomis, C.R., Hannun, Y.A. and Bell, R.M. (1986) Proc. Natl. Acad. Sci. U.S.A. 83, 1184-1188.
11. Wolf, M., LeVine, H. III, May, W.S. Jr., Cuatrecasas, P., and Sahyoun, N. (1985) Nature 317, 546-549.
12. Bazzi, M.D. and Nelsestuen G.L. (1987) Biochemistry 26, 115-122.

13. Dennis, E.A. (1983) The Enzymes (Boyer, P.D., ed.) Vol. 16, 307-353, Academic Press, New York.
14. Walker, J.M. and Sando, J.J. (1988) J. Biol. Chem. 263, 4537-4540.
15. Walker, J.M. and Sando, J.J., submitted.
16. Ohnishi, S. and Ito, T. (1974) Biochemistry 13, 881-887.
17. Das, S. and Rand, R.P. (1986) Biochemistry 25, 2882-2889.
18. Nishizuka, Y. (1988) Nature 334, 661-665.
19. Huang, K.P., Nakabayashi, H., and Huang, F.L. (1986) Proc. Natl. Acad. Sci. U.S.A. 83, 8535-8539.
20. Jaken, S. and Kiley, S.C. (1987) Proc. Natl. Acad. Sci. U.S.A. 84, 4418-4422.
21. Sekiguchi, K., Tsukuda, M., Ogita, K., Kikkawa, U., and Nishizuka, Y. (1987) Biochem. Biophys. Res. Commun. 145, 797-802.
22. Kikkawa, U., Takai, Y., Minakuchi, R., Inohara, S., and Nishizuka, Y. (1982) J. Biol. Chem. 257, 13341-13348.
23. Bazzi, M.D. and Nelsestuen G.L. (1987) Biochemistry 26, 1974-1982.
24. Bazzi, M.D. and Nelsestuen G.L. (1987) Biochemistry 26, 5002-5008.
25. Newton, A.C., and Koshland, D.E. (1987) J. Biol. Chem. 262, 10185-10188.
26. Huang, K.P., Chan, K.F.J., Singh, T.J., Nakabayashi, H., and Huang, F.L. (1986) J. Biol. Chem. 261, 12134-12140.
27. Tausk, R.J.M., Van Esch, J., Karmiggelt, J., Voordouw, G., and Overbeek J.Th.G. (1974) Biophys. Chem. 1, 184-203.
28. Hoshijima, M., Kikuchi, A., Tanimoto, T., Kaibuchi, K., and Takai, Y. (1986) Cancer Res. 46, 3000-3004.
29. Lee, M. and Bell, R.M. (1986) J. Biol. Chem. 261, 14867-14870.
30. Tanford, C. (1980) The Hydrophobic Effect: Formation of Micelles and Biological Membranes, John Wiley & Sons, New York.

INOSITOL PHOSPHATE METABOLISM AND CELLULAR SIGNAL TRANSDUCTION

James W. Putney, Jr., Arlene R. Hughes, Debra A. Horstman, and Haruo Takemura

Calcium Regulation Section, Laboratory of Cellular and Molecular Pharmacology, National Institute of Environmental Health Sciences, National Institutes of Health P. O. Box 12233, Research Triangle Park, NC 27709, USA

INTRODUCTION

Over thirty years have now past since the original report by Hokin and Hokin of receptor-stimulated turnover of inositol lipids[1]. Today, the impact of this phenomenon on a multitude of important biological systems is widely appreciated. Since the first papers by the Hokins, the phosphoinositides have enjoyed periods of interest, neglect, controversy and finally acceptance as important precursors for biological signals in a variety of systems. The important contributions which lead to our current understanding of this system came from a number of different laboratories. Michell's[2] hypothesis that the phosphoinositides somehow served to couple receptors to cellular calcium mobilization provoked considerable research and criticism. Progress in understanding the exact role of inositol lipid turnover in receptor mechanisms was hindered by lack of knowledge of the biochemical pathways involved. By the early 1980's the experimental evidence began to indicate that the initial reaction in stimulated phosphoinositide turnover was the breakdown of not the major known inositide, phosphatidylinositol, but rather a minor phosphorylated derivative, phosphatidylinositol 4,5-bisphosphate[3,4]. Berridge[5] suggested that the water soluble product of this reaction, inositol 1,4,5-trisphosphate [(1,4,5)IP_3], might act as a second messenger to activate the release of Ca^{2+} from intracellular stores. Subsequently, it was demonstrated that this molecule indeed had the predicted biological activity; (1,4,5)IP_3 in micromolar concentrations rapidly released Ca^{2+} from a non-mitochondrial store in permeabilized pancreatic acinar cells[6]. This result was quickly confirmed in a number of different laboratories[7]. It now seems likely that (1,4,5)IP_3 is indeed the second messenger signalling the release of intracellular Ca^{2+} which occurs on activation of Ca^{2+} mobilizing receptors.

This chapter summarizes recent work from the authors' laboratory on the metabolism of (1,4,5)IP_3 in stimulated cells, and the possible relationship of that metabolism to cellular Ca^{2+} signalling. In addition, recent studies on the nature of the coupling of this release to Ca^{2+} entry at the level of the plasma membrane will be discussed.

Just four years ago, the formation of inositol tris- and bisphosphates in response to the activation of surface membrane receptors was demonstrated[8]. At this time a rather simple biochemical pathway was envisioned involving the sequential dephosphorylation of (1,4,5)IP_3 to (1,4)IP_2, to (1)IP and finally to free inositol by a lithium-sensitive inositol 1-phosphatase. Today, the picture of inositol phosphate metabolism is extremely complex[9,10]. This is attributable largely to the power of HPLC analytical procedures that can separate inositol phosphates with only subtle structural differences. (1,4,5)IP_3 is dephosphorylated by an extremely active 5-phosphatase to (1,4)IP_2[11], and (1,4)IP_2 is believed to be dephosphorylated almost exclusively to (4)IP by a relatively non-specific inositol polyphosphate 1-phosphatase[12]. In addition to the dephosphorylation of (1,4,5)IP_3 by the 5-phosphatase, there exists in most tissues thus far examined a 3-kinase which transfers a phosphate from ATP to the 3 position of (1,4,5)IP_3 to form (1,3,4,5)IP_4[13]. This molecule is then dephosphorylated by, probably, the same 5-phosphatase which degrades (1,4,5)IP_3 to form an isomeric inositol trisphosphate, (1,3,4)IP_3. (1,3,4)IP_3 is then dephosphorylated by the inositol polyphosphate 1-phosphatase to (3,4)IP_2[12], and to (1,3)IP_2[14] by an enzyme that has not been as well characterized. These bisphosphates are then dephosphorylated primarily to a mixture of (1)IP and (3)IP, which are stereoisomers, and not resolved by conventional HPLC.

The complexity of this metabolic pathway suggests that some biological function may be regulated by one of these metabolites. For example, there is evidence for biological activity of (1,3,4,5)IP_4 in either potentiating the action of (1,4,5)IP_3, or in modulating Ca^{2+} entry[15,16]. However, recent studies from our laboratory cast doubt on an obligatory role for (1,3,4,5)IP_4 in cellular Ca^{2+} signalling. When AR42J pancreatoma cells were stimulated with substance P, there was a rapid increase in the cellular level of (1,4,5)IP_3[17], but no increase in (1,3,4,5)IP_4 (Fig.1). In addition, metabolites of (1,3,4,5)IP_4, specifically (1,3,4)IP_3, and (1)IP or (3)IP were not increased, even in the presence of the monophosphatase inhibitor, LiCl. Homogenates of the AR42J cells could metabolize exogenous (1,3,4,5)IP_4 to all of the same products seen in other cells, but could not convert (1,4,5)IP_3 to (1,3,4,5)IP_4. It was concluded that AR42J cells, under these specific culture conditions, had no apparent (1,4,5)IP_3 3-kinase activity and could not therefore convert (1,4,5)IP_3 to (1,3,4,5)IP_4 on stimulation[17]. However, in studies using the intracellular Ca^{2+} indicator, fura-2, substance P provoked a substantial mobilization of cellular Ca^{2+}. This rise in cytosolic [Ca^{2+}] was partially dependent on extracellular Ca^{2+} (Fig. 2), and was partially inhibited by the inorganic Ca^{2+} antagonist, La^{3+}. Thus, these cells appear fully capable of responding to agonist activation by release of intracellular Ca^{2+}, and by increased entry of Ca^{2+} from the extracellular space, despite the inability to phosphorylate (1,4,5)IP_3 to (1,3,4,5)IP_4[17].

Another potential regulator of cytoplasmic Ca^{2+} levels is inositol cyclic 1:2,4,5-trisphosphate ((cl:2,4,5)IP_3). Majerus and his coworkers have demonstrated that purified phospholipase C-catalyzed hydrolysis of PIP_2 results in the formation of (1,4,5)IP_3 and (cl:2,4,5)IP_3 (with the 1-phosphate cyclized to the 2-hydroxyl[18,19]. (cl:2,4,5)IP_3 elicits $^{45}Ca^{2+}$ release from detergent-permeabilized platelets[20] and Swiss 3T3 cells[21] with a potency similar to that of (1,4,5)IP_3 and appears more potent than (1,4,5)IP_3 when injected into _Limulus_ photoreceptors[20]. However, (cl:2,4,5)IP_3 is degraded by the 5-phosphatase more slowly than (1,4,5)IP_3 (which may account for its increased apparent potency[22]) and is not phosphorylated by the 3-kinase to a cyclic derivative of (1,3,4,5)IP_4 (23). The slower inactivation of (cl:2,4,5)IP_3 suggests

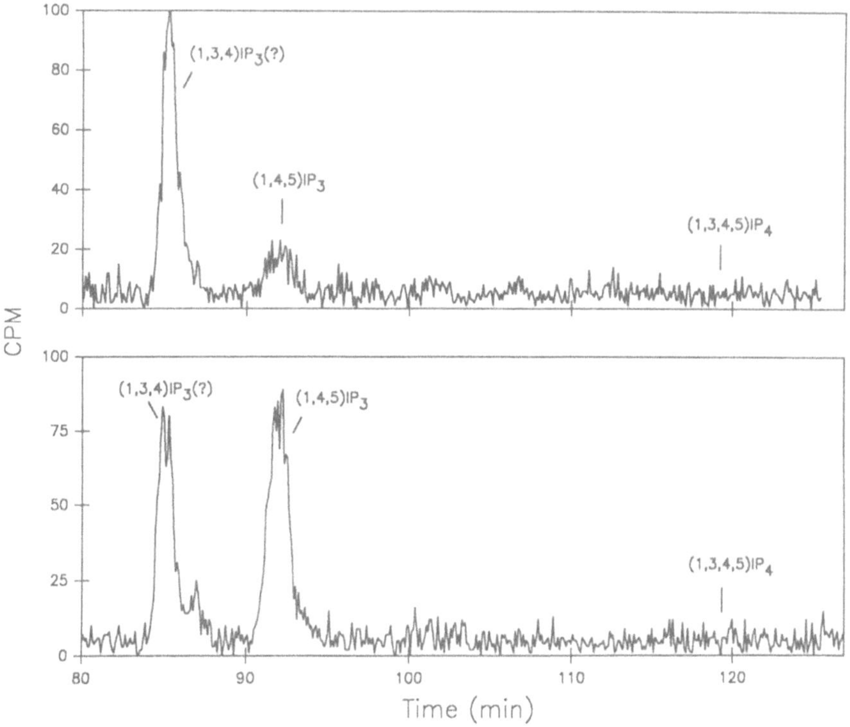

Fig. 1. HPLC analysis of inositol phosphates in substance P-stimulated AR42J cells.

that if it is formed in cells, its effects on Ca^{2+} mobilization might be more sustained than the effects of (1,4,5)IP_3. Thus, it is important to determine precisely which inositol trisphosphates are formed <u>in vivo</u> as well as their role(s) in the Ca^{2+} response. In this regard, several laboratories have reported receptor stimulated (cl:2,4,5)IP_3 formation in platelets[24,25], pancreas[26,27], and parotid minilobules[28]. In pancreatic minilobules, Dixon and Hokin (26) determined that, upon muscarinic receptor stimulation, the levels of (1,4,5)IP_3 rapidly increased, then fell to a new elevated steady state whereas (cl:2,4,5)IP_3 slowly increased. Furthermore, they suggested that (1,4,5)IP_3 could be responsible for intracellular Ca^{2+} release at short times of stimulation whereas both (1,4,5)IP_3 and (cl:2,4,5)IP_3 could equally contribute to Ca^{2+} release during prolonged stimulation.

We have further investigated the metabolism of (cl:2,4,5)IP_3 in parotid acinar cells, with attention to the possible relationship of its metabolism to a possible role in Ca^{2+} signalling[29]. Thus, the turnover rates of the cyclic and noncyclic inositol trisphosphates were estimated in cholinergically stimulated parotid cells. In these experiments, parotid acinar cells were stimulated with the muscarinic cholinergic receptor agonist, methacholine, for 40 min, by which time inositol phosphate levels are in steady state[30]. Then, inositol phosphate formation was terminated by the addition of a pharmacological excess of the muscarinic receptor antagonist, atropine (100 uM) and the decline in the inositol phosphates was monitored with time. Figure 3 illustrates the time course of decline in $[^3H]$(cl:2,4,5)IP_3, $[^3H]$(1,4,5)IP_3, as well as $[^3H]$(1,3,4,5)IP_4 following the addition of atropine. [3H](1,3,4,5)IP_3

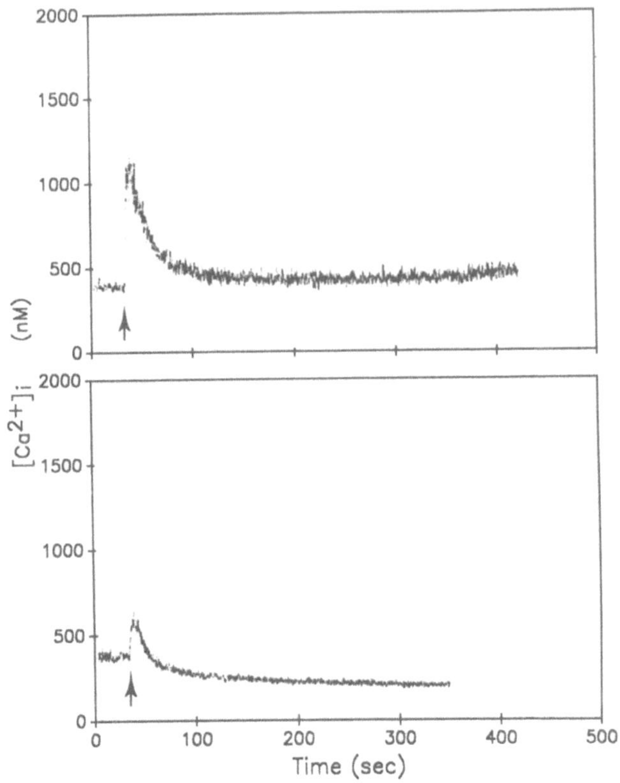

Fig. 2. Time course of intracellular Ca^{2+} in fura-2-loaded AR42J cells. In both panels, the cells were stimulated with 1 uM substance P. In the lower panel 3 mM EGTA was added to the cells together with substance P. Partial inhibition of the Ca^{2+} signal by EGTA is indicative of a contribution of Ca^{2+} entry to the response.

disappears rapidly, presumably due to its metabolism to $[^3H](1,3,4,5)IP_4$ and $[^3H](1,4)IP_2$. Following a short lag period (probably due to continued conversion of $[^3H](1,4,5)IP_3$ to $[^3H](1,3,4,5)IP_4$), $[^3H](1,3,4,5)$-IP_4 also rapidly declines, although more slowly than $[^3H](1,4,5)IP_3$. In contrast, $[^3H](cl{:}2,4,5)IP_3$ decays very slowly. Rate constants for the metabolism of $(1,4,5)IP_3$, $(cl{:}2,4,5)IP_3$, and $(1,3,4,5)IP_4$ were calculated and $t_{1/2}$ values determined. $(1,4,5)IP_3$ is metabolized with a $t_{1/2}$ = 7.6 sec; $(1,3,4,5)IP_4$ with a $t_{1/2}$ of 43.0 sec; and $(cl{:}2,4,5)IP_3$ metabolized with a $t_{1/2}$ value of 9.7 min. Furthermore, the calculated rates of flux through the various inositol phosphates indicate that $(1,4,5)IP_3$ formation occurs at about 100 times the rate of $(cl{:}2,4,5)IP_3$ formation.

The Ca^{2+} response of parotid cells also was determined under similar experimental protocols, but with cells which had been exposed to methacholine for 5 or 40 min in order to determine differences in Ca^{2+} metabolism when different proportions of $(1,4,5)IP_3$ and $(cl{:}2,4,5)IP_3$ were present. Figure 4 illustrates the change in cytosolic $[Ca^{2+}]$ in rat parotid acinar cells in response to a 5 min exposure to methacholine. Previous studies have shown that a transient, intracellular release of Ca^{2+} is responsible for the Ca^{2+} response during the first 1-2 min, while by 5 min, the response is due to extracellular Ca^{2+} entry[31]. Addition of atropine, as in the inositol phosphate experiments, causes a rapid

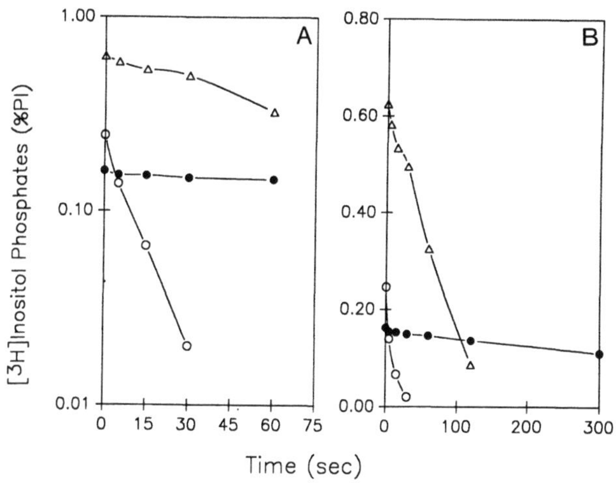

Fig. 3. Time course of decline in $[^3H]$inositol phosphate levels following muscarinic receptor blockade. Rat parotid acinar cells were stimulated with 100 uM methacholine for 40 min. 100 uM atropine was added at 40 min and 100 ml aliquots of cells were removed into boiling water at the indicated times. The inositol phosphates were separated on HPLC Partisil SAX 10 columns and the integrated peaks of radioactivity expressed as a percentage of the radioactivity in the phosphoinositides (%PI). Panel A. The decay in the radioactivity associated with (1,4,5)IP_3 (open circles), (cl:2,4,5)IP_3 (closed circles), and (1,3,4,5)IP_4 (open triangles) is plotted on a logarithmic scale versus time. Panel B. A more extended time course of the degradation of (1,4,5)IP_3, (cl:2,4,5)IP_3 and (1,3,4,5)IP_4 is plotted on an arithmetic scale versus time. The data are representative of results obtained in four experiments, each performed in duplicate.

decline in cytosolic $[Ca^{2+}]$ to basal levels. Subsequently, the status of the agonist-sensitive Ca^{2+} pool was assessed by the addition of substance P, an agonist capable of releasing this same pool[32]. This experiment also was carried out with cells which had been exposed to methacholine for 40 min prior to the addition of atropine.

Figure 5 summarizes data for the time course of restoration of cytosolic $[Ca^{2+}]$ from several such experiments. Following either the 5 or 40 min exposure to methacholine, cytosolic $[Ca^{2+}]$ declined over a matter of seconds, and was much faster than the decline in cellular $[^3H]$(cl:2,4,5)IP_3. Also, in cells treated with methacholine for 5 or 40 min, the responsiveness to substance P (and by inference the agonist-sensitive Ca^{2+} pool) was rapidly restored; again, the rate was much more rapid than the observed rate of decline of cellular $[^3H]$(cl:2,4,5)IP_3.

These data present an apparent paradox. Thus if, as has been shown for other systems, (cl:2,4,5)IP_3 has similar Ca^{2+} releasing activity to (1,4,5)IP_3, then at the proportions obtained at 40 min, there should be more than enough (cl:2,4,5)IP_3 present to saturate the Ca^{2+} mobilizing mechanism. The simplest explanation is that, unlike those systems in which the Ca^{2+} mobilizing activity of (cl:2,4,5)IP_3 has been previously demonstrated, in the rat parotid gland, the cyclic compound may not be able to activate the (1,4,5)IP_3 receptor to promote Ca^{2+} release. In fact, our preliminary data suggest that this is true. If confirmed, this would be somewhat surprising, since other inositol phosphates with vicinal 4 and 5 phosphates are capable of releasing Ca^{2+} from the

$(1,4,5)IP_3$-sensitive pool[33]. Perhaps the cyclic 1:2 phosphodiester bond imparts steric constraints on the molecule altering the relative positions of the critical phosphates at 4 and 5 positions.

The known pathways for metabolism of inositol phosphates that arise as a result of phospholipase activation are illustrated in Figure 6. Inositol pentakisphosphate (IP_5) and hexakisphosphate (IP_6; phytic acid) are also present in most mammalian cells, but their levels do not change noticeably on stimulation[34]. This would suggest that they are not formed by phosphorylation of the known IP_3's or IP_4. Their source and physiological function are at present unknown.

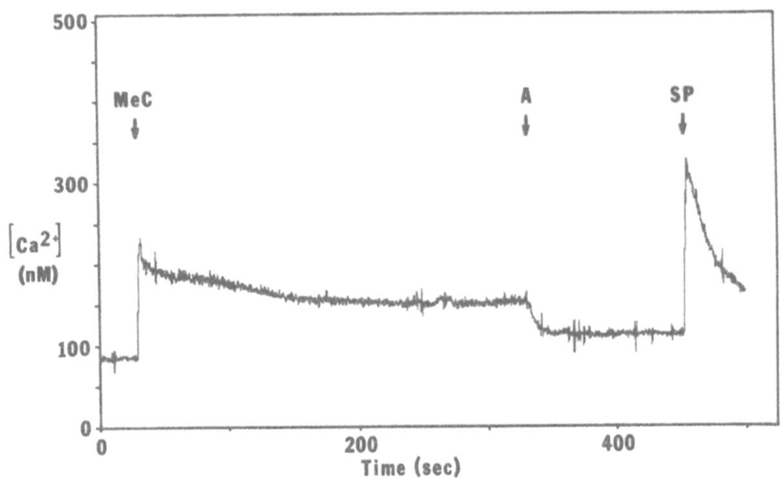

Fig. 4. Cytoplasmic Ca^{2+} concentration of rat parotid acinar cells. Rat parotid acinar cells were incubated with 2 uM fura-2/AM for 45 min at 37°C and washed immediately prior to use. Cells were stimulated with the muscarinic receptor agonist, methacholine (MeC, 100 uM) for 5 min (in the experiment in the figure) or 40 min at which time a supramaximal concentration of the muscarinic receptor antagonist, atropine (A, 100 uM) was added. In this experiment, 0.1 uM substance P (SP) was added to the cells two min after the addition of atropine. $[Ca^{2+}]i$ was calculated from the measurements of the ratio of fluorescence intensities as described by Grynkiewicz et al. (1986).

CALCIUM ENTRY

By comparison to calcium release, the regulation of calcium entry, or the second phase of calcium mobilization, is poorly understood[35]. There is strong albeit indirect evidence that inositol lipid metabolism is important for calcium entry[36], and recently, the injection of inositol phosphates into sea urchin eggs and lacrimal gland cells produced a response pattern suggestive of an activation of both intracellular calcium release and entry of calcium from the extracellular space[15,16,37,38]. However, the direct application of $(1,4,5)IP_3$ to plasma membranes does not increase their permeability to calcium[39,40]. Collectively these observations suggest that $(1,4,5)IP_3$ may activate calcium entry in cells, but not by acting directly on the plasma membrane.

Recently a mechanism which is consistent with this idea has been proposed. According to this hypothesis, the emptying of a (1,4,5)IP_3-sensitive pool secondarily signals calcium entry. A detailed discussion of the circumstantial evidence supporting this idea has been presented elsewhere[35]. The major evidence is derived from studies of the kinetics of emptying and refilling of a receptor-regulated calcium pool in the rat parotid gland. In the parotid gland under resting conditions, the intracellular calcium pool was resistant to depletion by extracellular chelating agents. However, when emptied by agonist stimulation, this pool could rapidly be filled from outside of the cell, even in the absence of agonists and (presumably) second messengers, such as inositol phosphates (41). These results suggest that the loss of calcium from this pool somehow activates a pathway for entry into the pool from the

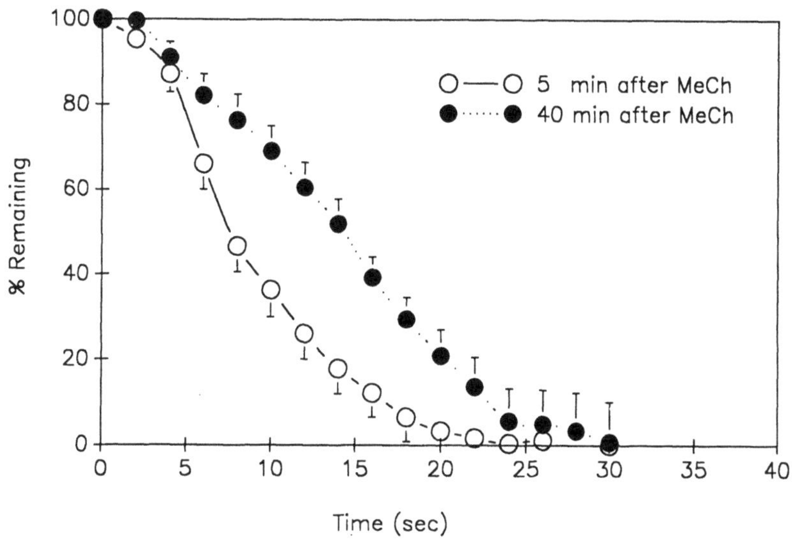

Fig 5. Time course of the decline in cytosolic $[Ca^{2+}]$ after atropine. Fura-2-loaded parotid acinar cells were stimulated with 100 uM methacholine and the $[Ca^{2+}]_i$ determined. 100 uM atropine was added to the cells either 5 min (open circles) or 40 min (closed circles) after methacholine stimulation, and the $[Ca^{2+}]_i$ was monitored with time. The results are plotted as the % decline in $[Ca^{2+}]_i$ versus time, with 100% representing the $[Ca^{2+}]_i$ immediately prior to the addition of atropine and 0% representing the baseline cytosolic $[Ca^{2+}]$. The results are the mean ± SE of 9-11 experiments.

extracellular space. Thus, in the continued presence of agonist, when (1,4,5)IP_3 levels are being continuously maintained, the pool would presumably be held empty, the pathway from the extracellular space would be open, and calcium would enter the pool and subsequently the cytosol through the (1,4,5)IP_3-activated channels.

In more recent experiments, the intracellular Ca^{2+} indicator, fura-2, has been used to monitor changes in cytosolic [Ca^{2+}] in parotid acinar cells[31,32]. With this more sensitive indicator, a brief elevation of cytosolic [Ca^{2+}] is detected during the reloading process which can be used to infer the influence of pool Ca^{2+} content on plasma membrane Ca^{2+} fluxes[42]. Thus, when parotid cells were incubated in a medium containing low [Ca^{2+}], and Ca^{2+} was restored to the physiological range, there was a small increase in cytosolic [Ca^{2+}] (Fig. 7). In another experiment shown in Figure 7, receptor activation was terminated prior to the addition of Ca^{2+} by the addition of a pharmacological excess of the muscarinic receptor antagonist, atropine. This resulted in a rapid ($t_{0.5}$ = 7.6 sec) decline in the cellular level of the Ca^{2+}-mobilizing second messenger, (1,4,5)IP_3[29]. However, the absence of extracellular Ca^{2+} premented the agonist-sensitive Ca^{2+} pool from refilling[41]. When extracellular Ca^{2+} was restored under these conditions, i.e., in the absence of receptor activation but when the agonist-sensitive Ca^{2+} pool had been depleted, the initial increase in cytosolic [Ca^{2+}] was faster and transiently larger than in the control cells with the pool intact.

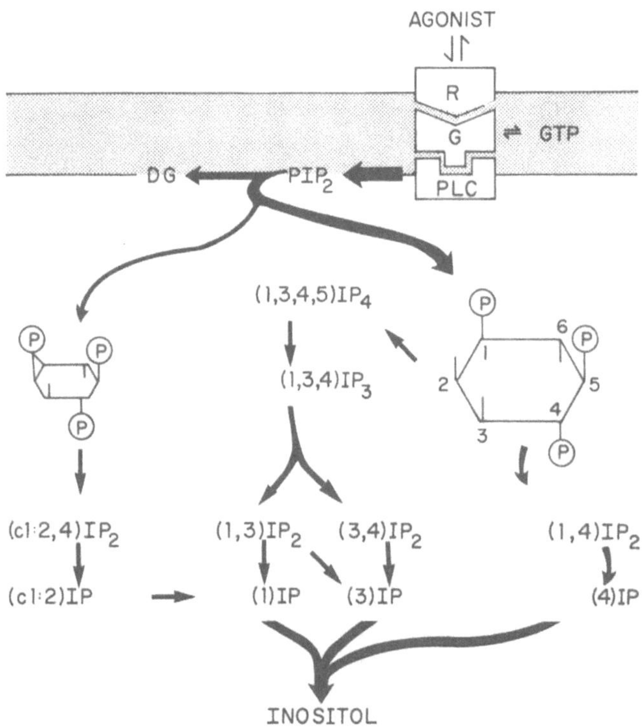

Fig. 6. Major (known) pathways for metabolism of (1,4,5)IP_3 and (cycli-cl:2,4,5)IP_3. See text for details.

In the experiments shown in figure 7, the addition of Ca^{2+} occurred one minute after the addition of atropine. Thus, it is unlikely that the additional flux can be attributed to persisting levels of (1,4,5)IP_3 which would decline with a half-time of 7.6 sec (29). However, other

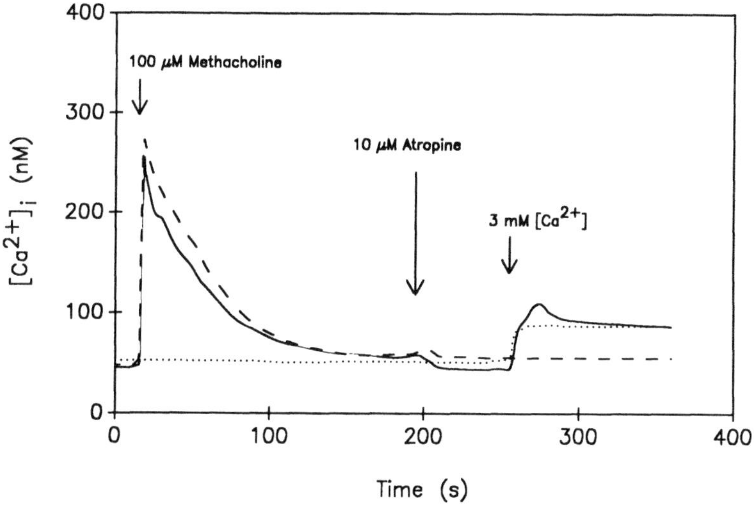

Fig. 7 The increase in $[Ca^{2+}]_i$ due to the addition of 3.0 mM Ca^{2+} to parotid acinar cells incubated in medium containing no added Ca^{2+} and 0.2 mM EGTA. $[Ca^{2+}]_i$ was determined in cells loaded with fura-2, as described by Takemura and Putney (in press). The Ca^{2+} was added either to control cells (not previously stimulated, dotted line), or to cells treated with 0.1 uM methacholine followed by 10 uM atropine (solid line, dashed line). In the experiment indicated by the dashed line, 0.3 mM $LaCl_3$ was added with the atropine ($[La^{3+}]$ of about 0.1 mM).

inositol phosphates formed in acinar cells decline with half-times of the order of minutes[29], and some of these have been suggested to have Ca^{2+} mobilizing or regulating actions (for example, inositol cyclic 1:2,4,5-trisphosphate[20], inositol 1,3,4-trisphosphate[21], inositol 1,3,4,5-tetrakisphosphate[16]). Two experimental protocols were utilized to demonstrate the independence of the pool-dependent Ca^{2+} influx on the presence or metabolism of inositol phosphates or other agonist-dependent messengers. First, the extra Ca^{2+} influx seen in pool-depleted cells persisted even when the interval between the addition of atropine and Ca^{2+} was increased from one to 20 min. In a second strategy, the pool was allowed to refill by adding atropine in the presence of extracellular Ca^{2+}. Extracellular Ca^{2+} was subsequently removed, and the rise in cytosolic $[Ca^{2+}]$ following application of 3 mM extracellular Ca^{2+} was determined. Under these conditions, the rise in cytosolic $[Ca^{2+}]$ following the addition of extracellular Ca^{2+} was not rapid, and resembled the increase seen in unstimulated cells. In other words, the extra Ca^{2+} influx could be observed as late as 20 min following the blockade of the muscarinic receptor as long as refilling of the agonist-sensitive pool during the intervening period was prevented by the absence of extracellular Ca^{2+}. Conversely, the extra influx was <u>not</u> observed 3 min after atropine treatment if the intracellular pool was first allowed to refill.

Taken together, these results demonstrate that when the agonist-sensitive Ca^{2+} pool is emptied by an agonist, Ca^{2+} influx across the plasma membrane is increased. This influx of Ca^{2+} occurs independently of the levels of inositol phosphates and probably of any second messengers linked directly to receptor activation. It appears rather to be a consequence of the empty state of the Ca^{2+} pool. It follows therefore that whenever the agonist-sensitive Ca^{2+} pool is emptied by agonist activation, the plasma membrane permeability to Ca^{2+} will be increased. In these experiments, this effect is short-lived because the test application of Ca^{2+} serves to restore the pool. However, in the continued presence of an agonist, (1,4,5)IP_3 would prevent the refilling of the pool, and a sustained influx of Ca^{2+} would result.

CONCLUSIONS

In the past few years, our understanding of the relationships between receptors, inositol lipids and calcium has advanced at an incredible rate. Indeed, only 5 years before the writing of this review, Streb et al. (1983) first demonstrated the biological activity of (1,4,5)IP_3; in the short interval since, this molecule has now gained general acceptance as an important and ubiquitous second messenger. It is apparent that this second messenger system is exceedingly complex and possesses many potential sites of regulation. As with most important biochemical pathways, our knowledge raises numerous challenging questions. These include the significance and regulation of the complex pathways (phosphorylation/dephosphorylation) of (1,4,5)IP_3 metabolism, and the mechanisms by which calcium entry is regulated. Indeed, we need to further pursue how the entire system is regulated, and how it interacts with other signalling systems such as those for cyclic nucleotides. Hopefully, the substantial progress we have made in the past few years will continue toward the resolution of these issues and to pose new questions as well.

REFERENCES

1. Hokin, M.R., and L.E. Hokin, Enzyme secretion and the incorporation of P32 into phospholipides of pancreas slices, J. Biol. Chem. 203:967 (1953).
2. Michell, R.H., Inositol phospholipids and cell surface receptor function, Biochim. Biophys. Acta 415:81 (1975).
3. Abdel-Latif, A.A., R. Akhtar, and J.N. Hawthorne, Acetylcholine increases the breakdown of triphosphoinositide of rabbit iris muscle prelabelled with [^{32}P]phosphate, Biochem. J. 162:61 (1977).
4. Kirk, C.J., J.A. Creba, C.P. Downes, and R.H. Michell, Hormone-stimulated metabolism of inositol lipids and its relationship to hepatic receptor function, Biochem. Soc. Trans. 9:377 (1981).
5. Berridge, M.J., Rapid accumulation of inositol trisphosphate reveals that agonists hydrolyse polyphosphoinositides instead of phosphatidylinositol, Biochem. J. 212:849 (1983).
6. Streb, H., R.F. Irvine, M.J. Berridge, and I. Schulz, Release of Ca^{2+} from a nonmitochondrial store in pancreatic cells by inositol-1,4,5-trisphosphate, Nature 306:67 (1983).
7. Berridge, M.J., Inositol phosphates as second messengers. in: Phosphoinositides and Receptor Mechanisms, Putney, J.W., Jr., ed., p. 25-46, Alan R. Liss, Inc., New York (1986).
8. Berridge, M.J., R.M. Dawson, C.P. Downes, J.P. Heslop, and R.F. Irvine, Changes in the levels of inositol phosphates after agonist-dependent hydrolysis of membrane phosphoinositides, Biochem. J. 212:473 (1983).
9. Majerus, P.W., T.M. Connolly, V.S. Bansal, R.C. Inhorn, T.S. Ross, and D.L. Lips, Inositol phosphates: Synthesis and degradation, J. Biol. Chem. 263:3051 (1988).

10. Putney, J.W., Jr., Calcium-mobilizing receptors, Trends Pharmacol. Sci. 8:481 (1987).
11. Downes, C.P., M.C. Mussat, and R.H. Michell, The inositol trisphosphate phosphomonoesterase of the human erythrocyte membrane, Biochem. J. 203:169 (1982).
12. Inhorn, R.C., V.S. Bansal, and P.W. Majerus, Pathway for 1,3,4-trisphosphate and 1,4-bisphosphate metabolism, Proc. Nat. Acad. Sci. USA 84:2170 (1987).
13. Irvine, R.F., A.J. Letcher, J.P. Heslop, and M.J. Berridge, The inositol tris/tetrakisphosphate pathway - demonstration of Ins(1,4,5)P3 3-kinase activity in animal tissues, Nature 320:631 (1986).
14. Bansal, V.S., R.C. Inhorn, and P.W. Majerus, The metabolism of inositol 1,3,4-trisphosphate to inositol 1,3-bisphosphate, J. Biol. Chem. 262:9444 (1987).
15. Irvine, R.F., and R.M. Moor, Micro-injection of inositol 1,3,4,5-tetrakisphosphate activates sea urchin eggs by a mechanism dependent on external Ca^{2+}, Biochem. J. 240:917 (1986).
16. Morris, A.P., D.V. Gallacher, R.F. Irvine, and O.H. Petersen, Synergism of inositol trisphosphate and tetrakisphosphate in activating Ca^{2+}-dependent K^{+} channels, Nature 330:653 (1987).
17. Horstman, D.A., H. Takemura, and J.W. Putney,Jr., Formation and metabolism of [^{3}H]inositol phosphates in AR42J pancreatoma cells: Substance P-induced Ca^{2+} mobilization in the apparent absence of inositol 1,4,5-trisphosphate 3-kinase activity, J. Biol. Chem. in press.
18. Wilson, D.B., T.E. Bross, S.L. Hofmann, and P.W. Majerus, Hydrolysis of polyphosphoinositides by purified sheep seminal vesicle phospholipase C enzymes, J. Biol. Chem. 259:11718 (1984).
19. Wilson, D.B., T.E. Bross, W.R. Sherman, R.A. Berger, and P.W. Majerus, Inositol cyclic phosphates are produced by cleavage of phosphatidylphosphoinositols (polyphosphoinositides) with purified sheep seminal vesicle phospholipase C enzymes, Proc. Nat. Acad. Sci. USA 82:4013 (1985).
20. Wilson, D.B., T.M. Connolly, T.E. Bross, P.W. Majerus, W.R. Sherman, A.N. Tyler, L.J. Rubin, and J.E. Brown, Isolation and characterization of the inositol cyclic phosphate products of polyphosphoinositide cleavage by phospholipase C. Physiological effects in permeabilized platelets and Limulus photoreceptor cells, J. Biol. Chem. 260:13496 (1985).
21. Irvine, R.F., A.J. Letcher, D.J. Lander, and M.J. Berridge, Specificity of inositol phosphate-stimulated Ca2+ mobilization from Swiss-mouse 3T3 cells, Biochem. J. 240:301 (1986).
22. Connolly, T.M., D.B. Wilson, T.E. Bross, and P.W. Majerus, Isolation and characterization of the inositol cyclic phosphate products of phosphoinositide cleavage by phospholipase C. Metabolism in cell-free extracts, J. Biol. Chem. 261:122 (1986).
23. Connolly, T.M., V.S. Bansal, T.E. Bross, R.F. Irvine, and P.W. Majerus, The metabolism of tris- and tetraphosphates of inositol by 5- phosphomonoesterase and 3-kinase enzymes, J. Biol. Chem. 262:2146 (1987).
24. Ishii, H., T.M. Connolly, T.E. Bross, and P.W. Majerus, Inositol cyclic trisphosphate (inositol 1:2-cyclic 4,5-trisphosphate) is formed upon thrombin stimulation of human platelets, Proc. Nat. Acad. Sci. USA 83:6397 (1986).
25. Tarver, A.P., W.G. King, and S.E. Rittenhouse, Inositol 1,4,5-trisphosphate and inositol 1,2-cyclic 4,5-trisphosphate are minor components of total mass of inositol trisphosphate in thrombin-stimulated platelets, J. Biol. Chem. 262:17268 (1987).
26. Dixon, J.F., and L.E. Hokin, Inositol 1,2-cyclic 4,5-trisphosphate concentration relative to inositol 1,4,5-trisphosphate in

pancreatic minilobules on stimulation with carbamylcholine in the absence of lithium. Possible role as a second messenger in long- but not short-term responses, J. Biol. Chem. 262:13892 (1987).
27. Sekar, M.C., J.F. Dixon, and L.E. Hokin, The formation of inositol 1,2-cyclic 4,5-trisphosphate and inositol 1,2-cyclic 4-bisphosphate on stimulation of mouse pancreatic minilobules with carbamylcholine, J. Biol. Chem. 262:340 (1987).
28. Dixon, J.F., and L.E. Hokin, Inositol 1,2-cyclic 4,5-trisphosphate is formed in the rat parotid gland on muscarinic stimulation, Biochem. Biophys. Res. Comm. 149:1208 (1987).
29. Hughes, A.R., H. Takemura, and J.W. Putney,Jr., Kinetics of inositol 1,4,5-trisphosphate and inositol cyclic 1:2,4,5-trisphosphate metabolism in intact rat parotid acinar cells: Relationship to calcium signalling, J. Biol. Chem. 263:10314 (1988).
30. Aub, D.L., and J.W. Putney,Jr., Metabolism of inositol phosphates in parotid cells: implications for the pathway of the phosphoinositide effect and for the possible messenger role of inositol trisphosphate, Life Sciences 34:1347 (1984).
31. Merritt, J.E., and T.J. Rink, Regulation of cytosolic free calcium in fura-2-loaded rat parotid acinar cells, J. Biol. Chem. 262: 17362 (1987).
32. Merritt, J.E., and T.J. Rink, The Effects of substance P and carbachol on inositol tris- and titrakisphosphate formation and cytosolic free calcium in rat parotid acinar cells. A correlation between inositol phosphate levels and calcium entry, J. Biol. Chem. 262:14912 (1987).
33. Aub, D.L., and J.W. Putney,Jr., Mobilization of intracellular calcium by methacholine and inositol 1,4,5-trisphosphate in rat parotid acinar cells, J. Dent. Res. 66:547 (1987).
34. Heslop, J.P., R.F. Irvine, A.H. Tashjian, and M.J. Berridge, Inositol tetrakis- and pentakisphosphates in GH_4 cells, J. Exp. Biol. 119:395 (1985).
35. Putney, J.W., Jr., A model for receptor-regulated calcium entry, Cell Calcium 7:1 (1986).
36. Berridge, M.J., and J.N. Fain, Inhibition of phosphatidylinositol synthesis and the inactivation of calcium entry after prolonged exposure of the blowfly salivary gland to 5-hydroxytryptamine, Biochem. J. 178:59 (1979).
37. Slack, B.E., J.E. Bell, and D.J. Benos, Inositol 1,4,5-trisphosphate injection mimics fertilization potentials in sea urchin eggs, Am. J. Physiol. 250:C340 (1986).
38. Irvine, R.F., and R.M. Moor, Inositol(1,3,4,5)tetrakisphosphate-induced activation of sea urchin eggs requires the presence of inositol trisphosphate, Biochem. Biophys. Res. Comm. 146:284 (1987).
39. Delfert, D.M., S. Hill, H.A. Pershadsingh, and W.R. Sherman, myo-Inositol 1,4,5-trisphosphate mobilizes Ca^{2+} from isolated adipocyte endoplasmic reticulum but not from plasma membranes, Biochem. J. 236:37 (1986).
40. Ueda, T., S.H. Church, M.W. Noel, and D.L. Gill, Influence of inositol 1,4,5-trisphosphate and guanine nucleotides on intracellular calcium release within the N1E-115 neuronal cell line, J. Biol. Chem. 261:3184 (1986).
41. Aub, D.L., J.S. McKinney, and J.W. Putney,Jr., Nature of the receptor-regulated calcium pool in the rat parotid gland, J. Physiol. (Lond.) 331:557 (1982).
42. Takemura, H., and J.W. Putney,Jr., Capacitative calcium entry in parotid acinar cells, Nature, submitted.
43. Grynkiewicz, G., M. Poenie, and R.Y. Tsien, A new generation of Ca^{2+} indicators with indicators with greatly improved fluorescence properties, J. Biol. Chem. 260:3440 (1986).

ROLE OF CALCIUM RELEASED FROM THE SARCOPLASMIC RETICULUM OF SMOOTH MUSCLE CELLS AS INDUCED BY INOSITOL PHOSPHATIDES

Kenji Kitamura and Hirosi Kuriyama

Department of Pharmacology, Faculty of Medicine

Kyushu University, Fukuoka 812, Japan

INTRODUCTION

Phosphatidylinositol 4,5-bisphosphate ($PI-P_2$), distributed at the sarcolemma in smooth muscles, as a phospholipid, is hydrolyzed into inositol 1,4,5-trisphosphate (below Ins-1,4,5-P_3) and diacylglycerol (DG) by activations of each receptor through actions of GTP binding protein (G-protein) and by those of phospholipase C. The latter is also synthesized from phosphatidyl inositide monophosphate (Berridge, 1984; Berridge and Irvine, 1984; Abdel-Latif, 1986). It is now clear that Ins-1,4,5-P_3 releases Ca from the sarcoplasmic reticulum (SR) of smooth muscle cells and increases free Ca concentration in the cytosol, thereby initiating contraction (Streb et al., 1983; Suematsu et al., 1984; Somlyo et al., 1985; Yamamoto and van Breemen, 1985; Hashimoto et al., 1986; Sasaguri et al., 1985). DG with phosphatidylserine (PS) activates protein kinase C and phosphorylates the proteins required to regulate physiological functions in the cells (Nishizuka, 1984; 1986). In smooth muscles, 12-O-tetradecanoylphorbol-13-acetate (TPA), a phorbol ester and a substitute of DG, together with PS increases the Ca sensitivity of contractile proteins and enhances the mechanical response evoked by high K (below 59 mM) in intact smooth muscles and also by Ca (below 0.5 μM) in skinned muscle tissues (Itoh et al. 1987). TPA also inhibits the hydrolysis of $PI-P_2$ triggered by agonists and reduces the synthesis of Ins-1,4,5-P_3 (Itoh et al., 1988).

We describe here in reviews our recent studies on the role of Ca released from the SR of smooth muscles. Attention is directed to the effects of released Ca on the membrane electrical activity, as measured using the whole cell voltage clamp method.

THE RELEASE OF CALCIUM FROM THE SARCOPLASMIC RETICULUM, AS INDUCED BY INOSITOL TRISPHOSPHATE

Pioneering work concerning inositol phospholipid metabolism in relation to Ca mobilization is briefly described. In 1953, Hokin and Hokin observed an accelerated incorparation of ^{32}P into phosophatidylinositol and phosphatidic acid in the pancreas slices stimulated by acetylcholine. In 1975, Michell presented the Ca gate theory in relation to phosphatidylinositol turnover, i.e. activations of PI metabolism trigger the opening of the Ca gate at the surface membrane and the influx of Ca is increased. In 1977, Abdel-Latif et al. reported that acetylcholine accelerates the hydrolysis of $PI-P_2$ and of 4-monophosphate. They also suggested that this hydrolysis requires the presence of extracellular Ca. On the other hand, Michell (1982) did experiments on vasopressin-treated liver slices and found that the hydrolysis of $PI-P_2$ precedes that of phosphatidylinositol and that this process does not require the extracellular Ca ions. Streb et al. (1983) reported that in rat pancreatic acinar cells, Ins-1,4,5-P_3 releases Ca stored in the cell, probably from the endoplasmic reticulum. In 1984, Berridge and Irvine examined the salivary gland of calliphora stimulated by 5-hydroxytryptamine and noted that the synthesis of Ins-1,4,5-P_3 preceded that of inositol monophosphate.

During the course of invesigations in various laboratories, the metabolism of inositol phosphatide was clarified and a putative cascade map was drawn up. Neurohumoral transmitters, hormones, autacoids and many putative peptides which activate the receptor are reported to hydrolyse $PI-P_2$ and the synthesis of Ins-1,4,5-P_3, Inositol-1,3,4,5-tetrakisphosphate, Ins-P_3, Ins-cyclic 1:2,4,5-P_3, Ins-2,4,5-P3, Ins-1,3,4,5,6-$_5$ and other metabolites (Downess and Michell, 1985; Lindon et al., 1986; Tomlinson and Ballou, 1961; Irvine et al., 1984 a & b, 1985, 1986 a and b; Burgess et al., 1985; Heslop et al. 1985). Accumulation of Ins-1,3,4,5-P_4 does not occur with a slower time course than that of Ins-1,4,5-P_3 and accumulation of Ins-P_3 is slower than that of Ins-1,4,5-P_3. Ins-1,3,4,5-P_4 is synthesized from Ins-1,4,5-P_3 by 3 kinase present in the cytosol and Ins-1,3,4,5-P_4 is thought to be dephosphorylated by 5-phosphatase (Irvine and Moor, 1986; Batty et al., 1985; Michell, 1986). The Ca releasd by inositol phosphatides from the SR has been mainly investigated using Ins-1,4,5-P_3 or Ins-1,3,4,5-P_4, the latter with little affect on the release of Ca from the SR but which does acts on the sarcolemma (Irvine et al., 1986 a and b; Higashida and Brown, 1986a,b). There are reports that Ins-1,3,4,5-P_4 requires the presence of Ins-1,4,5-P_3 or Ins-P_3 (Irvine et al., 1986a and b; Morris et al., 1987).

Ins-1,4,5-P_3 can release Ca from the SR but not from the sarcolemmal fraction of smooth muscle cells and the Km value, as measured from the SR fraction is 0.6 μM (Suematsu et al., 1984). This value was much the same as noted with macrophages (Hirata et al., 1984). The minimum Km value of 25 nM was noted in the case of insulinoma in the hamster (Joseph et al., 1984) and in skeletal muscles, the maximum Km value was reported to be 4-5 μM (Volpe et al., 1985).

The amount of Ca released from the SR by Ins-1,4,5-P_3 can be estimated from the contraction evoked in skinned muscle tissues. In vascular smooth muscles (porcine coronary and rabbit mesenteric arteries), after skinning smooth muscle strips with saponin, application of low concentrations of Ca solution containing EGTA led to an accumulation of Ca in the SR, concentration and time dependently, and the following application of Ins-1,4,5-P_3 or caffeine led to a contraction of the tissue (Hashimoto et al., 1986). When caffeine was applied before the application of Ins-1,4,5-P_3, the resulting contraction induced by Ins-1,4,5-P_3 was markedly inhibited, and vice versa. Concentration (Ca)- response relationships, observed after application of Ins-1,4,5-P_3, shifted to the left. Thus, Ins-1,4,5-P_3 accelerates the contraction as a consequence of additional increases in Ca concentrations in the cytosol. The amplitude of the Ins-1,4,5-P_3-induced contraction in skinned muscles depends on the concentration of EGTA in the relaxing solution (EGTA added Ca-free solution containing Mg-ATP). Therefore, the absolute amount of Ca released from the SR cannot be estimated from the amplitude of the Ins-1,4,5-P3-induced contraction. Recently, a caged Ins-1,4,5-P_3 was introduced to clarify mechanisms involved in release of Ca from the SR. The results supported previous observations that Ins-1,4,5-P_3 can release Ca from the SR (Walker et al., 1987).

The amount of Ca released by Ins-1,4,5-P_3 can be estimated from the Ca transient, using fura-2. After obtaining dispersed smooth muscle cells, enzymatically, these cells were permeabilized by saponin (chemically skinned dispersed cells), and fura-2 was added to the relaxing solution (Ca-free solution containing ATP). The dispersed cell suspension was then washed with the relaxing solution. Application of Ins-1,4,5-P_3 increased the intensity of the fluorescence (increases in the amount of Ca) as observed with the application of caffeine.

The site of action of Ins-1,4,5-P_3 in the Ca store site can be estimated using the photoaffinity labelling technique. Following incubation with Ins-1,4,5-P_3 coupled with para-azidobenzyoyl β-alanine, the Ins-1,4,5-P_3-induced Ca release was blocked after photo-irradiation of the saponin permeabilized dispersed smooth muscle cells. Therefore, there is a covalent binding of Ins-1,4,5-P_3-para-azidobenzyol β-alanine to the specific receptor on the store site, i.e. this complex irreversibly opens the Ca channel in the SR and reduces the Ins-1,4,5-P_3-releasable Ca.

THE IONIC CURRENT RECORDED FROM SINGLE SMOOTH MUSCLE CELLS IN RELATION TO MOBILIZATION OF CALCIUM

Since 1981, when Hamill et al. introduced improved patch clamp techniques, studies on biophysical features of the smooth muscle cell membrane rapidly progresed. From results with unitary current recording and whole-cell voltage clamp procedures (macroscopic current recording), numerous ion channels have been identified. Some features of the ionic current modified by intracellular Ca, namely the Ca dependent K current will be briefly discussed.

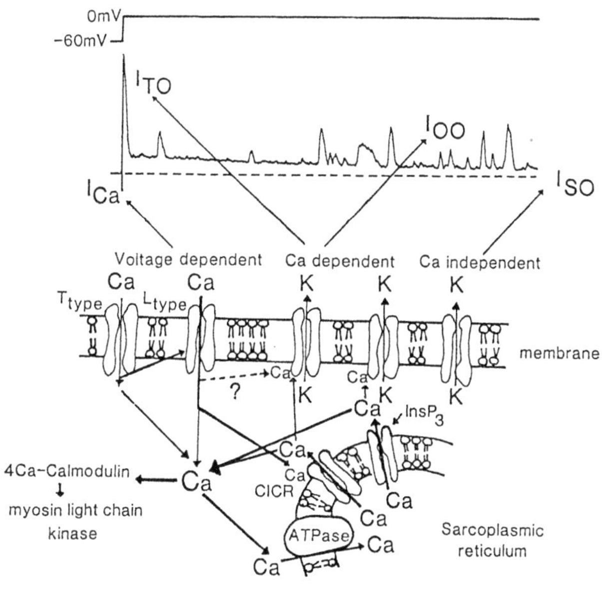

Fig. 1. Typical responses of smooth muscle cell membrane using the whole cell voltage clamp method. Low : schematic model of Ca mobilization in relation to K channel.
I_{Ca}; inward Ca current, I_{TO}; transient outward current, I_{OO}; oscillatory outward current, I_{SO}; sustained outward current.
T and L types; two voltage dependent Ca channels, CICR; Ca induced Ca release mechanism.

When collagenase dispersed single smooth muscle cells of the guinea-pig ileum were depolarized (0 mV) from the holding potential of -60 mV, using the whole cell voltage clamp method, evidence was obtained for an initial inward current, transient outward current (initial outward current), sustained outward current and oscillatory outward current (spontaneous transient outward current) superimposed on the sustained outward current. The initial inward current is a voltage dependent Ca inward current, the sustained outward current is apparently due to activation of the Ca independent K current and the transient and oscillatory outward currents seem to be due to activations of the Ca dependent K current (Benham and Bolton, 1986; Ohya et al., 1986; 1987). Figure 1 shows typical electrical responses of the smooth muscle cell membrane of the guinea-pig ileum recorded using the whole cell voltage clamp method (holding potential of -60 mV and command pulse to 0 mV). These latter two currents (transient and oscillatory outward currents) are closely related to intracellular as well as extracellular Ca. The nature of the Ca dependent K current from many smooth muscle cells observed using various procedures is summarized in Table 1.

Table 1. Ca dependent K current recorded from various smooth muscle cells of different species.

Ca dependent K channel

unitary current			
conductance	solution inside / outside	preparation	reference
200pS	126 / 126 mM KCl	rabbit jejunum mecenteric artery	Benham et al. (1985)
266pS	135 / 135 mM KCl	canine trachea	McCann & Welsh (1986)
273pS	142 / 142 mM KCl	rabbit portal vein	Inoue et al. (1985)
92pS	142 / 142 mM KCl	rabbit portal vein	Inoue et al. (1985)
250pS	130 / 130 mM KCl	toad stomach	Singer & Walsh (1987)
Macroscopic current			
transient outward current (peak outward current)		toad stomach	Walsh & Singer (1981)
		guinea-pig urinary bladder	Klöchner & Isenberg (1985)
		rabbit ileum	Ohya et al. (1987)
		rabbit portal vein	Ohya et al. (1988)
oscillartory outward current (spontaneous transient outward current)		rabbit jejunum	Benham & Bolton (1986)
		rabbit ear artery	
		rabbit ileum	Ohya et al. (1987)
		rabbit portal vein	Ohya et al. (1988)
extracellular Ca dependent K channel			
180pS	142 / 142 mM KCl	rabbit portal vein	Inoue et al. (1986)

In dispersed smooth muscle cells of the rabbit portal vein, the frequency of the oscillatory outward current observed with long depolarization was increased, in a voltage dependent (above -40 mV) manner and also in an extracellular Ca dependent manner. However, reduction in the concentration of free Ca to zero did not block generation of the oscillatory outward current but did block generation of the transient outward current. By reducing the extracellular Ca concentration, the inhibition of the small amplitude rather than that of the large amplitude of the oscillatory outward current occurred, as estimated from the amplitude histograms obtained with various concentrations of Ca. When the relationship between the relative amount of the oscillatory outward current (the area of oscillatory outward current generated in 1 min was divided by the value obtained from the same cells in physiological solution) and extracellular Ca were observed, the amount of oscillatory outward current was saturated in the concentration of above 1 mM (Ohya et al., 1986).

After application of low concentrations (1 -10 nM) of A23187, a Ca ionophore, both transient and oscillatory outward currents ceased, with no effect on the inward current in the presence of extracellular Ca. Such concentrations of A23187 acted more selectively on the SR rather than the sarcolemma and blocked functions of the SR. This would indicate that transient outward currents have a causal relation with the influx of Ca but prerequisites the preservation of ability to release Ca from the SR. On the other hand, generation of the oscillatory outward current does not require the influx of Ca (Ohya et al., 1986). Thus, it is postulated that the former may be closely related with the Ca-induced Ca release mechanism while the latter may be closely related with releases of Ca from the SR, with a slow time course. The physiological nature of the transient outward current is thought to be related with the falling phase of the action potential and after-hyperpolarization of the membrane. The latter may be related to the rectifying property of the membrane.

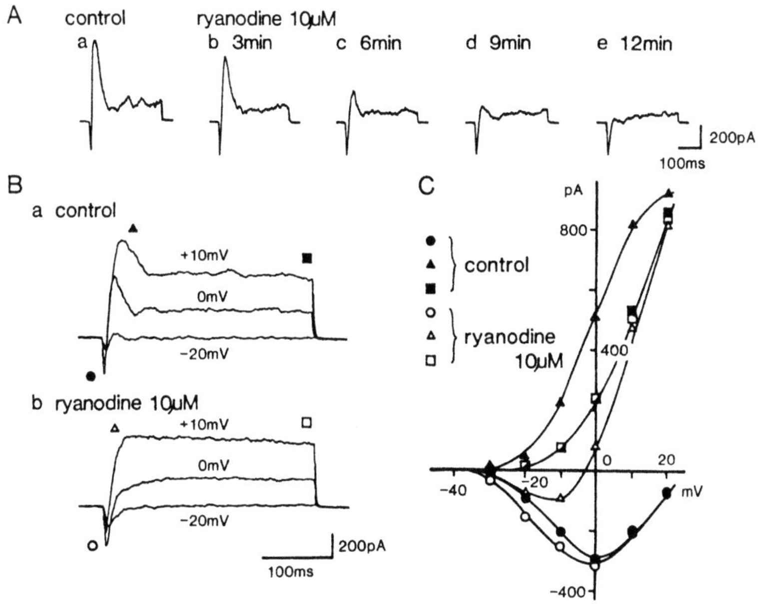

Fig. 2. Effects of ryanodine (10 μM) on the membrane current recorded from the guinea-pig portal vein using whole cell voltage clamp method. The effects of ryanodine were measured 10 min after application.

The events described above concerning the nature of the transient and oscillatory outward currents were confirmed by application of ryanodine, an insecticidal alkaloid from Ryania speciosa Vahl, to smooth muscle cell membranes of the rabbit ileum (longitudinal muscle layer). This drug inhibits or enhances the release of Ca from intracellular store sites

(Sutko and Kenyon, 1983; Hilgemann et al., 1983; Hunter et al., 1983; Sutko et al., 1985; Meissner, 1986; Rousseau et al., 1987; Ito et al., 1986; Hwang and vanBreemen, 1987; McLeod and Bers, 1987). The pharmacological actions of ryanodine on the action potential or voltage dependent Ca current observed in cardiac muscles were also controversial (Mitchell et al.,1984; Marben and Wier, 1985; Nishio et al., 1986; Shattock et al., 1987; Rasmussen, 1987). In dispersed smooth muscle cells (rabbit ileum), ryanodine (below 30 μM) neither modified the basal membrane current recorded at the holding potential (-60 mV) nor the sustained outward current but did inhibit the transient and oscillatory outward currents. However, the concentrations of ryanodine required to inhibit the generation of both currents to half (5.5 and 4.5 μM, respectively) were higher than that observed in skeletal muscles. This action of ryanodine differs from that observed with caffeine, i.e. the action of ryanodine was irreversible while that of caffeine was reversible. Under conditions of pretreatment with low concentrations of A23187, the transient and oscillatory outward currents ceased, and ryanodine slightly enhanced the amplitude of the voltage dependent Ca current evoked low depolarization and inhibited that evoked by higher depolarization. Ryanodine (below 30 μM) had no effects on the single Ca dependent K current recorded using patch clamp procedures but did enhance the amplitude of the single channel current recorded in the cell attached condition. When 1 μM Ca buffered with 10 mM EGTA was superfused on the intracellular membrane side, ryanodine (10 μM) had no effect, a finding indicating that ryanodine does not directly act on the Ca dependent K channel, in the sarcolemma. Figure 2 summarizes the effects of ryanodine (10 μM) on various currents recorded using the whole cell voltage clamp method (cell configuration). A shows changes in the current against the exposure time and B shows changes in the current evoked by three intensities of the depolarization pulse (after 10 min). C shons the current-voltage relationship observed in case of the inward and outward currents (voltage dependent Ca, transient outward and sustained outward currents). Figure 3 shows the effects of ryanodine on the oscillatory outward current (depolarization to 0 mV from the holding potential of -60 mV). Thus, ryanodine reversibly opens the Ca channel in the SR and inhibits the Ca dependent K channel following depletion of the stored Ca (Sakai et al., 1988).

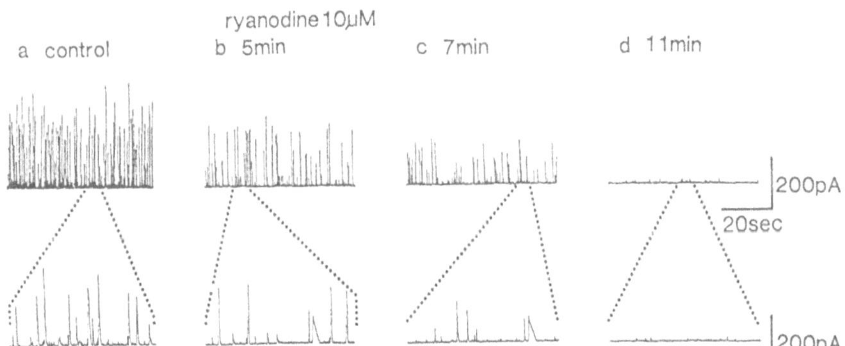

Fig. 3. The effects of ryanodine (10 μM) on the oscillatory outward current (I_{OO}; depolarized to 0 mV from the holding potential of -60 mV) recorded from the rabbit portal vein.

EFFECTS OF INOSITOLPHOSPHATIDES ON THE CALCIUM DEPENDENT POTASSIUM CURRENT

Inositol 1,4,5-trisphosphate

Following intracellular perfusion of Ins-1,4,5-P_3 through the recording pipette, the oscillatory outward current was generated at the holding potential of -60 mV and the frequency was gradually increased. When Ins-1,4,5-P_3 was applied at the membrane potential of -40 mV, the amplitude and frequency of the oscillatory current were further increased. However, basal currents at either -60 or -40 mV were not modified by 10 μM Ins-1,4,5-P_3 (Figure 4).

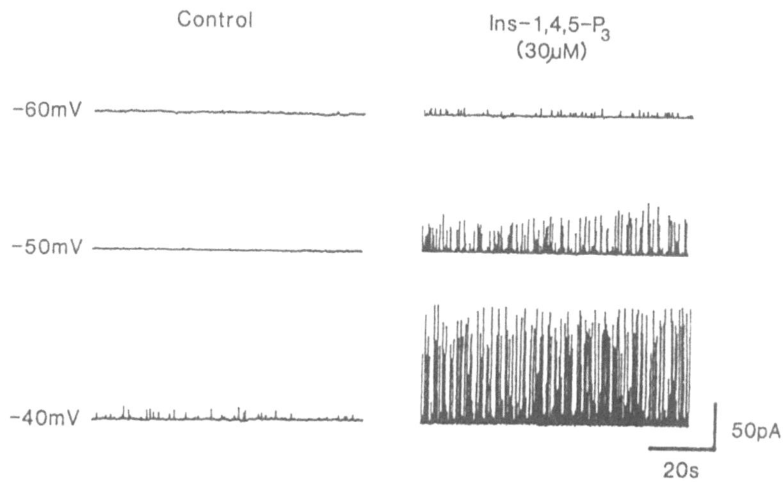

Fig. 4. Effects of $InsP_3$ (30 μM) on the oscillatory outward current (I_{OO}) recorded using the whole cell voltage clamp method at the membrane potential of -60 mV, -50 mV, or -40 mV of a dispersed smooth muscle cell of the guinea-pig portal vein.

The effects of Ins-1,4,5-P_3, (perfused into the cell), on the inward (voltage dependent Ca current) and outward currents (transient, oscillatory and sustained outward currents) were observed. The Ca inward current was slightly reduced but the frequency of the oscillatory outward current was enhanced, with no change in the sustained outward current. When the amplitude histogram of the oscillatory outward current was obtained under conditions of application of different concentrations of Ins-1,4,5-P_3, both small and large amplitudes of the oscillatory outward were increased, but the latter were more enhanced than the former by application of any given concentration of Ins-1,4,5-P_3 (10-30 μM). The oscillatory outward current ceased after application of low concentration of A23817, as was observed with application of various concentrations of Ca. Ins-1,4,5-P_3 (30 μM) was only slightly enhanced after 15 min perfusion of A23187, and after 30 min, Ins-1,4,5-P_3 did not generate the oscillatory outward current. From the histogram, the amplitude of oscillatory outward current could be resolved into more than one Gaussian distribution. When the concentration of Ca was reduced, the small amplitude of currents ceased faster than the large amplitude of the oscillatory outward current. The small amplitude seems to be more sensitive to extracellular Ca of the oscillatory outward current. Ins-1,4,5-P_3 enhanced the frequency more than the amplitude of the oscillatory outward current. Thus, the main action of Ins-1,4,5-P_3 on the smooth muscle cells of the rabbit portal vein is to increase the frequency of release of Ca from the SR rather than the amount of Ca released during each event (Ohya et al., 1988).

In Ca-free solution containing 3 mM EGTA, caffeine produced the oscillatory outward current but the following application of Ins-1,4,5-P_3 did not generate the oscillatory outward current. Much the same effects of caffeine (25 mM) were observed in the Ca free solution, and when caffeine was applied before the application of Ins-1,4,5-P_3, the subsequently applied Ins-1,4,5-P_3 no longer generated the oscillatory outward current and the response of the cells was reversed.

Inositol 1,3,4,5-tetrakisphosphate

When the effects of intracellularly perfused Ins-1,3,4,5-P_4 (10 μM) on dispersed smooth muscle cells of the guinea-pig portal vein were observed with regend to the voltge dependent Ca inward current, and Ca dependent and independent outward currents by depolarization of the membrane to 0 mV from the holding potential of -60 mV in a solution containing 0.5 mM Ca, none of the currents recorded were modified. The depolarization of the membrane to -40 mV, did generate the oscillatory outward current with no remarkable change in other currents. However, when concentrations of Ca were increased to 10 mM, the frequency of the oscillatory outward current was increased by increases in the concentration of Ins-1,3,4,5-P_4 up to 30 μM, and this oscillatory outward current occurred at -60 mV (Ohya et al., 1988).

Higashida and Brown (1986) reported that Ins-1,3,4,5-P_4 can modify the membrane current as that observed on the membrane of NG108-15 cell and that this agent has no potential to release Ca from the SR. However, in smooth muscles, Ins-1,3,4,5-P_4 initiated generations of the oscillatory outward current, in the precence of high concentrations of Ca. Irvine et al. (1986a and b) reported that Ins-1,3,4,5-P_4 is dephosphorylated by Ins-1,3,4,5-P_4 5 phosphatase into Ins-P_3 which is able to release Ca from the SR with a lower potency than that of Ins-1,4,5-P_3. Irvine and Moor (1986) and Morris et al. (1987) reported that Ins-1,3,4,5-P_4 alone does not release Ca from the stored site but together with Ins-1,4,5-P_3 Ca release does occur. In smooth muscle cells, Ins-1,3,4,5-P_4 alone releases Ca from the SR, in high Ca solution. Therefore, Ins-P_3, following dephosophorylation of Ins-1,3,4,5-P_4 by 5 phosphatase, may release Ca from the SR, albeit with a weak potency.

CONCLUSION

Ins-1,4,5-P_3 synthesized by hydrolysis of PI-P_2 releases Ca from the SR, as estimated from the amount of free Ca, the amount of Ca stored in the SR and the Ca-induced contraction. The Ca released by Ins-1,4,5-P_3 facilitates formation of the ternary complex of Ca-calmodulin-protein kinase C complex, phosphorylates 20Kd protein of myosin light chain and activates the cross bridge of actin and myosin. The released Ca also activates the Ca dependent K channel and forms transient and oscillatory outward currents. The physiological function of the oscillatory outward current is unknown. The measurements of the oscillatory outward current may also be used as a tool to estimate the release of Ca from the SR. Furthermore, Ins-1,3,4,5-P_4 synthesized by Ins-1,4,5-P_3 3 kinase may not directly act on the SR but Ins-P_3 dephosphorylated by Ins-1,3,4,5-P_4-5 phosphatase can release Ca from the SR, with a weaker potency than Ins-1,4,5-P_3. Thus, Ins-1,4,5-P_3 may be one main substance which leads to pharmaco-contraction coupling in smooth muscle cells.

ACKNOWLEDGEMENTS

We thank M. Ohara for reading the manuscript. This work was supported in part by a grant of the Ministry of Education Science and Culture, Japan.

REFERENCES

Abdel-Latif, A.A., Akhtar, R.A. and Hawthorne, J.N., 1977, Acetylcholine increases the breakdown of triphosphainositide of rabbit iris muscle prelabelled with [^{32}P] phosphate. Biochem. J., 162: 61-73

Adel-Latif, A.A., 1986, Calcium-mobilizing receptors, polyphosphoinositides, and the generation of second messengers. Pharmacol. Rev. 38: 227-272

Batty, I.R., Nahorski, S.R. and Irvine, R.F., 1985, Rapid formation of inositol 1,3,4,5-tetrakisphosphate following muscarinic receptor stimulation of rat cerebral cortical slices. Biochem. J. 232: 211-215

Benham, C.D. and Bolton, T.B., 1986, Spontaneous transient outward currents in single visceral and vascular smooth muscle cells of the rabbit. J. Physiol. 381:385-406

Benham, C.D., Bolton, T.B., Lang, R.J. and Takewaki, T., 1985, The mechanism of action of Ba2+ and TEA on single Ca2+-activated K+-channels in arterial and intestinal smooth muscle cell membranes. Pflügers Arch. 403: 120-127

Berridge, M. J., 1984, Inositol trisphosphate and diacylglycerol as second messengers. Biochem. J. 220: 345-360

Berridge, M.J. and Irvine, R.F., 1984, Inositol trisphosphate a novel second messenger in cellular signal transduction. Nature, 312:315-321

Burgess, G.M., McKinney, J.S., Irvine, R.F. and Putney, J.W. Jr., 1985, Inositol 1,4,5-trisphosphate and inositol 1,3,4-trisphosphate formation in Ca2+-mobilizing-hormone-activated cells. Biochem. J. 232:237-243

Downes, C.P. and Michell, R.H. Inositol phosphlipid breakdown as a receptor-controlled generator of second messengers. In Molecular Mechanisms of Transmembrane Signalling. (ed. P. Cohen and M.D. Houslay) pp 3-56, Amsterdam, Elsevier.

Hamill, O.P., Marty, A., Neher, E., Sakmam, B. and Sigworth, F.J., 1981, Improved Patch-Clamp Techniques for High-Resolution Current Recording from cells and Cell-Free Membrane Patches. Pflügers Arch. 391:85-100

Hashimoto, T., Hirata, M., Itoh, T., Kanmura, Y. and Kuriyama, H., 1986, Inositol 1,4,5-trisphosphate activates pharmacomechanical coupling in smooth muscle of the rabbit mesenteric artery. J. Physiol. 370:605-618

Heslop, J.P., Irvine, R.F., Tashjian, Jr. A.H. and Berridge, M.J., 1985, Short communication inositol tetrakis- and pentakisphosphates in GH4 cells. J. Exp. Biol. 119:395-401

Higashida, H. and Brown, D.A., 1986a, Membrane current responses to intracellular injections of inositol 1,3,4-trisphosphate in NG 108-15 hybrid cells. FEBS letters 208:283-286

Higashida, H. Brown, D.A., 1986b, Two polyphosphatide inositide metabolites control two K+ currents in a neuronal cell. Nature 323:333-335

Hilgemann, D.W., Delay, M. and Langer, G.A., 1983, Activation-dependent cumulative depletions of extracellular free calcium in guinea pig atrium measured with antipyralazo III and tetramethylmurexide. Circ. Res. 53:779-793

Hirata, M., Suematsu, E., Hashimoto, T., Hamachi, T. and Koga, T., 1984, Release of Ca2+ from a non-mitochondrial store site in peritoneal

macrophages treated with saponin by inositol 1,4,5-trisphosphate. Biochem. J. 223:229-236
Hokin, M.R. and Hokin, L.F., 1953, Enzyme secretion and the incorporation of P32 into phospholipids of pancreas slices. J. Biol. Chem. 203:967-977
Hunter, D.R., Haworth, R.A. and Berkoff, H.A., 1983, Modulation of cellular calcium stores in the perfused rat heart by isoproterenol and ryanodine. Circ. Res. 53:703-712
Hwang, K.S. and van Breemen, C., 1987, Ryanodine modulation of 45Ca efflnx and tension in rabbit aortic smooth muscle. Pflügers Arch. Eur. J. Physiol. 408:343-350
Inoue, R., Kitamura, K. and Kuriyama H., 1985, Two Ca-dependent K-channels classified by the application of tetraethylammonium distribute to smooth muscle membranes of the rabbit portal vein. Pflügers Arch., 405: 173-179
Inoue, R., Okabe, K., Kitamura, K. and Kuriyama, H., 1986, A newly identified Ca2+ dependent K+ channel in the smooth muscle membrane of single cells dispersed from the rabbit portal vein. Pflügers Arch., 406; 138-143
Irvine, R.F., Auggard, E.E., Letcher, A.J. and Downess, C.P., 1985, Metabolism of inositol 1,4,5-trisphosphate and inositol 1,3,4-trisphosphate in rat parotid glands. Biochem. J. 229 505-511
Irvine, R.F., Letcher, A.J., Heslop, J.P. and Berridge, M.J., 1986a, The inositol tris/tetrakis phosphate pathway-demonstration of Ins(1,4,5)P3 3 kinase activity in animal tissues. Nature 320:631-634
Irvine, R.F., Letcher, A.J., Lander, D.J. and Berridge M.J., 1986b, Specificity of inositol phosphate-stimulated Ca2+ mobilization from Swiss-mouse 3T3 cells. Biochem. J. 240: 301-304
Irvine, R.F., Letcher, A.J., Lander, D.J. and Downes, C.P., 1984, Inositol trisphosphates in carbachol-stimulated rat parotid glands. Biochem. J. 223:237-243
Irvine, R.F., and Moor, R.M., 1986, Micro-injection of inositol 1,3,4,5-tetrakisphosphate activates sea urchin eggs by a mechanism dependent on external Ca2+. Biochem. J. 240:917-920
Irvine, R.F. and Moor, R.M., 1987, Inositol (1,3,4,5) tetrakisphosphate-induced activation of sea urchin inositol trisphosphate. Biochem. Biophys. Res. Commun. 146:284-290
Ito, K., Takakura, S., Sato, K. and Sutko, J.L., 1986, Ryanodine inhibits the release of calcium from intracellular stores in guinea pig aortic smooth muscle. Circ. Res. 58:730-734
Itoh, T., Kubota, Y. and Kuriyama, H., 1988, Effects of a phorbol ester on acetylcholine-induced Ca2+ mobilization and contraction in the porcine coronary artery. J. Physiol. 397:401-419
Itoh, T., Kanmura, Y., Kuriyama, H. and Sumimoto, K., 1987, A phorbol ester has dual actions on the mechanical response in the rabbit mesenteric and porcine coronary arteries. J. Physiol. 375:515-534
Joseph, S.K., Williams, R.J., Corkey, B.E., Matschinsky, F.M. and Williamson, J.R., 1984, The effect of inositol trisphosphate on Ca2+ fluxes in insulin-secreting tumor cells. J. Biol. Chem. 259:12952-12955
Klöckner, U. and Isenberg, G., 1985, Action potentials and net membrane currents of isolated smooth muscle cells (urinary bladder of the guinea-pig). Pflügers Arch., 405: 329-339
Lindon, J.C., Baber, D.J., Farrant, R.D. and Williams, J. M., 1986, 1H, 13C and 31P n.m.r. spectra and molecular conformation of myo-inositol 1,4,5-trisphosphate. Biochem. J. 233:275-277
MacLeod, K.T. and Bers, D.M., 1987, Effects of rest duration and ryanodine on changes of extracellular [Ca] in cardiac muscle from rabbits. Am. J. Physiol. 253:C398-C407
Marben, E. and Wier, W.G., 1985, Ryanodine as a tool to determine the contributions of calcium entry and calcium release to the calcium

transient and contraction of cardiac purkinje fibers. Circ. Res. 56:133-138
McCann, J.D. and Welsh, M.J., 1986, Calcium-activated potassium channels in canine airway smooth muscle. J. Physiol., 372: 113-127
Meissner, G., 1986, Ryanodine activation and inhibition of the Ca2+ release channel of sarcoplasmic reticulum. (1986) J. Biol. Chem. 261:6300-6306
Michell, B., 1986, A second messenger function for inositol tetrakisphosphate. Nature 324:613
Michell, R.H., 1975, Inositol phospholipids and cell surface receptor function. Biochim. Biophys. Acta 415:81-147
Michell, R.H., 1982, Is phosphatidylinositol really out of the calcium gate? Nature 296:492-493
Mitchell, M.R., Powell, T., Terrar, D.A. and Twist, V.W., 1984, Ryanodine prolongs Ca currents while suppressing contraction in rat ventricular muscle cells. Br. J. Pharmacol. 81:13-15
Morris, A.P., Gallacher, D.V., Irvine, R.F. and Petersen, O.H., 1987, Synergism of inositol trisphosphate and tetrakisphosphate in activating Ca2+-dependent K+ channels. Nature 330:1653-1655
Nishio, M., Kigoshi, S. and Muramasu, I., 1986, Ryanodine has no effect on the Ca current ventricular cells of guinea-pig. Eur. J. Pharmacol. 124:353-356
Nishizuka, Y., 1984, The role of protein kinase C in cell surface signal transduction and tumor promotion. Nature (London) 308:693-698
Nishizuka, Y., 1986, Studies and perspectives of protein kinase C. Science 233:305-312
Ohya, Y., Kitamura, K. and Kuriyama, H., 1987, Cellular calcium regulates outward currents in rabbit intestinal smooth muscle cell. Am. J. Physiol., 252:C401-C410
Ohya, Y., Terada, K., Kitamura, K. and Kuriyama, H., 1986, Membrane currents recorded from a fragment of rabbit intestinal smooth muscle cell. Am. J. Physiol. 251:C335-C346
Ohya, Y., Terada, K., Yamaguchi, K., Inoue, R., Okabe, K., Kitamura, K., Hirata, M. and Kuriyama, H., 1988, Effects of inositol phosphates on the membrane activity of smooth muscle cells of the rabbit portal vein. Pflügers Arch. Eur. J. Physiol., (in press).
Rasmussen Jr., C.A.F., Sutko, J.L. and Barry, W.H., 1987, Effects of ryanodine and caffeine on contractility, membrane voltage, and calcium exchange in cultured heart cells. Circ. Res. 60:495-504
Rousseau, E., Smith, J.S. and Meissner, G., 1987, Ryanodine modifies conductance and gating behavior of single Ca2+ release channel. Am. J. Physiol. 253:C364-C368
Sakai, T., Terada, K., Kitamura, K. and Kuriyama, H., 1988, Ryanodine inhibits the Ca dependent K current after depletion of Ca stored in smooth muscle cells of the rabbit ileal longitudinal muscle. Br. J. Pharmacol. in press
Sasaguri, T., Hirata, M. and Kuriyama, H., 1985, Dependence on Ca2+ of the activities of phosphatidylinositol 4,5-bisphosphate phosphodiesterase and inositol 1,4,5-trisphosphate in smooth muscles of the porcine coronary artery. Biochem. J. 231:497-503
Shattock, M.J., Warner, K.C., Tidball, J.G. and Bers, D.M., 1987, Two different electrophysiological responses to ryanodine: Evidence for two populations of muscles isolated from the rabbit right ventricle. J. Mol. Cell. Cardiol. 19:751-762
Singer, J.J. and Walsh, Jr., J.V., 1987, Characterization of calcium-activated potassium channels in single smooth muscle cells using the patch-clamp technique. Phlügers Arch., 408: 98-111
Somlyo, A.V., Bond, M., Somlyo, A.P. and Scarpa, A., 1985, Inositol trisphosphate-induced calcium release and contraction in vascular smooth muscle. Circ. Res. 57:497-507
Streb, H., Irvine, R.F., Berridge, M.J. and Schulz, I., 1983, Release of

Ca2+ from a non mitochodrial intracellular store in pancreatic acinar cells by inositol-1,4,5-trisphosphate. Nature 306:67-69

Suematsu, E., Hirata, M., Hashimoto, T. and Kuriyama, H., 1984, Inositol 1,4,5-trisphosphate releases Ca^{2+} from intracellular store sites in skinned single cells of porcine coronary artery. Biochem. Biophys. Res. Commun. 120:481-485

Sutko, J.L., Ito, K. and Kenyon, J.L., 1985, Ryanodine: A modifier of sarcoplasmic reticulum calcium release in striated muscle. Fed. Proc. 44:2984-2988

Sutko, J.L. and Kenyon, J.L., 1983, Ryanodine modification of cardiac muscle responses to potassium-free solutions. J. Gen. Physiol. 82:385-404

Tomlinson, R.V. and Ballou, C.E., 1961, Complete characterization of the Myo-inositol polyphosphates from beef brain phosphoinositide. J. Biol. Chem. 236:1902-1906.

Volpe, P., Giomanni, S., Di Virgilio, F. and Pozzan, T., 1985, Inositol 1,4,5-trisphosphate induces calcium release from sarcoplasmic reticulum of skeletal muscle. Nature 316:347-349

Walker, J.W., Somlyo, A.V., Goldman, Y.E., Somlyo, A.P. and Trentham, D.R., 1987, Kinetics of smooth and skeletal muscle activation by laser pulse photolysis of caged inositol 1,4,5-trisphosphate. Nature 327:249-252

Walsh, Jr., J.V. and Singer, J.J., 1981, Voltage clamp of single freshly dissociated smooth muscle cells: Current-voltage relationships for three currents. Pflügers Arch., 390: 207-210

Yamamoto, H. and Van Breemen, C., 1985, Inositol-1,4,5-trisphosphate releases calcium from skinned cultured smooth muscle cells. Biochem. Biophys. Res. Commun. 130:270-274

THE CALCIUM PUMP OF THE PLASMA MEMBRANE: STRUCTURE-FUNCTION RELATIONSHIPS

Ernesto Carafoli*, Anil K. Verma**, Peter James*, Emanuel Strehler*, and John T. Penniston**

* Laboratory of Biochemistry
Swiss Federal Institute of Technology (ETH)
8092 Zurich (Switzerland)

** Department of Biochemistry and Molecular Biology
Mayo Clinic
Rochester, Minnesota 55905 (U.S.A.)

INTRODUCTION

The Ca pump of the plasma membrane is an ATPase of the P-class (1-2), i.e.; it forms a phosphorylated intermediate during the reaction cycle and is inhibited by low concentrations of vanadate (see 3-4 for comprehensive reviews). Calmodulin stimulates the ATPase by direct interaction, shifting the Ca affinity of the enzyme from the normal K_m value of between 10 and 20 µM to values around 0.5 µM. The direct interaction with calmodulin has been exploited to purify the enzyme to essential homogeneity on a calmodulin affinity chromatography column (5). The purified enzyme has been shown to be fully competent functionally: it has the expected high affinity for Ca in the presence of calmodulin, it is sensitive to vanadate, and can be reconstituted as an ATP-dependent Ca-transporting system in liposomes. Work on the purified enzyme has permitted to establish that its Ca/ATP-stoichiometry is 1, and that protons are obligatorily exchanged for Ca in the transport reaction. Table 1 offers a summary of the properties of the ATPase: A comprehensive review on the properties of the purified enzyme has appeared in 1982 (6). Most of the work on the pump has so far been carried out on erythrocytes, but the enzyme has been detected, with essentially the same properties, in all plasma membranes so far examined, with the possible exception of liver. One interesting property of the pump, first established on heart plasma membranes(7) but later extended to the enzyme purified from heart sarcolemma and from erythrocytes (8) is the activation by a phosphorylation reaction mediated by the cAMP-dependent protein kinase.

Efforts aimed at establishing the primary structure of the pump were initiated in our Laboratory about two years ago using a combination of DNA cloning and protein chemistry methods. Prior to this, extensive proteolytic work on the purified erythrocyte enzyme had led to the proposal of a model for the architecture of the protein in the membrane in which the calmodulin binding domain was tentatively placed in a C-terminal region of minimal M_r about 9,000 (9-10). Later work combining CNBr

TABLE 1

A summary of the properties of the plasma membrane Ca pumping ATPase

Pump type	P-class
Molecular weight	about 140,000
Vanadate sensitivity	half maximal inhibition 3 μM
Phosphorylated intermediate	aspartyl phosphate
Charge balance	proton exchanger. Electroneutral?
Ca-ATP stoichiometry	1:1
Ca affinity	10-20μM in low affinity state less than 0.5μM in high affinity state
Calmodulin	induces high affinity state
Purification	on calmodulin columns

cleavage and specific labelling of the calmodulin binding domain of the purified erythrocyte enzyme (11) has permitted its isolation as a peptide of about 30 amino acids and has confirmed its location near the C-terminus of the enzyme. The sequence of the labelled calmodulin binding peptide is reminiscent of that of the calmodulin binding domains of calmodulin modulated proteins and calmodulin binding venoms. The proteolytic work on the purified erythrocyte enzyme has also led to the isolation and sequencing of the fluorescein isothyocyanate (FITC) binding site, believed to be a part of the ATP-binding region, (1) and of the region containing the aspartic acid residue where the acylphosphate intermediate is formed (13).

A number of other small M_r tryptic fragments have also been isolated and sequenced. Suitable oligodeoxy-nucleotide probes have then been synthesized and used to screen a human teratoma λgt10 cDNA library, eventually resulting in the solution of the complete primary structure of the human pump (14). Independent cloning work using a rat brain cDNA library has in the meantime led others to deduce the amino acid sequences of two isoforms of the rat brain pump (15). In this report a succinct account will be presented of the work on the primary structure of the pump carried out in our Laboratory, and on the arrangement of some of the functional domains of the enzyme along its sequence.

PRIMARY STRUCTURE OF THE HUMAN PLASMA MEMBRANE Ca PUMP

The two tryptic sequences from the purified erythrocyte pump initially used to synthesize oligonucleotide probes were FAMGIA and MTHPEF. The 17mer oligonucleotides were 32P-end-labelled, mixed, and used to screen about 1 x 10^6 plaques of a human teratoma cDNA library constructed in λgt10 (kindly donated by Drs. Maxine Singer and Jacek Skowronski of the National Cancer Institute, Bethesda, MD, USA). The initial screening yielded three positive clones, but only one hybridized to both oligonucleotide probes. This clone (t19c) was purified and analyzed by restriction mapping and nucleotide sequencing: it was about 1.2 kb long and contained two stretches that matched with the oligonucleotide probes in the same (and only) open reading frame that encompassed its entire sequence (Figure 1). Subfragments of the clone insert were used to rescreen the cDNA library, yielding several additional clones, the longest being about 3.4kb long (t6). The reading frame was found to remain open up to the 5'-end of cDNA t6, indicating that the sequence coding for the N terminus of the ATPase had not been reached. The cDNA library was then rescreened with an oligonucleotide corresponding to the 5'-end sequence of cDNA clone t6 and with a mixture of oligonucleotides corresponding to a peptide of the erythrocyte pump (EGDFGC) not contained in the translated sequence of clone t6 and thus assumed to belong to the N-terminal portion of the pump. The rescreening produced clone t8.1, which overlapped with clone t6 at its 3'-end and extended it by more than one kb towards its 5'-end. It contained a region corresponding to the oligonucleotide probe made on the basis of the sequence EGDFGC in 14 of the first 15 positions (in the clone this region read DGDFGI). The combined nucleotide sequence of the three clones $t19_c$, t6, and t8.1 is presented in Figure 2. The reading frame as predetermined by the sequence of clone t6 continued into clone t8.1 and most likely began at the first in frame Met residue (nucleotide positions 1-3 in Figure 2). However, a second M-codon is also present downstream of the last stop in the 5'-untranslated region (Met-4), and in fact the nucleotide sequence around Met-4 (CATGG) resembles the consensus sequence for mRNA translational start sequences more closely than that around Met-1. Met-4 may thus well be the N-terminal residue of

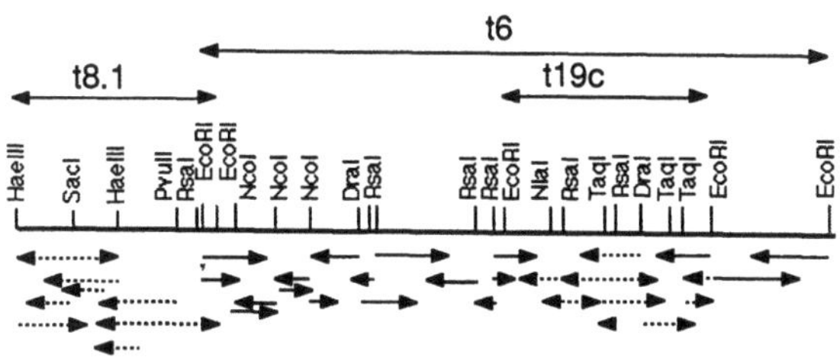

Fig. 1. Partial restriction map and sequencing strategy of cDNA clones t19c, t6, t8.1 (see text). The arrows indicated the direction and limits of inidividual sequence readings. The dashed arrows indicate sequencing on clones t19c and t8.1. The restriction sites indicated by a dotted line refer to EcoRI sites created by linkers (Taken from Ref. 14)

the pump. Attempts to establish the N-terminal sequence of the pump by chemical sequencing of two independent preparations of the erythrocyte pump after de-blocking the acetylated N-terminus with a N-acylaminoacyl peptide hydrolase yielded the sequence (M)ANGTVFI/LTVE, suggesting that the N-terminal residue of the pump may indeed by Met-4. Sequence analysis showed that several of the cDNA clones ended with a poly-A tail, indicating that they contained the 3'-terminal sequence of the corresponding RNA (see also Figure 2). The open reading frame ends at position 3660 with a TGA stop codon, followed by a 3'-untranslated stretch of 554 bp and a poly-A tail.

The difference between the deduced amino acid sequence of teratoma clone t8.1 and of the erythrocyte pump at the N-terminus shows that at least two isoforms of the pump exist in humans. In fact, although most of the sequenced peptides from the erythrocyte enzyme are identical to the corresponding deduced sequences in the teratoma clone(s), significant divergences also exist. Analysis of the sequence of about 200 amino acids from the erythrocyte enzyme shows only about 85% identity with the corresponding deduced sequences of the teratoma clone(s). However, the

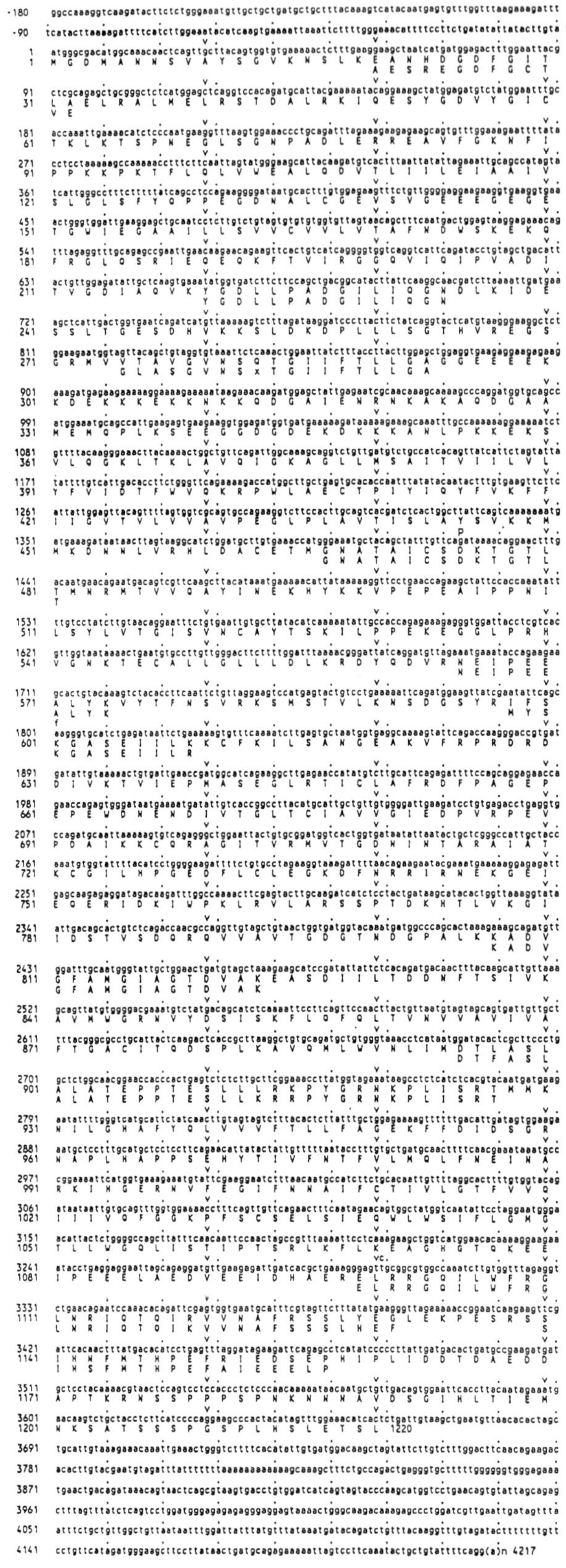

Fig. 2. Cloned nucleotide sequence of the plasma membrane Ca pump with translation and aligned peptides. Each segment of the sequence is presented in four rows: 1) pointers, 2) nucleotide sequence, 3) amino acid sequence, 4) aligned tryptic peptides from the purified erythrocyte pump. The pointers now contain a period at every 10th nucleotide and an arrowhead pointing downward every 10th amino acid. (p) indicates the site of phosphoenzyme formation, (f) indicates the FITC binding residue. The erythrocyte peptides obtained with trypsin are 120-234, 565-574, 598-609, 807-823, 1140-1157. Those produced by CNBr are 20-32, 275-293, 467-481, 895-927, and 1099-1131. (Taken from Ref. 14)

sequences around crucial domains of the pump (e.g., the ATP (FITC) binding domain, the site of aspartyl-phosphate formation, and the calmodulin binding domain) are identical, or nearly so. Figure 3 shows the translated amino acid sequence of the clone, and identifies residues and domains of critical importance to the ATPase protein. The hydrophobic domains spanning the membrane can be identified only tentatively: hydrophobicity plots using a window size of 18 and the Kyte-Doolittle algorithm (16) have identified ten, in analogy with the conclusion of MacLennan et.al. (17) on the sarcoplasmic reticulum Ca-ATPase. They correspond to residues 105-127, 156-173, 372-394, 418-438, 855-878, 885-904, 929-953, 970-990, 1004-1026, 1040-1060. Interestingly, on comparing the distribution of the putative hydrophobic membrane spanning domains in a number of other ion pumps of the P-class it appears evident that the domains tends to be "clustered" in the vicinity of the C-terminus and in the N-terminal region. A relatively large middle portion of the enzyme molecule is invariably free of hydrophobic domain (17-20). It may also be added in this connection that the overall sequence identity of the plasma membrane Ca pump with

```
   1 M G D M A N N S V A Y S G V K N S L K E A N H D G D F G I T L A E L R A L M E L
  41 R S T D A L R K I Q E S Y G D V Y G I C T K L K T S P N E G L S G N P A D L E R
  81 R E A V F K K N F I P P K K P K T F L Q L V W E A L Q D V T L I I L E I A A I V
 121 S L G L S F Y Q P P E G D N A L C G E V S V G E E E G E G E T G W I E G A A I L
 161 L S V V C V V L V T A F N D W S K E K Q F R G L Q S R I E Q E Q K F T V I R G G
 201 Q V I Q I P V A D I T V G D I A Q V K Y G D L L P A D G I L I Q G N D L K I D E
 244 S S L T G E S D H V K K S L D K D P L L L S G T H V R E G S G R M V V T A V G V
 281 N S Q T G I I F T L L G A G G E E E E K K D E K K K E K K N K K Q D G A I E N R
 321 N K A K A Q D G A A M E M Q P L K S E E G G D G D E K D K K K A N L P K K E K S
 361 V L Q G K L T K L A V Q I G K A G L L M S A I T V I I L V L Y F V I D T F W V Q
 401 K R P W L A E C T P I Y I Q Y F V K F F I I G V T V L V V A V P E G L P L A V T
 441 I S L A Y S V K K M M K D N N L V R H L D A C E T M G N A T A I C S D K T G T L
 481 T M N R M T V V Q A Y I N E K H Y K K V P E P E A I P P N I L S Y L V T G I S V
 521 N V A Y T S K I L P P E K E G G L P R H V G N K T E C A L L G L L L D L K R D Y
 561 Q D V R N E I P E E A L Y K V Y T F N S V R K S M S T V L K N S D G S Y R I F S
 601 K G A S E I I L K K C F K I L S A N G E A K V F R P R D R D D I V K T V I E P M
 641 A S E G L R T I C L A F R D F P A G E P E P E W D N E N D I V T G L T C I A V V
 681 G I E D P V R P E V P D A I K K C Q R A G I T V R M V T G D N I N T A R A I A T
 721 K C G I L H P G E D F L C L E G K D F N R R I R N E K G E I E Q E R I D K I W P
 761 K L R V L A R S S P T D K H T L V K G I I D S T V S D Q R Q V V A V T G D G T N
 801 D G P A L K K A D V G F A M G I A G T D V A K E A S D I I L T D D N F T S I V K
 841 A V M W G R N V Y D S I S K F L Q F Q L T V N V V A V I V A F T G A C I T Q D S
 881 P L K A V Q M L W V N L I M D T L A S L A L A T E P P T E S L L L R K P Y G R N
 921 K P L I S R T M M K N I L G H A F Y Q L V V V F T L L F A G E K F F D I D S G R
 961 N A P L H A P P S E H Y T I V F N T F V L M Q L F N E I N A R K I H G E R N V F
1001 E G I F N N A I F C T I V L G T F V V Q I I I V Q F G G K P F S C S E L S I E Q
1041 W L W S I F L G M G T L L W G Q L I S T I P T S R L K F L K E A G H G T Q K E E
1081 I P E E E L A E D V E E I D H A E R E L R R G Q I L W F R G L N R I Q T Q I R V
1121 V N A F R S S L Y E G L E K P E S R S S I H N F M T H P E F R I E D S E P H I P
1161 L I D D T D A E D D A P T K R N S S P P P S P N K N N N A V D S G I H L T I E M
1201 N K S A T S S S P G S P L H S L E T S L 1220
```

Fig. 3. Amino acid sequence of the plasma membrane Ca-pump. The boxed domains represent the putative trans-membrane helices. The aspartic acid where the phosphoenzyme is formed, and the lysine which binds the ATP analog FITC are circled. The calmodulin binding domain is underlined with a solid line, and the negatively charged domain N-terminal to it (the putative Ca-binding site) is underlined with a dashed line.

the other ion transporters of the P-type is rather low, but is striking in the regions connected with the catalytic function. It is, also, very high in the "hinge" region in the protruding hydrophilic portion of the ATPase which is assumed to permit intramolecular motion necessary to bring the ATP (FITC) binding site and the critical aspartic acid residue (normally spaced wide apart) close to each other.

The FITC (ATP) binding residue (12) is identified in Figure 3 as Lys 601, and the residue forming the phosphoenzyme as Asp 475. The Figure also shows that the calmodulin binding domain (11) is indeed located next to the C-terminus of the ATPase as tentatively proposed earlier (10, see above), and identifies, immediately N-terminal to the calmodulin binding domain, a 20 amino acid stretch which is particularly rich in glutamic and aspartic acid residues (1079-1099). It is tempting to propose that this acidic domain functions as a potential Ca binding site. In the absence of calmodulin it would, however, be Ca-free, and bound to the adjoining basic calmodulin binding domain. Calmodulin would complex its basically charged binding domain, removing it from the acidic sequence which would thus be free to bind Ca (Figure 4). Recent work in our Laboratory using synthetic peptides has provided some initial indications in favour of the model, i.e., it has shown that the synthetic basic calmodulin binding domain interacts with calmodulin in a Ca-dependent manner. It has further shown that the putative putative Ca binding domain interacts very strongly with the basic calmodulin binding domain in the absence of Ca. Work now in progress attempts to demonstrate that, a), the synthetic acidic domain

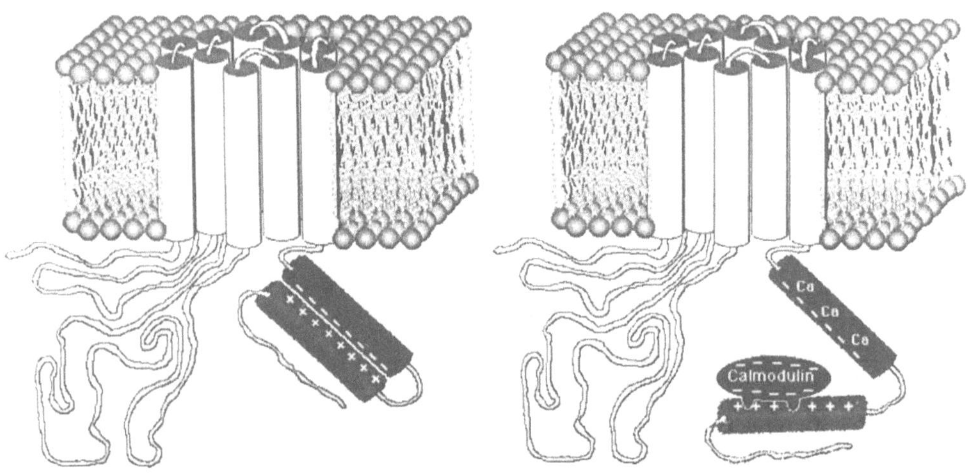

Fig. 4. A model of the interaction of calmodulin with the plasma membrane Ca pump, and of its effect on the binding of Ca. The hydrophilic protruding C-terminal domain of the ATPase contains the positively charged calmodulin binding domain (the cylinder identified with (+) signs) and, N-terminal to it, a domain rich in acidic residues (the cylinder identified with the (-)signs). The two domains interact in the non-activated enzyme (absence of Ca-calmodulin, left portion of the Figure). In the presence of Ca calmodulin swings the calmodulin binding domain away from the acidic sequence, which is now free to bind Ca. (Right protein of the Figure)

indeed binds Ca with the appropriate high affinity and, b), that in the presence of Ca calmodulin complexes the basic calmodulin binding domain away from the synthetic putative acidic Ca binding domain.

The matter of the binding of Ca by the ATPase has another aspect of potential interest. When the translated sequence of the entire teratoma clone was searched for similarities with an average Ca-binding loop such as existing in EF-hand structure of Ca-binding proteins (21) using the University of Wisconsin sequence analysis software programs (Table 2), two

TABLE 2

Using the University of Wisconsin Genetics Computer Group program PROFILE, an "average" EF-hand was constructed using 25 known EF-hans. This averaged EF-hand was then used to search the data base, and the plasma membrane ATPase sequence discussed in this contribution, for homologies. A representative set of proteins searched is given above for comparison to the plasma membrane ATPase. The proteins are given below.

Protein	Score
Calmodulin (a)	11.20
Calmodulin (b)	9.60
Calpain (c)	8.60
Ca^{2+}-ATPase of plasma membrane (residues 22-33)	7.60
Ca2+-ATPase of plasma membrane (residues 310-321)	7.40
Calcium binding protein (d)	5.60
Trypsin inhibitor (e)	4.20
Histone H3 (human)	2.10

(a)Human brain calmodulin (22), (b)Streptomyces erythraeus calmodulin-like protein (23), (c)Chicken Ca activated neutral protease (24), (d) Rat intestinal mucosa steroid-induced Ca binding protein (25),(e) Maize trypsin inhibitor (26),(f) Human H3 histone (27)

regions were identified (residues 22-33 and 310-321) which scored rather high, in fact even higher than some classical EF-hand proteins (14). They are visually represented in the scheme of Figure 5, which offers a general summary of the molecular architecture of the pump in the plasma membrane. Both regions show differences from the classical Ca-coordinating loops of EF-hand structures (see 14 for a discussion of the matter), and, in addition, the flanking sequences of the regions do not show the strong propensity for α helix formation expected of true EF-hand structures. For these reasons great caution is necessary at the moment in postulating a function in Ca-binding for the two EF-hand-type regions. The possibility that they may play a role, perhaps a regulatory one, in the process appears nevertheless stimulating.

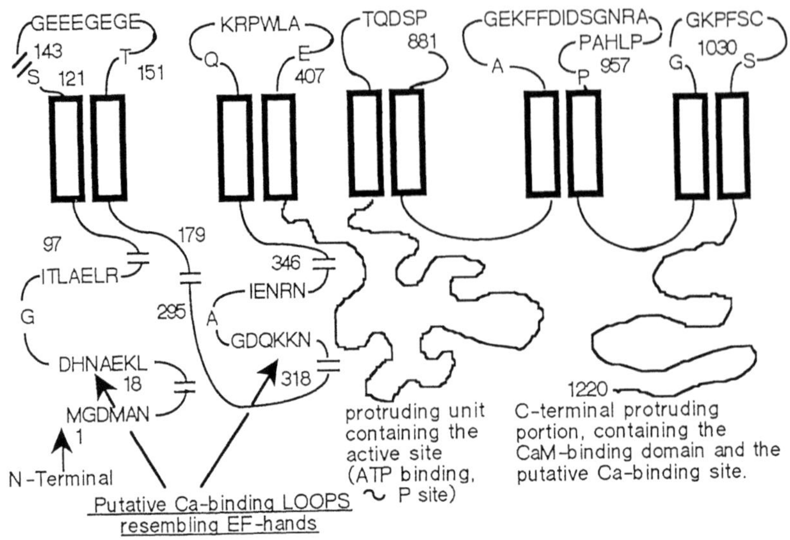

Fig. 5. A scheme of the architecture of the Ca pump in the plasma membrane. The boxes represent the ten putative transmembrane domains. The small stretches of protein protruding on the outside are shown in full, whereas the large hydrophilic domains protruding into the cytosolic space are not. The cytosolic domain between transmembrane helices 3 and 4 contains more that 400 residues, including the catalytic site(s) and the "hinge" region (see text). The two loops resembling Ca-binding EF-hands are formed by residues 22 through 33 and 310 through 321. The numbers in the Figure indicate the approximate position of the domains in the structure.

ACKNOWLEDGEMENTS

The original work described in this contribution has been made possible by the financial contribution of the Swiss Nationalfonds (85.159.0.84 and 3.531-0-86) and of the National Institutes of Health (GM28835).

REFERENCES

1. P. L. Pedersen and E. Carafoli, Ion-motive ATPases, Trends in Biochem. Sciences. 12:146(1987)
2. P. L. Pedersen and E. Carafoli, Ion-motive ATPases, Trends in Biochem. Sciences. 12:186 (1987)
3. H. J. Schatzmann, The calcium pump of erythrocytes and other animal cells, in: "Membrane Transport of Calcium," E. Carafoli, ed., Acad. Press, London (1982)
4. J. T. Penniston, Plasma membrane Ca^{2+} ATPase as active Ca^{2+} pumps, in: "Calcium and Cell Function," W. Y. Cheung, ed., Acad. Press, New York (1984)
5. V. Niggli, J. T. Penniston, and E. Carafoli, Purification of the (Ca^{2+} + Mg^{2+})-ATPase from human erythrocyte membranes using a calmodulin affinity column, J. Biol. Chem. 254:9955 (1979)
6. E. Carafoli and M. Zurini,, The Ca^{2+}-pumping ATPase of plasma membranes. Purification, reconstitution and properties, Biochim. Biophys. Acta. Rev. in Bioenerg. 683:279 (1982)
7. P. Caroni and E. Carafoli, Regulation of Ca^{2+}-pumping ATPase of heart

sarcolemma by a phosphorylation-dephosphorylation process, J. Biol. Chem. 256:9371 (1981)

8. L. Neyses, L. Reinlib, and E. Carafoli, Phosphorylation of the Ca^{2+} pumping ATPase of heart sarcolemma and erythrocyte plasma membrane by the cAMP-dependent protein kinase, J. Biol. Chem. 260:10283 (1985)

9. M. Zurini, J. Krebs, J. T. Penniston, and E. Carafoli, Controlled proteolysis of the purified Ca^{2+}-ATPase of erythrocyte membrane, J. Biol. Chem. 259:618 (1984)

10. G. Benaim, M. Zurini, and E. Carafoli, Different conformational statesof the purified Ca^{2+} ATPase of the erythrocyte plasma membrane, revealed by controlled Trypsin proteolysis, J. Biol. Chem. 259:8471 (1984)

11. P. James, M. Maeda, R. Fischer, A. K. Verma, J. Krebs, J. T. Penniston, and E. Carafoli, Identification and primary structure of a calmodulin binding domain of the Ca^{2+}-pump of human erythrocytes, J. Biol. Chem. 263:2905 (1988)

12. A. G. Filoteo, J. P. Gorski, and J. T. Penniston, The ATP-binding site of the erythrocyte membrane Ca^{2+}pump, J. Biol. Chem. 262:6526 (1987)

13. P. James, E. Zvaritch, M. Shakhparonov, J. T. Penniston, and E. Carafoli, The amino acid sequence of the phosphorylation domain of the erythrocyte Ca^{2+} ATPase, Biochem. Biophys. Res. Commun. 149:7 (1987)

14. A. K. Verma, A. G. Filoteo, D. R. Stanford, E. D. Wieben, J. T. Penniston, E. E. Strehler, R. Fischer, R. Heim, G. Vogel, S. Mathews, M. A. Strehler-Page, P. James, T. Vorherr, J. Krebs, and E. Carafoli, Complete primary structure of a human plasma membrane Ca^{2+} pump, J. Biol. Chem. 263:in press (1988)

15. G. E. Shull and J. Greeb, Molecular cloning of two isoforms of the plasma membrane Ca^{2+}-transporting ATPase from rat brain, J. Biol. Chem. 263:8646 (1988)

16. J. Kyte and R. F. Doolittle, A simple method for displaying the hydropathic character of a protein, J. Mol. Biol. 157:105 (1982)

17. D. H. MacLennan, C. J. Brandl, B. Korczak, and N. M. Green, Amino acid sequence of a Ca^{2+} + Mg^{2+}-dependent ATPase from rabbit muscle sarcoplasmic reticulum, deduced from its complementary DNA sequence, Nature. 316:696 (1985)

18. K. Kawakami, T. Ohta, H. Nojima, and K. Nagano, Primary structure of the α-subunit of human Na/K-ATPase deduced from cDNA sequence, J. Biochem. (Toyko). 100:389 (1986)

19. K. M. Hager, S. M. Mandala, J. W. Davenport, D. W. Speicher, E. J. Benz, and C.W. Slayman, Amino acid sequence of the plasma membrane ATPase or neurospora crassa:deduction from genomic and cDNA sequences, Proc. Nat. Acad. Sci (USA). 83:7693 (1986)

20. G. E. Shull and J. B. Lingrel, Molecular cloning of the rat stomach (H^{+} + K^{+}) ATPase. J. Biol. Chem. 261:16788 (1986)

21. R. H. Kretsinger, Structure and evolution of calcium modulated proteins, CRC Lit. Rev. Biochem. 8:119 (1980)

22. T. Sasagawa, L. H. Ericsson, K. A. Walsh, W.E. Schreiber, E.H. Fischer, and K. Titani, Complete amino acid sequence of human brain calmodulin, Biochemistry. 21:2565 (1982)

23. D. G. Swan, R. S. Hall, N. Dhillon, and P. F. Leadlay, A bacterial calcium binding protein homologous to calmodulin, Nature. 329:84 (1987)

24. S. Ohno, Y. Emori, S. Imajoh, H. Kawasaki, M. Kisargi, and K. Suzuki, Evolutionary origin of a calcium dependent protease by fusion of genes for a thiol protase and a calcium binding protein, Nature, 312:566 (1984)

25. J. P. McManus, D. C. Watson, and M. Yaguchi, The purification and complete amino acid sequence of the 9000- Mr Ca^{2+}-binding protein

from rat placenta, Biochem. J. 235:585 (1985)
26. K. Hochstrasser, K. Illchmann, and E. Werle, The amino acid sequence of maize trypsin inhibitor, Hoppe Seyler's Z. Physiol. Chemie. 351:721 (1970)
27. F. Marashi, S. Helms, A. Shiels, S. Silverstein, D. Greenspan, G. Stein, and J. Stein, The sequence of human histone 3 as deduced from the cDNA sequence, Biochem. Cell Biol. 64:277(1985)

IDENTIFICATION OF A SYNAPTIC MEMBRANE-LOCALIZED ISOFORM OF THE CALCIUM-PUMPING ATPASE

D.M. Mann, P. Brandt, B. Sisken[1], and T.C. Vanaman

University of Kentucky School of Medicine, Department of Biochemistry and [1]Department of Anatomy, Lexington, KY 40536

INTRODUCTION

Calcium-pumping membrane ATPases play the essential function of terminating Ca^{2+}-dependent responses. Two broad classes of Ca^{2+}-pumping ATPases are now known to exist in animal cells, those of internal membranes (100 kDa) and the plasma membrane enzymes (120-140 kDa). These two families of Ca^{2+}-pumping ATPases differ substantially in molecular, enzymatic and regulatory properties commensurate with their differiential localization and physiological roles. Recent studies in a number of laboratories have provided evidence that multiple isoforms of each type of Ca^{+2}-ATPase are produced in different cell types as products of either differential splicing events or different genes. These sets of isoforms are almost certain to have subtle differences in functional properties related to the requirements of a particular biological setting.

In this report, we present further analyses of an isoform of the plasma membrane enzyme which we have cloned from brain cDNA libraries (Brandt et al., 1988). It is clearly unique from 2 other isoforms also cloned from mammalian brain (Shull and Greeb, 1988) as well as from one cloned from a human teratoma library (Verma, et al., 1988). All four isoforms cloned to date share substantial sequence identity throughout most of their structures but differ prominently in the C-terminal region involved in regulation of activity. We report here studies with monospecific antibodies prepared to a synthetic peptide of our isoform which indicate its presence as a unique isoform in synaptic membranes, where it may be the primary Ca^{2+} pumping activity. Further comparison of its structure to other plasma membrane isoforms shows that, while all are highly related, the synaptic membrane isoform is clearly the product of a separate gene.

```
BCAATP4      1 eFGGKPFSCt kLtLsQWfWC lFIGiGELLW GQVISTIPTq sLKFLKEAGH
PMCA1     1025 QFGGKPFSCS ELSIEQWLWS IFLGMGTLLW GQLISTIPTS RLKFLKEAGH
PMCA2     1003 QFGGKPFSCS pLqLdQWmWC IFIGlGELvW GQVIaTIPTS RLKFLKEAGr
Verma     1025 QFGGKPFSCS ELSIEQWLWS IFLGMGTLLW GQLISTIPTS RLKFLKEAGH

BCAATP4     51 GTtKEEItkd --AEgldEID HAEmELRRGQ ILWFRGLNRI QTQIDViNtF
PMCA1     1075 GTQKEEIPEE ELAEDVEEID HAERELRRGQ ILWFRGLNRI QTQmDVVNAF
PMCA2     1053 lTQKEEIPEE ELnEDVEEID HAERELRRGQ ILWFRGLNRI QTQIRVVkAF
Verma     1075 GTQKEEIPEE ELAEDVEEID HAERELRRGQ ILWFRGLNRI QTQIRVVNAF

BCAATP4     99 QTGASFKGVL KRQTMGQHLD VKHVPSSSYV TVAPVKSPPT TSVAAAVSSP
PMCA1     1125 QSGG------ ---------- ---------- ---------- ----------
PMCA2          ---------- ---------- ---------- ---------- ----------
Verma          ---------- ---------- ---------- ---------- ----------

BCAATP4    149 TLGNQSGQSV SivSLhEsiq KPknqnSIHN FMTHPEFtId eegPrtPLlD
PMCA1     1129 ---------- ---------- ------SI-q galrrqpsIa -SQhH----D
PMCA2     1101 ---------- -RSSLYEGLE KPESRTSIHN FMaHPEFRIE DSQPHIPLID
Verma     1123 ---------- -RSSLYEGLE KPESRSSIHN FMTHPEFRIE DSePHIPLID

BCAATP4    199 eqe-EEifek VskpqtktSS LDGEVTPQTN KNNNtVDCcq qI--vaShSd
PMCA1     1147 vT-nvstpth VvfsSStaSt P--------v gypSg-eCis
PMCA2     1142 DTDlEEDAal -KqNSSPPSS L--------N KNNSAiDSGI nLTtdtSKSA
Verma     1164 DTDaEdDApt -KrNSSPPpS P--------N KNNNAVDSGI hLTiemnKSA

BCAATP4    246 ------SPLH SLETSv
PMCA2     1183 TSSSPGSPiH SLETSL
Verma     1205 TSSSPGSPLH SLETSL
```

Figure 1. Comparison of the deduced amino acid sequences of the carboxy-terminus of known plasma membrane Ca^{2+}-ATPases from bovine brain (BCAATP4), rat brain (PMCA1 and PMCA2, Shull and Greeb, 1988) and human teratoma (Verma, et al., 1988). The residue that is predominant at a position is in upper case letters. If only no residue is predominant all residues are lower case. Peptide I is underlined in the BCAATP4 sequence.

METHODS

Details of the experimental procedures used in molecular cloning, immunoblot analysis, enzyme purification and assay, etc., have been published previously in a number of different reports (Brandt et al., 1988; Mann and Vanaman, 1988) and will not be reproduced here. Our strategy for extending previously isolated bovine brain cDNAs encoding the synaptic membrane Ca^{+2}-ATPase was based on specifically primed cDNA synthesis from poly A(+) RNA to ensure that the correct isoform was cloned. Accordingly, the 15-mer oligodeoxynucleotide, 5'-TGACACGTCCGAGGG-3', representing position 451-465 in our previous clone CAATP1, was used to prime reverse transcription of 10 ug of purified poly A(+) RNA from bovine brain. Subsequent second strand synthesis, cloning, isolation and sequencing of cDNAs for this isoform were performed essentially as described by Neve, et al., (1986). A 640 bp fragment from the 5'-end of CAATP1 was used for library screening.

As shown in Figure 1, the deduced amino acid sequences for the four putative isoforms of the plasma membrane Ca^{2+} ATPase cloned to date differ in the C-terminal region. A synthetic peptide was prepared with the sequence LDEQEEEIFEKVSKPGTK (peptide I) present in BCAATP4 which differs significantly from this region encoded by other cloned isoforms. Antibodies were raised in rabbits to this peptide and affinity purified from whole serum on peptide Sepharose as previously described for the human erythrocyte Ca^{2+}ATPase by Brandt et al., (1988).

The antigens were detected following SDS-PAGE on 7.5 % gels (Laemmli, 1972) and electrophoretic transfer to nitrocellulose as described by Burnette (1981). Gels were silver stained by the procedure of Morrissey, (1981). For detection of calmodulin-binding activity, plant calmodulin biotinylated on Cys-27 was prepared as described by Mann and Vanaman (1989) and used as described by Billingsley and coworkers (1985). For detection of the antigens in bovine brain, synaptosomal membranes were prepared according to the procedure of Cotman (1972). Immunoblots were developed using biotin-conjugated goat anti-rabbit IgG (Cappel) followed by incubation with avidin-peroxidase (Boeringer-Mannheim) and color development with 4-chloronapthol.

Immunolocalization studies using affinity-purified antipeptide antibodies were performed essentially as previously described (Brandt et al., 1987). Whole brains from adult rats were fixed in 10 % (v/v) neutral formalin, and imbedded in paraffin. Immunohistochemical localization of the Ca^{2+}-ATPase was performed on 8 mu sections after removal of the paraffin, and rehydrating the sections. Application of the antipeptide I and secondary antibody to the sections, and development were as described above except 3-amino-9-ethylcarbazole (AEC) was used as the chromogen. The sections were rinsed in distilled water, air dried, covered with Crystal Mount and coverslipped with Permount.

RESULTS

Screening of 25,000 recombinant bacteriophage from the specifically primed bovine brain cDNA library resulted in five positive clones. Only three of the isolates were determined to have legitimate extensions of the Ca^{+2}-ATPase as judged by comparison of the deduced peptide sequence to published peptide sequence in this region (James, et al., 1988). The longest of these was designated ¶BCAATP4. Comparison of the deduced peptide sequence of BCAATP4 with the deduced peptide sequences for the two rat brain (Shull and Greeb, 1988) and human teratoma (Verma, et al., 1988) isoforms are shown in Figure 1. As can be seen, the N-termini of all the sequences are very homologous with rat brain PMCA1 and the human teratoma sequences being nearly identical. However, the homology drops off precipitously near the C-terminus in most isoforms with PMCA2 and the teratoma sequences retaining homology. The region thought to contain the calmodulin-binding domain (James, et al., 1987) (residues 72-105 in BCAATP4) the greatest degree of homology, but, non-homology commences immediately following. This loss of homology is also seen at the DNA level (data not shown).

The antibodies raised against peptide I, which represents a unique sequence in the carboxy-terminal coding region of BCAATP4, were used to screen for the presence of this antigen in preparations of Ca^{+2}-ATPase isolated from both human and porcine erythrocyte ghosts. Bovine synaptosomal membranes were also assayed for the presence of cross-reacting antigen. Purified Ca^{+2}-ATPase prepared from outdated bank blood is detected as a single band of molecular weight 138 kDa by biotinylated calmodulin, as shown in Figure 2, lane 1. In lane 2, biotinylated calmodulin was able to detect several bands present in bovine brain synaptosomal membranes. A faint band at a molecular weight slightly lower than 138 kDa is of particular interest in that a band migrating in this same position is detected by affinity purified anti-peptide I (lane 3) and anti-human ATPase (lane 4) IgGs.

Interestingly, only a lower molecular weight ATPase doublet was detected by anti-peptide I IgG in purified Ca^{+2}-ATPase prepared from both human (lane 5) and porcine (lane 6) fresh blood. When the Ca^{+2}-ATPase was purified from outdated bank blood, this doublet was not seen and the main (138 kDa) form of the enzyme was also not recognized by anti-peptide I (lane 7). In lane 8, only the 138 kDa form is present in Ca^{+2}-ATPase prepared from outdated bank blood, whereas both the 138 kDa

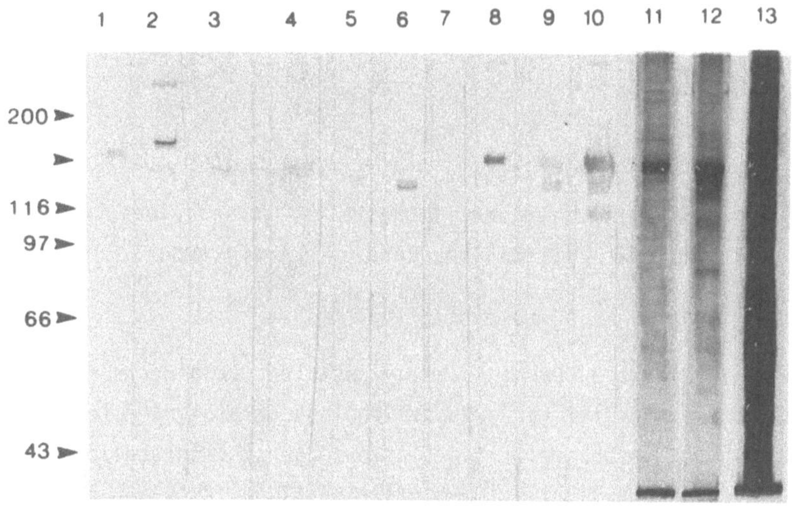

Figure 2. Immunoblotting of electrophoretically transferred proteins using anti-peptide I (lanes 3, 5, 6, and 7), polyclonal anti-ATPase (lanes 4, 8, and 9), and biotinylated calmodulin (lanes 1 and 10). The transferred proteins were: Ca^{+2}-ATPase purified from outdated bank blood (lanes 1, 7 and 8); Ca^{+2}-ATPase purified from fresh porcine blood (lanes 6, 9 and 10); and, Ca^{+2}-ATPase purified from fresh human blood (lane 5) bovine brain synaptosomes (lanes 2, 3, and 4). Lanes 11, 12, and 13 represent Ca^{+2}-ATPase purified from outdated bank blood, the ATPase prepared from fresh porcine blood, and bovine brain synaptosomal membranes, respectively, following silver staining of proteins.

form and the lower molecular weight doublet is observed in an immunoblot of the ATPase purified from fresh porcine blood (lane 9). In lane 10, biotinylated calmodulin is able to recognize both the higher and lower molecular weight forms. Lanes 11, 12, and 13 show silver stained proteins that were transferred to nitrocellulose for the immunoblotting shown in lanes 1-10.

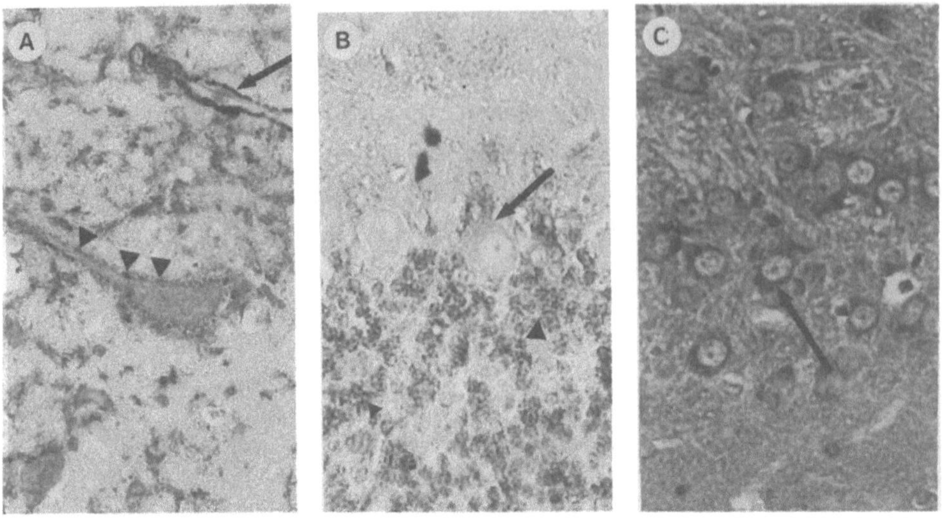

Figure 3. Immunolocalization studies with anti-peptide I in rat brain tissue. Samples were prepared and immunostained as described in METHODS. Panel A - pons; Panel B - cerebellum; Panel C - hippocampus.

As shown in Figure 3, Panel A, heavy positive staining is seen in axosomatic and axodendritic (arrowheads) endings of pontine motor neurons as well as in blood vessel walls (long arrow). In the cerebellum (Figure 4B), Purkinje cells are lightly immunopositive (long arrow) while granule cells are stained heterogeneously, with some cells darkly stained (arrowheads), or others showing no immunostaining. In the hippocampus, neurons of the CA4 region are homogenously immunopositive (arrows).

DISCUSSION

The plasma membrane Ca^{+2}-ATPase from human erythrocytes has been shown to contain a calmodulin regulatory domain at the C-terminal region of the protein. This domain includes both a calmodulin-binding region (James, et al., 1988) and an inhibitory domain (Benaim, et al., 1984). We have previously isolated a cDNA fragment, CAATP1, which is believed to contain the inhibitory domain of this form of the enzyme. It is clear that enzyme prepared from fresh blood sources contains a lower molecular weight form that is recognized by polyclonal anti-ATPase, anti-peptide I and biotinylated calmodulin.

Recently, three new cDNAs have been isolated for the plasma membrane Ca^{+2}-ATPase, two from a rat brain cDNA library (Shull and Greeb, 1988), and one from a human teratoma library (Verma, et al., 1988). None of these cDNAs contained a 3'-end homologous to CAATP1. To exclude the possibility that CAATP1 was not the 3'-end of a Ca^{+2}-ATPase message, we used primer extension to extend its sequence into the calmodulin-binding domain region. Comparison of the deduced peptide sequences of all of the isoforms shows that they are related at the amino-terminal region but not at the carboxyl-terminal region. This homology is also seen at the DNA level (data not shown). Based on the sequence homologies, it seems likely that BCAATP4 is the product of a unique gene. We have also isolated a homolog to BCAATP4 from a human brain cDNA library that is 95 % homologous at the DNA level and 93 % at the protein level, further suggesting that this is a unique gene product.

Using bovine brain synaptosomes, we have detected an immunoreactive band at an approximate molecular weight of 130 kDa with anti-peptide I, polyclonal anti-ATPase and biotinylated calmodulin. Since the sequence used to generate the anti-peptide I antibody is not present in either of the rat brain isoforms or the human teratoma isoform, it must represent a unique isoform. The immunolocalization studies with anti-peptide I IgGs indicate that this form of the ATPase is expressed prominently in several different regions of rat brain but largely restricted to nerve endings. This coupled with the low abundance of mRNA (0.01 % of total poly A+ RNA) for this isoform (data not shown) suggests that this isoform may play a unique role in Ca^{2+}-metabolism at the synapse which may be further restricted to specific types of neurones. Studies are currently in progress to isolate and characterize this specific isoform and to compare its properties and distrsibution to the others now known. Of particular importance is the elucidation of any differences in the regulation of their activities suggested by substantial differences in the structure of their putative regulatory domains.

REFERENCES

1. Benaim, G., Zurini, M., and Carafoli, E. (1984). J. Biol. Chem. 259: 8471-8477.

2. Billingsley, M.E., Pennypacker, K.R., Hoover, C.G., Brigati, D.J., and Kincaid, R.L. (1985). Proc. Nat'l. Acad. Sci. (USA). 82: 7585-7589.

3. Brandt, P., Zurini, M., Sisken, B.F., Rhoads, R.E. and Vanaman, T.C. (1987). in Calcium Binding Proteins In Health and Disease, eds. Norman, A.W., Means, A.R. and Vanaman, T.C. (Academic Press), pp. 544-554.

4. Brandt, P., Zurini, M., Neve, R.L., Rhoads, R.E., and Vanaman, T.C. (1988). Proc. Nat'l. Acad. Sci. (USA). 85: 2914-2918.

5. Burnette, W.N. (1981). Anal. Biochem. 112, 195-203.

6. Cotman, C.W. (1974). In: Methods in Enzymology 31, 445-452.

7. James, P., Maeda, M., Fisher, R., Verma, A.K., Krebs, J., Penniston, J.T., and Carafoli, E. (1988). J. Biol. Chem. 263: 2905-2910.

8. Laemmli, U. (1970). Nature (Lond.) 227, 680-685.

9. Mann, D.M., and Vanaman, T.C. (1989). J. Biol. Chem. (in press).

10. Morrissey, J.H. (1981). Anal. Biochem. 117, 307-310.

11. Neve, R.L., Harris, P., Kosik, K.S., Kurnit, D.M. and Donlon, T.A. (1986). Mol. Brain Res. 1, 271-280.

12. Niggli, V., Penniston, J.T., and Carafoli, E. (1979). J. Biol. Chem. 254, 9955-9958.

13. Shull, G.E., and Greeb, J. (1988). J. Biol. Chem. 263, 8646-8657.

14. Verma, A.K., Filoteo, A.G., Stanford, D.R., Wieben, E.D., Penniston, J.T., Strehler, E.E., Fischer, R., Heim, R., Vogel, G., Mathews, S., Strehler-Page, M.-A., James, P., Vorherr, T., Krebs, J., and Carafoli, E. (1988). J. Biol. Chem. 263, 14152-14159.

MOLECULAR STRUCTURE AND FUNCTION OF PHOSPHOLAMBAN:

THE REGULATORY PROTEIN OF CALCIUM PUMP IN CARDIAC SARCOPLASMIC RETICULUM

Michihiko Tada, Masaaki Kadoma, Junichi Fujii,
Yoshihiro Kimura, and Yoshiyuki Kijima

Departments of Medicine and Pathophysiology
Osaka University School of Medicine
Fukushima-ku, Osaka 553, Japan

INTRODUCTION

The excitation-contraction coupling of the myocardium represents a three-part process, involving three kinds of subcellular systems. These are sarcolemma, sarcoplasmic reticulum (SR), and myofibrillar proteins. Information transfer among these systems is exclusively carried out by Ca ions[1] in that both membranes of sarcolemma and SR exhibit bi-directional Ca fluxes, and the myofibrillar system contains Ca receptor protein troponin. It is important to note that all of these three subcellular systems provide phosphorylation sites for protein kinases and, in addition, such phosphorylation reactions are thought to accompany profound alterations in Ca-related events in these systems. Among these, phosphorylation of phospholamban, a membrane protein in cardiac SR, and its functional consequences are extensively defined[2,3], in that phospholamban presumably serves to modulate Ca pump ATPase of SR by augmenting the key elementary steps of ATPase. Phospholamban of cardiac SR was purified to near homogeneity and was sequenced by amino acid and cDNA sequencing, demonstrating a unique molecular properties. This paper defines the functional and structural characteristics of the phospholamban-ATPase system and attemps to propose a molecular model for the functional unit of phospholamban, which provides a basic understanding for the regulatory mechanism of ion transport and bioenergetic transduction across biomembrane.

STIMULATORY ACTION OF PHOSPHOLAMBAN ON Ca PUMP ATPase

Elementary Steps of Ca Pump ATPase Reaction

Ca pump of cardiac SR is energized by Ca^{2+}-dependent ATPase of E_1E_2 type. During translocation of Ca across the SR membrane, the ATPase undergoes a complex series of reactions in which phosphorylated intermediates (EP) are sequentially formed and degraded according to the following equation:

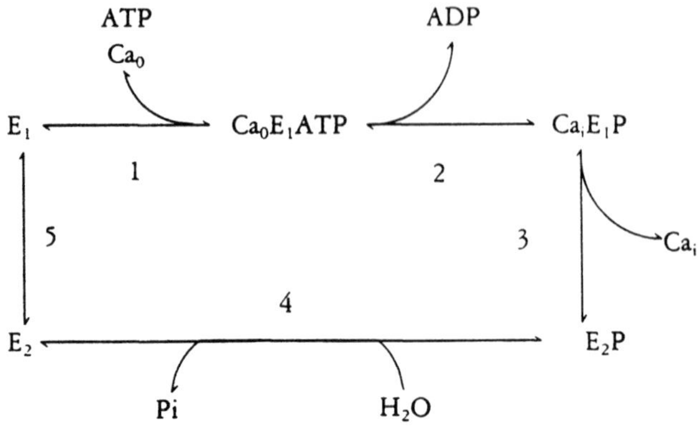

where E_1 and E_2 represent two different states of the ATPase, Ca^{2+}-bound and Ca^{2+}-free enzymes (see above); *i* and *o* indicate the inside and outside of SR membranes, respectively. E_1P is the phosphorylated intermediate which has a high affinity for Ca^{2+}, while E_2P has a low affinity for Ca^{2+}. E_1P can react with ADP to form ATP, and is thus termed ADP-sensitive phosphoenzyme, while E_2P could not form ATP upon addition of ADP, thus termed ADP-insensitive phosphoenzyme. As usually measured, EP represents the sum of E_1P and E_2P. Like the step from E_2 to E_1 (Step 5), the step at which E_1P is converted to E_2P (Step 3) is rate determining. Under the condition in which concentrations of Ca^{2+}, Mg^{2+} and K^+ are saturating, the phosphorylated enzyme is predominantly at E_1P form, since the decomposition of E_2P, but not that of E_1P, is accelerated under these conditions.

Effects of Phospholamban on Steady-state Kinetics of Ca Pump ATPase

Ca pump ATPase is originally found to be enhanced by phosphorylation of phospholamban[4,5] in that the Ca^{2+}-dependence profiles of ATPase and Ca transport at the steady state are significantly shifted toward low Ca^{2+} concentration, due to phospholamban phosphorylation. The ATPase activity increased in the presence of saturating concentrations of Ca^{2+} and ATP without alteration of EP level when phospholamban was phosphorylated, suggesting that enhancement of ATPase activity may be derived from augmentation of turnover rate of EP. When the rates of Pi liberation (v) and EP levels were simultaneously determined as a function of pCa, the value of v/[EP], which was independent of Ca^{2+}, was greatly enhanced by phospholamban phosphorylation, indicating that the rate-determining step of EP decomposition is markedly enhanced[6]. These results are consistent with the finding that the rate constant Kd of EP decomposition was enhanced. In the light of equation 1, enhancement of EP decomposition can be derived from stimulation of any of the Steps 3 and 4. Among these steps, Step 3 is probably accelerated since it is rate-limiting during EP decomposition and E_1P usually predominates under conditions where saturating concentrations of Ca^{2+}, Mg^2 and KCl are present.

Effects of Phospholamban on Presteady-state Kinetics of Ca Pump ATPase

The step at which EP is formed is an extremely rapid process, taking place within tens of milliseconds . Employing a rapid mixing device, the rate of EP formation was determined in reactions initiated at two different states of the ATPase, E_1 and E_2. In reactions initiated at Ca^{2+}-bound enzyme (E_1), the initial rates of EP formation were virtually unaltered by phospholamban phosphorylation, whereas EP levels were slightly augmented. In reactions initiated at Ca^{2+}-free enzyme (E_2), the initial rate of EP formation, which was much lower than that initiated at Ca^{2+}-bound enzyme, was highly enhanced by phospholamban phosphorylation, suggesting that phospholamban phosphorylation greatly accelerates Step 5, at which the affinity of the substrate-free enzyme for Ca^{2+} is greatly enhanced[7,8]. Transient state measurements of EP decomposition were also performed, suggesting that decay in ADP-sensitive EP (Step 3) was markedly accelerated by phospholamban phosphorylation[3].

Kinetical data indicated that both of the two key rate-determining steps (Steps 3 and 5) during the turnover of ATPase are enhanced when phospholamban is at the phosphorylated state. These are the major steps at which the affinity of the enzyme for Ca^{2+} is greatly altered, suggesting that phospholamban would probably exert its action by regulating the cation-mediated conformational change of this region. A direct protein-protein interaction has been proposed, with the assumption that the conformational state of a region of the ATPase molecule appears to be under direct control of phospholamban.

MOLECULAR CHARACTERISTICS OF PURIFIED PHOSPHOLAMBAN

Properties of Purified Phospholamban

Phospholamban was purified to near homogeneity with its inherent properties reasonably preserved, using non-ionic detergents like $C_{12}E_8$ and Zwittergent[9,10]. We overcame several difficulties by using $C_{12}E_8$ (octa-ethyleneglycol *n*-dodecylether) and obtained purified phospholamban. Judging from SDS-polyacrylamide gel electrophoresis and the extent of phosphorylation, our procedures yielded phospholamban with more than 99% purity. The amounts of phosphate incorporated were 3.73 mole/mole of 27,000-dalton phospholamban oligomer (see below). Suzuki and Wang[11] purified phospholamban with lower phosphorylatable property by monoclonal antibody immunoaffinity column chromatography. They also reported that the extent of phosphorylation was restored by the addition of phosphatidylinositol. Jakab and Kranias[12] reported that purified phospholamban, associated with endogenous phospholipids, was phosphorylated to the full extent.

The molecular weight of phospholamban was originally reported to be 22,000[13]. Employing purified phospholamban in unphosphorylated form, we demonstrated that the apparent molecular weight of phospholamban varied by varying the gel system for electrophoresis[9]. In the Weber and Osborn neutral gel system, the molecular weight of phospholamban was 22,000 as originally reported, against 27,000 in the Laemmli alkaline gel system, when either gel system consisted of 15% polyacrylamide and 0.1% SDS. Accordingly, Laemmli gel system is largely employed in the subsequent studies.

The purified phospholamban preparation exhibited unusual electrophoretic behavior[14]. While phospholamban migrated as 27,000-dalton component in the Laemmli alkaline gel system, the heat treatment (90 °C or above) of phospholamban preparation lowered the apparent molecular weight to 6,000-dalton component. We tentatively designated the 27,000-dalton form of phospholamban as PN_H and the 6,000-dalton form as PN_L.

The molecular weight of phospholamban monomer, calculated as 6,080 from the primary structure (see below), suggested the possibility that PN_H and PN_L represents an oligomer and a monomer. Further electrophoretic analyses, examining heat-modifiable transition between PN_H and PN_L and phosphorylation-induced stepwise mobility shift[15,16], suggested that phospholamban holoprotein would consist of four or five identical monomers. However, due largely to unusual properties of phospholamban, in which electrophoretic mobility varies with varying polyacrylamide concentration, the molecular weight of phospholamban holoprotein should be subjected to more accurate analysis, such as that we are pursuing now by low-angle laser light scattering photometry[17]. The other important point in this sense is the issue related to what kind of chemical interactions would serve to assemble phospholamban monomers, stabilizing the assembly of monomers in SDS.

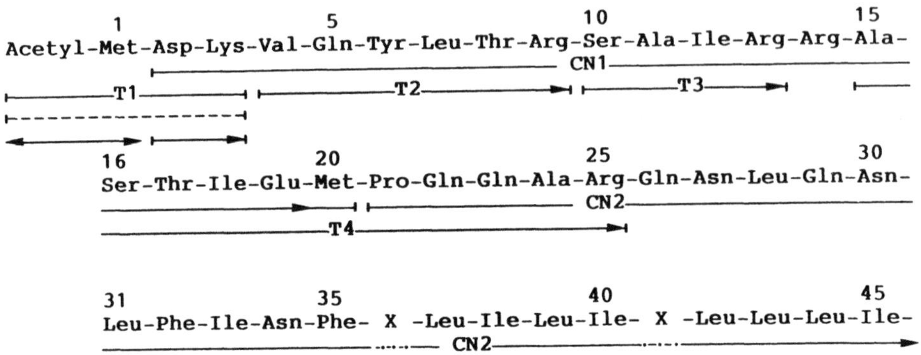

Fig. 1. **Partial amino acid sequence of phospholamban from canine cardiac SR.** Designations are; **CN**, cyanogen bromide cleaved peptide; **T**, tryptic peptide; **X**, unidentified residue; |—|, amino acid analysis; —►, automatic Edman degradation; |--|, fast atom bombardment mass spectrometry; ◄—►, reverse phase HPLC[14].

Primary Structure of Phospholamban

Purified phospholamban was subjected to amino acid sequencing[14]. By direct Edman degradation, we could not detect significant PTH-amino acid derivatives released from intact and S-aminoethylated phospholamban, indicating that the amino terminus of the protein is blocked. When the intact protein was cleaved by cyanogen bromide, two peptide fragments, CN1 and CN2, were obtained by fractionation through HPLC on a Phenyl 5PW-RP column. Edman degradation determined the first 18 amino acids of CN1, which are finally placed from Asp 2 to Glu 19 (Fig. 1). Although the C-terminal homoserine remained undetermined, Met 20 was placed at the C-terminus of CN1 from the sequence of T4 as mentioned later. Analysis of CN2 gave the unambiguous determination of the N-terminal 25 amino acids of which two of the residues were not identified.

The S-aminoethylated protein was digested with TPCK-trypsin and fractionated by Cosmosil columns, resulting in the separation of four tryptic fragments, T1 to T4 (Fig. 1). T1 was a tripeptide composed of Met, Asp, and Lys and could not generate the N-terminal PTH-amino acid, suggesting that T1 was originated from the blocked amino terminus of the protein. Fast atom bombardment mass spectrometry gave a major peak at M/Z 435.0 (Acetyl-Met-Asp-Lys-OH, MW = 434.5), indicating that T1 was N^{α}-acetylated. Cyanogen bromide-cleaved T1 showed sequence Asp-Lys and acetylhomoserine was determined by reverse phase HPLC. The N-terminus of T1 was concluded to be Acetyl-Met. T4 provided sequential overlap for CN1 and CN2.

Fujii et al.[18] reported the complete amino acid sequence of phospholamban deduced from cDNA. To isolate cDNA for phospholamban, a canine cardiac cDNA library was screened by hybridization with a mixture of 32 synthetic oligodeoxyribonucleotide probes predicted from a partial

```
                                                          5'-AGAAAACTTTCTAACTAAACAC -159

CGATAAGACTTCATACAACTCACAATACTTTATATTGTAATCATCACAAGAGCCAAGGCTACCTAAAAGAAGAGAGTGG  -80

TTGAGCTCACATTTGGCCGCCAGCTTTTTACCTTTCTCTTCACCATTTAAAACTTGAGACTTCCTGCTTTCCTGGGGTC   -1

 1                                      10                                      20
Met Asp Lys Val Gln Tyr Leu Thr Arg Ser Ala Ile Arg Arg Ala Ser Thr Ile Glu Met
ATG GAT AAA GTC CAA TAC CTC ACT CGC TCT GCT ATT AGA AGA GCT TCA ACC ATT GAA ATG   60

21                                      30                                      40
Pro Gln Gln Ala Arg Gln Asn Leu Gln Asn Leu Phe Ile Asn Phe Cys Leu Ile Leu Ile
CCT CAA CAA GCA CGT CAA AAT CTT CAG AAC CTA TTT ATA AAT TTC TGT CTC ATT TTA ATA  120

41                                      50      52
Cys Leu Leu Leu Ile Cys Ile Ile Val Met Leu Leu End
TGT CTC TTG TTG ATC TGC ATC ATT GTG ATG CTT CTC TGA AGTTCTGCTGCAATCTCCAGTGATGCA  187

ACTTGTCACCATCAACTTAATATCTGCCATCCCATGAAGAGGGGAAAATAATACTATATAACAGACCACTTCTAAGTAG  266

AAGATTTTACTTGTGAAAAGGTCAAGATTCAGAACAAAAGAAATTATTAACAAATGTCTTCATCTGTGGGATTTTGTAA  345

ACATGAAAAGAGCTTTATTTTCAAAAATTAACTTCAAAATGACTATAGGTGCGCATAATGTAATTGCTGAATTCCTCAA  424

CAAAGCTTGTAAAAGTTTCTATGCCAAATTTTTTCTGAGGGTAAAGTAGGAGTTTAGTTTTAAAACTGCTCTGCTAACC  503

AGTTCACTTCACATATAAAGCATTAGCTTCACTATTTGAGCTAAATATTTATATTGTACTGTAAATGCCTATGTAATGT  582

TTATTAAGATTTTTCAAGTCTCCGCTAAGTACGAAAATAATCATCCAAATGAAGTCATCATTTGAAATAGC-3'         652
```

Fig. 2. **Nucleotide sequence of pPLB1 cDNA and deduced amino acid sequence of canine cardiac phospholamban.** Nucleotide residues are numbered in the 5' to 3' direction, beginning with the first residue of ATG triplet encoding the initiator methionine. Nucleotides on the 5' side of residue 1 are indicated by negative numbers. A poly(A) tail on 3'-end is not shown. The predicted amino acid sequence of phospholamban is displayed above the nucleotide sequence with its residue number beginning with the initiator methionine. A box indicates the presumptive polyadenylation signal. The in-frame stop codon preceding the initiator codon is underlined[18].

amino acid sequence (Glu 19-Met-Pro-Gln-Gln-Ala 24) of phospholamban. Three hybridization-positive clones were isolated from about 3,000 transformants. All of these plasmids contained the same size insert of about 800 bases with the same restriction maps, suggesting that they were derived from the same mRNA. Fig. 2 depicts the complete nucleotide sequence of the cDNA insert of the plasmid, having 832 base pairs. Analysis of the cloned cDNA showed an open reading frame of 156 nucleotides (52 codons) starting with the ATG codon (position 1) and ending with the TGA stop codon (position 157). Fig. 2 also shows the amino acid sequence deduced from the 52 codons. In this reading frame, the sequence of nucleotide residues 1-105, encoding amino acids 1-35, corresponded precisely to the 35 N-terminal sequence from Acetyl-Met 1 to Phe 35 (Fig. 1). The two residues unidentified by Edman degradation for purified phospholamban, positions 36 and 41 (Fig. 1), were both deduced to be cysteine residues. Seven amino acids following Ile 45 would represent C-terminal residues ending at Leu 52, although protein sequencing is necessary to confirm these findings.

The cloned cDNA extended for 180 nucleotides to the 5'-end of the sequence coding for the N-terminus. Because this sequence contained an in-frame stop codon (TGA, position -78) and did not contain an ATG codon between the TGA and the Met 1, we conclude that phospholamban is synthesized without an N-terminal signal sequence. Therefore, it is likely that the information required for integration of phospholamban into the membrane is encoded in an uncleavable signal sequence in the C-terminus.

Simmerman et al.[19] reported a partial amino acid sequence and phosphorylation sites in phospholamban. They indicated 36 amino acid residues, which correspond precisely to those in the peptide from Ser 10 to Ile 45 in Fig. 1. Phosphorylation sites at Ser 7 and Thr 8 in their sequence were cAMP- and calmodulin-dependent and would correspond to Ser 16 and Thr 17 in our sequence, respectively.

MOLECULAR MODEL OF PHOSPHOLAMBAN-ATPase SYSTEM

The molecular weight of encoded protein was estimated to be 6,080 in good agreement with the apparent molecular weight of phospholamban monomer (6,000)[14,18]. The hydropathic profile of phospholamban indicates that the protein is amphipathic in nature. The protein is divided into two domains. The N-terminal domain from Met 1 to Asn 30 (domain I) was composed largely of hydrophilic amino acids, whereas the other domain representing 22 amino acids from Leu 31 to Leu 52 (domain II) was highly hydrophobic. Domain I, but not domain II, contained phosphorylatable serine (Ser 16) and threonine (Thr 17) residues, indicating that domain I faces cytoplasmic surface of SR membrane. Secondary structure prediction suggested that the majority of domains I and II forms an α-helix, which are being supported by circular dichroism analysis.

Phospholamban is characterized by two distinct molecular properties, when we assume the functional expression of this molecule. One is that domain I α-helix is likely to break into two portions (domain IA and IB) at around Pro 21, while the other is the possibility of forming an oligomeric complex. Fig. 3 illustrates the oligomeric model of phospholamban, in which each of phosphorylation site faces the cytoplasmic milieu with Pro 21 serving as a hinge[20]. The pentameric assembly is tentatively assumed to form this model, although the exact number of monomers in a holoprotein is not yet determined (see above). Domain IA probably assume the most important function, based on the following two potential reasons. One is the charge effect in which the phosphate incorporation into Ser 16 and/or Thr 17 would serve to neutralize the positive charge given by arginine residues (Arg 13, 14), which are situated just above these phosphorylation sites. The second feature would be the re-arragement of domain IA against IB, which would depend on the states of phosphorylation sites. Under phosphorylated state, domain IA could be rotated or get kinked with Pro 21 serving as a hinge. The hydrophobic side chains on domain II α-helix , consisting of 21 amino acids, allow a hydrophobic interaction with membrane lipids and could anchor phospholamban in SR membrane.

It is interesting to speculate which portion of phospholamban monomer would permit inter-molecular interaction to form the oligomeric assembly. Three cysteine residues in domain II are reported to exist as free sulfhydryl groups[19]. These cysteine residues play a unique role in oligomeric assembly of phospholamban*. A tryptic fragment of

*J. Fujii, personal communication.

phospholamban holoprotein, which was devoid of the phosphorylation site, remained oligomeric[21]. Since trypsinization under these conditions results in the removal of N-terminal residues up to Arg 25, the ability to maintain an oligomeric organization would reside in the C-terminal residues, possibly in domain II.

Simmerman et al.[19] proposed the pentameric model of phospholamban, based on their partial sequence, which corresponded precisely to the peptide from Ser 10 to Ile 45 in our sequence. Their model assumes that phospholamban holoprotein might serve as an ion channel. However, phospholamban would be unlikely to function as a channel in view of our model, in which the surface of membrane-penetrating domain II helix is considered to be covered with extremely hydrophobic side chains. It is more likely to assume, therefore, that phospholamban modulates Ca^{2+} transporting domains, representing the hallmark of ion-transport ATPase enzyme of SR[22].

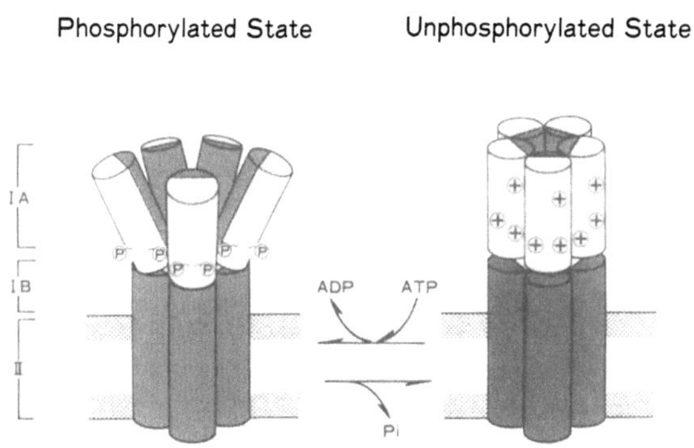

Fig. 3. **Molecular model of phospholamban.** When each monomer is assembled into oligomer the hydrophilic residues including phosphorylatable Ser 16 and Thr 17 face the cytoplasmic milieu. The positive charge of arginine residue in unphosphorylated phospholamban is indicated by ⊕, whereas the phosphate group covalently bound to the side chain of Ser 16 and/or Thr 17 is indicated by Ⓟ. The cross-hatched area indicates hydrophobic region of the α-helix. In domain **IA**, the helical wheel analysis indicates that the hydrophobic residues face the core of oligomer[20].

It is intriguing to speculate on the molecular mechanism by which phospholamban phosphorylation regulates Ca pump ATPase. The molecular structure of phospholamban, based upon amino acid sequence data, provides us with the view that domain IA of phospholamban could execute a certain influence over a key domain of Ca pump ATPase. Thus, the Ca^{2+} binding sites of the ATPase molecule are assumed to undergo a direct protein-protein interaction with the functional domain of phospholamban molecule.

Suzuki and Wang[23] observed that a monoclonal antibody directed against phospholamban blocked the phosphorylation and dephosphorylation of phospholamban, but markedly increased the ATP-dependent Ca pump activity by cardiac SR. Possibly the residues near phosphorylation sites in domain IA are responsible for exerting an influence over the Ca pump ATPase. MacLennan proposed the assumption that the putative Ca^{2+} binding sites in Ca pump ATPase, which are believed to be localized in the stalks or S region, just above the trans-membrane segments, are negatively charged due largely to the existence of glutamic and aspartic residues[22]. Although alternative molecular models are possible, the charge interaction between Ca^{2+} binding sites of ATPase (negative charge) and domain I of phospholamban (positive charge), which are modifiable by phosphorylation and de-phosphorylation induced by protein kinase, may play a critical role in regulation of Ca^{2+} transport across cardiac SR membrane.

PHYSIOLOGICAL RELEVANCE OF PHOSPHOLAMBAN-ATPase SYSTEM

The two mechanical effects of catecholamines on the myocardium are the increased contractility, i.e., positive inotropic effect, and the abbreviation of systole, i.e., accelerated rates of contraction and relaxation. Such effects of catecholamines are considered to be produced during the excitation-contraction coupling by altering Ca fluxes across the two principal membrane systems, SR and sarcolemma, of the myocardial cells[2,24]. The cAMP and phospholamban-ATPase system in SR could alter the rate of Ca uptake, subsequently changing the rate of Ca release (see below). Ca influx across the sarcolemmal membrane was also found to increase during β-adrenergic stimulation of the myocardial cells[25], possibly due to the phosphorylation of a channel protein by cAMP-dependent protein kinase. In the presence of cAMP, the acceleration of Ca uptake by the ATPase-phospholamban system may explain the acceleration of relaxation, because the increased rate of Ca uptake by SR would increase the rate at which Ca^{2+} is removed from troponin. This effect could eventually increase the amount of Ca^{2+} stored within the SR, for some of the Ca^{2+} remaining within the SR might otherwise be lost during diastole. Catecholamine-induced enhancement of Ca influx across sarcolemma would produce the following two effects on the SR: (1) enhancement of Ca-induced Ca release from the SR and (2) increased amounts of Ca loading on SR. Increased amounts of Ca accumulation into the SR, brought out by the increases in Ca uptake in SR and Ca influx in sarcolemma, could add to the amounts of Ca^{2+} available for delivery to the myofibrillar proteins in subsequent contractions, thus promoting myocardial contractility. Increased Ca influx across sarcolemma could increase the Ca release, due to the Ca-induced Ca release mechanism, thus partly contributing to the latter effect. Increased Ca release from the SR could increase both the rate and extent of myofibrillar contractions.

The in vivo evidence supporting these intracellular mechanisms was obtained by several investigators[26,27,28], who documented that the addition of isoproterenol to the isolated heart or the sliced heart tissue perfused with [^{32}P]Pi resulted in increased ^{32}P incorporation into phospholamban in situ, with the simultaneous increase in the rates of contraction and relaxation. Interestingly, cholinergic agonists are found to antagonize the isoproterenol-induced augmentation of phospholamban phosphorylation. Calmodulin inhibitor (fluphenazine) significantly reduced in vivo phosphorylation of phospholamban[27], although the physiological relevance of such an effect is not entirely clarified.

There are other evidences that are consistent with the mechanism by which Ca^{2+} fluxes of SR are controlled by the cAMP-phospholamban system. Employing a skinned cardiac cell, which exhibits cycles of phasic contractions upon addition of Ca^{2+}, Fabiato and Fabiato[29] demonstrated that a brief preincubation with cAMP results in an increased amplitude of contraction and faster rates of tension development and relaxation. The more direct evidence is obtained by Allen and Blinks[30] who measured intracellular Ca^{2+} by aequorin, a Ca^{2+}-sensitive bioluminescent protein. They found that isoproterenol augments the initial rate of Ca^{2+} release from SR during the early phase of contraction, with the simultaneous enhancement in the rate of Ca^{2+} reduction at the onset of relaxation.

Ca pump ATPase in SR consists of two distinct isozymes: type I in cardiac and slow-contracting skeletal muscle and type II in fast-contracting skeletal muscle[31,32]. Phospholamban was also found in SR of slow-contracting muscles[33,34]. This is in contrast to the finding that SR of fast-contracting skeletal muscle is devoid of phospholamban. It is intriguing to assume that Ca pump ATPase of type I muscle, but not type II muscle, requires phospholamban as a co-factor. This assumption finds supports from the work by Fujii et al.[35], who found that cDNA-derived primary structures of phospholamban from cardiac and slow-contracting skeletal muscle of rabbit were identical.

REFERENCES

1. M. Tada, T. Yamamoto, and Y. Tonomura, Molecular mechanism of active calcium transport by sarcoplasmic reticulum, Physiol. Rev. 58:1 (1978)
2. M. Tada and A. M. Katz, Phosphorylation of the sarcoplasmic reticulum and sarcolemma, Annu. Rev. Physiol. 44:401 (1982)
3. M. Tada, M. Yamada, M. Kadoma, M. Inui, and F. Ohmori, Calcium transport by cardiac sarcoplasmic reticulum and phosphorylation of phospholamban, Mol. Cell. Biochem. 46:73 (1982)
4. M. Tada, M. A. Kirchberger, D. I. Repke, and A. M. Katz, The stimulation of calcium transport in cardiac sarcoplasmic reticulum by adenosine 3':5'-monophosphate-dependent protein kinase, J. Biol. Chem. 249:6174 (1974)
5. M. A. Kirchberger, M. Tada, and A. M. Katz, Adenosine 3':5'-monophosphate-dependent protein kinase-catalyzed phosphorylation reaction and its relationship to calcium transport in cardiac sarcoplasmic reticulum, J. Biol. Chem. 249:6166 (1974)
6. M. Tada, F. Ohmori, M. Yamada, and H. Abe, Mechanism of the stimulation of Ca^{2+}-dependent ATPase of cardiac sarcoplasmic reticulum by adenosine 3':5'-monophosphate-dependent protein kinase, J. Biol. Chem. 254:319 (1979)
7. M. Tada, M. Yamada, F. Ohmori, T. Kuzuya, M. Inui, and H. Abe, Transient state kinetic studies of Ca^{2+}-dependent ATPase and calcium transport by cardiac sarcoplasmic reticulum, J. Biol. Chem. 255:1985 (1980)
8. E. G. Kranias, F. Mandel, T. Wang, and A. Schwartz, Mechanism of the stimulation of calcium ion dependent adenosine triphosphatase of cardiac sarcoplasmic reticulum by adenosine 3',5'-monophosphate dependent protein kinase, Biochemistry 19:5434 (1980)
9. M. Inui, M. Kadoma, and M. Tada, Purification and characterization of phospholamban from canine cardiac sarcoplasmic reticulum, J. Biol. Chem. 260:3708 (1985)
10. L. R. Jones, H. K. B. Simmerman, W. W. Wilson, F. R. N. Gurd, and A. D. Wegener, Purification and characterization of phospholamban from canine cardiac sarcoplasmic reticulum, J. Biol. Chem. 260:7721 (1985)
11. T. Suzuki and J. H. Wang, The phosphorylation of purified phospholamban by cyclic AMP-dependent protein kinase is stimulated by phosphatidylinositol, J. Biol. Chem. 262:3880 (1987)
12. G. Jakab and E. G. Kranias, Phosphorylation and dephosphorylation of

purified phospholamban and associated phosphatidylinositides, Biochemistry 27:3799 (1988)
13. M. Tada, M. A. Kirchberger, and A. M. Katz, Phosphorylation of a 22,000-dalton component of the cardiac sarcoplasmic reticulum by adenosine 3':5'-monophosphate-dependent protein kinase, J. Biol. Chem. 250:2640 (1975)
14. J. Fujii, M. Kadoma, M. Tada, M. Toda, and F. Sakiyama, Characterization of structural unit of phospholamban by amino acid sequencing and electrophoretic analysis, Biochem. Biophys. Res. Commun. 138:1044 (1986)
15. A. D. Wegener and L. R. Jones, Phosphorylation-induced mobility shift in phospholamban in sodium dodecyl sulfate-polyacrylamide gels, J. Biol. Chem. 259:1834 (1984)
16. T. Imagawa, T. Watanabe, and T. Nakamura, Subunit structure and multiple phosphorylation sites of phospholamban, J. Biochem. (Tokyo) 99:41 (1986)
17. T. Takagi, Determination of protein molecular weight by gel permeation chromatography equipped with low-angle laser light scattering photometer, Progress in HPLC 1:27 (1985)
18. J. Fujii, A. Ueno, K. Kitano, S. Tanaka, M. Kadoma, and M. Tada, Complete complementary DNA-derived amino acid sequence of canine cardiac phospholamban, J. Clin. Invest., 79:301 (1987)
19. H. K. B. Simmerman, J. H. Collins, J. L. Theibert, A. D. Wegener, and L. R. Jones, Sequence analysis of phospholamban: Identification of phosphorylation sites and two major structural domains, J. Biol. Chem. 261:13333 (1986)
20. M. Tada, M. Kadoma, M. Inui, and J. Fujii, Regulation of Ca^{2+}-pump from cardiac sarcoplasmic reticulum, Methods Enzymol. 157:107 (1988)
21. A. D. Wegener, H. K. B. Simmerman, J. Liepnieks, and L. R. Jones, Proteolytic cleavage of phospholamban purified from canine cardiac sarcoplasmic reticulum vesicles, J. Biol. Chem. 261:5154 (1986)
22. D. H. MacLennan, C. J. Brandl, B. Korczak, and N. M. Green, Calcium ATPases: Contribution of molecular genetics to our understanding of structure and function, in: "Proteins of Excitable Membranes, Society of General Physiologists Series" 41:287, B. Hille and D. M. Fambrough, eds, Raven Press, New York (1987)
23. T. Suzuki and J. H. Wang, Stimulation of bovine cardiac sarcoplasmic reticulum Ca^{2+}-pump and blocking of phospholamban phosphorylation and dephosphorylation by a phospholamban monoclonal antibody, J. Biol. Chem. 261:7018 (1986)
24. M. Tada and M. Inui, Regulation of Calcium transport by the ATPase-phospholamban system, J. Mol. Cell. Cardiol. 15:565 (1983)
25. W. Osterrieder, G. Brum, J. Hescheler, W. Trautwein, V. Flockerzi, and F. Hofmann, Injection of subunuits of cyclic AMP-dependent protein kinase into cardiac myocytes modurates Ca^{2+}-current, Nature 298:576 (1982)
26. Y. Iwasa and M. M. Hosey, Cholinergic antagonism of β-adrenergic stimulation of cardiac membrane phosphorylation in situ, J. Biol. Chem. 258:4571 (1983)
27. C. J. Le Peuch, J.-C. Guilleux, and J.G. Demaille, Phospholamban phosphorylation in the perfused rat heart is not solely dependent on β-adrenergic stimulation, FEBS Lett. 114:165 (1980)
28. J. P. Lindemann, L. R. Jones, D. R. Hathaway, B. G. Henry, and A. M. Watanabe, β-adrenergic stimulation of phospholamban phosphorylation and Ca^{2+}-ATPase activity in guinea pig ventricles, J. Biol. Chem. 258:464 (1983)
29. A. Fabiato and F. Fabiato, Relaxing and inotropic effects of cyclic AMP on skinned cardiac cells, Nature 253:556 (1975)
30. D. G. Allen and J. R. Blinks, Calcium transients in equorin-injected frog cardiac muscle, Nature 273:509 (1978)

31. D. H. MacLennan, C. J. Brandl, B. Korczak, and N. M. Green, Amino-acid sequence of a Ca^{2+}+Mg^{2+}-dependent ATPase from rabbit muscle sarcoplasmic reticulum, deduced from its complementary DNA sequence, Nature 316:696 (1985)
32. C. J. Brandl, N. M. Green, B. Korczak, and D. H. MacLennan, Two Ca^{2+} ATPase genes: Homologies and mechanistic implications of deduced amino acid sequences, Cell 44:597 (1986)
33. M. A. Kirchberger and M. Tada, Effects of adenosine 3':5'-monophosphate-dependent protein kinase on sarcoplasmic reticulum isolated from cardiac and slow and fast contracting skeletal muscles, J. Biol. Chem. 251:725 (1976)
34. A. O. Jorgensen and L. R. Jones, Localization of phospholamban in slow but not fast canine skeletal muscle fibers, J. Biol. Chem. 261:3775 (1986)
35. J. Fujii, J. Lytton, M. Tada, and D. H. MacLennan, Rabbit cardiac and slow-twitch muscle express the same phospholamban gene, FEBS Lett. 227:51 (1988)

SARCOLEMMAL (Ca^{2+} + Mg^{2+})-ATPase OF VASCULAR SMOOTH MUSCLE

Shoichi Imai, Yutaka Yoshida and Hsiao-Tung Sun

Department of Pharmacology, Niigata University School of Medicine, Niigata, Japan

INTRODUCTION

The existence of a plasmalemmal ATP-dependent Ca^{2+}-pump that extrudes Ca^{2+} out of the smooth muscle cells against its electrochemical gradient is well established and has been studied by measuring the ATP-dependent accumulation of ^{45}Ca by microsomal fractions of smooth muscles.

However, it proved more difficult to demonstrate the existence of a corresponding (Ca^{2+} + Mg^{2+})-ATPase. The main purpose of this study was to find out the microsomal fractions of the vascular smooth muscles with which the activities of the (Ca^{2+} + Mg^{2+})-ATPase could be studied. Since the enzyme activities could be detected in a fraction obtained from the digitonin-treated microsomes of the porcine aortic smooth muscle, preliminary experiments were conducted to study the effects of protein kinases on the enzyme activities.

METHODS

Preparation of the microsomal fractions

KCl-extracted microsomal fractions were prepared from the porcine aorta and the large extramural branches of the porcine coronary artery smooth muscle. Aortae and hearts were obtained from the slaughter house and brought to the laboratory in ice-cold modified Krebs solution buffered with Hepes at pH 7.4. The coronary artery was dissected out and the medial layer of the aorta and the coronary artery was isolated and frozen in liquid N_2. The frozen muscle was pulverized in a stainless-steel percussion mortar at a liquid N_2 temperature and stored until the time of preparation of microsome fractions, which was conducted using a slight modification of the method described by Furukawa and Nakamura (1984). In the case of the aorta the so-called F2 fraction was also prepared following the method of Suematsu et al. (1984). For comparison the sarcolemmal fraction was prepared from the porcine ventricular muscle with the method of Jones and Besch, Jr. (1984). (Ca^{2+} + Mg^{2+})-ATPase activities of these fractions were assayed with the malachite green method developed by Chan et al. (1986) in a reaction mixture containing 100 mM KCl, 0.98 mM $CaCl_2$, 2mM $MgCl_2$, 1 mM EGTA, 20 mM Tris-maleate, 2 mM ATP, 1 mM ouabain and 10 mM NaN_3 (pH 7.5).

Preparation of the digitonin-treated microsomes

Making use of the fact that digitonin mainly interacts with cholesterol, the density of the sarcolemmal fraction rich in cholesterol was increased by treating the KCl-extracted microsomes with digitonin using a modification of the method originally devised by Raeymaekers et al. (1985) to purify the endoplasmic reticulum fraction from visceral smooth muscles. Briefly, KCl-extracted microsomes were suspended in 0.25 M sucrose solution containing 0.6 M KCl, 20 mM Hepes-KOH (pH 7.4) and 1 mM dithiothreitol. Sixty mg of digitonin was added to 100 g of the starting material. After vigorous mixing the solution was left for 30 min on ice. Then the solution was layered on top of the sucrose gradient of 15, 28, 40 and 60 %, which additionally contained 0.6 M KCl, 1 mM dithiothreitol and 20 mM Hepes-KOH (pH 7.4). The three fractions were obtained after centrifugation with a Hitachi RP50-2 angle-type rotor for 2 hr at 100000 g at 4°C as shown in Fig. 2a and were designated as DF3, DF4 and DF5. Fig. 2a also depicts the distribution of Na^+, K^+-ATPase, a representative marker enzyme of the plasmalemma among the various fractions obtained with our sucrose density gradient method. On the left half of the figure, the distributions of the enzyme among the fractions obtained from the usual KCl-extracted microsomes are shown, while the right half depicts the distributions among the fractions obtained from digitonin-treated microsomes. As is evident from this figure, Na^+, K^+-ATPase is most enriched in DF3 fraction of digitonin-treated microsome, while it is most enriched in CF2 fraction of the ordinary microsomes. We, therefore, selected DF3 fraction as the fraction most enriched in plasmalemma. After dilution with 0.25 M sucrose solution containing 0.1 M KCl and 20 mM Hepes-KOH (pH 7.4), the fraction was centrifuged again at 100000 g for 40 min. The final pellet was resuspended in 0.25 M sucrose solution containing 0.1 M KCl, 20 mM Hepes-KOH (pH 7.4) and stored at -80°C until the time of the analysis.

Purification of the (Ca^{2+} + Mg^{2+})-ATPase

The purification of the (Ca^{2+} + Mg^{2+})-ATPase from the KCl-extracted microsomes was conducted using the calmodulin affinity chromatographic method as described by Furukawa and Nakamura (1984). Briefly, after repeated washings with alternately hypotonic and hypertonic solutions containing EGTA to remove calmodulin, the microsomes were solubilized with the non-ionic detergent Triton X-100 at a detergent / protein ratio (w/w) of 1:1 and put on the calmodulin sepharose 4B column. The elution of the enzyme was conducted with EDTA solution supplemented with phosphatidylcholine.

MATERIALS

The sources of materials used in this study were as follows: malachite green oxalate from Merck; (-) ouabain octahydrate from Aldrich Chemical; CNBr-activated Sepharose 4B from Pharmacia; Triton X-100 (polyoxyethylene (10) octyl phenylether), digitonin and EGTA from Wako Pure Chemical; calmodulin, saponin, L-α-phosphatidyl-L-serine, 1-oleoyl-2-acetyl-rac-glycerol (OAG), L-α-phosphatidylcholine and bovine heart 3',5'-cyclic AMP-dependent protein kinase (A-kinase) from Sigma Chemical; N-(6-aminohexyl)-5-chloro-1-naphthalene sulfonamide hydrochloride (W-7) and N-(6-aminohexyl)-1-naphthalene sulfonamine hydrocholoride (W-5) from Seikagaku Kogyo.

RESULTS AND DISCUSSION

Mg^{2+}-dependency of high affinity Ca^{2+}-ATPase of the KCl-extracted microsome fractions obtained from the porcine aorta and the coronary artery smooth muscles

Fig. 1a and b illustrate the Mg^{2+}-dependency of the high-affinity Ca^{2+}-ATPase activities of the KCl-extracted microsome fractions from the

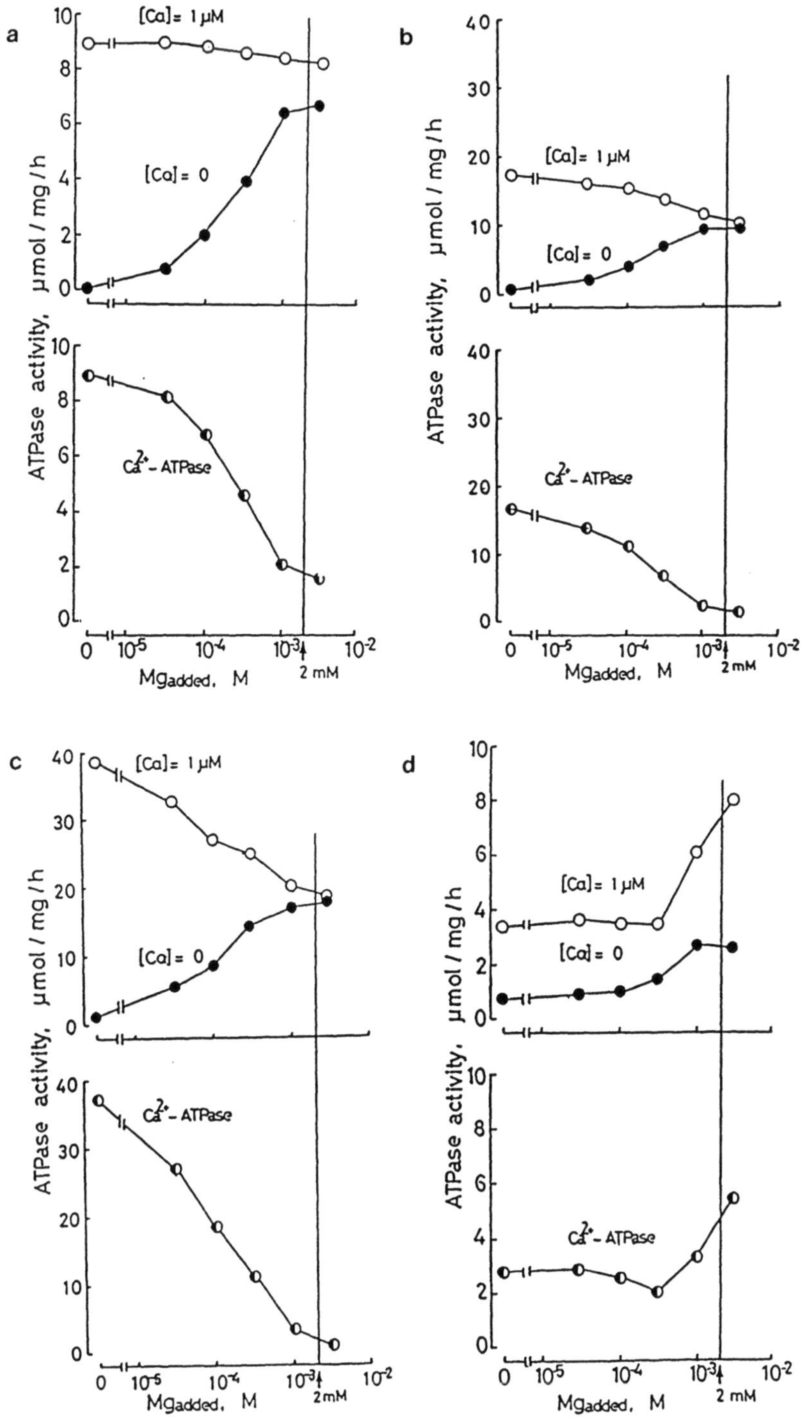

Fig. 1 Mg^{2+}-dependency of high affinity Ca^{2+}-ATPase of the KCl-extracted microsomes of the porcine aorta and coronary artery smooth muscle, (a and b), the plasmalemma-enriched F2 fraction of the porcine aorta (c) and the sarcolemmal fraction of the porcine ventricular muscle (d).

porcine aorta and coronary artery determined in the presence of 1 mM of ouabain and 10 mM of NaN_3. The closed circles depict the ATPase activities at zero Ca^{2+} concentration, while the open circles, the ATPase activities determined at a concentration of calcium of 1 μM. The difference between these two representing the extra activities elicited by calcium are plotted with half-closed circles. The extra activities attributable to calcium, a high affinity Ca^{2+}-ATPase activities, decreased as the concentrations of magnesium increased and only low activities of around 2 μmole/mg protein / hr remained at 2 mM Mg^{2+}. Furthermore, the extra activities were insensitive to stimulation by 5 μg/ml of calmodulin and to inhibition by vanadate (2 μM)(data not shown). Thus, the activities cannot be taken to represent the enzymatic basis of the sarcolemmal Ca^{2+}-pump (Ca^{2+}-pumping ATPase).

Mg^{2+}-dependency of high-affinity Ca^{2+}-ATPase of the microsome fraction enriched in plasma membrane (so-called F2 fraction) of the porcine aorta as compared with that of the sarcolemma fraction of the porcine ventricular muscle

As shown in Fig. 1c, high-affinity Ca^{2+}-ATPase activities of F2 fractions decreased as the concentrations of magnesium increased, leaving again only a very low activity of around 2 μmole/mg protein/hr at 2 mM Mg^{2+}. As was the case with the ordinary microsome fractions, the remaining activities were insensitive to calmodulin and vanadate. By contrast the Ca^{2+}-ATPase activities in the sarcolemmal fractions of the ventricular muscle showed a rise at Mg^{2+} concentrations of 0.3 mM and above amounting to the activity of around 4 μmole/mg protein/hr at 2 mM Mg^{2+} (Fig. 1d). Furthermore, this activity was stimulated with 5 μg/ml of calmodulin and was inhibited by 2 μM of sodium vanadate. From these studies it became clear that it is difficult if not impossible to detect the Ca^{2+}-pumping ATPase activities in KCl-extracted microsomes of the vascular smooth muscle preparations even when we prepared the plasmalemma-enriched F2 fraction.

(Ca^{2+} + Mg^{2+})-ATPase activities of the DF3 fraction of the porcine aortic smooth muscle microsomes

Fig. 2b depicts the Mg^{2+}-dependency of the Ca^{2+}-ATPase activities of DF3 fraction obtained from the digitonin-treated microsomes of the porcine aortic smooth muscle. In contrast to the activities of the ordinary KCl-extracted microsomes, the Ca^{2+}-ATPase activities showed a rise at Mg^{2+} concentrations of 0.3 mM and above, amounting to around 5 μmole/mg protein/ hr at 2 mM Mg^{2+}.

The enzyme activities were stimulated by 5 μg/ml of calmodulin by a factor of 1.3 as shown in Fig. 3a. Although this is only a moderate stimulation, it is still statistically significant. Furthermore, the ATPase activities were significantly inhibited by 2 μM of vanadate. As far as I am aware, this is the first demonstration of the presence of (Ca^{2+} + Mg^{2+})-ATPase activity in the microsome fractions of the vascular smooth muscle that is sensitive not only to stimulation by calmodulin but also to inhibition by vanadate. This portion of (Ca^{2+} + Mg^{2+})-ATPase activities could be taken to represent the Ca^{2+}-pumping ATPase, since the stimulation by calmodulin was completely reversed by vanadate, and the vanadate-insensitive portion of the activities was not stimulated by calmodulin.

As shown in Fig. 3b, the enzyme activity could be inhibited concentration-dependently by W-7, a calmodulin antagonist. Inhibition produced by 300 μM of W-7 corresponded to that produced by 2-3 μM of vanadate (Fig. 3c). W-5, a compound structurally related to W-7 but without calmodulin antagonistic activities, did not produce such an inhibition (Fig. 3b).

Fig. 4a depicts the effects of treatment of the DF3 fraction with saponin. The disruption of the microsome with saponin resulted in an unmasking of the latent activity; the enzyme activity was increased by a factor of 1.3. The activation by saponin was confined to that portion of

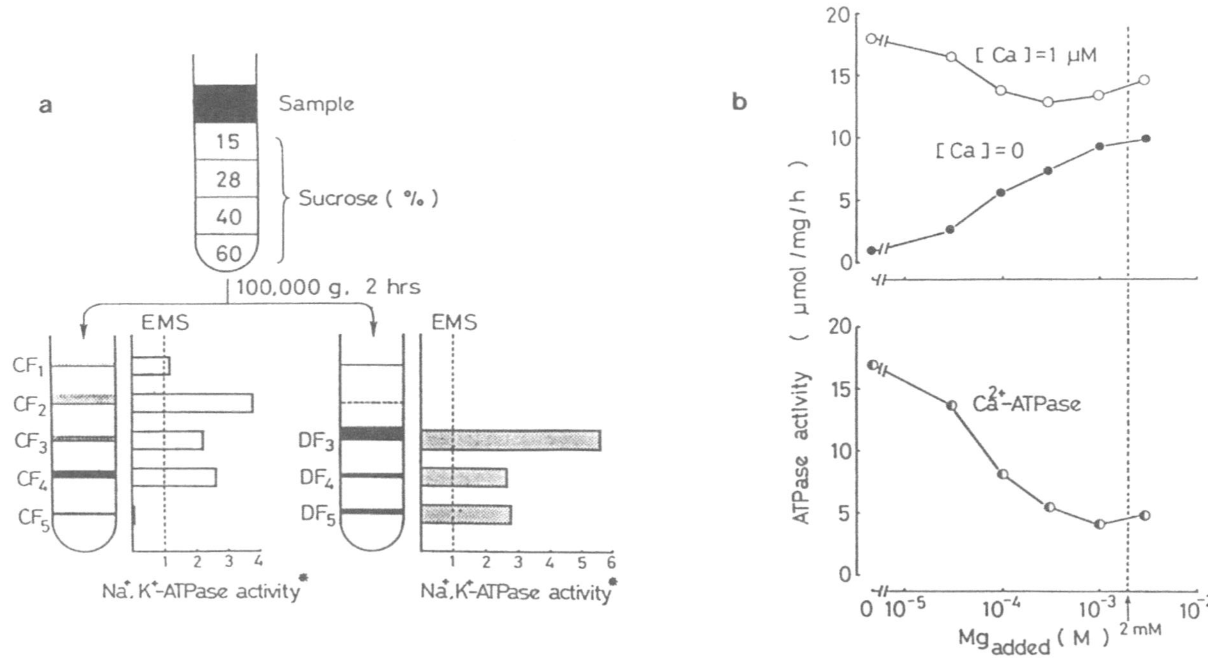

Fig. 2 Subfractionation of the digitonin-treated microsomes of the porcine aorta. a: Distribution of Na^+, K^+-ATPase among the various fractions obtained with a sucrose density gradient method. Fractions from the ordinary microsomes were designated as CF, while fractions from digitonin-treated microsomes, DF. b: Mg^{2+}-dependency of the high-affinity Ca^{2+}-ATPase of the DF3 fraction.

the enzyme activity which was sensitive to inhibition of a low concentration of vanadate and addition of vanadate reduced the saponin-stimulated activity as well as the basal activity to a same low level.

As demonstrated by DeSchutter, Wuytack and Casteels (1984) in porcine gastric antrum smooth muscles, it seems that saponin in some way affects the interaction of calmodulin with the enzyme, for the stimulation of the enzyme by calmodulin was specifically potentiated after saponin as shown in Fig. 4b; more than additive effects were observed and the activities almost doubling the basal was reached.

Effects of protein kinases on the vandate-sensitive portion of the (Ca^{2+} + Mg^{2+})-ATPase activities of DF3

Although there were reports indicating that cyclic GMP- or cyclic AMP-dependent protein kinase could stimulate the enzyme activities (Suematsu et al., 1984; Popescu et al., 1985; Furukawa and Nakamura, 1987; Rashatwar et al., 1987), we could not demonstrate the stimulation of the vanadate-sensitive portion of (Ca^{2+} + Mg^{2+})-ATPase activity of our DF3 fraction by the cyclic GMP-dependent kinase prepared with the method of Lincoln (1983) from the porcine lung (data not shown). The cyclic GMP-dependent protein kinase did not stimulate the purified (Ca^{2+} + Mg^{2+})-ATPase, either (data not shown). Cyclic AMP and cyclic AMP-dependent protein kinse also failed to stimulate the enzyme activity (data not shown). Instead, we found a significant stimulation of the vanadate-sensitive portion of the enzyme activity by Ca^{2+}- and phospholipid-dependent protein kinase (C-kinase). Fig. 5a depicts the stimulation of the enzyme by C-kinase prepared from the rat brain following the method of Kikkawa et al. (1986). Although the Ca^{2+}-ATPase activity was stimulated by a combined administration of phosphatidylserine (PS) and 1-oleoyl-2-acetyl-Sn-glycerol (OAG), the stimulation produced by administration of C-kinase, PS and OAG was significantly greater than that produced by PS plus OAG. Therefore, we regarded the activation of the enzyme produced by a combined administration of PS, OAG and C-kinase as representing the action of C-kinase. This, I believe, is the first clear-cut demonstration of the activation of the (Ca^{2+} + Mg^{2+})-ATPase of the vascular smooth muscle by C-kinase. Rashatwar et al. (1987) reported on the activation of a Mg^{2+}-dependent Ca^{2+}-activated ATPase activity of the microsomes obtained from the cultured aortic smooth muscle cells of the rat. However, whether the the enzyme originated from the sarcolemma or the sarcoplasmic reticulum was not ascertained in their experiments. Fig. 6 also depicts the effects of a combined administration of calmodulin and C-kinase. The effects of these two substances were found to be simply additive, indicating the difference in mechanism of action of these two substances.

Effects of C-kinase on the activity of purified (Ca^{2+} + Mg^{2+})-ATPase

Smallwood et al. (1988) reported recently on the direct stimulation of the purified (Ca^{2+} + Mg^{2+})-ATPase of the human erythrocyte by C-kinase. However, no such direct stimulation was observed with the (Ca^{2+} + Mg^{2+})-ATPase purified from the porcine aortic smooth muscle with calmodulin-sepharose 4B affinity column method (Fig. 7). Thus, we must postulate an indirect activation through intermediate effectors, e.g. alterations in ion fluxes.

Fig. 8 depicts the possible role of calmodulin and C-kinase in the regulation of Ca^{2+}-pump activity of the plasma membrane. When a certain contraction-producing agonists bind to the receptor coupled to phospholipase C, inositol 1,4,5-trisphosphate is liberated and induces the release of calcium from the intracellular store sites to produce an elevation of the intracellular Ca^{2+} concentrations. Calmodulin stimulates the pump activity under this condition. In addition diacylglycerol is produced not only from

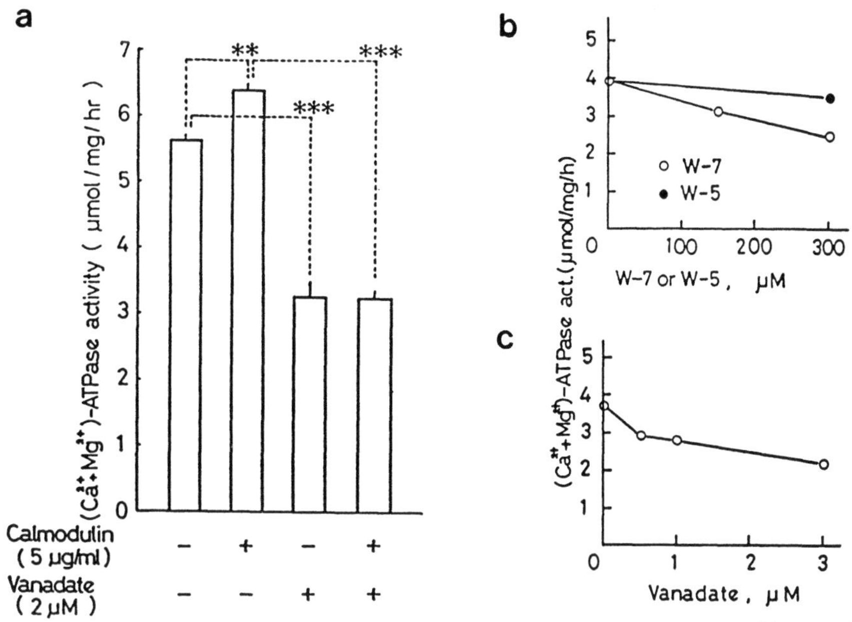

Fig. 3 Effects of calmodulin, vanadate, W-7 and W-5 on the (Ca^{2+} + Mg^{2+}) ATPase activities of DF3 fraction of the porcine aortic microsomes.

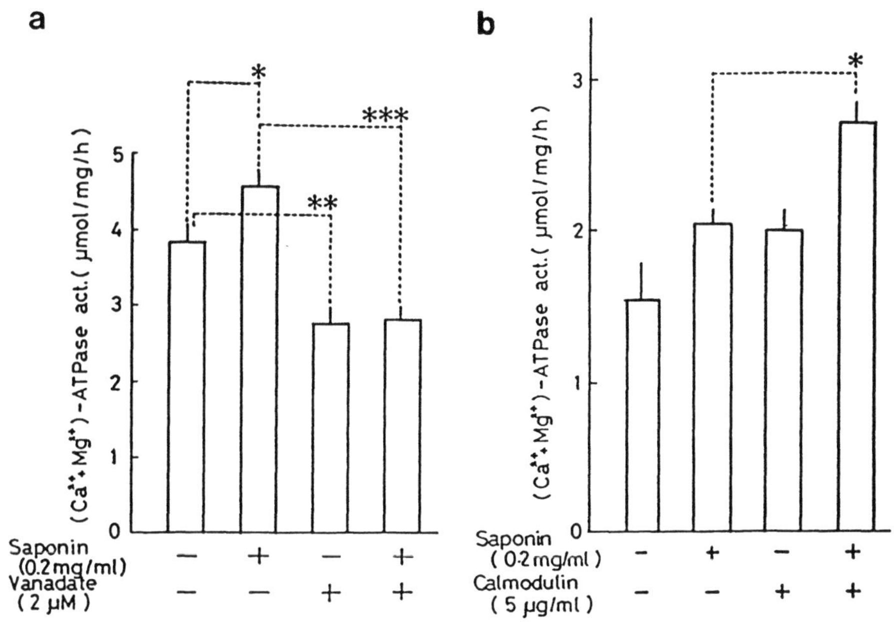

Fig. 4 Effects of saponin, vanadate and calmodulin on the (Ca^{2+} + Mg^{2+}) ATPase activities of DF3 fraction.

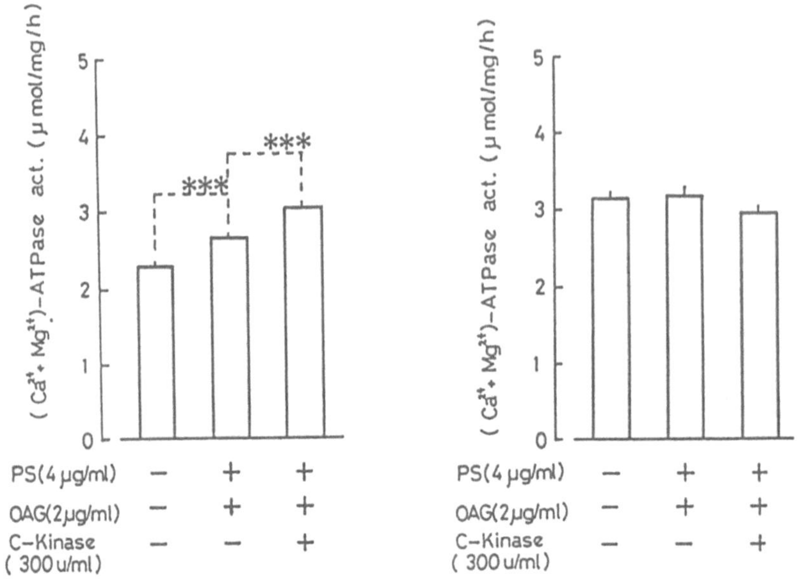

Fig. 5 a) Effects of C-kinase on the vanadate-sensitive portion of the $(Ca^{2+} + Mg^{2+})$-ATPase activities of DF3 fraction.

b) Effects on the vanadate-insensitive portion.

PS: phosphatidylserine, OAG: 1-oleoyl-2-acetyl-rac-glycerol

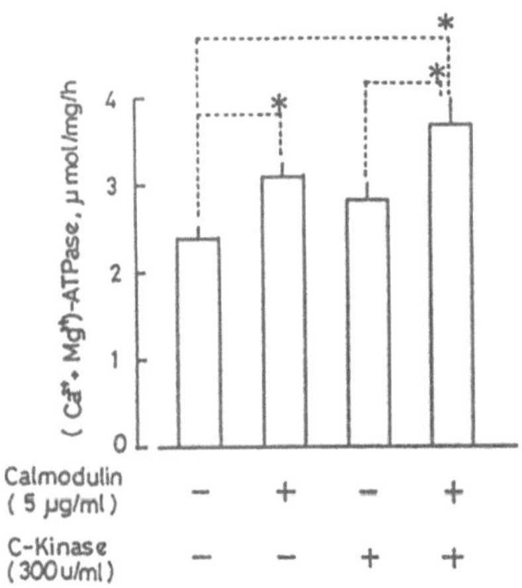

Fig. 6 Effects of C-kinase and calmodulin on the vanadate-sensitive portion of the $(Ca^{2+} + Mg^{2+})$-ATPase activities of DF3 fraction.

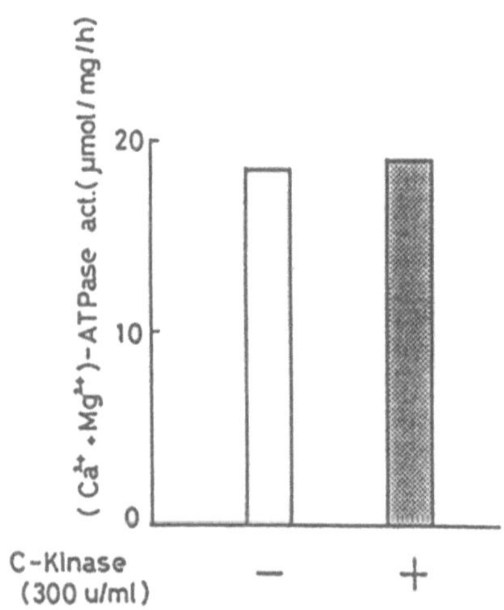

Fig. 7 Effects of C-kinase on the purified $(Ca^{2+} + Mg^{2+})$-ATPase of the porcine aorta.

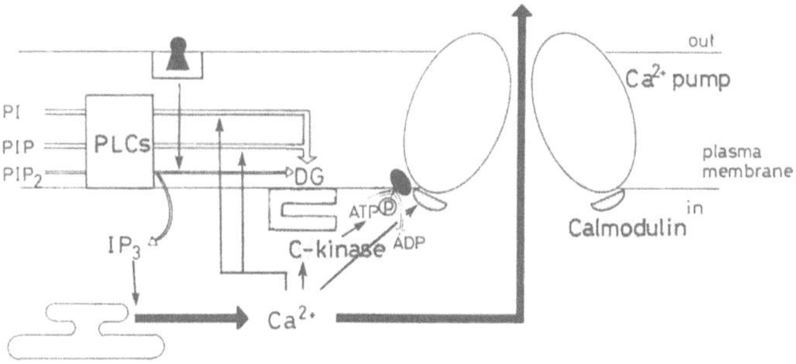

Fig. 8 Regulation of the Ca^{2+}-pumping ATPase by calmodulin and C-kinase. PI: phosphatidylinositol, PIP: phosphatidylinositol 4-phosphate, PIP_2: phosphatidylinositol 4,5-bisphosphate, IP_3: inositol 1,4,5-trisphosphate, PLC: phospholipase C, DG: diacylglycerol

PIP_2 but also from PIP or PI and in cooperation with calcium and phospholipid activates C-kinase. Activated C-kinase, in turn, produces a stimulation of the Ca^{2+}-pumping ATPase. The stimulation of the Ca^{2+}-pumping ATPase by C-kinase found in the present study can, therefore, be taken to denote the initiation of the compensatory mechanism for prevention of an excessive increase in the intracellular Ca^{2+} concentrations.

In conclusion, preparation of the digitonin-treated microsomes made it possible to determine the Ca^{2+}-pumping ATPase activities of the vascular smooth muscle in its nearly natural environment. It was found that the enzyme could be stimulated under this conditions by protein kinase C. The digitonin-treated microsome preparations would be useful for the study on the role of Ca^{2+}-pumping ATPase in the relaxant actions of various drugs acting on the vascular smooth muscle and on the mechanisms of intiation of various disease states.

Acknowledgements: The authors wish to express their thanks to Miss S. Tsuchida for help in preparation of the manuscript.

REFERENCES

Chan, K.-M., Delfert, D. and Junger, K.D., 1986, A direct colorimetric assay for Ca^{2+}-stimulated ATPase activity, Anal. Biochem., 157:375.
Deshutter, G., Wuytack F. and Casteels, R., 1984, Tissue levels and purification by affinity chromatography of the calmodlin-stimulated Ca^{2+}-transport ATPase in pig antrum smooth muscle, Biochim. Biophys. Acta, 773:1.
Furukawa, K. and Nakamura, H., 1984, Characterization of the (Ca^{2+} + Mg^{2+})ATPase purified by calmodulin-affinity chromatography from bovine aortic smooth muscle, J. Biochem., 96:1343.
Furukawa, K. and Nakamura, H., 1987, Cyclic GMP regulation of the plasma membrane (Ca^{2+}-Mg^{2+}) ATPase in vascular smooth muscle, J. Biochem., 101:287.
Jones, L. R. and Besch, H. R. Jr., 1984, Isolation of canine cardiac sarcolemmal vesicles, Methods Pharmacol., 5:1.
Kikkawa, U., Go, M., Koumoto, J. and Nishizuka, Y., 1986, Rapid purification of protein kinase C by high performance liquid chromatography, Biochem. Biophys. Res. Commun., 135:636.
Lincoln, T.M., 1983, cGMP-dependent protein kinase, Methods Enzymol., 99:62.
Popescu, L. M., Panoiu, C., Hinescu, M. and Nutu, O., The mechanism of

cGMP-induced relaxation in vascular smooth muscle, Eur. J. Pharmacol., 107:393
Raeymarkers, L., Wuytack, F. and Casteels, R., 1985, Subcellular fractionation of pig stomach smooth muscle. A study of the distribution of the $(Ca^{2+} + Mg^{2+})$-ATPase activity in plasmalemma and endoplasmic reticulum, Biochim. Biophys. Acta, 815:441.
Rashatwar, S. S., Cornwell, T. L. and Lincoln, T. M., 1987, Effects of 8-bromo-cGMP on Ca^{2+} levels in vascular smooth muscle cells: Possible regulation of Ca^{2+}-ATPase by cGMP-dependent protein kinase, Proc. Natl. Acad. Sci., 84:5685.
Smallwood, J. I., Gügi, B. and Rasmussen, H., 1988, Regulation of erythrocyte Ca^{2+} pump activity by protein kinase C, J. Biol. Chem., 263:2195.
Suematu, E., Hirata, M. and Kuriyama, H., 1984, Effects of cAMP- and cGMP-dependent protein kinase, and calmodulin on Ca^{2+} uptake by highly purified sarcolemmal vesicles of vascular smooth muscle, Biochim. Biophys. Acta, 773:83.

MOLECULAR PROPERTIES OF VOLTAGE-SENSITIVE CALCIUM CHANNELS

William A. Catterall, Michael J. Seagar, Masami Takahashi, and Kazuo Nunoki

Department of Pharmacology, SJ-30, University of Washington
Seattle, WA 98195

INTRODUCTION

Voltage sensitive calcium channels participate in action potential generation in electrically excitable cells and constitute an essential link between transient changes in membrane potential and a variety of cellular responses including secretion of neurotransmitters and hormones, initiation of contraction in cardiac and smooth muscle, and activation of second messenger responses in many cell types. Electrophysiological measurements have established the existence of multiple classes of calcium channels.[1-4] Although work on many cell systems has contributed to current understanding of calcium channel function, the molecular properties of the channel have been investigated most thoroughly in skeletal muscle which has a particularly high density of calcium channels. This chapter will therefore focus primarily on skeletal muscle calcium channels.

Vertebrate skeletal muscle fibers have substantial voltage- activated calcium currents which are restricted to the transverse tubule system in adult muscle.[5, 6] These calcium currents are mediated by two separate calcium channel subtypes.[7, 8] One subtype mediates a transient calcium current that is rapidly activated at membrane potentials more positive than -65 mV and is not sensitive to organic calcium channel blockers. The second mediates a long lasting calcium current that is activated at membrane potentials more positive than -20 mV and is blocked by organic calcium channel antagonists including the phenylalkylamine verapamil and the dihydropyridine nifedipine.[6, 8] The organic calcium channel antagonists have provided specific probes for studies of the molecular properties of these dihydropyridine-sensitive calcium channels.

The calcium antagonist receptor has three separate calcium antagonist binding sites. The dihydropyridine calcium channel antagonists nifedipine, nitrendipine, and PN200-110 bind with high affinity and specificity to a single class of receptor sites which are present in high concentration in transverse tubule membranes.[9, 10] Two other structurally distinct classes of calcium channel antagonists, phenylalkylamines such as verapamil and the benzothiazepines such as diltiazem, also inhibit dihydropyridine-sensitive calcium channels in skeletal muscle fibers[6] and bind to a similar number of high affinity receptor sites in transverse tubule membranes.[11, 12] Binding of diltiazem enhances the binding of dihydropyridines and slows their dissociation through allosteric interactions, while verapamil decreases the binding of dihydropyridines and enhances their dissociation through allosteric interactions.[11] Direct measurements of specific binding of verapamil (or its desmethoxy derivative) and diltiazem have shown that their binding is competitive.[11, 12] However, studies of the kinetics of ligand dissociation and the concentration dependence of competitive displacement of these two ligands in transverse tubules (11 and earlier references cited therein) and in cardiac plasma membranes[12] provide clear evidence that verapamil and diltiazem bind to two distinct sites that have negative allosteric interactions. These data support a model[8, 10] of the calcium antagonist receptor having three separate, allosterically linked, binding sites for calcium antagonist drugs.

Identification by photoaffinity labeling. One approach to determination of the size of the calcium antagonist receptor is specific covalent labeling by reactive derivatives of the drugs. Ferry et al[13] first succeeded in covalently labeling the dihydropyridine receptor site in transverse tubule membranes with [^{3}H]-azidopine, an arylazido derivative of a dihydropyridine calcium channel antagonist. A single polypeptide with an apparent size of approximately 145 kDa as analyzed by SDS-PAGE after reduction of disulfide bonds was specifically labeled. Galizzi et al[12] labeled the diltiazem and dihydropyridine binding sites of the calcium antagonist receptor by high intensity ultraviolet irradiation of transverse tubule membranes with bound drugs. Both drug classes reacted with multiple polypeptides, but only a single polypeptide with an apparent size of 170 kDa after reduction of disulfide bonds was labeled specifically as judged by inhibition of labeling in the presence of excess unlabeled drug. Streissnig et al[14] have recently described a photoreactive analog of verapamil which radioactively labels a polypeptide of 165 kDa in transverse tubule membranes. These results suggest that a single polypeptide of 145 to 170 kDa may contain all three receptor sites for calcium antagonist drugs which inhibit calcium channels in transverse tubule membranes.

Purification. Survey of a broad range of detergents led us to conclude that digitonin is suitable for solubilization of the calcium antagonist receptor with retention of binding activity for dihydropyridines and allosteric coupling with receptor sites for verapamil and diltiazem.[15, 16] Studies of the solubilized calcium antagonist receptor showed that it is a large glycoprotein which contains complex carbohydrate chains and retains all three calcium antagonist binding sites. Since skeletal muscle transverse tubules contain few glycoproteins, the finding that the calcium antagonist receptor is a glycoprotein suggested that affinity chromatography on immobilized lectin columns would be a powerful purification method. We therefore[16] isolated transverse tubule membranes from freshly dissected skeletal muscle, prelabeled calcium antagonist receptors by binding of a ^{3}H-labeled dihydropyridine and solubilized in digitonin. The resulting calcium antagonist/receptor complex was purified by lectin affinity chromatography, ion exchange chromatography, and sucrose gradient sedimentation. Correction for dissociation of the pre-bound dihydropyridine label indicated a calcium antagonist receptor concentration of 0.78 per 400 kDa, the approximate mass of the purified calcium channel complex (see below). Analysis of such preparations by SDS-PAGE without reduction of disulfide bonds revealed three noncovalently associated subunits which quantitatively comigrated with dihydropyridine binding activity: α of 162 kDa, β of 50 kDa, and γ of 33 kDa (Figure 1, lane 1).

Subunit structure of the dihydropyridine-sensitive calcium channel. The number, transmembrane arrangement, and possible functional roles of the polypeptide components of these dihydropyridine-sensitive calcium channels have been considerably clarified by affinity labeling studies. The heterogeneous α subunit band described in the initial purification of functional dihydropyridine-sensitive calcium channels[16] can be resolved after reduction of disulfide bonds to yield two polypeptides (Fig. 1, lane 2) which have distinctly different properties as revealed by specific labeling techniques.[14, 17-21] Because biochemical properties of these two different polypeptides had been previously ascribed to the "α" subunit, they have been designated α_1 and α_2.[17, 21] The α_1 subunit has an apparent size of approximately 175 kDa after reduction of disulfide bonds (Fig. 1, lane 2) and binds subunit specific antibodies that do not recognize the α_2 subunit (Fig. 1, lane 3). The α_1 subunit contains the receptor site for dihydropyridines as determined by photoaffinity labeling with [^{3}H]-azidopine (Fig. 1, lane 6) and the receptor site for verapamil as determined by photoaffinity labeling with [^{3}H]-LU49888.[14] The α_1 subunit is not detectably glycosylated as assessed by binding of the lectins concanavalin A or wheat germ agglutinin which are specific for high mannose or complex carbohydrate chains, respectively (Fig. 1, lanes 4 and 5), but it is heavily labeled by trifluoromethyl-iodophenyl diazirine (TID) (Fig. 1, lane 7), a hydrophobic photoaffinity probe that labels transmembrane segments in membrane proteins.[17] These characteristics mark α_1 as the principal functional subunit of the dihydropyridine-sensitive calcium channel since it has the receptor sites for calcium antagonist drugs and transmembrane segments required for formation of a transmembrane pore.

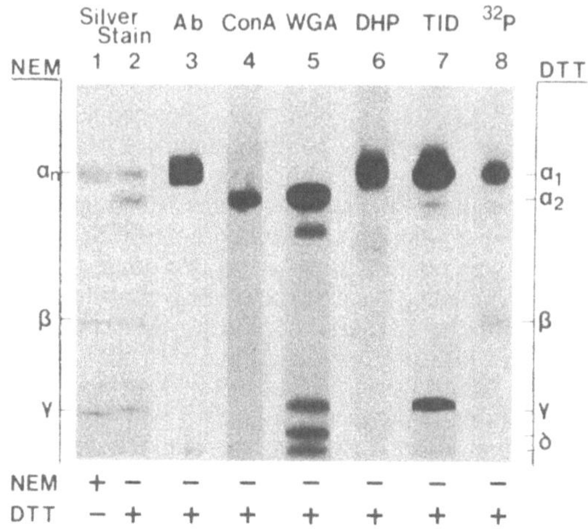

FIGURE 1

Differential labeling of calcium channel subunits. *Lanes 1 and 2*, purified calcium channels were analyzed by SDS-PAGE and silver staining with or without reduction of disulfide bonds as indicated beneath each lane. *Lane 3*, polypeptides separated by SDS-PAGE with reduction of disulfide bonds were electrophoretically transferred to nitrocellulose strips and immunolabeled by incubation without PAC-10, or followed by incubation with ^{125}I-labeled protein A, washing, and autoradiography. *Lane 4*, calcium channel subunits were transferred to a nitrocellulose sheet and labeled with [^{125}I]ConA. *Lane 5*, calcium chanel subunits were labeled directly in the polyacrylamide gel with [^{125}I]WGA. *Lane 6, photoaffinity labeling:* T-tubule membranes (0.4 mg/ml) in 25 mM HEPES, 1 mM CaCl, adjusted to pH 7.5 with Tris base, were incubated with 6 nM [^{3}H]azidopine and irradiated for 15 min at 4°C with a 30 watt UV source (λ_{max} 356 nm). The membranes were solubilized in 1% digitonin, 10 mM HEPES, 185 mM NaCl, 0.5 mM CaCl, 0.1 mM PMSF, 1 μM pepstatin A adjusted to pH 7.5 with Tris base, and calcium channels were partially purified by chromatography on WGA-Sepharose and analyzed by SDS-PAGE and fluorography. *Lane 7, hydrophobic labeling:* [^{125}I]TID (15 Ci/mmol) was prepared, and purified calcium channel was labeled with 100 μCi/ml [^{125}I]TID in a buffer containing 0.1% digitonin as previously described (17). *Lane 8, phosphorylation:* Purified calcium channel was incubated with 0.3 μM cAMP-dependent kinase catalytic subunit and 0.12 μM carrier free [γ-^{32}P]ATP for 15 min at 37°C as previously described (17).

In contrast to α_1, α_2 has an apparent size of 135 to 150 kDa after reduction of disulfide bonds (Fig. 1, lane 2) and is strongly labeled by both concanavalin A and wheat germ agglutinin indicating that it is heavily glycosylated (Fig. 1, lanes 4 and 5). Enzymatic removal of N-linked carbohydrate reduces its size to 105 kDa.[17] Moreover, it does not contain any of the calcium antagonist receptor sites that have been identified to date, and is only weakly labeled by hydrophobic photoaffinity probes indicating that it has few transmembrane segments (Fig. 1, lanes 6 and 7). α_2 is linked by disulfide bonds to a glycoprotein of 24 to 29 kDa, designated δ, increasing its apparent size to 175 kDa when disulfide bonds are intact (Fig. 1, lane 1). The δ subunit is clearly visualized by labeling with wheat germ agglutinin, but is not easily seen with protein stains (Fig. 1, lanes 2 and 5). Since the $\alpha_2\delta$ complex is heavily glycosylated but has few transmembrane segments, it is likely to be primarily exposed at the extracellular surface.

The remaining subunits of the purified calcium channel, β and γ, are peripherally associated with the α_1 subunit. β is not glycosylated and is not labeled by hydrophobic probes (Fig. 1, lanes 4, 5, and 7). In contrast, γ is heavily labeled by hydrophobic photoaffinity probes and is glycosylated (Fig. 1, lanes 4, 5, and 7). Enzymatic deglycosylation reduces its apparent size to 20 kDa.[17]

These results show that the α_1 subunit has the properties expected of the dihydropyridine-sensitive calcium channel: receptor sites for dihydropyridines and verapamil and numerous transmembrane segments appropriate for formation of an ion channel. The $\alpha_2\delta$, β, and γ polypeptides cannot now be assigned any functional role. Are they truly subunits of an oligomeric calcium channel complex? Several lines of evidence suggest that they are. All three size classes of polypeptides quantitatively comigrate with dihydropyridine binding activity during separation of polypeptides by sucrose gradient sedimentation.[16] Moreover, antibodies that are specific for the α_1 subunit (Fig. 1, lane 3) immunoprecipitate α_2, β and γ in digitonin or CHAPS where the calcium channel retains functional activity (Fig. 2, right, lanes 1 and 2), but not in Triton X-100 or SDS in which calcium channel function is unstable (Fig. 2, right, lanes 3 and 4). Lentil lectin, which is specific for high mannose carbohydrate carbohydrate chains like concanavalin A and therefore binds the α_2 subunit specifically (Fig. 1, lane 4), also binds α_1, β, and γ in digitonin or CHAPS (Fig. 2, left, lanes 1 and 2) but not in Triton X-100 or SDS (Fig. 2, left, lanes 3 and 4). These results show that all four of these calcium channel components are specifically associated with the α_1 subunit under conditions where the calcium channel retains functional activity.

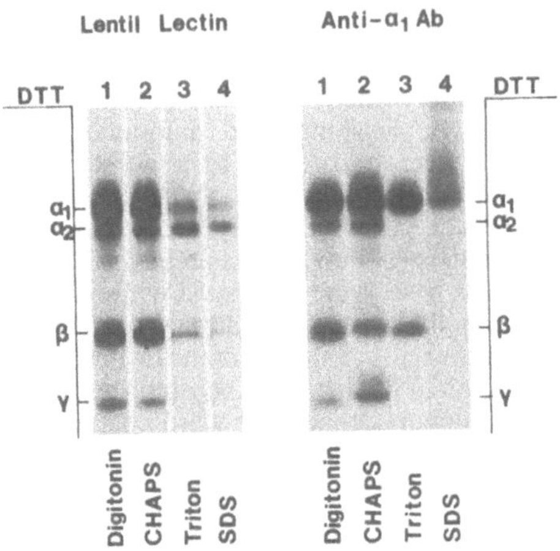

FIGURE 2

Noncovalent association of calcium channel subunits. *Lentil lectin:* One hundred nanograms of ^{125}I-labeled calcium channel in 150 μl of the detergents indicated below and 25 mM HEPES, 150 mM NaCl, 1 mM $CaCl_2$, 1 mM $MnCl_2$, 0.1 mM PMSF, 1 mM pepstatin A (adjusted to pH 7.5 with Tris base) was incubated for 90 min at 4°C with 50 μl of lentil lectin-agarose, equilibrated in the same buffer, under agitation. The resin was spun down and the flow-through collected. The resin was washed three times with 1 ml buffer and resuspended in 150 μl containing 0.2 M α-methylmannoside. The eluate was collected after a 90-min batch incubation. Specifically bound samples were analyzed by SDS-PAGE and autoradiography. *Lane 1*, 0.1% digitonin; *lane 2*, 0.1% CHAPS; *lane 3*, 0.5% Triton X-100; *lane 4*, samples treated with 1% SDS were exchanged into 0.5% Triton X-100 by filtration over a 2 ml Sephadex G-50 column and analyzed in 0.5% Triton X-100. *Anti-α_1 antibody:* ^{125}I-labeled calcium channel was immunoprecipitated using affinity purified anti-α_1 antibodies in an immunoassay buffer containing the indicated detergents: 0.5% digitonin, 0.1% CHAPS, and 1% Triton X-100, or 1% SDS for 2 min at 100°C followed by immunoassay in 0.5% digitonin detergent after exchange by gel filtration on a 2 ml Sephadex G-50 column.

On the basis of these results, and in analogy with current models of the structure of voltage-sensitive sodium channels,[22] we have proposed a model of the subunit structure of the skeletal muscle calcium channel (Fig. 3) based on a central ion channel-forming element interacting with three other noncovalently associated subunits. The α_1 subunit, which contains the calcium antagonist binding sites, cAMP-dependent phosphorylation sites (see below), and the largest hydrophobic domains, is proposed to be the central ion channel-forming component of the complex. Its apparent molecular weight of 175 kDa from $NaDodSO_4$/PAGE is likely to be a reasonable approximation of the true polypeptide molecular weight since no N-glycosylation was detected. This calcium channel subunit is therefore large enough to contain four homologous transmembrane domains analogous to those of the rat brain sodium channel α subunit, whose mRNA alone encodes a functional ion channel.[23,24] The β subunit is also a substrate for cAMP-dependent kinase (see below), hydrophobic labeling indicates that it does not interact with the membrane phase and it is not a glycoprotein. It is probably therefore tightly associated with an intracellular domain of α_1. The γ subunit interacts independently with α_1, contains hydrophobic segments, and consists of approximately 30% carbohydrate, so it is placed as a transmembrane component of the complex. The $\alpha_2\delta$ dimer interacts specifically with α_1, since the conditions necessary to achieve dissociation result in a loss of DHP binding activity. The 105 kDa core polypeptide of α_2 contains a heavily glycosylated extracellular domain but displays weak hydrophobic labeling, indicating a limited intramembrane domain. For this reason it seems unlikely that the ion channel is formed jointly by α_1 and α_2 at their zone of interaction. Thus, we believe that α_2 is also peripherally associated with the α_1 subunit in the native calcium channel complex as illustrated in Fig. 3.

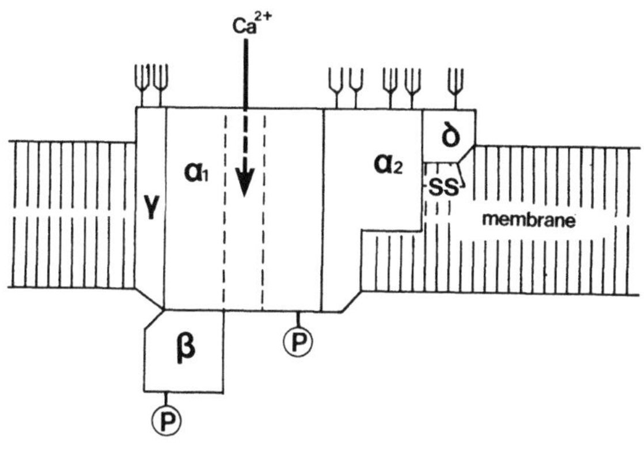

FIGURE 3

Proposed model for calcium channel structure. Sites of cAMP-dependent phosphorylation (P), glycosylation, and interaction with the membrane are illustrated.

The proposed model assumes a complex containing one of each subunit type. Our results suggest that each complex contains at least one α_1 and one α_2 subunit, but do not specify the stoichiometry of any subunits. α_1 and α_2 appear to be present in approximately equal amounts on silver stained gels, and the α_1 and β subunits incorporate approximately one mole of ^{32}P per mole of complex (see below). The oligomeric molecular weight of the skeletal muscle calcium channel is 380 kDa within reasonable range of the predicted size of the complex represented in Fig. 4 (416 kDa). Thus, an

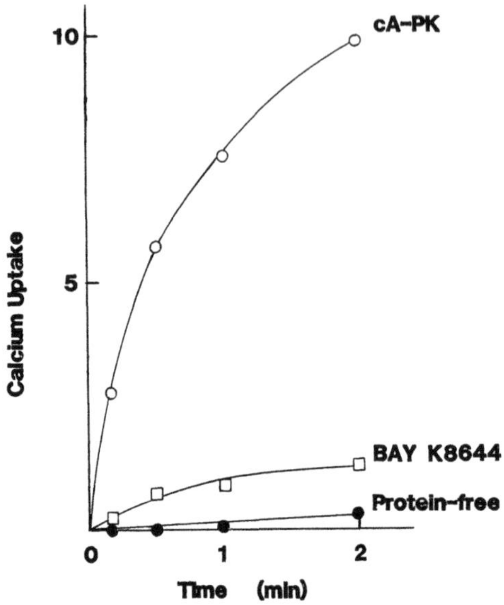

FIGURE 4

Activation of purified calcium channels by cAMP-dependent protein kinase. Calcium channels were radioactively labeled with PN200-110 at approximately 1% occupancy and saturated with Bay K 8644. The channels were solubilized, purified, and reconstituted into phosphatidylcholine vesicles. The reconstituted vesicles were incubated under control conditions or with 1 mM ATP and 2 μM cAMP-dependent protein kinase for 1 hr at 4°C and the initial rate of $^{45}Ca^{++}$ influx was measured as described previously.[26] Control experiments with protein-free vesicles were carried out in parallel.

assumption of one mole of each subunit in the complex is plausible, but requires direct experimental verification.

Functional reconstitution of skeletal muscle calcium channels and regulation by protein phosphorylation. After detergent solubilization and purification, calcium antagonist receptors retain binding activity for dihydropyridines and allosteric interactions among the receptor sites for dihydropyridines, verapamil, and diltiazem.[25-28] It is important to determine whether the purified calcium antagonist receptor can also function as a calcium channel. The development of dihydropyridines such as Bay K 8644 that activate calcium channels rather than inhibiting them[29] provides the necessary reagent for these studies. Bay K 8644 was bound to calcium antagonist receptors in transverse tubule membranes, and the receptors were solubilized, purified in digitonin, and incorporated into phospholipid vesicles.[26] Bay K 8644 increased the initial rate of influx of $^{45}Ca^{2+}$ into the vesicles 2- to 3-fold, and this increase was inhibited by PN200-110 and verapamil at concentrations that inhibit calcium currents in transverse tubule membranes. A close correlation was observed between the presence of the α, β, and γ size classes of subunits and the ability to mediate Bay K 8644-stimulated calcium influx, indicating that the purified calcium antagonist receptor was indeed capable of acting as a functional calcium channel. Fusion of similar reconstituted vesicles with phospholipid bilayers on the tips of patch micropipettes allowed recording single channel currents of approximately 20 pS (20 x 10^{-12} ohm^{-1}) mediated by purified calcium antagonist receptors.[27] The probability of channel activation was increased by Bay K 8644 and reduced by desmethoxyverapamil, but was not strongly affected by voltage. Thus, these results also support the conclusion that the purified calcium antagonist receptor can function as a calcium channel whose ion conductance activity is regulated by the same pharmacological agents that regulate dihydropyridine-sensitive calcium channels in situ.

Voltage sensitive calcium channels are subject to long-term modulation by a variety of second messengers and neurotransmitters. For example, the activation of dihydropyridine-sensitive calcium channels in the heart is enhanced by a pathway involving β-adrenergic receptors, adenylate cyclase, and cAMP-dependent protein kinase, and this pathway plays a critical role in modulation of beat rate and myocardial contractility by the sympathetic nervous system.[30-32] Activation of the calcium channels in skeletal muscle is also increased by cAMP and cAMP-dependent protein kinase.[33] Regulation of dihydropyridine-sensitive calcium channels by cAMP-dependent protein phosphorylation might involve phosphorylation of the calcium channel itself or phosphorylation of other proteins which in turn regulate channel activity. The catalytic subunit of cAMP-dependent protein kinase phosphorylates both the $\alpha 1$ and β subunits of the purified calcium channel (Fig. 1, lane 8), and can incorporate up to 0.9 mol of phosphate per mol of calcium antagonist receptor under optimal conditions.[17, 34] Thus, these two components of the calcium channel complex itself are probable sites of cAMP-dependent phosphorylation during regulation of the channel.

The functional effect of phosphorylation on purified calcium channels has been examined by both single channel recording methods[27] and ion flux methods. In reconstituted planar bilayers, the fraction of time that individual purified calcium channels spend in the active state at +165 mV was increased from 0.17 to 0.68 by phosphorylation with cAMP-dependent protein kinase after incorporation into the phospholipid membrane patch. In companion experiments, the α_1 and β subunits were phosphorylated under similar conditions.[27] In order to examine the effect of protein phosphorylation on the functional activity of the purified population of calcium channels as a whole, channels in reconstituted vesicles were phosphorylated by cAMP-dependent protein kinase and ATP and their initial rate of calcium flux after activation by Bay K8644 was measured (Fig. 4). Phosphorylation increased both the initial rate and the extent of calcium uptake 8- to 10-fold. Since there are fewer than one calcium channel per reconstituted vesicle in these preparations, these results indicate that both the probability of channel opening and the fraction of calcium channels that can be activated by Bay K8644 can be increased by cAMP-dependent phosphorylation. Evidently, the purified calcium channel complex is sufficient for physiological function, pharmacological modulation, and regulation by protein phosphorylation.

Structural homologies between the subunits of sodium and calcium channels. The sodium channel and the dihydropyridine-sensitive calcium channel are the first two voltage-gated ion channels to be purified and characterized. Tanabe et al[35] have isolated cDNA clones encoding the α_1 subunit of the calcium channel and deduced its primary structure. As previously observed for the α subunit of sodium channels,[36] the mRNA for the α_1 subunit of dihydropyridine-sensitive calcium channels encodes a polypeptide of 210 kDa containing four internally homologous domains which each has multiple (probably 6) hydrophobic sequences that are likely to be transmembrane segments. It is homologous to sodium channel α subunits in both amino acid sequence and overall structure. In particular, predicted transmembrane segments that are thought to play important functional roles in ion transport and voltage-dependent gating of the sodium channel[22, 36] are highly conserved in the calcium channel structure.[35] It seems likely that the models proposed for the functional architecture of the sodium channel[22, 36] also apply to the calcium channel,[35] and therefore that the α_1 subunit is the principal structural component of the transmembrane pore as diagrammed in Fig. 3.

Of the remaining calcium channel subunits, only the γ subunit appears to have a potential homolog among sodium channel subunits. The β1 subunits of the brain and skeletal muscle sodium channels are highly glycosylated, hydrophobic proteins with native sizes of 36 kDa and core polypeptides of 23 to 26 kDa.[22] In all these respects they are similar to the γ subunits of the calcium channel. In future experiments, it will be of interest to determine the function of this apparently homologous polypeptide as well as the other members of the calcium channel complex and to probe more thoroughly the relationship between the structure and function of the principal subunits of these two related voltage-sensitive ion channels.

REFERENCES

1. S. Hagiwara and L. Byerly, Calcium channel, Ann. Rev. Neurosci. 4:69 (1981).
2. C. M. Armstrong and D. R. Matteson, Two distinct populations of calcium channels in a clonal line of pituitary cells, Science 227:65 (1985).
3. E. Carbone and H. D. Lux, A low voltage-activated fully inactivating calcium channel in vertebrate sensory neurones, Nature 310:501 (1984).
4. M.C. Nowycky, A. P. Fox, and R. W. Tsien, Three types of neuronal calcium channel with different calcium agonist sensitivity, Nature 316:440 (1985).
5. J. Sanchez and E. Stefani, Inward calcium current in twitch muscle fibers of the frog, J. Physiol. (Lond.) 283:197 (1978).
6. W. Almers, E. McCleskey, and P. Palade, in: "Calcium in Biological Systems" R. P. Rubin, G. B. Weiss, J. W. Putney, Jr. eds., Plenum Press, New York, London. pp. 321-330 (1985).
7. G. Cota and E. Stefani, A fast-activated inward calcium current in twitch muscle fibres of the frog (*Rana montezume*,) J. Physiol. 370:151 (1986).
8. C. Cognard, M. Lazdunski, and G. Romey, Different types of Ca^{2+} channels in mammalian skeletal muscle cells in culture, Proc. Natl. Acad. Sci. USA 83:517 (1986).
9. H. Glossmann, D. R. Ferry, and C. B. Boschek, Purification of the putative calcium channel from skeletal muscle with the aid of [^{3}H]-nimodipine binding, Naunyn-Schmiedeberg's Arch. Pharmacol. 323:1 (1983).
10. M. Fosset, E. Jaimovich, E. Delpont, and M. Lazdunski, [^{3}H] nitrendipine receptors in skeletal muscle. Properties and preferential localization in transverse tubules, J. Biol. Chem. 258:6086 (1983).
11. H. Glossmann, R. Ferry, A. Goll, J. Striessnig, and M. Schober, Calcium channels: Basic properties as revealed by radioligand binding studies, J. Cardio. Pharmacol. 7:520 (1985).
12. J. Galizzi, M. Borsotto, J. Barhanin, M. Fosset, and M. Lazdunski, Characterization and photoaffinity labeling of receptor sites for the Ca^{2+} channel inhibitors d-cis-diltiazem, (+)-bepridil, desmethoxyverapamil and (+)PN200-110 in skeletal muscle transverse tubule membranes, J. Biol. Chem. 261:1393 (1986).
13. D. Ferry, M. Rombusch, A. Goll, and H. Glossmann, Photoaffinity labelling of Ca^{2+} channels with [3H]azidopine, FEBS Lett. 169:112 (1984).
14. J. Striessnig, H. -G. Knaus, M. Grabner, K. Moosburger, W. Seitz, H. Lietz, and H. Glossmann, Photoaffinity labelling of the phenylalkylamine receptor of the skeletal muscle transverse tubule calcium channel, FEBS Lett. 212:247 (1987).
15. B. M. Curtis and W. A. Catterall, Solubilization of the calcium antagonist receptor from rat brain, J. Biol. Chem. 258:7280 (1983).
16. B. M. Curtis and W. A. Catterall, Purification of the calcium antagonist receptor of the voltage-sensitive calcium channel from skeletal muscle transverse tubules, Biochemistry 23:2113 (1984).
17. M. Takahashi, M. Seagar, J. Jones, B.F.X. Reber, and W. A. Catterall, Subunit structure of dihydropyridine-sensitive calcium channels from skeletal muscle, Proc. Natl. Acad. Sci. USA 84:5478 (1987).
18. A. Leung, T. Imagawa, and K. Campbell, Structural characterization of the 1,4 dihydropyridine receptor of the voltage-dependent Ca^{2+} channel from rabbit skeletal muscle, J. Biol. Chem. 262:7943 (1987).
19. A. H. Sharp, T. Imagawa, A. T. Leung, and K. P. Campbell, Identification and characterization of the dihydropyridine binding subunit of the skeletal muscle dihydropyridine receptor, J. Biol. Chem. 262:12309 (1987).
20. M. E. Morton and S. C. Froehner, Monoclonal antibody identifies a 200-kDa subunit of the dihydropyridine-sensitive calcium channel, J. Biol. Chem. 262:11904 (1987).

21. P. L. Vaghy, J. Streissnig, K. Miwa, H. -G. Knaus, K. Itagaki, E. McKenna, H. Glossmann, and A. Schwartz, Identification of a novel 1,4-dihydropyridine- and phenylalkylamine-binding polypeptide in calcium channel preparations, J. Biol. Chem. 262:14337 (1987).
22. W. A. Catterall, Molecular properties of voltage-sensitive sodium channels, Ann. Rev. Biochem. 55:953 (1986).
23. A. Goldin, T. Smith, H. Lubbert, A. Dowsett, J. Marshall, V. Auld, W. Downey, L. C. Fritz, H. A. Lester, R. Dunn, W. A. Catterall, and N. Davidson, Messenger RNA coding for only the α subunit of the rat brain Na channel is sufficient for expression of functional channels in *Xenopus* oocytes, Proc. Natl. Acad. Sci. USA 83:7503 (1986).
24. M. Noda, T. Ikeda, T. Suzuki, H. Takeshima, Takahashi M., M. Kuno, and S. Numa, Expression of functional sodium channels from cloned cDNA, Nature 32:826 (1986).
25. J. Barhanin, T. Coppola, A. Schmid, M. Borsotto, and M. Lazdunski, The calcium channel antagonists receptor form rabbit skeletal muscle. Reconstitution after purification and subunit characterization, Eur. J. Biochem. 164:525 (1987).
26. B. M. Curtis and W. A. Catterall, Reconstitution of the voltage-sensitive calcium channel purified from skeletal muscle transverse tubules, Biochemistry 25:3077 (1986).
27. V. Flockerzi, H. -J. Oeken, F. Hofmann, D. Pelzer, A. Cavalie, and W. Trautwein, Purified dihydropyridine binding site from skeletal muscle T-tubules in a functional calcium channel, Nature (Lond.) 323:66 (1986).
28. J. Striessnig, A. Goll, K. Moosburger, and H. Glossmann, Purified calcium channels have three allosterically coupled drug receptors, FEBS Lett. 197:204 (1986).
29. M. Schramm, G. Thomas, R. Towart, and G. Franckowiak, Novel dihydropyridines with positive inotropic action through activation of Ca^{2+} channels, Nature 303:535 (1983).
30. H. Reuter, Properties of two inward membrane currents in the heart. Ann. Rev. Physiol. 41:413 (1974).
31. R. Tsien, B. Bean, P.Hess, J. Lansman, B. Nilius, and M. Nowycky, Mechanisms of calcium channel modulation by β-adrenergic agents and dihydropyridine calcium agonists, J. Mol. Cell Cardiol. 18:691 (1986).
32. G. Brum, V. Flockerzi, F. Hofmann, W. Osterrieder, and W. Trautwein, Injection of catalytic subunit of cAMP-dependent protein kinase into isolated cardiac myocytes, Pflugers Archiv. 398:147 (1983).
33. J. Arreola, J. Calvo, M. C. Garcia, and J. A. Sanchez, Modulation of calcium channels of twitch skeletal muscle of the frog by adrenaline and cyclic adenosine monophosphate fibres, J. Physiol. (Lond.) 393:307 (1987).
34. B. M. Curtis and W. A. Catterall, Phosphorylation of the calcium antagonist receptor of the voltage-sensitive calcium channel by cAMP-dependent protein kinase, Proc. Natl. Acad. Sci. USA 82:2528 (1985).
35. T .Tanabe, H. Takeshima, A. Mikami, V. Flockerzi, H. Takahashi, K. Kanagawa, M. Kojima, H. Matsuo, T. Hirose, S. Numa, Primary structure of dihydropyridine binding calcium channel from rabbit skeletal muscle, Nature 328:313 (1987).
36. M. Noda, T. Ikeda, T. Kayano, H. Suzuki, H. Takeshima, M. Kurasaki, H. Takahashi, S. and Numa, Existence of distinct sodium channel messenger RNAs in rat brain, Nature 320:188 (1986).

INTRACELLULAR MECHANISMS INVOLVED IN "RUN-DOWN" OF CALCIUM CHANNELS

M. Kameyama, A. Kameyama, M. Kaibara, T. Nakayama and H. Irisawa

National Institute for Physiological Sciences
Okazaki 444, Japan

INTRODUCTION

Voltage-dependent calcium channels in the cell membrane play important roles in cellular excitation related to rhythmic firing, synaptic transmission, muscle contraction and secretion (1-3). It is known that neural and hormonal signals modulate activities of L- and N-type Ca^{2+} channels by mechanisms involving protein phosphorylation or coupling of GTP-binding proteins to the channel molecules (3-5). In most electrophysiological studies, Ca^{2+}-channel currents were isolated from other membrane currents by replacing the cytoplasm with a simple solution using cell dialysis method or by single-channel recording using patch clamp technique. It has, however, been noticed, since pioneering work by Kostyuk and collaborators (6), that Ca^{2+} current becomes labile and diminished progressively when the intracellular side of the membrane is perfused with an artificial solution. Such a loss of channel activity during intercellular dialysis in whole-cell clamp or in cell-free patches is referred to as 'run-down' or 'wash-out' (1,6-8). Since L- and N- but not T-type Ca^{2+} channels are subject to run-down, it has been suggested that the run-down of Ca^{2+} channel may result from disturbance of cellular metabolism or of some regulatory mechanism of the Ca^{2+} channels (7-13). Thus, possible mechanisms of the run-down are 1) dephosphorylation of channel protein due to loss of protein kinases or activation of appropriate protein phosphatases; 2) proteolysis of channel molecule by activation of intrinsic proteases; 3) decoupling of GTP-binding protein from the channel; and 4) loss of other cytoplasmic factor(s) required for the channel activity.

In the present study we examined possible involvement of these mechanisms in the run-down of cardiac L-type Ca^{2+} channels in inside-out patches (14) and in reconstituted liposomes (15). The results suggest that the run-down is most likely caused by a loss of cytoplasmic protein(s) which maintains the Ca^{2+} channel to be available upon depolarization.

METHODS

Electrophysiology (14). Single cardiac myocytes were dispersed from the guinea-pig ventricle by collagenase. Ca^{2+}-channel currents was recorded at 31-35 °C by a patch clamp method with either cell-attached or inside-out configurations. The recording pipette contained (in mM) $BaCl_2$

50, choline Cl 70, Bay K 8644 0.003 (Ca^{2+}-channel agonist), and HEPES-Tris buffer 10, pH 7.4, and bathing solution contained (in mM) K aspartate 90, KCl 30, KH_2PO_4 10, EGTA 0.5-1, and HEPES-KOH buffer 5-10, pH 7.2-7.4 (basic internal solution). Depolarizing pulses from -80 to 0 mV applied at a rate of 0.5 Hz with a duration of 200-300 ms elicited eventual openings of L-type Ca^{2+} channels.

Reconstitution of Ca^{2+} channels into liposomes (15). Sarcolemmal fraction of bovine ventricle was separated by affinity purification using wheat germ agglutinin-sepharose gel, and membrane proteins were solubilized by CHAPS and reconstituted into soybean phospholipid liposomes by sonication. Then, proteoliposomes were loaded with 1 mM Ca^{2+} and 10 mM K^+ by freeze-thaw cycles. Ca^{2+} efflux from proteoliposomes was induced by diluting the liposome solution (2-5 µl) into 2.5 ml of test solution containing 10 mM KCl, 140 mM choline-Cl, 0.1 µM valinomycin, 0.5 µM fura-2, and 20 mM HEPES-Tris buffer 20 at pH 7.4. A resultant increase in Ca^{2+} level was monitored by fura-2 fluorescence with excitation and emission wave lengths of 340/380 and 500 nm, respectively.

Cardiac tissue extract. Guinea-pig or bovine ventricles were homogenized in 3 vol. (v/w) of the basic internal solution at 4 °C, centrifuged at 100,000 x g for 30 min and then filtered through a 0.22 µm Millipore filter. Protein content in the supernatant was adjusted to 5-10 mg/ml.

RESULTS AND DISCUSSION

It is known that cardiac L-type Ca^{2+} channel runs down very quickly in the inside-out than in the whole-cell clamp mode (13,16-18). In the cell-attached mode, the activity of Ca^{2+} channel, represented by a product of number of channels (N) times open probability of channels (P_0), is stable as long as hours although there exists sweep-to-sweep fluctuation arising from a slow gating mechanism of the channel. When, however, a membrane patch was excised into external solution to obtain an inside-out patch, the channel activity was progressively decreased and finally abolished within 2-3 min in most cases. This run-down of Ca^{2+} channel was not affected by 20 mM EGTA or 2 mM BAPTA in the external solution, suggesting that the run-down is not mediated by a Ca^{2+}-dependent process such as Ca^{2+}-mediated inactivation of the channel. Similarly, the run-down was not prevented by protease inhibitors, such as leupeptin (10 µM), benzamidine (1 mM) and phenylmethylsulphonyl fluoride (0.1 mM) or by activation of a GTP-binding protein by 0.1 mM GTPγS in the bath and 1 µM isoprenaline in the pipette. Thus, the results do not support the idea that the run-down of Ca^{2+} channel in the inside-out mode is caused by proteolysis of channel protein or decoupling of the channel from GTP-binding protein.

Then, we examined the effect of phosphorylation on the run-down of Ca^{2+} channel. Formation of inside-out patches in the presence of 2 µM catalytic subunit (C subunit) of cAMP-dependent protein kinase (PKA) or 120 mM phosphate, which may suppress activities of phosphatases, failed to maintain the channel activity. In 3 out of 12 experiment, however, a prolongation in the time course of run-down (>6 min) was observed. Application of C subunit after run-down of the channel did not restore channel activity. These results indicate that the run-down is not accounted for by a loss of PKA and resulting dephosphorylation of a phosphoprotein which regulates the activity of Ca^{2+} channel (see also ref. 13).

The idea that some soluble cytoplasmic factor(s) is required for the activity of Ca^{2+} channel and hence a loss of this factor leads to run-down of the channel has been proposed in earlier studies (1,7,8). Since the

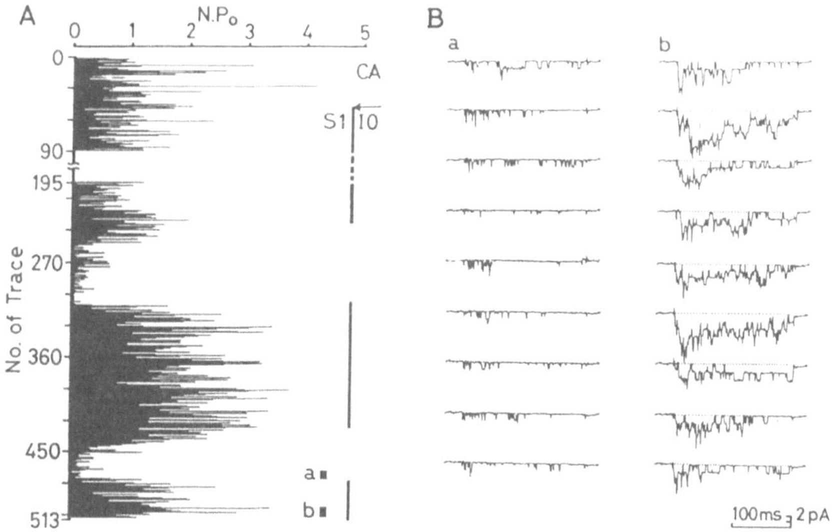

Fig. 1. Effects of cardiac tissue extract on the activity of Ca^{2+} channels. Ca^{2+}-channel current was elicited by depolarizing pulses from -80 to 0 mV at 0.5 Hz, with pipette solution containing 50 mM Ba^{2+} and 3 μM Bay K 8644. A. Channel activity, represented by the product of number of channels in the patch (N) x mean open-state probability during initial 100 ms of a test pulse (P_0), is plotted on a sweep-by-sweep basis: CA, cell-attached mode; IO, inside-out mode. The tissue extract (S1) together with 10 μM cAMP and 3 mM MgATP was present in the bathing solution during the times indicated by lines. Sweeps of No. 90-193 have been omitted. B. Example sweeps obtained in the absence (a, No. 467-475) or in the presence (b, No. 503-511) of the extract during the periods marked with thick bars in A. (from ref. 14)

results in the above experiments do not support PKA as a candidate, we examined the effect of cytoplasm on the Ca^{2+}-channel activity in inside-out patches. When a membrane patch was excised into the external solution containing the supernatant fraction of cardiac tissue homogenate (5-10 mg protein/ml), the activity of Ca^{2+} channel was maintained as shown in Fig. 1. Washing out the tissue extract resulted in a marked decrease in $N.P_0$ value, and reintroduction of the extract increased the channel activity again. Although the extract used in the experiment of Fig. 1 contained cAMP and MgATP, these nucleotides were not essential for the effect of the extract but seemed to produce an additional increase in $N.P_0$. In other experiments, the extract could induce channel openings in inside-out patches after run-down of Ca^{2+} channels. The extract, however, did not induce openings of a channel in silent patches, where Ca^{2+} channel activity was not observed in the preceding cell-attached mode. The channels activated by the extract was indistinguishable form the Ca^{2+} channel observed in the cell-attached mode in terms of the single channel conductance (20-25 pS), distribution of open times (time constants of ∿1 and ∿10 ms), and voltage-dependent gating. These results strongly suggest that the run-down of Ca^{2+} channel observed in the inside-out mode is not an irreversible process such as proteolysis but is caused by a loss of some factor(s) in the cytoplasm.

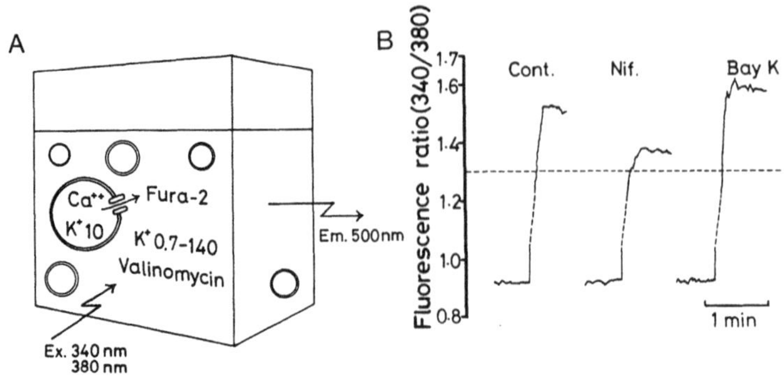

Fig. 2. Measurement of Ca^{2+} efflux through Ca^{2+} channels reconstituted into liposomes. A. Schematic illustration of Ca^{2+} efflux measurement with the fluorescent dye fura-2. Membrane potential of liposomes was controlled by K^+ gradient with K^+ ionophore valinomycin. Excitation wave length was 340 or 380 nm, and 500 nm emission was measured. B. Fluorescence signals induced by addition of liposomes into the cuvette of photometer. Membrane potential of liposomes was changed from 63 to 0 mV (note inside-out form of the channels), and fluorescence ratio at 340/380 nm excitation was measured. Broken line indicates the basal increase in Ca^{2+} level. The excess increase was suppressed by nifedipine (Nif, 1 μM) and enhanced by Bay K 8644 (Bay K, 1 μM).

The cytoplasmic factor involved in the activity of Ca^{2+} channel seems to be a protein because the effect of cytoplasm on the channel activity was abolished by heating (80 °C, 5 min) or trypsin digestion. Therefore, we attempted to characterize this factor by using chromatography and reconstituted Ca^{2+} channels as an assay system (15). Ca^{2+} channels from bovine heart were partially purified, solubilized and incorporated into soybean phospholipid liposomes. As illustrated in Fig. 2, when proteoliposomes, loaded with 1 mM Ca^{2+} and 10 mM K^+, were added into a solution in the cuvette of a fluorescent spectrophotometer, containing 10 mM K^+, 0.1 μM valinomycin and a fluorescent Ca^{2+} indicator, fura-2 (0.5 μM), the fluorescence intensity of fura-2 increased resulting from an increased Ca^{2+} level. This Ca^{2+} increase originated, in part from residual Ca^{2+} in the suspension medium of liposomes (basal increase). The excess increase was due to an efflux of Ca^{2+} from proteoliposomes through reconstituted Ca^{2+} channels because the fluorescence signal was 1) dependent on the membrane potential of liposomes developed by K^+ gradient; 2) suppressed by a Ca^{2+} channel blocker, nifedipine (1 μM), and enhanced by a Ca^{2+} channel agonist, Bay K 8644 (1 μM); and 3) enhanced 3-fold by pre-incubation of liposomes with C subunit (5 μM) and MgATP. Thus, most of functional Ca^{2+} channels in the liposomes may exist in an inside-out form. It should, however, be mentioned that a very small percentage of the channels seems to function in the liposemes. This notion is based on a finding that channel density of liposomes estimated by binding to a Ca^{2+} channel ligand, $[^3H]$(+)PN 200-110, is far greater than to account for the amount of Ca^{2+} efflux from liposomes. One plausible explanation for this disparity may be that most of the reconstituted channels might be in non-functional states, including a state of 'run-down'. Consequently, the cardiac tissue extract (0.2 mg protein/ml) increased Ca^{2+} efflux from liposomes 3-6 times.

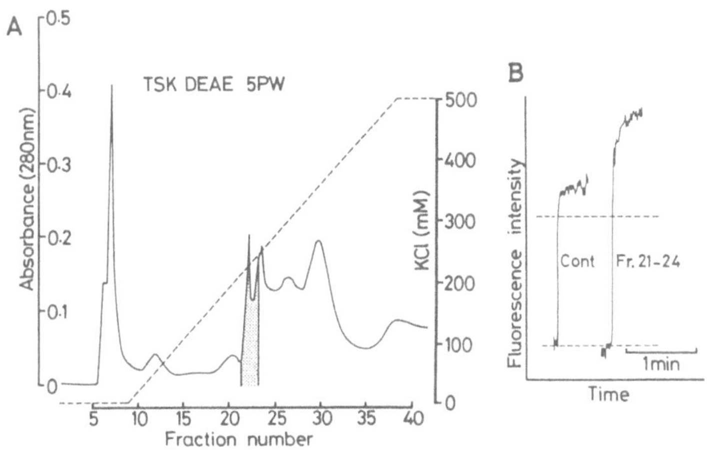

Fig. 3. A. Elution profile of the cardiac tissue extract on a DEAE-5PW gel. Fraction No. 21-22 (shaded area) was found to increase Ca^{2+} efflux from liposomes most effectively. B. Enhancement of Ca^{2+} efflux by fraction No. 21-24. Fluorescence signal was measured at 340 nm excitation. Total calcium level was 1 nM before addition of liposomes; 25 nM for basal increase (broken line); 32 and 70 nM in the absence and presence of the extract, respectively.

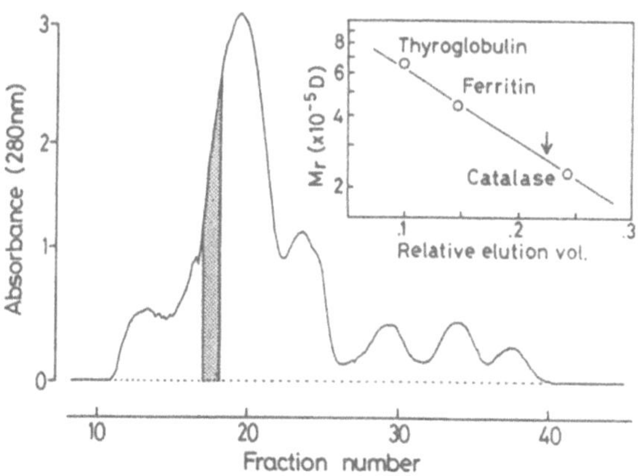

Fig. 4. Gel filtration of the cardiac tissue extract. The column (TSK HW-60 gel) was calibrated with thyroglobulin (670k), ferritin (440k), catalase (232k) and aldolase (160k). Fraction No. 17 (shaded area) was most effective in increasing Ca^{2+} efflux from liposomes as well as in inducing channel activity in inside-out patches. An apparent molecular weight for the factor of about 230×10^3 was estimated as indicated by the arrow in the inset. (from ref. 14)

Proteins in the cardiac extract were fractionated by chromatography using a DEAE-5PW gel and fractions are assayed by the liposome method for the ability to enhance Ca^{2+} efflux. Fig. 3-A. shows an elution pattern of the cardiac extract on DEAE-5PW column. We found that Ca^{2+} efflux from proteoliposomes was enhanced by fractions of No. 21-24 (around shaded area in the figure), which were eluted at about 250 mM KCl. As depicted in Fig. 3-B, Ca^{2+} efflux is enhanced 6-fold by this fraction. The cardiac extract was also fractionated by gel filtration and fractions were assayed by the liposome method.

Fig. 4. shows a gel filtration profile of the extract, in which fractions in the shaded area were found to enhance Ca^{2+} efflux. These fractions corresponded to an apparent molecular weight of 200-300 x 10^3, and were also able to restore Ca^{2+} channel activity in inside-out patches. Further characterization and purification are under investigation in this laboratory.

The results in the present study suggest that some cytoplasmic protein(s) in the heart maintains the activity of cardiac L-type Ca^{2+} channel and that run-down of the channel in inside-out patches results from a loss of this cytoplasmic protein. This protein, however, is not cAMP-dependent protein kinase because its C subunit does not recover the channel from the run-down (13) and the cytoplasm can induce channel activity in the absence of MgATP. Recently, Belles et al. (19,20) reported that the run-down of whole-cell Ca^{2+} current in cardiac cells is Ca^{2+}-dependent and, furthermore, that the run-down is accelerated by calpain, a Ca^{2+}-activated protease, and slowed by calpastatin, an inhibitor of calpain. In our experiment, the run-down in inside-out patches is not prevented by lowering Ca^{2+} level or by protease inhibitors such as leupeptin. These results suggest that mechanisms underlying run-down of the Ca^{2+} channel may vary under different conditions.

Acknowledgements The authors are grateful to Mr. O. Nagata and Mr. M. Ohara for their technical assistance. This work was supported by a Research Grant from the Ministry of Education, Science and Culture of Japan.

REFERENCES

1. S. Hagiwara and L. Byerly, The calcium channel, Trend. Neurosci. 6:189 (1983).
2. R. W. Tsien, Calcium channels in excitable cell membranes, Ann. Rev. Physiol. 45:341 (1983).
3. R. J. Miller, Multiple calcium channels and neuronal function, Science 235:46 (1987).
4. H. Reuter, Calcium channel modulation by neurotransmitters, enzymes and drugs, Nature 301:569 (1983).
5. F. Hofmann, W. Nastainczyk, A. Röhrkasten, T. Schneider and M. Sieber, Regulation of the L-type calcium channel, Trend. Pharmacol. Sci. 8:393 (1987).
6. P. G. Kostyuk, O. A. Krishtal and V. I. Pidoplichko, Calcium inward current and related charge movements in the membrane of snail neurones, J. Physiol. 310:403 (1981).
7. P. G. Kostyuk, Intracellular perfusion of nerve cells and its effects on membrane currents, Physiol. Rev. 64:435 (1984).
8. E. M. Fenwick, A. Marty and E. Neher, Sodium and Calcium channels in bovine chromaffin cells, J. Physiol. 331:599 (1982).

9. H. Chad, R. Eckert, An enzymatic mechanism for calcium current inactivation in dialyzed Helix neurones, J. Physiol. 378:31 (1986).
10. P. Forscher, and G. S. Oxford, Modulation of calcium channels by norepinephrine in internally dialyzed avian sensory neurons, J. Gen. Physiol. 85:743 (1985).
11. L. Byerly and B. Yazejian, Intracellular factor for the maintenance of calcium currents in perfused neurones from the snail, Lymnaea Stagnalis, J. Physiol. 370:631 (1986).
12. D. Armstrong and R. Eckert, Voltage-gated calcium channels that must be phosphorylated to respond to membrane depolarization, Proc. Natl. Acad. Sci. USA 84:2518 (1987).
13. A. Yatani, J. Codina, Y. Imoto, J. P. Reeves, L. Birnbaumer and A. M. Brown, A Gprotein directly regulates mammalian cardiac calcium channels, Science 238:1288 (1987).
14. M. Kameyama, A. Kameyama, T. Nakayama and M. Kaibara, Tissue extract recovers cardiac calcium channels from 'run-down', Pflügers Arch. 412:328 (1988).
15. A. Kameyama and T. Nakayama, Calcium efflux through cardiac calcium channels reconstituted into liposomes - flux measurement with fura-2, Biochem. Biophys. Res. Comm. 154:1067 (1988).
16. A. Cavalié, R. Ochi, D. Pelzer and W. Trautwein, Elementary currents through Ca^{2+} channels in guinea pig myocytes, Pflügers Arch. 398:284 (1983).
17. B. Nilius, P. Hess, J. B. Lansman and R. W. Tsien, A novel type of cardiac calcium channel in ventricular cells, Nature 316:443 (1985).
18. M. Kaibara and M. Kameyama, Inhibition of the Ca^{2+} channel by intracellular proton in single ventricular myocytes of guinea-pig, J. Physiol. 403:621 (1988).
19. B. Belles, C. O. Malécot, J. Hescheler and Trautwein, "Run-down" of the Ca current during long whole-cell recordings in guinea pig heart cells: role of phosphorylation and intracellular calcium, Pflügers Arch. 411:353 (1988).
20. B. Belles, J. Hescheler and Trautwein, K. Blomgren and J. O. Karlsson, A possible physiological role of the Ca-dependent protease calpain and its inhibitor calpastatin on the Ca current in guinea pig myocytes, Pflügers Arch. 412:554 (1988).

A COMPARATIVE STUDY ABOUT VOLTAGE-DEPENDENT Ca CURRENTS IN SMOOTH MUSCLE CELLS ISOLATED FROM SEVERAL TISSUES

M.Watanabe, Y. Imaizumi, K. Muraki and M. Takeda

Department of Chemical Pharmacology
Faculty of Pharmaceutical Sciences
Nagoya City University
467 Nagoya, Japan

INTRODUCTION

It is well known that many characteristics of smooth muscles vary significantly depending on the source tissues or organs. Therefore, smooth muscles have been classified into two or three groups with regard to the innervation pattern of autonomic nerves, membrane excitability, automaticity, cell to cell coupling etc.[1,2] The physiology and pharmacology of each group have been often described separately or comparatively according to this basis of the classification.

Recent development of techniques in electrophysiology and cell isolation allows measurement of whole cell or single channel current under whole cell clamp or patch clamp[3], respectively. Characteristics of macroscopic[4-9] and single channel currents[10-14] through voltage dependent Ca channels have been reported in several kinds of smooth muscle cells. However, experimental conditions varied markedly from study to study and were often far from the physiological ones. Therefore, information for direct comparison of characteristics between Ca currents in different smooth muscle cells still is not available. Our main goal in this study was to examine the characteristics of Ca channel current in various types of smooth muscle cells under the same experimental conditions. Quantitative comparison of Ca currents and counteracting K currents in various smooth muscle cells suggests an explanation for the observed differences in membrane excitability. A noninactivating Ca current which may be responsible for Ca influx during sustained depolarization in high K^+ solution was also examined and compared[15] in several types of smooth muscle cells.

METHODS

Cell-Isolation and Electrical Recordings Single smooth muscle cells were isolated using collagenase and elastase from following tissues; taenia caecum, urinary bladder, ureter, vas deferens of the guinea pig, portal vein, pulmonary artery, femoral artery, aorta, iris sphincter of the rabbit and trachea of the dog. The procedures of cell-isolation were the same as described previously[16,17]. The cells were placed in a bath having a volume of 0.5ml and mounted on a stage of phase-contrast microscope. Membrane currents were recorded from single cells under the voltage-clamp using a single suction micropipette method[3]. The recording system and details of data storage, aquisition and analysis were the same as described previously[17].

Application of Test Substances A test substance was applied by changing the perfusion solution to one which contained the substance at a certain concentration. This required a delay of about 30sec to reach the final concentration because of the replacement of the medium in the bath. In some experiments, therefore, test substances were applied by puff of the solution from a micropipette placed just beside the target cell[15]. The response to this application occurred after very short delay (<10msec), but the exact concentration at the site of effects was not known. Concentration-clamp was occasionally performed by the method introduced by Hering et al.[18] and was used to avoid the disadvantages of puff application.

Solutions The composition of normal medium was as follows (mM); NaCl 137, KCl 5.9, $MgCl_2$ 1.2, $CaCl_2$ 2.2 or 5.0, glucose 14.0, HEPES 10.0. The pH of the medium was adjusted to 7.2 by NaOH. In experiments to record I_{Ca}, most of NaCl (120mM) in the medium was replaced with equimolar tetraethylammonium chloride (TEA solution). The pipette-filling solution (pipette solution) contained (mM): K aspartate 110, KCl 30, $MgCl_2$ 4.0, HEPES 10, ATP 2Na 5, EGTA 0.5. The pH of the pipette solution was adjusted to 7.2 by KOH. When I_{Ca} was recorded, 110mM K aspartate and 30mM KCl were replaced with equimolar CsCl, and EGTA concentration was elevated to 5mM in the pipette solution. Most of the experiments were done at $30 \pm 1^{o}C$.

RESULTS and DISCUSSION

A Classification of Smooth Muscles

For comparison, smooth muscles were roughly classified into three groups with regard to the membrane excitability as shown in Table 1. This table is neither exhaustive nor completely accurate but is rather shown to make the comparison easy in the following paragraphs. Class A includes highly excitable smooth muscles having myogenic spontaneous activities under physiological conditions. Class B includes smooth muscles which elicit action potentials only when activated by nerve or direct electrical stimulation. Smooth muscles having low excitability were listed in Class C. Neither nerve nor direct stimulation elicit action potentials in these muscles. Moreover, some agonists including transmitters induce contraction without action potentials or even significant depolarization by pharmaco-mechanical coupling.

Table 1 A Classification of Smooth Muscles with Regard to Membrane Excitability

A: Spontaneous Action Potentials
intestinal smooth muscles (duodenum, jejunum, ileum, taenia cecum*, etc.), portal vein*, urinary bladder*, pyloric antrum, myometrium, mesenteric vein, etc.

B: Evoked Action Potentials
ureter*, vas deferens*, mesenteric artery, fundus ventriculi, etc.

C: No Action Potentials
large arteries (aorta*, pulmonary artery*, femoral artery*), trachea*, iris sphincter*, etc.

*: Tissues from which single cells were isolated and used in this study

Amplitude of Ca currents in Various Smooth Muscle Cells

Ca currents (I_{Ca}s) were recorded from single cells under the same experimental conditions to compare the amplitude. Fig. 1 shows I_{Ca}s recorded from 5 types of smooth muscle cells. Outward currents were blocked by Cs^{+} in the pipette solution. The medium contained 2.2mM Ca^{++}. Large I_{Ca}s were recorded in cells from urinary bladder (Fig.1A), portal vein (B) and vas deferens (C). In aortic and iris sphincter cells, I_{Ca}s were extremely small; approximately 20pA and 10pA at the peak, respectively. Application of Ca channel agonist, Bay K 8644, or replacement of extracellular Ca with 30mM Ba^{++} significantly increased the amplitude of inward currents even in cells from aorta or iris sphincter (Fig. 1D and E). Fig.2 illustrates the summary of the maximum amplitude of I_{Ca} in the presence of 5mM Ca^{++}. To minimize outward currents and Ca-induced I_{Ca} inactivation, the pipette

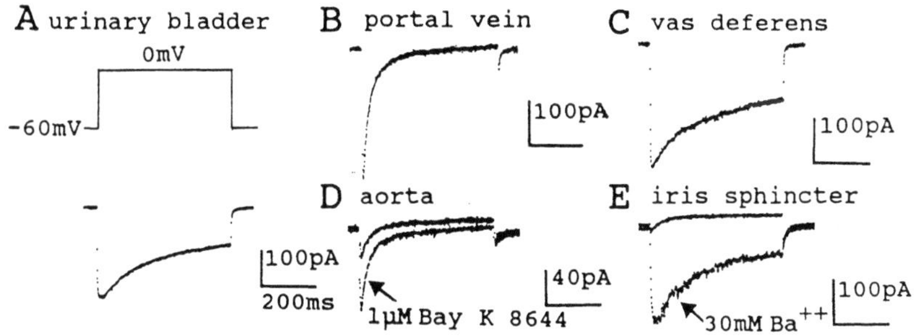

Fig. 1 I_{Ca}s were recorded from 5 types of cells under voltage clamp. Upper trace in A shows the applied pulse. $[Ca^{++}]_o$=2.2mM. Pipette solution contained 140mM CsCl. In D and E, two traces were superimposed.

solution contained 140mM CsCl and 5mM EGTA[4], respectively. Since fast Na current has been found in some smooth muscle cells[19,20], extracellular Na^+ was replaced with tetraethylammonium (TEA) to avoid the possible contamination of Na currents in these cells. Current-voltage relationships of peak I_{Ca} in these cells (not shown) indicate that inward currents were recorded in the voltage range from about -30 to +40mV and that the maximum amplitude was obtained between -10 and 0mV in portal vein and between 0 and +10mV in other cells examined. The maximum amplitude of I_{Ca}s was large in smooth muscles which show spontaneous or evoked action potentials under physiological conditions. Cells isolated from low or not excitable tissues had small I_{Ca}s (Fig.2). Therefore, roughly speaking, excitability of smooth muscles well correlates with the amplitude of I_{Ca} upon depolarization. Of interest is that canine tracheal smooth muscle cells, however, have substantial amplitude of I_{Ca} (cf. ref. 21) which is close to those in cells from taenia caecum or ureter of the guinea pig.

Factors Deciding Membrane Excitability in Smooth Muscle Cells

To elucidate the mechanism of the difference in excitability, other membrane currents were also compared in cells from vas deferens and trachea. An action potential was easily evoked under normal conditions in vas deferens (Fig. 3A; cf. ref. 22). In tracheal cells, however, action potentials are observed only in the presence of TEA and/or 4-AP (Fig.3B). Depolarization from -60 to 0mV elicited transient inward (arrow in Fig.4A, left) and following outward currents in cells from vas deferens. It is clear that this transient inward current is responsible for the initial depolarizing phase in action potential.

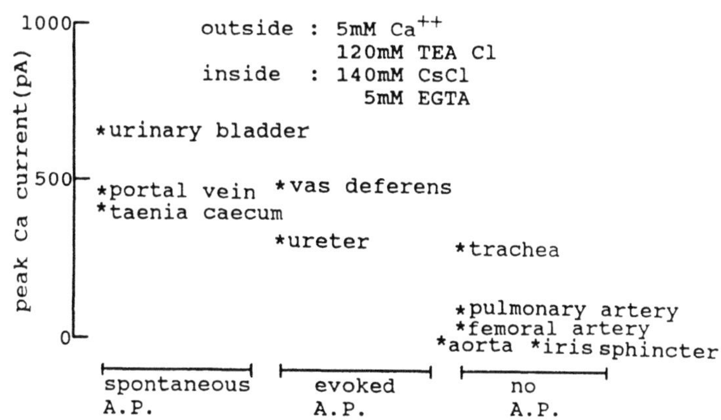

Fig.2 The summarized data of maximum I_{Ca} amplitude in ten types of cells measured under the same conditions shown at the top. V_H = -60mV.

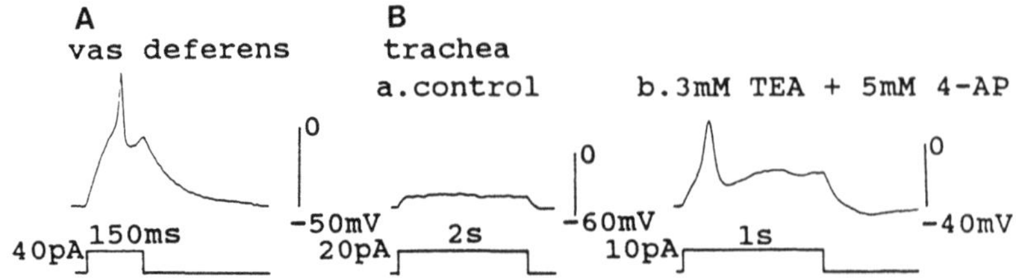

Fig. 3 Action potentials in cells from vas deferens (A) and trachea. (B) Depolarizing currents were applied through the recording pipettes as shown in lower traces under current clamp mode.

In tracheal cells, the inward current upon depolarization was not detected or extremely small (Fig.4A; right). When outward currents were partly blocked by K channel blockers, TEA and 4-AP, the inward currents were unmasked (but not completely) in both cell types (Fig.4B). This observation explains the decrease in the strong outward rectification and the occur-rence of action potential in tracheal cells in the presence of TEA and 4-AP (Fig. 3B). The currents in Fig. 4C illustrate the net outward currents obtained by subtraction of the traces in B from the corresponding traces in A. It is clear that the net outward current in tracheal cells includes a sustained component while the initial transient component is predominant in vas deferens. The following were common observations in smooth muscle cells obtained from low excitable tissues such as large arteries, trachea and iris sphincter. 1) The inward current upon depolarization was not detected unless the counteracting outward currents were decreased by TEA, 4-AP and/or Cs. 2) The outward current contains a large sustained component which is not due to Ca-activated K current ($I_{K(Ca)}$).

At present, existence of 5 types of K currents have been suggested in smooth muscle cells. Table 2 shows four types of K currents and whether the currents exist or not in the cells examined. The other K channel current has been reported in relation to slow pacemaker potentials in intestinal smooth muscle[23] (not shown in Table 2). In addition, some mammalian visceral smooth muscle cells might have M type K current. In all types of smooth muscles examined, $I_{K(Ca)}$ was the largest outward current and

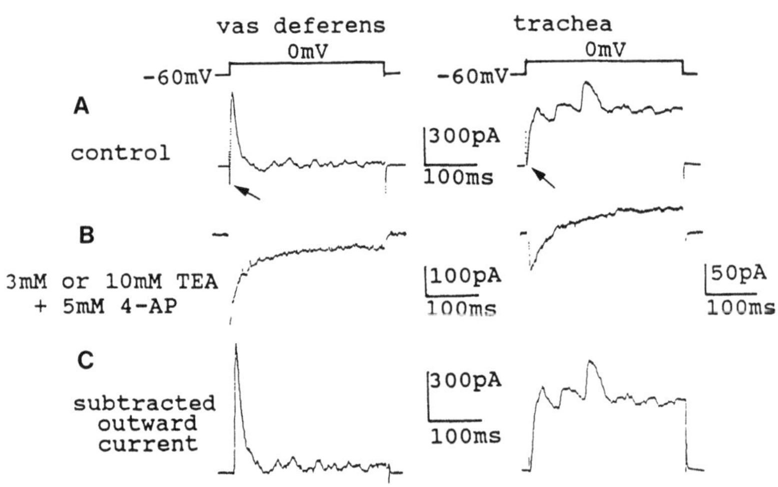

Fig. 4 A: total membrane currents in cells from vas deferens (left) and trachea (right). The pipette solution contains high K-aspartate. B: TEA and 4-AP were simultaneously applied. C: Subtraction of traces in A from those in B.

selectively blocked by low concentration of TEA (<2mM). Although three types of channels underlying $I_{K(Ca)}$ have been reported[24,25], the channel which has large conductance (about 200pS under high K^+ symmetry[26-29]) and is quite sensitive to extracellular TEA[24,28] appeared to be dominant[30-32]. There seems to be two kinds of voltage and time dependent K currents, so called delayed rectifier K currents (I_K). One of them is sensitive to 4-AP as shown in pulmonary artery[33]. This type of I_K was found in cells from trachea, femoral artery and iris sphincter. The I_K observed in cells from taenia caecum[6] and urinary bladder was blocked by 20mM TEA but not by 5mM 4-AP. The onset of I_Ks in smooth muscle cells examined here was faster than that in cardiac cells and slower than that in nerve cells. Ca independent transient outward current (I_A; cf. ref.34) and corresponding single channel K current were recorded in cells from ureter[17], vas deferens and portal vein[36]. Although distinct single channel currents underlying these two types of I_K have not been recorded, they are tentatively shown as separate in Table 2.

Table 2 Ca currents and K currents in several smooth muscle cells

Tissues	$I_{Ca}{}^a$ (nA)	Total[b] (nA)	I_A	$I_{K(Ca)}$	Delayed rectifier I_K sensitive to 4-AP	TEA
Urinary Bladder	0.3	1	−	++	−	+
Vas Deferens	0.25	0.7	+	++	−	+
Taenia Caecum	0.3	0.7	−	++	−	++
Portal Vein	0.3	0.6	+	++	−(?)	+(?)
Ureter	0.2	0.3	+	++	−	−
Trachea	0.2	0.4	−	++	++	+
Femoral Artery	20pA	0.4	−	++	+	+
Pulmonary Artery	60pA	0.4	−	++	++	+(?)
Iris Sphincter	<10pA	0.2	+	++	++	+

Cells were depolarized from -60 to 0mV, $[Ca^{++}]_o$=2.2mM. Pipette sol. contained 140mM Cs^+ and 5mM EGTA (a); 140mM K^+ and 0.5mM EGTA (b). $I_{Ca}{}^a$ and total[b] indicate the peak amplitude of I_{Ca} and outward current, respectively. The species of animals used were the same as described in Methods.

The excitability of cell membrane is regulated mainly by couterbalance of inward and outward currents. The low excitability in the smooth muscle cells from large arteries and iris sphincter is due to 1) tiny I_{Ca}s which are about one tenth or less of those in highly excitable cells and also 2) substantial outward currents which are markedly larger than I_{Ca} in amplitude. Although tiny Ca current may result in smaller activation of $I_{K(Ca)}$, both I_Ks are substantial and contribute to the strong rectification of the membrane. This does not allow the inward current to dominate at any time even when the membrane is depolarized artificially under current clamp mode.

Differences in Ca current inactivation in various smooth muscle cells

The differences in I_{Ca} in various smooth muscle cells include not only differences in amplitude but also in shape during depolarization pulse. The inactivation of I_{Ca} activated by depolarization from -80 to +10mV (Fig.5 A) in portal vein is quite fast, while in ureter and taenia caecum cells it is slow and intermediate, respectively, under the same experimental conditions. The half inactivation time is markedly different in these cells. I_{Ca} in ureter showed very small decay in comparison with that in portal vein. Two possibilities can be postulated from these findings. First, L type Ca channels which are activated and inactivated at less negative potentials and underlie the long lasting I_{Ca}[9,13,14,37] may be predominant in ureter cells, while T type channel may significantly contribute to macroscopic I_{Ca} in portal vein[35]. Second, I_{Ca} in portal vein is strongly inactivated by elevation of intracellular Ca concentration[38-41] while this mechanism may be much smaller in ureter[17] than portal vein. The result obtained from the experiment changing holding potentials (Fig. 5C) indicate that T type channels may not contribute to macroscopic I_{Ca} in ureter cells[17]. Although these hypotheses are quite interesting to

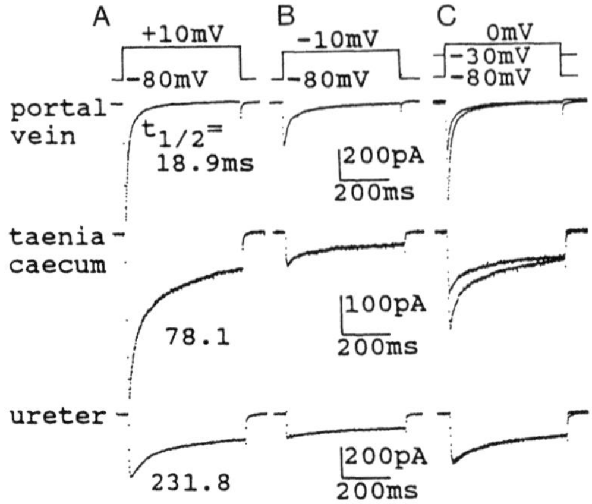

Fig. 5 Three types of cells were depolarized from -80 (A and B) or -30mV (C) to +10 (A), -10 (B) or 0mV (C).

examine, this study was focused on the functional meaning of the difference in inactivation time course between cell types. In ureter cell, a large I_{Ca} remains at the end of 500msec pulse, while that in portal vein is quite small. It has been widely accepted that the tonic contraction induced by high K^+ solution in smooth muscle is due to sustained Ca influx through voltage dependent Ca channel[42,43]. If this is true, noninactivating component of Ca current must remain during prolonged depolarization. This component might be smaller in the cells from portal vein than those from ureter.

Noninactivating Ca currents in various smooth muscle cells

Fig.6 A shows the effect of puff application of Cd^{++} (5mM in puff solution) on I_{Ca} elicited by depolarization from -60 to +10mV in ureter cell. Three traces obtained before and during the application of Cd^{++} and after washout are superimposed. Cd^{++} blocked the inward current quickly and the effect was washed out within an interval of the pulses (5sec). Although puff application occasionally produced small inhibitory artifact on I_{Ca}, the shown effects in this study did not include significant artifacts. In Fig.6B, Cd^{++} puffed about 1.5 min after the start of depolarization blocked the sustained I_{Ca} of about 10pA. This strongly suggests that a small but noninactivating I_{Ca} ($I_{Ca(ni)}$) exists under these conditions[15]. In Fig.6C, a solution containing 10μM Bay K 8644 was puffed about 4 min after

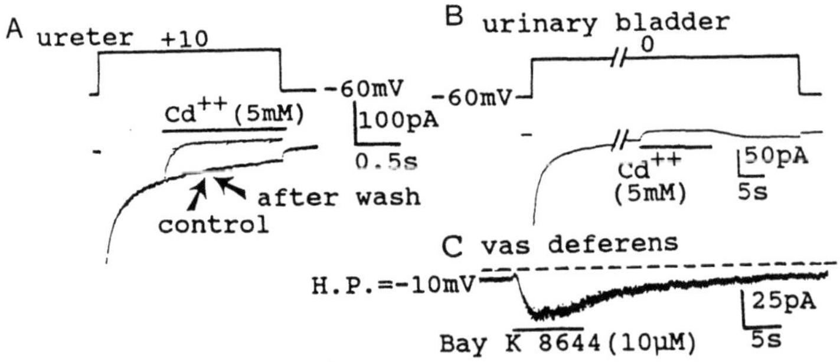

Fig. 6 Effects of Cd^{++} (A and B) and Bay K 8644 (C) on I_{Ca} were examined by puff-application after about 0.7s (A), 1.5min (B) and 4min (C) after the start of depolarization in cells from ureter (A), urinary bladder (B) and vas deferens (C), respectively.

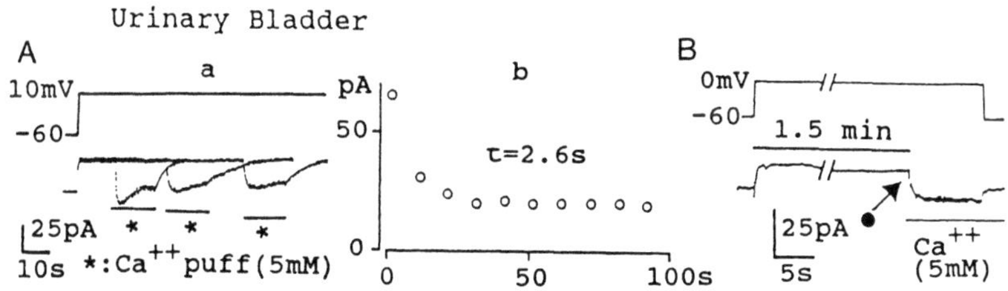

Fig. 7 Measurement of $I_{Ca(ni)}$ by Ca^{++}-application. Aa:Ca^{++} (5mM in the pipette sol.) was applied by puff in Ca-free sol. Three traces were superimposed. Ab: The peak amplitude of I_{Ca} elicited by Ca puff were plotted against the time when Ca^{++} was applied after the start of depolarization. B: $[Ca^{++}]_o$ was elevated by concentration-clamp from 0.1 to 5mM at the time indicated by a circle and an arrow.

the vas deferens cell was depolarized to -10mV. The $I_{Ca(ni)}$ was markedly enhanced by Bay K 8644 (cf. ref.44). Application of 1-10μM nifedipine in a similar manner decreased I_{Ca} slowly but did not potentiated it at any potentials (cf. ref.44 & 45). Fig.7A shows the effects of repetitive puff application of 5mM Ca^{++} solution to an urinary bladder cell in Ca-free solution. The depolarization did not elicit inward current in the urinary bladder cell. When Ca was puffed after a certain period of depolarization, an inward current was induced. Three traces were superimposed in Fig.7A. Fig.7Ab illustrates a relationship between time of Ca^{++}-puff after the start of depolarization and the amplitude of induced I_{Ca}. This time course was fitted with single exponential function (τ=2.6S) and reached steady value at about 30sec. The amplitude of steady level which corresponds to that of $I_{Ca(ni)}$ was about 18pA. Another way to measure $I_{Ca(ni)}$ involved the concentration clamp method introduced by Hering et al. About 1.5min after the start of depolarization, extracellular Ca concentration was elevated from 0.1 to 5mM by concentration clamp (Fig.7B). $I_{Ca(ni)}$ of about 20pA was observed. Fig.8 illustrates the voltage-dependency of the $I_{Ca(ni)}$ in urinary bladder cells measured by three methods; shown as means of several measurements. The peaks of I-V relationships were obtained at -10mV in both Cd^{++}-puff and Ca^{++}-puff, while $I_{Ca(ni)}$ measured by Ca^{++}-puff was larger than that by Cd^{++}. These observations were consistent in all types of cells examined. The Ca-induced inactivation of I_{Ca} may be responsible for the smaller amplitude of $I_{Ca(ni)}$ measured by Cd^{++}-puff than that by Ca^{++}-puff.

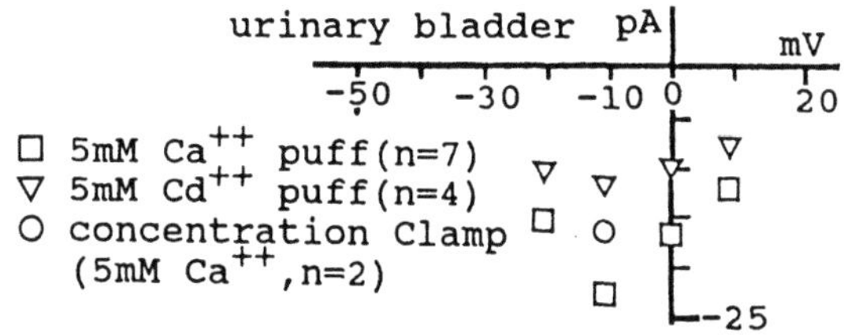

Fig. 8 Current-voltage relationship of $I_{Ca(ni)}$ measured by three means (see text and Fig. 6 and 7) in urinary bladder cells.

Steady-state activation and inactivation of Ca current in various smooth muscle cells

The relationship between voltage and the steady-state inactivation was determined in several types of smooth muscle cells using conventional double-pulse protocols (Fig.9A). During the conditioning pulse for 3 sec (P_1), the cell was clamped at various voltages and the inactivation of Ca channel reached steady state. Thereafter, a second pulse (P_2) activates the available Ca channels. The steady-state inactivation curves show the ratio of available and unavailable Ca channels as a function of voltage. The data were fitted with Boltzman function (solid line calculated using equation I). The curves for portal vein and femoral artery are about 10mV more negative than those for ureter and urinary bladder cells (Fig.9C). Moreover, the curves for urinary bladder and ureter cells do not reach zero even at very positive potentials (as indicated by * in Fig.9A; see ref.44).

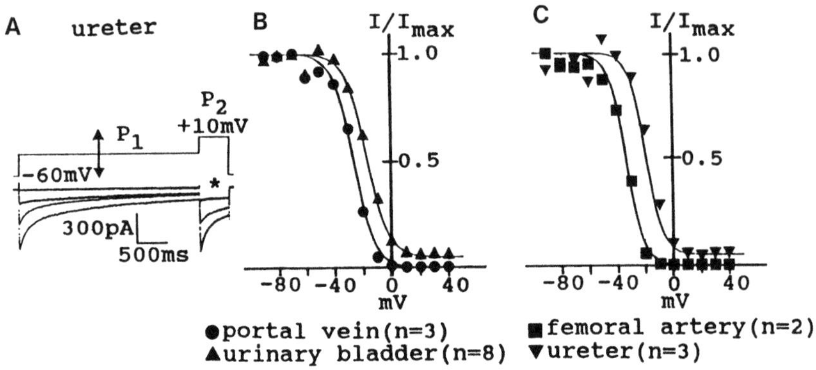

Fig. 9 A: The protocol and current recordings to determine steady state inactivation. B: Inactivation curves for portal vein ($V_{1/2}$=-26.3mV; S=-7.2mV; see equ I) and urinary bladder (-17.4;-7.5). C: femoral artery (-33.6;-6.2) and ureter (-19.7;-5.9).

Simulation of noninactivating Ca currents

The area under both steady-state activation (Fig.10A circles) and inactivation curves (triangles) suggests the presence of a noninactivating current. Cardiac electro-physiologists call this window current[46]. The window current was calculated using equation III. The g_{max} was obtained as maximum chord conductance assuming E_{Ca} of +60mV (Fig.10B). The steady state activation of I_{Ca} was not directly measured but obtained from I-V curve of peak I_{Ca} (ref.4 and Fig.10B). The activation curve was also fitted with Boltzman equation (equation II). Fig.10C illustrates the simulated curves assuming the power of d in equ II as 1, 2 or 3, respectively. The simulated I-V curve using d_{∞}^1 shows a maximum near -10mV and is similar to that of $I_{Ca(ni)}$ measured directly (Fig.8). The calculation using d_{∞}^2 or d_{∞}^3 resulted in quite different shapes of I-V curves. It was common in all types of cells examined.

$$f_{\infty} = (1-C)\,\{1+\mathrm{EXP}[(V_{1/2}-V_m)/S]\}^{-1}+C \quad \text{(I)}$$

$$d_{\infty} = \{1+\mathrm{EXP}[(V_{1/2}-V_m)/S]\}^{-1} \quad \text{(II)}$$

$$I_{Ca} = g_{max}\cdot(d_{\infty})^n\cdot f_{\infty}\cdot(E_m-E_{Ca}) \quad \text{(III)}$$

where f_{∞} and d_{∞} are inactivation and activation factors, respectively. $V_{1/2}$ and S are half activation (II) or inactivation voltage (I) and slope factor, respectively. E_{Ca} is reversal potential of I_{Ca}. C is constant for steady component.

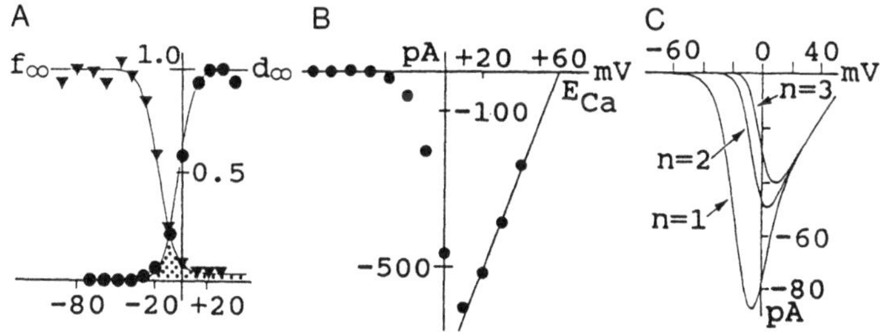

Fig. 10 A:Steady state activation and inactivation curves of I_{Ca} in urinary bladder cells. B:I-V relationship of I_{Ca} and measurement of chord conductance. C:I-V curves obtained by simulation using equ III.

Table 3 summarizes the measured $I_{Ca(ni)}$ and the window currents calculated assuming d_{∞}^{1} at -10mV. The amplitude of $I_{Ca(ni)}$ was smaller than the window current in most types of cells. Of interest is that cells from ureter have smaller I_{Ca} but larger $I_{Ca(ni)}$ than those from portal vein. This indicates that the amplitude of $I_{Ca(ni)}$ has a marked correlation with the ratio of I_{Ca} amplitude measured at the peak and the end of 500msec pulse (I_e/I_p). L type Ca channel may mainly contribute to $I_{Ca(ni)}$. In femoral artery, $I_{Ca(ni)}$ and I_e were too small to be measured. Although this simulation of Ca window current includes some uncertainties, it may give theoretical insight[15] into the $I_{Ca(ni)}$ and the resulting sustained Ca-influx through voltage dependent Ca channel(s) during tonic contraction in high K solution.

Table 3. Comparison of $I_{Ca(ni)}$ and window currents

Tissues	measured (pA) a	measured (pA) b	calculated (pA)	I_{Ca}* (pA)	I_e/I_p#	
Ureter	22.9 (5)		31.8 (3)	320	0.51	(5)
Urinary Bladder	21.5 (7)	10.5 (4)	84.5 (10)	650	0.18	(11)
Taenia Caecum	8.0 (4)		28.8 (6)	410	0.16	(5)
Vas Deferens		7.2 (3)	41.7 (7)	480	0.40	(4)
Portal Vein	3.2 (3)		22.2 (3)	460	0.09	(5)
Femoral Artery	<1.0 (3)		0.54 (3)	40	0	(2)

Measured $I_{Ca(ni)}$s are those measured by 5mM Ca^{++}-puff after 60s depolarization (a), and by 5mM Cd^{++}-puff (b), respectively.
Calculated $I_{Ca(ni)}$s were obtained by the simulation of window current (see text).
*: peak amplitude of I_{Ca}s elicited by deplarization from -60 to 0mV (5mM Ca^{++}).
#: I_e and I_p were I_{Ca} measured at the end of 500ms pulse and the peak, respectively, in cells which were depolarized from -60 to -10mV (5mM Ca^{++}).

From these results, the elevation of $[Ca^{++}]_i$ by $I_{Ca(ni)}$ can be calculated in a simplest manner on the basis of following assumptions. 1) The volume of the cell is 10pl. 2) The influx of Ca^{++} as the $I_{Ca(ni)}$ of 10pA in amplitude continues for 10sec. 3) Other cellular functions for Ca metabolism such as uptake by intracellular store sites, extrusion by Ca-pump and/or Na-Ca exchange are ignored. The calculation gave a value of 50μM, which may be higher than the concentration required for maximum contraction. Although the third assumption is apparently not completely valid, this calculation may help to account for the sustained contraction in high K solution. Even when the $I_{Ca(ni)}$ is less than lpA, it could contribute to keep the $[Ca^{++}]_i$ high during a tonic phase of high K induced contraction. In addition, it is worth emphasizing that Ca extrusion by Na-Ca exchange mechanism does not work in high K solution if Na is replaced with K in the solution.

ACKNOWLEDGEMENTS

We thank Dr.Wayne Giles (University of Calgary, Canada) for his valuable comments on this manuscript and his kindness in giving us the data-acquisition and -analysis programs for the IBM-AT. Fig.6,7,8,10 and Table 3 in this manuscript are reproduced with permission of Am.J. Physiol. (ref.15). A part of this study was supported by grant from the Japanese Ministry of Education, Science and Culture.

REFERENCES

1 E.Bozler Experientia 4:213 (1948).
2 G.Burnstock, Structure of smooth muscle and its innervation, in: "Smooth Muscle," Bülbring et al. ed. Arnold (1970).
3 O.P.Hamill, A.Marty, E.Neher, B.Sakmann,B. & F.J.Sigworth Pflügers Arch. 391:85 (1981).
4 U.Klöckner & G.Isenberg, Pflügers Arch. 405:340 (1985).
5 R.Mitra & M.Morad, Sciences 229:269 (1985).
6 V.Y.Ganitkevich, M.F.Shuba & A.J.Smirnou, J.Physiol. 380:1 (1986).
7 G.Droogmans & G.Callewaert, Pflügers Arch. 406:259 (1986).
8 T.Amedee, C.Mironneau & J.Mironneau, J.Physiol. 392:253 (1987).
9 B.P.Bean, M.Sturek, M.Puga & K.Hermsmeyer, Cir.Res. 59:229 (1986).
10 M.Yoshino & H.Yabu, Pflügers Arch. 404:285 (1985).
11 J.M.Caffrey, I.R.Josephson & A.M.Brown, Biophys.J. 49:1237 (1986).
12 J.F.Worley III, J.W.Deitmer & M.T.Nelson, Proc.Natl.Acad.Sci.USA. 83:5746 (1986).
13 A.Yatani, C.L.Seidel, J.Allen & A.M.Brown, Cir.Res. 60:523 (1987).
14 C.D.Benham, P.Hess & R.Tsien, Cir.Res. 61(suppl I):I-10 (1987).
15 Y.Imaizumi, K.Muraki, M.Takeda & M.Watanabe, Am.J.Physiol. in press.
16 K.Momose & Y.Gomi, Chem.Pharmaceut.Bull.(Tokyo) 25:2449 (1977).
17 Y.Imaizumi, K.Muraki & M.Watanabe, J.Physiol. in press.
18 S.Hering, D.J.Beech & T.B.Bolton, Pflügers Arch. 410:335(1987).
19 M.Sturek & K.Hermsmeyer, Sciences, 233:475 (1986).
20 K.Okabe, K.Kitamura, & H. Kuriyama, Pflügers Arch. 411:423 (1988).
21 M.I.Kotlikoff, Am.J.Physiol. 254:C793 (1988).
22 K.Nakazawa, N.Matsuki, K.Shigenobu & Y.Kasuya, Pflügers Arch. 408:112 (1987).
23 C.D.Benham & T.B.Bolton, J.Physiol. 340:469 (1983).
24 R.Inoue, K.Kitamura & H.Kuriyama, Pflügers Arch. 405:173 (1985).
25 R.Inoue, K.Okabe, K.Kitamura & H.Kuriyama, Pflügers Arch. 406:138 (1986).
26 C.D.Benham, T.B.Bolton, R.J.Lang & T.Takewaki, J.Physiol. 371:45 (1986).
27 W.Berger, R.Grygorcyk, W.Schwarz, Pflügers Arch. 402:18 (1984).
28 J.D.McCann & M.J.Welsh, J.Physiol. 372:113 (1986).
29 J.J.Singer & J.V.Walsh, Pflügers Arch. 408:98 (1987).
30 U.Klöckner & G.Isenberg, Pflügers Arch. 405:329 (1985).
31 Y.Ohya, K.Terada, K.Kitamura & H.Kuriyama, Am.J.Physiol. 251:C335 (1986).
32 J.V.Walsh & J.J.Singer, Pflügers Arch. 408:83 (1987).
33 K.Okabe, K.Kitamura & H.Kuriyama, Pflügers Arch. 409:561(1987).
34 M.A.Rogawski, Trends in Neurosci. 8:214 (1985).
35 G.Loirand, P.Pacaud, C.Mironneau & J.Mironneau, Pflügers Arch. 407:566 (1986).
36 D.J.Beech & T.B.Bolton, Pflügers Arch. 411:Suppl.1,R200 (1988).
37 M.Yoshino, T.Someya, A.Nishio & H.Yabu, Pflügers Arch. 411:229 (1988).
38 R.W.Tsien, Ann. Rev. Physiol. 45:341 (1983).
39 V.Y..Ganitkevich, M.F.Shuba & S.V.Smirnov, J.Physiol. 392:431 (1987).
40 K.Nakazawa, H.Saito & N.Matsuki, J.Membrane Biol. 100:13 (1987).
41 Y.Ohya, K.Kitamura & H.Kuriyama, Cir.Res. 62:375 (1988).
42 G.B.Weiss, Stimulation with high potassium, in: "Methods in Pharmacology", Vol.3, E.E.Daniel & D.M.Paton, ed., New York and London, Plenum Press. (1975).
43 T.B.Bolton, Physiol. Rev. 59; 606 (1979).
44 S.Hering, D.J.Beech, T.B.Bolton & S.P.Lim, Pflügers Arch. 411:590 (1988).
45 P.I.Aaronson, T.B.Bolton, R.J.Lang & I.MaCkenzie, J.Physiol. 405:57 (1988).
46 D.Atwell, I.Cohen, D.Eisner, M.Ohba & C.Ojeda, Pflügers Arch. 379:137 (1979).

TWO TYPES OF CA CHANNELS IN SMOOTH MUSCLE CELLS ISOLATED FROM GUINEA-PIG TAENIA COLI

Hideyo Yabu, Masami Yoshino, Tetsufumi Someya and Morio Totsuka

Department of Physiology, Sapporo Medical College
Sapporo, 060, Japan

INTRODUCTION

Ca influx through Ca channels in smooth muscle cells plays a crucial role in both membrane excitation and muscle contraction. In 1985, we demonstrated that a single Ca channel currents, with Ba ions acting as charge carriers, could be recorded by the cell attached patch clamp method in enzymatically isolated smooth muscle cells of guinea pig taenia coli (Yoshino and Yabu, 1985). Recent patch clamp studies has provided evidence that multiple types of Ca channels exist in many excitable cells. It has been suggested that each channel type contributes to different physiological functions, such as a source of Ca ions during excitation-contraction coupling, the generation of pacemaker activity and the release of transmitter from nerve terminals. Two different types of Ca channels has also been reported for smooth muscle cell membrane, mainly vascular muscles, from single channel recordings(Aaronson et al., 1986 ; Benham et al., 1987) and whole cell recording (Sturek and Hermsmeyer, 1986 ; Bean et al., 1986). We examined whether different types of Ca channels in mamalian intestinal smooth muscles cells from guinea-pig taenia coli (Yoshino et al., 1987 ; Yoshino et al., 1988). Then, we concluded that both channel types represent fast and slow inactivating voltage dependent Ca channels exist in this cell.

METHODS

Single guinea-pig taenia coli cells were isolated with a combination of 0.3% collagenase, 0.6% trypsin inhibitor and 1% bovine serum albumin in Ca-free solution as per the method of Obara and Yabu(1982). The freshly isolated single cells were then incubated in the KB medium (Isenberg and Klockner, 1982) at 4°C for 1hr. This procedure made it possible to obtain Ca tolerant cells analogous to the cardiac myocytes prepared by Calalie et al., (1983).

Whole-cell currents were recorded under voltage clamp conditions similar to that described by Hamil et al.,(1981). Ca channel currents were isolated by the suppression of potassium currents using patch pipettes filled with 140mMCsCl, 10mM glucose, 4mM EGTA, 5mM HEPES and 2mM Na_2ATP. Cells were bathed in a solution containing 50mM $BaCl_2$, 75mM CholineCl and 5mM HEPES-Tris(pH=7.4). Experiments were performed on small smooth muscle cells (cell capacitance of 33.1 ± 9.4 pF, N=6) to facilitate voltage control and internal perfusion. The currents were measured with a List LM/EPC7 patch-clamp amplifier with cancellation of capacitive transients and compensation of series resistance. Patch clamp recordings of single Ca channel activity were made from cell-attached patches with pipettes filled with 50 mM BaCl , 75 mM CholineCl and 5mM HEPES-Tris(pH=7.4) as described previously(Yoshino and Yabu, 1985). The cell resting potential was zeroed with an external solution containing 145 mM KCl, 1 mM $MgCl_2$, 10 mM glucose and 5 mM HEPES (pH=7.4). Current signals were stored on a PCM data recorder(model RP-880). Stored signals were low pass filtered (cut-off frequently, 1.0 kHz) and sampled at 5 kHz. Digitized signals were further analyzed by a Nihon Kohden ATAC-450. The leak and capacitive currents were subtracted by adding the averaged hyperpolarizing currents on each current. Seals of resistance of 10-1000 GΩwere obtained by positioning the tip of the pipettes against the cell surface and then applying gentle suction (tip resistance was 2-10 MΩ). All experiments were carried out at room temperature(25-30°C).

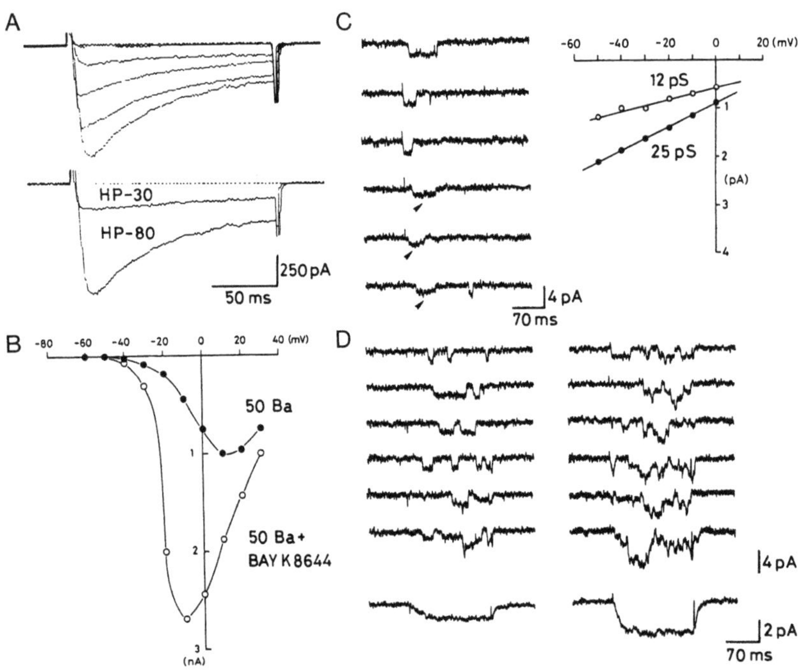

Fig.1A Upper, Superimposed traces of whole-cell Ba currents in 50 mM Ba. The cell was depolarized to -60, -20, +40, +20 and 0mV (from upper to lower traces) from a holding potential of -80mV. A lower, Inward Ba currents elicited by depolarization to 0mV from two different holding potentials (HP -80 and -30mV). B) Current voltage curves for peak whole-cell Ba currents in the absence (closed circles) and presence (open circles of 1 μ M Bay K 8644 in the external solution. C-left) Two types of unitary Ba currents elicited by depolarization to -40 mV from a holding potential of -60 mV using patch pipettes filled with 50 mM Ba and 1 μM Bay K 8644. The arrows indicate unitary currentswith small conductance. C-right) The current voltage relationship in 50 mM Ba gives single channel slope conductances of 12 pS and 25 pS. D) Representative current traces obtained from the same membrane patch shown in Fig. 1C when the membrane patch was depolarized to -30 mV, (left) and -20 mV, (right) from a holding potential of -60 mV. The lowest trace of each column shows the mean current from 58 sweeps.

RESULTS AND DISCUSSION

The superimposed traces of whole cell Ba currents when the cell membrane was depolarized from a holding potential of -80mV was showed in Fig.1A. The activation threshold of the Ba currents was about -40 mV and the current reached its peak value at +10 mV(Fig.1B). The Ba currents recorded from a holding potential of -80 mV to a test potential of 0 mV was composed of fast and slow inactivating phase (Fig.1A., lower). When the holding potential was set at -30 mV, only the slow inactivating component persisted(Fig.1A, lower trace). Thus the whole cell experiments indicated the possible existence of two types of Ca channels with different inactivation kinetics. Next, we tried to record the currents through single Ca channel using the cell-attached patch clamp technique. To resolve the single Ca channel activity, BAY K 8644 was added to the patch solution. Two different sizes of unitary Ba currents were recorded when the patch membrane was depolarized to -40mV from a holding potential of -60mV (Fig.1C). Both channel types currents appeared near -40mV and increased their opening probability with membrane depolarization. The current-voltage relationship showed that the large and small unitary currents had slope conductances of 25pS and 12pS respectively(Fig.1C). Fig.1D shows the unitary currents records when the membrane patch was depolarized to -30mV(Fig.1D left) and to -20mV(Fig.1D right). The lower traces of each column show the mean currents obtained by averaging individual current trace which contain both types of channel activities. Both mean currents had no decay. The amplitude of the mean current increased with further depolarizations and reached its peak at -10mV. The threshold potential of 12pS channel was similar to that of 25pS. 12pS channel is a Ca channel since the unit currents had extrapolated reversal potential more positive than -60mV and were not blocked by 30uM TTX.

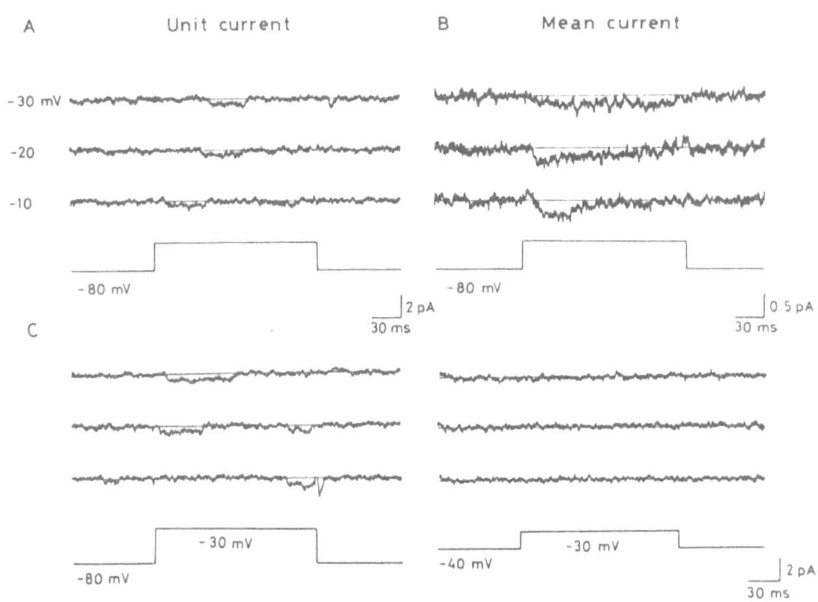

Fig.2A Representative current traces through a single Ca channel with small conductance elicited by depolarization to the indicated levels from a holding potential of -80mV. B) The mean currents. C) Unitary currents with small conductances elicited by depolarization to -30mV from holding potentials of -80 and -40mV. Note that the unitary currents disappeared with a holding potentail of -40mV.

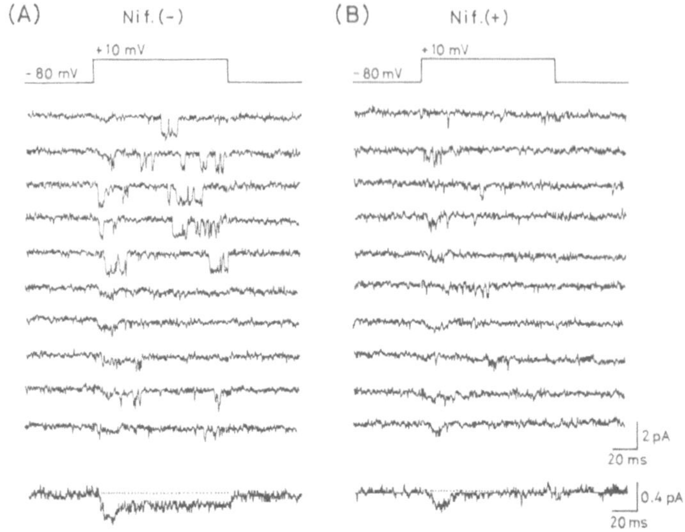

Fig. 3. Different effects of the DHP Ca antagonist nifedipine on the large and small conductance channels. The cell was depolarized to 10 mV from a holding potential of -30 mV in the absence (A) and presence (B) of 2 μM nifedipine in the external solution. Lowest traces in each column show averaged currents from 69 (A) and 47 (B) sweeps. Note that the large conductance channel activity disappeared in the presence of nifedipine(B).

Fig.2A shows the current traces which contain only the small conductance channel at different potentials without BAY K 8644 in the pipette solution. The mean current of this channel activity inactivated completely during 150msec (Fig.2B). The peak value for the mean current was obtained at -10mV. Fig. 2C shows the small unitary currents when the patch membrane was depolarized to -30mV from -80mV (left) and from -40mV (right). With a holding potential of -40mV, small unitary currents completely disappeared. Further depolarizations of up to -30mV from the holding potential of -40mV never induced small unitary currents.

The small conductance Ca channels showed a tendency to open at the beginning of depolarization clamp pulses and voltage dependent inactivation. In contrast to the large conductance Ca channels, the small one was insensitive to high dose of the dihydropyridine derivative, Nifedipine(Fig. 3., Yoshino et al., in preparation). Cadmium ions(10μM in patch pipette) inhibited the large conductance Ca channel currents in flickering manner (Fig.4A right., Yoshino et al., in preparation), though the small conductance Ca channel currents were not blocked (Fig.4B, right). These properties of the small conductance channel appear to be similar to those of the T type Ca channel that has been identified in neuronal and cardiac cells (Nilius et al., 1985; Nowycky et al., 1985). However, the activation threshold of both channel types is close to -40mV so that the separation of both channel activity by changing the holding potential or the activation voltage is hard. Recently, Hagiwara et al.,(1988) reported that in sino-atrial node cells a low concentration of tetramethrin (0.1 μM) abolished transient type Ca current without affecting long-lasting Ca current. We found that this pyrethroid insecticide diminished the unitary Ca channel currents elicited by a test pulse of depolarization to -20mV from a holding potential of -80mV (Someya et al., in preparation). Tetramethrin should be a good tool to separate the two types of Ca channels pharmacologically and to explain the physiological function of the existence of two types of Ca channels in smooth muscle cells of guinea pig taenia coli.

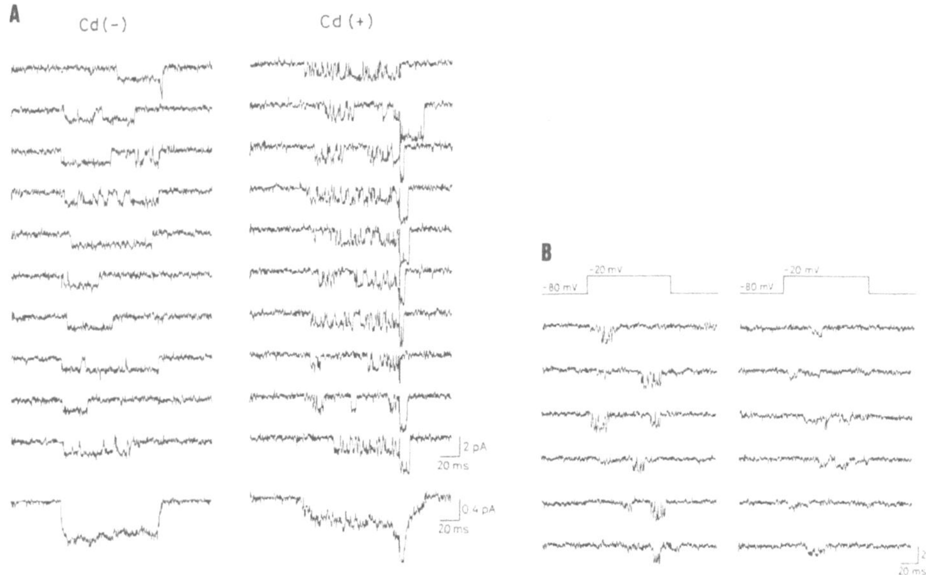

Fig. 4A Effect of Cd^{++} on the large conductance channel. Selected current traces were shown in the absence (left) and presence (right) of 10 μM Cd in the patch pipettes. The external solution contained 2 μM Bay K 8644 to prolong the open times of the large conductance channel. The cell was depolarized to -10mV from a holding potential of -80mV with 50 mM Ba as the charge carrier.

Fig. 4B Effects of Cd^{++} on the large (left) and small (right) conductance channels. Each current trace was selected from the same cell. The cell was depolarized to -20mV from a holding potential of -80mV in the absence of Bay K 8644.

SUMMARY

1) Two types of voltage dependent Ca channels with different conductances and inactivation kinetics were identified from cell-attached patch clamp recordings. One type, with a larger conductance of 25pS, had a threshold of activation near -40mV and the mean current inactivated slowly. A second type of Ca channel, with a smaller conductance of 12pS channel, but the averaged mean current inactivated rapidly.

2) Cadmium ions inhibited the large conductance Ca channel currents, while the small conductance Ca channel currents was not blocked.

3) Large conductance Ca channel current selectively inhibited by dihydropyridine derivative, Nifedipine. Small conductance Ca channel current is abolished in dose dependent manner by pyrethroid insecticide, tetranethrine.

REFERENCES

Aaronson, P. I., Benham, C. D., Bolton, T. B., Hess, P., Lang, R. J. and Tsien, R. W. 1986, Two types of single channel and whole-cell Calcium or barium currents in single smooth muscle cells of rabbit ear artery and the effects of noradrenaline, J. Physiol., 377:36P.

Bean, B. P., Sturek, M., Puga, A. and Hermsmeyer, K., 1986, Calcium

channels in muscle cells isolated from rat mesentric arteries: Modulation by dihydropyridine drugs, Circ. Res., 59:229-235.
Benham, C. D., Hess, P. and Tsien, R. W., 1987, Two types of Calcium channels in single smooth muscle cells from rabbit ear artery studied with whole-cell and single-channel recordings, Circ. Res., 61:10-16.
Caralie, A., Ochi, R., Pelzer, D. and Trautwein, W., 1982, Elementary currents through Ca^{2+} channels in guinea pig myocytes. Pflugers Archiv 398:284-297.
Hamil, O. P., Marty, A., Neher, Sakman and Sigworth, F. J., 1981, Improved patch-clamp techniques for high-resolution current recording from cells and cell-attached membrane patches, pfluger Arch., 391:85-100
Hagiwara, N., Irisawa, H. and Kameyama, M., 1988, Contribution of two types of Calcium currents to the pacemaker potentials of rabbit sino-atrial node cells, J. Physiol., 395:233-253.
Isenberg, G., and Klockner, V., 1982, Calcium tolerant ventricular myocytes prepared by preincubation in a "KB medium", Pflugers Arch., 395:6-18.
Nilius, B., Hess, P., Lansman and Tsien, R. W., 1985, A novel type of cardiac calcium channel in ventricular cells, Nature., 316:433-446.
Nowycky, M. C., Fox, and Tsien, R. W., 1985, Three types of neuronal Calcium channel with different calcium agonist sensitivity., Nature., 316:440-443.
Obara, K., and Yabu, H., 1982, Effects of chlorotetracycline fluorescence in single smooth muscle fluorescence in single smooth muscle cells of guinea pig taenia coli., Jpn. J. Physiol., 32:1002-1006.
Sturek, M. and Hermsmeyer, K., 1986, Calcium and sodium channels in spontaneously contracting vascular muscle cells., Science., 233:475-478.
Yoshino, M., and Yabu, H., 1985, Single Ca channel currents in mammalian visceral smooth muscle cell., Pflugers Arch., 404:285-286.
Yoshino, M., Nishio, A. and Yabu, H., 1987, Calcium channel modulation by the dihydropyridine derivative BAY K 8644 in mammalian visceral smooth muscle cells, Comp. Biochem. Physiol. C., 89:39-43.
Yoshino, M., Someya, A., Nishio, A. and Yabu, H., 1988, Whole-cell and unitary Ca channel in mammalian intestinal smooth muscle cells: evidence for the existence of two types of Ca channels, Pflugers Arch., 411:228-231.

MOLECULAR ANALYSIS OF CALMODULIN AND SMOOTH MUSCLE MYOSIN LIGHT CHAIN KINASE

Anthony R. Means, Mark F.A. VanBerkum, Samuel E. George and Indrani C. Bagchi

Department of Cell Biology
Baylor College of Medicine
Houston, TX 77030

INTRODUCTION

Of all the known members of the superfamily of proteins that utilize the EF-hand helix-loop-helix configuration to bind Ca^{++}, calmodulin is unique. This intracellular receptor is ubiquitous in eukaryotes and is highly conserved at the primary amino acid sequence level. In vertebrates only a single conservative amino acid substitution exists between fish and humans [1]. Even between primitive eukaryotes such as yeasts and higher vertebrate species, the proteins show at least 80% amino acid identity. In addition calmodulin serves as the obligatory Ca^{++}-dependent activator of a variety of enzymes, exists in enzyme and organelle complexes in the Ca^{++} free state and is associated with several intracellular structural proteins. These characteristics are in sharp contrast to other members of this superfamily such as troponin C, calbindins, S-100, calretinin, calcineurin B and myosin light chains. The function of these proteins, when known, tends to be highly specific. Distribution is largely restricted to vertebrates and even within members of this phylum, is found only in selected cell types. The one exception to this generalization is calcineurin B which enjoys a much broader species distribution. However this breadth may also be related to calmodulin since calcineurin is the only known Ca^{++}/calmodulin-dependent protein phosphatase. Indeed some calmodulin-dependent enzymes are more widely distributed among phyla and between cell types of a given organism than are the other members of the calmodulin superfamily [2].

A single gene encodes calmodulin in classes of mycophyta such as Saccharomyces cervisiae [3], Schizosaccharomyces pombe [4], and Aspergillus nidulans (unpublished, A.R.M., C.D.R.). In each case disruption of the gene is lethal demonstrating that calmodulin is essential for survival of the organism. In classes of vertebrates multiple calmodulin genes that differ considerably at the nucleic acid level, encode identical proteins. Rats and humans contain 3 calmodulin genes whereas two genes occur in chickens (5-11). In the mammalian orders the genes are located on different chromosomes. Perhaps this unusual gene duplication is to insure that calmodulin exists in appropriate concentrations necessary for function and proliferation of individual cells. It is known that overproduction of calmodulin in mammalian cells shortens the cell generation times by specifically decreasing the duration of the prereplicative G_1

period (12). Conversely decreased levels of calmodulin arrests cell growth at the G_1/S boundary of the cell cycle and in mitosis [13,14]. Even in S. cerevisiae, whose cell cycle is markedly different than that of mammalian cells, calmodulin has been shown to be essential for both nuclear and cell division [15,16].

A major concern of investigators in the broad field of calmodulin research is how the protein can execute such a diverse array of functions many of which occur within the same cell. Insight into the structure/function relationships of this Ca^{++} receptor was gained by solving the 3 dimensional structure [17]. The crystal structure of calmodulin reveals a 148 amino acid oblong molecule with two homologous globular domains separated by a 8 turn solvent exposed α-helix. Each globular domain contains two classic EF-hands that retain their ternary structure and calcium binding capabilities when separated from the rest of the molecule as tryptic fragments of amino acids 1-75 and 77-148 [18]. However, neither of the two tryptic fragments is capable of activating CaM dependent enzymes [19,20]. This suggests that the intact molecule is required for proper protein-protein interactions during CaM activation of its several target enzymes. Upon calcium ion binding, CaM undergoes a conformational change leading to an increase in both α-helical content (and/or orientation) and hydrophobicity [21,21a]. The two halves of CaM appear to communicate during calcium induced conformational changes and target enzyme binding [22,23]. This communication is presumably mediated via the connecting central helix. Lysines 75 and 77 of the central helix also become more susceptible to trace acetylation upon calcium binding [24-26] but lys 75 remains preferentially protected in the presence of smooth muscle myosin light chain kinase [26]. The acidic charge cluster (glu 82-84) has also been implicated in CaM binding of the basic amphiphillic helices present in most putative CaM binding domains of target enzymes [27]. Small angle x-ray scattering studies of CaM in solution suggest that while the globular domains retain a structure quite similar to the crystal structure, the central helix may be bent [28]. Only a slight bend in the central helix is observed in the crystal structure [29]. Therefore, the orientation and flexibility of the central helix may be important for CaM interactions with its target enzymes.

Four EF-hand Ca^{++} binding proteins have now been crystallized and the structures have been resolved to less than 3Å [11-15,17,30-33]. Calmodulin and skeletal muscle troponin C both interact with specific target proteins and share a similar structure in which the two pairs of Ca^{++} binding sites are separated by a multiturn central helix (helix IV) [17,32,33]. Parvalbumin and calbindin do not interact with other proteins and do not contain this central helix [30,31]. These facts along with a number of others have led to the postulate that the central helical region of calmodulin and troponin C is important for target protein recognition. Both chemical modifications [25,26,34-38] and genetically engineered mutants [39-43] have been utilized to emphasize the importance of this central region for enzyme recognition. The remarkable fact is that no single calmodulin variant affects all enzymes the same.

PRODUCTION OF CALMODULIN MUTANTS

We have utilized M13 phage-based site-specific mutagenesis to alter the central helix region [41]. Our first mutants contained a single Pro for Thr^{79} (PM) and an insert in which Pro-Ser-Thr-Asp occurred between Asp^{78} and Pro^{79} (IM). As summarized in Table 1 both PM and IM could maximally stimulate cGMP-PDE but the concentration required for half-maximal activation (Kapp) was increased 2 fold for CaMPM and 8-10 fold for CaMIM. It was unclear whether the shift in Kapp values observed for

Table 1. cGMP-Phosphodiesterase Activation by Calmodulin Central Helix Mutants

Protein	Partial Amino Acid Sequence	K_{act}(nM)
	75 80 85	
CaM23	MARKMKDTDSEEEI	1.2
CAMPM	MARKMKDPDSEEEI	2.6
CAMIM	MARKMKDPSTDPDSEEEI	9.3
CAMDM(-2)	MARKMKD--SEEEI	34.0
CAMDM(-3)	MARKMKD---EEEI	5.8
CAMDM(-5)	MARKI-----EEEI	322
CAMDM(-8)	MAR--------EEI	>710

Phosphodiesterase assays were performed using 50 pM cGMP-PDE,0.15 Ci 8-[^{3}H]-cGMP (final concentration of 0.03 mM cGMP) and varying concentrations of bacterial expressed calmodulin or its mutants. The amino acid sequence of the central helix region affected by the mutation is illustrated. Altered amino acid sequences are underlined, and deleted amino acids are represented by a dashed line.

these mutant proteins was related to the addition of one helical turn in the central helix or the disruption of this helix by two closely spaced Pro residues. Therefore, site-directed mutagenesis was again used to convert the two Pro residues in CaMIM to Lys, Gln or Thr. The mutants analyzed have either the two Pro replaced by Thr and Gln (CaMIM-TQ) or Lys and Gln (CaMIM-KQ) and one additional mutant have only the first Pro replaced with Thr (CaMIM-T). These mutants were all capable of maximally stimulating cGMP-PDE in a manner similar to CaMIM (Table 1). These results suggest the length of the central helix is important for proper target enzyme recognition and activation. This hypothesis has been strengthened by using deletion mutants that have removed 2,3,5 and 8 amino acids from the central helix (Table 1). Deletion of 3 amino acids ^{79,80,81} causes a 3-fold decrease in the K_{act} for PDE whereas the 2 amino acid deletion ^{79,80} was 15-fold less active than calmodulin. Removal of 5 $^{77-81}$ or 8 $^{75-82}$ residues results in a requirement of >300 nM CaM to achieve half-maximal PDE activation. These experiments suggest that, in addition to the length of the central helix, the orientation of the two globular portions of calmodulin is important.

PRODUCTION OF CALMODULIN/TROPONIN C HYBRID PROTEINS

As mentioned earlier calmodulin and troponin C are structurally similar. These proteins also demonstrate 55% direct amino acid identity although troponin C will not activate the majority of calmodulin-dependent enzymes. The fast skeletal muscle troponin C has been crystallized and is known to bind four Ca^{++} [32,33]. The cardiac form of troponin C (cTnC) is also very similar to calmodulin at the primary amino acid sequence level but does not bind Ca^{++} in the first EF-hand [44]. This inability to bind Ca^{++} is likely due to a single amino acid insertion in the loop.

Other differences between calmodulin and cTnC include an 8 amino acid extension at the NH_2-terminus, a 3 amino acid extension in the central helix and a 1 amino acid extension at the COOH-terminus. Whereas CaM regulates a variety of enzymes and physiological processes, cTnC regulates only actomyosin ATPase activity in cardiac and slow skeletal muscle cells. To investigate the molecular basis for the specificity of cTnC, we have cloned and sequenced this cDNA from rat muscle [45]. Bacterial expression vectors were constructed for cTnC and CaM. In addition vectors were prepared that would produce mRNA encoding 3 hybrid proteins. These vectors were TaM which contains the first domain of cTnC and the last 3 domains of CaM, and CaT which contains the first domain of CaM and the last 3 domains of cTnC. Finally aM was constructed to produce only the last 3 domains of CaM. All of the proteins were produced in and isolated from bacterial cells. The proteins were then examined for ability to activate 2 CaM-dependent enzymes, bovine brain cGMP phosphodiesterase (PDE) and rat brain calmodulin kinase II (CaMKII). The results are summarized in Table 2. Calmodulin and TaM activate PDE with identical kinetics whereas TaM only activates CaMKII to <20% of the value obtained with CaM as an agonist. Even this degree of activation requires 10 times more TaM (5 μM) than CaM (0.5 μM). cTnC and CaT require 1000 times higher concentration than CaM to activate PDE. Whereas cTnC is a poor activator of CaMKII, CaT may be only slightly less effective than CaM. Finally aM does not activate PDE and is a relatively poor activator of CaMKII. In the latter case more than a 10-fold excess of aM, as compared to CaM, is required to achieve a 40% activation of CaMKII. These results indicate that different portions of the CaM molecule must be important for the activation of PDE compared to CaMKII. Neither the inactive first Ca^{++} domain nor the NH_2-terminal extension are responsible for the inability of cTnC to activate PDE. A more likely reason is the addition of 3 amino acids in the central helix. On the other hand, substitution of the first domain of cTnC for its homolog of CaM results in inability to activate CaMKII, whereas the extra turn of the central helix

Table 2. Activation of cGMP Phosphodiesterase and Ca-Calmodulin Dependent Protein Kinase II by Calmodulin-Troponin C Chimeric Proteins

Activator Protein	PDE K_{act} (nM)	PDE % of Maximal Activation	CaMKII K_{act} (nM)	CaMKII % of Maximal Activation
CaM	1.4±0.2	100	285±25	100
TaM	1.4±0.2	100±2	2600±400	19±3
TnC	1450±200	92±5	-	0
CaT	1400±150	97±3	250±20*	90±5
aM	-	0	1600±200	37±5

Phosphodiesterase assays were done as described in the legend to Table 1. Ca-CaM dependent protein kinase assays were done in 50 1 volumes containing 10 mM HEPES pH 7.5, 0.5 mM DTT, 0.5 mM EGTA, 1.5 mM $CaCl_2$, 5 mM $MgCl_2$, 25 μM ATP (50 μCi γ-^{32}PATP). Reactions were incubated at 30°C for 2 minutes and stopped by the addition of 12 μl of 5X SDS sample buffer. Samples were run on 10% SDS acrylamide gels. Bands corresponding to the 50 kDa and 60 kDa subunits were localized by autoradiography, excised and counted. Results presented are the mean of six determinations, ± standard error of estimate. *These results are very preliminary and must be verified using additional CaT protein preparations.

appears to be less important. Thus the molecular basis for the specificity of cTnC is complex and must differ relative to the structure of the target protein. These studies support the contention that different portions of calmodulin must be involved in activation of the various calmodulin-dependent enzymes.

STUDIES ON THE STRUCTURE AND FUNCTION OF SMOOTH MUSCLE MLCK

The primary sequence is now known for several calmodulin-dependent enzymes [46-53]. Synthetic peptides have been utilized to define a high affinity calmodulin binding site [50,51,54-56]. Whereas these 20 amino acid peptides show less than 40% sequence identity, they are not distinguishable in their ability to bind calmodulin [50]. A portion of the smooth muscle myosin light chain kinase (MLCK) binding region was found to serve as potent substrate antagonist of this enzyme [56]. Peptide analogs modeled on this region would specifically inhibit the constitutive form of MLCK [57]. Similar studies have now been carried out with skeletal muscle MLCK [58], calmodulin kinase II [59,60], Ca^{++} ATPase [51] and calcineurin ([61] and C. Klee, personal communication). In all cases similar results have been obtained. Therefore calmodulin derepresses rather than activates many of its target enzymes and the specificity of the repression is built directly into the primary sequence of each enzyme. The corollary to these facts is that calmodulin binding to the protein should be distinguishable from activation. In theory it should be possible to design a mutant calmodulin that would bind to but not activate a given enzyme. Alternatively it should be possible to design peptide analogs that would activate the enzyme in the absence of calmodulin. Finally, since the substrate antagonist region of an enzyme is unique to that enzyme, specific inhibitors of the enzyme should be able to be developed based on a knowledge of the interaction of the pseudo-substrate region with the active site. However such information must be specifically designed for the enzyme in question. This knowledge helps to explain why limited proteolysis activates so many calmodulin dependent enzymes and why calmodulin binding drugs are so notoriously nonspecific. The remainder of this chapter will be devoted to a discussion of our efforts to understand how calmodulin regulates the chicken smooth muscle form of MLCK.

Isolation of cDNA clones for MLCK was undertaken to provide information on the sequence and functional domains of the enzyme. A cDNA clone of 2.1 kb (MK 2.1) was isolated from a λgt11 expression library made from chicken gizzard mRNA [47]. This clone was sequenced and contains an open reading frame coding for 655 amino acids. Analysis of the DNA sequence revealed a region containing the consensus sequence for the ATP binding site and sites phosphorylated by cAMP dependent protein kinase. Using a ^{125}I-calmodulin overlay procedure it was determined that the proteins produced by the phage contained the CaM binding region of MLCK. To define the CaM binding domain smaller cDNAs from the 2.1 kb clone were subcloned into a plasmid expression vector. These additional constructs defined the region encoding the CaM binding activity to a 200 bp piece 5' to the area coding for the phosphorylation site and 3' to that of the ATP binding site. Within this 64 amino acid region is a sequence with homology to the corresponding sequence reported for the CaM binding region of skeletal muscle MLCK [55]. Inspection of these sequences revealed that they share a similar number and spatial arrangement of basic residues with those present in the myosin light chain substrate. We synthesized a 22-residue peptide corresponding to amino acids 480-501 (determined from the cDNA) of the smooth muscle MLCK. This peptide inhibited CaM-dependent activation of MLCK with an IC_{50} of 46 nM [56]. At saturating concentration of CaM the 22-residue peptide competitively inhibited myosin light chain and synthetic peptide substrate phosphorylation with IC_{50} values of 2.7 μM and 0.9 μM respectively. An 11-residue synthetic peptide analog, corresponding to part of the CaM-binding sequence in skeletal muscle MLCK also competitively inhibited substrate phosphorylation with a K_i of 1 μM. The competitive inhibitory activity of the CaM-binding region is similar to the apparent K_m of 2.7 μM for phosphorylation of the 23-residue peptide analog of the smooth muscle myosin light chain and raised the possibility that the CaM binding region of MLCK may act as a pseudosubstrate inhibitor of the enzyme.

Proteolysis by trypsin of MLCK in the absence of Ca^{2+}-CaM causes a biphasic effect on kinase activity. During the initial phase of proteolysis, kinase activity is reduced over 1000-fold. Further proteolysis causes an increase in activity that is independent of Ca^{2+}-CaM. Loss of activity is associated with the formation of a 64 kDa fragment. CaM-independent activity is associated with the formation of a 61 kDa fragment. Ikebe et al. [57] developed procedures for the isolation of each fragment. Tryptic hydrolysis of the isolated 64 kDa peptide generates the 61 kDa peptide and increases CaM-independent activity. K_m values for ATP and light chains for the native kinase and two fragments are the same, i.e. approximately 100 and 5 μM, respectively. Synthetic peptides corresponding to the CaM-binding regions of the smooth and skeletal muscle kinases are potent inhibitors of the 61 kDa fragment. These data confirm the existence of an inhibitory region located between the active site and the CaM-binding site. Microsequencing and amino acid analysis of peptides generated from the 61 and 64 kDa fragments by proteolysis were used to determine the aminoterminal and carboxyterminal sequence of the 64 kDa fragment (62). As shown diagrammatically in Figure 1, the 64 kDa fragment results from cleavage in the CaM-binding region at Arg^{505}, is catalytically inactive and incapable of binding CaM. Further digestion removes the pseudosubstrate sequence from Ser^{484}-Arg^{505} resulting in a constitutively active CaM-independent 61 kDa fragment. This fragment is inhibited by peptide analogs representing the A^{480}-L^{510} region. Both fragments have the same aminoterminal sequences. The best estimate for the COOH-terminus of the 61 kDa enzyme is between K^{445} and R^{459}. However the precise extent of the pseudosubstrate area (the difference between the 64 and 61 kDa enzyme) is unknown. Nevertheless, these results provide direct support for the concept that CaM regulates MLCK by modulating the intramolecular interaction of a pseudosubstrate structure with the active site. It seems reasonable that pseudosubstrates may be utilized in analogous ways by a number of other allosterically regulated enzymes.

To facilitate studies of structure-function relationships in MLCK, we wished to express a portion of the cDNA in bacteria. A 1400 nucleotide cDNA fragment encoding the proposed catalytic and CaM binding domains (Asp^{147}-Glu^{610}) was inserted into a pOTS plasmid vector designed to express foreign gene products efficiently in E. Coli. This vector, designated POTS-MK1.4, is under the control of a λ promoter inducible by nalidixic acid. When induced, E. Coli cells containing POTS-MK1.4 synthesized a protein of about 50 kDa. This protein, produced and isolated at 2mg/l, had the following characteristics: (1) It reacted with antibody to smooth muscle MLCK by a Western blot analysis; (2) It specifically bound CaM in the presence of Ca^{++} in a gel overlay assay; (3) It phosphorylated both native gizzard myosin and isolated light chains in a Ca^{++}-CaM dependent manner; (4) The specific activity and kinetics of activation are similar to native chicken gizzard MLCK. Truncation of this fragment to Met^{513} at the COOH-end did not alter any of the enzymatic properties (Fig. 1). The enzyme that ends at M^{513} is about 40 kDa in size. This size is extremely similar to the 30kDa form

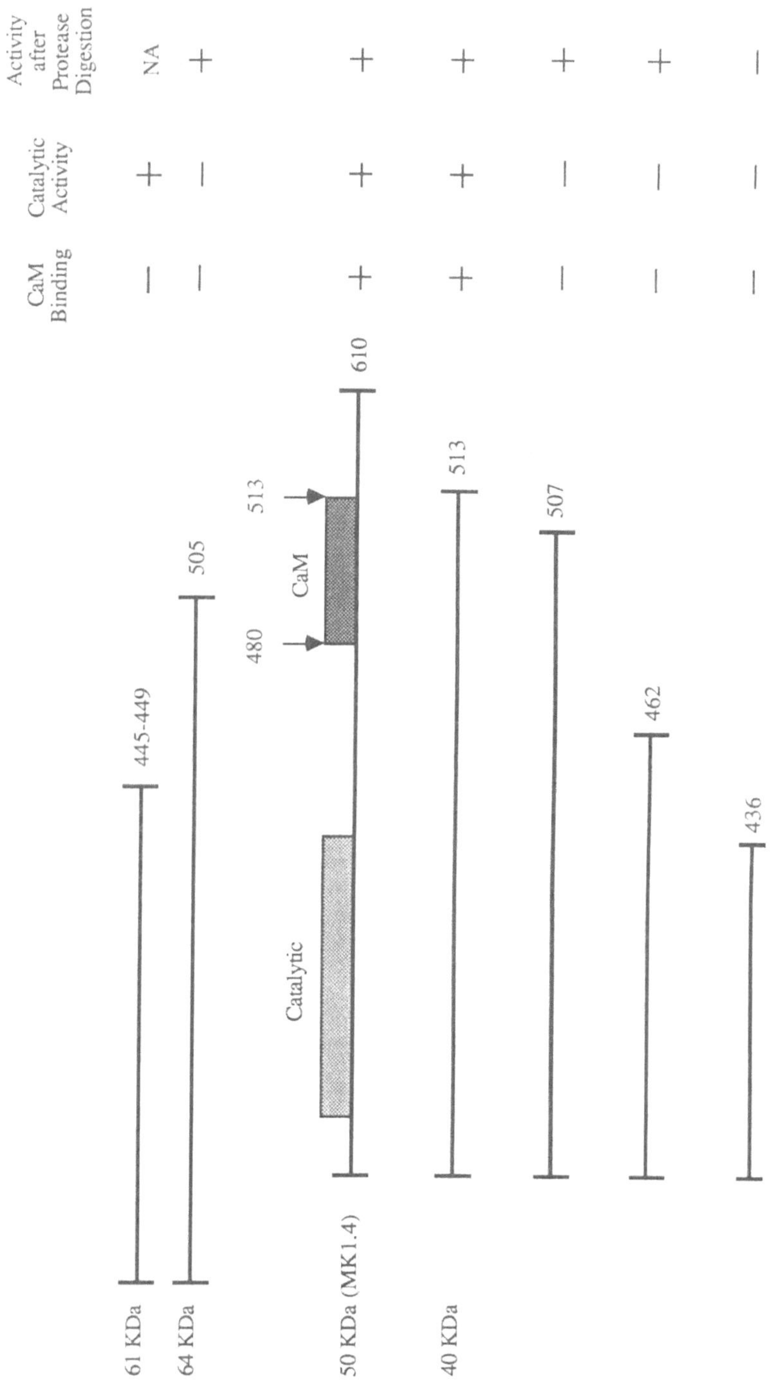

Activity after Protease Digestion
Catalytic Activity
CaM Binding
61 KDa
445-449
64 KDa
505
50 KDa (MK1.4)
Catalytic
480
CaM
513
610
40 KDa
513
507
462
436
NA

Figure 1

of the enzyme present in Limulus [63] and represents the smallest enzyme so far demonstrated to retain calmodulin dependency, substrate specificity and maximum velocity. The smallest enzyme purified to date is from Dictyostelium [64]. However this 35 kDa protein is not calmodulin-dependent. We suggest that it may have lost the calmodulin binding and substrate inhibitory regions.

Truncation of the 40 kDa chicken gizzard MLCK fragment to Ala^{506} resulted in a completely inactive enzyme that could not be activated by CaM (Fig. 1). These data confirmed the proteolysis studies [57,62]. They also revealed that residues critical for binding of and regulation by calmodulin must be between A^{506} and S^{512}. To determine the limits of this region precisely, the restriction endonuclease NcoI was utilized to cleave the cDNA at the codon for M^{513}. A series of deletion mutants were prepared by incubating the cDNA for various times in the presence of exonuclease III. Individual clones were obtained, sequenced and the protein products analyzed for both enzyme activity and calmodulin dependency (Fig. 1). Truncation of the enzyme to L^{510} resulted in no change from the properties of the fragment that terminates in M^{513} (our data and refs. [54,62]). However if the enzyme ends at I^{507} it is neither active nor does it bind calmodulin. This reveals that G^{508} and/or R^{509} are absolutely critical for calmodulin-dependent activation of MLCK activity. Site-specific mutagenesis is currently being used to determine the importance of the specific amino acid at position 509.

Proteins have been produced in the MLCK bacterial expression system that do not exhibit kinase activity. These kin^- mutants have resulted from conversion of K^{245} of the ATP binding site to A, deleting a region between L^{370} and E^{420} and by deleting a region between E^{420} and M^{492}. Exonuclease III has also been used to delete the enzyme NH_2-terminal to A^{506}. Deletion to T^{463} does not restore activity. However enzyme fragments that terminate between T^{463} and G^{508} can be activated by limited proteolysis with trypsin, and the active forms can be inhibited by the peptide analogs to the pseudosubstrate region. Deletion to W^{436} results in complete loss of activity, and proteolysis does not produce an active enzyme. Therefore residues between W^{436} and T^{463} are critical for kinase activity. This 33 amino acid stretch is 65% identical to the homologous region of the rabbit skeletal muscle enzyme whereas the two enzymes are 54% identical over 340 amino acids. A common R residue exists in both enzymes that is at position 459 in smooth muscle MLCK, and an R residue is always found at the COOH-terminus of the catalytic region of protein kinases. Since proteolysis experiments define the COOH-terminus of the active 61 kDa enzyme between K^{445} and R^{459}, we suspect the R^{459} is the likely COOH-terminal boundary of the constitutive enzyme. Further deletion from T^{463} through R^{459} will be required to answer this question.

CONCLUDING REMARKS

Determining the structure of the constitutive form of MLCK together with a peptide analog of its structure antagonist region could eventually lead to useful applications. Phosphorylation of myosin on the light chains is rate-limiting for smooth muscle contraction [65]. Therefore by inhibiting MLCK, one would effectively achieve both vasodilation and decreased vascular tone. Several classes of pharmacologic agents such as direct vasodilators, calcium entry blockers, and angiotensin converting enzyme inhibitors are currently available to decrease systemic vascular resistance (SVR) [66]. Each class has proved to have uses and limitations in disease states characterized by elevated SVR. Specifically designed inhibitors of MLCK would provide an entirely new class of drug for the treatment of hypertension and congestive heart failure. Such agents would have the potential to be as significant an advance in therapy as angiotensin-converting enzyme inhibitors. In addition, there is reason to believe that this new class of drugs will be potent antianginal agents. Calcium entry blocking agents have been found to be effective in the treatment of chronic stable angina [67,68] and in the prevention of recurrent myocardial infarction after non Q wave myocardial infarction [69]. These agents have been shown to be effective in causing relaxation of vascular smooth muscle in both the systemic arterial and coronary arterial beds [70,72]. It is known that transient elevation in intracellular Ca^{++} results in phosphorylation of the myosin regulatory light chain via the MLCK system. By decreasing the magnitude of this transient Ca^{++} current, the Ca^{++} entry blockers may act by decreasing the extent of myosin light chain phosphorylation [71,72]. The availability of specific inhibitors of MLCK could provide a new, more specific means of decreasing coronary vascular tone. We would expect these agents to be synergistic with calcium entry blockers in the treatment of coronary heart disease. Finally, these new agents would not be expected to depress myocardial contractility and/or interfere with conduction of the cardiac action potential at the atrioventricular node, as do some calcium entry blockers [71,72].

ACKNOWLEDGEMENTS

The authors thank Ms. Patricia Donnelly for excellent technical assistance, Dr. John A. Putkey for initial constructs and helpful advice, and Dr. Tomio Ono who purified TaM and CaT initially and carried out phosphodiesterase assays. Dr. Richard G. Cook provided amino acid composition and sequence data for the regulatory proteins. Finally we owe a great debt to Dr. Bruce E. Kemp (St. Vincent's Institute of Medical Research) who is a long-time and continuing collaborator on the MLCK project. The work described was supported in part by research grants to ARM from the NIH (GM-33976 and HD-07503); the Robert A. Welch Foundation (Q-611); and the Jerry Lewis Muscular Disease Research Center. Ms. Lisa Gamble showed great patience in the preparation of the typescript.

REFERENCES

1. L. Lagace', T. Chandra, S.L.C. Woo, and A.R. Means, J. Biol. Chem. 258:1684 (1983).
2. D. Bartlet, S. Fidel, L.H. Farber, D.J. Wolff, and R.L. Hammell, Proc. Natl. Acad. Sci. USA 85:3284 (1983).
3. T.N. Davis, M.S. Urdea, F.R. Masiarz, and J. Thorner, Cell 47:423 (1986).
4. T. Takeda and M. Yamamoto, Proc. Natl. Acad. USA 84:3580 (1987).
5. J.A. Putkey, K.F. Ts'ui, T. Tanaka, L. Lagace', J.P. Stein, E.C. Lai, and A.R. Means, J. Biol. Chem. 258:11864 (1983).
6. D.J. Mangelsdorf, B.S. Komm, D.P. McDonnell, J.W. Pike, and M.R. Haussler, Biochemistry 26:8332 (1987).
7. H. Nojima, K. Kishi, and H. Sokabe, Mol. Cell. Biol. 7:1873 (1987).
8. A. Sherbany, A.S. Parent, and J. Brosius, DNA 6:267 (1987).
9. E.J. Wawrzynczak and R.N. Perham, Biochem. Int. 9:177 (1984).
10. B. Sen Gupta, F. Friedberg, and S.D. Detera-Wadleigh, J. Biol. Chem. 262:16663 (1987).
11. R. Fischer, M. Koller, M. Flura, S. Mathews, M.-A.Strehler-Page, J. Krebs, J.T. Penniston, E. Carafoli, and E.E. Strehler, J. Biol. Chem. (in press).
12. C.D. Rasmussen and A.R. Means, EMBO J. 6:3961 (1987).

13. J.G. Chafouleas, W.E. Bolton, H. Hidaka, A.E. Boyd, and A.R. Means, Cell 28:41 (1982).
14. Y. Sasaki and H. Hidaka, Biochem. Biophys. Res. Commun. 104:451 (1982).
15. Y. Eilam and D. Chernichovsky, J. Gen. Microbiol. 134:1063 (1988).
16. Y. Ohya and Y. Anraku, Nature (in press).
17. Y.S. Babu, J.S. Sack, T.J. Greenhough, C.E. Bugg, A.R. Means, and W.J. Cook, Nature 315:37 (1985).
18. O. Minowa and K. Yagi, J. Biochem. 96:1175 (1984).
19. D.L. Newton, M.D. Oldewurted, M.H. Krinks, J. Shiloach, and C. B. Klee, J. Biol. Chem. 259:4419 (1984).
20. O. Minowa, M. Yazawa, K. Sobue, K. Ito, and K. Yagi, J. Biochem. 103:531 (1988).
21. J.R. Dedman, J.D. Potter, R.L. Jackson, J.D. Johnson, and A.R Means, J. Biol. Chem. 252:8415 (1977).
21a.E.W. Small, and S.R. Anderson, Biochemistry 27:419 (1988).
22. R.E. Klevit, D.K. Blumenthal, D.E. Wemmer, and E.G. Krebs, Biochemistry 24:8152 (1985).
23. S.H. Seeholzer, M. Cohn, J.A. Putkey, A.R. Means, and H.L. Crespi, Proc. Natl. Acad. Sci. USA 83:3634 (1986).
24. A.E. Jackson, K.L. Carraway, D. Puett, and K. Brew, J. Biol. Chem. 261:12226 (1986).
25. A.E. Jackson, K.L. Carraway III, M.E. Payne, A.R. Means, D. Puett, and K. Brew, Proteins 2:202 (1987).
26. M.A. Winkler, V.A. Fried, D.L. Merat, and W.Y. Cheung, J. Biol. Chem. 262:15466 (1987).
27. K.T. O'Neil and W.F. Delgrado, Proc. Natl. Acad. Sci. USA 82:4954 (1985).
28. D.B. Heidorn and J. Trewhella, Biochemistry 27:909 (1988).
29. Y.S. Babu, C.E. Bugg, and W.J. Cook, (1988) J. Mol. Biol. (in press).
30. R.H. Kretsinger and C.E. Nockolds, J. Biol. Chem. 248:3313 (1973).
31. D.M.E. Szebenyi, S.K. Obendorf, and K. Moffat, Nature 294:327 (1981).
32. O. Herzberg and M.N.G. James, Nature 313:653 (1985).
33. K.A. Satyshur, S.T. Rao, D. Pyzalska, W. Drendel, M. Greaser, and M. Sundaralingam, J. Biol. Chem. 263:1628 (1988).
34. J.A. Cox, Biochem. J. 249:621 (1988).
35. D. Guerini, J. Krebs and E. Carafoli, Eur. J. Biochem. 170:35 (1987).
36. A.S. Manalan and C.B. Klee, Biochemistry 26:1382 (1987).
37. D. Mann and T.C. Vanaman, Meth. Enzymol. 139:417 (1987).
38. E.W. Small and S.R. Anderson, Biochemistry 27:419 (1988).
39. J.A. Putkey, G.R. Slaughter, and A.R. Means, J. Biol. Chem. 260:4704 (1985).
40. J.A. Putkey, G.F. Draetta, G.R. Slaughter, C.B. Klee, P. Cohen, J.T. Stull, and A.R. Means, J. Biol. Chem. 261:9896 (1986).
41. J.A. Putkey, T. Ono, M.F.A. VanBerkum, and A.R. Means, J. Biol. Chem. (in press).
42. T.A. Craig, D.M. Watterson, F.G. Prendergast, J. Haiech, and D.M. Roberts, J. Biol. Chem. 262:3278 (1987).
43. A. Persechi, D.O. Hardy, D.K. Blumenthal, H.W. Jarrett, and R.H. Kretsinger, Biophys. J. 53:A252 (1988).
44. J.D. Johnson, J.H. Collins, and J.D. Potter, J. Biol. Chem. 253:6451 (1978).
45. J.A. Putkey, S.L. Carroll, and A.R. Means, Mol. Cell. Biol. 7:1549 (1987).
46. K. Takio, D.K. Blumenthal, A.M. Edelman, K.A. Walsh, E.G. Krebs, and K. Titani, Biochemistry 24:6028 (1985).
47. V. Guerriero, M.A. Russo, N.J. Olson, J.A. Putkey, and A.R. Means, Biochemistry 30:8372 (1986).
48. M.K. Bennett and M.B. Kennedy, Proc. Natl. Acad. Sci. USA 84:1794 (1987).

49. C.R. Lin, M.S. Kapiloff, S. Durgerian, K. Tatemoto, A.F. Russo, P. Hanson, H. Schulman, and M.G. Rosenfeld, Proc. Natl. Acad. Sci. USA 84:5962 (1987).
50. R.M. Hanley, A.R. Means, T. Ono, B.E. Kemp, K.E. Burgin, N. Waxham, and P.T. Kelly, Science 237:293 (1987).
51. P. James, M. Maeda, R. Fischer, A.K. Verma, J. Krebs, J.T. Penniston and E. Carafoli, J. Biol. Chem. 263:2905 (1988).
52. E.M. Reimann, K. Titani, L.H. Ericsson, R.D. Wade, E.H. Fischer, and K.A. Walsh, (1984) Biochemistry 23:4185 (1984).
53. P. Glaser, D. Ladant, O. Sezer, F. Pichot, A. Ullmann, and A. Danchin, Mol. Microbiol. 2:19 (1988).
54. T.J. Lukas, W.H. Burgess, F.G. Prendergast, W. Lau, and D.M. Watterson, 25:1458 (1986).
55. D.K. Blumenthal, K. Takio, A.M. Edelman, H. Charbonneau, K. Titani, K.A. Walsh, and E.G. Krebs, Proc. Natl. Acad. Sci. USA 82:3187 (1985).
56. B.E. Kemp, R.B. Pearson, V. Guerriero, I.C. Bagchi, and A.R. Means, J. Biol. Chem. 262:2542 (1987).
57. M. Ikebe, S. Malgorzata, B.E. Kemp, A.R. Means, and D.J. Hartshorne, J. Biol. Chem. 260:12828 (1987).
58. P.J. Kennelly, A.M. Edelman, D.K. Blumenthal, and E.G. Krebs, J. Biol. Chem. 262:11958 (1987).
59. M.E. Payne, Y.-L. Fong, T. Ono, R.J. Colbran, B.E. Kemp, T.R. Soderling, and A.R. Means, J. Biol. Chem. 263:7190 (1988).
60. R.M. Hanley, A.R. Means, B.E. Kemp, and S. Shenolikar, Biochem. Biophys. Res. Commun. 152:122 (1988).
61. E.A. Tallant, L.M. Brumley and R.W. Wallace, 27:2205 (1988).
62. R.B. Pearson, R.E.H. Wettenhall, A.R. Means, D.J. Hartshorne, and B.E. Kemp, Science 241:970 (1988).
63. J.R. Sellers and E.V. Harvey, Biochemistry 23:5821 (1984).
64. L.M. Griffith, S.M. Downs and J.A. Spudich, J. Cell. Biol. 104:1309 (1987).
65. K.E. Kamm and J.T. Stull, Ann. Rev. Pharmacol. Toxicol. 25:593 (1985).
66. N.M. Kaplan, Systemic Hypertension: Therapy in: "Heart Disease: A Textbook of Cardiovascular Medicine," E. Braunwald, W.B. Saunders & Co., pp. 819-861 (1988).
67. P.H. Stone, Z. Turi, J.E. Muller, E. Geltman, A. Jaffe, E. Braunwald, J. Am. Coll. Cardiol. 1:596 (1983).
68. V.B. Subramanian, N.S. Khurmi, M.J. Bowles, M. O'Hara and E.B. Raferty, J. Am. Coll. Cardiol. 1:1144 (1983).
69. R.S. Gibson, W.E. Boden, P. Theroux, et al. N. Engl. J. Med. 315:423 (1986).
70. C.Y.C. Chew, B.G. Brown, B.W. Single, M.M. Wong, C. Pierce, R. Peterson, Am. J. Cardiol. 51:699 (1983).
71. E. Braunwald, N. Engl. J. Med. 307:1618 (1983).
72. L.H. Opie, Cardiovascular Drug and Therapy 1:411 (1987).

INTERACTION BETWEEN CALMODULIN AND TARGET PROTEINS

Koichi Yagi, Michio Yazawa,
Mitsuhiko Ikura* and Kunio Hikichi*

Department of Chemistry and High Resolution
NMR Laboratory*, Faculty of Science
Hokkaido University, Sapporo 060, Japan

INTRODUCTION

When myosin light chain kinase (MLCK) activity was measured at various different free Ca^{2+} concentrations, the activity attained abruptively the maximum value within a very narrow concentration range of Ca^{2+}. A result of this phenomenon is shown in Fig.1 (curve e), accompanied with the Ca^{2+} binding curve of calmodulin (curve a) (1). Since the Ca^{2+}-calmodulin is known to work as an activator of MLCK, a large difference between the enzyme activation curve and the Ca^{2+} binding curve of calmodulin indicates that Ca^{2+} effect on the MLCK activity can not directly be explained by the comparison of these two curves. Olwin et al. have already shown that the affinity of Ca^{2+} to calmodulin was largely enhanced by the addition of MLCK (2). It is expected that the analysis of enhancing effect of MLCK to the Ca^{2+} binding to calmodulin may give an answer how the enzyme attains the maximum activity within such a narrow concentration range of Ca^{2+} and how the Ca^{2+} concentration is fixed to this range. The Ca^{2+} binding to calmodulin was measured in the presence of mastoparan (curve c in Fig. 1) (3). Mastoparan is a small peptide composed of 14 amino acid residues, and Malencik and Anderson have reported a strong binding constant of mastoparan to calmodulin (4). The Ca^{2+} binding curve of calmodulin in the presence of mastoparan was almost identical with the result of Olwin et al. obtained in the presence of MLCK. Both of the Ca^{2+} binding curves shift to the lower Ca^{2+} concentration range and to the same extent from the curve obtained in their absence. This result suggests that mastoparan can be used as a model compound of MLCK. The measurements of Ca^{2+} binding to calmodulin by equilibrium dialysis or flow dialysis in the presence of MLCK (target enzymes) were difficult experiments because of their high molecular weights. The detailed study of ^{1}H-NMR of calmodulin in the presence of MLCK was almost impossible. Therefore, a small molecular weight model compound is useful for the extensive investigations of the effect of MLCK to calmodulin.

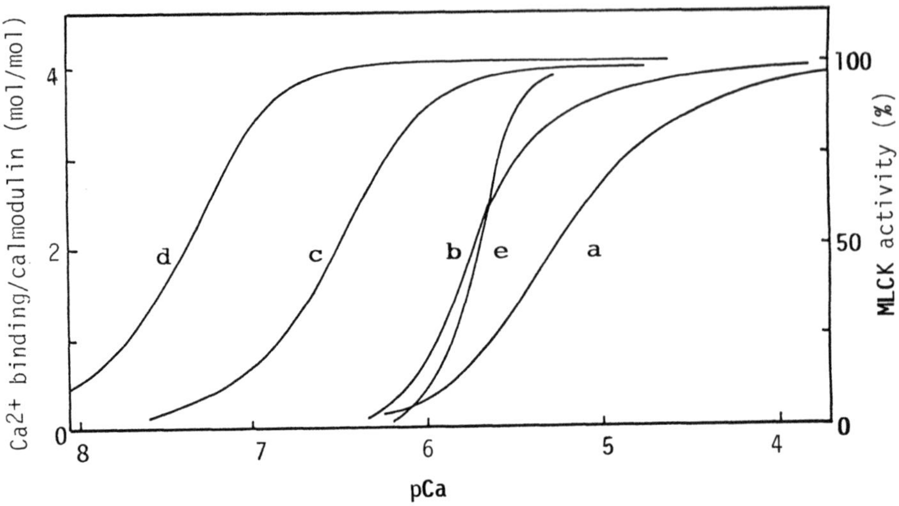

Fig. 1 Ca^{2+} binding to calmodulin and Ca^{2+} dependence of the MLCK activity. Ca^{2+} binding to scallop testis calmodulin was measured by the flow-dialysis method in 0.1 M NaCl, 20 mM MOPS-NaOH (pH 7.0) at 25 °C. Concentration of calmodulin was 16.2 μM. a, calmodulin alone. b, +28.5 μM 40K fragment of caldesmon. c, +60∿300 μM mastoparan. d, +43.9 μM M13. The Ca^{2+} binding curves are the calculated best fit curves with four macroscopic binding constants. Chicken gizzard MLCK activity was measured in 0.1 M NaCl, 20 mM MOPS-NaOH (pH 7.0), 10 mM $MgCl_2$, 2.56 nM MLCK, 6.06 nM chicken gizzard calmodulin, 58.5 μM 20K light chain and 1 mM [γ^{32}P]ATP at 25 °C. e, MLCK activity and the maximum activity was 14.9 S^{-1}.

CALMODULIN MOLECULE

As clearly shown by X-ray crystallography, N-terminal domain and the C-terminal domain are separated by a polypeptide chain composed of 10 to 15 amino acid residues (5). When the calmodulin is exposed to trypsin, it is cleaved at the middle of the connecting polypeptide. This procedure was first discovered by the late Dr. W. Drabikowski and his group (6). They demonstrated that a similar result can be obtained with troponin C and also with calmodulin. N-terminal and C-terminal half-fragments can be isolated by this procedure of tryptic digestion and we designated them F12 and F34, respectively. Interaction between the two domains of calmodulin in the native state was investigated and we could prove that they were separated independently (1,7).

^{1}H-NMR spectrum of low magnetic field of calmodulin is shown in Fig. 2. The ^{1}H-NMR spectra of F12,F34 and a mixture containing an equivalent amount of F12 and F34 are also shown in the same figure (8). Sum of the spectra of F12 and F34 was identical with the spectrum of mixture. Moreover, the spectrum of mixture was identical with that of calmodulin (bottom of the figure). Although this result

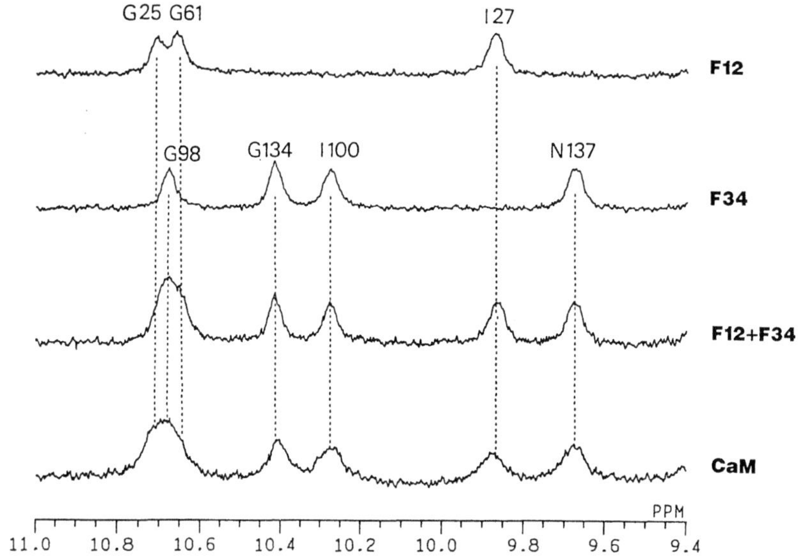

Fig. 2 Low-field amide region of ^{1}H-NMR spectra of F12, F34, a mixture of F12 and F34, and calmodulin (CaM) in the presence of Ca^{2+}. NMR spectra were measured at 30 °C on a JEOL JNM-GX500 spectrometer (^{1}H 500 MHz). Samples are 3.0 mM in 0.2 M KCl. pH was adjusted to 6.4 - 6.6 by NaOH.

represents a part of calmodulin structure, it strongly suggests that the conformation of amino acid residues in calmodulin did not change by the tryptic cleavage at the middle of calmodulin, that is, F12 and F34 maintained the same conformations as those of the N-domain and C-domain, respectively, in the calmodulin.

The sizes and the amino acid sequences of the two domains of calmodulin are similar to each other. However, there is a distinct difference between the Ca^{2+} binding properties of the two domains. As shown in Fig. 3 (9), the change of UV difference spectrum at 287 nm, which is caused by the conformational change around tyrosin-138 in C-domain, is observed by the addition of Ca^{2+}. The maximal change was attained at 1.7 mol Ca^{2+} added per mol of calmodulin. It suggests that the conformational change in C-domain occurs by the first 2 mol of Ca^{2+} bindings to calmodulin. Since no interaction can be assumed between the two domains, it is likely that the conformational change in C-domain occured by the first 2 Ca^{2+} binding to the C-domain.

The change in the chemical shift of ^{13}C-NMR of calmodulin was observed by the addition of Ca^{2+} (9). ^{13}C was introduced into cysteine-27 of plant calmodulin using DTNB and $K^{13}CN$. The side chain $-CH_2-SH$ group of the cysteine was converted to $-CH_2-S^{13}CN$. Since the cysteine-27 is in the N-domain, the downfield shift around 114-115 ppm observed by the addition of Ca^{2+} can be regarded as the result of the conformational change in the N-domain. The change occured at 2-3.5 mol Ca^{2+} per mol of calmodulin. It indicates that 2 mol of Ca^{2+} bind to the Ca^{2+} binding sites

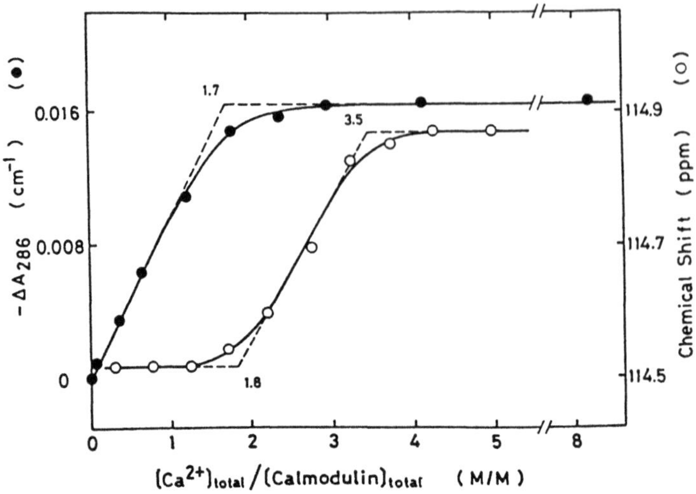

Fig. 3 Changes in the chemical shift (○) plotted against the ratio of total molar concentration of Ca^{2+} to that of calmodulin. NMR spectrum was recorded with 3 mM ^{13}CN-calmodulin, 0.25 M KCl, 10 mM K-phosphate (pH 7.0) in D_2O at 25 °C. The Ca^{2+} dependence of ΔA_{286} of the Ca^{2+}-induced UV absorption difference spectrum of calmodulin was measured with 85.3 μM of ^{13}CN-calmodulin, 0.2 M KCl, 20 mM MOPS (pH 7.0) at 25 °C.

of N-domain after the two sites in C-domain were saturated by Ca^{2+}.

These results show that the high affinity sites are in C-domain and low affinity sites are in N-domain. Each of the two domains showed the positive cooperative two calcium bindings, but the Ca^{2+}-bindings to N-domain and C-domain were independent from each other (1,3).

EFFECT OF M13

As shown in Fig. 1, Ca^{2+} bindings to calmodulin were enhanced by the addition of caldesmon or mastoparan (3). A result of Ca^{2+} binding to calmodulin is also shown in the presence of a peptide designated by Blumenthal et al. as M13 which corresponds to a calmodulin binding portion of MLCK (10). The Ca^{2+} binding was enhanced greatly by M13. The mid-point of binding curve was 0.04 μM and it was even smaller than the mid-point obtained by the addition of mastoparan, which is a model compound of MLCK. In order to investigate the effect of M13 in detail, we measured ^{113}Cd-NMR of Cd^{2+}-calmodulin solution (11). Forsen and his group have already investigated ^{113}Cd-NMR spectrum in the presence of calmodulin (12). They obtained two chemical shifts of ^{113}Cd from the ^{113}Cd-calmodulin. We confirmed their results which were obtained in the range of Cd^{2+} concentration from 0.52 to 4.16 mol/mol Cd^{2+}-calmodulin. The peak heights become the maximum when the amount of Cd^{2+} arrived at 2 mol/mol. The interaction of Cd^{2+} with N-domain could not be observed probably because of the high exchange rates of Cd^{2+} in N-domain.

When Cd^{+2} was added to M13-calmodulin complex, 4

chemical shifts were clearly observed at 1.04 mol Cd^{2+} per mol of calmodulin in the range between -100 and -122 ppm. The 4 peak heights increased at the same rate accompanied with the increase in the Cd^{2+} concentration and they attained the maximum together, as shown in Fig. 4. If the effect of MLCK might be the same as that of M13, the MLCK activity is ought to coincide with the Ca^{2+} binding curve of calmodulin in the presence of M13. Unfortunately, the Ca^{2+} concentration necessary for the MLCK activation was much higher than the Ca^{2+} binding obtained in the presence of M13. Therefore, it is difficult to correlate the result of Cd^{2+}-binding obtained in the presence of M13 to the MLCK activation.

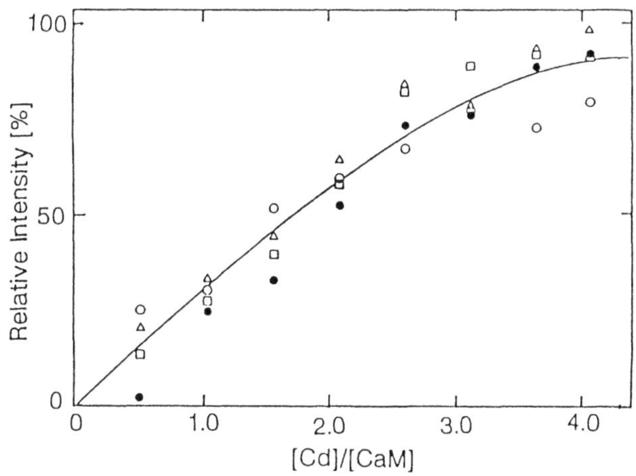

Fig. 4 Plots of the intensity of 4 peaks (○,□,△,●) of ^{113}Cd-NMR spectra against the molar ratio of total Cd^{2+} to calmodulin (CaM) in the presence of M13. Conditions are 1.2 mM calmodulin, 1.2 mM M13, 0.2 M KCl, pH 6.9 - 7.3, at 25 °C. The four peaks were observed at -101.86, -113.27, -117.67 and -122.07 ppm.

As we have shown previously in the absence of any target proteins or peptides, the strong positive cooperative Ca^{2+} binding was observed in C-domain and the weak cooperativity was also observed in N-domain (1). However, the cooperativity between the two domains was not indicated. In the presence of mastoparan, we still observed discrimination among the binding sites, that is, stronger two Ca^{2+} bindings to C-domain and weaker two bindings to N-domain (3). However, an apparently complete positive cooperativity was observed for the 4 Cd^{2+} bindings to calmodulin in the presence of M13.

EFFECT OF MLCK

The Ca^{2+} binding was not enhanced by MLCK itself so strong as the effect by M13 (Fig. 1). We then assume that MLCK contains the other domain which participates to suppress the calmodulin binding to MLCK. The combination of the calmodulin binding region with this suppressive region of MLCK may determine the interaction of Ca^{2+}, calmodulin and MLCK.

In order to investigate the relation of the effects of MLCK for the Cd^{2+} binding to those for the Ca^{2+} binding, we measured ^{113}Cd-NMR of Cd^{2+}-calmodulin solution in the presence of mastoparan, a favorable model compound of MLCK. The ^{113}Cd spectra of Cd^{2+}-calmodulin the presence of mastoparan were not so simple as the ^{113}Cd-calmodulin NMR spectra obtained in the presence of M13 (11). At 0.52 mol Cd^{2+} per mol of calmodulin three small chemical shifts were observed in the range from -95 to -120 ppm. With increase in the Cd^{2+} concentrations to 1.04 - 1.56 mol/mol, the chemical shift at -95.18 ppm disappeared and the peaks at -100.23 and -119.23 ppm became larger. They were not observed at 2.60 mol/mol and new peaks at -102.84, -113.13 and -120.60 ppm became apparent. They attained the maximum at 3.64 or 4.16 mol of Cd^{2+} per mol of calmodulin. These results are summarized in Fig. 5. One curve soon disappeared and two disappeared later. The finally appeared three curves were in parallel and became the maximum together at nearly 4 mol Cd^{2+} per mol. At this Cd^{2+} concentration, the 4 Ca^{2+} binding sites of calmodulin were probably occupied by the 4 Cd^{2+}. We assume at the moment that this complicated changes of the chemical shifts represent the manner of Ca^{2+} bindings to calmodulin in the presence of mastoparan.

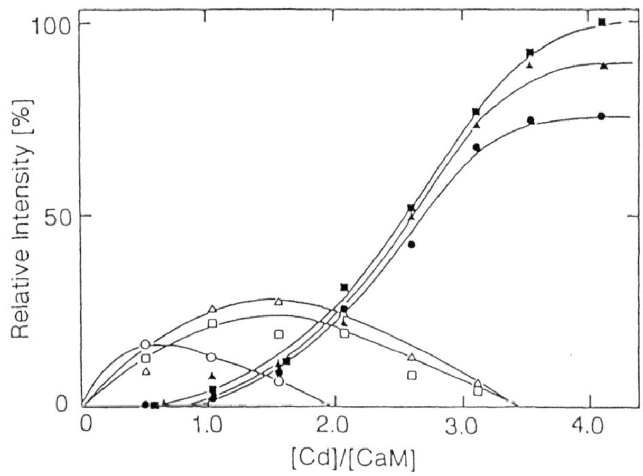

Fig. 5 Plots of the intensity of peaks of ^{113}Cd-NMR spectra against the molar ratio of total Cd^{2+} to calmodulin in the presence of mastoparan. Conditions are 1.2 mM calmodulin, 1.2 mM mastoparan, 0.2 M KCl, pH 7.1 - 7.4 at 25 °C. The peaks were observed at -95.18 (○), -100.23 (□), -119.23 (△), -102.84 (●), -113.13 (two species together, ■) and -120.60 ppm (▲).

COMMUNICATION BETWEEN THE TWO DOMAINS

The location of histidine-107 is in C-domain. We measured ^{1}H-NMR spectrum of histidine in order to investigate the conformational change of C-domain of calmodulin in relation to the amount of bound Ca^{2+} (3). Only two chemical shifts were observed in the absence of mastoparan. They are the chemical shifts one of which corresponds to the conformation of Ca^{2+}-free calmodulin and the other to the conformation of Ca^{2+}-saturated calmodulin. The chemical shift corresponding to the calcium-saturated calmodulin increased the peak height by the increase in the added Ca^{2+} concentration until 2 mol Ca^{2+} per mol. It appears to become the maximum by the addition of 2 mol Ca^{2+} per mol of calmodulin.

On the other hand, three states were observed in the presence of mastoparan. They may correspond to the conformations of Ca^{2+}-free, intermediate and Ca^{2+}-saturated calmodulins. We strongly emphasize for this result that the peak height of Ca^{2+}-saturated calmodulin increased by the addition of Ca^{2+} more than 2 mol per mol. It is known that the first 2 Ca^{2+} bindings occur at the C-domain of calmodulin and the third and fourth Ca^{2+} bindings at N-domain. It, therefore, leads to a view that the Ca^{2+} binding to N-domain induces the conformational change in C-domain. This change was not observed in the absence of mastoparan. Mastoparan is a model compound of MLCK and the second conformational change in C-domain, which is induced by a signal from N-domain, may directly concern to the enzyme activation.

The affinity of Ca^{2+} to calmodulin was enhanced by target proteins or peptides (Fig. 1). The shape of the curve became steep by their addition, which was clearly indicated by the calculation of Hill coefficient of the curves. The target proteins and peptides may introduce a strong positive cooperativity for the Ca^{2+} binding to N-domain and it may be a reason to give the steep Ca^{2+} binding curves of calmodulin.

The Hill coefficient of Ca^{2+} binding curve enhanced by caldesmon was very large, but the displacement of the mid-point of curve was not so large as the displacement by mastoparan. Caldesmon is one of the calmodulin-dependent cytoskeletal proteins and the smaller displacement than MLCK might be a character of the cytoskeletal proteins.

CONCLUSION

MLCK activity increased to the maximum value abruptively from the minimum at the Ca^{2+} concentration where the calmodulin was nearly saturated by Ca^{2+} (Fig 1). The binding of mastoparan to calmodulin was indicated in the absence of Ca^{2+} by NMR experiments (3). Therefore, the Ca^{2+} binding to calmodulin in the presence of mastoparan (and MLCK) can occur at the lower Ca^{2+} concentrations than in the absence. Moreover, the positive cooperative Ca^{2+} binding to N-domain can be intensified by mastoparan (and MLCK), which may result in the abruptive MLCK activation.

It was suggested from the experiments using M13 that MLCK contains the suppressive region which suppresses the interaction of Ca^{2+}, calmodulin and MLCK by yet unknown manner. On the other hand, the calmodulin binding region has already been isolated as M13 (10). The combination of the two domains may control the Ca^{2+} binding to calmodulin and fix an appropriate Ca^{2+} concentration for the MLCK activation.

REFERENCES

1. O. Minowa and K. Yagi, Calcium binding to tryptic fragments of calmodulin, J. Biochem. 96:1175 (1984)
2. B.B. Olwin, A.M. Edelman, E.G. Krebs, and D.R. Storm, Quantitation of energy coupling between Ca^{2+}, calmodulin, skeletal muscle myosin light chain kinase, and kinase substrates, J. Biol. Chem. 259:10949 (1984)
3. M. Yazawa, M. Ikura, K. Hikichi, Y. Luan and K. Yagi, Communication between two globular domains of calmodlulin in the presence of mastoparan or caldesmon fragment. Ca^{2+} binding and ^{1}H NMR, J. Biol. Chem. 262:10951 (1987)
4. D.A. Malencik and S.R. Anderson, High affinity binding of the mastoparans by calmodulin, Biochem. Biophys. Res. Commun. 114:50 (1983)
5. Y.S. Babu, J.S. Sack, T.J. Greenhough, C.E. Bugg, A.R. Means and W.J. Cook, Three-dimensional structure of calmoldulin, Nature 315:37 (1985)
6. W. Drabikowski, H. Brzeska and S.Y. Venyaminov, Tryptic fragments of calmodulin. Ca^{2+}- and Mg^{2+}-induced conformational changes, J. Biol. Chem. 257:11584 (1982)
7. M. Ikura, T. Hiraoki, K. Hikichi, O. Minowa, H. Yamaguchi, M. Yazawa and K. Yagi, Nuclear magnetic resonance studies on calmodulin: Ca^{2+}-dependent spectral change of proteolytic fragments, Biochemistry 23:3124 (1984)
8. M. Ikura, O. Minowa, M. Yazawa, K. Yagi and K. Hikichi, Sequence-specific assignments of downfield-shifted amide proton resonances of calmodulin, FEBS Lett. 219:17 (1987)
9. M. Yazawa, E. Kawamura, O. Minowa, K. Yagi, M. Ikura and K. Hikichi, N-terminal region (Domain 1) of calmodulin is the low affinity site for Ca^{2+}. A ^{13}C NMR study of S-cyanocalmodulin, J. Biochem. 95:443 (1984)
10. D.K. Blumenthal, K. Takio, A.M. Edelman, H. Charbonneau, K. Titani, K.A. Walsh and E.G. Krebs, Identification of the calmodulin-binding domain of skeletal muscle myosin light chain kinase, Proc. Natl. Acad. Sci. USA 82:3187 (1985)
11. M. Ikura, N. Hasegawa, K. Hikichi, S. Aimoto, M. Yazawa and K. Yagi, ^{113}Cd-NMR Evidence for cooperative interaction between the amino- and carboxyl-terminal domains of calmodulin, in preparation
12. S. Forsén, E. Thulin, T. Drakenberg, J. Krebs and K. Seamon, A ^{113}Cd NMR study of calmodulin and its interaction with calcium, magnesium and trifluoperazine, FEBS Lett. 117:189 (1980)

ACTIVATION OF RABBIT SKELETAL MUSCLE MYOSIN LIGHT CHAIN KINASE BY CALMODULIN- A MECHANISTIC OVERVIEW

Peter J. Kennelly[1], Melissa Starovasnik[2], and Edwin G. Krebs[1,2,3]

Howard Hughes Medical Institute[1] and Departments of Biochemistry[2] and Pharmacology[3], University of Washington Seattle, Washington 98195, U. S. A.

BACKGROUND: WHAT IS MYOSIN LIGHT CHAIN KINASE?

The Myosin Light Chain Kinase Isozyme Family

The myosin light chain kinases [MLCK, E.C. 2.7.1.37] are a family of tissue-specific isozymes which share as their substrate the phosphoryl- or P-light chain component of myosin. Distinct forms of MLCK have been isolated from smooth, skeletal, and cardiac muscle. The enzyme appears to be ubiquitous, with non-muscle MLCK envisaged as being very similar, if not identical, to the smooth muscle isozyme (For a review see (43)).

Curiously, although the myosin light chain kinases all phosphorylate the P-light chain component of myosin *in vivo*, the physiological consequences of this event vary dramatically from tissue to tissue. In smooth muscle, phosphorylation of myosin P-light chains by MLCK is an obligatory step in the initiation of contraction (15). In striated muscle, P-light chain phosphorylation plays a more subtle modulatory role, regulating the contractile machinery rather than forming an integral part of it (22,43).

These functional differences are mirrored by the physical differences between the smooth and skeletal muscle isozymes of MLCK, the two most intensely studied to date. They differ from one another in molecular weight (32,43) and number of phosphorylation sites for cAMP-dependent protein kinase (4,8,42). Antibodies to the smooth and skeletal muscle forms display no cross-reactivity (31,32), and a cDNA for rat skeletal muscle MLCK does not hybridize to MLCK mRNAs from other tissues during Northern transfer analysis(40). Comparison of the protein and nucleotide sequences of skeletal muscle MLCK with a partial clone from a smooth muscle source shows them to possess areas of significant, but not overwhelming, homology (12,40,45). Catalytically, although skeletal muscle MLCK will phosphorylate P-light chains from all tissue sources with high efficiency *in vitro*, smooth muscle MLCK will only phosphorylate smooth muscle myosin P-light chains (43).

Despite the above emphasis on areas of difference, it should be pointed out that these isozymes also share a great deal in common. One of these common properties is the ability to bind to and be activated by calcium/calmodulin (CaM). CaM is absolutely required for expression of MLCK activity (6,13,29,46). This CaM-dependence allows cells to modulate MLCK activity through fluctuations in the intracellular second messenger calcium. It is the mechanism by which CaM-binding tranforms MLCK from an inert to a catalytically active entity that is the subject of this article.

Domain Organization of Skeletal Muscle MLCK

Through the efforts of several laboratories a substantial body of data concerning the structural organization of skeletal muscle MLCK has been accumulated. This body of data includes:

a) The complete amino acid sequence of the rabbit skeletal muscle enzyme (45).
b) The complete cDNA sequence of the rat skeletal muscle enzyme (40).
c) Identification of the tail and catalytic domains by partial proteolysis (9,27).
d) Identification of the CaM-binding site (3).

The picture which has emerged from these studies is shown in Figure 1, below.

MECHANISM OF CaM ACTIVATION: THEORETICAL MODELS VERSUS EXPERIMENTAL OBSERVATIONS

The Models

Two main models have been suggested for the activation of MLCK, the energy transfer hypothesis and the autoinhibitor (or pseudosubstrate) hypothesis. As its name implies, the energy transfer hypothesis focuses upon the fate of the free energy released when CaM binds MLCK. It is proposed that the CaM-binding domain is physically coupled to the remainder of the protein such that a portion of the free energy of binding is transmitted into the protein's structure. Here, it evokes a conformational change that reconfigures MLCK into a catalytically competent state.

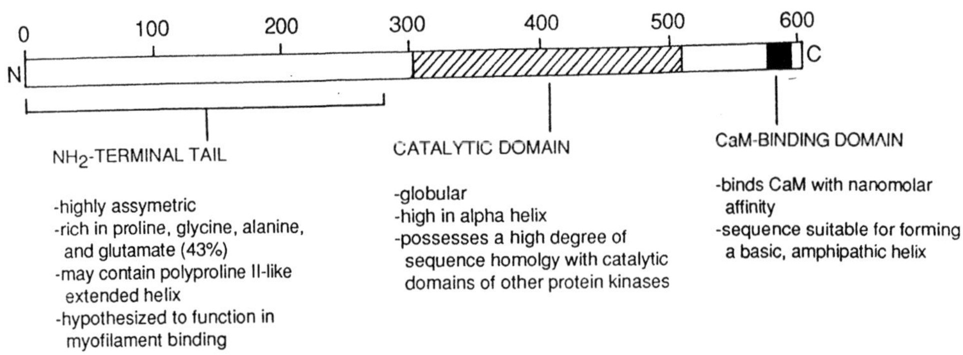

FIGURE 1. Domain organization of rabbit skeletal MLCK.

The autoinhibitor hypothesis focuses upon the similarities between the amino acid sequences of the CaM-binding domain of MLCK and its substrate proteins (Fig. 2). Although largely limited to the presence and spacing of clusters of basic amino acids, these positive clusters are known to be the major determinants of the enzyme's substrate specificity (16,28). Substitutions of the remaining largely neutral amino acids with others of similar character are well-tolerated by MLCK. The autoinhibitor hypothesis suggests that the CaM-binding domain of MLCK is bifunctional, able to bind either to the active site as a substrate analogue or to CaM, but not to both simultaneously. In the absence of CaM, binding to the active site blocks substrate binding and hence catalysis. The enzyme is then considered to be self- or auto- inhibited. Removal of this autoinhibitor upon its binding to CaM opens the active site to substrates and permits catalysis to proceed.

MLCK CaM-binding domain (2)
Skeletal P-light chain (37)

MLCK CaM-binding domain (2)
Smooth P-light chain (35)

FIGURE 2. Sequence comparison of MLCK CaM-binding domain and the phosphorylation sites (*) in myosin P-light chains.

On paper, each model offers certain inherent advantages. For energy-transfer the strong suit is flexibility. By invoking a sufficiently complex or comprehensive conformational change one can readily account for dramatic and diverse changes in an enzyme's physical and catalytic properties. The virtue of the autoinhibitor hypothesis is its simplicity. By making the CaM-binding domain a bifunctional chameleon of sorts, a regulatory mechanism which is straightforward and extremely economical results. In addition, by tailoring the sequence appropriately, autoinhibitors for a large number of CaM-regulated enzymes could be constructed. Each of these models makes several predictions concerning the properties of MLCK and its CaM complex. Some of those which distinguish the energy-transfer hypothesis from the autoinhibitor hypothesis are listed below:

a) The energy transfer hypothesis predicts that MLCK undergoes a global conformational transformation upon CaM binding. The autoinhibitor hypothesis suggests that the conformational effects of CaM binding are strictly localized to the CaM-binding domain.

b) The autoinihibitor hypothesis suggests that the CaM-binding domain is bifunctional, capable of acting as either a CaM-antagonist or a substrate-antagonist. The energy transfer hypothesis maintains that any substrate-antagonist capability would be strictly fortuitous in nature.

c) The energy-transfer hypothesis predicts that the CaM-binding domain of MLCK is tightly coupled to the rest of the protein. Changes in its status are transmitted to and affect other parts of the enzyme. The autoinhibitor hypothesis maintains that the CaM binding domain is only loosely coupled, physically attached to the protein but essentially free to move about as circumstances dictate.

Experimental Observations

Several lines of investigation bear on the question of which model best describes the CaM activation of MLCK. Unfortunately, little direct information exists concerning the three dimensional structure of MLCK or its CaM complex. Circular dichroism measurements detected a small increase in the net content of alpha helix when MLCK binds CaM (27). Since CaM already contains a very high proportion of alpha helix, it is thought that this increase must occur in MLCK itself. It has been suggested, based upon the marked propensity of peptide models to adopt a helical conformation upon binding CaM, that this increase in helix content represents the local transition of the CaM-binding domain from an extended to a helical state (21). Such a local change in secondary structure neither precludes nor predicts more extensive changes in tertiary structure, however.

Partial proteolysis has been successfully used to activate MLCK in a CaM-independent fashion. Using chymotrypsin, an active catalytic core can be isolated which neither binds to nor is stimulated by CaM (9,42). Careful mapping of the various cleavage sites in fragments of protease-treated MLCK revealled that removal of 9-10 amino acids in the C-terminal half of the CaM-binding domain were responsible for the appearance of CaM-independent activity (ref. 9 , Fig. 3):

Such behavior suggests two important things concerning the activation of MLCK. First of all, CaM activation involves the release of some negative constraint on the protein. The catalytic domain either exists in, or has the propensity to fold into, a catalytically competent configuration. The constraint could take the form of an active site-directed inhibitor, as suggested by the autoinhibitor hypothesis, or of some structure which is coupled to the catalytic center in a manner which distorts it from an active configuration, as suggested by energy transfer. Second, and perhaps more intriguingly, the location of the activating cleavage event (Figure 3) indicates that the CaM-binding domain itself is or forms an integral part of the constraining/ inhibitory domain (9). It is noteworthy that such behavior exactly corresponds to that predicted by the autoinhibitor hypothesis.

Further evidence that the CaM-binding domain might function as the inhibitory domain as well comes from studies with synthetic peptides. Peptides modelled after the CaM-binding domain of MLCK have long been known to bind CaM with high affinity and block the activation of MLCK thereby (3). Using an

----KRRWKKNFIAVSAANRFKKISSSGALM-COOH

↑ (CaM-independent, constituitively active) ↑↑ (Activity still CaM-dependent)

FIGURE 3. Proteolytic cleavage sites in the carboxyl terminus of MLCK and their effect on activity. The underlined portion is the CaM-binding domain. Arrows indicate the sites of proteolytic cleavage (adapted from (9)).

active chymotryptic fragment of MLCK it was possible to study the effects of CaM-binding domain peptides on catalysis independent of their CaM-antagonist properties. It was observed that such peptides are potent inhibitors of MLCK activity (Ki = 3μM) and that inhibition was competitive with respect to peptide substrates (18). Introduction of an appropriate serine within their sequence resulted in a peptide which was phosphorylated by MLCK. Taken together, these results suggest that inhibition results from the direct binding of these peptides to MLCK's myosin P-light chain binding site. In the presence of CaM, moreover, both the inhibitory potency and phosphorylation of these peptides were abolished (18). Thus, the sequence of the CaM-binding domain confers upon it the potential to act as a CaM-regulated inhibitor of MLCK. The bifunctional nature of model peptides, together with the striking correlation between proteolytic cleavage in the CaM-binding domain and activation, represents the principal evidence in support of the autoinhibitor hypothesis.

Despite the ability of the autoinhibitor hypothesis to neatly account for both the proteolytic activation of MLCK and the properties of CaM-binding domain peptides, other investigations strongly hint at both CaM-induced conformational changes in MLCK and the tight coupling of its catalytic and CaM-binding sites. Binding of CaM to MLCK results in an enhancement of the latters tryptophan fluorecence (CaM does not contain tryptophan). Examination of the time course of this increase revealled that it was slow relative to the presumed rate of MLCK-CaM complex formation, independent of either MLCK or CaM concentration, and biphasic (14). These observations suggest the occurrence of slow conformational changes in the environment surrounding one or more of MLCK's four tryptophanyl residues following CaM binding (14). Presumably some of these changes involve tryptophan 580 within the CaM-binding domain, but they are not necessarily limited to it.

Using a fluorescent derivative of CaM, it is possible to measure its binding to MLCK. When the binding constant was measured in the presence of myosin P-light chains and the ATP-analogue AMPPNP, it was observed that affinity for CaM increased 3-5 fold over that of MLCK alone (33). In our laboratory we have observed that binding to CaM renders MLCK unstable at a temperature, 30°C, at which MLCK alone is quite content (P. J. Kennelly, M. Starovasnik, and A. M. Edelman, unpublished observations). This inactivation was first order over a wide range of MLCK concentrations and absolutely dependent upon the binding of MLCK to CaM, indicating that the instability was intrinsic to the complex and not the manifestation of some contaminating factor. The addition of either peptide substrates or MgATP effectively inhibited inactivation, indicating that binding events at the active site can counteract an effect of binding at the CaM site. Together with the ability of substrates to influence CaM binding affinity, this suggests that the CaM-binding domain and active site are coupled, with changes in the status of one directly affecting the other. Such behavior matches that predicted by the energy transfer hypothesis.

CaM Activation of MLCK- A Pluralistic Mechanism?

As can be seen from the brief overview above, a case can be made for either the energy-transfer hypothesis or the autoinhibitor hypothesis as the best model for describing MLCK activation. Both neatly account for certain experimental behaviors, but are unable to convincingly account for others. Both have ample precedent elsewhere in nature. For energy transfer there is the widespread phenomenom of allosterism, where small ligands modulate enzyme activity through binding-induced conformational changes. For autoinhibition there is the example of the cyclic nucleotide-dependent protein kinases, whose activities are modulated by the controlled binding of substrate-like inhibitory domains (1).

In trying to choose the hypothesis which best fits the data, it would appear that the most satisfactory model would contain elements of each. Such an idea is not as unrealistic as our rather cut-and-dried explanations of these models imply. Autoinhibition and energy-transfer represent discrete, and purposefully extreme, points in a continuum of mechansitic possibilities. In reality, both mechanisms share important features in common, such as the necessity for some form of conformational change. They differ in the degree or scope of the change, but not its occurrence. Autoinhibition suggests that the change is strictly local to the CaM-binding domain itself (its opening out from the active site), while energy transfer suggests it is widespread. There is much room for maneuver between these two possibilities, and no clear delineation between where "local" ends and "widespread" begins.

If one allows the binding of the autoinhibitor domain to the active site to precipitate a sufficiently high degree of conformational change, one obtains a hybrid model that accommodates both the self-inhibitory and conformational components suggested for the CaM activation process. Such conformational coupling could be accomplished through either direct physical linkage via the polypeptide backbone (a push-pull mechanism) or through a process of induced fit at the active site (a hand-in-glove mechanism). Under such a conformationally coupled variant of the autoinhibitor model, one could envisage activation as proceeding via the following steps:

1. CaM binds MLCK.
2. CaM elicits local conformational changes in the CaM binding domain, such as the formation of alpha helix.
3. The CaM-complexed domain swings out of the active site.
4. The catalytic domain relaxes into its catalytically optimal configuration.

The order of steps 2 and 3 is arbitrary. For a pictoral representation see Figure 4 on the next page.

OTHER CaM-DEPENDENT ENZYMES OF THE PHOSPHOTRANSFERASE FAMILY- A SEARCH FOR PARALLELS

Since the discovery of calmodulin, much attention has centered upon the questions of whether there is a single target interaction site on CaM, a single type of CaM-binding site on its target proteins, or a single common mechanism of activation. At present, it appears safe to say that there is no universal mode of binding or single mechanism of activation. Variants of CaM have been produced in the laboratory which possess differing affinities for, and induce differing effects in, CaM's target proteins (5,11,30,39). However, within enzyme families there is still some likelihood that related proteins may display commonality of mechanism. In the family of proteins involved in

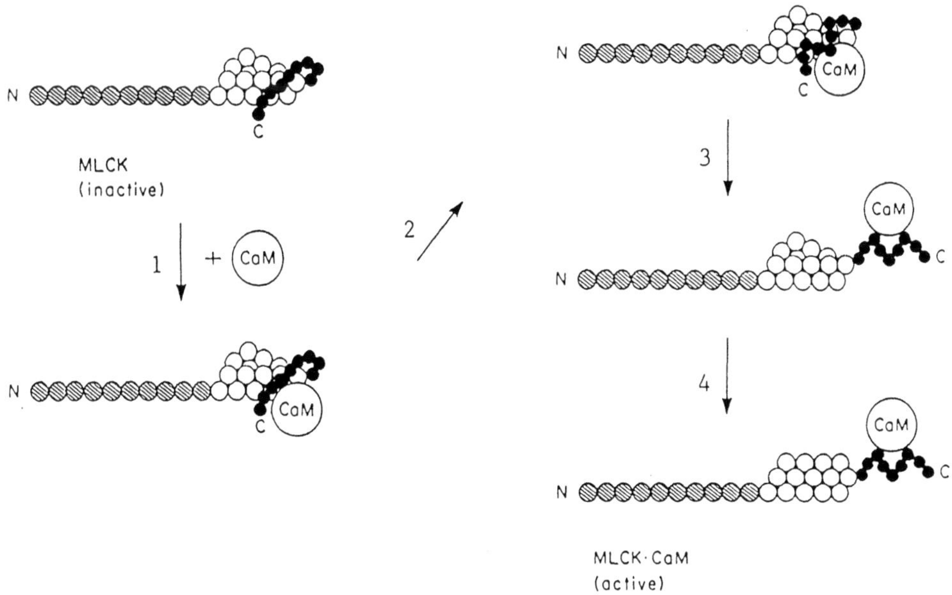

FIGURE 4. The conformationally-coupled autoinhibitor model for activation of skeletal muscle MLCK. The tail, catalytic, and CaM-binding domains of MLCK are represented by the hatched, open, and filled circles respectively. The numbered steps correspond to those given in the text on the previous page.

phosphotransferase reactions, the CaM-dependent protein kinases and protein phosphatases, several members have been subject to extensive study: skeletal muscle MLCK, smooth muscle MLCK, phosphorylase kinase, the multifunctional CaM-dependent protein kinase (CaM kinase II), and the calcium/calmodulin-dependent protein phosphatase (calcineurin).

All of these proteins can be activated by partial proteolysis in lieu of calcium and calmodulin. While the MLCKs are made up of a single polypeptide chain, the latter three proteins are multimeric, with phosphorylase kinase and calcineurin containing integral regulatory subunits nearly identical to calmodulin. These subunits bind Ca^{2+} and upon so doing activate the enzymes (20,38). In addition, both proteins bind to and are further stimulated by exogenous CaM. Both enzymes bind other divalent metal ions and this also affects their catalytic properties. Such a complex, multi-effector/variable response system of regulation is best suited to an energy transfer model. The ability to vary conformation is flexible enough to accomodate a wide range of responses and the stepwise gradation of response, whereas autoinhibition represents a simple on-off switch. Experimental evidence of effector-induced conformational changes reinforces this view. For phosphorylase kinase, chemical cross-linking reagents suggest that subunit interrelationships undergo slow changes upon binding Ca^{2+} and Mg^{2+} (10). A monoclonal antibody to calcineurin displays different affinities for the enzyme's inactive and Ni^{2+}-activated forms (26). In addition, the enzyme undergoes a slow inactivation upon binding CaM (19), a behavior closely analogous to that of skeletal muscle MLCK. Unlike MLCK, substrates and competitive inhibitors of catalysis accelerate the inactivation rather than retard it, while certain divalent metals such as Mn^{2+} and Ni^{2+} can reverse it. Thus, both phosphorylase kinase and calcineurin show evidence of the conformational changes and site-site coupling suggested by the energy transfer hypothesis. Although certain promising candidates are currently under investigation, the CaM-binding sites of neither enzyme have been identified with certainty. Therefore, it has not been possible to determine if there is any contribution of a self-inhibitory component to activation.

CaM kinase II is intermediate between MLCK and phosphorlyase kinase/calcineurin in structural complexity. It has multiple subunits, generally 10-12, but they are made up of three isoforms of a single subunit type (2). Like MLCK, the enzyme requires CaM for activation- calcium alone is insufficient. However, unlike MLCK it can attain a state marked by CaM-independent activity following autophosphorylation (24,41). Moreover, it requires the autophosphorylation of only 2-3 subunits per multimer to trigger complete activation (23). This last observation strongly suggests conformational or other coupling between the enzyme's active sites and/or regulatory domains and the points of subunit contact. However, certain other behaviors, notably those of CaM-binding domain peptides, suggest an element of autoinhibition as well. Model peptides have revealled the existence of an area with substrate-antagonist properties overlapping the amino-terminal half of the CaM-binding domain (34). This inhibitory region was, in turn, immediately adjacent to the threonine residue (thr 286) whose phosphorylation apparently renders the enzyme CaM-independent (34). This raises the possibility that CaM activation involves the removal of this inhibitory sequence from a location in the active site, and that phosphorylation of threonine 286, by blocking reassociation with the active site, maintains the "non-inhibited state" even after the removal of CaM.

Smooth muscle MLCK shares many common features with its skeletal muscle counterpart, including the very strong suggestion of an autoinhibitor-type mechanism. Its CaM-binding domain overlaps a region which bears strong resemblance to its substrate, and studies with model peptides have confirmed this inhibitory potential (17). Moreover, careful analysis of the sites of protease cleavage of the enzyme revealled that removal of this presumed autoinhibitor domain correlates with the appearance of CaM-independent activity (36). As with skeletal muscle MLCK, there does appear to be some possibility of site-site coupling as CaM binding increases the affinity of the enzyme for the fluorescent probe 9-anthroylcholine, an ATP analogue (25).

To summarize, there exist several other CaM-dependent enzymes in the phosphotransferase family. Examination of their CaM-activation suggests that for the multimeric members of the family, strong evidence for conformational changes and site-site coupling exists. Elements of active-site directed autoinhibition may play a role in their activation but, with the exception of CaM-kinase II, no direct evidence consistent with such a mechanism has appeared. Smooth muscle MLCK, like its skeletal muscle counterpart, displays behaviors which very strongly suggest that autoinhibition plays a major role in the modulation of its activity.

As mentioned earlier, autoinhibition and energy transfer represent discrete possibilities in a continuum of possible mechanisms. Assuming for a moment that some element of each plays a role in the regulation of this family of CaM-regulated phosphotransfer enzymes, a correlation appears to exist between the complexity of the target and the balance between the relative contributions of these two mechanisms. Where the target is simple and the response a straightforward on-off event (smooth and skeletal muscle MLCK), the evidence suggests a very strong role for autoinhibition. Where the target is structurally complex and the response a graded one (phosphorylase kinase and calcineurin), the evidence suggests a very strong energy transfer component. CaM kinase II, intermediate in structural and regulatory complexity, shows suggestions of each.

REFERENCES

1. Beebe, S. J., and J. D. Corbin (1986) In "The Enzymes: Enzyme Control by Phosphorylation" (Krebs, E. G., and P. Boyer, eds.) ,vol. 17, pp. 43-111, Academic Press, New York.
2. Bennett, M. K., N. E. Erondu, and M. B. Kennedy (1983) J. Biol. Chem. 258, 12735-12744.
3. Blumenthal, D. K., K. Takio, A. M. Edelman, H. Charbonneau, K,. Titani, K. A. Walsh, and E. G. Krebs (1985) Proc. Natl. Acad. Sci. USA 82, 3187-3191.
4. Conti, M. A., and R. S. Adelstein (1981) J. Biol. Chem. 256, 3178-3181.
5. Craig, T. A., D. M. Watterson, F. G. Prendergast, J. Haeich, and D. M. Roberts (1987) J. Biol. Chem. 262, 3278-3284.
6. Dabrowska, R., J. M. F. Sherry, D. K. Aromatorio, and D. J. Hartshorne (1978) Biochemistry 17, 253-258.
7. Daniel, J. L., and R. S. Adelstein (1976) Biochemistry 15, 2370-2377.
8. Edelman, A. M., and E. G. Krebs (1982) FEBS Lett. 138, 293-298.
9. Edelman, A. M., K. Takio, D. K. Blumenthal, R. S. Hansen, K. A. Walsh, K. Titani, and E. G. Krebs (1985) J. Biol. Chem. 260, 11275-11285.
10. Fitzgerald, T. J., and G. M. Carlson (1984) J. Biol. Chem. 259, 3266-3274.
11. Guerini, D., J. Krebs, and E. Carafoli (1987) Eur. J. Biochem. 170, 35-42.
12. Guerriero, V., Jr., M. A. Russo, N. J. Olson, J. A. Putney, and A. R. Means (1986) Biochemistry 25, 8372-8381.
13. Hathaway, D. R., and R. S. Adelstein (1979) Proc. Natl. Acad. Sci. USA 76, 1653-1657.
14. Johnson, J. D., M. J. Holyrode, T. H. Crouch, R. J. Solaro, and J. D. Potter (1981) J. Biol. Chem. 256, 12194-12198.
15. Kamm, K. E., and J. T. Stull (1985) Ann. Rev. Pharmacol. Toxicol. 25, 593-620.
16. Kemp, B. E., and R. B. Pearson (1985) J. Biol. Chem. 260, 3355-3359.
17. Kemp, B. E., R. B. Pearson, V. Guerriero, Jr., I. C. Bagchi, and A. R. Means (1987) J. Biol. Chem. 262, 2542-2548.
18. Kennelly, P. J., A. M. Edelman, D. K. Blumenthal, and E.G. Krebs (1987) J. Biol. Chem. 262, 11958-11963.
19. King, M. M., and C. Y. Huang (1984) J. Biol. Chem. 259, 8847-8856.
20. Klee, C. B., M. H. Krinks, and A. S. Manalan (1983) In

"Calcium Binding Proteins 1983" (DeBernard, B., ed.), pp. 103-112, Elsevier Science Publishing, Amsterdam
21. Klevit, R. E., D. K. Blumenthal, D. E. Wemmer, and E. G. Krebs (1985) Biochemistry 24, 8152-8157.
22. Klug, G. A., B. R. Botterman, and J. T. Stull (1982) J. Biol. Chem. 257, 4688-4690.
23. Kwiatkowski, A. P., D. J. Shell, and M. M. King (1988) J. Biol. Chem. 263, 6484-6486.
24. Lou, L. L., S. J. Lloyd, and H. Schulman (1986) Proc. Natl. Acad. Sci. USA 83, 9497-9501.
25. Malencik, D. A., and S. R. Anderson (1983) Biochem. Biophys. Res. Commun. 114, 50-56.
26. Matsui, H., C. J. Pallen, A. Adachi, J. H. Wang, and H. Lam (1985) J. Biol. Chem. 260, 4174-4179.
27. Mayr, G. W., and L. M. G. Heilemyer, Jr. (1983) Biochemistry 22, 4316-4326.
28. Michnoff, C. H., B. E. Kemp, and J. T. Stull (1986) J. Biol. Chem. 261, 8320-8326.
29. Nairn, A. C., and S. V. Perry (1979) Biochem. J. 179, 151-164.
30. Newton, D. L., M. D. Oldenwurtel, M. H. Krinks, J. Shiloach, and C. B. Klee (1984) J. Biol. Chem. 259, 4419-4426.
31. Nunnally, M. H., S. B. Rybicki, and J. T. Stull (1985) J. Biol. Chem. 260, 1020-1026.
32. Nunnally, M. H., and J. T. Stull (1984) J. Biol. Chem. 259, 1776-1780.
33. Olwin, B. B., A. M. Edelman, E. G. Krebs, and D. R. Storm (1984), J. Biol. Chem. 259, 10949-10955.
34. Payne, M. E., Y. Fong, T. Ono, R. J. Colbran, B. E. Kemp, T. R. Soderling, and A. R. Means (1988) J. Biol. Chem. 263, 7190-7195.
35. Pearson, R. B., R. Jakes, M. John, J. Kendrick-Jones, and B. E. Kemp (1984) FEBS Lett. 168, 108-112.
36. Pearson, R. B., R. E. H. Wettenhall, A. R. Means, D. J. Hartshorne, and B. E. Kemp (1988) Science 241, 970-973.
37. Perrie, W. T., L. B. Smillie, and S. V. Perry (1973) Biochem. J. 135, 151-164.
38. Pickett-Gies, C. A., and D. A. Walsh (1986) In "The Enzymes: Enzyme Control by Phosphorylation" (Krebs, E. G., and P. Boyer, eds.), pp. 395-459, Academic Press, New York.
39. Putkey, J. A., G. F. Draetta, G. R. Slaughter, C. B. Klee, P. Cohen, J. T. Stull, and A. R. Means (1986) J. Biol. Chem. 261, 9896-9903.
40. Roush, C. L., P. J. Kennelly, M. B. Glaccum, D. M. Helfman, J. D. Scott, and E. G. Krebs (1988) J. Biol. Chem. 263, 10510-10516.
41. Schworer, C. M., R. J. Colbran, and T. R. Soderling (1986) J. Biol. Chem. 261, 8581-8584.
42. Srivastava, S., and D. J. Hartshorne (1983) Biochem. Biophys. Res. Commun. 110, 701-708.
43. Stull, J. T., M. H. Nunnally, and C. M. Michnoff (1986) In "The Enzymes: Enzyme Control by Phosphorylation" (Krebs, E. G., and P. Boyer, eds.), Vol. 18, pp. 114-166, Academic Press, New York.
44. Stull, J. T., P. J. Silver, J. R. Miller, D. K. Blumenthal, B. R. Botterman, and G. A. Klug (1983) Fed. Proc. 42, 21-26.
45. Takio, K., D. K. Blumenthal, K. A. Walsh, K. Titani, and E. G. Krebs (1986) Biochemistry 25, 8049-8057.
46. Yagi, K., M. Yazawa, S. Kakiuchi, M. Ohshima, and K. Uenishi (1978) J. Biol. Chem. 253, 1338-1340.

CALCIUM SIGNALING OF CALCIUM-BINDING PROTEINS AND DRUG ACTIONS

Toshio Tanaka and Hiroyoshi Hidaka

Department of Molecular and Cellular Pharmacology
Mie University School of Medicine, Edobashi, Tsu 514, and
Department of Pharmacology, Nagoya University School of
Medicine, Showa-ku, Nagoya 466, Japan

INTRODUCTION

Intracellular calcium ion plays a critical role in cell activation in response to various agonists and calmodulin is a major intracellular receptor for calcium ion. At first, there is the formation of a calcium-calmodulin complex. Subsequently, the calcium-calmodulin complex interact with target proteins. Then, the target protein becomes active and transfer calcium signal to the next molecule. Recently, we have found many calmodulin antagonists with various chemical structures (Figure 1). These compounds selectively inhibit the process of interaction of the calcium-calmodulin complex with target proteins.

In this paper, we have reported recent development in our research on calmodulin antagonists. We focused on the following points, 1) their pharmacological actions at molecular and cellular levels, 2) different binding sites of various compounds on calmodulin, 3) physicochemical properties of their binding sites, 4) interaction with other calcium binding proteins.

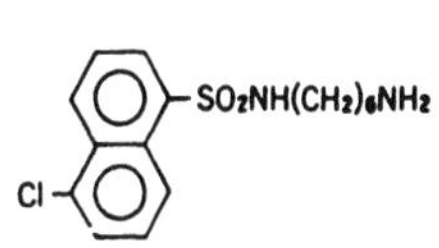

W-7

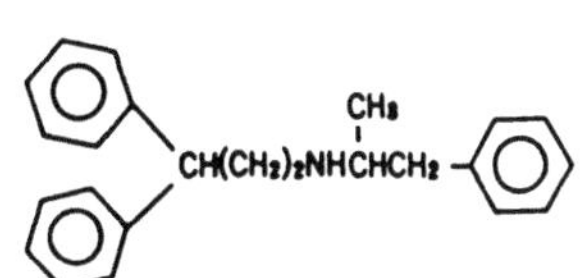

Prenylamine

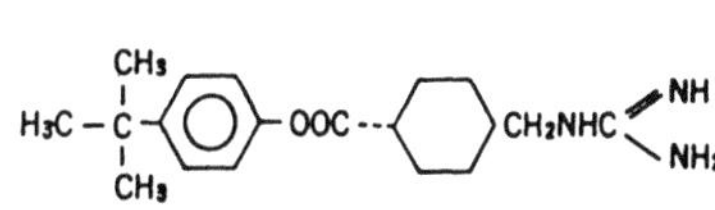

GMCHA

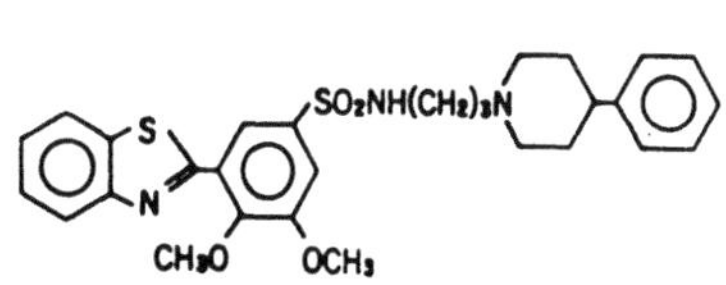

HT-74

Fig. 1. Chemical structure of various calmodulin antagonists.

In clinical medicine, calcium entry blockers are estblished as most effective drugs in the treatment for hypertension, angina pectoris, and other cardiovascular diseases[1]. Calcium entry blockers are defined as drugs that inhibit the entry of calcium ion. These compounds show great variety in chemical structure and pharmacological properties. We found that some calcium entry blockers such as prenylamine[2] and bepridil[3] have calmodulin antagonistic actions.

Figure 2 shows chemical structures and pharmacological actions of various calcium entry blockers. Nifedipin, verapamil and diltiazem are selective for calcium channel and almost no effects on sodium channels and calmodulin function. On the other hand, prenylamine and bepridil are called as nonselective calcium entry blockers because they act at similar concentrations on calcium channels and sodium channels. Moreover, we found prenylamine[2] and bepridil[3] have calmodulin antagonistic actions. From these results, it is suggested that only nonselective calcium entry blockers may inhibit calmodulin function.

	Selective	Nonselective
Effects on Fast Sodium Channels	(—)	(+)
Effects on Calmodulin Function	(—)	(+)

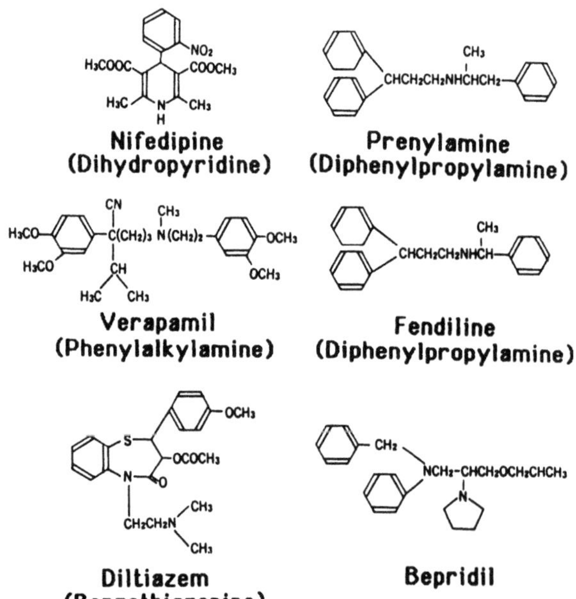

Fig. 2. Chemical structures and actions of calcium entry blockers.

It is reported that dihydropyridines such as nifedipine have no calmodulin antagonistic action. However, we found that some derivatives of dihydropyridine such as CV-159 are potent calmodulin antagonists (Figure 3). Then, it is suggested that absence of calmodulin antagonistic action is not essential property of dihydropyridine type calcium entry blockers. It was found that the dihydropyridine derivative, CV-159 inhibited myosin light chain kinase. The addition of CV-159 to the reaction mixture resulted in a marked inhibition of the calcium, calmodulin-dependent myosin light chain kinase activity. On the other hand, this compound had little effect on the trypsin-treated calmodulin-independent myosin light chain kinase activity. The CV-159-induced inhibition of myosin light chain kinase activity was restored by the addition of excess calmodulin. Kinetic analysis of CV-159-induced inhibition of myosin light chain kinase by Dixon plots suggests that CV-159 inhibited myosin light chain kinase competitively with respect to calmodulin and its Ki value is 0.8μM. These results suggest that some dihydropyridine derivatives can inhibit calmodulin activity[4].

Phenylalkylamines such as verapamil were reported to be weak calmodulin antagonists. It was found that some modification of verapamil in chemical structure enable to increase in affinity for calmodulin.

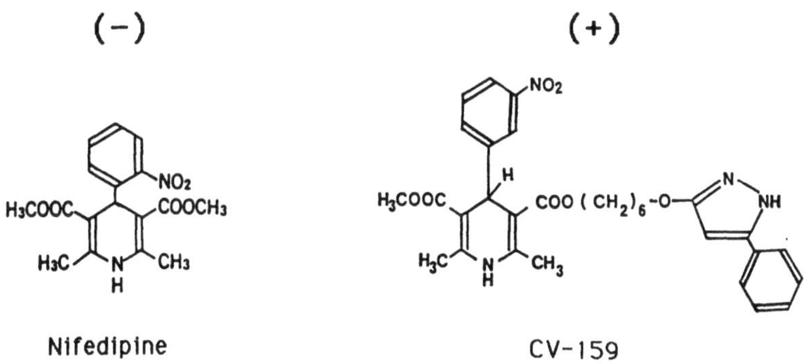

Fig. 3. Calmodulin antagonistic action of dihydropyridines.

CALMODULIN ANTAGONISTIC ACTION OF AN ANTIALLERGIC AGENT

Numerous antiallergic agents prevent the release of histamine from mast cells. Disodium cromoglycate inhibits the antigen-antibody-mediated influx of calcium ion into mast cells[5] and modulates the phosphorylation of mast cell proteins[6]. Several other antiallergic compounds seem to act in a similar manner and it was reported that all tested have no significant effects on the functions of calmodulin[7]. However, we found that this new antiallergic agent trans-4-guanidinomethylcyclohexanecarboxylic acid p-tert-butylphenylester (GMCHA). GMCHA interacts with calcium-calmodulin complex, suppresses calcium, calmodulin-dependent protein phosphorylation and inhibits platelet secretion[8].

The effect of GMCHA on platelet ATP secretion and vascular smooth muscle contraction was investigated. GMCHA inhibited ATP and serotonin secretion of human platelets dose-dependently. This compound produced a significant relaxation of vascular strips which contracted by KCl in a concentration of 20mM and the ED_{50} value in rabbit mesenteric arteries was 3.0μM. It was found that GMCHA inhibited myosin light chain phosphorylation of human platelets stimulated by thrombin. ^{32}P-labeled platelets were activated by thrombin in the presence or absence of GMCHA. When the stimulus-induced phosphorylation of platelet proteins was determined by autoradiography following SDS-polyacrylamide gel electrophoresis, the 20K dalton myosin light chain was phosphorylated time-dependently. GMCHA inhibited the phosphorylation of myosin light chain dose-dependently.

The effective concentrations of GMCHA in interaction with the calcium-calmodulin complex, inhibition of calcium, calmodulin-dependent protein phosphorylation, suppression of calcium-dependent secretion of mast cells and platelets, and relaxation of vascular strips were investigated. The affinities of this compound for calmodulin determined by several procedures, are similar to the concentractions which inhibit release reactions from mast cells and platelets. These results suggest that GMCHA inhibits these calcium-dependent cell functions through inhibitory effects on calmodulin.

DIFFERENT BINDING SITES ON CALMODULIN

Recently, we found many calmodulin antagonists with various chemical structure. Among those calmodulin antagonists, it was found that these four compounds have different binding sites on calmodulin (Figure 1).

W-7 was found as a platelet inhibitor[9], prenylamine as a nonselective calcium entry blocker[2], GMCHA as an antiallergic agent[8], HT-74 as a vascular relaxant[10]. Moreover, these compounds inhibited W-7 binding to calmodulin non-competitively. We found the different effects of these four compounds on the three kinds of fluorescence and calcium binding to calmodulin. First one is the fluorescence of dansylated calmodulin. Second fluorescence is that of hydrophobic probe ANS with calcium-calmodulin complex. Third one is the fluorescence of thiol-specific fluorescent probe ANM bound to the twenty-sixth cystein of wheat germ calmodulin. The data indicate that these four kinds of drugs bind to different sites on calmodulin and there may be multiple binding sites on calmodulin of target enzymes.

Moreover, we found the specific binding of HT-74 to calmodulin, compared with other calcium binding proteins such as troponin-C, S-100 protein and parvalbumin. On the other hand, W-7 interact with calmodulin selectively but also can bind to troponin-C and S-100 protein with lower affinity (Figure 4). These difference may be related to their different binding sites on calmodulin.

BINDING SITES OF W-7 ON CALMODULIN

In order to determine the primary structure for drug binding sites, we tried covalent binding of W-7 to calmodulin. We used water soluble carbodiimide EDC. Reaction mixture of calmodulin, radiolabeled W-7 and EDC in the presence or absence of calcium ion was applied onto the PD-10 column. The results suggest calcium-dependent cross-linking of tritium-labeled W-7 to calmodulin. We tried to separate W-7-calmodulin complex by W-7 Sepharose affinity column chromatography. It was found that W-7-calmodulin complex could not be absorbed to W-7 Sepharose in the presence of calcium ion. Uncrosslinked calmodulin bound to W-7 Sepharose in the presence of calcium ion, and was eluted with the buffer containing EGTA. The stoichiometry of the W-7-calmodulin complex is one mole of W-7 per mole of calmodulin. Limited proteolysis of the complex with trypsin in the presence of calcium ion suggests that crosslinked W-7 locates in the C-terminal half of calmodulin. More detailed study of W-7 binding site on calmodulin is now on the way.

	W-7	HT-74
Calmodulin	+++	+++
Troponin C	++	−
S-100 Protein	+	−
Parvalbumin	−	−

Fig. 4. Affinity of W-7 or HT-74 for calcium-binding proteins.

HYDROPHOBIC REGIONS OF CALCIUM BINDING PROTEINS

There is similarity in chemical structure of a calmodulin antagonist W-7 and a fluorescent hydrophobic probe 1-NPN. In fact, NPN has potent calmodulin antagonistic action and has same binding sites on calmodulin as W-7. Then, we tried to characterize the binding sites of W-7 on calmodulin by using the fluoresent hydrophobic probe, NPN. Moreover, we have done comparative study of NPN fluoresce with calcium binding proteins and various alcohols. The emission characteristics of NPN dissolved in various alcohols vary markedly with the polarity of the solvent. As the polarity of the alcohol decreases, the quantum yield of NPN emission increases, and the fluorescence maximum shifts to the blue. We compared the degree of blue shift in fluorescence of NPN with various calcium binding proteins and alcohols. The emission of NPN bound to calmodulin is the furthest shifted to the blue. Troponin C and S-100 protein produced moderately the blue shift of NPN fluorescence. There is almost no blue shift in the fluorescence with parvalbumin and myosin light chain.

The EF hand structure was found not only in calmodulin but also in other calcium-binding proteins such as troponin C, S-100 protein, myosin light chain and parvalbumin, and proposed to be responsible for the calcium binding function. However, the intramolecular structure and how these proteins interact with target proteins or drugs remain obscure. We have proposed that calcium binding produces conformational changes in calmodulin that expose hydrophobic regions on the surface of the molecule, and these regions may act as active sites for interaction with its target proteins or calmodulin antagonists[11, 12]. Moreover, calcium-dependent hydrophobic regions of troponin C and S-100 protein were demonstrated by using fluorescent hydro-phobic probes[13] and hydrophobic chromatography[14]. However, these calcium-binding proteins cannot activated calmodulin-dependent enzymes. Then, we investigated comparatively the properties of their calcium-dependent hydrophobic regions by NPN and demonstrated the binding sites for this compound on calmodulin are extremely hydrophobic, and might be responsible for calmodulin specific function such as enzyme activation. Troponin C and S-100 protein produced moderately the blue shift of the emission maximum. These results suggest that some degree of hydrophobicity may be necessary for common function of these proteins such as microtubules disassembly[15]. At present time, molecular mechanisms of calcium signaling of myosin light chain and parvalbumin are not known.

Acknowledgements

We are indebted to our collaborators for carrying out the experiments presented in this chapter and Miss N. Wada for manuscript preparation.

This investigation was supported in part by Grants-In-Aid for Scientific Research from the Ministry of Education, Science and Culture, Japan.

REFERENCE

1. T. Godfraind, R. Miller, and M. Wibo, Calcium antagonism and calcium entry blockade., Pharmacological Reviews, 38(4):321 (1986).
2. H. Hidaka, T. Yamaki, M. Naka, T. Tanaka, H. Haigawara, and R. Kobayashi, Calcium-regulated modulator protein interacting agents inhibit muscle calcium-stimulated protein kinase and ATPase., Mol. Pharmacol., 17:66 (1980).
3. H. Itoh, T. Tanaka, Y. Mitani, and H. Hidaka, The binding of the calcium channel blocker, bepridil, to calmodulin., Biochem. Pharmacol., 35(2):217 (1986).
4. H. Umekawa, K. Yamakawa, K. Nunoki, N. Taira, T. Tanaka, and H. Hidaka, Inhibition of calmodulin function by CV-159, a novel dihydropyridine compound., Biochem. Pharmacol., 37(18):3377 (1988).
5. J. C. Foreman, M. B. Hallett, and J. L. Mongar, Site of action of the antiallergic drugs cromoglycate and doxantrazole., Br. J. Pharmacol., 59:473 (1977).
6. T. C. Theoharides, W. Sieghart, P. Greengard, and T. C. Douglas, Antiallergic drug cromolyn may inhibit histamine secretion by regulating phosphorylation of a mast cell protein., Science, 207:80 (1980).
7. A. F. Welton, and G. Wiggan, Antiallergic drugs do not act through antagonism of calmodulin., Fedn Proc. Fedn Am. Socs exp. Biol., 42:569 (1983).
8. T. Tanaka, M. Saitoh, M. Ito, T. Shin, M. Naka, K. Endo, and H. Hidaka, Inhibition of human platelet secretion and of Ca^{2+}, calmodulin-dependent protein phosphorylation by the antiallergic agent, GMCHA., Biochem. Pharmacol., 37(13):2537 (1988).
9. M. Nishikawa, T. Tanaka, and H. Hidaka, Ca^{2+}-calmodulin dependent

phosphorylation and platelet secretion., Nature, 287:863 (1980).
10. T. Tanaka, H. Umekawa, M. Saitoh, T. Ishikawa, T. Shin, M. Ito, H. Itoh, Y. Kawamotsu, H. Sugihara, and H. Hidaka, Modulation of calmodulin function and of Ca^{2+}-induced smooth muscle contraction by the calmodulin antagonist, HT-74., Mol. Pharmacol., 29:264 (1986).
11. T. Tanaka, and H. Hidaka, Hydrophobic regions function in calmodulin-enzymes interactions., J. Biol. Chem., 255(23):11078 (1980).
12. T. Tanaka, T. Ohmura, and H. Hidaka, Hydrophobic interaction of the Ca^{2+}-calmodulin complex with calmodulin antagonists: naphthalenesulfonamide derivatives., Mol. Pharmacol., 22:403 (1982).
13. T. Tanaka, and H. Hidaka, Hydrophobic regions of calcium-binding proteins exposed by calcium., Biochem. Int., 2:71 (1981).
14. T. Tanaka, H. Umekawa, T. Ohmura, and H. Hidaka, Calcium-dependent hydrophobic chromatography of calmodulin, S-100 protein and troponin-C., Biochim. Biophys. Acta., 787:158 (1984).
15. J. M. Marcum, J. R. Dedman, B. R. Brinkley, and A. R. Means, Control of microtubule assembly-disassembly by calcium-dependent regulator protein., Proc. Natl. Acad. Sci. USA., 75:3771 (1978).

ANALYSIS OF CALCIUM-BINDING SITES IN CALCIUM-ACTIVATED NEUTRAL PROTEASE

Koichi Suzuki, Yasufumi Minami, Yasufumi Emori, Shinobu Imajoh, and Hiroshi Kawasaki

Department of Molecular Biology, Tokoyo Metropolitan Institute of Medical Science, Honkomagome, Bunkyo-ku Tokyo-113, Japan

INTRODUCTION

Calcium-activated neutral protease (CANP) is a typical intracellular protease responsible for the processing and turnover of various intracellular proteins.[1-5] Uncontrolled, CANP can destroy various cellular components, leading to various disease processes like muscular dystrophy.[3] Hence the activity must be tightly regulated. Because CANP absolutely requires calcium for activity, calcium plays a pivotal role in its regulation. Although many of the important steps of CANP activity regulation have been clarified, most studies have not provided any insight into the binding of calcium to CANP[6]. Structural analyses of CANP have revealed the existence of a calmodulin-like domain in the C-terminal region of each of the two constituent subunits (80K and 30K subunits).[7-9] These domains are presumed to bind calcium and determine the calcium sensitivity of CANP. Nevertheless, the following questions remain unsolved: whether the calmodulin domains indeed bind calcium; whether calcium-binding sites exist only in the calmodulin domains; and, if so, how many calcium ions bind to each domain and which EF hand structures in the calmodulin domains are functional calcium-binding sites.

Calcium-binding studies are hindered by the rapid autolysis of CANP that occurs in the presence of calcium, making it impossible to use native CANP in these studies. In this report, CANP calmodulin domains, prepared by expression in E. coli of corresponding cDNA fragments, were examined to investigate the binding of calcium to CANP.

MATERIALS AND METHODS

Calcium Binding Assay

Calcium binding activity was determined quantitatively by the microscale filter assay.[10] Samples containing a fixed amount of protein (1-2 µg) were preincubated with various concentrations of $^{45}CaCl_2$ (specific activity: 0.1-0.5 pmol/cpm) in 50 mM Tris-HCl, 5 mM 2-mercapto-ethanol, pH 7.5, filtered through a nitrocellulose membrane (PH79, Scheicher & Schuell) and the redioactivity retained on the filter was counted in a scintillation mixture. Calcium gel overlay, a qualitative assay for calcium binding activity, was performed as described by Maruyama et al.[11] In brief, samples separated by SDS polyacrylamide gel electrophoresis were blotted onto a nitrocellulose membrane. The membrane was incubated with $^{45}CaCl_2$ (1 μCi/ml) and autoradiographed after brief washing.

Preparation of Samples for Calcium Binding Assay

cDNA fragments corresponding to the calmodulin domains in the rabbit μ and m,[12] and chicken[7] 80K subunits and rabbit 30K subunit[8] were inserted into an expression vector (pUC8 or 18) and expressed in *E. coli* strain Y21 by induction with isopropyl-1-thio-beta-D-galactoside.[13] Expressed products (U80, M80, Ch80, and DG30, respectively) were purified by DEAE-Sephacel chromatography and their structures confirmed by N and C-terminal analyses. Ch80, U80, M80 and DG30 are C-terminal fragments composed of 207, 218, 216, and 204 residues, respectively.[13] The cDNA fragment of the calmodulin domain in the 30K subunit[8] of rabbit CANP was truncated by digestion with Bal31 and the resulting fragment was ligated with a termination linker, inserted into pUC18 and expressed in *E. coli* as described above. Expressed proteins were purified similarly.[14]

Site-Directed Mutagenesis

Site-directed mutagenesis using a single stranded cDNA for the rabbit 30K subunit calmodulin domain was carried out using an *in vitro* mutagenesis system obtained from Amersham.[14] Oligonucleotides of 24mer with base changes at their near central positions were synthesized by a DNA synthesizer (Applied Biosystems 380A) and used for mutagenesis. All amino acid replacements could be achieved by a single base change. Three mutant clones, pDR31, pDR32, and pDR34, corresponding to mutants for the 1st, 2nd, and 4th EF hand structures, respectively, were obtained and expressed in *E. coli*. The sonicated bacterial extract was purified successively by DEAE-Sephacel chromatography, ammmonium sulfate fractionation and Superose 6 column chromatography.

Dissociation and Reconstitution of CANP

CANP was dissociated by treatment with 1 M NaSCN and the subunits were separated by gel filtration on NaSCN. Reconstitution of CANP from the large and small subunits or their variants was peformed as described before.[15,16]

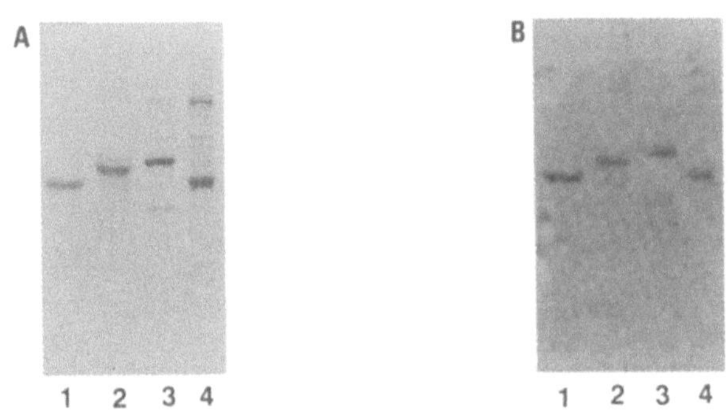

Fig. 1. SDS polyacrylamide gel electrophoresis (A) and calcium-gel overlay (B) of expressed calmodulin domains in CANP. 1 to 4 are calmodulin domains in chicken (Ch80), rabbit m and μ 80K (M80 and u80), and rabbit 30K (DG30) subunits, respectively.

Table 1. Calcium Binding Activity of Expressed Calmodulin Domains in CANP

Calmodulin Domain	Number of Calcium Binding Site	Kd (μM)	Ka (μM)[a]
DG 30	2.1 (2)	150	-
U 80	1.6 (2)	35	50
M 80	3.3 (3)	79	700
Ch 80	2.1 (2)	61	150

[a]Calcium concentration for 50% activity.

Other Methods

Protein concentrations were determined by the method of Lowry using bovine serum albumin as a standard.[17] SDS-polyacryamide gel electrophoresis was performed according to Laemmli[18] and protease activity was measured using casein as a substrate.[6]

RESULTS AND DISCUSSION

Binding of Calcium to Calmodulin Domains in CANP

Figure 1 shows SDS gel electrophoresis of expressed and partially %purified calmodulin domains in CANP together with the result of calcium gel overlay. Clearly, all expressed calmodulin domains can bind calcium. Quantitative analysis of calcium-binding to these domains as analyzed by double reciprocal plot is summarized in Table 1. M80 bound 3 moles of calcium, whereas the other three domains each bound 2 moles. Thus rabbit μ and mCANPs presumably bind 4 and 5 moles of calcium, respectively. Previously we reported that carboxymethylated chicken CANP binds 5-6 moles of calcium with an apparent average dissociation constant of 75 μM.[16] Taken together, these results are the first direct evidence that the calmodulin domains in CANP bind calcium and that the calcium binding sites found in CANP can be explained by the binding of calcium to calmodulin domains. Calcium binding sites in CANP exist only in the calmodulin domains; other domains do not bind calcium. The order of the apparent calcium dissociation constants for the 80K subunits shown in Table 1 roughly corresponds to that of the Ka values, the calcium concentration for 50% activity: i.e., μCANP has the smallest Kd value, and mCANP the highest Kd and Ka values. This indicates that the calcium sensitivity of CANP is determined essentially by the binding of calcium to the calmodulin domain of the large subunit, as predicted previously.[19] Assuming that the EF hand structures are the actual calcium binding sites, there are 8 potential calcium binding sites in CANP, four in each of the two calmodulin domains. The above results indicate that not all these sites are functional calcium binding sites. To confirm that the EF hand structures are actual calcium binding sites and to identify functional calcium binding sites among the 4 potential sites, we selected the calmodulin domain in the 30K subunit as an example and performed site-directed mutagenesis.

C-Terminal truncation of the calmodulin domain in the 30K subunit (DG30)

Two derivatives of DG30 (TR2 and TR22) were obtained by expression of plasmids containing truncated cDNAs. As shown in Fig. 2, TR2 and TR22 are shorter than the original sequence by 2 and 22 residues, respectively, and

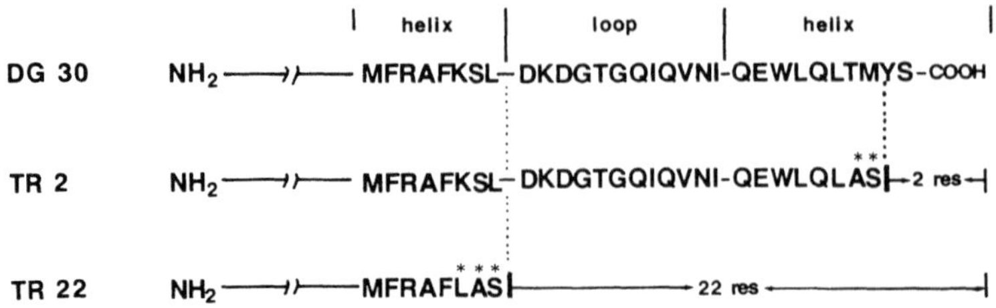

Fig. 2. C-terminal truncation of DG30. C-terminal sequences of the original DG30 and two truncated derivatives, TR2 and TR22, are shown. Residues marked by asterisks are those derived from the termination linker and are different from the original sequence.

their C-terminal 2 or 3 residues are different from those in DG30. TR22 lacks the C-terminal 4th EF hand structure and retains the remaining 3 potential calcium binding sites. Calcium binding activity of these two samples was measured after purification. From the double reciprocal plot analyses, the number of calcium binding sites and the apparent average dissociation constant were calculated (Table 2). The number of calcium binding sites in TR2 was the same as in DG30 and the Kd values were roughly the same, indicating that removal of the C-terminal 2 residues has no significant effect on the calcium binding activity of DG30. In contrast, loss of the 4th EF hand structure by removal of a further 20 residues from the C-terminus eliminated one calcium binding site. A simple interpretation is that the 4th EF hand structure is one of the two calcium binding sites in DG30. The Kd of TR22 represents the binding of calcium to the remaining site assuming that the structural changes which might occur on deletion of the C-terminal region do not affect the calcium-binding properties.

Site-Directed mutagenesis The above experiments suggest that the 4th EF hand structure is one of two calcium binding sites in DG30. We used site-directed mutagenesis to confirm this and to identify the other calcium binding site. The correlation between the amino acid sequence of the EF hand structure and calcium binding ability is, however, not clear

Table 2. Binding of Calcium to Derivatives of Calmodulin Domains in CANP

Calmodulin Domain	Number of Calcium Bound (mole/mole)	Apparent Kd (µM)
DG 30	2.3 (2)	150
TR 2	2.3 (2)	190
TR 22	1.3 (1)	110
M 80	3.3 (3)	79
M 1	2.6 (3)	88
M 2	3.4 (3)	65
M 3	3.1 (3)	93
M 4	3.3 (3)	40
DR 31	1.1 (1)	63
DR 32	2.4 (2)	52
DR 34	1.3 (1)	95

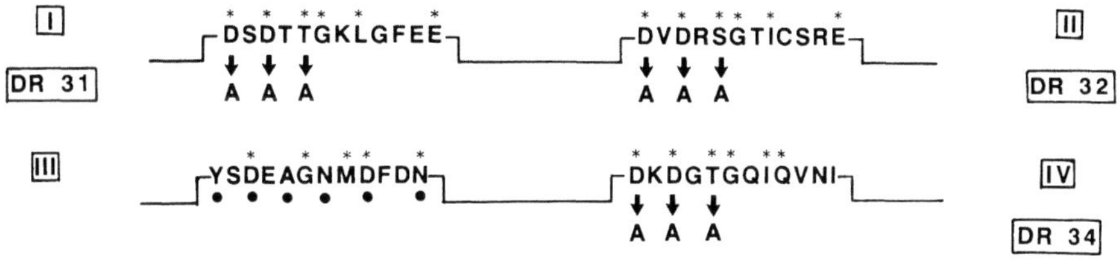

Fig. 3. Site-directed mutagenesis of the EF hand structures of DG30. Four sequences correspond to the loop regions of the EF hand structures (I-IV). Asterisks indicate residues matching the consensus calcium binding sequence.[21] Dotted residues shown in the 3rd EF hand structure as an example are the 6 calcium coordination positions, corresponding to X, Y, Z, -Y, -X, and -Z from the N-terminal end.

at present. Reinach et al.[20] reported that the replacement of the aspartic acid residues at the 1st (X) and 6th (-Z) calcium coordination sites of the EF hand loop region[21] by uncharged alanine abolished the calcium binding ability of the scallop myosin regulatory light chain. In a similar experiment, we replaced aspartic acids at the 2nd (Y) calcium coordination site of each EF hand structure of M80 with valine, but the calcium binding ability was unchanged. As shown in Table 2, 4 mutant proteins, M1 to M4, had nearly the same number of calcium binding sites and Kd values. Furthermore, changes in the calcium binding ability of DG30 were not abolished after replacement of the first calcium coordination site (X) by alanine. These results indicate that substitution of a single acidic amino acid at the calcium coordination position by a neutral residue is not enough to abolish the calcium binding ability of the EF hand structure in the case of CANP. Therefore, three residues at the calcium coordination positions were changed simultaneously. Aspartic acids and the other oxygen containing amino acid residue at the 1st (X), 2nd (Y) and 3rd (Z) calcium coordination sites were all replaced by alanines. We modified only the 1st, 2nd and 4th EF hand structures of

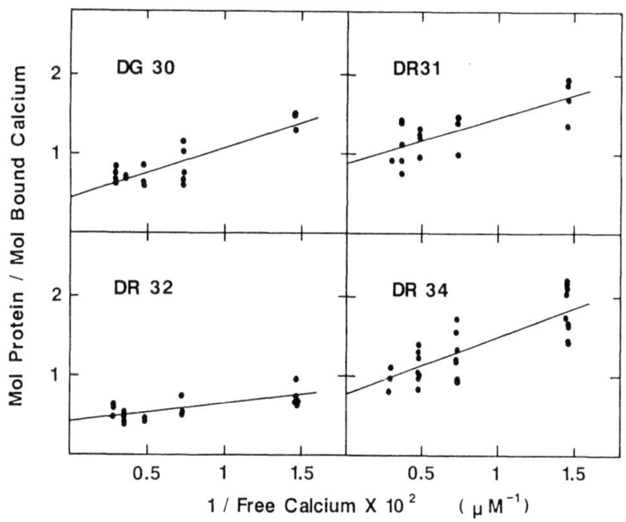

Fig. 4. Double reciprocal plot of calcium binding to the mutated calmodulin domains in DG30.

DG30 as shown in Fig. 3. The 3rd EF hand structure is presumed to be an inactive calcium binding site, because the loop region of the 3rd EF hand structure shows the lowest sequence homology to the consensus sequence and the residues at two terminal calcium coordination sites (X and -Z) are not acidic.

Three mutant clones encoding DR31, 32 and 34 were isolated and their proteins expressed in <u>E. coli</u> were purified. DR31, 32 and 34 have mutations only at the 1st, 2nd or 4th EF hand structures, respectively. Double reciprocal plots of calcium binding to the mutated DG30s are shown in Fig. 4 and the results are summarized in Table 2.

DG30 binds 2 moles of calcium, but after modification of the 1st or 4th site (DR31 or DR34), one mole is bound. Modification of the 2nd site had no effect on the number of calcium ions bound. These results together with the C-terminal truncation study indicate that the 1st and 4th EF hand structures are the functional calcium binding sites in DG30. The Kd values of DR31 and DR34 correspond to the binding of calcium to the remaining 4th or 1st EF hand structures, respectively. The results are confirmed by the fact that TR22, which lacks the 4th EF hand structure by deletion, has a Kd value similar to that of DR34, which has the same functional site (1st site). Removal of the 4th EF hand structure does not affect the structure of the 1st.

Interestingly, when one of the functional EF hand structures is modified, the apparent Kd values decrease. A similar tendency is also observed upon removal of the 4th EF hand structure (TR22). The Kd value of DG30 is distinctly higher than the average Kd values for the 1st and 4th sites. Thus, the two functional calcium binding sites may not be independent but interact negatively. Since modification of the non-functional 2nd site increases the affinity for calcium binding, it may participate in the interaction as well.

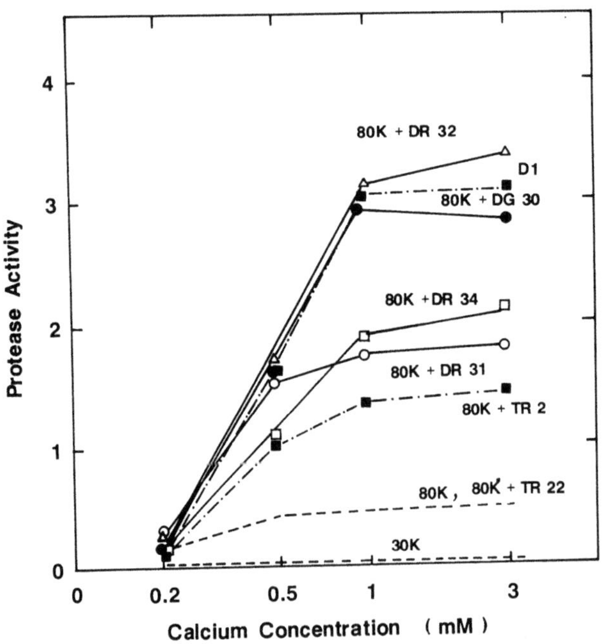

Fig. 5. Reconstitution of CANP from the native 80K subunit and derivatives of DG30.

Reconstitution of CANP from Native 80K and Derivatives of Calmodulin Domain of 30K subunit

Various derivatives of DG30 were used for reconstitution with the native 80K subunit to analyze the role of the small subunit in the regulation and activation of CANP protease activity. The 30K subunit is composed of two domains, the N-terminal glycine-rich hydrophobic domain and C-terminal calmodulin domain[8,9]. The N-terminal domain is removed during activation of CANP at the membrane, indicating that the N-terminal domain is not essential for the interaction with the 80K subunit, at least once the complex is formed, nor for the expression of protease activity.[22] The 30K subunit or DG30 have no protease activity but the 80K subunit expresses distinct, albeit low, protease activity as shown in Fig. 5. Reconstitution of the 30K subunit with the 80K subunit enhances significantly the recovered protease activity.[15,16] When the native 30K subunit is replaced by DG30, the recovered activity decreases to about 50%. Modification of the inactive 2nd site (DR32) has no significant effect on the recovered protease activity. Similarly, substitution of valine for a single aspartic acid residue at the 1st calcium coordination site in the 1st (D1) EF hand structure produces no appreciable change in the protease activity. In contrast, modifying the active calcium binding sites, i.e. DR31 and DR34, reduces the maximum activation to about half. The binding of 2 moles of calcium is essential for expression of full protease activity, but enhanced protease activity is also be obtained upon the binding of one mole. Activation may proceed in a stepwise manner, because about 50% activation was observed when calcium binds to one of the two calcium binding sites. The maximum levels of protease activity recovered with various DG30 derivatives reflect at least partly the efficiency of complex formation with the 80K subunit. TR22 does not stimulate the activity of the 80K subunit, even though it binds one mole of calcium. Complex formation as determined by polyacrylamide disc gel electrophoresis was not observed in this case. The maximum activity obtained with TR2, which binds 2 moles of calcium, was about 50% of that obtained with DG30. Since the C-terminal region of the 30K subunit is indispensable for complex formation with the 80K subunit,[15] the decrease in the maximum level of activation with TR2 and TR22 can be ascribed mainly to the change in the C-terminal region. The 30K subunit and its derivatives are potent enhancers of the protease activity of the 80K subunit which alone exerts very low activity. A double mutant of DG30 modified at the 1st and 4th EF hand structures so that it no longer binds calcium must be analyzed to clarify further the role of the binding of calcium to the 30K subunit in activity regulation.

Another important point is that modification of the calcium binding sites in the 30K subunit does not significantly change the calcium sensitivity of the protease activity of reconstituted CANP. This is consistent with the fact that the protease activity calcium sensitivity is determined essentially by the binding of calcium to the calmodulin domain of the 80K subunit.[16,19]

Calcium Binding Sites in CANP

In this study we analyzed the calcium binding sites of the 30K subunit by site-directed mutagenesis and verified that the 1st and 4th EF hand structures are the functional calcium binding sites. This is the first instance where active calcium binding sites have been mapped successfully by site-directed mutagenesis. Since the calmodulin domains in the 30K and 80K subunits have evolved from the same cognate ancestral gene,[23,24] their sequences are highly homologous (ca. 50% homology). Accordingly, it is possible to deduce the calcium binding sites in the 80K subunit on the basis of the results obtained with DG30 as shown in Fig. 6.

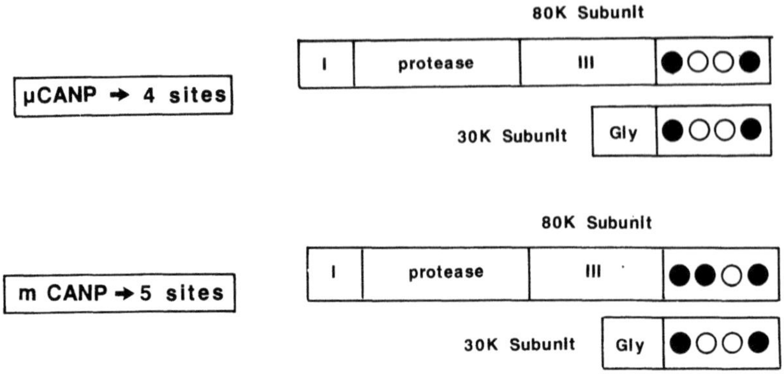

Fig. 6. Predicted calcium binding sites in rabbit mCANP and μCANP. Filled and open circles are EF hand structures correponding to functional and nonfunctional calcium binding sites, respectively.

U80 and M80 bind 2 and 3 moles of calcium. Two calcium binding sites should be the 1st and 4th EF hand structures as in the case of DG30. The third site in M80 should be the 2nd EF hand structure, because the 3rd EF hand structure will be a functional binding site for the same reasons discussed above for DG30. However, why the 2nd EF hand structure of the 80K subunit binds calcium, while the corresponding sites in U80 and DG30 do not, is not apparent from the sequence. Summarizing, μCANP binds 4 moles and mCANP binds 5 moles of calcium per mole as shown in Fig. 6. Similar site-directed mutagenesis of the calmodulin domain in the 80K subunit is required to confirm the predicted calcium binding sites in the 80K subunit.

Calcium Binding Phenomena observed in Vitro and Presumed in Vivo

For expression of protease activity purified CANP requires a calcium concentration which can not be reached under physiological conditions. This implies that the calcium sensitivity of CANP must be increased *in vivo* to become active at physiological calcium concentrations. This activation mechanism is a critical point for understanding the expression of CANP activity *in vivo*. Recent studies on CANP have revealed the activation mechanism summarized in Fig. 7.[25,26] According to this model, CANP is an inactive proenzyme that translocates to the membrane upon binding of calcium. Activation, i.e. removal of the N-terminal region of the 80K subunit, occurs autocatalytically at the membrane in the presence of calcium and phospholipid. The active enzyme functions at the membrane and in the cytosol after liberation from the membrane. This model predicts the presence of at least two kinds of calcium binding sites, one which is independent of protease activity and induces conformational changes upon calcium binding to expose hydrophobic regions required for interaction with the biological membrane, and another which has direct correlation with the expression of protease activity. Both sites must bind calcium under physiological conditions, indicating that their Kd values are around μM order. Further, the former site must be exposed to the surface.

When purified CANP is incubated with plasma membranes at low temperature, CANP binds to the membrane only in the presence of calcium. This binding is observed at a calcium concentration as low as 1 or 2 μM. Upon removal of calcium by addition of EDTA, native CANP is recovered without modification. When similar experiments are performed at higher calcium concentration or at room temperature, CANP binds to the membrane, but simultaneous cleavage of the 80K subunit is observed. This autocatalytic activation is inhibited by lowering the temperature or by adding a protease inhibitor, leupeptin. Activated CANP hydrolyzes various substrates at μM levels of calcium. These results support the existence of two types of calcium binding sites with Kd values of a μM oder. In the absence of biological membranes, autolysis requires unphysiologically high calcium concentrations.[26,27] The glycine-rich hydrophobic domain in the 30K subunit interacts with membrane phospholipids and reduces the calcium concentration for autolysis.[26] Two kinds of calcium binding sites reported previously[6] presumably correspond to the above two types of calcium binding. These two types of binding sites must be explained in terms of the calcium binding sites revealed in the present study. However, the Kd values of the calcium binding sites measured here are one or two orders of magnitude higher than predicted above. In the present study, we used the calmodulin domain rather than native CANP. This may be a reason we did not find a calcium binding site of μM order.

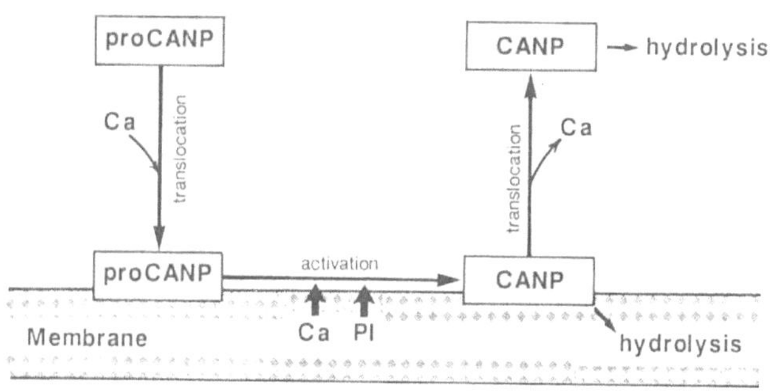

Fig. 7. Activation mechanism of CANP at the membrane.

To understand the activation and regulatory mechanism of CANP we must analyze calcium binding sites in both native and activated CANPs. As discussed, native active CANP can not be used for calcium binding studies unless a sophisticated method is devised. It may be possible to use chemically modified inactive CANP.[6] The active site SH group of CANP is buried and can be modified only in the presence of calcium.[6,28] Thus, simultaneous autolysis during modification is inevitable. Modification of CANP in the presence of calcium produces a mixture of modified native and autolyzed CANPs which makes precise calcium binding studies very difficult. Calcium-binding stuyies can be done in the presence of protease inhibitor. However, even a high concentration of leupeptin does not inhibit autolysis completely. Previously, we reported the binding of calcium to CANP after inactivation by carboxymethylation.[6] No other studies have described the actual binding of calcium to CANP. Although the CANP preparation used in our study was a mixture of native and activated CANPs, the coincidence in the number of calcium binding sites reported previously[6] and in the present paper suggests that the calcium binding sites identified here are functional in native and/or activated CANP. The Kd values, however, do not correspond to those presumed for native CANP. It is likely that some sites are buried and become exposed only after conformational changes induced upon binding of calcium to exposed sites. Further, analyses of the calcium binding sites are essential to understand the function of the calmodulin domains and calcium ions in the regulation of CANP activity.

REFERENCES

1. K. Suzuki, Calcium-activated neutral protease: domain structure and activity regulation, Trend. Biol. Sci., 12:103 (1987).
2. S. Pontremoli and E. Melloni, Extralysosomal degradation, Ann. Rev. Biochem., 55:455 (1986).
3. K. Suzuki, S. Ohno, Y. Emori, S. Imajoh, and H. Kawasaki, Calcium-activated neutral protease (CANP) and its biological and medical implications, Prog. Clin. Biochem. Med., 5:43 (1987).
4. T. Murachi, M. Hatanaka, and T. Hamakubo, Calpains and neuropeptide metabolism, in "Neuropeptides and their peptidases," Turner, ed., Ellis Horwood Ltd., Chichester (1987).
5. R. Mellgren, Calcium-dependent proteases: an enzyme system active at cellular membrane? FASEB J., 2:110 (1988).
6. K. Suzuki and S. Ishiura, Effect of metal ions on the activity of calcium-activated neutral protease (CANP), J. Biochem., 93:1463 (1983).
7. S. Ohno, Y. Emori, S. Imajoh, H. Kawasaki, M. Kisaragi, and K. Suzuki, Evolutionary origin of calcium dependent protease by fusion of genes for a thiol protease and a calcium binding protein? Nature, 312:566 (1984).
8. Y. Emori, H. Kawasaki, S. Imajoh, S. Kawashima, and K. Suzuki, Isolation and sequence analysis of cDNA clones for the small subunit of rabbit calcium-dependent protease, J. Biol. Chem., 261:9472 (1986).
9. T. Sakihama, H. Kakidani, K. Zenita, N. Yumoto, T. Kikuchi, T. Sasaki, R. Knnagi, S. Nakanishi, M. Ohmori, K. Takio, K. Titani, and T. Murachi, A putative Ca^{2+}-binding protein: structure of the light subunit of porcine calpain elucidated by molecular cloning and protein sequence analysis, Proc. Natl. Acad. Sci.USA, 82:6075 (1985).
10. H. Kawasaki, H. Kasai, and T. Okuyama, Protein analyses and reagents: microscale assay of calcium-binding activity of proteins and peptides using a nitrocellulose membrane, Anal. Biochem., 148:297 (1985).
11. K. Maruyama, T. Mikawa, and S. Ebashi, Detection of calcium binding proteins by ^{45}Ca autoradiography on nitrocellulose membrane after sodium dodecylsulfate gel electrophoresis, J. Biochem., 95:511 (1984).
12. Y. Emori, H. Kawasaki, H. Sugihara, S. Imajoh, S. Kawashima, and K. Suzuki, Isolation and sequence analyses of cDNA clones for the large subunits of two isozymes of rabbit calcium-dependent protease, J. Biol. Chem., 261:9472 (1986).
13. Y. Minami, Y. Emori, H. Kawasaki, and K. Suzuki, E-F hand structure domain of calcium-activated neutral protease (CANP) can bind Ca^{2+} ions, J. Biochem., 101:889 (1987).
14. Y. Minami, Y. Emori, S. Imajoh, H. Kawasaki, and K. Suzuki, Carboxy-terminal truncation and site-directed mutagenesis of the small subunit of rabbit calcium-dependent protease, J. Biochem., 103: in press (1988).
15. S. Imajoh, H. Kawasaki, and K. Suzuki, The COOH-terminal EF hand structure of calcium activated neutral protease (CANP) is important for the association of subunits and resulting protease activity, J. Biochem., 101:447 (1987).
16. T. Kikuchi, N. Yumoto, T. Sasaki, and T. Murachi, Reconstitution of calpain I and calpain II from their subunits. Interchangeability of the light subunits, Arch. Biophys. Biochem., 234:639 (1984).
17. O.H. Lowry, N.I. Rosenbrough, A.L. Farr, and R.J. Randall, Protein measurement with the Folin-phenol reagent, J. Biol. Chem., 193: 265 (1955).
18. U.K. Laemmli, Cleavage of structural proteins during the assembly of the head of bacteriophage T4, Nature, 227:680 (1970).
19. H. Kawasaki, S. Imajoh, S. Kawashima, H. Hayashi, and K. Suzuki, The small subunit of calcium dependent protease with different calcium sensitivities are identical, J. Biochem., 99:1525 (1986).

20. F.C. Reinach, K. Nagai, and J. Kendrick-Jones, Site-directed mutagenesis of the regulatory light chain Ca^{2+}/Mg^{2+} binding site and its role in hybrid myosins, Nature, 322:80 (1986).
21. R.M. Tufty, R.H. Kretsinger, Troponin and parvalbumin calcium binding regions predicted in myosin light chain and T4 lysozyme, Science, 187:167 (1975).
22. S. Imajoh, H. Kawasaki, and K. Suzuki, Limited autolysis of calcium-activated neutral protease (CANP): reduction of the Ca^{2+}-requirement is due to the N-terminal processing of the large subunit, J. Biochem., 100:633 (1986).
23. Y. Emori, S. Ohno, M. Tobita, and K. Suzuki, Gene structure of calcium-dependent protease retains the ancestral organization of the calcium-binding protein gene, FEBS Letters, 194:249 (1986).
24. S. Miyake, Y. Emori, and K. Suzuki, Gene organization of the small subunit of human calcium-activated neutral protease, Nucleic Acids Res., 14:8805 (1986).
25. K. Suzuki, S. Imajoh, Y. Emori, H. Kawasaki, Y. Minami, and S. Ohno, Calcium-activated neutral protease and its endogenous inhibitor. Activationat the cell membrane and biological function, FEBS Letters, 220:271 (1987).
26. K. Suzuki, S. Imajoh, Y. Emori, H. Kawasaki, Y. Minami, and S. Ohno, Regulation of activity of calcium-activated neutral protease, Adv. Enzyme Regul., 27:153 (1988).
27. K. Suzuki, S. Tsuji, K. Kubota, Y. Kimura, and K. Suzuki, Limited autolysis of Ca^{2+}-activated neutral protease changes its sensitivity to Ca^{2+} ions, J. Biochem., 90:275 (1981).
28. K. Suzuki, Reaction of calcium-activated neutral protease (CANP) with an epoxysuccinyl derivative (E64c) and iodoacetic acid, J. Biochem., 93: 1305 (1983).

NEUTRALIZATION OF SURFACE CHARGES MARKEDLY AFFECTS THE PROPERTIES OF BOVINE CALBINDIN D_{9k}

S. Forsén, M. Akke, P. Brodin*, P. Bayley**, T. Drakenberg, T. Grundström*, C. Johansson, S. Linse, S. Martin**, and E. Thulin

Physical Chemistry2, Chemical Centre, P.O. Box 124
Lund University, S-221 00 Lund, Sweden

*Unit of Appl. Cell and Molecular Biology Umeå University
S-901 87 Umeå, Sweden

**Div. of Physical Biochemistry, Natl. Institute for Medical Research, Mill Hill, London NW7 1AA, England

INTRODUCTION

Calbindin D_{9k} constitutes an attractive model system for exploring the relationships between structure, dynamics and function in the EF-hand family of proteins. It is the smallest protein known ($M_r \approx 8{,}500$; 75 a.a:s) with a pair of EF-hand Ca^{2+}-binding sites and its structure resembles closely the globular domains of the homologous proteins calmodulin, parvalbumin and skeletal muscle troponin C. [1,7]. The schematic structure of calbindin is shown in Fig. 1.

Some of the many current questions concerning this family of proteins are:

- How is the Ca^{2+}-affinity of the individual sites related to the nature and spatial arrangement of the ligands?
- How do the Ca^{2+}-sites in pairs of EF-hands interact and what is the free energy of interaction?
- What are the rates of Ca^{2+}-binding and release?
- What changes in tertiary structure accompanies the binding or release of Ca^{2+}-ions?

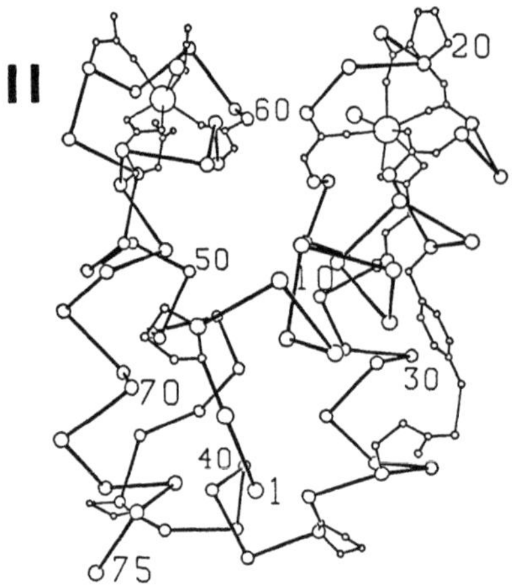

Figure 1. Schematic structure of bovine calbindin D_{9k} in the Ca^{2+}-loaded form according to Szebenyi and Moffat [7]. The C-terminal Ca^{2+}-site (site II) has the same peptide fold as in the archetypal EF-hand whereas the N-terminal site (site I) constitutes a variant hand in which the Ca^{2+}-binding loop has two additional amino acids as compared with the normal EF-hand.

During the last few years we have attempted to address some of these questions using a combination of biophysical techniques and the powerful new method of site directed mutagenesis. The gene of bovine calbindin D_{9k} (the minor A form for which the X-ray structure [7] is known) has been synthesized and expressed in E. coli [2]. Some 30 mutant calbindins with single, double and triple amino acid substitutions and/or deletions have been prepared to date. The initial set of mutants were confined to the N-terminal Ca^{2+}-binding site (site I)which has some unusual features as compared to a normal EF-hand. Also the intramolecular hydrogen bond between Tyr-13 and Glu-35, which links the helices flanking the N-terminal Ca^{2+}-site, has been removed through the change Tyr 13 → Phe. Results obtained on a first set of 4 to 5 mutants have recently been presented elsewhere [1, 3-6]. These studies have shown among other things that the modifications within the Ca^{2+}-binding loop of site I cause only localized structural changes and that the global structure is hardly affected. The reduction in the Ca^{2+}-affinity of the site modified is primarily due to an increased rate of Ca^{2+} dissociation. Furthermore it has been possible to establish that the Ca^{2+}-binding in the wild-type protein, and in the Tyr-13 → Phe mutant, is cooperative and that the free energy of interaction between the sites, $\Delta\Delta G$, is -5.1 ± 0.4 kJ/mole in the former.

In the present work we will address the role which charged amino acid residues at the surface of the protein play in the Ca^{2+}-binding properties of calbindin D_{9k}. The distribution of surface charges on calbindin D_{9k} is non-uniform. A cluster of negatively charged amino acid side-chains is present around the Ca^{2+} sites [7].

The disposition of the carboxylate groups and some relevant inter-residue distances are schematically shown in Fig. 2. In the present work we will concentrate on three of the surface residues: Glu-17, Asp-19 and Glu-26. Note that the carboxylate groups of these are not directly liganded to the Ca^{2+}-ions.

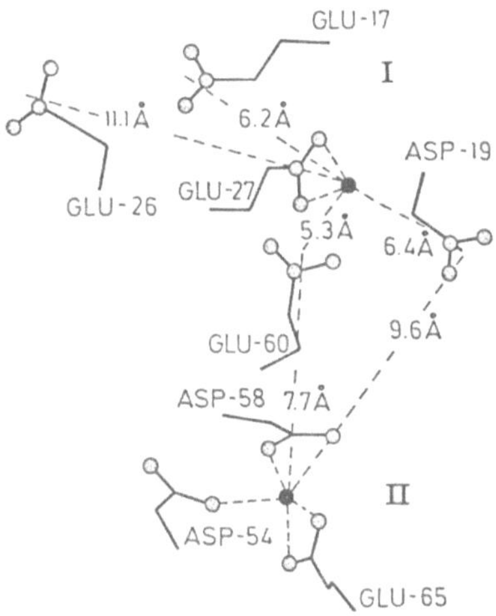

Figure 2. Schematic picture of the cluster of negative surface residues around the two sites of calbindin D_{9k}. X-ray data show that only the carboxylate groups of Glu 27, Asp 54, Asp 58 and Glu 65 are part of the ligand spheres of sites I and II. Distances between the Ca^{2+}-ions (•) and the non-liganded carboxylate groups (midpoint between the oxygens) are indicated. Mutant forms of the protein prepared and studied in this work are:

M5: Glu 17 → Gln
M6: Asp 19 → Asn
M7: Glu 17 → Gln; Asp 19 → Asn
M8: Glu 26 → Gln
M9: Glu 26 → Gln, Glu 17 → Gln
M10: Glu 26 → Gln; Asp 19 → Asn
M11: Glu 17 → Gln; Asp 19 → Asn; Glu 26 → Gln

Choice of mutants for biophysical studies

The contributions of the surface residues Glu-17, Asp-19 and Glu-26 to the free energy of binding of two Ca^{2+}-ion, to the cooperativity of binding and the stability towards unfolding of calbindin D_{9k} has been assessed through a study of seven mutant proteins. The charges have been neutralized with a minimal change in side chain structure and volume requirement by the substitutions Asp → Asn and Glu → Gln. All possible single, double and triple mutants have been constructed - the designation of the individual mutant proteins is given in the legend to Fig. 2. The wild-type will in the following be denoted "M0". The proteins have been produced in E. coli using synthetic genes and have been purified according to methods detailed elsewhere [1-3]. ^{1}H and ^{113}Cd NMR studies of mutants M5 through M11 strongly indicate that their tertiary structure is virtually identical to that of M0.

RESULTS AND DISCUSSION

Ca^{2+}-binding properties

The macroscopic (stoichiometric) Ca^{2+}-binding constants have been obtained by titrating the mutant proteins with Ca^{2+} in the presence of the chromophoric chelator 5,5'-Br_2BAPTA in the manner previously described for other mutants using the analogous chelator Quin 2. The Ca^{2+}-binding constant of 5,5'-Br_2 BAPTA was separately determined as $1 \cdot 10^{7}$ M^{-1} under the conditions of our experiments (low salt, pH 7.0). The results are summarized in Table I. It is evident that the neutralization of the three surface charges has a profound influence on the Ca^{2}-affinity of calbindin D_{9k}. The changes in affinity are perhaps best reflected in the total free energy of Ca^{2+}-binding, $\Delta G^{Ca^{2+}}_{tot} = -RT \ln(K_1 K_2)$, where K_1 and K_2 are two macroscopic binding constants. In the wild-type (M0) $\Delta G^{Ca^{2+}}_{tot}$ has the value -97 kJ/mol.

Table I. Equilibrium constants (K_1 and K_2), total free energy of Ca^{2+} binding $\Delta G^{Ca^{2+}}_{tot}$ (= $-RT \ln(K_1 \cdot K_2)$) and lower limit of the free energy of interaction between the two sites, $|\Delta\Delta G|_{min}$, for wild-type and seven mutant calbindins D_{9k}.

Protein	Side-chains neutralized	log K_1	log K_2	$\Delta G^{Ca^{2+}}_{tot}$ kJ/mol	$\|\Delta\Delta G\|_{min}$ kJ/mol
M0	-	8.3	8.6	-97	5.1
M5	Glu 17	7.4	8.1	-89	7.7
M6	Asp 19	7.6	8.0	-89	5.8
M7	Glu 17, Asp 19	7.4	7.1	-83	1.7
M8	Glu 26	7.8	8.4	-93	6.4
M9	Glu 17, Glu 26	6.8	7.8	-83	9.2
M10	Asp 19, Glu 26	6.9	7.6	-83	7.8
M11	Glu 17, Asp 19, Glu 26	6.8	6.7	-77	2.9

Neutralization of one charge, as in M5, M6 and M8, changes $\Delta G_{tot}^{Ca^{2+}}$ to about -90 kJ/mol with some variation depending on which individual charge is neutralized. Neutralization of two charges, as in M7, M9 and M10, results in a $\Delta G_{tot}^{Ca^{2+}}$-value around -83 kJ/mol and when all three surface charges are neutralized the corresponding value is -77 kJ/mol. On the average each surface charge contributes -7 kJ/mol to $\Delta G_{tot}^{Ca^{2+}}$ and in addition it is seen that the $\Delta G_{tot}^{Ca^{2+}}$-values are additive within the experimental error. The present results demonstrate unambigously that charged amino acids not directly liganded to the Ca^{2+}-ions have a large influence on the Ca^{2+}-affinity of the protein. Attempts to rationalize the Ca^{2+} affinity of EF-hand proteins entirely in terms of the nature and spatial arrangement of groups directly involved as Ca^{2+}-ligands will, therefore, very likely be unsuccessful.

The equilibrium constant data provide information also about another parameter of some importance: the free energy of interaction between the two Ca^{2+}-sites, $\Delta\Delta G$. We have previously shown that $\Delta\Delta G$ may be expressed as

$$\Delta\Delta G = - RT \ln \left[\frac{K_2}{K_1} \frac{(1+\eta)^2}{\eta}\right] \qquad (1)$$

where $\eta = K_{II}/K_I$ (K_I and K_{II} are the microscopic bidning constants of Ca^{2+} to sites I and II respectively). Since $|\Delta\Delta G|$ has a minimum for $\eta = 1$ a lower limit of $|\Delta\Delta G|$ can be obtained solely from the macroscopic constants as

$$|\Delta\Delta G|_{min} = RT \ln \left[\frac{4K_2}{K_1}\right] \qquad (2)$$

Calculated values of $|\Delta\Delta G|_{min}$ for mutants M5 through M11 are listed in the last column of Table I. Positive cooperativity ($\Delta\Delta G < 0$) is observed for the Ca^{2+} bidning process in all mutant proteins studied. For M0 the free energy of site-site interaction has previously been determined as $\Delta\Delta G = -5.1 \pm 0.4$ kJ/mol [3]. It is obvious that the site-site interaction in mutant calbindins where one or two surface charges are neutralized - M7 excluded - is increased relative to that in the wild-type. This unexpected result may be the consequence of a reduced electrostatic repulsion between the two halves of the protein.

Apart from the three negative surface sites studied in the present work, which are all located in the N-terminal half, there are additional surface sites on the adjacent Ca^{2+} binding loop of the C-terminal half (cf. Figure 2). Neutralization of surface charges very likely facilitates an approach of the two Ca^{2+}-sites. Consequently, transmission of conformational effects of Ca^{2+}-binding in one site to the other site would accordingly become more efficient. Further discussion as to the reason for the low values of $|\Delta\Delta G|_{min}$ found in M7 and M11 will have to await ^1H NMR studies of the relative Ca^{2+} affinities of the two individual binding sites.

Rates of Ca^{2+}-dissociation and -association

Stopped-flow kinetic studies of the rate of Ca^{2+}-dissociation from the mutant proteins M5 through M11 have been performed. The fluorescent chelator Quin 2 was used to monitor the release of Ca^{2+} from the initially Ca^{2+}-loaded proteins in the manner described elsewhere [6]. Table II summarizes the set of reaction rates and amplitudes obtained at 20° under essentially the same conditions as used in determinations of Ca^{2+} binding constatns. For comparison the values obtained for the

Table II. Ca^{2+}-dissociation rates (k) and amplitudes (A) obtained from stopped-flow experiments on wild-type (M0) and seven mutant calbindins (for amino acids modifed cf text to Figure 2). Experimental conditions: 20 mM Pipes/KOH, pH 7.0, t=20°C.

Protein	k_{slow} [s^{-1}]	A^*_{slow}	k_{fast} [s^{-1}]	A^*_{fast}
M0	3.5±0.5	1.8±0.3	-	-
M5	4.3±0.8	1.7±0.2	-	-
M6	8.3±1.5	0.8±0.1	21.5±2.3	0.7±0.2
M7	6.3±1.2	0.43±0.2	15.3±3.5	0.8±0.2
M8	2.4±0.6	0.8±0.2	12.4±4.2	0.7±0.15
M9	3.9±1.2	0.7±0.2	18.3±3.6	0.8±0.2
M10	4.8±1.3	1.6±0.2	-	-
M11	4.5±0.8	1.15±0.25	-	-

*) Amplitudes in terms of number of Ca^{2+} ions released based on the amplitude of a calibration reaction (dissociation of Ca^{2+} from calmodulin).

wild-type protein are also listed. We note that single dissociation steps, corresponding to the release of two Ca^{2+} ions, are observed for M0 and the mutant proteins M5, M10 and M11. At higher ionic strength conditions (0.1 M KCl) the single steps can be resolved into two one-Ca^{2+} steps. For the other mutants two dissociation steps, each approximately corresponding to the release of one Ca^{2+}-ion, are observed. An extensive analysis of the observed finer variations in the rate parameters is beyond the scope of this brief communication; a more detailed analysis is under preparation [S.R. Martin et al.to be published]. We may, however, note that the observed nearly 100-fold reduction in Ca^{2+} affinity as successive surface charges are neutralized is not reflected in a corresponding increase in Ca^{2+} dissociation rate. In fact rate parameters observed for the mutant where all three surface charges have been neutralized (mutant M11) are virtually the same as for the wild-type protein. We may conclude therefore that the observed reduction in Ca^{2+}-affinity must primarily be due to a reduced rate of Ca^{2+}-association. The kinetic results imply that the dynamic properties of the protein molecule, and the large scale fluctuations that must accompany the loss of a Ca^{2+}-ion, are not much affected by the disappearing surface charges. By contrast the surface charges around the two Ca^{2+}-sites play an important role in the series of molecular events leading to the capture of an incoming Ca^{2+}-ion - perhaps by enhancing the affinity of an initial encounter complex.

Stability towards unfolding

Previous studies in our laboratories have shown that the stability towards unfolding of native calbindin D_{9k} is considerable. The Ca^{2+} loaded form is not significantly denatured by heating to 90°C or by adding 8M urea [1,5]. In the Ca^{2+} free, (apo) form thermal unfolding is only half complete at 90°C while denaturation is complete in 7-8 M urea [1,5]. Preliminary data on the urea denaturation of mutants M5 through M11 are given in Table III. Unfolding was followed by CD measurements at 223 nm as described earlier [5]. The unfolding process was treated using the two-state equilibrium model. The fraction of native (f_N) and denatured (f_D) protein in the transition region defines the apparent free energy difference, ΔG_D, of native and denatured forms as $\Delta G_D = -RT \ln(f_D/f_N)$. ΔG_D was to a first approximation assumed to depend linearly on urea concentration, i.e. $\Delta G_D = \Delta G_D^{H_2O} + A[U]$, where $\Delta G_D^{H_2O}$ is the free-energy difference between native and denaturated protein in the absence of urea, A is a constant and [U] the urea concentration in mol/lit.

Table III. Preliminary values for wild-type calbindin D_{9k} (MO) and mutants M5 through M11 of stabilization free energies of unfolding ($\Delta G_D^{H_2O}$), urea concentrations at transition midpoints (C_m) and the coefficient A multiplying the linear concentration term in the equation used for extrapolating the experimental data to zero urea concentration (cf. text).

Protein	$\Delta G_D^{H_2O}$ [kJ/mol]	C_m [M]	A [kJ/mol, M]
MO	20±1	5.2	-3.8
M5	23.3	5.5	-4.1
M6	25.1	5.0	-4.2
M7	27.7	6.0	-4.6
M8	24.1	5.2	-4.7
M9	22.1	5.4	-4.1
M10	26.3	5.7	-4.6
M11	27.4	6.0	-4.5

We note from Table III that neutralization of surface charges on calbindin results in an increased stability towards unfolding. The general tendency is that the stability increases with the number of charges neutralized - M9 being a notable exception. The tendency is manifested not only in the value of $\Delta G_D^{H_2O}$ but also in the value of C_m (the urea concentration at the midpoint of the unfolding transition). The latter parameter is much less sensitive than $\Delta G_D^{H_2O}$ to the particular analytical expression used to extrapolate ΔG_D to [U] = 0 [5].

How can we rationalize the present results? We are inclined to attribute the observed changes mainly to an increased stability of the native form of the proteins. Although the tertiary structure of native apocalbindin is not yet known in detail it appears from ^{1}H NMR studies that no major structural changes have occurred in comparison with the Ca^{2+} loaded form. The mini β-sheet between amino acids Leu 23 and Val 61 observed in the Ca^{2+} form is however no longer seen which suggests that the two halves of the protein have moved further apart in the region of the Ca^{2+} sites. It thus seems likely that Glu 17, Asp 19 and Glu 26, like other negatively charged amino acids in this region (cf. figure 2), are still located fairly close together at the surface of the protein. Electrostatic interactions between these charged groups would tend to lower the stability of the native structure as compared to the unfolded form. Accordingly neutralizations of some of the negative charges, as in mutants M5 through M11 would increase the stability of the native structure.

CONCLUSION

The present study has highlighted the important role played by surface charges in bovine calbindin D_{9k}. Not only do they have a profound influence on the Ca^{2+}-affinity and kinetics of binding, they also affect the cooperativity between the two sites and contribute largely to the stability of the protein towards unfolding. In one sense the result is not unexpected - coulombic interactions are long range with a dependence of $(\text{distance})^{-1}$. The results obtained on calbindin can very likely be generalized to other proteins.

Finally some speculations about a biological role of the negative surface charges on bovine calbindin. A proposed function for this protein is to facilitate Ca^{2+}-transport through the intestinal epithelial cells [8,9]. A Ca^{2+}-ATPase in the enterocyte basolateral membrane (BLM) pumps Ca^{2+} out of these cells. Interaction between calbindin and the BLM Ca^{2+}-ATPase has been observed by Walters [10,11] and might occur through complementary surface charges. Such an interaction could then play a double role. First the proteins would be docked in a favourable position for Ca^{2+}-ions to be transferred from calbindin to the ATPase. Secondly, the surface charges of both proteins would be partly neutralized as a result of the close approach of opposite charges. The Ca^{2+}-affinity of calbindin would then decrease as shown in this work. In addition the Ca^{2+}-affinity of the ATPase might increase. The net effect would favour Ca^{2+}-transport from calbindin to the Ca^{2+}-pump.

REFERENCES

1. S. Forsén, T. Drakenberg, S. Linse, P. Brodin, P. Sellers, C. Johansson, E. Thulin and T. Grundström. "Site Directed Mutagenesis and Structure-Function Relations in EF-hand Ca^{2+}-Binding Proteins" in "Metal Ions in Biology" (G. Sigel ed.) Vol. **25**. M. Dekker Inc. New York (1988/89) In press.

2. P. Brodin, T. Grundström, T. Hofmann, T. Drakenberg, E. Thulin and S. Forsén. Biochemistry **25**, 5371-5377 (1986).

3. S. Linse, P. Brodin, T. Drakenberg, E. Thulin, P. Sellers, K. Elmdén, T. Grundström and S. Forsén. Biochemistry **26**, 6723-6735 (1987).

4. S. Forsén, T. Drakenberg, E. Thulin, P. Brodin, T. Grundström, S. Linse, P. Sellers and K. Elmdén in "Calcium Binding Proteins in Health and Disease" (A.W. Norman, T.C. Vanaman adn A.R. Means, eds). Academic Press Inc. New York, p. 348-359 (1987).

5. B. Lindegård, T. Hofmann, S.R. Martin, P. Bayley, P. Brodin, T. Grundström, S. Linse and S. Forsén. Eur. J. Biochem. In press (1988).

6. S. Forsén, S. Linse, E. Thulin, B. Lindegård, S.R. Martin, P. Bayley, P. Brodin and T. Grundström. Eur.J. Biochem. In press. (1988).

7. D.M.E. Szebenyi and K. Moffat J. Biol. Chem. **261**, 8761-8777 (1986).

8. B.A. Levine and R.J.P. Williams in "Calcium and the Cell Function" (W.Y. CHeung, ed). Vol. II. Acad. Press New York, 1982.

9. R.J.P. Williams in "Calcium and the Cell", CIBA Foundation Symposium 122. J. Wiley, Chichester, 1986, pp. 145-159.

10. J.R.F. Walters in "Calcium Binding Proteins in Health and Disease" (A.W. Norman et al., eds.). Acad. Press New York, 1987. pp. 122-124.

11. J.R.F. Walters, personal communication.

CONFORMATION OF A TROPONIN-I PEPTIDE BOUND TO TROPONIN-C AS DETERMINED BY ^1H NMR

A. Patricia Campbell and Brian D. Sykes

Department of Biochemistry, and MRC Group in Protein Structure and Function, University of Alberta
Edmonton, Alberta, Canada, T6G 2H7, Canada

I INTRODUCTION

The Ca^{+2}-mediated interaction between troponin-I (Tn-I) and troponin-C (Tn-C) in the thin filament of skeletal muscle is an important element in the regulation of the actomyosin ATPase. It has been proposed that the structural changes produced by Ca^{+2} binding result in the presentation of regions on the surface of Tn-C capable of binding Tn-I with high affinity (1,2,3). In particular, it appears that residues 84-115[1] on Tn-C (which represent the N-terminal side of Ca^{+2}-binding site III) interact with residues 97-117 of Tn-I (3,5). A synthetic analog of the inhibitory region of Tn-I, Nα-acetyl-Tn-I(104-115)amide, has been made by solid phase synthesis. This region represents the minimum sequence necessary for inhibition of actomyosin ATPase activity and is found to bind to Tn-C in a Ca^{+2}-dependent fashion, with a binding constant of $1 \times 10^5 M^{-1}$ (6,7). The following body of work primarily addresses the question of secondary structure adopted by the Tn-I peptide when bound to Tn-C. In order to probe the structure of the bound Tn-I peptide, high field two-dimensional Nuclear Magnetic Resonance (NMR) techniques are used. Possible binding sites on the Tn-C molecule are also considered in light of recent findings from cross-linking studies between rabbit s-Tn-C, to which a maleimide cross-linker molecule has been attached at Cys98, and residues on whole Tn-I (8).

With the recent development of high-field spectrometers and new two-dimensional (2D) NMR techniques (9), it has become possible to determine the three-dimensional structure of small proteins and peptides in solution by NMR methods (10). A widely used protocol that can be applied for

1. The amino acid numbering from rabbit s-Tn-C is used here, so that amino acids 84-115 in rabbit s-Tn-C correspond to amino acids 87-115 in chicken s-Tn-C, and Cys98 in rabbit s-Tn-C corresponds to Cys101 in chicken s-Tn-C(4).

structure determination of biomolecules by NMR involves three basic steps: 1) assignment of proton resonances using 2D correlated spectroscopy (COSY) 2) determination of proton-proton distance restraints from 2D nuclear Overhauser spectroscopy (NOESY), and 3) calculation of a family of structures using the experimentally derived distance restraints and the covalent structure.

The assignment of proton resonances may be accomplished via the sequential resonance assignment technique outlined by Wüthrich (9), which employs information obtained from complementary COSY and NOESY data sets. The COSY spectrum consists of peaks along the diagonal and cross-peaks off the diagonal. Off-diagonal cross-peaks arise only between protons connected through J-coupling networks. Thus, the COSY spectrum identifies isolated spin systems characteristic of residue type since every amino acid side chain has a characteristic spin system and cross-peak connectivity pattern.

The determination of proton-proton distances is accomplished via the NOESY experiment which connects protons that are spatially in close proximity. The NOESY spectrum also consists of diagonal and off-diagonal cross-peaks. The origin of the cross-peak is dipolar cross-relaxation between protons. Because of the r^{-6} distance dependence of the effect, the cross-peak intensity is measureable only between protons that are 5Å or less in separation. The NOESY spectrum allows identification of sequential residues (i,i+1), since in any type of secondary structure, the αCH of residue i is never more than 5Å separated from the NH of residue i+1. Thus the presence of a cross-peak between residues i and i+1 in the αCH-NH region of the spectrum allows assignment of residues i and i+1 to sequential position in the polymer. This sequential information plus the information of residue type from the COSY allows unequivocal assignment in a nonrepetitive sequence. Other useful short distances involving backbone protons that are manifested in NOESY spectra are those between the αCH, βCH and NH of residue i and the NH of residue i+1 ($d\alpha N$, $d\beta N$, dNN, respectively). The presence of these cross-peaks can be diagnostic of secondary structural elements such as α-helix or β-turns.

The generation of a structure using the NMR distances and geometric constraints can be accomplished with restrained energy minimization (REM) and restrained molecular dynamics (RMD). The RMD calculation involves solving Newton's equations of motion:

$$\mathbf{F}_i = m_i \delta^2 \mathbf{r}_i / \delta t^2 \qquad \text{Eqn.1}$$

where $\mathbf{F}_i$ = -dE/d$\mathbf{r}_i$ = force on atom i
m_i = mass of atom i
t = time step
$\mathbf{r}_i$ = coordinates of atom i

E is the sum of energy potentials used for both the REM and RMD calculations which are comprised of terms for bond lengths, bond angles, proper and improper dihedrals, and Van der Waals and electrostatic non-bonding terms:

$$E = E_{BOND} + E_{ANGLE} + E_{DIHEDR} + E_{VdW} + E_{COULOMB} \qquad \text{Eqn.2}$$

A potential energy term representing the interproton distance restraints $E_{nOe}(r_{ij})$ is added to the total energy of the system in the form of a biharmonic potential:

$$E_{nOe}(r_{ij}) = CDIS\ (\ r_{ij} - r_{ij}^{o}\ )^2 \qquad \text{Eqn.3}$$

where r_{ij} and r_{ij}^{o} are the calculated and target interproton distances, respectively, and where CDIS is the force constant. The RMD calculation is usually preceded by REM to reduce unfavourable contacts, bond lengths and bond angles. REM usually changes the conformation only slightly and cannot take it out of local minima. RMD is able to overcome barriers of the order of kT because of the kinetic energy of the system, and has a much larger radius of convergence.

II MATERIALS AND METHODS

Tn-C was prepared from turkey skeletal muscle according the procedure outlined by McCubbin et al. (11). Turkey skeletal muscle was used since the crystal structure for the turkey skeletal Tn-C has been recently solved by Herzberg et al (12). There are only three differences (13) in the amino acid sequence of the turkey skeletal Tn-C and that of the chicken sequence as reported by Wilkinson (14). The Nα-acetyl Tn-I(104-115)-amide inhibitory peptide was synthesized by using the general procedures for solid-phase peptide synthesis (15) on a Beckman Model 990 synthesizer. The crude synthetic material was purified by high-performance liquid chromatography techniques using a SynChropak CM-300 weak cation-exchange resin and a SynChropak RP-P C18 reversed-phase resin.

All two-dimensional NMR spectra were collected on a Varian 500 MHz VXR-500 NMR spectrometer with a spectral width of 5000 Hz and a relaxation delay of 1.6 s. The 2048 point free-induction decays were collected at 256 t_1 values for the Double Quantum-Filtered COSY (DQFCOSY) data sets, and 512 t_1 values for the Transferred-NOESY (TRNOESY) data sets, and both sets of experiments zero-filled to 2048 points in the t_1 dimension. For the TRNOESY experiments, a mixing time, τ_m , of 500 ms was used for assignment purposes and 150 ms for distance measurements. The TRNOESY and DQFCOSY experiments were performed in a phase-sensitive mode according to the method of States et al. (16).

The sample used for both the DQFCOSY and TRNOESY experiments was 10mM inhibitory Tn-I(104-115) peptide, 1mM turkey s-Tn-C, 4mM $CaCl_2$ and 50mM KCl, dissolved in a 80%H_2O/20%D_2O solvent mixture at pH 6.5. Figure 1 shows a 1D spectrum of the NH resonances of the Tn-I peptide in the presence and absence of Tn-C (Ca^{+2} saturated). Figure 2 shows the upfield portion of the TRNOESY spectrum ; and figure 3 shows the αCH-NH portion of the TRNOESY spectrum.

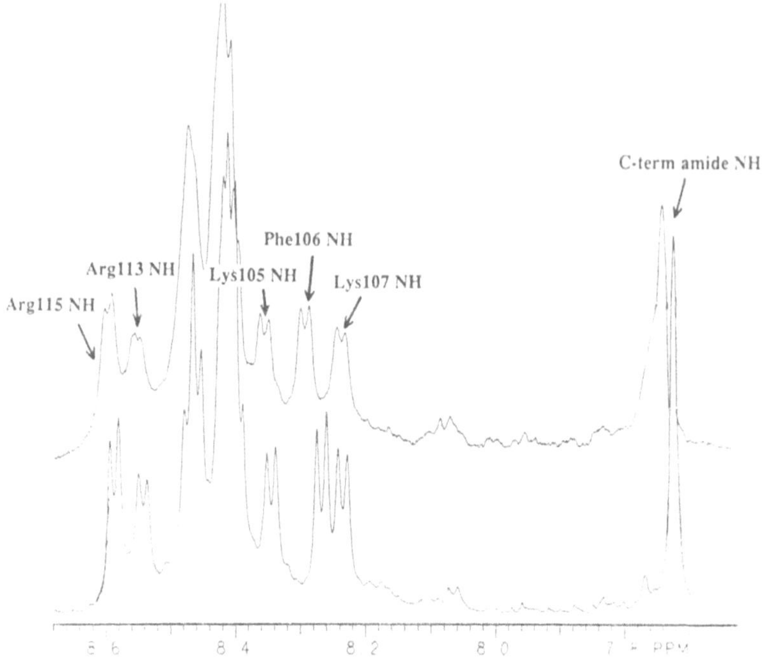

Figure 1 500 MHz ^{1}H NMR Spectrum of the Tn-I Peptide in the Presence (top) and Absence (bottom) of 0.1 Molar Equivalents of Tn-C (Ca^{+2} saturated)

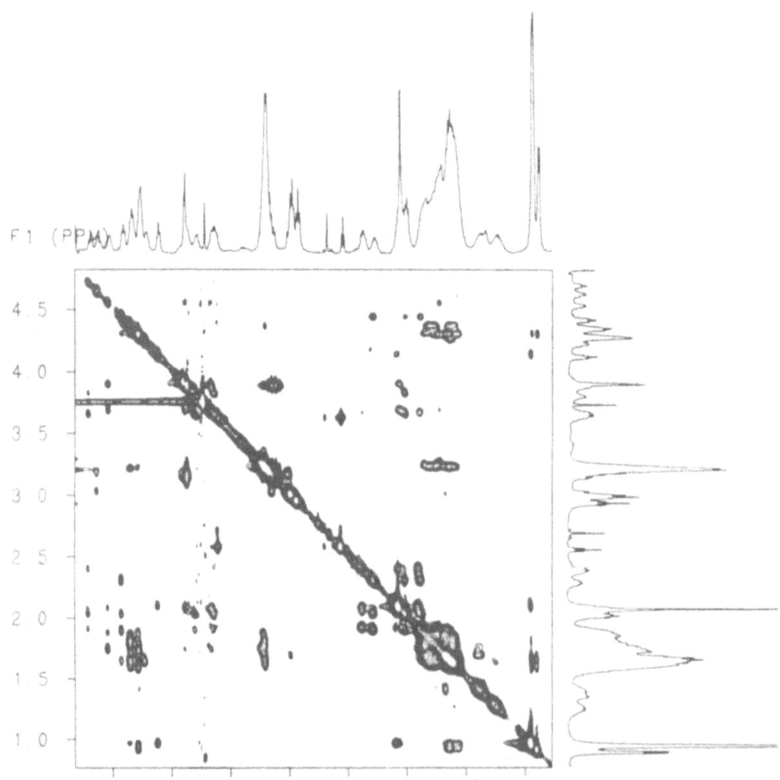

Figure 2 Upfield Portion of TRNOESY Spectrum

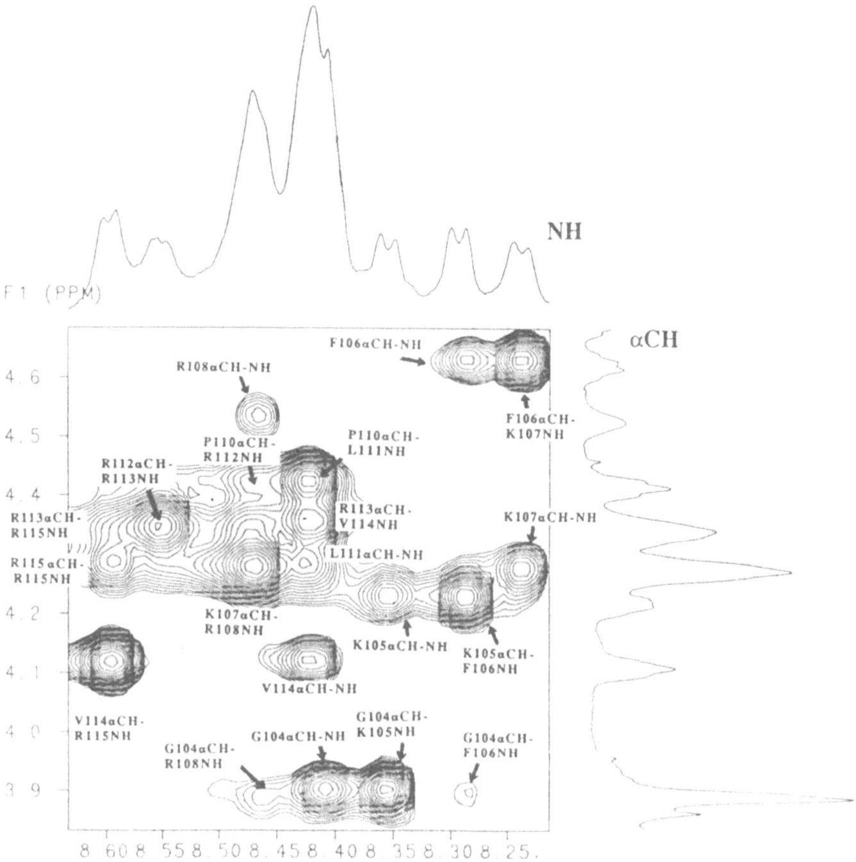

Figure 3 αCH-NH Portion of TRNOESY Spectrum

III STRUCTURE OF THE TN-I PEPTIDE BOUND TO TNC

a) Transferred Nuclear Overhauser Enhancement Two-Dimensional NMR Spectra (TRNOESY)

As previously stated, the nuclear Overhauser effect (nOe) is a result of short-range dipolar (through-space) interactions, leading to cross-relaxation between nuclei i and j if i and j are within 5Å of each other in space. The nOe is defined as the fractional change in intensity of one NMR resonance (corresponding to nucleus i) when another resonance (corresponding to nucleus j) is selectively perturbed. For short τm[2,3], the magnitude of the nOe (N_{ij}) between i and j is given by:

$$N_{ij} = (\sigma_{ij}) \tau_m = K (r_{ij})^{1/6} \tau_m \qquad \text{Eqn.4}$$

where σ_{ij} is the cross-relaxation rate between i and j, and K is a constant dependent on the correlation time of the internuclear vector.

2. τm is the 'mixing time' of the experiment during which cross-relaxation is allowed to evolve between spins
3. The adoption of a 2 spin approximation is justified by the use of short τm. For longer τm, a full multispin relaxation matrix treatment is necessary (17).

Conformational changes in the peptide induced by the formation of the peptide-Tn-C (Ca^{+2} saturated) complex were followed by the study of the transferred nOe (trnOe). In the unbound form, the peptide tumbles rapidly in solution ($\omega\tau_c$ = 1: extreme narrowing limit) and the peptide proton-proton nOes are small and either positive or negative; but when the peptide is bound to Tn-C, it tumbles more slowly with the protein ($\omega\tau_c$ > 1 :spin diffusion limit) and the nOes become large and negative. The trnOe involves the transfer of magnetization from the bound state to the free state by chemical exchange. Thus, in the presence of Tn-C, negative nOes are measure on the easily detected free ligand resonances , providing information on the proximity in space of two bound ligand protons (18,19). Consider the equilibrium:

$$\text{Tn-I(104-115)} + \text{Tn-C} \underset{k_{-1}}{\overset{k_1[\text{Tn-C}]}{\rightleftharpoons}} \text{Tn-I(104-115). Tn-C} \qquad \text{Eqn.5}$$

-free ligand (F) -bound ligand (B)

Given fast exchange on the NMR timescale ($\tau_m^{-2} >> \Delta\omega_m^2 >> (T_{2m}\tau_m)^{-1}$; where τ_m = life-time of a bound ligand nucleus, $\Delta\omega_m$ = chemical shift between the free and bound states, and T_{2m} = spin-spin relaxation time in the bound environment (21)),the average chemical shifts δ^{av}, and linewidths , $\Delta\nu_{1/2}^{av}$ are dominated by the free ligand population, since 1-a >> a (where a is the fraction of bound peptide) :

$$\delta^{av} = (1-a)\ \delta^F + a\ \delta^B \qquad \text{Eqn.6}$$

$$\Delta\nu_{1/2}^{av} = (1-a)\ \Delta\nu_{1/2}^F + a\ \Delta\nu_{1/2}^B \qquad \text{Eqn.7}$$

Conversely, the TRNOESY[4] cross-peak intensity , N_{ij}, is dominated by the bound ligand population, since $\sigma_{ij}^B >> \sigma_{ij}^F$:

$$N_{ij} = \{ a\sigma_{ij}^B + (1-a)\ \sigma_{ij}^F \} \cong a\ \sigma_{ij}^B \qquad \text{Eqn.8}$$

σ_{ij}^B yields interproton distances in the bound peptide :

$$\sigma_{ij}^B \propto \langle 1/r_{ij}^{B\ 6} \rangle \qquad \text{Eqn.9}$$

From Eqns. 8 and 9, it can be seen that for any two pairs of bound proton resonances, i and j, r_{ij}^B will be simply given by Eqn10, using a suitable reference distance :

$$r_{ij}^B = \{ (N_{REF}/ N_{ij}) \times r_{REF}^{\ 6} \}^{1/6} \qquad \text{Eqn.10}$$

The experimentally derived r_{ij}^B values may be used as distance restraints in REM and RMD programs to generate a bound Tn-I peptide conformation.

In summary, 232 interproton distance restraints, r_{ij}'s, were derived

4.The TRNOESY and the NOESY experiments are equivalent to the ID transient trnOe and nOe experiments, respectively, where enhancements develop in the absence of irradiation during τ_m, following selective pre-irradiation during the preparation period (20).

from the TRNOESY data of Nα-acetyl-Tn-I(104-115)amide bound to Tn-C(Ca^{+2})$_4$. Of these, 102 were intraresidue (i,i), 80 were sequential (i,i+1), 37 were medium range (i,i+2-i+4), and 13 were long range (i,i+5-i+8). The most important sequential range rij's were between Arg108αCH - Pro109 δCH2 and Pro109αCH - Pro110δCH2. These defined a trans Arg108-Pro109 and Pro109-Pro110 peptide bond orientation. The most important medium range r_{ij}'s involved the backbone αCH and NH protons of two separate regions; the region spanned by Gly104, Lys105, Phe106, Lys107 and Arg108 and the separate region spanned by Arg112, Arg113, Val114 and Arg115. These medium range distances defined turns involving residues 104 to 108 and 112 to 115. Another important set of medium range r_{ij}'s were found between the Phe106 ring protons and the Pro109 and Pro110γCH2 protons which defined a hydrophobic face of the peptide. Finally, the 13 long range r_{ij}'s were found between the Phe106 ring protons and the Leu111(δCH3)2 and the Val114(γCH3)2 methyl protons, defining a hydrophobic core in the peptide capable of interacting, along with the hydrophobic face spanned by Pro109-Pro110, with a hydrophobic patch on Tn-C.

b) Energy Minimization and Molecular Dynamics Programs

The REM and RMD programs in the GROMOS library (22) involved minimization of the potential energy function, comprised of the energy terms listed in equations 2 and 3. Solvent molecules were not included in the calculations. In the RMD program, initial velocities were assigned to a Maxwellian distribution at the appropriate temperature.

The structure calculations used an extended β-strand structure for the initial coordinates of the Tn-I peptide. Standard geometry was used to generate the extended β-strand elements using φ,ψ values of 113° and 180°, respectively. This structure was then energy minimized with 500 cycles of unrestrained EM and the energy-minimized β-strand coordinates used as input for the structure determination stage. The structure determination state proceeded in the following phases with the inclusion of the distance restraints derived from the TRNOESY measurements, the output coordinates of which corresponded to the bound Tn-I peptide conformation: 1) 500 cycles of REM (CDIS = 400 KJ/mol/nm) 2) 1.4 ps RMD with intervening REM to cool the structure (CDIS = 400 KJ/mol/nm) 3) another 0.6 ps RMD with intervening REM (CDIS = 800 KJ/mol/nm) 4) a final 6.6 ps RMD with intervening REM (CDIS =1600 KJ/mol/nm) 5) 200 cycles of REM (CDIS =1600 KJ/mol/nm) to anneal the structure.

c) Structure of the Tn-I Peptide Bound to Tn-C

The bound Tn-I peptide conformation is illustrated in Figure 5. Two 'bends' in the structure around residues Gly104-Arg108 and Arg112-Arg115 bring the aromatic benzene ring into close proximity with the aliphatic side chains of Leu111 and Val114, thereby defining a hydrophobic core in the peptide and presenting a hydrophobic face spanned by Pro109-Pro110 capable of interacting with a hydrophobic patch on the surface of Tn-C. The structure contains 4 arginines (Arg108, Arg112, Arg113 and Arg115) and 2 lysines (Lys105 and Lys107), and therefore is

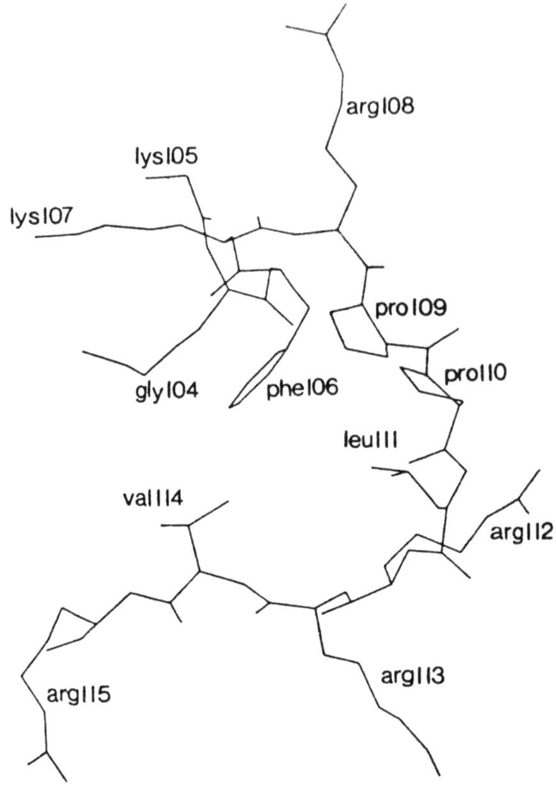

Figure 4 Conformation of Tn-I Peptide Bound to Tn-C (Ca^{+2} saturated)

highly basic. These bulky and charged side chains extend out from the core of the structure and may either interact with the solvent or with acidic residues on Tn-C. (The Helix-Loop-Helix region around Site III on turkey s-Tn-C contains 5 glutamic acids and 4 glutamines[5] : Glu96, Glu97, Gln100, Gln106, Gln110, Gln114, Glu116, Glu117 and Glu120 (23).)

All peptide bonds are trans in the bound structure. The ϕ and ψ angles define a collagen-type structure in dihedral space. There are steric constraints on the conformation of the proline rings. The first proline in the sequence, Pro109, displays a Cβ-endo ring pucker ($\chi^1 = -42^o$, $\phi = -18^o$) , whereas the second proline, Pro110, displays a Cβ-exo ring pucker ($\chi^1 = 34^o$, $\phi = 105^o$).

IV POSITION OF THE TN-I PEPTIDE BOUND TO TNC

The question remains as to the location of binding of the Tn-I peptide spanning residues 104-115 to Tn-C(Ca^{+2})$_4$. Leszyk et al. (8) have used the photoactive reagent 4-maleimido-benzophenone to identify the residues in]Tn-I that are close to Cys98 in Tn-C. They determined that the residues

5. The amino acid numbering corresponds to the chicken s-Tn-C sequence, which was used to fit the electron density map of the crystallized turkey s-Tn-C (12).

in Tn-I that appeared to be most highly cross-linked to Cys98 were Arg108 and Pro110, and to a lesser extent Arg103 and Lys107. These data suggest that the cross-linked residues were within 10Å of Cys98, the length of the extended maleimide cross-linker molecule. This would place the Tn-I peptide on or near the N-terminal side of Ca^{+2}-binding site III.

The use of paramagnetic lanthanides as Ca^{+2} analogs in sites III and IV to determine distances to assigned resonances in the bound Tn-I peptide is being currently studied in the lab. The paramagnetic lanthanide, Gd^{+3}, which contains 7 unpaired 4f electrons, is being used to determine the distances, r_{iM}'s, between individually assigned protons on the peptide and Gd^{+3}-filled sites III and IV on Tn-C. The unpaired 4f electrons on the Gd^{+3} metal increase the spin-lattice relaxation rates ($1/T_{1M}$) of nearby nuclei. Increases in $1/T_{1M}$ have an r^{-6} distance dependence according to the following equation:

$$1/T_{1M} = K \, (r_{iM})^{1/6} \qquad \text{Eqn.11}$$

where r_{iM} is the distance between the Gd^{+3} metal and the nucleus in question, and K is a constant dependent on the correlation time of the internuclear vector. Thus, by measuring increases in $1/T_{1M}$ for assigned nuclei on the Tn-I peptide, one can determine the distance between these specific nuclei and sites III and IV on Tn-C. The results of these experiments in addition to the results of the Leszyk cross-linking studies and possible spin-label attachment to Cys98 should produce a feasible model of the attachment site of Nα-acetyl-Tn-I(104-115)amide to Tn-C.

Acknowledgements The authors thank J. Van Eyk and Dr. R.S. Hodges for providing the peptide for initial stages of the project and for helpful discussions, and N. Strynadka for assistance with the computer graphics.

REFERENCES

1. D.C. LaPorte, B.M. Wierman and D.R. Storm, Biochemistry 19: 3814 (1980)
2. J. Gariepy and R.S. Hodges FEBS Lett. 160:1 (1983)
3. P.J. Cachia, J. Gariepy and R.S. Hodges, in: "Calmodulin Antagonists and Cellular Physiology," Hidaka, H., and Hartshorne, D.J., eds., pp 63-88, Acadmic Press, New York (1985)
4. O. Herzberg and M.N.G. James, CIBA Found. Symp. 122: 120 (1986)
5. S.V. Perry, "The Regulation of Contractile Activity in Muscle," Bioc. Soc. Trans. 7: 573 (1977)
6. D.A. Malencik and S.R. Anderson Biochemistry 23: 2420 (1984)
7. P.J. Cachia, B.D. Sykes and R.S. Hodges Biochemistry 22: 4145 (1983)
8. J. Leszyk, J.H. Collins , P.C. Leavis and T. Tao, Biochemistry 26: 7042 (1987)
9. R.R. Ernst, G. Bodenhausen and A. Wokaun, "Principles of Nuclear Magnetic Resonance in One and Two Dimensions," Clarendon Press, Oxford (1987)
10. K. Wüthrich, "NMR of Proteins and Nucleic Acids," Wiley, New York (1986)

11. W.D. McCubbin, K. Oikawa, B.D. Sykes and C.M. Kay, Biochemistry 21: 5948 (1982)
12. O. Herzberg and M.N.G James, Nature 313: 365(1985)
13. K. Golosinska, M.J. Carpenter and L.B. Smillie, personal communications
14. J.M. Wilkinson, FEBS Lett. 70: 254 (1976)
15. B.W. Erickson and R.B. Merrifield, in: "The Proteins," Neurath, H. and Hill, R.H., eds., VolII, pp225-527, Academic Press, New York (1976)
16. D.J. States, R.A. Haberkorn and D.J. Ruben, J. Magn. Reson. 48: 286 (1982)
17. S. Macura and R.R. Ernst, Molec. Phys. 41: 95 (1980)
18. G.M. Clore and A.M. Gronenborn, J. Magn. Reson. 48:402 (1982)
19. G.M. Clore and A.M. Gronenborn, J. Magn. Reson. 53: 423 (1983)
20. M.P. Williamson and D. Newhaus, J. Magn. Reson. 72:369 (1987)
21. T.J. Swift and R.E. Connick, J. Chem. Phys. 37: 307 (1962)
22. W.F. van Gunsteren and H.J.C. Berendsen, Laboratory of Physical Chemistry, University of Groningen, Nijenborgh16, 9747 Ag Groningen, The Netherlands
23. O. Herzberg and M.N.G. James, Biochemistry 24: 5298 (1985)

COOPERATIVITY IN CALCIUM BINDING AND CALCIUM DEPENDENT REACTIONS

Yasuo Ogawa

Department of Pharmacology
Juntendo University School of Medicine
Bunkyo-ku, Tokyo 113

INTRODUCTION

Among various Ca^{2+}-binding proteins, calmodulin (CaM) and troponin C (Tn C) or troponin (Tn) are well established as Ca^{2+}-receptor proteins which transmit intracellular Ca^{2+}-signaling, resulting in regulation of Ca^{2+}-dependent reactions. While Ca^{2+}-binding to these Ca^{2+}-receptor proteins can be considered non-cooperative, the resulting reactions show steep Ca^{2+}- dependence and high positive cooperativity.[1] This would indicate that those Ca^{2+}-receptor proteins incorporated into the complex might show positive cooperative Ca^{2+}-binding. In this paper, we will discuss whether the non-cooperative Ca^{2+}-binding can explain the steep Ca^{2+}-dependent reactions. Since Ebashi and Ogawa[1] have already explained a relevant matter with the CaM-system in detail, only an outline of that discussion will be given here, and then we will focus on the troponin-system, i.e., the relation between Ca^{2+}-binding and tension by striated muscle fibers. We would like to point out that the Ca^{2+}-signaling is a change of intracellular Ca^{2+} from sub-μM (e.g., 0.1 μM) to μM order <u>in the presence of mM order Mg^{2+}</u> (1 mM is assumed to be the intracellular Mg^{2+} concentration).

CALMODULIN SYSTEM

First of all, let us consider the meaning of the results of Ca^{2+}-binding. As an example, Ca^{2+}-binding to 3 CaM molecules is considered as illustrated in Fig. 1, where there are 12 binding sites as a whole. The binding sites are assumed to be independent and homogeneous. A binding probability of 0.5 means that half the total binding sites are occupied by Ca^{2+}. If three site occupation in a CaM molecule is the requirement for activation, there is no activating CaM molecule ($Ca_n \cdot CaM^*$) in case A. In case B, however, two CaM molecules are activating. In each case from C to E, the number of $Ca_n \cdot CaM^*$ is one. Overall, the probability of activating CaM molecules ($Ca_n \cdot CaM^*$, $n \geq 3$) is calculated to be 0.313 according to equation $\Sigma\, {}_4C_n p^n (1-p)^{4-n}$ where p stands for a binding probability. It should be noted that the probability is not p^3.

Figure 2 shows the calculated concentrations of activating calmodulin ($Ca_n \cdot CaM^*$) in the presence of various Ca^{2+} concentrations of physiological relevance when the total amount of CaM is 100 μM. The affinity of CaM for Ca^{2+} which is adopted in this calculation is $1.2 \times 10^5\ M^{-1}$ which corre-

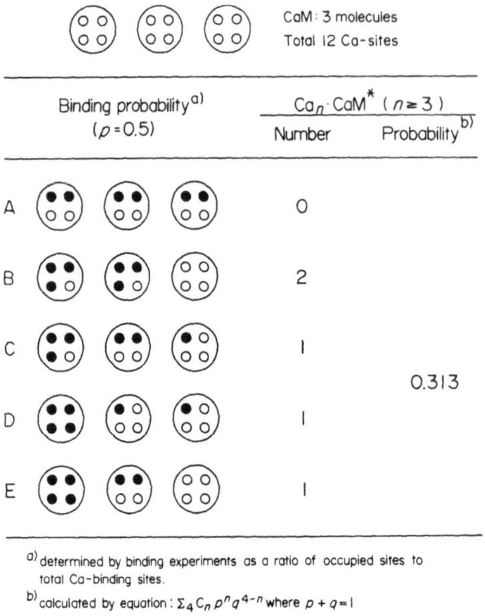

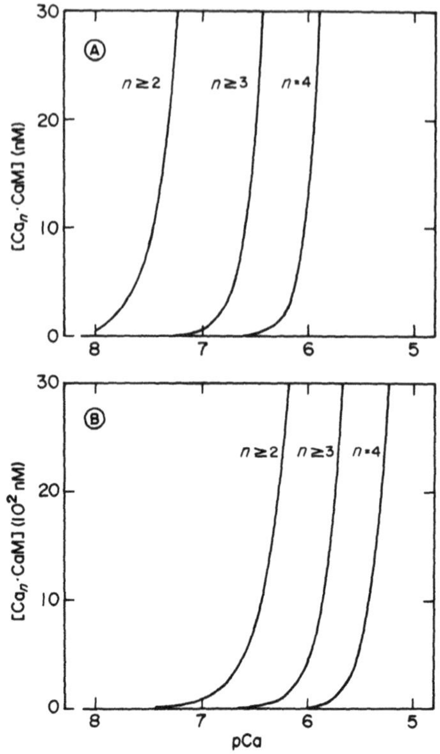

Fig. 1. (Left-hand side) Ca^{2+}-binding to calmodulin and probability of site-occupancy.

Fig. 2. (Right-hand side) Calculated amount of activating CaM ($Ca_n \cdot CaM$) in the presence of 100 μM total CaM. For details, refer to the text.

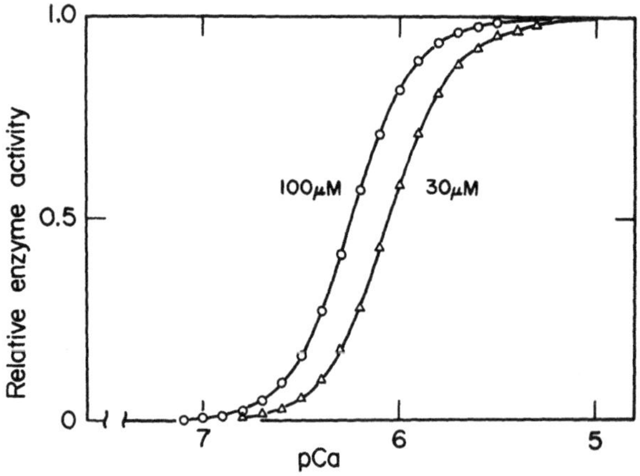

Fig. 3. Calculated relative activity of a CaM-dependent enzyme. circles, 100 μM total CaM; triangles, 30 μM total CaM; $\underline{K}_E = 10^7\ M^{-1}$; $\underline{n} \geq 3$

sponds to the affinity under a condition simulating the cytoplasm. Two site occupation ($\underline{n}\geqq 2$), three site occupation ($\underline{n}\geqq 3$) and full occupation ($\underline{n}=4$), respectively, are assumed to be the requirement for activation. Note difference in the scales of the ordinates between panels A and B.

A CaM-dependent enzyme (Enz) is known to be activated according to the two-step model shown as follows:

$$\underline{n}\ Ca^{2+} + CaM \rightleftharpoons Ca_{\underline{n}} \cdot CaM^{*} \qquad \text{——(1)}$$

$$Ca_{\underline{n}} \cdot CaM^{*} + Enz \rightleftharpoons Ca_{\underline{n}} \cdot CaM^{*} \cdot Enz^{\S} \qquad \text{——(2)}$$

$$\frac{[\ Ca_{\underline{n}} \cdot CaM^{*} \cdot Enz^{\S}\ ]}{[\ Ca_{\underline{n}} \cdot CaM^{*}\ ][\ Enz\ ]} = \underline{K}_E$$

$$\frac{[\ Ca_{\underline{n}} \cdot CaM^{*} \cdot Enz^{\S}\ ]}{\text{Total Enzyme}} = \frac{\underline{K}_E \cdot [\ Ca_{\underline{n}} \cdot CaM^{*}\ ]}{1 + \underline{K}_E \cdot [\ Ca_{\underline{n}} \cdot CaM^{*}\ ]} \qquad \text{——(3)}$$

where $\underline{K}_E$ represents the affinity for activating CaM ($Ca_n \cdot CaM^{*}$) of a target enzyme (Enz), and $Ca_n \cdot CaM^{*} \cdot Enz^{\S}$ stands for the activated enzyme. The intracellular concentration of CaM amounts to as much as 60 μM in such organs of high content as brain and testis, while the concentration of a target enzyme is of a micromolar order.[2] Therefore, the equilibrium of equation(1) is little interfered with by the presence of a target enzyme (equation(2)). For the first approximation, we can use the curves shown in Fig. 2 for [$Ca_n \cdot CaM^{*}$] in equation(3); then the fraction of the activated enzyme can be calculated. The values for $\underline{K}_E$ are reported to be 10^{7} - 10^{10} M^{-1}.[1,2] Figure 3 shows an example of such calculations in the presence of 30 and 100 μM CaM under the assumption of $\underline{K}_E=10^{7}$ M^{-1} and $\underline{n}\geqq 3$. Hill coefficients for these curves are about 2.8. As shown, this explanation means that an apparent affinity for Ca^{2+} of a CaM-activated reaction should depend upon the concentration of CaM used in the experiment. As the amount of CaM decreases, the curve is shifted to the right along the abscissa of pCa. This has already been reported with myosin light chain kinase activity by Blumenthal and Stull[3], with Ca^{2+}-induced tension by skinned smooth muscle by Sparrow et al.[4] and with other systems (for details, see Ref. 1). The finding[5] that most CaM in contracted smooth muscle is free of Ca^{2+} is also consistent with our explanation. The value of $\underline{K}_E$ is also important, and will determine the range of Ca^{2+} where the target enzyme is activated. Panels A and B in Fig.2 can be useful if $\underline{K}_E$ is in the range of 10^{9}-10^{8} M^{-1} and 10^{7}-10^{6} M^{-1}, respectively. It is not determined whether $\underline{n}$ should be fixed for a target enzyme or is variable for each one. Even if fixed, the value is controversial among investigators.

It is of interest to consider the deactivation process in relation to the explanation for activation. In addition to the reverse reaction of equation(2), Ca^{2+} release from the complex is also possible :

$$Ca_{\underline{n}} \cdot CaM^{*} \cdot Enz^{\S} \rightleftharpoons Ca^{2+} + Ca_{\underline{n-1}} \cdot CaM \cdot Enz \qquad \text{——(4)}$$

The CaM-Enz complex generally has a higher affinity for Ca^{2+} than uncomplexed free CaM, the factor being usually about 7 to 26.[1] The association rate constant of free CaM for Ca^{2+} is about 10^{7} M^{-1} s^{-1} per Ca^{2+}-binding site.[6] Trifluoperazine increased affinity for Ca^{2+} of CaM, mainly by reducing the dissociation rate constant;[6] this may also be the case with a natural target enzyme. Then the dissociation rate constant of Ca^{2+} in equation(4) can be estimated to be at least 1 - 2 s^{-1}. On the other hand, $\underline{K}_E$ is reported to be 10^{7} - 10^{10} M^{-1}. Association rate constants for diffusion-limited reactions are reported to be 10^{6} - 10^{9} M^{-1} s^{-1}; therefore, a dissociation rate constant for equation(2) might be as small as 10^{-4} s^{-1}. Wang et al.[7] reported that a dissociation rate constant of equation(2) for phosphodiesterase was $3 \times 10^{-3} s^{-1}$, the association rate constant being $4 \times 10^{6}\ M^{-1}s^{-1}$. The dissociation rate constant is too small to explain the rate of inhibition of the enzyme activity upon addition of EGTA which was reported by Klee and Haiech.[8] The latter may rather be consistent with the result by Johnson et al.[9] that the fluorescence increase produced by calmodulin binding to myosin light chain kinase was completely reversed at $2\ s^{-1}$ by EGTA, which would probably correspond to the value for equation(4).

TROPONIN SYSTEM

It is well known that both troponin C and troponin have two independent classes of Ca^{2+}-binding sites, namely high and low affinity sites. Each class is also composed of two independent and homogeneous binding sites. The affinities for Ca^{2+} and Mg^{2+} shown in Table 1 were determined in our laboratory,[10] and are similar to the results of other investigators.[11,12] Concerning Ca^{2+}-binding to myofibrils, on the other hand, various results have been reported. The most recent,[13,14] however, indicate that Ca^{2+}-binding to myofibrils is similar to that to troponin, at least under a condition free of ATP. The relative amounts of Ca bound to the thin filaments of skinned fibers which were determined by means of electron probe microanalysis[15] (squares in Fig. 4) fall on the Ca^{2+}-binding probability curve for troponin (smooth curve in Fig. 4) in the presence of 1 mM Mg^{2+} under an ionic strength similar to the myoplasmic one (1.19×10^{7} and $2.43 \times 10^{6}\ M^{-1}$ are assumed for apparent affinities for high and low affinity sites, respectively). In contrast, Ca^{2+}-induced tensions[16] under similar conditions show much steeper Ca^{2+}-dependence as shown in Fig. 4. Our concern is to what extent the steep tension-pCa relationship can be explained by the Ca^{2+}-binding curve.

As in the case of CaM, the probability of activating Tn molecules can be calculated on the basis of the Ca^{2+}-binding curve ("(H+L)/2" in Fig. 5A) if the requirement is specified. The curve labeled "(H*L)" stands for the case where the requirement for activation is simultaneous one site occupation of both the high and low affinity sites. The curve (H2*L) corresponds to the case where full site occupation of high affinity sites and one site occupation of low affinity sites are required. The curve (H2*L2) represents the case where all Ca sites of Tn are saturated. The probability curve for each case is that for any one of the Tn molecules which are aligned on a thin filament at regular intervals of 38 nm.[17] In the presence of a saturating Ca^{2+} concentration which fully activates thin filaments, the maximum tension is known to be proportional to the length of the thin filament overlapped by the thick filament.[18] The following discussion will concern thin filament activation in the presence of intermediate Ca^{2+} concentrations.

At the one extreme where the interaction between the adjacent troponin-tropomyosin complexes is so strong that the activation of any one Tn molecule on a thin filament is sufficient for the activation of its entire length, then the probability of activation of the thin filament corresponds to that of a single Tn molecule (i.e., each curve in Fig. 5A). In contrast, at the other extreme where each troponin-tropomyosin complex is entirely independent and where all the n troponin molecules in the overlapped region on the thin filament must be activated, then the probability of activation for the thin filament would be the nth power of that for a single Tn molecule. If the interaction is intermediate in strength, the power would be a value between 1 and n. Recently, Kishino and Yanagida[19] reported results which indicate very weak interaction, if any, between adjacent tropomyosin molecules. Figure 5B-D shows the probability curves for the thin filament activation in correspondence to the requirements for Tn activation, (H*L), (H2*L) and (H2*L2), respectively. The thin filament activation curve is shifted to the right along the abscissa of pCa and becomes steeper as the power increases. The numerical increase of the

Table 1. Ca^{2+}- and Mg^{2+}-binding properties of skeletal muscle troponin C and troponin

	High affinity site		Low affinity site	
	K_{Ca} (M^{-1})	K_{Mg} (M^{-1})	K_{Ca} (M^{-1})	K_{Mg} (M^{-1})
Troponin C	4.5×10^6	1×10^3	6.4×10^4	520
Troponin	2.5×10^8	2×10^4	3.5×10^6	440

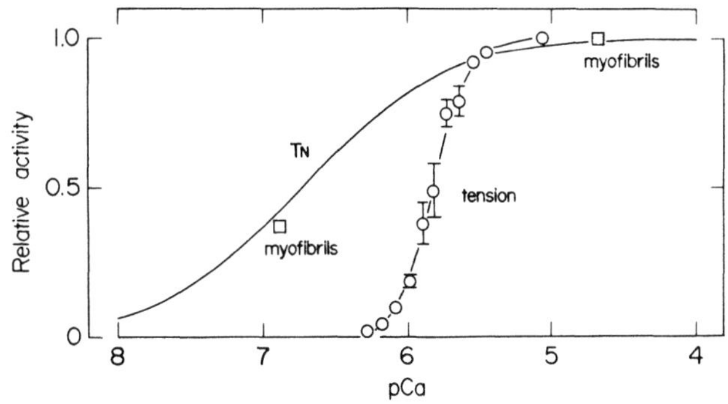

Fig. 4. Ca^{2+} binding curve to troponin and tension-pCa relationship. The smooth curve shows Ca^{2+}-binding to Tn. Squares represent relative amounts of Ca^{2+} bound to thin filaments of skinned fibers.[12] Tension-pCa curve was adopted from Kurebayashi and Ogawa.[13]

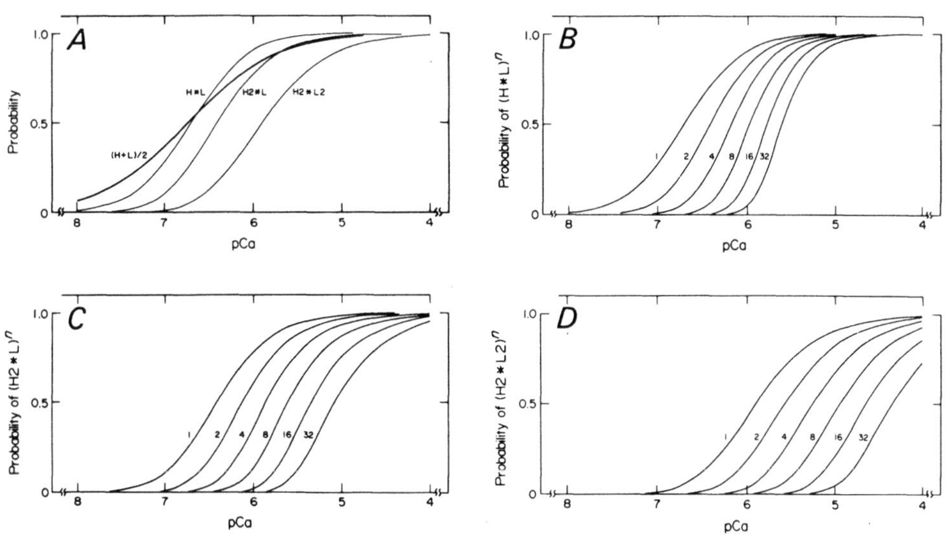

Fig. 5. A. Ca^{2+} binding curve to Tn and probability curves for Tn activation. B-D. Probability curves for thin filament activation when the requirements for Tn activation are (H∗L), (H2∗L) and (H2∗L2), respectively. For details, see the text.

power stands for not only weaker interaction between the troponin-tropomyosin complexes at a constant sarcomere length, but also a shorter sarcomere length under no (or constant) interaction. Each counterpart of the curves in panels B-D is located in a higher Ca^{2+} concentration range and becomes less steep moving from panel B to D. In respect to the steepness of a curve, the (H*L) model corresponding to panel B is preferable. The actual Ca^{2+}-induced tensions by guinea pig psoas muscle fibers (glycerol-treated followed by chemically skinned) at a sarcomere length of 2.8 μm are compared with the curve calculated to the model (H*L) with 12 for n, because 12 Tn molecules in the linear alignment on the thin filament would be overlapped by the thick filament at the sarcomere length of 2.8 μm[17] (Fig. 6). Good agreement was obtained as shown in Fig. 6. The experimental results are consistent with the indication of no interaction between adjacent troponin-tropomyosin complexes. It should be pointed out that the model calculation with n=12 shows a curvilinear Hill plot similar to that reported by Moss et al.[20] (Fig. 6B). If the results up to 0.75 of the maximum tension were adopted, an apparent slope of the Hill plot would be about 3.8. If the results near and above half the maximum tension were adopted, on the other hand, it would be about 2.0. The model calculation with a smaller value for n approaches a rectilinear Hill plot with a shallower slope (a limiting value of 1.4).

As shown in Fig. 5B, a probability curve of thin filament activation is shifted to the left along the abscissa of pCa as the value for n decreases. This may indicate that a tension-pCa curve depends on the sarcomere length on which determinations are carried out. In Fig. 7, the pCa value which would give a probability of 0.5 (pCa_{50}) is plotted against the sarcomere length calculated on the basis of dimensions of the Tn arrangement. The change in pCa_{50} would not be linear to the change in the sarcomere length. As summarized in Table 2, many investigators have reported a leftward shift of tension-pCa curve with a longer sarcomere length using skeletal and cardiac muscle fibers. The claim of a larger shift for cardiac and slow-twitch skeletal muscle[22,25] may be ascribable to different properties of Ca^{2+}-binding of their Tns.[1,11] These results are consistent with the value expected from our model calculation presented here. However, it must be pointed out that Endo[26] and Moss et al.[20] reported different results. They showed that the increase in Ca^{2+}-sensitivity was obvious only at a low Ca^{2+} concentration range where tensions were far below the maximum. We would like to point out that the maximum Ca^{2+}-induced tension is reduced with a longer sarcomere length and that the normalized tension-pCa curve in the descending limb of the tension-sarcomere length relationship may be subjected to a greater variation.

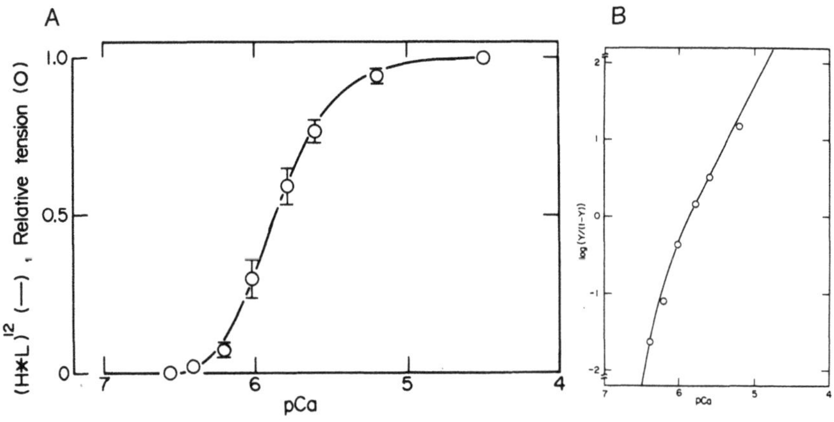

Fig. 6. Tension-pCa relation (A) and Hill plot (B) for guinea pig psoas fibers coincide with model calculation according to $(H{*}L)^{12}$. sarcomere length, 2.8 μm.

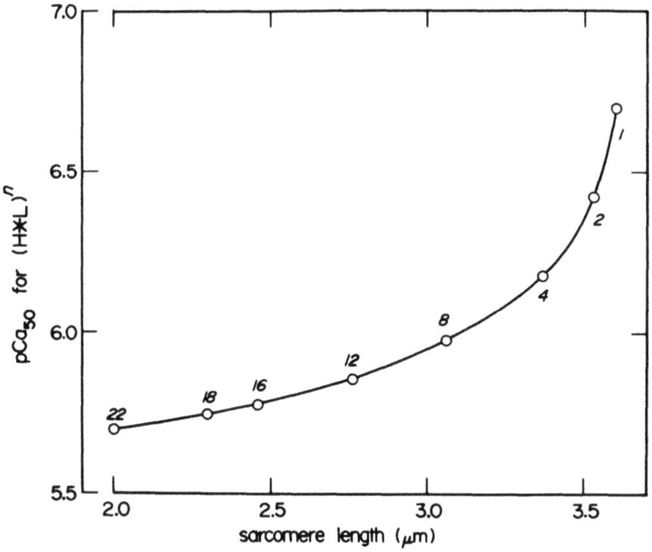

Fig. 7. Sarcomere length-pCa_{50} relationship deduced from Fig. 5B. Italicized numerals represent n, which, in turn, corresponds to sarcomere length.

Table 2. Experimental results which report a leftward shift of the tension-pCa curve with longer sarcomere length

Material	Change in sarcomere length (μm)	ΔpCa_{50}	Reference
Frog skeletal muscle	2.2→2.6 2.6→2.8	0.1 0	21
Rat fast-twitch	2.3→2.8	0.1	22
Rat slow-twitch	2.3→2.8	0.2	22
Frog ventricle	2.3→2.7 2.7→3.1	0.17 0.33	23
Rat ventricle	2.0→2.4	0.21	24
Hamster ventricle	1.9→2.4	0.13	25
This model	2.3→2.8 2.5→3.1	0.12 0.21	

In the experiments where Ca^{2+}-induced tension and myofibrillar ATPase activity were simultaneously determined, the normalized ATPase activity-pCa curve was located in a lower Ca^{2+} concentration range and was less steep than the normalized tension-pCa curve. The dissociation between the two activities is obvious in the Ca^{2+} concentration range which gives less than half the maximum tension.[27] The result is similar to that of Levy _et al_.[28] and is not incompatible with a recent report by Brenner.[29] In the model explaining Ca^{2+}-dependence of tension development, we take into account only thin filament activation, which may be an oversimplification. We also assume that all Tn molecules in the overlapped region on the thin filament must be activated for tension development; for ATPase activity, however, this may not be necessary. ATPase activity may be graduated, depending on the number of activated Tn molecules in the overlapped region on the thin filament. This would suggest that the presence of only one non-activating Tn molecule in the overlapped region on the thin filament may interact with myosin, sufficiently preventing the development of tension. This would be consistent with the role of Ca^{2+} in the troponin system being de-repression, i.e. removal of inhibition due to troponin I through complex formation of Tn C and Ca^{2+}. However, such molecular interactions between thick and thin filaments remains to be demonstrated.

CONCLUSION

A steep pCa-activity relationship does not necessarily mean positive cooperative Ca^{2+}-binding to a Ca^{2+}-receptor protein. Although further evidence is required for confirmation of the explanations proposed here, there are some viewpoints which should be taken into account: an excess amount of CaM over its target enzymes and the high affinities of these enzymes for $Ca_n \cdot CaM^*$ for the CaM-system, and the regular alignment of independent Tn molecules (or troponin-tropomyosin complex) on the thin filament for the Tn-system, respectively.

ACKNOWLEDGEMENTS

The author thanks Dr. N. Kurebayashi for allowing him to use her unpublished results of tension determinations shown in Fig. 6.

REFERENCES

1. S. Ebashi and Y. Ogawa, Troponin C and calmodulin as calcium receptors: mode of action and sensitivity to drugs, _in_: "Handbook of Experimental Pharmacology, Vol.83. Calcium in Drug Actions," P.F. Baker, ed. Springer-Verlag, Berlin, Heidelberg, New York, London, Paris and Tokyo (1988)
2. C. B. Klee and T. C. Vanaman, Calmodulin, Adv. Protein Chem. 35 : 213 (1982)
3. D. K. Blumenthal and J. T. Stull, Activation of skeletal muscle myosin light chain kinase by calcium (2+) and calmodulin, _Biochemistry_ 19 : 5608 (1980)
4. M. P. Sparrow, U. Mrwa, F. Hofmann and J. C. Ruegg, Calmodulin is essential for smooth muscle contraction, _FEBS Letters_ 125 : 141 (1981)
5. J. C. Ruegg, G. Pfitzer, M. Zimmer and F. Hoffman, The calmodulin fraction responsible for contraction in an intestinal smooth muscle, _FEBS Letters_ 170 : 383 (1984)

6. Y. Ogawa and M. Tanokura, Calcium binding to calmodulin and its modification by calmodulin-ligands in view of the regulatory role in vivo of the calmodulin system, in: "Cellular Regulation and Malignant Growth," S. Ebashi, ed. Japan Sci. Soc. Press, Tokyo / Springer-Verlag, Berlin, Heidelberg, New York, Tokyo (1985)
7. J. H. Wang, R. K. Sharma, C. Y. Huang, V. Chau and P. B. Chock, On the mechanism of activation of cyclic nucleotidephosphodiesterase by calmodulin, Ann. N. Y. Acad. Sci. 356 : 190 (1980)
8. C. B. Klee and J. Haiech, Concerted role of calmodulin and calcineurin in calcium regulation, Ann. N. Y. Acad. Sci. 356 : 43 (1980)
9. J. D. Johnson, M. J. Holroyde, T. H. Crouch, R. J. Solaro and J. D. Potter, Fluorescence studies of the interaction of calmodulin with myosin light chain kinase, J. Biol. Chem. 256 : 12194 (1981)
10. Y. Ogawa, Calcium binding to troponin C and troponin: effect of Mg^{2+}, ionic strength and pH, J. Biochem. 97 : 1011 (1985)
11. J. D. Potter and J. D. Johnson, Troponin, in: "Calcium and Cell Function, Vol.2," W.Y. Cheung, ed. Academic Press, New York (1982)
12. A. S. Zot and J. D. Potter, The effect of $[Mg^{2+}]$ on the Ca^{2+} dependence of ATPase and tension development of fast skeletal muscle: the role of the Ca^{2+}-specific sites of troponin C, J. Biol. Chem. 262 : 1966 (1987)
13. H. G. Zot and J. D. Potter, Calcium binding and fluorescence measurements of dansylaziridine-labelled troponin C in reconstituted thin filaments, J. Muscle Res. Cell Motil. 8 : 428 (1987)
14. K. Guth and J. D. Potter, Effect of rigor and cycling cross-bridges on the structure of troponin C and on the Ca^{2+} affinity of the Ca^{2+}-specific regulatory sites in skinned rabbit psoas fibers, J. Biol. Chem. 262 : 13627 (1987)
15. T. Kitazawa, H. Shuman and A. P. Somlyo, Calcium and magnesium binding to thin and thick filaments in skinned muscle fibres: electron probe analysis, J. Muscle. Res. Cell Motil. 3 : 437 (1982)
16. N. Kurebayashi and Y. Ogawa, Increase by trifluoperazine in calcium sensitivity of myofibrils in a skinned fibre from frog skeletal muscle, J. Physiol. 403 : 407 (1988)
17. I. Ohtsuki, Localization of troponin in thin filament and tropomyosin paracrystal, J. Biochem. 75 : 753 (1974)
18. A. M. Gordon, A. F. Huxley and F. J. Julian, The variation in isometric tension with sarcomere length in vertebrate muscle fibres, J. Physiol. 184 : 170 (1966)
19. A. Kishino and T. Yanagida, Force measurements by micromanipulation of a single actin filament by glass needles, Nature 334 : 74 (1988)
20. R. L. Moss, A. E. Swinford and M. L. Greaser, Alterations in the Ca^{2+} sensitivity of tension development by single skeletal muscle fibres at stretched length, Biophys. J. 43 : 115 (1983)
21. D. G. Moisescu and R. Thieleczek, Sarcomere length effects on the Sr^{2+}- and Ca^{2+}-activation curves in skinned frog muscle fibres, Biochim. Biophys. Acta 546 : 64 (1979)
22. D. G. Stephenson and D. A. Williams, Effects of sarcomere length on the force-pCa relation in fast- and slow-twitch skinned muscle fibres from the rat, J. Physiol. 333 : 637 (1982)
23. A. Fabiato and F. Fabiato, Myofilament-generated tension oscillations during partial calcium activation and activation dependence of the sarcomere length-tension relation of skinned cardiac cells, J. Gen. Physiol. 72 : 667 (1978)
24. M. G. Hibberd and B. R. Jewell, Calcium- and length-dependent force production in rat ventricular muscle, J. Physiol, 329 : 527 (1982)
25. A. Babu, E. Sonnenblick and J. Gulati, Molecular basis for the influence of muscle length on myocardial performance, Science 240 : 74 (1988)
26. M. Endo, Stretch-induced increase in activation of skinned muscle fibres by calcium, Nature New Biol. 237 : 211 (1972)
27. Y. Ogawa, N. Kurebayashi, Modulation by drugs of the relationship

between calcium binding to troponin C and tension, in: "Muscle Energetics," R. Paul, K. Yamada, G. Elzinga, eds., Alan R. Liss, Inc., New York (1988)
28. R. M. Levy, Y. Umazume and M. J. Kushmerick, Ca^{2+} dependence of tension and ADP production in segments of chemically skinned muscle fibres, Biochim. Biophys. Acta 430 : 352 (1976)
29. B. Brenner, Effect of Ca^{2+} on cross-bridge turnover kinetics in skinned single rabbit psoas fibers: implications for regulation of muscle contraction, Proc. Natl. Acad. Sci. USA 85 : 3265 (1988)

STRUCTURE AND FUNCTION OF PARVALBUMIN

Claus W. Heizmann and Urs Kägi

Department of Pediatrics, Division of Clinical Chemistry, University of Zürich, Steinwiesstr. 75 CH-8032 Zürich, Switzerland

INTRODUCTION

Parvalbumins were first isolated from muscles of lower vertebrates and their primary structures and biochemical properties have been investigated intensively (for review see 1). Later, parvalbumins were also detected in skeletal muscles of higher vertebrates, and the rabbit and rat parvalbumin have been sequenced (for reviews see 2-4). Fish fast muscles contain up to five different iso-parvalbumins, whereas in mammalian muscles only one parvalbumin has been detected so far.

Recently the structural organization of the rat parvalbumin gene has been elucidated[5]. The rat parvalbumin transcription unit is 15.5 kilobases in length and contains five exons interrupted by four introns. The last three exons encode most of the protein and are similar in size and organization to the coding region of genes encoding four-domain calcium-binding proteins (such as calmodulin and myosin light chains). From these data it was concluded that parvalbumin gene would maintain the structural pattern of the four domain calcium-binding protein gene family though it would exhibit deletion of one domain.

The single copy parvalbumin gene has been assigned to human chromosome 22, from pter to q11.2. This segment is deleted in some patients with DiGeorge syndrome, a metachromatic leukodystrophy. The parvalbumin gene may therefore serve as an useful marker in this immunodeficiency disease. Recently a loss of genes on chromosome 22 has been reported in tumorigenesis of human acoustic neuroma[6]. A loss of regions of chromosome 22 - possibly containing a tumor suppressor gene - is associated with several tumor types, including Schwannomas, neurofibromas, meningiomas and astrocytomas[7].

A human parvalbumin clone was isolated from a chromosome 22 specific library[5,8]. The coding region of human parvalbumin has 92% sequence homology to the rat protein, 9 conservative amino acid replacements were predicted from exon sequencing. A part of the human promotor (-79 to -110) is identical to the rat promotor. This region has a high homology to the promotor of the gene for myosin light chain 3F. Since both genes are expressed in fast-twitch (type II b) muscle fibers, it was suggested that these promotor regions may be important for their tissue and cell specific expression.

In the last few years we have focussed our interest on the distribution and possible roles of parvalbumin and other Ca^{2+}-binding proteins in the brain of various mammalian species including man[3,4]. Many processes in the brain depend on calcium and various Ca^{2+}-binding proteins such as calbindin D-28K, calcineurin, S-100 proteins, calretinin and novel Ca^{2+}-binding proteins have been detected in this tissue[4].

Impairment of neuronal Ca^{2+}-homeostasis has been implicated in several degenerative disorders such as Parkinson and Alzheimer diseases. For example, a significant loss of parvalbumin- immunoreactive neurons was found in the frontal and temporal cortex of Alzheimer type dementia patients. This finding suggested that parvalbumin-positive neurons in the cortex may be especially affected in Alzheimer type dementia[9]. Preliminary data also indicate that the distribution of parvalbumin in the human brain is very similar to that in rat brain. This is different from skeletal muscle where much higher concentrations of parvalbumin were found in fast twitch muscles of small mammals (correlated with its higher contraction/relaxation speed) compared to human muscle. The immunocytochemical distribution of parvalbumin has been described recently[3,4] and compared with the distribution of calbindin D-28k, calmodulin, calcineurin, calretinin and S-100 proteins in selected systems and areas of mammalian and avian brains.

Parvalbumin was found to be a marker for a distinct subpopulation of neurons in the central nervous system of the rat, cat, zebra finch, monkey and man[4].

In the cerebellum, Purkinje cells are immunoreactive for parvalbumin, calbindin D-28K, calcineurin and calmodulin. The presence of parvalbumin[10] and calbindin D-28K[11] -mRNA's in Purkinje cells has also been demonstrated by in situ hybridization techniques. The immuno- and mRNA signals in the cerebellum remained unchanged in rachitic animals indicating that the expression of both proteins is independent of the vitamin D status of the animal. Parvalbumin is so far the only Ca^{2+}-binding protein which is also found in cerebellar basket and stellate cells. The function of all these Ca^{2+}-binding proteins in Purkinje cells is unknown. Purkinje cells exhibit characteristic Ca^{2+}-spikes, which may spread over large portions of the dendritic tree or can be restricted to small segments of dentritic arborisations and spines. Parvalbumin may be an intracellular Ca^{2+}-buffer in such cells with such specialized electrophysiological properties[4].

The mammalian hippocampus contains several neuronal populations which are characterized by their immunoreactivity for parvalbumin, calbindin -28K, calcineurin and calmodulin. Parvalbumin-positive neurons are concentrated in the granule cell layer and hilus of the gyrus dendatus and in the stratum pyramidale and oriens in the CA_1 and CA_3 regions of the hippocampus and include non-pyramidal and basket cells[12]. In the hippocampus, like in other areas of the brain, a colocalization of parvalbumin and the inhibitory neurotransmitter γ-aminobutyric acid (GABA) or glutamic acid decarboxylase (GAD) immunoreactivity, respectively, has been found. While almost all parvalbumin-positive neurons were also immunoreactive to antiserum against GAD, not all GAD-positive neurons contained parvalbumin (1-50% depending on their location)[12].

In mammals, a colocalization of parvalbumin and GABA-immunoreactivity has been found in the visual cortex[13], in the dorsal lateral geniculate nucleus[14] and in horizontal cells of the retina[15]. There are a few exceptions, however, where parvalbumin-positive neurons do not exhibit GABA-immunoreactivity, e.g. the nucleus rotundus of zebra finches[16] and the ganglion cells in rat retina[15]. This suggests that parvalbumin may not be directly involved in the metabolism of GABA.

Generally, parvalbumin is associated with neurons that have a high firing rate and a high oxidative metabolism. The presence of parvalbumin in fast spiking cells in rat hippocampus (CA_1 region) has been directly demonstrated by intracellular injection of Lucifer yellow *in vitro* in combination with postembedding parvalbumin immunohistochemistry[17].

Many of the GABAergic and parvalbumin-containing interneurons in brain tend to respond to various kinds of different stimulation with burst discharges of high frequency. These cells have to cope with transient but large surges of intracellular calcium. Parvalbumin might thus play an important role in supressing these transient rises of calcium during repetitive firing and blocking the adaptation due to a calcium-activated potassium conductance in fast spiking cells[14,17].

An interesting observation has been made by Katharina Braun (TH-Darmstadt, FRG)[4]. Male song birds such as zebra finches and canaries have specialized brain areas which are involved in early song learning. These nuclei undergo dramatic morphological alterations during development and related to song learning. Nuclei of the auditory, visual and vocal motor systems are characterized by a high density of parvalbumin- and calbindin D-28K-immunoreactivity correlated with a high concentration of ligands for voltage dependent Ca^{2+}-channels (Jodipine and desmethoxyverapamil)[4]. This correlation especially in those areas involved in learning processes may suggest a possible interaction of Ca^{2+}-channels and intracellular Ca^{2+}-binding proteins. This interaction may not only regulate fast changes of neuronal activity (as described above) but also be involved in various long term alterations of cell morphology which accompany learning and neuronal functions[4].

Parvalbumin-immunoreactive neurons in the hippocampal CA_1 region have been shown to belong to the class of fast-spiking neurons[17]. In addition, the number of parvalbumin-positive neurons increases about 20% after kindling stimulation[18]. These results imply that changes in parvalbumin-immunoreactivity are related to seizure activity rather than to a long-term increase in seizure sensitivity in kindled animals. It is known that a large influx of Ca^{2+} plays a crucial role in epileptogenesis. The increase of intracellular parvalbumin-content might be a protective reaction of the neurons to cope with the massive Ca^{2+}-influx produced by kindling stimulation and therefore prevent Ca^{2+}-overload.

PARVALBUMIN IN ENDOCRINE GLANDS

Calcium is involved in the production and secretion of hormones in endocrine glands. Parvalbumin-immunoreactivity has been detected in the adrenal gland and ovary, but so far a detailed investigation has only been performed in rat testis and more recently in human testis (Fig 1). Parvalbumin was isolated from rat testis and shown to be indistinguishable from its muscle and brain counterparts. The expression of parvalbumin during rat testis development was followed by RIA, Northern blot analysis using a parvalbumin cDNA clone and by immunohistochemistry[19]. High parvalbumin- immunoreactivity in the Leydig cells was found only during the time of highest testosterone production, e.g. at late fetal stages until birth, and again starting with puberty around postnatal day 21 to adulthood. At stages of low Leydig cell activity, parvalbumin levels also remained below detection level. These data are in agreement with our general observation that the concentration of parvalbumin seems to be correlated with a high cellular activity.

To investigate further a possible involvement of parvalbumin in the Ca^{2+}-dependent production of testosterone in Leydig cells we performed the following *in vivo* experiments[20]:

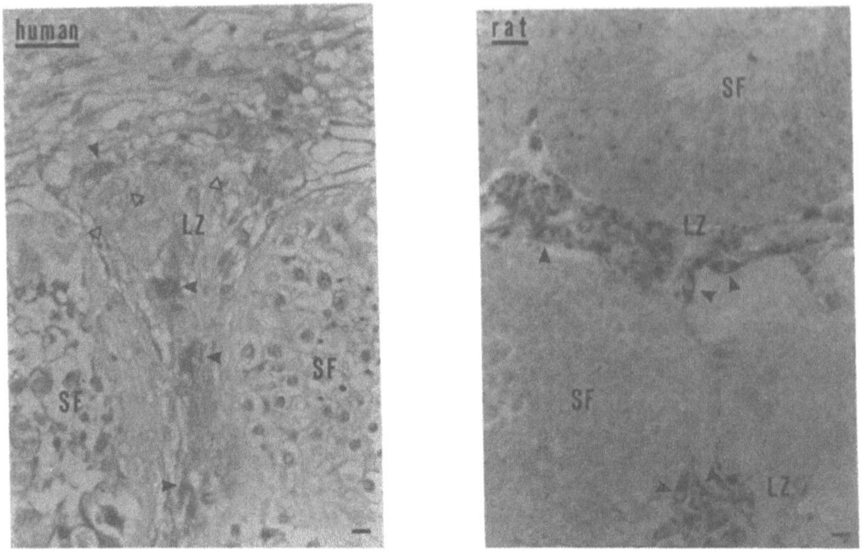

Fig. 1 Localization of parvalbumin in the testis of a 20 year-old man and a 55 day-old rat. Parvalbumin-immunoreactivity was seen in all Leydig cells in the rat, but only in some of them in man. Staining was performed by the peroxidase-anti-peroxidase (PAP)-technique.
LZ, Leydig cells (▶, stained; ▷, unstained); SF, seminiferous tubules; Bar 10 µm (magnification x490).

(1) Rats were hypophysectomized at mid-puberty (day 30-35), when Leydig cells were strongly parvalbumin-immunoreactive and able to synthesize testosterone. Parvalbumin-immunoreactivity in Leydig cells and hormone levels in the serum (measured by RIA) were followed after administration of FSH or LH or both in combination (Fig 2A).

(2) Adult rats were treated with a single dose of ethylene dimethane sulfonate (EDS). This drug selectively destroys the active Leydig cell population, leading to a decrease of hormone levels in the serum[21]. Parvalbumin-reexpression and resynthesis of testosterone in Leydig cells were followed during the regeneration period (Fig 2B). In this experiment, in contrast to those illustrated in Fig 2A, the pituitary gland remained intact and its direct or indirect hormonal effects on the testis were maintained.

Parvalbumin-immunostaining in Leydig cells completely vanished after hypophysectomy at mid-puberty (Fig 2A) and testosterone levels in sera dropped from approx 4nM to undetectable. The number of Leydig cells remained constant. Administration of either FSH or LH restored neither parvalbumin-immunoreaction in Leydig cells nor hormone levels in sera. However, injection of FSH for one week followed by FSH in combination with LH for a second week resulted in the reappearance of parvalbumin-immunoreaction in most Leydig cells and of testosterone in the serum (Fig 2A). This suggests that these hormones, together with other intratesticular factors, may influence the expression of parvalbumin in Leydig cells. The close correlation of parvalbumin expression in Leydig cells with hormone levels in the serum supports the view that parvalbumin may play a role in the Ca^{2+}-dependent production of testosterone in Leydig cells.

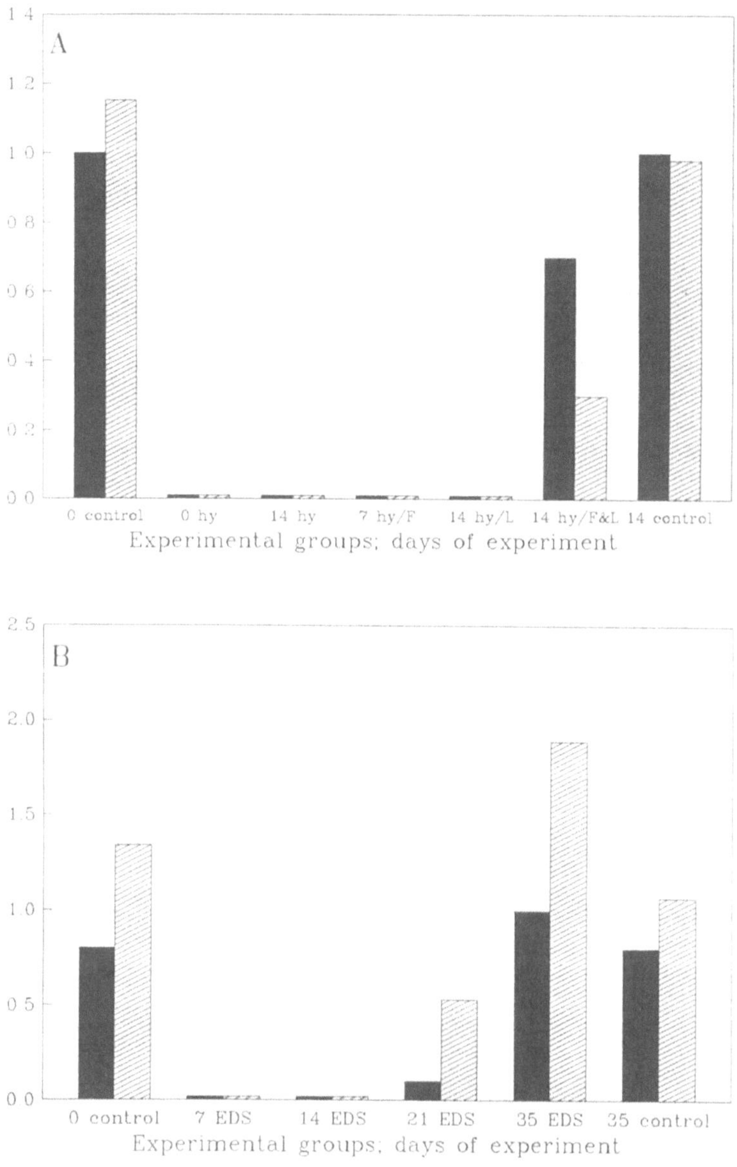

Fig. 2 Parvalbumin-staining in Leydig cells correlates with testosterone levels in sera. A, hypophysectomy followed by treatment with FSH or LH or both in combination. B, treatment with EDS.
parvalbumin-immunoreactivity (x100%); testosterone levels (ng/ml). Values are standarized to one and represent means of 5 animals per experimental group.
hy, hypophysectomized; F, FSH; L, LH. EDS, ethylene dimethane sulfonate.

This is further demonstrated by the EDS experiment (Fig 2B), where during the regeneration period and the appearance of new Leydig cells, a parvalbumin-signal was only found in cells of the mature phenotype.

Parvalbumin-immunoreactivity was also found in the seminiferous tubules[22] but only in maturing spermatids and not in Sertoli cells (Fig 3). In maturing spermatids, from rats aged 26 days postnatal and older, parvalbumin-labelling was observed at sites where the acrosome is formed (spermatid stages 1 to 15). This suggests that parvalbumin may be needed in certain processes of spermatogenesis.

Table 1. Immunohistochemical localization of parvalbumin with the location of calbindin D-28K, S-100 proteins and calmodulin in seminiferous tubules during testicular development.
FG, foetal spermatogonia; S, spermatids; SC, spermatocytes; acro, acrosome region in spermatids.

Age	cell type	Par-valbumin	Cal-bindin	S-100	Cal modulin
−3 d	FG	−	−	+ + some	−
14 d	SC	−	−	+ + some	+ + + some
21 d	SC	−	−	−	+ + +
26 d	SC	−	+ + +	−	+ + +
	S	+ + + acro	−	−	+ + +
30 d	SC	−	+ + +	−	+ + +
	S	+ + + acro	−	−	+ + +
45 d	SC	−	+ + +	−	+ + +
	S	+ + + acro	−	−	+ + +
50 d	SC	−	+ + +	−	+ + +
	S	+ + + acro	−	−	+ + +
60 d	SC	−	+ + +	−	+ + +
	S	+ + + acro	−	−	+ + +

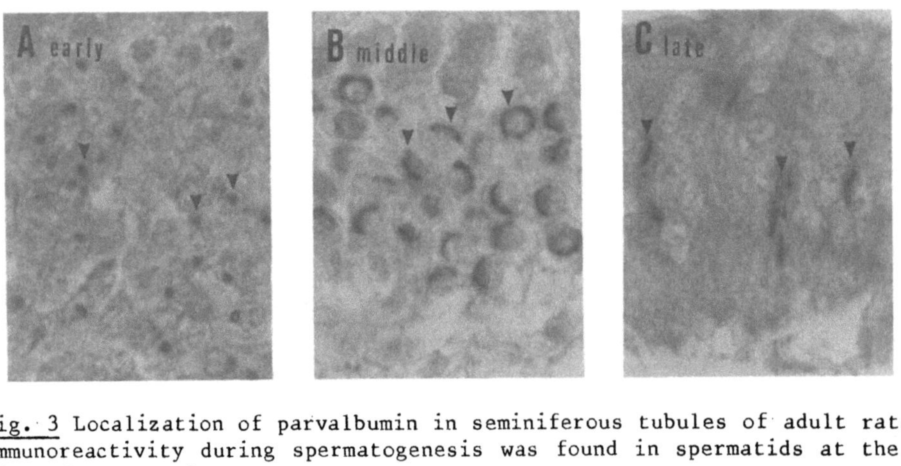

Fig. 3 Localization of parvalbumin in seminiferous tubules of adult rats. Immunoreactivity during spermatogenesis was found in spermatids at the site of the growing acrosome.
Spermatid stages: A, stages 1-3; B, 6-8; C, 10-14. Arrowheads point to areas of strong parvalbumin-immunoreactivity.

Hypophysectomy of animals at mid-puberty led to an arrest of spermatogenesis with all stages of sperm maturation still present (not shown). Parvalbumin-immunoreactivity in spermatids at equivalent stages was unchanged by the operation. Rats operated before puberty accumulated late spermatocytes in the seminiferous tubules. These cells showed a strong parvalbumin-staining. This suggests that parvalbumin expression might be regulated differently in germ-line- and Leydig cells.

In seminiferous tubules, parvalbumin co-exists with calbindin D-28K, S-100 proteins and calmodulin (Table 1); however, their distinct localization and developmental appearance suggests different regulatory roles during spermatogenesis.

ACKNOWLEDGEMENT

We are grateful to Dr. A. Rowlerson for critical reading of this manuscript. This study was supported by the Swiss National Science Foundation (No 3.139-0.88), EMDO-Stiftung Zürich, Jubiläumsspende für die Universität Zürich, and the Wilhelm Sander-Stiftung (FRG).

REFERENCES

1. Ch. Gerday, Soluble calcium binding proteins in vertebrate and invertebrate muscles in: Calcium and Calcium Binding Proteins. Molecular and Functional Aspects, Ch. Gerday, L. Bolis, R. Gilles, eds., pp23, Springer Verlag, Berlin, Heidelberg (1988).

2. C.W. Heizmann and M.W. Berchtold, Expression of parvalbumin and other calcium-binding proteins in normal and tumor cells: a topical review. Cell Calcium 8:1 (1987).

3. C.W. Heizmann and M.R. Celio, Immunolocalization of parvalbumin, Methods. Enzymol. 139:552 (1987).

4. C.W. Heizmann and K. Braun, Calcium-binding proteins. Molecular and functional aspects, CRC Press, The role of calcium in biological systems, L.J. Anghileri ed., Vol V, in press.

5. M.W. Berchtold, P. Epstein, A.L. Beaudet, M.E. Payne, C.W. Heizmann and A.R. Means, Structural organization and chromosomal assignement of the parvalbumin gene, J. Biol. Chem. 262:8696 (1987).

6. B.R. Seizinger, R.L. Martuza and J.F. Gusella, Loss of genes on chromosome 22 in tumorigenesis of human acoustic neuroma, Nature London 322:644 (1986).

7. R.L. Martuza, B.R. Seizinger, L.B. Jacoby, G.A. Rouleau and J.F. Gusella, The molecular biology of human glial tumors, TINS 11:22 (1988).

8. M.W. Berchtold, Evolution of the gene encoding the Ca^{2+}-binding parvalbumin, Experientia, Basel 44:A20 (1988).

9. H. Arai, P.C. Emson, C.Q. Mountjoy, L.H. Carassco and C.W. Heizmann, Loss of parvalbumin-immunoreactive neurons from cortex in Alzheimer-type dementia, Brain. Res. 418:164 (1987).

10. A. Seto-Ohshima, P.C. Emson, M.W. Berchtold and C.W. Heizmann, Localization of parvalbumin mRNA in rat brain by in situ hybridization histochemistry, Mol. Brain. Res., in press.

11. J.-M. Sequier, W. Hunziker and G. Richards, Localization of calbindin D28K mRNA in rat tissues by in situ hybridization, Neuroscience Letters 86:155 (1988).

12. T. Kosaka, H. Katsumaru, K. Hama, J.-Y. Wu and C.W. Heizmann, GABAergic neurons containing the calcium-binding protein parvalbumin in the rat hippocampus and dentate gyrus, Brain. Res. 419:119 (1987).

13. T. Kosaka, C.W. Heizmann, K. Tatushi, Y. Hamaoka and K. Hama, An aspect of the organizational principle of γ-aminobutyric acidergic system in the cerebral cortex, Brain Res. 409:403 (1987).

14. C.C. Stichel, W. Singer and C.W. Heizmann, Light and electron microscopic immunocytochemical localization of parvalbumin in the dorsal lateral geniculate nucleus of the cat: evidence for coexistence with GABA, J. Comp. Neurol. 268:29 (1988).

15. J. Röhrenbeck, H. Wässle and C.W. Heizmann, Immunocytochemical labelling of horizontal cells in mammalian retina using antibodies against calcium-binding proteins, Neuroscience Letters 77:255 (1987).

16. K. Braun, H. Scheich, C.W. Heizmann and C. Matute. Postnatal development of parvalbumin-, calbindin- and adult GABA-immunoreactivity in two visual nuclei of zebra finches, Brain Res., submitted.

17. Y. Kawaguchi, H. Katsumaru, T. Kosaka, C.W. Heizmann and K. Hama, Fast spiking cells in rat hippocampus (CA_1 region) contain the calcium-binding protein parvalbumin, Brain Res. 416:369 (1987).

18. W. Kamphuis, E. Huisman, W.J. Wadman, C.W. Heizmann and F.H. Lopes da Silva, Kindling induced changes in parvalbumin-immunoreactivity in rat hippocampus and its relation to long term decrease in GABA-immunoreactivity, Brain. Res., submitted.

19. U. Kägi, M.W. Berchtold and C.W. Heizmann, Ca^{2+}-binding parvalbumin in rat testis. Characterization, localization and expression during development, J. Biol. Chem. 262:7314 (1987).

20. U. Kägi, M. de Gasparo and C.W. Heizmann, Parvalbumin levels in Leydig cells of rat testis correlate with testosterone production in hypophysectomized, FSH/LH replaced and ethylene dimethane sulfonate treated rats, submitted.

21. R. Molenaar, D.G. de Rooij, F.F.G. Rommerts and H.J. van der Molen, Repopulation of Leydig Cells in mature rats after selective destruction of the existent Leydig Cells with ethylene dimethane sulfonate is dependent on luteinizing hormone and not follicle-stimulating hormone, Endocrinology 118:2546 (1986).

22. U. Kägi, J.G. Chafouleas, A.W. Norman and C.W. Heizmann, Developmental appearance of the Ca^{2+}-binding proteins parvalbumin, calbindin D-28K, S-100 proteins and calmodulin during testicular development in the rat, Cell Tissue Res. 252:359 (1988).

STRUCTURAL ORGANIZATION OF CALMODULIN GENES IN THE RAT GENOME

Hiroshi Nojima and Hirofumi Sokabe

Department of Pharmacology
Jichi Medical School
Minamikawachi-machi, Tochigi-ken 329-04, Japan

INTRODUCTION

Calmodulin plays multifunctional roles in living cells through the regulation of calcium interactions with a variety of fundamental enzymes. In order to know how many genes for calmodulin should exist to realize its mutlifunctional roles, we have investigated the structural organizations of calmodulin genes in the rat genome[1-8].

CLONING AND SEQUENCING OF RAT CALMODULIN GENES

From a rat brain cDNA library, we have cloned and sequenced five distinct calmodulin cDNAs, which are classified into three groups based on the nucleotide sequences of noncoding and coding regions. We call them pRCM1, pRCM3 & pRCM4. They encode calmodulins with the same amino acid sequences. The sizes and the nucleotide sequences of the 5' and the 3' noncoding regions are quite distinct among these cDNAs, thus we conclude that these cDNAs are derived from three distinct <u>bona fide</u> calmodulin genes. We named them CaM I, CaM II and CaM III genes (Fig.1). One of

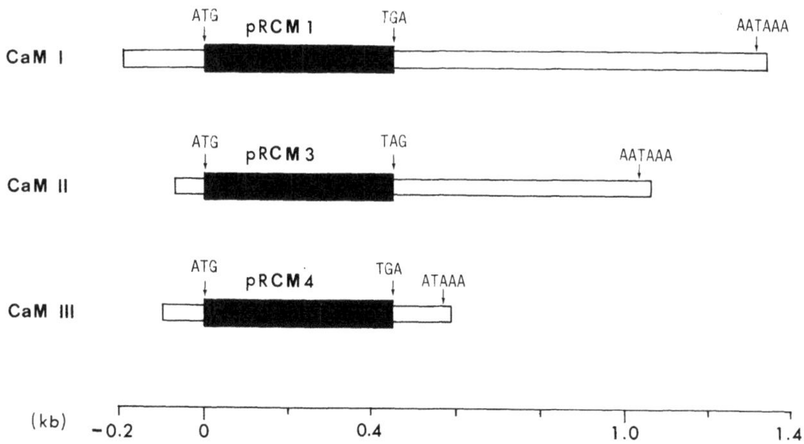

Fig.1. Structure of tree distinct calmodulin cDNAs cloned from a rat brain cDNA library. Filled boxes denote the coding regions. They are derived from three distinct calmodulin genes (CaM I, CaM II and CaM III).

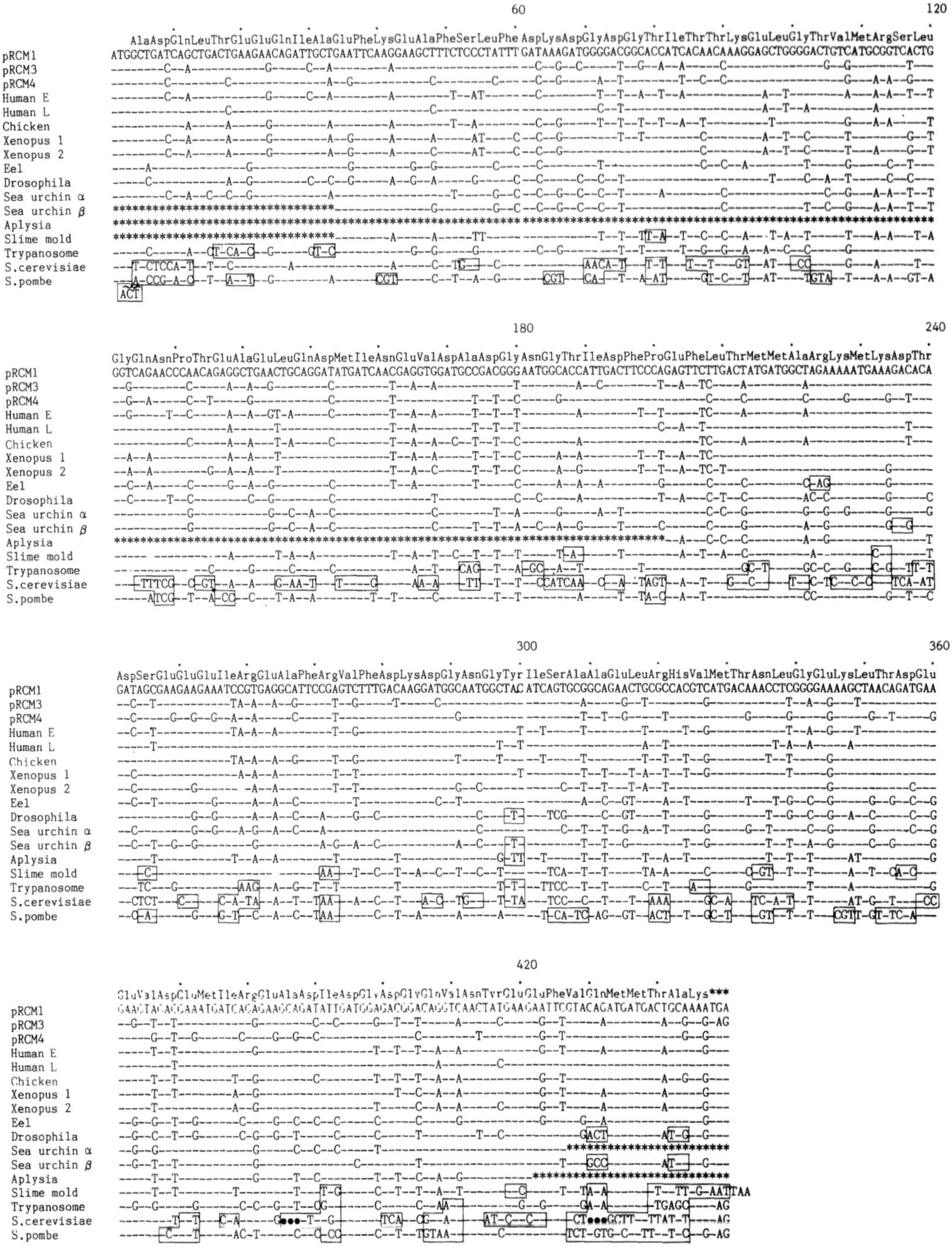

Fig.2. Alignment of nucleotide sequences of calmodulin cDNAs[10-22] reported heretofore to show maximum homology. Horizontal lines indicate nucleotide sequences identical to those of pRCM1. Only the substituted nucleotides are shown. Asterisks denote undetermined regions. Codons for substituted amino acids are encircled.

these cDNA clones (pRCM4) was found to be the same as the calmodulin cDNA cloned from a rat brain cDNA library by Sherbany et al.[9] except that the size of the 5' noncoding region of pRCM4 is longer by 92 bp.

The nucleotide sequences in the coding regions of calmodulin cDNAs of rat and other species reported heretofore[10-22] are aligned in Fig.2. Three rat calmodulin cDNAs show distinct nucleotide sequences within the limit of codon degeneracy. They exhibit about 15 % to 18 % difference in the nucleotide sequences in the coding region, namely, pRCM1 and pRCM3 differ at 73 sites, pRCM3 and pRCM4 differ at 81 sites and pRCM1 and pRCM4 differ at 66 sites out of 447 bp of the coding region plus termination codon. There is no apparent rule in the nucleotide displacements, that is, the sites of nucleotide substitutions are dispersed homogeneously throughout the coding region. It is noteworthy that pRCM1 is more homologous to human L (derived from liver) calmodulin cDNA than pRCM3 and pRCM4, i.e., pRCM1 and human L calmodulin cDNAs differ at only 31 sites. On the other hand, pRCM3 is more homologous to human E (derived from embryo) and chicken calmodulin cDNAs, viz., they differ at 41 and 53 sites respectively. It is surprising that chicken and human E calmodulin cDNAs differ only at 33 positions. These results indicate that human L calmodulin cDNA is pRCM1 (CaM I) type while human E and chicken calmodulin cDNAs are pRCM3 (CaM II) type. Homologies among these cDNAs can be displayed more apparently if the DNA sequences are compared to three rat cDNAs at the site where nucleotide substitutions occurred among pRCM1, pRCM3 and pRCM4. There are 104 such positions in the coding region plus termination codon. The results are given in Table 1. Assignment of human L cDNA to pRCM1 and human E and chicken cDNAs to pRCM3 are obvious again. Eel cDNA seems to be pRCM4(CaM III) type, while slime mold cDNA may be pRCM1(CaM I) type. As for calmodulin cDNAs of other species, it is difficult to classify them in this way.

Alignment of calmodulin cDNA sequences unveils interesting features on the evolution of calmodulin. Firstly, highly conserved amino acid sequences of calmodulin is reflected on the conserved codon usages of amino acids which are dispersed throughout the molecule. For example,

Table 1. Number of nucleotide sequences for calmodulin cDNAs of various species listed in Fig.2 that are identical to either pRCM1, pRCM3 and pRCM3 at the sites where DNA sequences among these three cDNAs are distinct.

	pRCM1	pRCM3	pRCM4	Reference
Human E (Embryo)	41	78	34	10
Human L (Liver)	83	33	40	11
Chicken	45	65	44	12
Xenopus 1	51	55	45	13
Xenopus 2	49	45	48	13
Eel	41	36	57	14
Drosophila	44	42	42	15,16
Sea urchin α	43	38	48	17
Sea urchin β	42	40	45	17
Aplysia	26	31	18	18
Slime mold	49	36	28	19
Trypanosome	41	30	44	20
S.cerevisae	44	39	34	21
S.pombe	38	37	36	22

Phe-12 codon is conserved in all species, Gln-8 and Asp-131 codon is substituted only in one species and Lys-94 and Ile-52 codon is changed only in two species. Conserved codon usages are also observed in Glu-139, Leu-4, Gln-49 and Asp-64. Even amino acid substitutions are the results of one nucleotide sequence. For example, in case of Ile-52, one nucleotide change is resulted in an amino acid substitution (Ile to Met). Secondly, comparisons of nucleotide sequences instead of amino acid sequences could unveil novel evolutional pathways for calmodulin. In case of Tyr-99, for example, replacement of Tyr to Phe or Leu was initiated by the exchange of the second codon from A to T or vice versa. Only in case of S.cerevisae the third codon was furthermore exchanged from C to A resulting in the amino acid replacement from Tyr to Leu.

Total genomic Southern hybridizations were performed using pRCM1, pRCM3 and pRCM4 as probes. Among various kinds of restriction enzymes, Pvu II was found to be most useful to assign each band to each gene. As shown in Fig.3, pRCM1 probe detected five bands at 6.0 kilobase (kb), 5.2 kb, 4.2 kb, 3.8 kb and 2.9 kb, three of which (middle three bands) were also detected by pRCM3 probe. On the other hand, pRCM4 detected novel two bands of smaller sizes at 2.3 kband 1.3 kb. By screening rat genomic libraries with these probes, we have hitherto cloned more than thirty independent clones. By restriction enzyme mappings and partial DNA sequencings, they are found to be classified into seven independent genomic fragments. They can be classified into three distinct bona fide genes, that is, CaM I, CaM II and CaM III and four processed retropseudogenes such as λSC8. λSC9, λSC19 and λSC27. Restriction enzyme mapping of each clone enabled us to assign each band on total genomic Southern blots to each clone as shown at the right side of Fig.3.

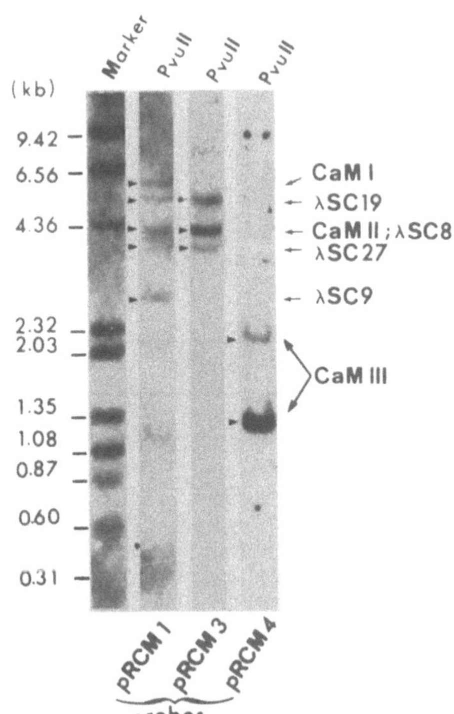

Fig.3. Total genomic Southern blots for rat DNA probed by three distinct calmodulin cDNAs (pRCM1, pRCM3 and pRCM4). Assignment of each band to each cloned gene are given at the right side of the figure.

DNA sequencing of cDNAs and genomic clones revealed genomic structures and intron/exon organizations of three bona fide calmodulin genes, CaM I, CaM II and CaM III. All intron exon junctions follow the GT/AG rule. CaM I spans about 9 kb and consisted of six exons interrupted by introns of various sizes. The first intron separates the initiatin codon (ATG) from the coding region of the protein. In the third intron, a segment homologous to the identifier sequence, a rat specific repetitive sequences, was observed. A TATA box and GC boxes which are important DNA element for transcriptional regulation are observed at the 5' upstream regulatory region of the gene. CaM II spans about 10 kb and consisted of five exons. One of the exons of CaM I was lost and exon 3 and 4 in CaM I was fused to one exon, exon 3 in CaM II gene. In chicken, the structure of a calmodulin gene was reported[23]. By comparing the nucleotide sequences in the coding region as described before (Fig.2 and Table 1) and the intron/exon organizations, it is found to be a CaM II type gene. CaM III is smaller than CaM I and CaM II. It spans about 7 kb. Especially the last four exons including 93 % of the coding region are found within 1 kb. Intron/exon organizations are the same with that of CaM I. A segment of sequence identical with the identifier sequence is found at intron 2 of the gene. Neither TATA box nor CAAT box is found at the 5' upstream regulatory region. Instead, the DNA sequences of this region is highly GC rich and three GC boxes are detected.

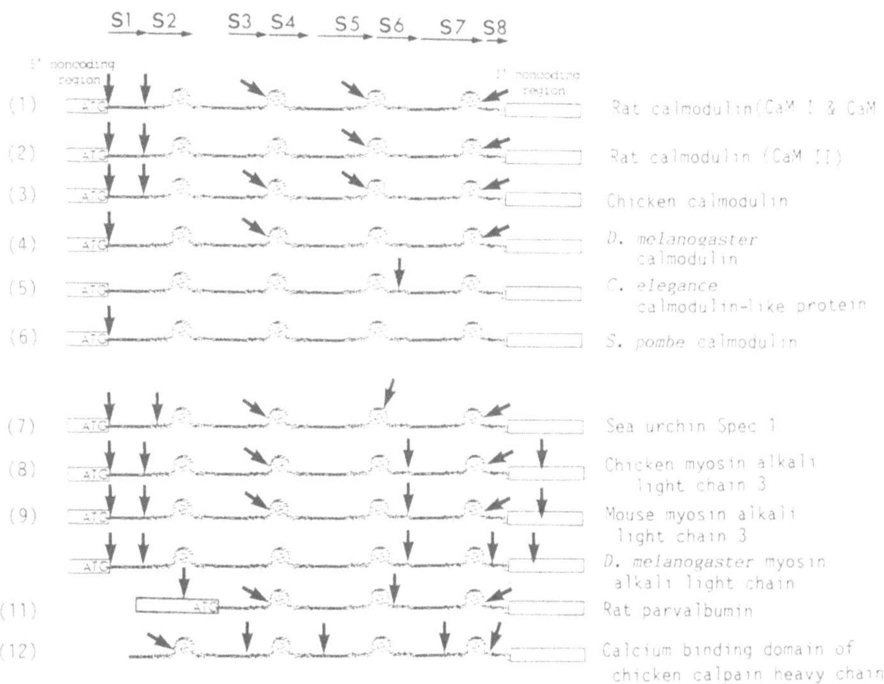

Fig.4. Comparisons of relative intron positions of calmodulin genes from various species and genes for several calcium binding proteins (See Table 1. for references). Vertical thick arrows denote intron positions. Horizontal thin arrows indicate homologous segments (S1-S8) observed in the coding regions of calmodulin genes[6]. Noncoding regions are shown by empty boxes. Four bumps in the coding sequences display putative calcium binding subdomains.

EVOLUTION OF CALMODULIN GENES

In Fig.4, we compared the positions of intron/exon junctions in the EF-hand domains of calmodulins from various species[4,8,15,16,22-24] and other Ca^{2+}-binding proteins such as sea urchin Spec 1[25], chicken[26], mouse[27] and drosophila[28] myosin alkali light chain, rat parvalbumin[29] and calcium binding domain of chicken calpain heavy chain[30], whose genomic structures have been characterized. The bumps or loops in Fig.4 indicate putative calcium-binding sites. As for calcium binding proteins, from (7) to (12) in Fig.4, the interval sizes between calcium binding loops are not exactly shown but only to compare the relative positions with those of calmodulin. It is evident from this figure that they all share a common ancestral origin. The separation of the first exon from other parts at right after the initiation codon ATG is commonly characteristic among these genes. Exon fusion seems to be popular among calmodulin genes.

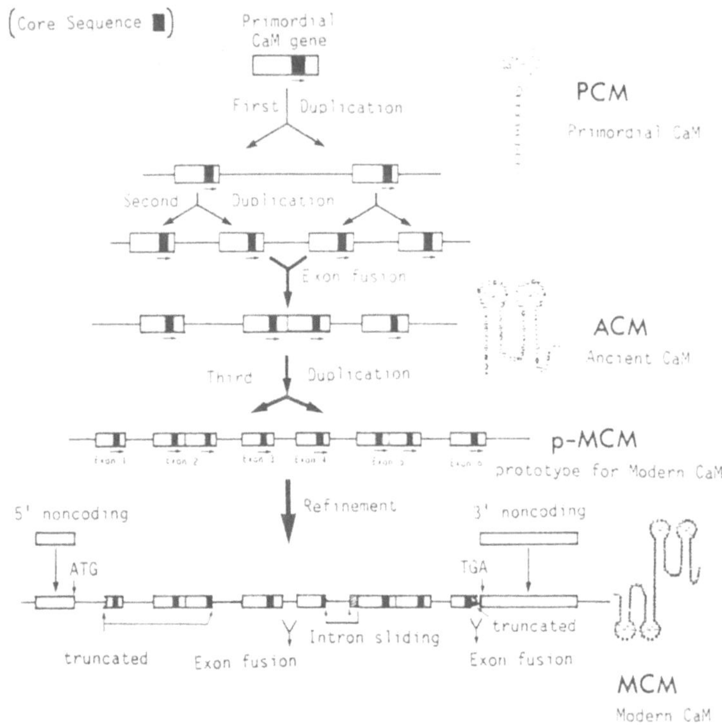

Fig.5. A possible model for the genesis of the calmodulin gene[6].

Rat CaM II lost one intron. Drosophila and yeast (S. pombe) calmodulin genes lost two and four introns respectively. Yeast (S.cerevisae) calmodulin possesses no intron at all. Nematode calmodulin like protein seems to be quite distinct from calmodulin. It is rather a novel type of calcium binding protein connecting calmodulin and myosin alkali light chain since the single intron of this gene resides at a similar position to myosin alkali light chain gene although the amino acid sequence of this gene resesmbles calmodulin.

It is noteworthy that introns reside in the middle of calcium binding loops instead of sitting in the linker domais correlating with the quarterly divided intramolecular homology of calmodulin proteins. It is against the exon-shuffling model of Gilbert[31] if calmodulin gene is evolved from an primitive gene encoding a peptide with one calcium-binding domain by twice duplication of the gene to create modern calmodulin with four calcium-binding subdomains[6].

In order to explain this enigma, We investigated the nucleotide sequences of calmodulin genes and found that it consisted of eight homologous segments (S1-S8)[6] as shown at the top of Fig.4. The junctions of each segment well correlates the intron positions except for segment 5 where the last part of the segment is engulfed into the next exon probably because of the intron sliding. Based on this eightfold intramolecular homology, One of the authors previously proposed a model for the genesis of modern calmodulin gene (Fig.5) which at least accomodate the exon-shuffling model with the enigmatic intron positions of calmodulin genes[6]. A primordial calmodulin (PCM) gene with 51 bp encoding 17 amino acids with one inefficient calcium-binding activity was firstly duplicated twice. At the end of the second duplication, the middle two exons were fused to form a double-sized exon. These processes create an ancient calmodulin (ACM) with two calcium-binding subdomains. ACM gene was furthermore duplicated to produce a prototype of modern calmodulin (p-MCM). Before the generation of modern calmodulin (MCM) gene, a refinement procedure took place. At this step, the 5' and the 3' noncoding segments that are unrelated to PCM were introduced into the gene. Some exon fusions occurred to create modern calmodulin (MCM) gene.

RETROPSEUDOGENES

From rat genomic libraries, we have cloned four other calmodulin related genes. After DNA sequencing they are found to be intronless retropseudogenes derived from either CaM I or CaM II genes through mRNA mediated process of insertions. Hypothetical products encoded in these retropseudogenes are shown in Fig.6. DNA sequences of λSC8, λ19 and λ27 clones are closely homologous to pRCM3, a cDNA form of CaM II gene. Nucleotide substitutions and deletions are observed in the coding and noncoding regions of these genes[8]. In the case of λSC19, the initiation codon ATG is changed to ATA, so that it is apparently an inactivated gene.

```
             1        10        20        30        40        50
CaM          ADQLTEEQIAEFKEAFSLFDKDGDGTITTKELGTVMRSLGQNPTEAELQD
λSC8         ----------------P---R---Q-V------M--G-------------
λSC19        V---S--H----------L---▲-------R--M------▲---------
λSC27(λWC1)  ------*--N▲-----------RY---K--------G--Q--H---D---
λSC9         --------------------------------------Q-------▲---

                       60        70        80        90       100
             MINEVDADGNGTIDFPEFLTMMARKMKDTDSEEEIREAFRVFDKDGNGYI
             --S----------N-------------G--V---------------••-T
             --R--HVG----V----S-M------------K-T--GSC-----D--S-
             ▲-----V------------------------K---K-V-C-YS-N-----
             --------------S----------------K-----------------F

                      110       120       130       140
             SAAELRHVMTNLGEKLTDEEVDEMIREADIDGDGQVNYEEFVQMMTAK
             ----F-----T-----------G------M------I-------V---
             ---••••••••••••••---GEKV------I-----------------
             --VQ-C-----------Y--------K-N-V---EL-SS---------
             -------------------DI-------------▲-----------S-
```

▲ sites for frameshift mutations ; ● deletions ; * stop codon

Fig.6. Hypothetical products encoded in rat calmodulin retropseudogenes.

Nucleotide deletions in the coding region of these genes caused frameshift mutations resulting in premature appearance of stop codons to make the gene inactive. The absence of transcrips of these genes are confirmed by S1 nuclease mapping. Only the λSC8 harbors no frameshift mutation. However, S1 nuclease mapping also denied the transcription of the gene[5]. These retropseudogenes are common in the genome of eukaryotes, however, it is surpizing that there are as many as four or more retropseudogenes for a single protein such as calmodulin.

TISSUE-SPECIFIC TRANSCRIPTION

By Northern blotting, transcriptions of CaM I, CaM II and CaM III are investigated using pRCM1, pRCM3 and pRCM4 as probes[5,8]. CaM I is transcribed into 1.7 kb and 4.0 kilo base mRNAs in similar amounts in liver, muscle and brain. 4.0 kb mRNA is observed primarily in skeletal muscle and almost undetectable in other tissues examined. By S1 nuclease mapping, we confirmed that this is generated from CaM I by transcribing extra amount of the 3' noncoding region[5]. CaM II is transcribed into 1.4 kb mRNA mainly in brain and faintly in other tissues. CaM III is transcribed primarily into 2.2 kb and faintly into 0.9 kb in brain and skeletal muscle but not in other tissues.

Table 2. Classification of plasmid and phage clones obtained from a brain cDNA library of Sprague-Dawley rats (SD) and the genomic libraries of spontaneously hypertensive rats (SHR) and Wistar-Kyoto rats (WKY).

	Bona fide gene	cDNA	Retropseudogene	mRNA (kb)	Transcribed mainly in
CaM I	λSC2,4,6,11 12,13,14	pRCM1 pRCM5	λSC1,3,5,7,9, 30,31	1.7 4.0	liver, brain muscle etc.
CaM II	λSC15,21 λWC40	pRCM3	λSC8,17,18,20,22 23,24,25,26 λSC19 λSC27 (= λWC1)	1.4	brain
CaM III	λSC16	pRCM4		0.9 2.2	brain, muscle
Strain	SHR(WKY)	SD	SHR (WKY)		

SUMMARY

In summary, we present a list of phage clones we have obtained from rat genomic libraries (from λSC1 to λSC31, λWC1 and λWC40) together with cDNA clones we have obtained from a rat brain cDNA library (PRCM1,5,3 and 4). pRCM5 corresponds to 4.0 kb mRNA species observed primarily in skeletal muscle. These clones can be classified into three groups. They belong to three bona fide calmodulin genes with five to six exons called CaM I, CaM II and CaM III and four intronless retropseudogenes, one derived from CaM I and three derived from CaM II. We have not obtained retropseudogenes for CaM III so far. These three bona fide genes are transcribed into multiple sized mRNA species in a tissue-specific manner, that is, CaM I is ubiquitous, CaM II is transcribed mainly in brain and CaM III is transcribed primarily in brain and skeletal muscle. Four retropseudogenes do not appear to be transcribed. They are probably relics of inactivated genes. The physiological meanings of multiple calmodulin mRNA species and mechanisms of transcriptional regulation of these three bona fide genes will be the main subjects of our future experiments.

REFERENCES

(1) H.Nojima and H.Sokabe, Structure of rat calmodulin processed genes with implications for a mRNA-mediated process of insertion. J. Mol. Biol. 190:391(1986).

(2) H.Nojma, K.Kishi and H.Sokabe, Organization of calmodulin genes in the spontaneously hypertensive rat. J. Hypertens. 4(Suppl 3): S275(1986).

(3) H.Nojima and H.Sokabe, A peculiar repetitive sequence in the rat genome. FEBS Lett 207:227(1986).

(4) H.Nojima and H.Sokabe, Structure of a gene for rat calmodulin. J. Mol. Biol. 193:439(1987).

(5) H.Nojima, K.Kishi and H.Sokabe, Multiple calmodulin mRNA species are derived from two distinct genes. Mol. Cell. Biol. 7:1873(1987).

(6) H.Nojima, Molecular evolution of calmodulin gene. FEBS Lett 217:187(1987)

(7) H.Nojima and H.Sokabe, Genes and pseudogenes for calmodulin in the spontaneously hypertensive rat. J. Hypertens. 6(suppl 6): in press.

(8) H.Nojima, Tissue-specific transcription of three distinct rat calmodulin genes. Submitted for publication.

(9) A.A.Sherbany, A.S.Parent and J.Brosius, Rat calmodulin cDNA. DNA.6:267(1987).

(10) B.SenGupta, F.Friedberg and S.D.Detera-Waldleigh, Molecular analysis of human and rat calmodulin cDNA clones. J. Biol. Chem. 262: 16663(1987).

(11) E.J.Warwrynczak and R.N.Perham, Isolation and nucleotide sequence of cDNA encoding human calmodulin. Biochem. Int. 9:177(1984).

(12) J.A.Putkey, K.F Ts'ui, T.Tanaka, L.Lagace, J.P.Stein, E.C.Lai and A.R.Means, Chicken calmodulin genes. J. Biol. Chem.258:11864(1983).

(13) Y.Chien and I.B.Dawid, Isolation and characterization of calmodulin genes from Xenopus laevis. Mol. Cell. Biol. 4:507(1984).

(14) L.Lagace, T.Chandra, S.L.C.Woo and A.R.Means, Identification of multiple species of calmodulin mRNA using a full length cDNA. J. Biol. Chem. 258:1684(1983).

(15) V.L.Smith, K.E.Doyle, J.F.Maune, R.P.Munjaal and K.Beckingham, Structure and Sequence of the Drosophila melanogaster calmodulin gene. J. Mol.Biol. 196:471(1987).

(16) M.K.Yamanaka, J.A.Saugstad, O.Hanson-painton, B.J.McCarthy and S.L.Tobin, Structure and expression of the Drosophila calmodulin gene. Nuc. Acids Res. 15:3335(1987).

(17) D.O.Hardy, P.K.Bender and R.H.Kretsinger, Two calmodulin genes are expressed in Arbacia punctulata. J. Mol. Biol. 199:223(1988).

(18) J.H.Schwartz and M.E.Swanson, Dissection of tissues for characterizing nucleic acids from Aplysia. Methods Enzymol. 139:277(1987).

(19) H.Goldhagen and M.Clarke, Identification of the single gene for calmodulin in Dictyostelium discoidium. Mol. Cell.Biol. 6: 1851(1986).

(20) C.Tschudi, A.S.Young, L.Ruben, C.L.Patton and F.F.Richards, Calmodulin genes in trypanosomes are tandemly repeated and produce multiple mRNAs with a common 5' leader sequence.Proc. Natl. Acad. Sci. U.S.A. 82:3998(1985).

(21) T.N.Davis, M.S.Urdea, F.R.Masiarz and J.Thorner, Isolation of the yeast calmodulin gene. Cell 47:423(1986).

(22) T.Takeda and M.Yamamoto, Analysis and in vivo disruption of the gene coding for calmodulin in Schizoaccharomyces pombe. Proc. Natl. Acad. Sci. U.S.A. 84:3580(1987).

(23) R.C.M.Simmen, T.Tanaka, K.F.Ts'ui, J.A.Putkey, M.J.Scott, E.C.Lai and A.R.Means, The structural organization of the chicken calmodulin gene. J. Biol. Chem. 260: 907(1987).

(24) M.Salvato, J.Sulston, D.Abertson and S.Brenner, A novel calmodulin-

like gene from the nematode Caenorhabditis elegance. J. Mol. Biol. 190:281(1986).

(25) S.H.Hardin, C.D.Carpenter, P.E.Hardin, A.M.Bruskin and W.H.Klein, Structure of the Spec 1 gene encoding a major calcium-binding protein in the embryonic ectoderm of the sea urchin, Strongylocentrotus purpuratus. J. Mol. Biol. 186:243(1985).

(26) Y.Nabeshima, Y.Fujii-Kuriyama, M.Muramatsu and K.Ogata, Alternative transcription and two modes of splicing result in two myosin light chains from one gene. Nature 308:333(1984).

(27) B.Robert, P.Daubas, M.Akimenko,A.Cohen, I.Garner, J.Guenet and M.Buckingham, A single locus in the mouse encodes both myosin light chains 1 and 3, a second locus corresponds to a related pseudogene. Cell 39:129(1984).

(28) A.Falkenthal, V.P.Parker and N.Davidson, Developmental variations in the splicing pattern of transcripts from the Drosophila gene encoding myosin alkali light chain result in different carboxyl-terminal amino acid sequences. Proc. Natl. Acad. Sci. U.S.A. 82:449(1985).

(29) M.W.Berchtold, P.Epstein, A.L.Beaudet, M.E.Payne, C.W.Heinzman and A.R.Means, Structural organization and chromosomal assignment of the parvalbumin gene. J. Biol.Chem. 262:8696(1987).

(30) Y.Emori, S.Ohno, M.Tobita and K.Suzuki, Gene structure of calcium-dependent protease retains the ancestral organization of the calcium-binding protein gene. FEBS Lett. 194:249(1986).

(31) W.Gilbert, Why genes in pieces?, Nature 271:501(1978)

THE HUMAN CALBINDINS: CDNA AND GENE CLONING

Marc Parmentier

I.R.I.B.H.N.
ULB Campus ERASME
808 route de Lennik
1070 Bruxelles

INTRODUCTION

Wasserman and Taylor described in 1966 that the major effect of vitamin D on chick intestine, besides the increase of calcium absorption, was the synthesis of a 28,000 Da protein. This protein was shown to bind calcium, and was named vitamin D-dependent calcium-binding protein or D-CaBP. A protein of 10,000 Da, similarly induced by vitamin D, was found in the intestine of rat and other mammals (Kallfeltz et al., 1967). No crossimmunoreactivity could be detected between these 28,000 and 10,000 Da calcium-binding proteins and this latter protein (ICaBP) will not be discussed further in this paper. The 28,000 Da protein was described by immunochemical methods in a variety of organs, mainly the distal convoluted tubule of the kidney (Taylor and Wasserman, 1967), the pancreas islets (Christakos et al., 1979) and the brain (Taylor, 1974), both in the chick and the rat. It was found later that D-CaBP was actually not dependent upon vitamin D in its brain location (Taylor, 1974, Thomasset et al., 1982). For this reason, the protein was renamed Calbindin.

We describe here the cloning of human cDNAs and genes for calbindin 27 kDa (corresponding to the 28 kDa protein of the chick intestine) and for a related protein, calbindin 29 kDa. Troughout this paper, the apparent molecular weights (from SDS gel migration) will be used to characterize the proteins. We think however that the real molecular weights (from cDNA and protein sequencing) should be used instead, in order to clarify a rather confusing nomenclature.

TWO CALBINDINS IN THE BRAIN

In previous works, by using the Western blotting technique as a control of immunohistochemical localization of calbindin in the brain and other organs of the rat, we found that the antibody directed against chick intestinal calbindin was labelling two bands on blots from brain tissues, corresponding to proteins of 27 and 29 kDa (Pochet et al., 1985). The same antibody was recognizing a single protein in the rat kidney (27 kDa) and in the chick intestine (27.5 kDa). Incubation of the samples for up to 16 hours at 37°C allowed to demonstrate that partial degradation was not responsible for the appearance of two bands in the brain. Both proteins were stable for hours even in the absence of protease inhibitors. In vitro translation of messenger RNA extracted from different tissues allowed to confirm these results. Two proteins (27 and 29 kDa) were synthesized from rat brain mRNA, while one protein was synthesized from kidney mRNA. From these experiments, it was concluded that two populations of mRNA were present in the brain cells (Pochet et al., 1985). It was not clear at that time if the second protein was related to calbindin or if they were completely different proteins accidentally sharing a common epitope. Affinity chromatography allowed to obtain enriched fractions for both proteins from rat brain. Calcium-binding assays on Western blots showed that the 29 kDa as well as the 27 kDa proteins were able to bind calcium (Pochet et al., unpublished). It was therefore postulated that the two proteins were related. They were called calbindin 27 kDa and calbindin 29 kDa. It was unknown however if their respective mRNA were derived from a single gene by a brain specific alternative splicing, or if a second gene was specifically expressed in the nervous system.

A phylogenetic study of calbindins in lower vertebrates by immunohistochemistry and Western blot (Parmentier et al., 1987a) showed that only one band was present in the fish brain while two bands were detected in the brain of amphibia, reptiles, birds and mammals. More interestingly, the affinity of the antibody for the 27,000 Da band was clearly decreasing from birds (source of the antigen) to amphibia, while the affinity for the 29,000 Da protein was almost not affected. This was the first indication that calbindin 29 kDa could have evolved independantly from the calbindin 27 kDa, and consequently be encoded by a different gene. To test this hypothesis, we decided to clone the cDNA coding for each of these proteins.

```
                           EL--LL--L*-*-*G-L*--*L--LL--L
27kDa-1      MAESHLQSSLITASQFFEIWLHFDADGSGYLEGKELQNLIQELQQARKKA
                  *     ***** *** ****** ** ***** *  ***  ***
29kDa-1 MAGPQQQPPYLHLAELTASQFLEIWKHFDADGNGYIEGKELENFFQELEKARKGS

                           EL--LL--L*-*-*G-L*--*L--LL--L
27kDa-2         GLE-----LSPEMKTFVDQYGQRDDGKIGIVELAHVLPTEENFLLLF
                *           ** *   *    ***    ***  ********* *
29kDa-2         GMMSKSDNFGEKMKEFMQKYDKNSDGKIEMAELAQILPTEENFLLCF

                           EL--LL--L*-*-*G-L*--*L--LL--L
27kDa-3           RCQQLKSCEEFMKTWRKYDTDHSGFIETEELKNFLKDLLEKANKT
                  * *   *  ***   ****** ** **  *** ** *** ***
29kDa-3           R-QHVGSSAEFMEAWRKYDTDRSGYIEANELKGFLSDLLKKANRP

                           EL--LL--L*-*-*G-L*--*L--LL--L
27kDa-4            VDDTKLAEYTDLMLKLFDSNNDGKLELTEMARLLPVQENFLL
                    *  ** ***   *  ** * **** * ** ***********
29kDa-4            YDEPKLQEYTQTILRMFDLNGDGKLGLSEMSRLLPVQENFLL

                           EL--LL--L*-*-*G-L*--*L--LL--L
27kDa-5          KFQGIKMCGKEFNKAFELYDQDGNGYIDENELDALLKDLCEKNKQDL
                 **** *    ***  *  ** *  ***** ********* ****
29kDa-5          KFQGMKLTSEEFNAIFTFYDKDRSGYIDEHELDALLKDLYEKNKKEI

                           EL--LL--L*-*-*G-L*--*L--LL--L
27kDa-6                DINNITTYKKNIMALSDGGKLYRTDLALILCAGDN
                        *   * * *  * *   ***** **   **
29kDa-6                NIQQLTNYRKSVMSLAEAGKLYRKDLEIVLCSEPPM
```

Fig. 1. Alignment of the 6 EF-hand domains of calbindins 27 and 29 kDa. The consensus of the EF-hand domain is represented on top (E = glutamine, L = hydrophobic residue, G = glycine, * = oxygen-containing residue, - = any residue).

CLONING OF HUMAN CDNA FOR CALBINDINS 27 AND 29 KDA

Two cDNA libraries (human brain stem and human cerebellum) constructed in the expression vector lambda gt11, were screened using antibodies directed against chick intestinal calbindin (Benton and Davis, 1977). Out of 1.1 10^6 recombinant clones, 105 were positive in the first screening. 44 clones were purified to homogeneity. Most of these clones contained internal EcoRI sites and cross-hydridized perfectly. The longest of these clones were subloned in M13 phages and sequenced by the dideoxy nucleotides chain-termination method of Sanger et al. (1977). The sequence (Parmentier et al., 1987b) revealed a single open reading frame of 261 amino-acids corresponding to a protein 97.7% identical to the bovine calbindin 27 kDa (Takagi et al. 1986) and 79.3% identical to the chicken calbindin (Wilson et al. 1985).

Amongst the remaining clones isolated by antibody screening, another population of 7 clones exhibited a completely different restriction fragment pattern and little affinity for the calbindin 27 kDa probes. A partial sequence revealed extensive homologies with calbindin 27 kDa and putative calcium-binding sites. Additional cDNA clones for this new calcium-binding protein were isolated by screening of the libraries, and the full sequence was determined. An open reading frame coding for a protein of 271 amino-acids was deduced. This protein was aligned with calbindin 27 kDa and was showed to contain 58.5% identical residues. The molecular weigths were calculated from the sequences. We obtained 30,300 Da for calbindin 27 kDa and 31,760 Da for the putative calbindin 29 kDa. Although these molecular weigths are higher than what can be estimated from SDS gel electrophoresis, the difference between these calculated values is compatible with the difference in gel mobility between the proteins. Calbindin 29 kDa was found homologous to the partial sequence obtained by Rodgers (1987) for the chicken retinal protein Calretinin. The amino-acids sequences are 87.8% identical. Therefore, Calbindin 29 kDa and Caretinin are probably to be considered as orthologous proteins.

The structure of both calbindins was analysed (fig. 1). Both contain 6 EF-hand domains as described by Kretsinger (1980). Domains 1,3,4 and 5 of calbindin 27 kDa and domains 1,2,3,4 and 5 of calbindin 29 kDa are in good agreement with the consensus, and presumably active. Domain 2 of calbindin 27 kDa and domain 6 of both proteins are deeply modified. It was showed previouly that calbindin 27 kDa was able to bind four Ca^{++} ions. It can be postulated that calbindin 29 kDa has a capacity of 5 Ca^{++} ions. The hydrophobicity profile is very similar for both proteins and indicates a soluble nature.

DISTRIBUTION OF CALBINDINS IN THE NERVOUS SYSTEM

The distribution of calbindin 27 kDa in the nervous system of various animals is well documented (Jande et al., 1981, Roth et al., 1981, Feldman and Christakos, 1983, Fournet et al., 1986). The cross reactivity between calbindins 27 and 29 kDa does not allow to exclude that both proteins were codetected in some of these studies. For this reason, we started to study the distribution of both protein in the rat brain, using an in situ hybridization technique and two cRNA probes that do not cross hybridize with each other. Preliminary data indicate that calbindins 27 and 29 kDa have a wide but distinct distribution in the central nervous system of the rat. This selective distribution of two closely related proteins reinforce the hypothesis that binding calcium is not their unique function in the neuron (V. Pohl, unpublished).

Considering human Calbindin 29 kDa and calretinin as orthologous proteins, evolutionary rates were calculated for different calcium-binding proteins of the troponin C superfamily and reference proteins (Kimura, 1983). Interestingly, calbindin 27 and 29 kDa appear, with evolutionary rates of 0.27 and 0.30 10^{-9} aa^{-1} $year^{-1}$ respectively, as strongly conserved in evolution (table 1). Proteins with low evolutionary rates are generally considered as essential proteins having extensive interactions with other macromolecules. The biological function of calbindins, usually described as calcium buffers (Jande et al., 1981, McBurney and Neering, 1987), could therefore be underestimated and regulatory actions should be considered and more actively searched for.

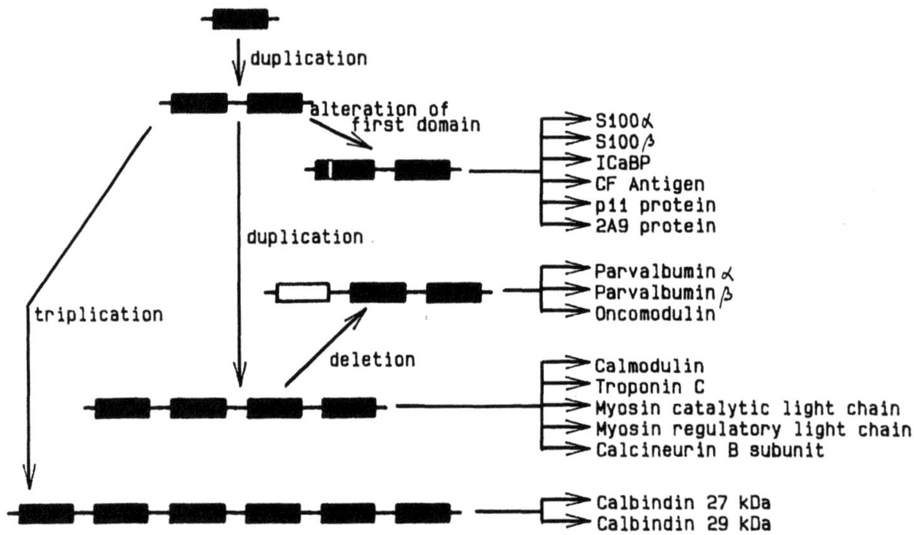

Fig. 2. Proposed evolutionary tree for the calcium-binding proteins of the Troponin C superfamily. Modified from Parmentier et al. (1987b)

The position of calbindins in the evolutionary tree of E-F hand domains containing proteins was analysed by comparing the homology between the individual domains of the various calcium-binding proteins. It appeared clearly that both calbindins were derived from a common six domained ancestor. An evolutionary tree was constructed and it was suggested that all members of the troponin C superfamily derive from a common two-domained ancestor, but that duplications leading to calbindins and to the four-domained calcium-binding proteins took place independantly on different branches of the evolutionary tree (fig. 2).

Table 1. Evolutionary rates of calbindin, calcium-binding proteins of the troponin C superfamily, and some reference proteins (Kimura, 1983).

The rates are expressed per amino acid site per year taking 10^{-9} as the unit.

Histone H4	0.01
Calmodulin	0.04
Troponin C	0.14
S100β chain	0.21
Calbindin 29 kDa	0.27
Myosin regulatory Nbs_2 light chain	0.27
Calbindin 27 kDa	0.30
Cytochrome C	0.30
Insulin	0.44
Myosin catalytic light chain	0.48
Parvalbumin β	0.62
Parvalbumin α	0.68
Myoglobin	0.89
Hemoglobin α	1.20
ICaBP	1.30
Lysozyme	2.00
Pancreatic ribonuclease	2.10
Fibrinopeptides	8.30

CLONING OF THE HUMAN GENE FOR CALBINDIN 27 KDA

A human genomic library constructed in the phage lambda Charon 4A, was screened with cDNA probes. The 5' and 3' regions of the gene were sequenced, including the first two and the last two exons. The four mapped introns were found at the same location as in the chicken gene (Wilson et al. 1988). The exons and introns of the human gene are therefore numbered by analogy with the chicken gene structure. The exon sequences confirm the corresponding cDNA sequence, so that there is not a single base mutation, even in the long 3' untranslated region. Several interspersed repeated elements are found in the 5' and 3' regions. Two Alu repeats lie 1200 bp ahead of the TATA box and 270 bp downstream of the polyadenylation site, respectively. A L1 element is located in intron I. An ATAAATA sequence which satisfies the TATA box consensus (Bucher and Trifonof, 1986, Nussinov et al., 1986) is found 30 nucleotides ahead of the putative cap site. No sequence close to the consensus of CAAT box could be found at reasonnable proximity of the TATA box. Little homology was found with the promoter region of the chicken gene although important regulatory elements would be expected to be conserved

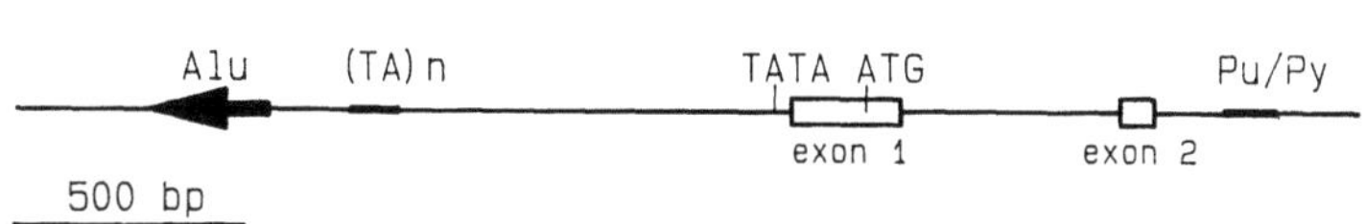

Fig. 3. Structure of the 5' region of the calbindin 27 kDa gene. The first two exons, the Alu repeat, the alternating purine-pyrimidine element and the homopurine-homopyrimidine stretch are indicated.

during the evolution of the gene. However, the evolutionary distance between the two species and the different pattern of expression could explain the lack of homology between the promoter regions. A potential SP1 binding site was found 57 bp upstream from the TATA box. Similar elements can be found in the GC-rich region of the chick promoter, but their significance is unknown.

In the 5' region (fig. 3), two sequence features are found that have been reported to be regularly associated with promoter regions: an alternating purine-pyrimidine element located 900 bp upstream from the TATA box, and a 120 bp long homopurine-homopyrimidine element in the second intron. Alternating $(A\text{-}T)_n$ sequences have been found frequently in the upstream region of eukaryotic genes (Mc Clellan et al 1986). They are thought to be subjected to easy tortional deformation and this flexibility is likely to be connected with a potential biological role. Homopurine-homopyrimidine tracts are also frequently encountered in the promoter of eukaryotic genes and are responsible for the S1 nuclease sensitivity often described within regulatory regions (Nickol and Felsenberg, 1983, Evans et al, 1984)

The calbindin 27 kDa gene was assigned to chromosome 8 using human-rodent hybrid cell lines, and a HindIII restriction fragment length polymorphism was detected at the calbindin locus, using a cDNA probe (Parmentier et al., 1988).

CLONING OF THE HUMAN GENE FOR CALBINDIN 29 KDA

Screening of the human genomic library with cDNA probes and oligonucleotide probes from the 5'end of the cDNA allowed to isolate six clones for the calbindin 29 kDa gene. Three overlapping clones contain most of the exonic sequence to the exception of the first exon. The last three clones overlap over a region that contains the first exon and the upstream sequences. The length of the first intron is not known. The junctions of the first four introns were sequenced, and their position appeared to be identical in the calbindin 29 and 27 kDa genes.

CONCLUSIONS

We have demonstrated the existence of two calbindins in the brain and determined their primary structure. These two proteins are the only calcium-binding proteins that harbor 6 EF-hand domains and constitute therefore a subfamily of the troponin C superfamily. Both appear to be highly conserved in evolution and could be considered as essential proteins with potential interactions with other macromolecules. An evolutionary tree of the Troponin c superfamily was constructed. The genes for both proteins were cloned and extensive regions of the calbindin 27 kDa gene were sequenced. We are now working on the sequencing of the 29 kDa gene promoter. Future work will focus on the regulation of the expression of these two calbindin genes and on their mapping in the human genome.

ACKNOWLEDGMENTS

This work was partially supported by the Belgian Fonds de la Recherche Scientifique Médicale, the Ministère de la Politique Scientifique (Sciences de la Vie), the European Economic Community (Stimulation program), the British Council, Solvay and the Association Recherche Biomédicale et Diagnostic. We thank D.E.M. Lawson for the gift of anti-calbindin serum, L. Muir for her participation in the sequencing work on the calbindin 27 kDa gene, J. De Vijlder and A. Geurts van Kessel for their collaboration in the chromosomal assignment, V. Pohl for the preliminary in situ hybridization study, and G. Vassart and R. Pochet for helpful discussions. M.P. is Chargé de Recherche du Fonds National de la Recherche Scientifique.

REFERENCES

Benton, W.D.and Davis, R.W. (1977). Screening lgt recombinant clones by hybridization to single plaques in situ. Science 196: 180-182.

Bucher, P. and Trifonov, E.N. (1986). Compilation and analysis of eukaryotic Pol II promoter sequences. Nucl. Acids Res. 14: 10009-10026.

Christakos, S., Friedlander, E.J., Frandsen, B.R. and Norman, A.W. (1979). Studies on the mode of action of calciferol. XIII. Development of a radioimmunoassay for vitamin D-dependent chick intestinal calcium-binding protein and tissue distribution. Endocrinology 104: 1495-1503.

Evans, T., Schon, E., Gora-Maslak, G., Patterson, J. and Efstratiadis, A. (1984). Sl-hypersensitive sites in eukaryotic promoter regions. Nucl. Acids Res. 12: 8043-8058.

Feldman, S.C. and Christakos, S. (1983). Vitamin D-dependent calcium-binding protein in rat brain: biochemical and immunocytochemical characterization. Endocrinology 112: 290-302.

Fournet, N., Garcia-Segura, L.M., Norman, A.W. & Orci, L. (1986). Selective localization of calcium-binding protein in human brain stem, cerebellum and spinal cord. Brain Res. 399: 310-316.

Jande, S.S., Maler, L. & Lawson, D.E.M. (1981). Immunohistochemical mapping of vitamin D-dependent calcium-binding protein in brain. Nature (London) 294: 765-767.

Kallfelz, F.A., Taylor, A.N. and Wasserman, R.H. (1967). Vitamin D-induced calcium-binding factor in rat intestinal mucosa. Proc. Soc. Exp. Biol. Med. 125: 54-58.

Kimura, M. (1983). The Neutral Theory of Evolution, Cambridge University Press, Cambridge.

Kretsinger, R.H. (1980). Structure and evolution of calcium-modulated proteins. C.R.C. Crit. Rev. Biochem. 8: 119-174.

McBurney, R.N. & Neering, I.R. (1987). Neuronal calcium homeostasis. Trends Neurosc. 10: 164-169.

McClellan, J.A., Palacek, E. and Lilley, D.M.J. (1986). $(A\text{-}T)_n$ tracts embedded in random sequence DNA - Formation of a structure which is chemically reactive and torsionally deformable. Nucl. Acids Res. 14: 9291-9309.

Nickol, J.M. and Felsenfeld, G. (1983). DNA conformation at the 5' end of the chicken adult β-globin gene. Cell 35: 467-477.

Nussinov, R., Owens, J. and Maizel, J.V. (1986). Sequence signals in eukaryotic upstream regions. Bioch. Bioph. Acta 866: 109-119.

Parmentier, M., Lawson, D.E.M. and Vassart, G. (1987a). Human 27-kDa calbindin complementary DNA sequence: Evolutionary and functional implications. Eur. J. Biochem. 170: 207-215.

Parmentier, M., Ghysels, M., Rypens, F., Lawson, D.E.M., Pasteels, J.L. & Pochet, R. (1987b). Calbindin in vertebrate classes: Immunohistchemical localization and Western blot analysis. Gen. Comp. Endocr. 65: 399-407.

Parmentier, M. and Vassart, G. (1988). HindIII RFLP on chromosome 8 detected with a Calbindin 27kDa cDNA probe, HBSC21. Nucl. Acids Res. (in press)

Pochet, R., Parmentier, M., Lawson, D.E.M. & Pasteels, J.L. (1985). Rat brain synthesizes two 'Vitamin D-dependent' calcium binding proteins. Brain Res. 345: 251-256.

Rodgers, J. (1987). Calretinin: a gene for a novel calcium-binding protein expressed principally in neurons. J. Cell Biol., 105: 1343-1353.
Roth, J., Baetens, D., Norman, A.W. and Garcia-Segura, L.-M. (1981). Specific neurons in chick central nervous system stain with an antibody against chick intestinal vitamin D-dependent calcium-binding protein . Brain Res. 222: 452-457.
Sanger, F., Nicklen, S., Coulson, A.R. (1977). DNA sequencing with chain-terminating inhibitors. Proc. Natl. Acad. Sci. USA 74: 5463-5467.
Takagi, T., Nojiri, M., Konishi, K., Maruyama, K. and Nonomura, Y. (1986). Amino acid sequence of vitamin D -dependent calcium-binding protein from bovine brain. FEBS Lett. 201: 41-45.
Taylor, A.N. (1974). Chick brain cerebellar CaBP: comparison with intestinal vitamin D-induced CaBP. Arch. Biochem. Biophys., 161: 100-108.
Taylor, A.N. and Wasserman, R.H. (1967). Vitamin D3-induced calcium-binding protein: Partial purification, electrophoretic visualization and tissue distribution. Arch. Biochem. Biophys. 119: 536-540.
Thomasset, M., Parkes, C.O. & Cuisinier-Gleizes, P. (1982). Rat calcium-binding protein: distribution, development and vitamin D dependence. Am. J. Physiol. 243: E483-E488.
Wasserman, R.H., and Taylor, A.N. (1966). Vitamin D3-induced calcium-binding protein in chick intestinal mucosa. Science 152: 791-793.
Wilson, P.W., Harding, M. and Lawson, D.E.M. (1985). Putative amino acid sequence of chick calcium-binding protein deduced from a complementary DNA sequence. Nucl. Acids Res. 13: 8867-8881.
Wilson, P.W., Rodgers, J., Harding, M., Pohl, V., Pattyn, G. and Lawson, D.E.M. (1988). Structure of chick chromosomal genes for calbindin and calretinin. J. Mol. Biol. 200: 615-625.

STRUCTURE OF THE RAT VITAMIN D-INDUCED CALBINDIN-D9K GENE AND EVOLUTION OF THE EF-HAND CALCIUM-BINDING PROTEIN FAMILY

Christine Perret, Nourredine Lomri and Monique Thomasset

INSERM U.120
44, Chemin de Ronde
78110 Le Vésinet, France

INTRODUCTION

Based upon their number of calcium sites, two intracellular vitamin D-regulated calcium-binding proteins, CaBPs, (Calbindin-D9K and 28K) are expressed in specific target organs of the rat (1,2). CaBP9K was first located in the absorptive cells of the duodenum and has two calcium-binding sites. CaBP28K, which is concentrated in the distal tubules of kidney and in specific neurons of the nervous system, has four functional calcium-binding sites. Only the CaBP28K is present in birds and is expressed in the intestine, the kidney and specific neurons of the nervous system. Both CaBP9K and 28K belong to the calciprotein family which includes calmodulin, troponin C, parvalbumin (3). In this family, each Ca-binding site is organized into a consensus structure called by Kretsinger an EF-hand which consists of a helix, a loop and a helix. Calcium is bound to the loop region (3).

The members of this gene family are all believed to be derived from a common ancestor, which, itself, was the product of two successive gene duplications (4). This is apparent from analysis of the internal similarities of calmodulin which is the protein closest to the common ancestor (4). There are considerably more similarities between sites I and III and between sites II and IV, than between all the other sites (fig. 1).

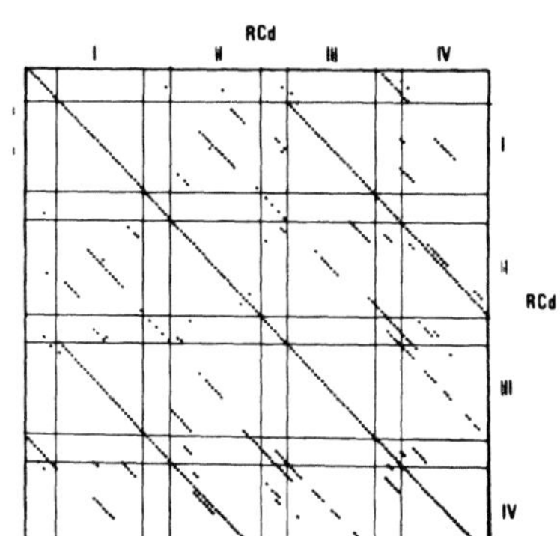

Fig. 1. Internal similarities of calmodulin. Dot matrix analysis of Staden of the sequence of rat calmodulin (RCd) versus RCd.

In this study, we compare and analyze the structures of the recently described genes coding for these calciproteins and propose an hypothesis for the evolution of the "EF hand" protein family (5).

ANALYSIS OF CaBP9K STRUCTURE

The complete sequence of the rat CaBP9K has been deduced from a cloned cDNA plus a 5' primer extension (6,7). The rat CaBP9K contains two calcium-binding sites composed of -helix-loop-helix. Site I is modified by the insertion of two aminoacids and site II has the classical EF hand structure with 29 aminoacids (8). S100 protein is also a protein with two calcium-binding sites. Site I is a modified EF-hand with 31 aminoacid residues and site II has a classical EF-hand structure (8). Several other two-site calcium-binding proteins have recently been described : calcyclin 2A9 protein (9), P_{10} or P_{11} the tyrosine kinase substrate (10), CFAg the cystic fibrosis antigen (11), MRP-8 and MRP-14 the macrophage rheumatoïd proteins (12) and p9Ka protein (13). The sequence of these proteins are compared in fig. 2.

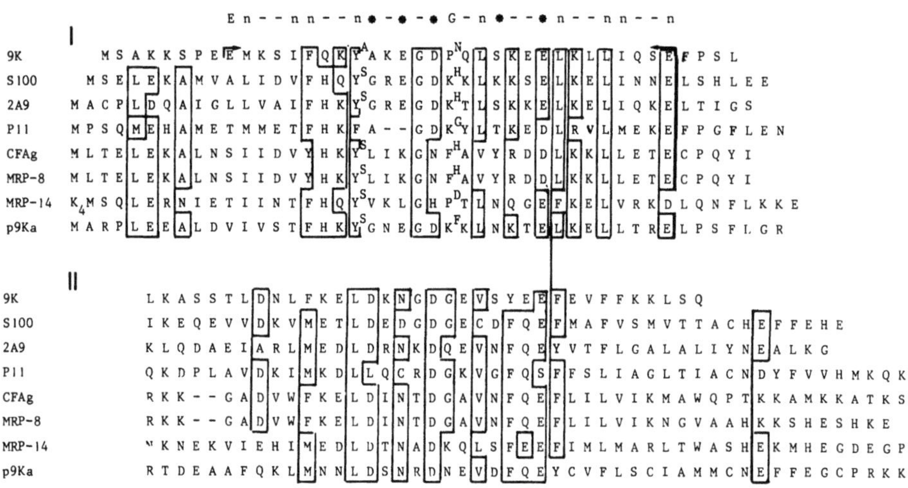

Fig. 2. Comparison of the aminoacid sequences of the two-site calcium-binding proteins with a modified site I. Aminoacids that are similar in more than 50% of the sequence are boxed. The above sequence represents the consensus EF hand sequence described by Kretsinger (3).

ANALYSIS OF CaBP9K GENE

The structure of CaBP9K gene is now known (7) and is shown in fig. 3. This gene contains three exons interrupted by two introns. The first exon contains almost the entire 5' non coding region. The second exon codes for the calcium-binding site I. The third exon codes for calcium site II and the 3' untranslated region. Therefore each calcium domain is encoded by a separate exon.

```
                                 cacccacccccattttgtatt -365
tttaaaagccatgtatactatccctctgtgacgcatttttttttctgagtag -313
gaaaattattgtcaagcctaaatctcgaattccattgcttttagaaatgtgg -261
actgtgcgcaaggtcaaaattctatactgaagggaaaatatgctaaaggaga -209
agaaaagccaggttttacccttgggactctgaaccattaagcaattgttgaa -157
gtagctctctattggagcttgaccttaattgagagtcttaagcttggtctca -105
gaaaccattaatcattacccttaaatagtaaacaaagggaaatttcataatc  -53
agggtggtgtgtctgtaaagac[tataaaa]gagctcctcgtcggctcttcatc   -1
  AGACCTCACCTGTTCCTGTCTGACTCTGGCAGCACTCACTGACAGCAAG   49
 +1
CAGgtcag.........                           .........tacag   359
CAC AGA AAA ATG AGC GCT AAG AAA TCT CCC GAA GAA ATG   398
            Met Ser Ala Lys Lys Ser Pro Glu Glu Met
AAG AGC ATT TTT CAA AAA TAT GCA GCC AAA GAA GGC GAT   437
Lys Ser Ile Phe Gln Lys Tyr Ala Ala Lys Glu Gly Asp
CCA AAC CAG CTG TCC AAG GAG GAG CTG AAG CTG CTG ATT   476
Pro Asn Gln Leu Ser Lys Glu Glu Leu Lys Leu Leu Ile
CAG TCA GAG TTC CCC AGC CTC CTG AAG gtgagt.........
Gln Ser Glu Phe Pro Ser Leu Leu Lys
        .........gaagcag GCT TCA AGT ACT CTA GAC AAT  2331
                         Ala Ser Ser Thr Leu Asp Asn
CTC TTT AAA GAG CTG GAT AAG AAC GGT GAT GGA GAA GTT  2370
Leu Phe Lys Glu Leu Asp Lys Asn Gly Asp Gly Glu Val
AGC TAT GAA GAA TTC GAA GTT TTC TTC AAA AAG TTA TCA  2409
Ser Tyr Glu Glu Phe Glu Val Phe Phe Lys Lys Leu Ser
CAA TGAAGCCAGAAGAAGGAGCTCCGACACCACCTACTGATTGAATCCTAT 2460
Gln
CCAATCCCAAAGATCTAGCTGTGAGAGCAAGATACTGTTAATAAAGCAAATT 2512
CTGAGAC   atgtctctcttgtgaagtactgactcttgattacttagattt 2561
cagaggaatcttagtctgatgtgcctatttgatttgtctgaaatttcctttg 2613
ttttacattcatttatttatgtggtgtatgtattgcacatatgtacat     2661
```

Fig. 3. Nucleotide sequence of rat CaBP 9K gene. The cap site (+1) is indicated. The 5' flanking region has been numbered negatively. GT/AG at splice junctions are double-underlined. Putative CAAT boxes and the polyadenylation signal are underlined. The TATA box is boxed.

Analysis of the known structure of the genes that encode other two -site calcium-binding proteins that are 2 A9 (14) and p9Ka (11) proteins, similar to CaBP9K, show similar structure (fig. 4). The intron position in the interdomain between each calcium-binding site is conserved together with the type of phase (phase 0).

There is one intron in the 5' non-coding region of the 2A9 gene. This structure (3 exons, 2 introns with one intron in the 5' non-coding region and one intron of phase 0 in the interdomain) is also found in the MRP-8 and MRP-14 genes recently described (15). The structure of the 5' region of the p9Ka gene is not yet known(13).

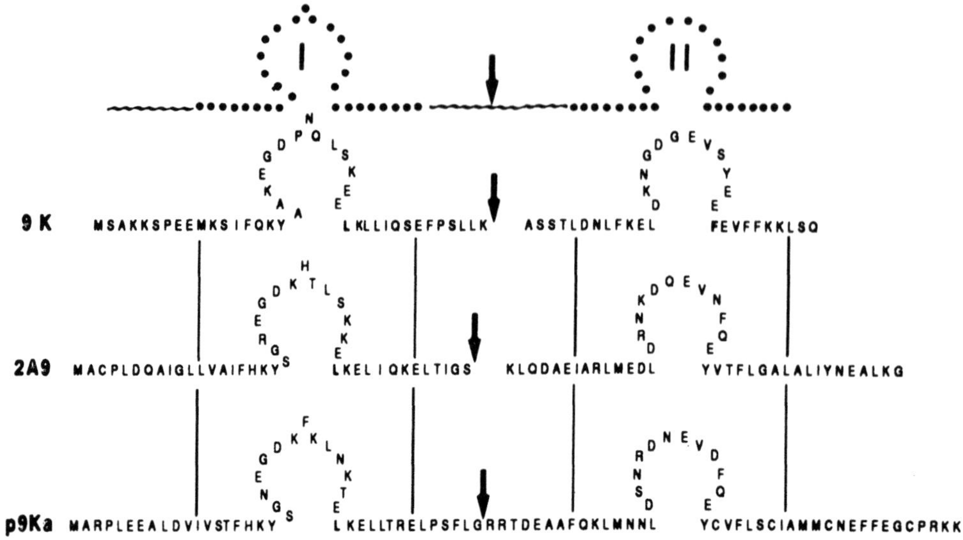

Fig. 4. Analysis of the intron position in relation to the calcium-binding sites of the two-calcium site proteins. Each calcium-binding site (I, II) is represented by a loop which indicates putative Ca^{++} binding sites and by 8 aminoacid residues adjacent to the loop which correspond to the putative regions of the helices. One intron of phase 0 (↓) is present between the 2 sites, in the interdomain region. The supposed common ancestor to these genes is represented in the upper figure.

EVOLUTION OF THE TWO-SITE CALCIUM-BINDING PROTEIN GENES

The above observations support the hypothesis that these genes are derived from a common ancestor which is represented in fig. 4. The differences in the intron positions in the interdomain region can be explained by a sliding mechanism (16).

Analysis of the internal sequence homologies of the CaBP9K cDNA showed similarities between site I sequence and the 3' sequence (called site III-like) this could suggest that CaBP9K is derived from a four-site protein by early termination (6). The most obvious four-site ancestor would be CaBP28K, as this is the only CaBP in birds.

We have looked for a cross-hybridization between the messengers to test this hypothesis. The CaBP9K cDNA hybridizes to a single mRNA species 600 nucleotides long from rat duodenal mRNA. No larger mRNA containing about two thousand nucleotides extracted from kidney and cerebellum and coding for CaBP28K hybridizes to the cDNA probe under either high and or low stringency conditions (17). The sequence of CaBP28K is now known it contains six EF hand structures, four functional calcium sites. No modified 31 aminoacid calcium site is present (18).

Moreover, comparison of CaBP9K, CaBP28K and S100 protein sequences using the dot matrix analysis (fig. 5) shows that there are more similarities between CaBP 9K and S100 than between CaBP9K and CaBP28K. Therefore CaBP28K is not the ancestor of CaBP9K and there is no direct filiation between the two proteins.

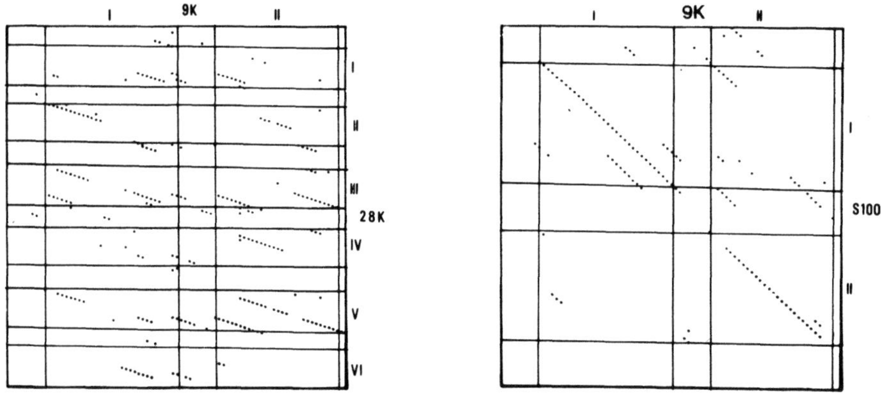

Fig. 5. Comparison between the protein sequences of rat CaBP9K, CaBP28K and S100 protein using the Staden's dot-matrix analysis. Calcium sites are noted by roman figure. 9K : rat CaBP 9K. 28K : rat CaBP 28K (19) - S100 : rat S100 protein (20).

EVOLUTION OF THE EF-HAND CALCIUM-BINDING PROTEIN FAMILY

Several other genes coding for the intracellular calcium-binding proteins have been recently reported as shown in fig. 6, calmodulin, Spec I protein, myosin alkali light chain, parvalbumin, myosin regulatory light chain, calpain genes (see ref 6). If the members of this family are all derived from a common ancestor that itself was the product of two successive gene duplications, these duplications should be recognizable in the structure of the genes. Introns should be present between the domains, or if introns interrupt domains, the position of the introns should be symmetric in relation to the domains. But this is not readily apparent from the known structures of these genes (fig. 6). Therefore several authors have proposed that introns are not directly involved in the duplication, although this is contrary to the exon shuffling theory of Gilbert (21). We have carefully reexamine the intron position and aminoacid conservation in relation to the domains so as recognize any symmetry.

The data indicate that, firstly, the aminoacid after the "EF-hand" domains I and III is highly conserved and is very often a glycine residue. These 2 Gly residues are also present after site I and site III in other proteins, such as troponin C and calcineurin B (references in 5). Secondly the first aminoacid of the C terminal helix is very often a leucine (L) in sites I and III and a phenylalanine (F) in sites II and IV. This L F L F alternance is also present in other proteins whose gene structure is not yet known, such as troponin C and calcineurin B. This aminoacid substitution probably took place after the first duplication and before the second duplication as shown in fig. 7. It is an argument in favour of the formation of the four-site ancestral gene resulting from a two successive duplications.

Finally when an intron is present inside the conserved codon glycine it always has the same phase (phase 1). There is such a phase 1 intron in MLC_3 and MLC_2 genes after site III and site I respectively (fig. 6).

The high degree of conservation of this glycine residue after sites I and III, and the presence of an intron of the same phase, when it exists, in this codon is intriguing. It therefore seems likely that these introns are the vestigial introns from the first duplication which have also been duplicated. They are labeled I_1 and $I_{1'}$ in fig. 7. Independent insertion

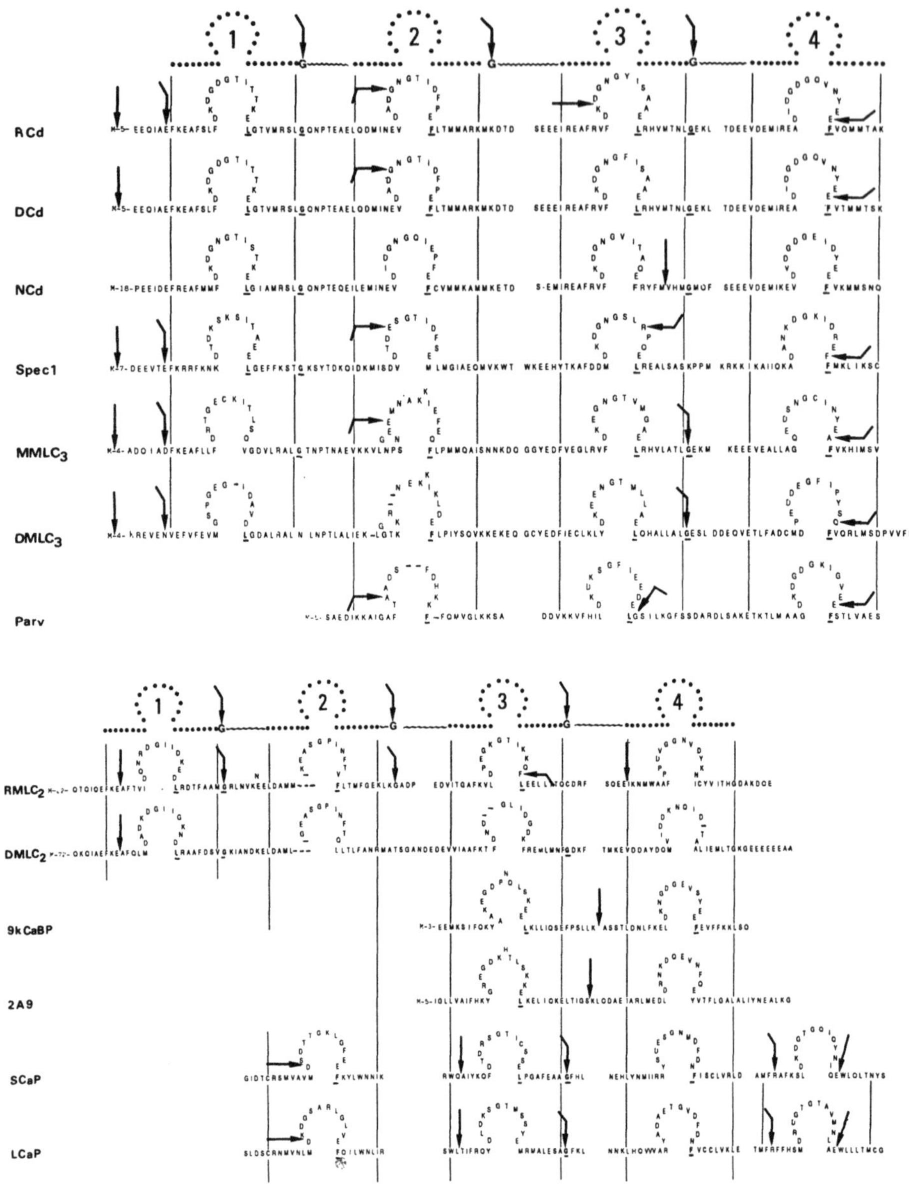

Fig. 6. Analysis of intron positions in relation to calcium-binding sites : the three classes of introns are indicated by the arrows : type 0 (↓), type 1 (↘), type 2 (↙). The representation is explained in fig. 3. The conserved aminoacid residues : glycine (G) after the C-terminal helix, and the first residue of the C-terminal helix (when it is L or F) are underlined.
RCD, DCd, NCd : Rat, Drosophila, Nematode Caenorhabditis elegans calmodulin - Spec I : Sea Urchin Spec I - MMLC3, DMLC3 : Mouse, Drosophila myosin alkali light chain - RMLC2, DMLC2 : Rat, Drosophila myosin regulatory light chain-9KCaBP : Calbindin-D9K - 2A9 : Calcyclin - SCaP : Small calpain - LCaP Large calpain (References in 5). Reproduced, by permission of Springer-Verlag N.Y. Inc. from Perret *et al* (5).

events cannot be excluded, but it seems more likely that these introns are the relics of orginal interdomain introns since, of the 9 introns present in the interdomains, 6 are of the same type (phase 1 intron in conserved glycine codon) and at homologous positions relative to the EF hand domains.

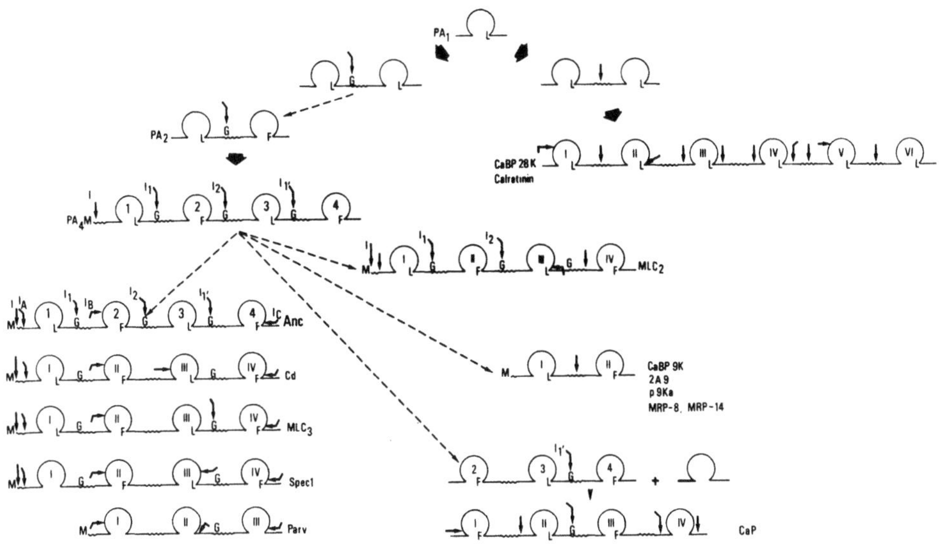

Fig. 7. Schematic outline of the evolution of the lineages in the calcium-binding protein family. Each calcium-binding site is reprented by a loop flanked by two straight lines for the two helices. The linker regions between each site are represented by a wavy line. Duplication or triplication events are noted by large dotted arrows. The shapes of the thin arrows indicate the type of the intron (see fig. 6). Roman numerals indicate the sites of the number in the present day proteins, the arabic numerals are for the proposed ancestral proteins.
PA1 : one-site primordial ancestor - PA2 : two-site primordial ancestor. - PA4 : four-site primordial ancestor.
M : first aminoacid of the protein - G : glycine - L : leucine - F : phenylalanine.

MLC$_2$ genes possess the same type of introns as I_1 and $I_{1'}$ between sites II and III (fig. 6) and shows the duplication events of the LFLF type described above. If we assume this intron called I_2 on fig. 7 is a homologous intron to I_1 and $I_{1'}$, it can be speculated that the 4 site primordial ancestor gene (PA4 in fig. 7) produced by two successive duplications had introns which flanked each domain. These introns would have permitted duplication. Such an arrangement is in full agreement with the proposal of Gilbert (21) on the role of introns. The distinct branches within this family of proteins can be interrelated if it is assumed that divergence took place by remodelling the structure of this ancestral gene and principally loss and insertion of introns. Each lineage of proteins will be analyzed by applying this hypothesis to explain their origin.

Evolution of the calmodulin-Spec I-MLC3-parvalbumin group

The structures of the genes of all the members of this group are remarkably similar (fig. 6). The positions of the three introns (labeled IA, IB and IC in fig. 7) are completely conserved, as is the intron phase. This structure confirms their common evolutionary origin. The 4-site primordial ancestor (PA4) probably evolved by insertion of the three introns IA, IB and IC to give rise to the common ancestor of this group, labeled Anc in fig. 7.

Calmodulin genes would arise from this gene Anc by loss of introns I1, I2 and I1' as well as Spec I, MLC3 by loss of introns I1 and I2, and parvalbumin by loss of site I and introns I2 and I1'.

Evolution of the myosin regulatory chain (MLC_2) genes

MLC_2 genes would arise from PA4 by loss of intron I1' and insertion of introns specific to MLC_2.

Evolution of the Calpain genes (CaP)

The calpains are calcium-dependent intracellular cystein proteases. They are heterodimers, made up of an 80 kDa large subunit (LCaP) and a 30 kDa small subunit (SCaP). The LCaP contains one protease domain and one domain that binds four calcium atoms represented in fig. 6 with four EF hand structures. The SCaP also has a Ca binding region with four EF hands. These two genes respect the same structure (fig. 6). The analysis of the structure of these genes is interesting since it is an argument in favour of our hypothesis.

Also, closer examination of the aminoacid residue shows that, instead of the alternating pattern I/L, II/F, III/L, IV/F, the pattern is I/F, II/L, III/F, IV/X, where X is not conserved (fig. 6).

This arrangement would suggest that the three first sites in CaP correspond to the last three sites of calmodulin so producing the alignement shown in fig. 6 resulting from the sliding, the intron between sites II and III is at exactly the same position and in the same phase as the I_1 intron from the initial duplication.

The evidence of this sliding hypothesis should be verified by the alignement of -site I of CaP with site II of the common ancestor and site II with site III. Indeed it is apparent, if we compare SCaP and Calmodulin, the protein closest to the common ancestor, site I of SCaP is aligned with site II of Cd and site II of SCaP is aligned with site III (fig. 8).

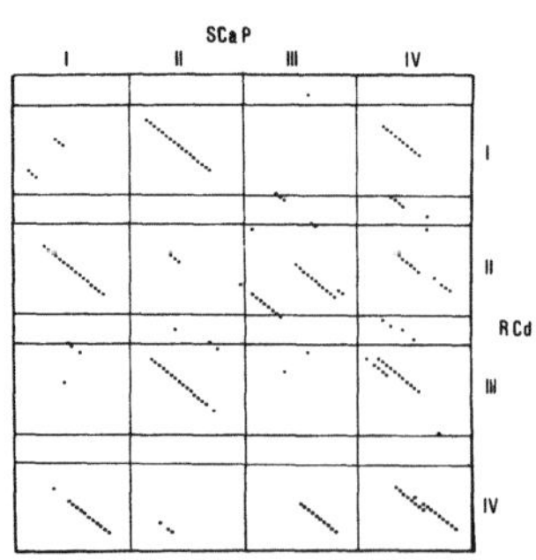

Fig. 8. Staden's dot matrix analysis of SCaP versus calmodulin. SCaP : small calpain. RCd : rat calmodulin.

Evolution of the two site calcium-binding proteins similar to CaBP9K

The evolution of the two site CaBP has been already analyzed. The ancestral gene would be derived from the four-site gene PA_4 by loss of two sites and genomic rearrangements that lead to a modified site I (fig. 7).

Evolution of the CaBP28K genes

The structure of the chicken CaBP28K gene is shown in fig. 9 and derived from (12).

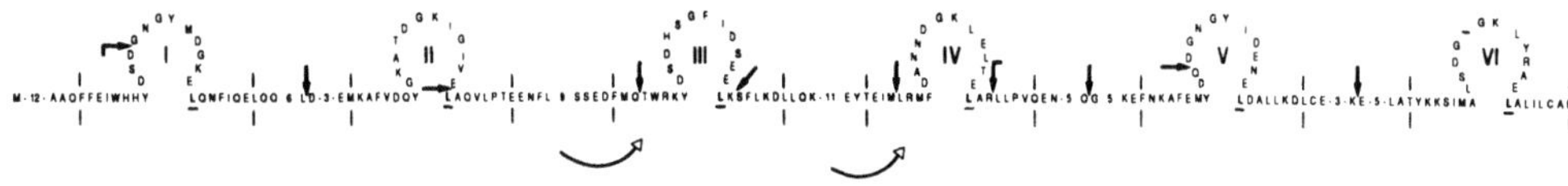

Fig. 9 Structure of the chicken CaBP28K gene. Analysis of intron positions in relation to the calcium binding sites. The representation is similar to that of figs 3, 4 and 6. The arrows represent a possible intron sliding.

We note that firstly the first aminoacid of the C terminal helix is always a leucine (L) and there is no L-F alternating sequence as seen in all the members. This is also true for CaBP28K of other species (5).

Secondly the Gly residue immediately following sites I and III which is highly conserved in the other members is absent from CaBP28K. This is also true for CaBP28K of other species.

At last, one intron in phase 0 is present in the interdomain after sites I, IV and V, two other phase 0 introns are in the N-terminal helix of sites III and IV and could have moved by a sliding mechanism (16) from the interdomain junctions as the sequences in between are not strongly conserved.

Moreover if we analyze the internal similarities in the CaBP28K, it is apparent that these proteins derives by a triplication event of a two-domain ancestor (fig. 10).

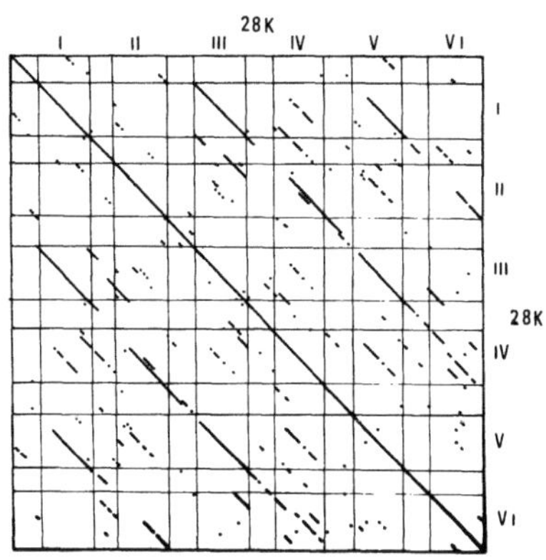

Fig. 10 Internal similarities in the CaBP28K. Staden's dot matrix analysis of the sequence of rat CaBP28K (19) versus rat CaBP28K.

Therefore, the evolutionary origin of CaBP28K is the most divergent from the rest of the calcium-binding protein, since the alternating event LFLF does not exist. The divergence would have taken place at the first duplication of the gene encoding one calcium-binding domain. An intron of phase 0 would have permitted this duplication event. Then it would have been followed by a triplication event leading to a six "EF-hand" structure. The five introns of phase 0 would have permitted the triplication. Introns would then be inserted. This evolutionary hypothesis could account for the CaBP28K gene and is shown in fig. 7.

CONCLUSIONS

The ancestral gene coding for one calcium-binding site was duplicated and a second duplication gave rise to the four domain ancestor gene.

The majority of the calciprotein genes are derived from this four-domain ancestor. However the CaBP28K gene would have arisen from another product of duplication of the ancestral gene coding for one calcium-binding site, followed by a triplication. All the duplication or triplication events would have involved introns present in the interdomain junctions according to the exon shuffling theory of Gilbert (21). Each distinct lineage evolved from these ancestors by genomic rearrangement, in which insertion of introns was a prominent feature.

Acknowledgments We thank L. Castera and INSERM SC6 personnel for their help in the preparation of the manuscript.

REFERENCES

1. M. Thomasset, C.O. Parkes, P. Cuisinier-Gleizes. Am. J. Physiol. 243:E483 (1982).
2. R.H. Wassermann, C.S. Fullmer. In Calcium and Cell Function (Cheung W.Y. ed.) Academic Press, New York, vol. 2 p. 175 (1982).
3. R.H. Kretsinger. Critical Review in Biochemistry :119 (1980).
4. M. Goodman, J.F. Pechere, J. Haiech, J.G. Demaille. J. Mol. Evol. 13:331 (1979).
5. C. Perret, N. Lomri, M. Thomasset. J. Mol. Evol. 27:351 (1988).
6. C. Desplan, O. Heidmann, J.W. Lillie, C. Auffray, M. Thomasset. J. Biol. Chem. 258:13502 (1983).
7. C. Perret, N. Lomri, N. Gouhier, C. Auffray, M. Thomasset. Eur. J. Biochem. 172:43 (1988).
8. D. Szebenyi, K. Moffat. J. Biol. Chem. 261:8761 (1986).
9. B. Calabretta et al. J. Biol. Chem. 261:12628 (1986).
10. V. Gerke, K. Weber. EMBO J. 4:2917 (1985).
11. J.R. Dorin et al. Nature 326:614 (1987).
12. K. Odink et al. Nature 330:80 (1987).
13. R. Barraclough et al. J. Mol. Biol. 198:13 (1987).
14. S. Ferrari et al. J. Biol. Chem. 267:8325 (1987).
15. E. Lagasse, R.G. Clerc. Mol. & Cell Biol. 8:2402 (1988).
16. C.S. Craik, W.J. Rutter, R. Flettrick. Science 220:1125 (1983).
17. C. Perret, C. Desplan, M. Thomasset. Eur. J. Biochem. 150:211 (1985).
18. Hunziker. Proc. Natl. Acad. Sci USA 83:7578 (1986).
19. Yamakuni et al. Nucl. Ac. Res. 14:6768 (1986).
20. Kuwano et al. Med. Acad. Res. 12:7455 (1984).
21. W. Gilbert. Nature 271:501 (1978).
22. P.W. Wilson, M. Hardin, V. Potel, G. Pattyn, D.E.M. Lawson. J. Mol. Biol. 200:615 (1988).

STRUCTURAL ORGANIZATION OF THE HUMAN PARVALBUMIN GENE

Martin W. Berchtold

Institute for Pharmacology and Biochemistry
University of Zürich-Irchel
8057 Zürich, Switzerland

INTRODUCTION

Several genes encoding Ca2+-binding proteins of the EF-hand type have been isolated from different species (for a recent review see[1]). Many structural features such as splice site positions with respect to the homologous amino acid sequences, are common among these genes. The original hypothesis put forward by Kretsinger[2] that EF-hand type Ca2+-binding proteins evolved from a common ancestor by gene duplication could be confirmed by molecular genetic analysis and a better understanding of the genetic diversity was the result of this approach. Based on the intron/exon splice position of the genes for parvalbumin, calmodulin, myosin light chain 1 and 3, Spec1, calbindin and calretinin an evolutionary tree has been recently proposed[3].

Parvalbumin is a three domain Ca2+-binding protein which most likely diverged relatively late during evolution from a four domain ancestor by a deletion of the N-terminal Ca2+-binding domain (for a review article see[4]) Comparison of rat parvalbumin cDNA with calmodulin cDNA revealed a region of high homology around the parvalbumin ATG translation start codon extending into the 5'noncoding region with the first Ca2+-binding domain of calmodulin[5].

There were two major rationales underlying this investigation. Firstly, analysis of the human parvalbumin gene. This should enable a deeper insight into the evolution of parvalbumins as so far only the genetic organization of rat parvalbumin has been elucidated[6]. Secondly, analysis of potentially conserved sequences in the parvalbumin promoter could lead to an understanding of mechanisms important for the transcriptional regulation of parvalbumin gene expression. The identification of a homologous promoter region in the parvalbumin gene and the gene for myosin light chain 3F[7], which is expressed in the same muscle fibre type (IIb) as parvalbumin was a starting point for a detailed study of cis- and trans-acting regulation important for the tight tissue specific and development dependent expression of parvalbumin[8,9,10].

CLONING AND ANALYSIS OF THE HUMAN PARVALBUMIN GENE

The human chromosomal assignment for the parvalbumin gene has been determined recently using the method of somatic cell hybridization with a Chinese hamster X human hybrid panel[6]. Lowest discordance was observed between the presence or absence of parvalbumin sequences and human chromosome 22. Based on these findings a chromosome 22 specific <u>Eco</u> R1 library was screened with the rat parvalbumin cDNA 9f[9] for the presence of the gene for human parvalbumin. A 7.5Kb Eco R1 fragment (termed λh7.5) was identified as part of the natural human parvalbumin gene based on the the fact that Eco R1 digested total human DNA showed a single 7.5Kb fragment which hybridized to parvalbumin sequences and labeled the same bands as rat parvalbumin cDNA on Northern blots with rat muscle RNA.

The human parvalbumin clone λh7.5 was then characterized in more detail. The result of Southern blot analysis using a variety of restriction enzymes and different exon specific probes from either full-length rat parvalbumin cDNA[5], or genomic DNA[6] is shown in Fig.1. Four out of the predicted five exons (in analogy to the rat parvalbumin) were identified in this clone. The 3'most exon was not present in λh7.5. Since in the rat parvalbumin the corresponding sequence is divided by an intron of a length of 9Kb it is possible that also in the human genome parvalbumin exons four and five are far apart. By using a rat cDNA fragment encoding exon five it was not possible to find positive clones in the chromosome 22 specific Eco R1 and Hind III libraries. This might be due to the fact that these two libraries contain inserts of restricted length of only up to approximatly 9Kb. The size of the analyzed part of the human parvalbumin gene is much shorter compared to the rat equivalent. All exons and the intron/exon

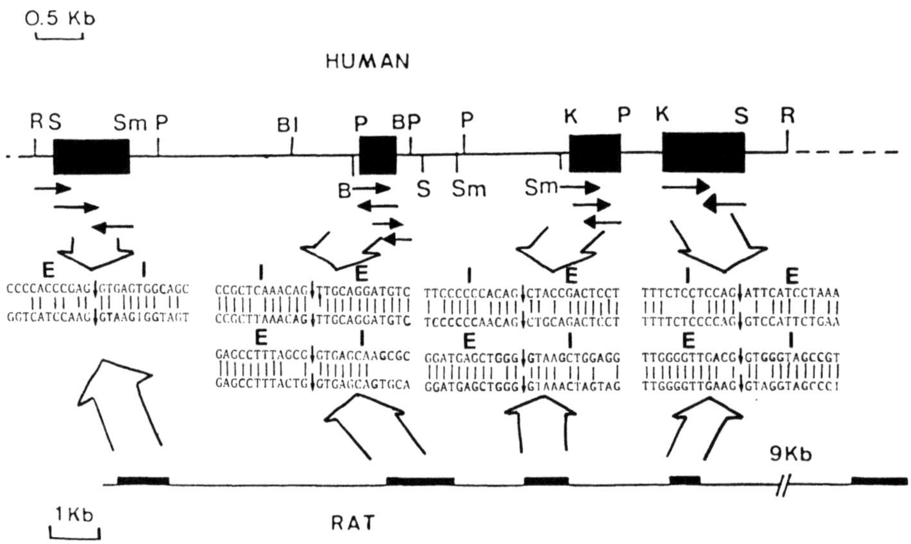

Fig.1. Intron/Exon organization of the human parvalbumin gene compared to the rat parvalbumin gene. Mapping was carried out by Southern blot analysis using combinations of different restriction enzymes: Restriction enzymes abbreviations are: B, Bam H1; Bl, Bgl II; K, Kpn 1; P, Pst 1; R, Eco R1; S, Sst 1; Sm, Sma 1. CDNA and genomic rat parvalbumin sequences were used as probes. Boxes indicate exons since they hybridized to fractions of the rat cDNA which are encoded by only one exon. Arrows below the gene map indicate the sequencing strategy used to analyze exons and intron/exon boundaries. Sequences around splice sites are shown between the representations of the human and rat parvalbumin. Exon sequences are indicated by E and introns by I. Sequence identities between the human and rat gene are shown by vertical dashes (I). Small arrows indicate splice site positions and large arrows show the rat or human origins of the boundary sequences.

junctions were sequenced mostly on both strands by subcloning fragments into M13 phages followed by standard dideoxy sequencing. Splice site positions with respect to the amino-acid sequences were identical in the rat and human parvalbumin gene. Splice site bordering sequences either in exons or introns showed a high degree of homology. This indicates that the gene structure was conserved in a stringent way during evolution.

DEDUCED AMINO-ACID SEQUENCE OF HUMAN PARVALBUMIN COMPARED TO OTHER MAMMALIAN AND CARP 4.25 PARVALBUMINS

The four analyzed exons in the human parvalbumin encode for amino-acids 1-100. Since the so far investigated mammalian parvalbumins[11,12] contain 109 amino-acids it is assumed that 9 amino-acids are missing in the characterized part of the human parvalbumin gene. Fig.2 shows a comparison between the human (this study), rabbit[11], rat[12] and carp 4.25[13] parvalbumin amino-acid sequences. In addition, the sequence of rat oncomodulin[14], a tumor specific parvalbumin-like protein is compared to human parvalbumin.

Only conservative amino-acid replacements between the three investigated mammalian parvalbumins were detected and all the amino-acids

```
                      A                   B             C

                 10         20         30         40         50
human   SMTDLLNAED IKKAVGAFSA TDSFDHKKFF QMVGLKKKSA DDVKKVFHML
rat           S        I   T  A                             I
rabbit  A  E           I   A  AE                  T E       I

carp    AFAGV  DA   AA LE CK  A    N  A   AK   TS         A AII
rat om   I  I S     AA LQECQD P T QPQ     TS  S M   SQ  DI RFI

                               D                E

                 60         70         80         90        100
human   DKDKSGFIEE DELGFILKGF SPDARDLSAK ETKMLMAAGD KDGDGKIGVD
rat                    S       S            T                E
rabbit             E                  V     T               A

carp     Q            KLF QN  KA   A TDG   TFLK    S
rat om   N Q  YLDG    KYF QK  QS   E TES   S  D A  N       A
```

Fig.2. Alignment of amino-acid sequences of mammalian parvalbumins (this study),[11,12], carp 4.25 parvalbumin[13] and oncomodulin[14] (om). Only the first 100 amino-acids were compared since the C-terminal of the human parvalbumin has not been sequenced so far. Horizontal lines above the protein sequences indicate α-helical regions A/B, C/D, E and triangles are putative Ca^{2+}-binding sites although loop 1 is non functional due to a deletion. Arrows indicate intron positions determined for the rat and human parvalbumin gene and the form of the arrow shows the position of the intron within the codon: ↴ ,splice sites after the 1st bp of the codon; ↲,splice sites after the 2nd bp of the codon.

potentially involved in Ca2+-chelating are identical with the carp 4.25 parvalbumin. The human parvalbumin amino-acid sequence 1-100 differs in 9 and 13 positions from the rat and rabbit counterparts. In addition also mouse parvalbumin (Ch.Zühlke and F.Schöffl, personel communication) is very similar to the other mammalian parvalbumins and differs in 14 positions from human parvalbumin. However carp 4.25 parvalbumin[13] and the parvalbumin-like oncomodulin[14] vary quite markedly in that 41 and 49 amino-acids respectively are replaced. The region with the highest sequence variation is around the first half of the loop from the AB domain of mammalian parvalbumins. This domain has lost its Ca2+-binding capacity due to a deletion of two amino-acids and therefore I conclude that less pressure to maintain the original structure was exerted on this part of the molecule during evolution. The predicted protein structure was almost identical for the four known mammalian parvalbumins when the computer aided method of Garnier-Robson[15] was applied. This fits quite well with the 3-dimensional structure obtained by X-ray diffraction analysis of carp 4.25 parvalbumin[16]. These combined results indicate that although several amino-acids vary among mammalian parvalbumins and even more when compared to fish parvalbumin the over all structural features and probably also the Ca2+-binding properties have been maintained

```
                      -220                  -240
                  .          .         .         .         .
HUMAN       GCCCCGTACC CTGGCGTCTA ATTTGGGTGC TGGAGGGAAG CAGGGGCCAA
            G  CC T C    G C T         TGGGTGC TG AGGGA G   AG GGCCAA
RAT         GTGCCCTGCT TGGACCTTAT TGCTGGGTGC TGTAGGGAGG TAGAGGCCAA

          -180                -160                     -140
              .          .         .         .         .
HUMAN       -GAGTTATTA ATAGTCTTGC CTGATGGGCC CAGGGAGGC- -TGAATGTGA
             GA T ATTA ATAG C TG   CT A    G C C GG  GG    TGAATGTGA
RAT         GGATTAATTA ATAGGCCTGG CTCAATAGGC CTGGATGGGG GTGAATGTGA

                         -120                -100
              .          .         .         .         .
HUMAN       TACAGACACC CAGCACCACG GTTGGGGAGT ACCTGACACC GGAAGGGGAG
            TACAGACA C CAGCACC   G G TGGGGAGT ACCTGACACC GGAAGGGGAG
RAT         TACAGACA-C CAGCACCGTG GCTGGGGAGT ACCTGACACC GGAAGGGGAG

             -80                 -60                        -40
              .          .         .         .         .
HUMAN PV    GGGGTCCGGG GCTACGGGGA GTGCCACCTC CCAAAATAGC CGA-GCAGAA
            GGGG C GGG GCT   GGG A G GCCACCTC CCAAAATAGC C A GC G A
RAT PV      GGGGCCAGGG GCTGGGGG-A GCGCCACCTC CCAAAATAGC CAAGGCTGTA

                          -20              +1
              .          .         .         .
HUMAN PV    GCCTATATA GGTGGCCATC CCTCCTCCAG GCTCA
            G CTAT TA  GTG C   TC CC CC C AG  CTCA
RAT PV      GACTATTTA AGTGACTGTC CCACCACAAG TCTCA
```

Fig. 3. Comparison of the human parvalbumin promoter with the rat parvalbumin promoter. 230 bp of the putative human parvalbumin promoter are aligned with the rat counterpart. Numbering is based on the putative transcription start site (+1). Some deletions (-) are inserted in both sequences to optimize homologies. General transcription regulatory sequences (TATA and CAAT boxes) are shown by faint boxes. The bold box has a 79% identity to the myosin light chain 3F promoter[7].

during evolution. In contrast, oncomodulin was shown to exhibit different Ca^{2+}/Mg^{2+} binding properties and it was concluded that this protein could therefore be functionally different from parvalbumins[17].

THE PARVALBUMIN PROMOTER

Sequencing of a 0.7Kb Sst 1-Sma 1 fragment of the human parvalbumin genomic clone λh7.5 (Fig.1) revealed the presence of putative transcription regulatory elements such as a TATA box and a CAAT box. 230 bp of the promoter region were aligned with the rat parvalbumin promoter[6] (Fig.3). The most interesting observation was that there exists regions of high similarity between the human and rat parvalbumin promoter. One identical region is located at position -81 to -112 from the transcription start site. This is the sequence where a 79% homology to the myosin light chain 3F promoter[7] was found. The latter gene is active in the same muscle fibre type (IIb) as the parvalbumin gene. To investigate if this highly homologous promoter sequence could be a common element important for the regulation of tissue specificity we are currently performing gel retardation experiments[18]. We are using a synthetic oligonucleotide probe for the parvalbumin promoter region -81 to -112 and other promoter sequences to identify putative tissue specific trans-acting factors in total cellular or nuclear extracts.

ACKNOWLEDGMENT

I thank M.Berger for expert technical work. The Swiss National Foundation (grant 3.634-0.87) and Krebsliga des Kantons Zürich are acknowledged for financial support.

REFERENCES

(1) A.R.Means, J.A.Putkey and P. Epstein, Organization and evolution of genes for calmodulin and other calcium binding proteins. in: "Calmodulin", P.Cohen and C.B.Klee eds.,pp 17-33 (1988)
(2) R.H.Kretsinger, Structure and evolution of calcium-modulated proteins. CRC Crit.Rev.Biochem.8:119 (1980)
(3) P.W.Wilson, J.Rogers, M.Harding, V.Pohl, G.Pattyn and D.E.M.Lawson, Structure of chick chromosomal genes for calbindin and calretinin. J.Mol.Biol. 200:615 (1988)
(4) C.W.Heizmann and M.W.Berchtold, Expression of parvalbumin and other Ca2+-binding proteins in normal and tumor cells: a topical review. Cell Calcium 8:1 (1987)
(5) P.Epstein, A.R.Means and M.W.Berchtold, Isolation and characterization of a rat parvalbumin gene and full length cDNA. J.Biol.Chem. 261:5886 (1986)
(6) M.W.Berchtold, P.Epstein, A.L.Beaudet, M.E.Payne, C.W.Heizmann and A.R.Means, Structural organization and chromosomal assignment of the parvalbumin gene. J.Biol.Chem. 262:8696 (1987)
(7) Y.Nabeshima, Y.Fujii-Kuriyama, M.Muramatsu and K.Ogata, Alternative transcription and two modes of splicing result in two myosin light chains from one gene. Nature 308:333 (1984)
(8) M.W.Berchtold, M.R.Celio and C.W.Heizmann, Parvalbumin in nonmuscle tissues of the rat. Quantitation and immunohistochemical localization. J.Biol.Chem. 259:5189 (1984)
(9) M.W.Berchtold and A.R.Means, The Ca2+-binding protein parvalbumin: Molecular cloning and developmental regulation of mRNA abundance. Proc.Natl.Acad.Sci.U.S.A. 82:1414 (1985)

(10) M.W.Berchtold, Cloning of the rat parvalbumin gene. Methods in Enzymol. 139:317 (1987)
(11) D.L.Enfield, L.H.Ericsson, H.E.Blum, E.H.Fischer and H.Neurath, Amino-acid sequence of parvalbumin from rabbit skeletal muscle. Proc.Natl.Acad.Sci.USA 72:1309 (1975)
(12) M.W.Berchtold, C.W.Heizmann, and K.J.Wilson, Primary structure of parvalbumin from rat skeletal muscle. Eur.J.Biochem. 127:381 (1982)
(13) C.J.Coffee, and R.A.Bradshaw, Carp muscle calcium-binding protein.I. Characterization of the tryptic peptides and the complete amino acid sequence.J.Biol.Chem. 248:3305 (1973)
(14) J.P.MacManus, D.C.Watson and M.Yaguchi, The complete amino acid sequence of oncomodulin - a parvalbumin like calcium-binding protein from Morris Hepatoma 6123 tc. Eur. J. Biochem. 136:9 (1983)
(15) J.Garnier, D.J. Osguthorpe and B.Robson, Analysis of the accuracy and implications of simple methods for predicting the secondary structure of globular proteins. J.Mol.Biol. 120:97 (1978)
(16) R.H.Kretsinger and C.E.Nockolds, Carp muscle calcium-binding protein.II. Structure determination and general description. J.Biol.Chem. 248:3313 (1973)
(17) T.C.Williams, D.C.Corson, B.D.Sykes and J.P.MacManus, Oncomodulin, 1H NMR and optical stopped-flow spectroscopic studies of its solution conformation and metal-binding properties. J.Biol.Chem. 262:6248 (1987)
(18) R.Schneider, I.Gander, U.Müller, R.Mertz and E.L.Winnacker, A sensitive and rapid gel retention assay for nuclear factor I and other DNA-binding proteins in crude nuclear extracts. Nucleic Acid Res 14: 1303 (1986)

GENE EXPRESSION OF A RAT CEREBELLAR CA-BINDING PROTEIN, SPOT 35 PROTEIN

Y. Takahashi, T. Yamakuni, R. Kuwano, T. Kumanishi*, and E. Ohama**

Departments of Neuropharmacology, *Neuropathology, and **Experimental Neuropathology, Brain Research Institute Niigata University, Niigata 951, Japan

INTRODUCTION

About ten years ago we discovered a cerebellar protein, spot 35 protein, during the systematic analysis of the rat brain protein by using two dimensional gel-electrophoresis.[1] Then we found that this spot 35 protein is an acidic Ca-binding protein by using ^{45}Ca-binding assay procedure and is localized in Purkinje cells, their dendrite and axons using immunohistochemical examination.[2,3] Translational experiment of cerebellar mRNA suggested the transcriptional regulation of gene expression of spot 35 protein.[3] Next immunohistochemical studies revealed the developmental changes of Purkinje cells in the rat and human cerebella.[4] Furthermore, our neurophysiological study showed involvement of spot 35 protein in modulation of Purkinje cell activity of rat cerebellum.[5] Recently, we succeeded in cloning the cDNA encoding this protein.[6,7] This cDNA was used to analyze the developmental and regional changes of mRNA coding for spot 35 protein in the rat brain, to probe the in situ hybridization of this mRNA in the rat cerebellum and to isolate the spot 35 gene in order to analyze its structure.

RESULTS AND DISCUSSION

The procedures for cloning and nucleotide sequence determination of spot 35 cDNA were previously described.[6,7] The sequence of our spot 35 cDNA clone contained 1714 bp which included the 783 bp of the complete coding region, encoding 261 amino acids. The molecular weight of the protein was calculated to be 29,996 dalton. This protein contains five Ca-binding domains and a domain which had lost its Ca-binding activity. The 5'-noncoding region contains 130 bp. The 3'-noncoding region is 801 bases long and contains a poly(A) signal. A considerably high homology was found between spot 35 and chick intestine calbindin (amino acids 79% and nucleotides 77%), although no high homology was seen in the noncoding region.[8] Very recently, Parmentier et al. published cDNA cloning of human 27 KDA calbindin from λgt11 brain library.[9] There is 91% sequence homology at the nucleotide level between rat spot 35 protein and their human brain calbindin. With this spot 35 cDNA probe, we analyzed the developmental and regional changes of mRNA coding spot 35 protein in the rat brains.[10] We also carried out the in situ hybridization of the mRNA

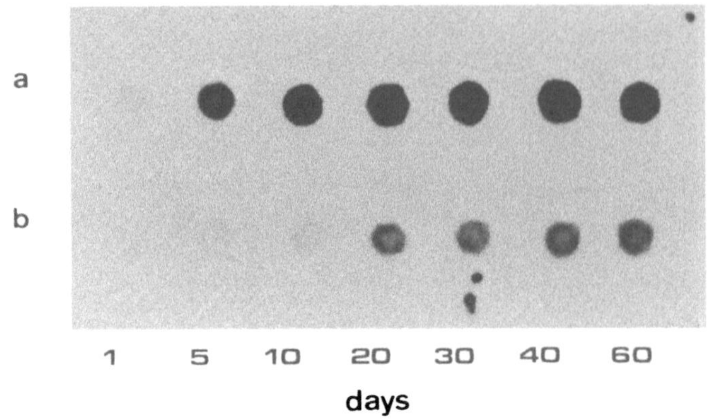

Fig. 1. Developmental changes of spot 35 mRNA in rat cerebella. Cytoplasmic poly(A) RNA of developing rat cerebella (postnatal, 1-60 days) was spotted onto a nitrocellulose membrane and hybridized with [^{32}P]cDNA. The RNA concentrations were 1 μg (a) and 0.3 μg (b).

with this probe. Furthermore, we attempted to isolate the spot 35 gene.

Fig. 1 shows the developmental changes of mRNA in rat cerebella by a dot-blot hybridization of cytoplasmic RNA from rat cerebella using [^{32}P]-cDNA. A rapid increase at 1-20 days was found. This result is in agreement with a developmental change of translatable spot 35 mRNA in a cell-free translation system and with that of immunohistochemical staining of Purkinje cells in the rat cerebellar sections. All data indicate that the expression of spot 35 protein is regulated mainly at the level of transcription and mRNA stability.

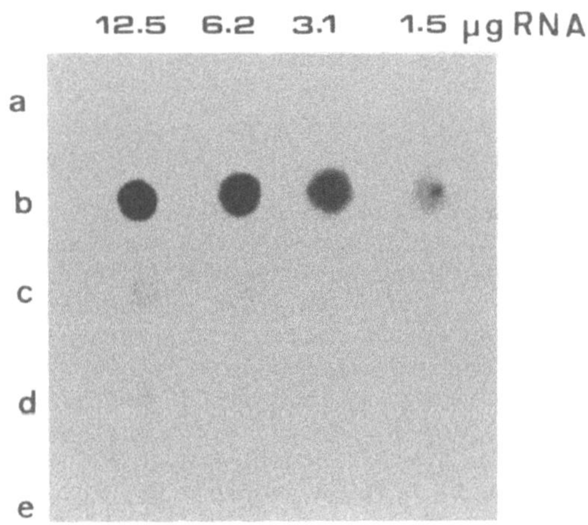

Fig. 2. Regional distribution of spot 35 mRNA in rat brains. The cytoplasmic RNA was analyzed by dot-blot hybridization using spot 35 [^{32}P]cDNA. a, Cerebral cortex; b, cerebellum; c, hippocampus; d, striatum; and e, liver rRNA.

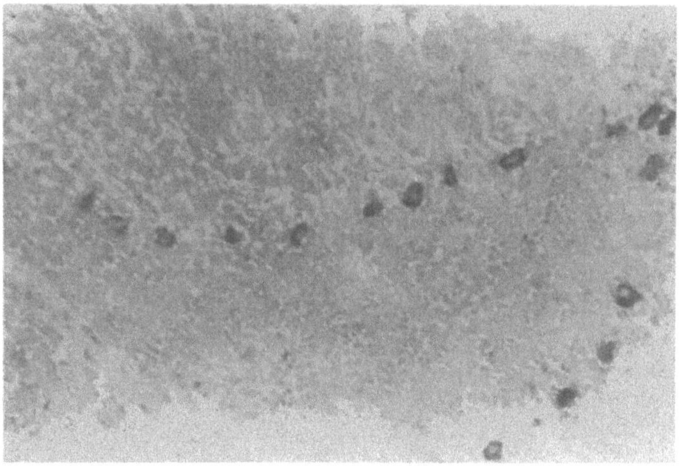

Fig. 3. In situ hybridization of spot 35 mRNA. The biotin-labeled pRS35-7 cDNA was hybridized to the fixed rat cerebellar tissue sections.

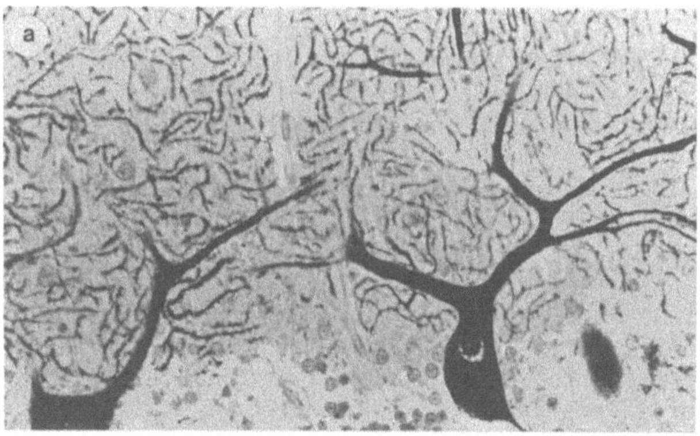

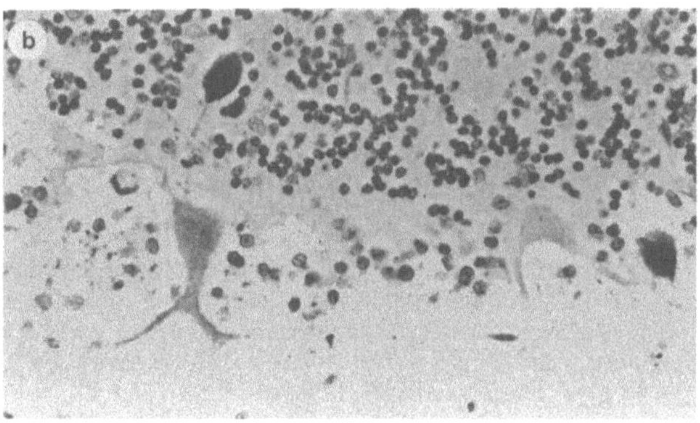

Fig. 4. a, The Purkinje cells, immunostained with the spot 35, antiserum, in a normal man, X 200. b, The Purkinje cells in the cerebellar cortex of a man with olivo-ponto-cerebellar atrophy, X 200. The spot 35 protein is completely or nearly completely undetectable not only in the remaining main dendritic trunks but also in the cell body.

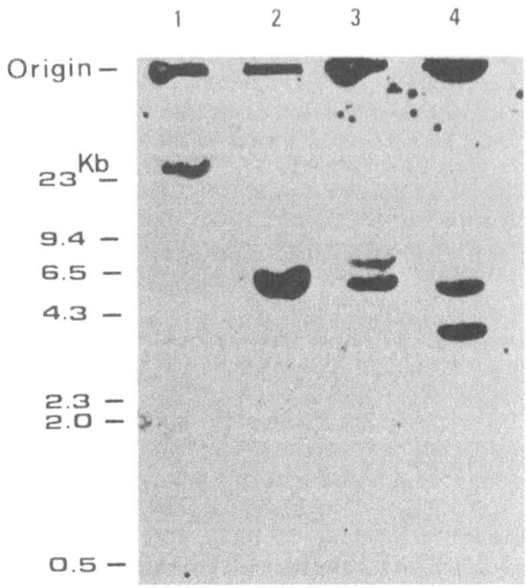

Fig. 5. Southern blot analysis of rat genomic DNA. Total genomic DNA (10 μg) of rat was digested with BamHI (1), EcoRI (2), HindIII (3) and PstI (4). The digested DNA fragments were electrophoresed on a 0.5% agarose gel and hybridized with [^{32}P] spot 35 cDNA.

Fig. 2 shows the distribution of spot 35 mRNA in various brain regions by a dot-blot hybridization. Spot 35 mRNA is localized in the cerebellum and is negligible in cerebral cortex, striatum and hippocampus. These data are also in agreement with the data on two-dimensional electrophoretic analysis of soluble proteins from various brain areas and the distribution pattern of translatable mRNA in a cell-free translation system. Similar findings on the regional distribution of spot 35 protein were also observed by the immunohistochemical study.

As the next step, we attempted to carry out in situ hybridization of spot 35 mRNA in the rat cerebellar sections using biotin-labeled cDNA as a probe.[11] As shown in Fig. 3, positive staining for spot 35 mRNA was observed in Purkinje cells in the rat cerebellar sections hybridized with the biotin-labeled cDNA probe (pRS 35-7). Fine granular, purplish-blue staining was observed in the cell bodies of almost all Purkinje cells. However, their nuclei, dendrites and axons were unstained. Similar findings were also obtained in the sections hybridized with the 3'-noncoding region probe. In each case, staining was not observed in nerve cells of the internal granule cell layer, the cerebellar nucleus and the brain stem, nor in glial cells. No positive staining was observed in the control section reacted without spot 35 cDNA probes or with vector DNA probes. These findings were different from those with peroxidase-antiperoxidase staining using spot 35 antiserum, which showed positive staining of the dendrites and axons of Purkinje cells. These results suggest that this spot 35 protein and its mRNA may be used as the useful markers for the further study on the cerebellar protein in spinocerebellar degeneration and mutant mice. In fact, as shown in Fig. 4a and b, our immunohistochemical study of human cerebella showed greatly decreased staining of spot 35 protein in the Purkinje cells of the spinocerebellar degeneration, suggesting that loss of this protein precedes loss of the neuronal structure.

As the first step to isolate the spot 35 gene, rat genomic DNA was analyzed by Southern blot analysis. [^{32}P]spot 35 cDNA was hybridized to a single or two bands in each DNA digest. These data show the existence of a sequence corresponding to spot 35 cDNA in total genomic DNA, indicating the presence of a single copy of spot 35 gene per haploid genome (Fig. 5).

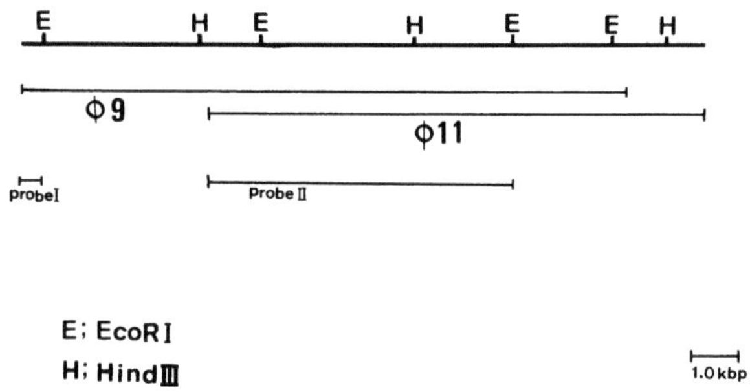

Fig. 6. The restriction map of the spot 35 protein gene prepared from analysis of the overlapping clones. The size and relative position of the two genomic clones (φ9 and φ11) are shown.

Then we started to screen the spot 35 gene from a rat genomic library by plaque hybridization. A rat genomic library was provided by Dr. J. Bonner. Eleven positive clones were isolated by using cDNA probe I (AvaI-EcoRI) derived from the 5' end of cDNA, probe II (EcoRI-PvuII) containing the 3' half of the coding region and a part of 3'-noncoding region and probe III (PvuII-AccI) derived from the 3' end of cDNA. DNA fragments of the longest clones (φ9 and φ11) were subcloned into pUC plasmids (Fig. 6).

We are currently analyzing their nucleotide sequence to examine the exon-intron organization of spot 35 gene. Very recently Wilson et al. published the structure of chick intestine calbindin gene.[12] Soon we can compare the gene structure between our protein and chick intestine calbindin.

REFERENCES

1. Y. Yoshida and Y. Takahashi, Compositional changes in soluble proteins of cerebral mantle, cerebellum and brain stem of rat brain during development: a two-dimensional gel electrophoretic analysis, Neurochem. Res. 5:81 (1980).
2. T. Yamakuni, H. Usui, T. Iwanaga, H. Kondo, S. Odani, and Y. Takahashi, Isolation and immunohistochemical localization of a cerebellar protein, Neurosci. Lett. 45:235 (1984).

3. T. Yamakuni, K. Araki, and Y. Takahashi, The developmental changes of mRNA levels for a cerebellar protein (spot 35 protein) in the rat brain, FEBS Lett. 188:127 (1985).
4. H. Takahashi-Iwanaga, H. Kondo, T. Yamakuni, and Y. Takahashi, An immunohistochemical study on the ontogeny of cell immunoreactive for spot 35 protein, a novel Purkinje cell-specific protein, in the rat cerebella, Dev. Brain Res. 29:225 (1986).
5. S. Maruyama, G. Zhang, Y. Tamura, T. Yamakuni, and Y. Takahashi, Involvement of spot 35 protein, a cerebellar protein in modulation of Purkinje cell activity of the rat cerebellum, Eur. J. Pharmacol. 108:309 (1985).
6. T. Yamakuni, R. Kuwano, S. Odani, N. Miki, Y. Yamaguchi, and Y. Takahashi, Nucleotide sequence of cDNA to mRNA for a cerebellar Ca-binding protein, spot 35 protein, Nucleic Acids Res. 14:6768 (1986).
7. T. Yamakuni, R. Kuwano, S. Odani, N. Miki, K. Yamaguchi, and Y. Takahashi, Molecular cloning of cDNA to mRNA for a cerebellar spot 35 protein, J. Neurochem. 48:1590 (1987).
8. P. W. Wilson, M. Harding, and D. E. M. Lawson, Putative amino acid sequence of chick calcium-binding protein deduced from a complementary DNA sequence, Nucleic Acids Res. 13:8867 (1985).
9. M. Parmentier, D. E. M. Lawson, and G. Vassart, Human 27-KDA calbindin complementary DNA sequence, evolutionary and functional implications, Eur. J. Biochem. 170:207 (1987).
10. T. Yamakuni, R. Kuwano, K. Araki, H. Usui, Y. Inoue, and Y. Takahashi, Developmental and regional changes of mRNA for a cerebellar protein (spot 35) in the rat brain, J. Neurochem. 50:282 (1988).
11. J. J. Leary, D. J. Brigati, and D. C. Ward, Rapid and sensitive colorimetric method for visualizing biotin-labeled DNA probes hybridized to DNA or RNA immobilized on nitrocellulose: Bio-blots, Proc. Natl. Acad. Sci. U.S.A. 80:4045 (1983).
12. P. W. Wilson, J. Rogers, M. Harding, V. Pohl, G. Pattyn, and D. E. M. Lawson, Structure of chick chromosomal genes for calbindin and calretinin, J. Mol. Biol. 200:615 (1988).

CALSPERMIN IS A TESTIS SPECIFIC CALMODULIN-BINDING PROTEIN CLOSELY RELATED TO Ca^{2+}/CALMODULIN-DEPENDENT PROTEIN KINASES

Tomio Ono and Anthony R. Means

Department of Cell Biology
Baylor College of Medicine
Houston, Texas 77030, USA

INTRODUCTION AND SUMMARY

Calmodulin regulates a wide variety of cellular functions by interacting with its binding proteins or enzymes [1]. Based on studies with a number of naturally occuring peptides that bind calmodulin with high affinity, it has been predicted that both ionic and hydrophobic interactions are involved in the association and that a general characteristic of a calmodulin-binding peptide is the ability to form a basic, amphiphilic helix [2-5]. Indeed as the sequence of calmodulin-binding proteins have become known, the basic, amphiphilic helix has been shown to be the one common feature of the calmodulin-binding domain [6-16].

Calspermin is unique among Ca^{2+}-dependent calmodulin-binding proteins [17-21]. It is a heat- and acid-stable protein first identified in rat testis that specifically binds calmodulin with low nM affinity. It exhibits an apparent Mr of 32,000 assessed by SDS polyacrylamide gel electrophoresis. Since it is the most acidic (pI=3.9) of the calmodulin-binding proteins so far reported, the primary structure of this protein and identification of its calmodulin-binding domain were of particular interest to our laboratory.

In order to elucidate the primary structure of calspermin and also to examine tissue/cell-specific expression of its mRNA, we have cloned and sequenced a full-length calspermin cDNA from an adult rat testis λgt11 library. From the deduced amino acid sequence, the Mr is determined to be 18,735. The putative calmodulin-binding domain is very close to the NH_2-terminus of the protein. Analysis of the deduced amino acid sequence revealed that the calmodulin-binding region was similar to that present in both subunits of rat brain CaM kinase II. This similarity is continued in the 5'-nontranslated region of calspermin where a 57.6% direct identity exists between this deduced sequence and the COOH-terminal region of the CaM kinase II enzyme. The data suggest that calspermin must be derived from a portion of a gene that also encodes a

The abbreviations used are IPTG, isopropyl-thio-b-D-galactopylanoside; Kb, kilobases; nt, nucleotides; TPCK-trypsine, N-tosyl-L-phenylalanine chromomethyl ketone-treated trypsine; CaM kinase II, Ca^{2+}/calmodulin-dependent protein kinase II.

novel Ca^{2+}/calmodulin-dependent protein kinase. The distribution of calspermin mRNA is severely restricted and even in testis is expressed predominantly in round spermatids. Such observations would be compatible with calspermin mRNA being a specialized transcript of the haploid genome.

Molecular cloning and sequence analysis of calspermin cDNA

In order to clone calspermin cDNA, purified calspermin from rat testis was subjected to amino acid sequence analysis in order to construct an oligonucleotide probe. The NH_2-terminus of calspermin was found to be blocked. Therefore the protein was partially digested with TPCK-trypsin, and Mr=23,000 and Mr=15,000 fragments were purified by electroelution following separation on a SDS-polyacrylamide gel. These two fragments had identical NH_2-termini and the first 29 amino acids were identified by automated sequence analysis. This sequence is underscored in Fig. 1 beginning with Leu30 and ending with Asp58. As an oligonucleotide was being prepared to this region, Sikela and Hahn [22] published the sequence of a cDNA for a calmodulin-binding protein. (λICM-1) that had been obtained by screening a mouse brain expression library with ^{125}I-calmodulin. The amino acid sequence of λICM-1 and calspermin were identical from Leu30 to Thr38. These data suggested that mouse λICM-1 and rat calspermin might be homologs. Therefore we synthesized a 27-mer oligonucleotide corresponding to the λICM-1 nucleotide sequence between Leu30 and Thr38. This oligonucleotide (oligo #2) was used to screen the rat testis library. Screening 150,000 recombinants of the library with oligo #2 yielded 9 positive plaques. Of these clones, clone 4A, which has the longest insert, was subcloned into pGEM-3Zf(+) and sequenced by Sanger's dideoxynucleotide method. The sequence began at an internal EcoRI site. The first 438 bp of clone 4A were in open-reading frame and ended with a TAA termination codon. This was followed by 148 bp of 3'-untranslated sequence and finally a stretch of about 70A residues. Authenticity of this clone was confirmed by finding that the deduced amino acid sequence contained the identical 29 amino acids determined from the tryptic peptide of rat calspermin (Fig. 1, Leu30 - Asp58). A β-galactosidase fusion protein was prepared from the phage harboring clone 4A. This fusion protein recognized anti-calspermin antiserum and also bound ^{125}I-calmodulin in a Ca^{2+}-dependent manner. These results further supported the contention that clone 4A contained a rat calspermin cDNA insert. In order to obtain the entire calspermin cDNA nucleotide sequence oligo#2 was used to extend rat testis poly(A^+)RNA. A unique oligonucleotide was synthesized corresponding to a portion of the extended sequence. This oligonucleotide and a 5' fragment of clone 4A were used to rescreen the rat testis library. Eleven positive clones were identified and plaque purified. The insert from clone R1, which had the longest 5' extension relative to clone 4A, was subcloned into pGEM-3Zf(-) and sequenced. Clone R1 contains 232 bp of 5'-nontranslated sequence, the entire open reading frame (510 bp) and 117 bp of 3'-nontranslated sequence. The sequence of the large overlap between clone 4A and R1 were identical. The overall cDNA consisted of 890 bp (minus the poly A stretch). The cDNA contained 232 bp of 5'-nontranslated region, 510 bp of open reading frame, and 148 bp of 3'-nontranslated region. The cDNA hybridized to a 1.1 kb mRNA present in rat testis. Assuming a poly A tail of 200 bp, this cDNA represents the full length mature transcript of the calspermin gene. The mouse λICM-1 begins at the internal EcoRI site and terminates in the A-rich stretch in the 3'-nontranslated region which corresponds to the calspermin sequence 850-859. At the nucleotide level the two clones show 82% identity in the coding region and 90% identity in the 3'-nontranslated region. These data reveal that the mouse and rat clones must represent related genes. Southern blotting analysis of genomic DNA from rat liver showed that calspermin is the product of a unique gene and the structure of the gene between the internal EcoRI site at nt 264 and HinfI site at nt 835 will be relatively simple, if not a single exon.

Calspermin cDNA encodes a 169 residue protein with a calculated Mr of 18,735. This value is considerably different than Mr=32,000 estimated for the rat testis protein by SDS polyacrylamide gel electrophoresis but is in reasonable agreement to the Mr=23,000 calculated from the stokes radius and sedimentation coefficient [18]. The putative calmodulin-binding domain is very close to the NH_2-terminus of the protein. This region shows 46% identity with the calmodulin-binding region of rat brain CaM kinase II [10] and 32% identity with the equivalent region of chicken smooth muscle myosin light chain kinase [9]. The putative calmodulin binding domains of λICM-1 and that of calspermin (Ala13 to Ser32) are identical (Fig. 1). From Glu10 to Tyr169 these latter two proteins are 70% identical providing that gaps of 4 amino acids are introduced into the λICM-1 sequence. Distal to the end of the calmodulin binding domain (Ser 32) the identity of the remaining 137 amino acid decreases to 60%. Within this region 17 Glu residues exist in calspermin but not in λICM-1. It is likely that this Glu-rich region contributes to the aberrent behavior of calspermin on polyacrylamide gels. Hydropathy analysis [23] of calspermin suggests that this protein is primarily hydrophilic. Analysis of amino acid sequence of calspermin by the algorithm of Garnier et al. [24] predicts 81% α-helix and 14% random coil. This prediction is consistent with an extended rod-like structure of calspermin which was initially predicted from the fact that biochemical analysis revealed a large fractional ratio [18].

```
1                  20                  40
MDTAQKKLQEFNARRKLKAAVKAVVASSRLGSASSSHTNI
.........-----------------------------S-

                   60                  80
QESNKASSEAQPAQDGKDKTDPLENKMQAGDHEAAKAAAD
--NH----DPPST--A--S--L-GK---EE-Q-EDQVE-E

                  100                 120
ETMKLQSEEVEEEEGVKEEEEEEEEEEETSRMVPQEPEDR
....ASAD-MRKLQSEEV-KDAGVK-----S----D---E

                  140                 160
LETDDQEMKRNSEETLKSVEEEMDPKAEEEAAAVGLGVPP
-----P----D---K----------MT----PDA----.-

        169
QQDAILPEY
-----Q---
```

Fig. 1 Amino acid sequene of calspermin. Deduced amino acid sequence of calspermin (upper line) and that of mouse brain λICM-1 (bottom line) [22] are aligned. the bars (-) indicate identical amino acids and the dots (•) represent gaps required for optimal alignment of the sequence. The box indicates the sequence of the putative calmodulin-binding domain. The contiguous line shows the amino acid sequence obtained from the tryptic fragment generated from authentic rat testis calspermin.

As the NH_2-terminus of the calspermin sequence showed significant homology to CaM kinase II, we have translated the 5'-nontranslated region into putative amino acid sequence and compared this region to the protein sequence of CaM kinase II. The amino acid sequence of the 5'-nontranslated region of calspermin cDNA together with the sequence of NH_2-terminus of calspermin shows significant homology to the COOH-end of the catalytic domain of CaM kinase II (Fig. 2). The overall homology between calspermin (residue -33 to 33) and CaM kinase II (residue 248 to 315) was 52.9%. The homology upstream of the initiator methionine (residue -33 to -1) was 57.6%. Since the sequence 5' to the initiation codon is not translated, calspermin does not contain any of the protein kinase catalytic region. Calspermin cannot be directly related to the enzyme known as CaM kinase II since the sequence of all subunits of this enzyme present in rat brain have been determined [10-12]. Instead, the remarkable similarity of the 5'-nontranslated region with known calmodulin dependent protein kinases [25] suggests the presence of a novel

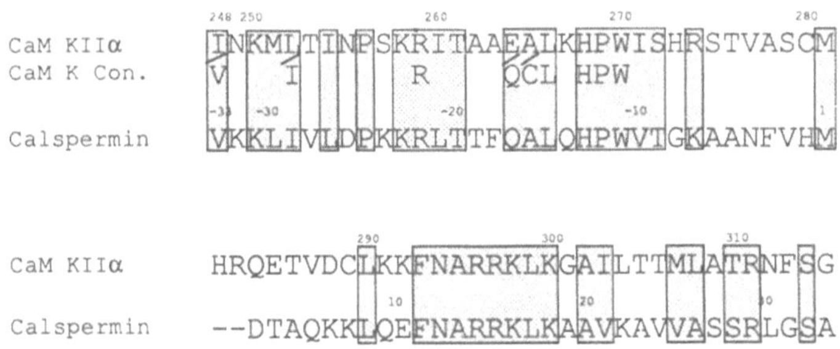

Fig. 2 Amino acid sequence alignment of calspermin and CaM kinase II. Calspermin cDNA sequence was translated to amino acid sequence assuming that 5'-nontranslated region was encoded in the same open reading frame as the protein coding region. The numbers above the amino acids indicate the position of the residues. The identical and/or conservative amino acid residues are boxed. Bars (-) represent gaps required to obtain optimal alignment. CaMKIIα, α-subunit of CaM kinase II in mouse brain [10]; CaMK Con., the consensus residues present in this region of all calmodulin-dependent protein kinases [25].

such enzyme in brain. We have some preliminary data in support of this suggestion. First, analysis of brain proteins by immunoblotting with anticalspermin antibody, reveals the presence of a 57 kDa protein. Second the antibody reacts positively with frozen sections of cerebellum and pituitary (as well as testis) but not with spleen or liver. Together our observations support the presence of a calmodulin-dependent protein kinase in brain that is related to calspermin. We predict that the same gene encodes both proteins and the calspermin transcript is generated from the protein kinase gene by tissue specific alternative splicing. It follows that the rat brain homology to mouse λICM-1 would have an identical sequence to that of rat calspermin in the COOH-half of the larger protein. Analysis of genomic DNA will be required to confirm or deny this prediction.

Tissue specificity and developmental change of calspermin mRNA

Northern blot of total RNA preparations isolated from various rat tissues and from the testes of various other vertebrates showed that calspermin cDNA hybridized predominantly to a 1.1 kb mRNA present only in testis. This signal was species specific to rat and was not seen in testicular RNA from adult mouse, rabbit, pig, bovine, or baboon. In rat testis RNA a strong signal was seen at 1.1 kb and a weak signal was observed at 2.0 kb. Calspermin mRNA was also examined as a function of testis development. Calspermin mRNA could not be detected before day 29. At 29 days of age signals at 1.1 and 2.0 kb were observed only in the poly(A^+)RNA. Relative signal intensity of both mRNA species reached a maximum at 40 days of age and remained high at day 61. The age-dependent changes in the appearance and concentration of calspermin mRNA raised the possibility that this species was predominantly found in post-meiotic germ cells. This possibility was examined by separating the individual cell types present in testis and isolating total RNA from each cell population. An intense signal was observed at 1.1 kb in RNA isolated from round spermatid. No calspermin mRNA could be detected in Sertoli cells, interstitial cells, nor in the mitotically proliferative germ cells. The specific developmental expression of calspermin suggests that the protein may be a product of the haploid genome and have a role in spermiogenesis and/or sperm function.

If our prediction concerning calspermin and a rat brain calmodulin-dependent protein kinase is correct, then mRNA encoding the latter protein must be present in rat brain RNA. Our inability to detect an RNA species hybridizing to calspermin cDNA is likely due to the use of total RNA rather than poly(A^+)RNA from rat brain. Precedent exists for this possibility since even though CaM kinase II is present in liver and skeletal muscle, signals cannot be seen in total RNA preparation. Even with poly(A^+)RNA hybridizing signals are weak. These results suggest that the mRNA for the rat brain homolog of calspermin is either a rare species or is restricted to specific anatomical regions of the brain.

REFERENCES

1. A.R. Means, Rec. Prog. Horm. Res. 44:223 (1988).
2. J.A. Cox, M. Comte, J.E. Fitton, and W.F. DeGrado, J. Biol. Chem. 260:2527 (1985).
3. L. McDowell, G. Sanyal, and F.G. Prendergast, Biochemistry 24:2979 (1985).
4. D.P. Geidroc, T.M. Keravis, J.V. Staros, N. Ling, J.N. Wells, and D. Puett, Biochemistry 24:1203 (1985).
5. D.A. Malencik, and S.R. Anderson, Biochem. Biophys. Res. Commun. 114:50 (1983).
6. W.F. DeGrado, S. Erickson-Viitanen, H.R. Wolfe, Jr. and K.T. O'Neil, Proteins 2:20 (1987).
7. D.K. Blumenthal, K. Takio, A.M. Edelman, H. Charbonneau, K. Titani, K.A. Walsh, and E.G. Krebs, Proc. Natl. Acad. Sci. USA 82:3187 (1985).
8. V. Guerriero, Jr., M.A. Russo, N.J. Olson, J.A. Putkey, and A.R. Means, Biochemistry 25:8372 (1986).
9. T.J. Lukas, W.H. Burgess, F.G. Prendergast, W. Lau, and D.M. Watterson, Biochemistry 25:1458 (1986).
10. R.M. Hanley, A.R. Means, T. Ono, B.E. Kemp, K.E. Burgin, N. Waxham, and P.T. Kelly, Science 237:293 (1987).
11. C.R. Lin, M.S. Kapiloff, S. Durgerian, K. Tatemoto, A.F. Russo, P. Hanson, H. Schulman, and M.G. Rosenfeld, Proc. Natl. Acad. Sci. USA 84:5962 (1987).

12. M.K. Bennett, and M.B. Kennedy, Proc. Natl. Acad. Sci. USA 84:1794 (1987).
13. B. Buschmeier, H.E. Meyer, and G.W. Mayr, J. Biol. Chem. 262:9454 (1987).
14. P. James, M. Maeda, R. Fischer, A.K. Verma, J. Krebs, J.T. Penniston and E. Carafoli, J. Biol. Chem. 263:2905 (1988).
15. G.E. Shull, and J. Greeb, J. Biol. Chem. 263:8646 (1988).
16. K.A. Alexander, B.T. Wakim, G.S. Doyle, K.A. Walsh, and D.R. Storm, J. Biol. Chem. 263:7544 (1988).
17. T. Ono, Y. Koide, Y. Arai, and K. Yamashita, J. Biol. Chem. 259:9011 (1984).
18. T. Ono, Y. Koide, Y. Arai and K. Yamashita, J. Biochem. (Tokyo) 98:1455 (1985).
19. Y. Koide, T. Ono, and K. Yamashita, Cell Calcium 7:329 (1986).
20. T. Ono, Y. Koide, Y. Arai and K. Yamashita, Arch. Biochem. Biophys. 255:102 (1987).
21. Y. Koide, T. Ono, C. Ishinami, and K. Yamashita, Amer. J. Physiol. 250:C299 (1986).
22. J.M. Sikela, and W.E. Hahn, Proc. Natl. Acad. Sci. USA 84:3038 (1987).
23. J. Kyte, and R.F. Doolittle, J. Mol. Biol. 157:105 (1982).
24. J. Garnier, D.J. Osquthorpe and B. Robson, J. Mol. Biol. 120:97 (1978).
25. S.K. Hanks, A.M. Quinn and T. Hunter, Science 241:42 (1988).

MYOSIN AND CONTRACTILE ACTIVITY IN SMOOTH MUSCLE

D. J. Hartshorne[1], M. Ito[1] and M. Ikebe[2]

1 Muscle Biology Group
University of Arizona
Tucson, AZ 85721

2 Dept. of Physiology and Biophysics
Case Western Reserve University
Cleveland, OH 44106

INTRODUCTION

In relaxed smooth muscle the contractile apparatus is dormant and the initiation of contraction requires an activation. It is generally agreed that the activation is achieved by phosphorylation of the two 20,000-dalton light chains of myosin (LC20) and usually Serine 19 is phosphorylated. This process is catalyzed by myosin light chain kinase (MLCK) and the Ca^{2+}-dependence of the system is due to the activation of the MLCK apoenzyme by the $Ca^{2+}{}_4$ calmodulin (CaM) complex. Dephosphorylation and inactivation of the contractile apparatus, reflects the activity of a light chain phosphatase. The latter, however, has not been well characterized and it is not known, for example, if one or more phosphatases are involved or, if the phosphatase activity is regulated.

The initial observation that prompted the phosphorylation hypothesis was that the actin-activated ATPase of myosin was activated by phosphorylation (review 1) and thus should promote contraction in smooth muscle. Several laboratories have confirmed this finding and have demonstrated a correlation between the level of ATPase and of phosphorylation. Whether the relationship is linear or indicative of cooperative interaction between the two myosin heads remains controversial. From these biochemical data the simplest interpretation of the phosphorylation hypothesis is that the system is either active (i.e. phosphorylated) or inactive (i.e. dephosphorylated) and that any variation is due to the level of phosphorylation, i.e. the number of heads recruited. If this is extrapolated to an ideal situation in terms of the mechanics of smooth muscle fibers this would allow for an increase in the levels of tension development, corresponding to the extent of phosphorylation, but at a constant velocity. As will be discussed below this is not consistent with the physiological evidence obtained with skinned or intact fibers.

Following the <u>in-vitro</u> discoveries attempts were made to correlate tension development with myosin phosphorylation in intact and skinned smooth muscle fibers. In general it was found that the initiation of contraction required phosphorylation of myosin and subsequent studies have shown that in the early stages of the contraction cycle there is a reasonable correlation between myosin phosphorylation, intracellular Ca^{2+} and tension development. (In general the MLCK activity tracks closely the Ca^{2+} concentration at all stages of the contraction-relaxation cycle.) However, several investigators have shown deviations from this simplest scenario using a variety of smooth muscles. It was reported that during K^+ stimulation of intact hog carotid artery, the myosin phosphorylation level rose initially and then fell to close to resting values, but the load-bearing capacity increased and remained high and did not follow the phosphorylation transient (2). On the basis of these results it was suggested that dephosphorylation of myosin can generate an attached, noncycling cross bridge (termed a latch bridge) that is responsible for the force maintenance. Although it is evident that the correlation between myosin phosphorylation and isometric force varies with both the agonist employed and the duration of simulation (3-5), it has also been confirmed that in the continued presence of stimulus, decreased levels of myosin phosphorylation are observed at constant levels of isometric force.

On the basis of these results it was suggested that a second Ca^{2+}-dependent mechanism (at resting Ca^{2+} levels the latch bridges are dissociated) must exist that controls the formation (and dissociation) of latch bridges (6). Since this occurs at low Ca^{2+} levels, presumably below the Ca^{2+}-CaM threshold since MLCK is not active, the unknown regulatory mechanism should involve a Ca^{2+}-binding site with higher affinity than CaM· To date this putative regulatory mechanism is unknown and a Ca^{2+} target with appropriate binding constants has not been identified. Despite much research on this topic the alternative, or complementary (to phosphorylation) mechanism that would account for latch bridge formation remains speculation.

If such a second regulatory mechanism does not exist, how can we account for an alteration in the rate of cross bridge cycling? Perhaps the myosin phosphorylation mechanism in fact is adequate for this purpose? That is the viewpoint of Hai and Murphy (7) and Driska (8) who based on models of actin attachment relative to the cross bridge cycle feel that a second regulatory mechanism is not required. However, this implies that the phosphorylation scheme is not simply an "on-off" switch (as initially thought) and it becomes imperative to understand at a molecular level the effects of phosphorylation.

EFFECTS OF PHOSPHORYLATION

There are initially two processes that any proposed molecular mechanism should explain, these are: the activation of myosin ATPase by actin as a consequence of phosphorylation and the formation of latch bridges. For the latter a more specific objective might be to rationalize a slow detachment rate of dephosphorylated myosin from actin (an important component in the two above models). Our current understanding of the changes involved in the myosin molecule as a result of phosphorylation are not adequate to satisfy either question, but there are some interesting preliminary results that might lead to a more complete appreciation.

Conformation of Smooth Muscle Myosin

Up to a few years ago it was assumed that smooth muscle myosin, as with other myosin types, existed only as an asymmetric extended molecule, approximately 160 nm in length. However, it was found that smooth muscle myosin and myosin from many non-muscle sources, could form a folded structure under appropriate ionic conditions. This transition was discovered by Prof. S. Watanabe and colleagues who found that filaments of dephosphorylated myosin in 0.15 M KCl disassembled on addition of ATP and formed a component of 10S (9). At higher ionic strength the filaments also dissociated, but formed a component sedimenting more slowly, i.e. 6S. After some initial confusion, it was discovered that both the 10S and 6S forms were monomers of myosin and represented folded (10S) and extended (6S) conformations. The dramatic change in hydrodynamic properties was shown by electron microscopy to be due to a folding of the rod portion of the molecule. Bending occurs at two "hinge" regions, approximately 60 and 100 nm from the C-terminus of the myosin heavy chains, and the loop that is formed appears to attach to a site close to the head-neck (or S1 - S2) junction.

The change in asymmetry that is involved in the 6S - 10S transition is detected easily by viscosity measurements and this offers a convenient method to monitor the conformational transition. Using viscosity measurements Ikebe et al. (10) showed that phosphorylation of myosin shifted the transition towards the 6S state. The formation of 10S occurs as the ionic strength is reduced and with phosphorylated myosin 10S is formed at lower KCl concentrations than for dephosphorylated myosin. It was shown also that the effect of phosphorylation on conformation required that both serine 19 light chain sites were phosphorylated (10). However the most significant finding came when enzymatic activity was compared to conformation. Mg^{2+}, $Ca^{2+}$and K^{+} EDTA-activated ATPase activities of gizzard myosin were measured and compared to the viscosities obtained over a range of ionic strengths. For both the phosphorylated and dephosphorylated myosin the formation of 10S myosin was accompanied by a change in enzymatic activity. With Ca^{2+}- and Mg^{2+}-ATPase activities the 10S conformer had a lower activity than the 6S and the reverse was true for the K^{+} EDTA-ATPase activity. From these results it was proposed that the conformation of myosin determines ATPase activity and the function of phosphorylation is to alter conformation. This simple hypothesis predicts that the conversion of 10S to 6S alters the enzymatic properties of myosin and that any factor that achieves this transition will also alter ATPase activity. Obviously, this can be tested and with only a limited number of available examples the predictions appears to be valid. One example is the effect of varying Mg^{2+} concentrations. It has been known for several years that with <u>in vitro</u> assays of smooth muscle actomyosin, relatively high concentrations of Mg^{2+} are required for maximum activity (review 1). It was shown (11) that this unusual Mg^{2+} dependence was not due to MLCK but was more likely to be a feature of smooth muscle myosin. Another distinctive feature that is observed with smooth muscle is the development of tension in skinned fibers at relatively high concentrations of Mg^{2+} in the absence of Ca^{2+} (review 1). We examined these effects to determine if the influence of Mg^{2+} could be correlated with a change in myosin conformation. For the actin-activated ATPase activity of phosphorylated gizzard myosin, it was found that at low concentrations of Mg^{2+} (on the order of 1 mM free Mg^{2+}), the activity was low and increased as the Mg^{2+} concentration was increased. At the lower concentrations of Mg^{2+}, the phosphorylated myosin existed as the 10S species and was converted to 6S myosin as the Mg^{2+}level was raised (12). In fact there was inverse relationship between the concentration of 10S myosin and the level of actin-activated ATPase activity. Using dephosphorylated myosin the transition from 10 to 6S could also be induced by increasing the $MgCl_2$ concentration, and again this transition was accompanied by an increase in the actin-activated ATPase activity. With skinned fibers from chicken gizzard, it was then shown (12) that an increase in Mg^{2+} caused tension development without a concomitant increase in the level of myosin phosphorylation. In addition, it was found that in the presence of inosine triphosphate, which is not a substrate for MLCK (13), tension can still be developed by increasing the Mg^{2+} concentration.

Another example where a change in conformation can be induced by a mechanism other than phosphorylation is by chemical modification of myosin. Reaction of thiol residues (14) shifts the 6S-10S equilibrium and also modifies ATPase activity. Similar effects are observed following modification of lysine residues by trinitrobenzene sulfonate (15).

Thus the basic premise is that the biological properties of myosin are directly linked to conformation and that the critical conformational changes occur as part of the 10S-6S transition. Obviously, the idea must remain speculative until future research confirms, modifies or reputes the hypothesis. One point that must be resolved is whether phosphorylation itself induces changes in the head region that might be too subtle to be detected by hydrodynamic monitors of the 6S-10S transition. For example, is the phosphorylated form of 10S myosin different from the dephosphorylated 10S state? This is an important question to be resolved in the future.

Measurement of Conformational Change

Most of the procedures used rely on the large change in hydrodynamic properties that occurs during the 10S-6S transition. These include measurements of sedimentation velocity and viscosity. The dramatic change is also easily observed by electron microscopy of rotary shadowed preparations. A very useful technique to probe for conformational changes in smooth muscle myosin has been limited proteolysis. Using papain it was shown that 10S myosin is more resistant to proteolysis than 6S myosin (16, 17) and one of the main regions involved in the change of proteolytic sensitivity was the S1-S2 junction. In contrast, hydrolysis at the junction between LMM and HMM was not sensitive to myosin conformation. Similar results were obtained with *Staph. aureus* protease (18). Further studies with different proteases showed that in the S1-S2 region of the molecule two major sites are affected by the 10S-6S transition. These are: a) the site at the S1-S2 junction (discussed above) which was termed site B (18). This site is cleaved more rapidly in the 6S conformation by papain and *Staph. aureus* protease, and b) a site 68 kDa from the N-terminus of the myosin heavy chain close to the junction of the C-terminal and central S1 domains, termed site A. This site is cleaved by papain, *Staph. aureus* protease and actinidin, is more resistent in the 10S conformation and also is protected by the binding of actin to 6S myosin. Based on the influence of actin it was proposed (19) that site A is part of, or close to, the actin-binding site. The data of Marianne-Pepin et al. (20) support this idea and they showed that actin was cross-linked to the 26-kDa C-terminal domain and an adduct of actin with the central 50-kDa domain was not obtained. Our contention is that actin induces changes at the 50-26 junction, but protection at this site by actin was not observed in the cross-linking studies. With regard to the location of the actin-binding site considerably more is known about S1 from skeletal muscle myosin. (The domains in skeletal S1 are 26-50-20-kDa, in sequence from the N-terminus of the heavy chain). It has been suggested that the interaction with actin involves both the central and C-terminal domains (see 19 for references). A two-site interaction with actin also was proposed by Morita and co-workers and involved a weak-binding site (J-site) close to the 50-20 junction and a stronger site (S-site) on the 20-kDa fragment between the SH1 and SH2 groups (21). In addition it was found that the isolated 20-kDa domain binds actin (22). Similar to the smooth muscle situation the junction between the central and C-terminal fragments of skeletal S1 also is protected from proteolysis by the binding of actin (see 19 for references). In addition it was shown that as the 50-20 junction is cleaved by trypsin, the actin-activated ATPase activity is lost (see 19 for references). This is probably due to a decrease in actin-binding affinity since at higher concentrations of actin the ATPase activity is not reduced. It was later suggested (23) that the loss of ATPase activity was due to the removal of a 2-kDa linker region between the two domains and if this linker is retained following proteolysis, actin-activation of ATPase activity is not altered. How this relates to smooth muscle S1 is not established, but it does appear that cleavage at site A by itself does not result in a reduced affinity for actin. It is unlikely that a linker region (analogous to the 2-kDa peptide) is present in smooth muscle S1 since further proteolysis of S1, involving degradation of the C-terminal domain does not eliminate actin-activated ATPase activity.

As mentioned above, one interpretation for the protection of site A against proteolysis by the binding of actin is that site A forms part of, or is close to, the actin-binding site. Since proteolysis of site A is retarded for 10S myosin, it might be inferred that the formation of the folded conformation masks or hinders accessibility of the proteases to the putative actin-binding site. If this is true, then it would be expected that the actin-binding properties of 6S and 10S myosins are different. The simplest interpretation is that the actin-binding site is exposed or available in the 6S conformation but is partially blocked in the 10S conformation. This hypothesis was tested (19) by measuring actin-binding affinity in the two conformations, 6S and 10S. For these experiments AMPPNP was used and it was shown that the binding of actin to the 10S conformation was weaker than to the 6S conformation. For the 6S phosphorylated myosin at 0.17 and 0.35 MKCl the Ka's (actin dissociation constants) were similar and were 0.31 and 1.33 μM, respectively. In contrast, Ka for 10S dephosphorylated myosin (0.17 MKCl) was 42.5 μM, compared to 2.4 μM for 6S dephosphorylated myosin (0.35 MKCl). The finding that myosin in the 6S conformation binds actin more strongly than 10S myosin is consistent with the results of Ikebe et al. (24) who reported that phosphorylation of gizzard HMM increased actin-affinity about 10-fold. The use of AMPPNP as a suitable analog of ATP may be questionable, and before these results can be fully accepted the experiments should be repeated.

The conclusions from the limited proteolysis studies indicate that discrete regions of the myosin molecule are affected during the 6S-10S transition. This situation, however, is probably over simplified and it is likely that many changes in local conformation occur. One procedure that detects more global changes is ^{1}H-NMR. We have recently investigated the spin echo ^{1}H-NMR resonances for myosin and HMM in the 10S and 6S, or equivalent 8.6 and 7S conformations of HMM, for phosphorylated and dephosphorylated species (Sommerville, Henry, Sykes and Hartshorne, unpublished observations). The only resonances detected by this technique are the more mobile ones, and when applied to myosin or HMM the resonances observed are those of S1. Those associated with S2 or LMM are too broad and are not detected. In the transition from 6S to 10S many of the S1 resonances are reduced in intensity and this is interpreted as a decrease in mobility of these groups as the 10S conformation is formed. This is consistent with the theory that in the 10S state the myosin heads are more constrained. Obviously those resonances that alter during the 10S-6S transition must be identified and associated with specific residues in the S1 structure and at the moment all that can be said is that several regions do change. Using this technique it will also be interesting to determine more precisely the effects of phosphorylated. An important question is whether 10S dephosphorylated myosin is identical to 10S phosphorylated myosin? There are some indications that the two species are different, but a detailed description is lacking.

It is surprising in view of the dramatic hydrodynamic changes involved in the 10S-6S transition that the α-helix content is the same for each conformation. For both the 10S and 6S conformation the content of α-helix is 80-81% (Sano, Tsuchiya and Hartshorne, unpublished observations). Presumably the folding of the myosin heavy chain to form the loop of 10S myosin does not involve an extensive disruption of the coiled-coil structure. It is possible that the hinge regions are not structured into the rigid α-helical coiled-coil, but may exist in a more flexible structure. Such a localized disruption of the α-helix should be reflected by discrete and unique portions of the sequence, i.e. by the appearance of non-helix forming or helix-disruption amino acids.

Other Monitors of Conformational Change

The most direct way to monitor gross conformational changes is by electron microscopy. These studies established that the myosin rod is folded in the 10S state, and also more recently it has been shown that in HMM the head orientation differs for the active and inactive forms (25). Our results also showed that for phosphorylated HMM the head position was generally extended away from the myosin rod region, whereas in the dephosphorylated state the heads were bent back towards the backbone of the molecule. In agreement with Suzuki et al. (25) we believe that this conformational change is important in determining enzymatic activity of smooth muscle myosin.

Fluorescence measurements were carried out on smooth muscle myosin in the 6 and 10S states (26). The addition of ATP enhanced tryptophan fluorescence and since this occurred at stoichiometric levels of ATP and myosin it was assumed that only high affinity ATP sites were involved, i.e. the ATP hydrolysis sites. It was shown that the tryptophan fluorescence was different for the 10S and 6S states and therefore it was argued that the conformation around the active sites differed in each conformation. This is reasonable since the ATP hydrolysis rates are also quite different for each conformer.

Recently monoclonal antibodies have been used to probe the structure-function relationship of smooth muscle myosin. Higashihara and Ikebe (27) generated an antibody to S1 of smooth muscle myosin. This stabilized the 6S formation and blocked transition to 10S. We have developed a monoclonal activity to gizzard myosin S2 (Ito and Hartshorne, unpublished observations) that has similar properties. It activates Ca^{2+} and Mg^{2+}-ATPase under conditions where the 10S conformation is formed and influences the rate of phosphorylation and dephosphorylation. Our interpretation of these results also is that the 6S conformation is stabilized by this antibody. Thus it appears that antibodies to different parts of the molecule can produce similar effects and it is predicted that the disturbance of the conformational equilibrium is due to binding within a general area of the molecule. An example of this would be binding within a region that would prevent the formation of the 10S head orientation.

Possible In-vivo Role of Conformational Changes

It is assumed that myosin in both contracted and relaxed muscle fibers exists in filamentous form (28) and it is therefore unlikely that the entire transition, involving folding of the molecule, is allowed. The parts of the transition that we believe are critical in determining the biological properties of myosin are: 1) the change in the flexibility of the S1-S2 (head-neck) junction; 2) the change at the actin-binding site, and 3) the change at the active site. We assume that the latter occurs, but there is no detailed information available at this time and the location of this site is not established. The following model will therefore concentrate on the first two changes.

We propose that the change at the S1-S2 junction leads to altered flexibility of the myosin heads. In the phosphorylated state, or 6S, the heads are more mobile and can interact freely with actin, leading to a higher rate of ATP hydrolysis. Dephosphorylation induces a conformational change at this site that causes the heads to become more rigid. In the intact fiber the ordered structure would amplify any effect due to movement of the heads away from the thin filament interaction zone. In solution studies, however, actin is freely accessible to the myosin heads and should have similar access in both 6S and 10S conformations. It is proposed, therefore, that this conformational change involves more than merely a reorientation of the heads and the effect that we term a constraint also inhibits ATP hydrolysis, possible by an effect on the active site. This effect requires the presence of the neck region, S2, since the isolated heads show no regulation by phosphorylation.

The second effect involves changes at the actin-binding site. In its simplest terms this means that actin-binding to 10S myosin is reduced, and although this is controversial , it is not an unreasonable assertion. A more interesting, but highly speculative, model can be proposed if these two effects are considered together and compared to the recent latch bridge models mentioned earlier (7, 8). Starting with relaxed muscle, the myosin is dephosphorylated, the heads are in a constrained position, the actin-binding site is masked and the cross bridges are detached. Phosphorylation alters both sites, and allows cross bridge cycling to occur. This would continue as long as the Ca^{2+} levels are sufficient to maintain phosphorylation. When the Ca^{2+} level is reduced, MLCK is inactivated and myosin phosphatase activity is manifested. The potential substrates for the phosphatase are myosin either in the detached or attached (to actin) states. Dephosphorylation of detached myosin allows the return of both sites to the starting position, i.e. the heads become constrained and the actin-binding site is masked. However, dephosphorylation of the attached cross bridge could lead to a different state. The change at the S1-S2 junction could still occur and this might impose a "stiffness" to the attached cross bridge. The change at the actin-binding site would not occur, since actin is bound and blocks masking of this site. If the detachment of dephosphorylated myosin from actin is slow this could then constitute a significant population of attached, dephosphorylated slowly-cycling cross bridges, i.e. the latch state. The proportion of cross bridges that occur as rapidly- or slowly-cycling would also be dependent on the ratio of kinase to phosphatase activities. This model is speculative, but several of its component parts can be challenged experimentally.

ACKNOWLEDGMENTS

Authors are supported by grants from the National Institutes of Health. HL 23615 and HL 20984 (to DJH) and AR 38431 (to M.I.).

REFERENCES

1. D. J. Hartshorne, Biochemistry of the contractile process in smooth muscle, in "Physiology of the Gastrointestinal Tract" 2nd. Ed. L. R. Johnson, ed., Raven Press, N.Y. pp. 423-482 (1987).
2. P. F. Dillon, M. O. Aksoy, S. P. Driska, and R. A. Murphy, Myosin phosphorylation and the cross bridge cycle in arterial smooth muscle, Science, 211:495-497 (1981).
3. M. O. Aksoy, S. Mras, K. E. Kamm, and R. A. Murphy, Ca^{++}, cAMP, and changes in myosin phosphorylation during contraction of smooth muscle, Am. J. Physiol., 245:C255-C270 (1983).
4. P. J. Silver, and J. T. Stull, Regulation of myosin light chain and phosphorylase phosphorylation in tracheal smooth muscle, J. Biol. Chem., 257:6145-6150 (1982).
5. P. J. Silver, and J. T. Stull, Phosphorylation of myosin light chain and phosphorylase in tracheal smooth muscle in response to KCl and carbachol, Mol. Pharmacol., 25:267-274 (1984).
6. R. A. Murphy, Myosin phosphorylation and cross bridge regulation in arterial smooth muscle, State-of-the-art review, Hypertension, 4(Suppl. 2):3-7 (1982).
7. C-M. Hai, and R. A. Murphy, Cross bridge phosphorylation and regulation of latch state in smooth muscle, Am. J. Physiol. 254:C99-C106 (1988).
8. S. P. Driska, High myosin light chain phosphatase activity in arterial smooth muscle: can it explain the latch phenomenon? in "Smooth Muscle Contraction", M. J. Siegman, ed., A. R. Liss, Inc., N.Y. 387-398 (1987).
9. H. Suzuki, H. Onishi, K. Takahashi, and S. Watanabe, Structure and function of chicken gizzard myosin, J. Biochem (Tokyo), 84:1529-1542 (1978).
10. M. Ikebe, S. Hinkins, and D. J. Hartshorne, Correlation of enzymatic properties and conformation of smooth muscle myosin, Biochemistry, 22:4580-4587 (1983).
11. D. J. Hartshorne, R. F. Siemankowski, and O. M. Aksoy, Ca regulation in smooth muscle and phosphorylation: Some properties of the myosin light chain kinase. In: Regulatory Mechanism of Muscle Contraction, S. Ebashi, K. Maruyama, and M. Endo, ed., Japan Society for the Promotion of Science and Fujihara Foundation of Science, Tokyo, 287-301, (1980).
12. M. Ikebe, R. J. Barsotti, S. Hinkins, and D. J. Hartshorne, Effects of magnesium chloride on smooth muscle actomyosin adenosine-5'-triphosphatase activity, myosin conformation and tension development in glycerinated smooth muscle fibers, Biochemistry 23:5062-5068 (1984).
13. E. M. V. Pires, and S. V. Perry, Purification and properties of myosin light-chain kinase from fast skeletal muscle, Biochem. J., 167:137-146 (1977).
14. T. S. Chandra, N. Nath, H. Suzuki, and J. C. Seidel, Modification of thiols of gizzard myosin alters ATPase activity, stability of myosin filaments, and the 6-10S conformational transition, J. Biol. Chem. 260:202-207 (1985).
15. S. Srivastava, M. Ikebe, and D. J. Hartshorne, Trinitrophenylation of smooth muscle myosin, Biochem. Biophys. Res. Commun., 126:748-755 (1985).
16. H. Onishi, and S. Watanabe, Correlation between the papain digestibility and the conformation of 10S-myosin from chicken gizzard, J. Biochem. (Tokyo), 95:899-902 (1984).
17. M. Ikebe, and D. J. Hartshorne, Conformation-dependent proteolysis of smooth-muscle myosin, J. Biol. Chem. 259:11639-11642 (1984).
18. M. Ikebe, and D. J. Hartshorne, Proteolysis of smooth muscle myosin by Staphylococcus aureus protease: Preparation of heavy

meromyosin and subfragment 1 with intact 20,000-dalton light chains, Biochemistry, 24:2380-2386 (1985).
19. M. Ikebe, and D. J. Hartshorne, Proteolysis and actin-binding properties of 10S and 6S smooth muscle myosin, Identification of a site protected from proteolysis in the 10S conformation and by the binding of actin, Biochemistry, 25:6177-6185 (1986).
20. T. Marianne-Pepin, D. Mornet, R. Bertrand, J-P. Labbe, and R. Kassab, Interaction of the heavy chain of gizzard myosin heads with skeletal F-actin, Biochemistry, 24:3024-3029 (1985).
21. T. Katoh, and F. Morita, Interaction between myosin and F-actin. Correlation with actin-binding sites on subfragment-1, J. Biochem. (Tokyo), 96:-1223-1230 (1984).
22. A. Muhlrad, and M. F. Morales, Isolation and partial renaturation of proteolytic fragments of the myosin head, Proc. Natl. Acad. Sci. USA, 81:1003-1007 (1984).
23. D. Mornet, R. Bertrand, P. Pantel, E. Audemard, and R. Kassab, Structure of the actin-myosin interface, Nature 292:301-306 (1981).
24. M. Ikebe, Y. Tonomura, H. Onishi, and S. Watanabe, Elementary steps in the F-actin activated Mg^{2+}-ATPase reaction of gizzard H-mero myosin:effects of phosphorylation of the light-chain subunit. J. Biochem. (Tokyo), 90:61-77 (1981).
25. H. Suzuki, W. F. Stafford III, H. S. Slayter, and J. C. Seidel, A conformational transition in gizzard heavy meromyosin involving the head-tail junction resulting in changes in sedimentation coefficient, ATPase activity, and orientation of heads, J. Biol. Chem., 260:14810-14817 (1985).
26. M. Ikebe, S. Hinkins, and D. J. Hartshorne, Correlation of intrinsic fluorescence and conformation of smooth muscle myosin, J. Biol. Chem., 258:14770-14773 (1983).
27. M. Higashihara, and M. Ikebe, Inhibition of 10S-6S conformational change of smooth muscle myosin by a monoclonal antibody, Biophys. J., 53:578a (1988).
28. A. V. Somlyo, T. M. Butler, M. Bond, and A. P. Somlyo, Myosin filaments have non-phosphorylated light chains in relaxed smooth muscle, Nature, 294:567-569 (1981).

SECOND MESSENGER EFFECTS ON THE MYOSIN PHOSPHORYLATION SYSTEM IN SMOOTH MUSCLE

J.T. Stull, B.F. Bowman, J.C. Colburn, L.-C. Hsu, C.H. Michnoff, D.A. Taylor, and K.E. Kamm

Department of Physiology and Moss Heart Center
University of Texas Southwestern Medical Center at Dallas
Dallas, Texas

INTRODUCTION

All smooth muscle cells contain the contractile proteins actin, myosin, and tropomyosin (Hartshorne, 1987). The enzyme myosin is the primary protein of the thick filament in smooth muscle and is composed of two high molecular weight subunits, or heavy chains, and two each of two types of low molecular weight subunits, or light chains. The molecular mass of each heavy chain subunit is about 200 kDa whereas the light chain subunits are 20 and 17 kDa respectively. The native hexameric form of myosin is configured as an intertwined coiled-tail region of two heavy chains embedded in the thick filament and two globular head regions that protrude at regular intervals to form cross bridges. These head regions contain the actin-binding domain, the catalytic site for ATP hydrolysis, and the associated light chain subunits. According to the sliding filament theory of muscle contraction, thick (myosin) and thin (actin and tropomyosin) filaments move past one another. This process is related to the binding of cross bridges to actin and to the hydrolysis of ATP. The sliding filament theory has been developed primarily from detailed investigations of skeletal muscle, but the general organization of thin and thick filaments in smooth muscle is consistent with a similar mechanism of contraction.

Ca^{2+} activation of contractile elements in vertebrate skeletal and cardiac muscle is mediated by a thin-filament Ca^{2+}-regulatory system, troponin-tropomyosin. The regulation of actin-myosin interactions in smooth muscle by Ca^{2+} is more complex, involving primarily a thick filament regulatory process (Kamm and Stull, 1985b). Ca^{2+} binds to calmodulin and the Ca^{2+}/calmodulin complex subsequently binds to and activates myosin light chain kinase. Activation of this protein kinase results in the phosphorylation of the 20 kDa light chain subunit of myosin. This phosphorylation is associated with stimulation of actin-activated MgATPase activity of smooth muscle myosin and contraction. There may be other secondary Ca^{2+} regulatory mechanisms, but these have not been defined completely (Kamm and Stull, 1989).

TABLE 1

MYOSIN CONTENT OF TRACHEAL TISSUE AND CELLS

Myosin content was calculated from percent total Coomassie stained area migrating at 200 kDa relative to total protein electrophoresed on gradient SDS-PAGE (3-20%). Following pyrophosphate-PAGE, myosin content was calculated by comparison with purified gizzard smooth muscle myosin standards electrophoresed on the same slab gel. Data represent the mean ± SEM of six to ten determinations.

	SDS-PAGE	Pyrophosphate-PAGE
	μg myosin/mg protein	
Tracheal Tissue	63 ± 1.0	89 ± 4.0
Tracheal Cells	37 ± 1.0	55 ± 12
Skin Fibroblasts	---	10 ± 01

REGULATION OF MYOSIN LIGHT CHAIN PHOSPHORYLATION BY CYTOPLASMIC CA^{2+}

Although the cellular and molecular mechanisms through which Ca^{2+} regulates smooth muscle contraction are still not completely understood, it is generally accepted that a primary step involves phosphorylation of the 20 kDa light chain of myosin. We have used primary cultures of adult bovine tracheal smooth muscle cells and examined by immunological and biochemical criteria, components of the Ca^{2+}-dependent myosin light chain phosphorylation system (Taylor and Stull, 1988). Our first objective in this study was to determine if bovine tracheal cells provide a useful model for studying specific Ca^{2+}-dependent processes such as myosin light chain phosphorylation in smooth muscle. The primary objective was to examine the relationship between cytosolic free Ca^{2+} concentration and the extent of myosin light chain phosphorylation in tracheal smooth muscle cells.

Properties of tracheal smooth muscle cells in culture

Myosin content Biochemical and immunological properties of the Ca^{2+}-dependent myosin light chain phosphorylation system were examined in bovine tracheal cells grown to confluence. We determined the myosin content of tracheal cells by 2 electrophoretic techniques, SDS-PAGE and pyrophosphate-PAGE. Tracheal tissue had about 75 μg myosin/mg protein (Table 1). Tracheal cells had 60% of the myosin demonstrated in tracheal tissue. Human skin fibroblasts, however, contained 10 μg myosin/mg protein or approximately 10% of the myosin measured in tracheal tissue.

Myosin light chain Antibodies raised against the 20 kDa light chain purified from bovine trachealis were used to compare immunoreactive light chain in homogenate from tracheal tissue, tracheal cells, and fibroblasts after SDS-PAGE. There was one immunoreactive form of the light chain from tracheal tissue and cells which migrates at 20 kDa. Skin fibroblasts show a second immunoreactive band which migrates slightly faster than tracheal tissue or cell light chain. The faster migrating band accounts for approximately 7% of the total immunoreactivity seen in fibroblasts. Thus, tracheal cells retain the phosphorylatable light chain of myosin at confluence in culture.

TABLE 2

MYOSIN LIGHT CHAIN KINASE ACTIVITY IN MUSCLE TISSUE AND CELLS

Myosin light chain kinase activity was determined as described (Taylor and Stull, 1988). Data represent the mean ± SEM of five experiments.

	Kinase Activity	Per Cent
	(nmol/min/mg protein)	
Tracheal Tissue	12.8 ± 0.8	100%
Tracheal Cells	9.5 ± 0.6	74%
3T3-L1 Cells	0.73 ± 0.11	6%

Myosin light chain kinase Ca^{2+}-dependent myosin light chain kinase activity was measured in tissue and cell extracts (Table 2). Tracheal cells had 74% of the myosin light chain kinase activity measured in tracheal tissue, whereas in nonmuscle cells only 6% of activity measured in tissue was present. Greater than 95% of the activity in tracheal cell and tissue extracts was abolished by the addition of EGTA to the enzyme reaction mixture. Thus the myosin light chain kinase expressed in tracheal cells in culture is Ca^{2+}-dependent.

Ca^{2+}-Dependent Myosin Light Chain Phosphorylation

Cell Measurements Cells were plated at 5 x 10^5 cells/60-mm dish containing four glass coverslips (9 x 35 mm) and used for experiments within 48 h of confluence. Coverslips were transferred to a 60-mm culture dish containing fura-2/AM for loading. Cells were then washed and transferred to a temperature-controlled cuvette for fluorescence measurements. To decrease specular reflection, the coverslip was placed in a Teflon frame to hold the coverslip at a 30° angle to the incident light in the cuvette. Addition of stimulating agent or trichloroacetic acid was accomplished through polyethylene tubing placed 1 cm from the bottom of the cuvette. All measurements were performed while stirring the buffer with an electronically controlled paddle to ensure rapid mixing of the added agents ($t_{1/2}$ of mixing = 0.25 s).

The samples were excited at 340 and 380 nm with a fixed emission wavelength of 510 nm. Autofluorescence of the cells and intrinsic fluorescence of all buffers were measured and subtracted as necessary. Cytosolic Ca^{2+} was calculated according to the ratio method of Grynkiewicz *et al.* (1985) using the formula $[Ca^{2+}]_i=K_d\ (R - R_{min})/(R_{max}-R)$ (Sf2/Sb2) where K_d=224 nM; R = observed ratio of 340/380 fluorescence; R_{max} = ratio of 340/380 fluorescence at a Ca^{2+} concentration (millimolar) that saturates fura-2; R_{min} = ratio of 340/380 fluorescence at a Ca^{2+} concentration (nanomolar) that results in no significant Ca^{2+} binding to fura-2; Sf2 = 380 fluorescence of fura-2 at a nanomolar Ca^{2+}; Sb2 = 380 fluorescence of fura-2 at millimolar Ca^{2+}.

Incompletely de-esterified fura-2/AM can contribute a Ca^{2+}-insensitive component to the fluorescence at 380 nm thereby decreasing the observed 340/380 ratio, R, and the calculated Ca^{2+} concentration

(Scanlon *et al.*, 1987). To alleviate this problem, control measurements were performed with each set of cells. Autofluorescence of buffers and cells at 340 and 380 nm was subtracted from each fluorescence value. Additionally extracellular fluorescence, determined by adding 5 mM EGTA and 12 mM Tris to the cuvette and observing rapid changes in 340 and 380 fluorescence, was subtracted from each signal. However, the most relevant control seemed to be the determination of R_{max}, R_{min}, Sf2, and Sb2 for each group of coverslips, which were then used to calculate Ca^{2+} concentrations for only the cells in that experiment. To determine R_{max}, cells were treated with 10 μM ionomycin in the presence of 1.9 mM Ca^{2+} to obtain a maximal fluorescence signal. EGTA (5 mM) and Tris base (12 mM) were added to determine a fluorescent signal in the absence of Ca^{2+} (R_{min}). Autofluorescence values of cells and buffer were subtracted from the maximal and minimal fluorescence values to obtain a corrected R_{max} and R_{min}, respectively. Sb2 and Sf2 were determined by subtracting autofluorescence and extracellular fluorescence from the signals determined at 380 nm in the presence and absence of Ca^{2+} bound to fura-2.

Since accurate measurements at both 340 and 380 nm excitation wavelengths were not possible at times less than 10 s for our instrument, cytosolic Ca^{2+} concentrations were also calculated from fluorescence measurements obtained by exciting at 340 nm according to the method of Tsien *et al.* (1982): $[Ca^{2+}]_i = K_d\ (F - F_{min}/F_{max} - F)$ where K_d = 224 nM; F = the observed 340 fluorescence; F_{max} = the 340 fluorescence at a Ca^{2+} concentration (millimolar) that saturates fura-2; F_{min} = the 340 fluorescence at a Ca^{2+} concentration (nanomolar) that results in no significant binding of Ca^{2+} to fura-2. Ca^{2+}concentrations calculated with each method at times greater than 10 s after a stimulus differed by less than 5%.

Addition of agents at times less than 10 s was accomplished with the use of a rapid injection device modeled after that of Eccleston *et al.* (1980). Gas pressure, controlled by electronically operated solenoid valves set at fixed time intervals, was used to eject solutions from microsyringes into the stirred solution within the cuvette. Time required for complete mixing of the injected solution was less than 0.5 s.

Cell proteins were precipitated by the rapid addition of 600 μl of 50% trichloracetic acid and 25 mM dithiothreitol to 2.5 ml of fluorescence buffer containing cells on a coverslip. After incubation on ice, protein was collected by scraping into 100 μl of 10% trichloroacetic acid. Precipitated material was collected by centrifugation and prepared for electrophoresis as described by Persechini *et al.* (1986).

The extent of myosin light chain phosphorylation in cells was measured by separation of nonphosphorylated, monophosphorylated, and diphosphorylated forms of the light chain by glycerol PAGE followed by electrophoretic transfer of the protein to nitrocellulose paper and quantitation of the relative amounts of each form by an immunoblot procedure. Immunoblot analysis was performed with antiserum raised to the phosphorylatable light chain of purified bovine tracheal myosin (Persechini *et al.*, 1986).

<u>Cell stimulation</u> Tracheal cells in culture containing fura-2 displayed an increase in cytosolic Ca^{2+} concentration which was dependent upon the concentration of applied ionomycin, a Ca^{2+} ionophore. Ten micromolar ionomycin elicited a maximal increase in cytosolic free Ca^{2+} concentration. Cytosolic Ca^{2+} concentration prior to cell

stimulation was 165 ± 13 nM. Upon stimulation, the Ca^{2+} concentration rapidly increased to about 4000 nM by 30 s and gradually declined to approximately 600 nM at 120 s, which was significantly greater than the basal concentration. Upon stimulation of the cells, myosin light chain phosphorylation also increased rapidly for up to 6 s when it reached a maximal value of 67%. Over 60 s, the percent total myosin light chain phosphorylation remained relatively constant despite a transient increase in cytosolic Ca^{2+} concentration. The percent monophosphorylated and diphosphorylated forms of the light chain, however, changed during this time period. The extent of monophosphorylated light chain increased for up to 6 s to a maximal value of 57% and decreased slightly during the remaining measurements presumably due to the conversion of monophosphorylated light chain to diphosphorylated light chain. The diphosphorylated form of myosin light chain increased slowly for up to 30 s and then decreased. Diphosphorylated light chain was detectable (9%) by 6 s when free cytosolic Ca^{2+} had risen above 400 nM, reached a maximal value of 19% by 30 s and remained significantly above basal values at 90 s (11%).

Exposure of cells to 10 μM carbachol resulted in an increase in cytosolic Ca^{2+} concentration to 492 ± 90 nM at 12 s which declined to 300 nM by 35 s where it remained for up to 60 s. Light chain phosphorylation followed the same relative time course, reaching a maximal value of approximately 50% at 12 s and declining significantly to 30% by 60 s. No diphosphorylated light chain was observed during carbachol stimulation.

Other pharmacological agents also elicited responses in tracheal smooth muscle cells in culture. Serotonin and histamine rapidly increased cytosolic free Ca^{2+} concentrations and the extent of myosin phosphorylation. The extent of myosin light chain phosphorylation was maximal by 6 to 15 s and remained significantly elevated above the basal value of 9 ± 2% for 40 to 60 s. Similarly, cytosolic free Ca^{2+} concentrations remained significantly greater than the basal value of 150 ± 10 nM for 40 to 60 s with these 2 agents. When extracellular Ca^{2+} was decreased by the addition of 5 mM EGTA for 15 s prior to stimulation, the ability of serotonin, histamine and ionomycin to increase cytosolic free Ca^{2+} concentrations and the extent of light chain phosphorylation was significantly diminished. The response of the cells to carbachol, however, was unaltered.

Stimulation of bovine tracheal smooth muscle cells with histamine, serotonin, carbachol or ionomycin yielded a single relationship between cytosolic free Ca^{2+} concentrations and the extent of myosin phosphorylation. The relationship between free cytosolic Ca^{2+} concentration and the extent of light chain phosphorylation showed 67% maximal phosphorylation and 250 nM Ca^{2+} required for half-maximal phosphorylation. This relationship appeared to be independent of the source of the Ca^{2+}, since the responses to carbachol were similar to the responses obtained with the other agents. Since the carbachol response was not inhibited by EGTA, it is probably due to Ca^{2+} release from internal stores, presumably through phosphoinositide hydrolysis and formation of inositol trisphosphate (Bitar *et al.*, 1986; Brading and Sneddon, 1980). EGTA significantly decreased the elevation in cytosolic free Ca^{2+} and the extent of myosin light chain phosphorylation in response to serotonin or histamine. This observation is consistent with the idea that serotonin and histamine stimulate Ca^{2+} influx across the plasma membrane (Brown *et al.*, 1986; Kanaide *et al.*, 1987). The cytosolic free Ca^{2+} concentration response to ionomycin is inhibited 70% in the presence of EGTA implying that approximately 30% of the Ca^{2+} signal may depend upon Ca^{2+} release from intracellular storage sites.

Williams et al. (1985) have reported that ionomycin can mobilize Ca^{2+} from the sarcoplasmic reticulum or nucleus in isolated smooth muscle cells. Thus, ionomycin may allow Ca^{2+} to move across the plasma membrane in the presence of high extracellular Ca^{2+} in addition to mobilizing Ca^{2+} from intracellular storage sites.

In summary, the relationship between cytosolic free Ca^{2+} concentration and the extent of myosin light chain phosphorylation in bovine tracheal smooth muscle cells is consistent with a tightly regulated biological system where small changes in cytosolic Ca^{2+} concentration result in activation of myosin light chain kinase. The relationship between Ca^{2+} concentration and the extent of light chain phosphorylation is independent of the source of free intracellular Ca^{2+} or the agent used to stimulate the cells. Furthermore, this relationship is not altered by previous exposure to high concentrations of free Ca^{2+}. Thus, no evidence was obtained for Ca^{2+} activation of myosin light chain phosphatase activity.

PHOSPHORYLATION OF MYOSIN IN CONTRACTING SMOOTH MUSCLE

Myosin Light Chain Phosphorylation

Smooth muscle myosin light chain kinase readily catalyzes the incorporation of one mol phosphate per mol myosin light chain into serine 19 which increases the actin-activated MgATPase activity of myosin (Kamm and Stull, 1985b). However, it was subsequently demonstrated that gizzard smooth muscle myosin light chain kinase phosphorylated threonine 18 in purified gizzard myosin light chain. However 32-fold more myosin light chain kinase was required for this phosphorylation as compared to the amount required for phosphorylation of serine 19 (Ikebe and Hartshorne, 1985; Ikebe *et al.*, 1986). Phosphorylation of the threonine further increased the actin-activated MgATPase activity of myosin which was associated with changes in conformation and other biochemical properties (Ikebe *et al.*, 1988).

The Ca^{2+}-activated, phospholipid-dependent protein kinase C also phosphorylates smooth muscle gizzard myosin light chain at multiple sites and these have recently been identified as threonine 9, serine 1 and serine 2 (Ikebe *et al.*, 1987; Bengur *et al.*, 1987). Phosphorylation at one or more of these sites decreased the actin-activated MgATPase activity of myosin that had been previously phosphorylated by myosin light chain kinase (Nishikawa *et al.*, 1984) and prevented a conformational change produced by myosin light chain kinase (Umekawa *et al.*, 1985). In addition, prior phosphorylation of light chain by protein kinase C decreased the subsequent rate of phosphorylation by myosin light chain kinase (Nishikawa *et al.*, 1984). These investigations with purified proteins demonstrate as many as five potential phosphorylation sites in smooth muscle myosin light chain that affect the actin-activation properties of myosin MgATPase activity.

At this time it is not clear whether there are multiple sites of phosphorylation on myosin light chain in intact contracting smooth muscle and if so, whether these are the result of both myosin light chain kinase and protein kinase C activities. Therefore, we compared the stoichiometry and sites of phosphorylation on myosin light chain in contracting tracheal smooth muscle during the initial and tonic phases of contraction (Colburn *et al.*, 1988).

<u>Phosphorylation of Purified Myosin Light Chain</u> Tracheal smooth muscle myosin light chain was phosphorylated to 1.0 or 2.0 mol ^{32}P per mol of light chain by myosin light chain kinase. The two-dimensional

tryptic-peptide maps of both monophosphorylated and diphosphorylated myosin light chain demonstrated a single major spot. The single major spot for diphosphorylated light chain indicates that the two phosphorylation sites are on the same peptide. The monophosphorylated peptide consistently migrated farther in both the electrophoretic and chromatographic directions.

The two-dimensional tryptic peptide map of tracheal smooth muscle myosin light chain diphosphorylated by protein kinase C had a pattern similar to that reported by Bengur *et al.* (1987b) for gizzard myosin light chain. This peptide map demonstrated the presence of two distinct phosphopeptides, both migrating much farther in the electrophoretic direction and less in the chromatographic direction than the single phosphopeptides observed in the maps obtained from digests of tracheal muscle myosin light chain phosphorylated by myosin light chain kinase.

Phosphorylation of Myosin Light Chain in Tracheal Tissue Tracheal smooth muscle strips were incubated with 0.1 mM ^{32}P-phosphate for 4 h and quick-frozen after various treatments. The immunoreactivity and radioactivity of tracheal smooth muscle monophosphorylated and diphosphorylated myosin light chains were compared between tissue homogenates and purified protein (Colburn *et al.*, 1988). Increased amounts of phosphate incorporated into purified tracheal myosin light chain resulted in increased protein mobility in urea/glycerol PAGE. Monophosphorylated and diphosphorylated myosin light chains from tracheal tissue homogenates have electrophoretic mobilities identical to those of the respective purified light chains as shown by Western blots and autoradiograms. The relative amounts of monophosphorylated and diphosphorylated myosin light chains from tracheal smooth muscles were quantitated by densitometric measurements from the Western blots (Table 3). The amount of both monophosphorylated and diphosphorylated myosin light chains increased after 1 min of carbachol stimulation of muscle tissue which correlated to an increase in force. After 30 min of stimulation, both phosphorylated forms decreased relative to the 1 min values although force had increased.

In order to identify phosphorylated amino acids in myosin light chain from stimulated trachealis, the proteins were isolated by two-dimensional gel electrophoresis (IEF and SDS-PAGE). Monophosphorylated and diphosphorylated myosin light chains were subjected to partial acid hydrolysis and high voltage paper electrophoresis at pH 1.9.

TABLE 3

PHOSPHORYLATION OF TRACHEAL LIGHT CHAIN IN RESPONSE TO CARBACHOL

Tracheal tissues were quick-frozen and analyzed for extent of light chain (LC) phosphorylation by the immunoblot procedure (Persechini *et al.*, 1986). The percentages of light chain in the monophosphorylated and diphosphorylated forms were determined from densitometric measurements of immunostained light chains. The results are means ± SEM for 5-6 values.

10 μM Carbachol	Monophosphorylated LC	Diphosphorylated LC	Force (g)
0	13 ± 7%	0	0
1 min	66 ± 5%	11 ± 2%	8.2 ± 0.3
30 min	47 ± 4%	5 ± 2%	11.7 ± 0.4

Monophosphorylated myosin light chain contained only phosphoserine. However, both phosphoserine and phosphothreonine were present in diphosphorylated myosin light chain. Thus, the specific phosphoamino acids were the same for both monophosphorylated and diphosphorylated myosin light chains phosphorylated *in vivo* and for light chains phosphorylated by myosin light chain kinase *in vitro*.

The isolated monophosphorylated and diphosphorylated light chains were proteolyzed with trypsin (Colburn *et al.*, 1988). Two-dimensional tryptic peptide maps of monophosphorylated myosin light chains from stimulated trachealis had mobilities similar to those mobilities from the peptide maps derived from purified tracheal muscle myosin light chain monophosphorylated by myosin light chain kinase. The peptide maps of the diphosphorylated myosin light chain from tissue were also similar to the maps obtained with purified light chain. There were no radioactive spots that comigrated with ^{32}P-peptides obtained from purified tracheal light chain phosphorylated by protein kinase C. Peptide maps of tryptic peptides of myosin light chain from stimulated trachealis were compared with light chain standards prepared by phosphorylation of purified tracheal light chain by known kinases. The conclusion was reached that phosphorylation of myosin light chain in tracheal smooth muscle is probably catalyzed by myosin light chain kinase; evidence for phosphorylation by protein kinase C was not observed. While it appears that myosin light chain may be phosphorylated by protein kinase C in nonmuscle cells (Ludowyke *et al.*, 1987; Bengur *et al.*, 1987; Naka *et al.*, 1988), we find no evidence for protein kinase C phosphorylation of myosin light chain in tracheal smooth muscle during the initial or tonic phases of contraction.

In tissue, diphosphorylated myosin light chain is formed in the presence of a high concentration of agonist (10 μM carbachol). Maximal levels of diphosphorylated light chain (11% of total light chain) probably occur within the first minute following the initiation of stimulation, as is found for monophosphorylated light chain. The physiological significance of these low levels of diphosphorylated light chain is not apparent.

Myosin Heavy Chain Phosphorylation

Control and carbachol stimulated tracheal smooth muscles were homogenized in SDS buffer and subjected to electrophoresis in 3-20% gradient SDS-PAGE. A number of proteins were ^{32}P-labeled in control muscles. Stimulation of tracheal muscles with carbachol for 1 to 30 min resulted in the additional phosphorylation of relatively few proteins as measured by one dimension SDS-PAGE. After 1 min stimulation with carbachol, the phosphorylation of myosin light chain was the most apparent. However, additional proteins appear to be phosphorylated at 30 min carbachol stimulation; the apparent molecular masses of these proteins are 28 kDa, 64 kDa, and 240 kDa.

We found that only minor amounts of phosphate were incorporated in myosin heavy chain in bovine tracheal smooth muscle as determined by single dimension SDS-PAGE and autoradiography. Because Kawamoto and Adelstein (1988) recently showed that myosin heavy chain was phosphorylated in aortic cells in culture, we examined this possibility more carefully. Myosin was immunoprecipitated from tissue homogenates (Kawamoto and Adelstein, 1988) and was analyzed by SDS-PAGE. Myosin heavy chain was phosphorylated, but the amount of ^{32}P incorporation was substantially less than ^{32}P incorporated into myosin light chain. Based on the measured radiospecific activity of myosin light chain, the stoichiometry of myosin heavy chain phosphorylation at rest was estimated at 0.06 ± 0.01 mol ^{32}P per mol of heavy chain. The level of phosphorylation of heavy chain after carbachol treatment for 1 min or 30 min was not significantly different than that in control tissues.

Kawamoto and Adelstein (1988) reported that myosin heavy chain in aortic smooth muscle cells in culture had 0.7 mol ^{32}P incorporated per mol heavy chain. Although ^{32}P incorporation was demonstrated for myosin heavy chain isolated from aortic tissue, the stoichiometry was not established. It is not clear whether these differences between tracheal and aortic smooth muscle are due to experimental techniques, cell types, or cells in culture.

In summary, we have found that with initial and tonic contractions stimulated by a cholinergic muscarinic agonist in tracheal smooth muscle, the extent of myosin heavy chain phosphorylation remains low. Stimulation of tracheal smooth muscle with a high concentration of carbachol results in formation of both monophosphorylated and diphosphorylated myosin light chain although the amount of diphosphorylated light chain is substantially less than monophosphorylated light chain. It appears that phosphorylation of the single serine site on myosin light chain by myosin light chain kinase is the primary phosphorylation that is associated with activation of smooth muscle contraction.

REFERENCES

Bengur, A.R., Robinson, E.A., Apella, E., and Sellers, J.R., 1987, Sequence of the sites phosphorylated by protein kinase C in the smooth muscle myosin light chain, *J. Biol. Chem.*, 262:7613-7616.

Bitar, K.N., Bradford, P., Putney, Jr., J.W., and Makhloug, G.M., 1986, Cytosolic calcium during contraction of isolated mammalian gastric muscle cells, *Science*, 232:1143-1145.

Brading, A.F. and Sneddon, P., 1980, Evidence for multiple sources of calcium for activation of the contractile mechanism of guinea-pig taenia coli on stimulation with carbachol, *Br. J. Pharmacol.*, 70:229-240.

Brown, R.D., Prendiville, P., and Cain, C., 1986, Alpha 1-adrenergic and H1-histamine receptor control of intracellular Ca^{2+} in a muscle cell line: the influence of prior agonist exposure on receptor responsiveness, *Molec. Pharmacol.*, 29:531-539.

Colburn, J.C., Michnoff, C.H., Hsu, L., Slaughter, C.A., Kamm, K.E., and Stull, J.T., 1988, Sites phosphorylated in myosin light chain in contracting smooth muscle, *J. Biol. Chem.*, in press.

Eccleston, J.F., Messerschmidt, R.G., and Yates, D.W., 1980, A simple rapid mixing device, *Anal. Biochem.*, 10673-77.

Grynkiewicz, G., Poenie, M., and Tsien, R.Y., 1985, A new generation of Ca^{2+} indicators with greatly improved fluorescence properties, *J. Biol. Chem.*, 260:3440-3450.

Hartshorne, D.J., 1987, Biochemistry of the contractile process in smooth muscle, in: "Physiology of the Gastrointestinal Tract, Second Edition, L.R. Johnson, ed., Raven Press, New York.

Ikebe, M., and Hartshorne, D.J., 1985, Phosphorylation of smooth muscle myosin at two distinct sites by myosin light chain kinase, *J. Biol. Chem.*, 260:10027-10031.

Ikebe, M., Hartshorne, D.J., and Elzinga, M., 1986, Identification, phosphorylation, and dephosphorylation of a second site for myosin light chain kinase on the 20,000 dalton light chain of smooth muscle myosin, *J. Biol. Chem.*, 261:36-39.

Ikebe, M., Hartshorne, D.J., and Elzinga, M., 1987, Phosphorylation of the 20,000-dalton light chain of smooth muscle myosin by the calcium-

activated, phospholipid-dependent protein kinase, *J. Biol. Chem.*, 262:9569-9573.
Ikebe, M., Koretz, J., and Hartshorne, D.J., 1988, Effects of phosphorylation of light chain residues threonine 18 and serine 19 on the properties and conformation of smooth muscle myosin, *J. Biol. Chem.*, 263:6432-6437.
Kamm, K.E., and Stull, J.T., 1985b, Myosin phosphorylation, force and maximal shortening velocity in neurally stimulated tracheal smooth muscle, *Amer. J. Physiol.*, 249:C238-C247.
Kamm, K.E., and Stull, J.T., 1989, Regulation of smooth muscle contractile elements by second messengers, *Annu. Rev. Physiol.*, in press.
Kanaide, H., Hasegawa, M., Kobayashi, S., and Nakamura, M., 1987, Serotonin-induced cytosolic free calcium transients in cultured vascular smooth muscle cells, *Biochem. Biophys. Res. Comm.*, 143:532-538.
Kawamoto, S. and Adelstein, R.S.,1988, The heavy chain of smooth muscle myosin is phosphorylated in aorta cells, *J. Biol. Chem.*, 263:1099-1102.
Ludowyke, R.I., Peleg, I., Beaven, M.A., and Adelstein, R.S., 1987, Analysis of phosphorylation of myosin heavy and light chains during antigen stimulation of basophil leukemic cells, *J. Cell. Biol.*, 105:118a.
Naka, M., Saitoh, M., and Hidaka, H., 1988, Two phosphorylated forms of myosin in thrombin-stimulated platelets, *Arch. of Biochem. and Biophysics*, 261:235-240.
Nishikawa, M., Sellers, J.R., Adelstein, R.S., and Hidaka, H., 1984, Protein kinase C modulates *in vitro* phosphorylation of the smooth muscle heavy meromyosin by myosin light chain kinase, *J. Biol. Chem.*, 259:8808-8814.
Persechini, A., Kamm, K.E., and Stull, J.T., 1986, Different phosphorylated forms of myosin in contracting tracheal smooth muscle, *J. Biol. Chem.*, 261:6293-6299.
Scanlon, M., Williams, D.A., and Fay, F.S., 1987, A Ca^{2+}-insensitive form of fura-2 associated with polymorphonuclear leukocytes, *J. Biol Chem.*, 262:6308-6312
Taylor, D.A., and Stull, J.R., 1988, Calcium dependence of myosin light chain phosphorylation in smooth muscle cells, *J. Biol. Chem.*, 263:14456-14462.
Tsien, R.Y., Pozzan, T., and Rink, T.J., 1982, Calcium homeostasis in intact lymphocytes: Cytoplasmic free calcium monitored with a new, intracellularly trapped fluorescent indicator, *J. Cell Biol.*, 94:325-334.
Umekawa, H., Naka, M., Inagaki, M., Onishi, H., Wakabayashi, T., and Hidaka, H., 1985, Conformational studies of myosin phosphorylated by protein kinase C, *J. Biol. Chem.*, 260:9833-9837.
Williams, D.A., Fogarty, K.E., Tsien, R.Y., and Fay, F.S., 1985, Calcium gradients in single smooth muscle cells revealed by the digital imaging microscope using fura-2, *Nature*, 318:558-561.

IN SITU PHOSPHORYLATION OF HUMAN PLATELET AND RAT BASOPHILIC LEUKEMIA CELL (RBL-2H3) MYOSIN HEAVY CHAIN AND LIGHT CHAIN

Robert S. Adelstein, Michael A. Beaven, A. Resai Bengur, Sachiyo Kawamoto, Russell I. Ludowyke, Itzhak Peleg, and James R. Sellers

National Heart, Lung, and Blood Institute
National Institutes of Health
Bethesda, MD 20892 USA

INTRODUCTION

Cytoplasmic myosin has been isolated and characterized from numerous vertebrate nonmuscle cells, including platelets, fibroblasts, intestinal brush border and macrophages (for reviews see refs. 1 and 2). Similar to smooth muscle myosin, vertebrate cytoplasmic (also referred to as nonmuscle) myosin is composed of a pair of 200 kDa heavy chains and two pairs of light chains (20 kDa and 17 kDa).

The 20 kDa myosin light chain is a substrate for the calcium-calmodulin-dependent enzyme, myosin light chain kinase, which catalyzes phosphorylation of a serine residue, the equivalent residue to serine-19 in the avian smooth muscle myosin light chain. The result of this phosphorylation is a marked increase in the actin-activated MgATPase activity concomitant with an increase in myosin filament formation[1,2].

In addition to acting as a substrate for myosin light chain kinase the 20 kDa light chain of platelet myosin can also serve as a substrate for protein kinase C[3]. Previous studies with the 20 kDa avian gizzard light chain identified three sites of protein kinase C phosphorylation *in vitro*: serine-1, serine-2 and threonine-9[4,5]. In this paper we compare the sites phosphorylated on the 20 kDa platelet myosin light chain by protein kinase C under *in vitro* and *in situ* conditions[6]. We also studied the sites phosphorylated on the 20 kDa light chain of cultured rat basophilic leukemia cells (RBL-2H3 cells) following antigenic stimulation, which results in activation of protein kinase C[7].

In the course of these experiments we discovered that, in addition to phosphorylating the 20 kDa light chain of vertebrate nonmuscle myosin, protein kinase C also catalyzed the phosphorylation of the myosin heavy chain. We therefore, also report on myosin heavy chain phosphorylation under *in vitro* and *in situ* conditions[6,7]. Finally, we show that the same stimulus that results in the activation of protein kinase C and subsequent phosphorylation of the myosin heavy chain and light chain in RBL-2H3 cells also results in the release of histamine from these cells[7].

EXPERIMENTAL PROCEDURES

The techniques used in the in vitro and in situ phosphorylation of human platelet myosin are outlined in detail in Kawamoto et al.[6].

In Vitro Phosphorylation of Platelet Myosin Human platelet myosin was purified as described[8]. Protein kinase C was purified from rat brain as described by Huang et al.[9] except the final hydroxyapatite column was omitted. Platelet myosin (1-2 mg/ml) was incubated at 30°C in 5 mM NaCl, 20 mM Tris-HCl (pH 7.5), 6 mM $MgCl_2$, 0.7 mM $CaCl_2$, 80 µg/ml phosphatidylserine, 1 µg/ml diolein, 0.5 mM dithiothreitol, 1 mM [γ-^{32}P]ATP (100 Ci/mol) and 1 µg/ml protein kinase C. The extent of total myosin phosphorylation was determined by applying 10 µl of reaction mixture to Whatman No. 3MM filter discs which were then immersed in 10% TCA and 8% Na-pyrophosphate on ice, washed in 10% TCA and 2% Na-pyrophosphate and counted in a scintillation counter. Another 10 µl of the same reaction mixture was added to SDS-sample buffer[10] in boiling water and analyzed by SDS-polyacrylamide gel electrophoresis (SDS-PAGE).

Labeling of Platelets with ^{32}P and RBL-2H3 Cells with ^{32}P and [^{35}S]-methionine Human platelet concentrates were sedimented at 1200 x g for 8 min, and the platelet pellets were suspended in phosphate-free Medium 199 with 5 mM EDTA and 2 mM EGTA. The platelets were labeled with [^{32}P]orthophosphoric acid (0.2-0.5 mCi/ml) for 90 min at 37°C, separated from unincorporated [^{32}P]orthophosphate by centrifugation and resuspended in the above medium. RBL-2H3 cells were maintained as monolayer cultures and harvested by trypsin treatment as described previously[11]. The cells (2.5×10^6) were incubated at 37° for 2-3 h in 2.5 ml of 119 mM NaCl; 5 mM KCl; 5.6 mM glucose; 0.4 mM $MgCl_2$; 1 mM $CaCl_2$; 0.1% BSA and 25 mM PIPES-NaOH (pH 7.2) in the presence of 0.1 mCi/ml [^{32}P]orthophosphoric acid. Labeling with [^{35}S]methionine, 0.07 mCi/ml, was carried out overnight at 37° in growth medium.

Stimulation of Platelets with TPA Platelets were incubated with 162 nM 12-O-tetradecanoylphorbol 13-acetate (TPA) at 37° and the reaction stopped by addition of TCA to a final concentration of 10%. TCA-precipitated proteins were washed in water and/or acetone and dissolved in SDS-sample buffer[10] or isoelectric focusing (IEF)-sample buffer[12].

Stimulation of RBL-2H3 Cells Cells were sensitized by addition of 0.5 µg/ml of DNP-specific IgE and incubated overnight at 37° with 5% CO_2. Cells were stimulated with 10 ng/ml $DNP_{24}BSA$ (dinitrophenylated bovine serum albumin, DNP-BSA) at 37° for various periods of time. Histamine secretion was measured as previously described[13].

Miscellaneous Methods Myosin was immunoprecipitated as outlined previously[14]. Tryptic peptide mapping of the phosphorylated myosin light chain was carried out as previously described[15] as was tryptic peptide mapping of the myosin heavy chain[14]. The stoichiometry of phosphorylation using [^{35}S]methionine labeled myosin was determined by modification of the method used previously[14]. The stoichiometry of the platelet myosin light chain was determined using two-dimensional IEF/SDS-PAGE at pH 4-6 and staining the gel with Coomassie brilliant blue to identify the various forms of the unphosphorylated and phosphorylated myosin light chain[6].

RESULTS

In Vitro Phosphorylation of Human Platelet Myosin When human platelet myosin is incubated with protein kinase C purified from rat brains in the presence of [γ-^{32}P]ATP both the myosin heavy chain and the 20 kDa myosin light chain incorporate radioactive phosphate. Two-dimensional tryptic peptide maps of the myosin light chains show that phosphate is incorporated into peptides containing residues equivalent to threonine-9 (Fig. 1, peptide D) and serine-1 or serine-2 (Fig. 1, peptides B and C) in the avian gizzard myosin light chain. Two-dimensional tryptic peptide mapping of the phosphorylated myosin heavy chain results in a single major phosphoserine containing peptide (Fig. 2, left). The exact location of the phosphorylated serine residue has yet to be determined, but preliminary experiments (JRS, unpublished results) suggest that it is located in the rod portion of the heavy chain.

In Situ Phosphorylation of Human Platelet Myosin Treatment of intact human platelets, which have previously been equilibrated with radioactive phosphoric acid, with 162 nM TPA results in the phosphorylation of a number of proteins including platelet myosin (see Fig. 3). Both the 20 kDa light chain and the 200 kDa myosin heavy chain incorporate phosphate following treatment with TPA.

Analysis of the sites of phosphorylation on the myosin heavy chain and light chain shows both similarities and differences between *in vitro* phosphorylation using protein kinase C and *in situ* phosphorylation following treatment with TPA. In the case of the myosin heavy chain, two-dimensional tryptic peptide mapping of the *in situ* phosphorylated myosin resulted in the same major phosphoserine-containing peptide seen following *in vitro* phosphorylation with protein kinase C (Fig. 2, right, compare to Fig. 2, left).

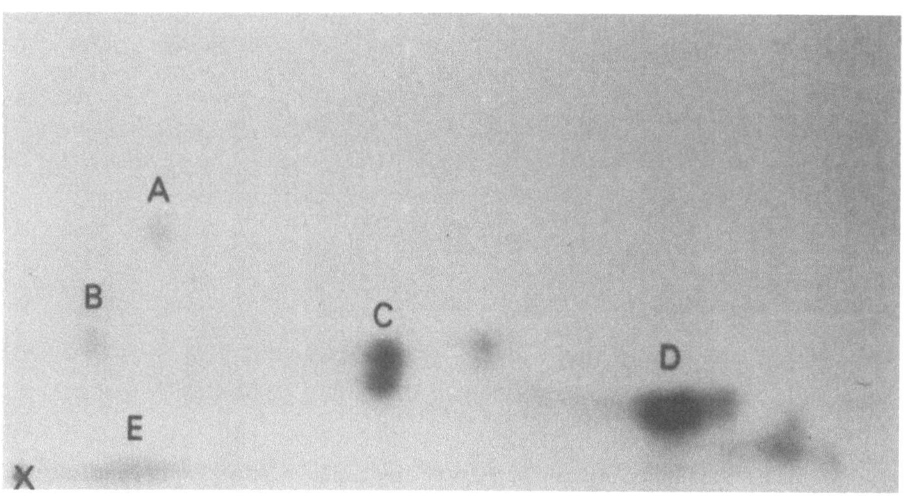

Fig. 1. Autoradiogram of a tryptic peptide map of the phosphorylated platelet 20 kDa myosin light chain following *in vitro* phosphorylation by protein kinase C. Peptides B and C are phosphoserine-containing peptides corresponding to peptides containing phosphorylated serine-1 or -2 in the gizzard myosin light chain[4,5]. Peptide D is a phosphothreonine-containing peptide corresponding to a peptide containing threonine-9[4,5]. X marks the origin. This figure is reproduced from reference 6.

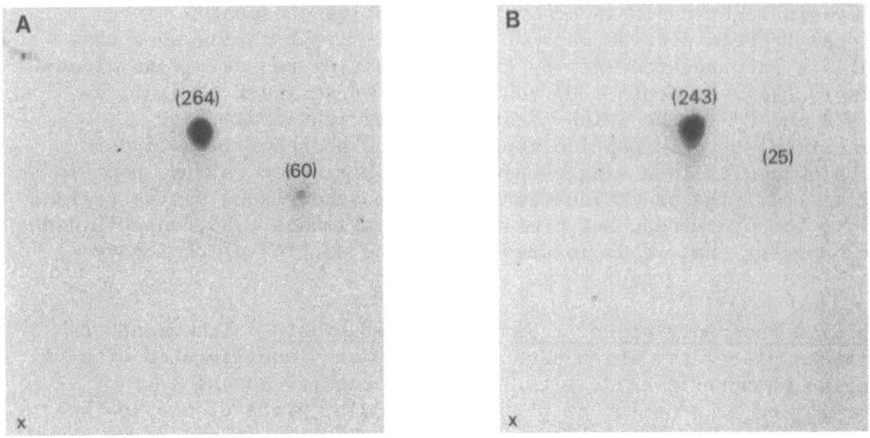

Fig. 2. Identification of the phosphorylated myosin heavy chain tryptic peptides following protein kinase C phosphorylation _in vitro_ (A) and _in situ_ (B). The autoradiogram on the left is from a phosphopeptide map from purified human platelet myosin which has been phosphorylated using purified protein kinase C _in vitro_. The one on the right is from a tryptic phosphopeptide map from intact platelets which have been treated with TPA (162 nM, 15 min). Both myosin heavy chains were isolated by SDS-PAGE using 5% polyacrylamide gels. X marks the origin. This figure is reproduced from reference 6.

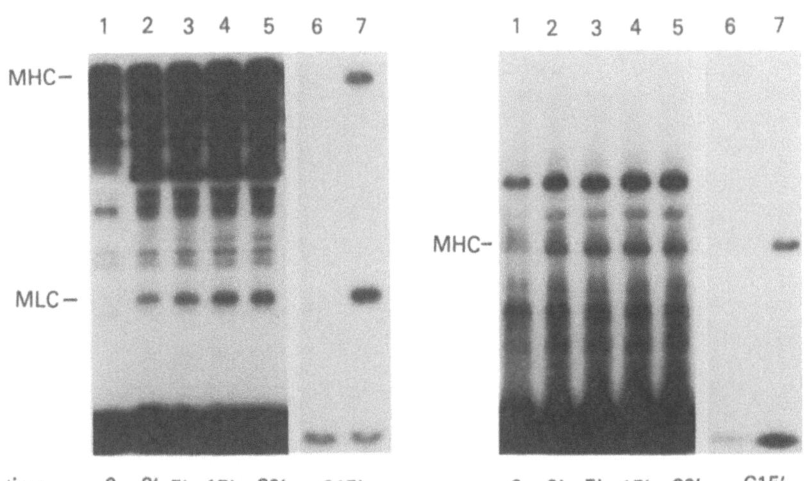

Fig. 3. Myosin phosphorylation in TPA-treated platelets. ^{32}P-labeled human platelets were treated with 162 nM TPA for the time (min) indicated. TCA-precipitated total proteins (lanes 1-5) or immunoprecipitates following addition of anti-platelet myosin antibodies (lanes 6 and 7) were separated by SDS-PAGE using a 13% polyacrylamide gel (left) or 5% polyacrylamide gel (right). ^{32}P autoradiograms are shown. C, is a control sample that was incubated for 15 min in the absence of TPA; MHC, myosin heavy chain; MLC, myosin light chain.

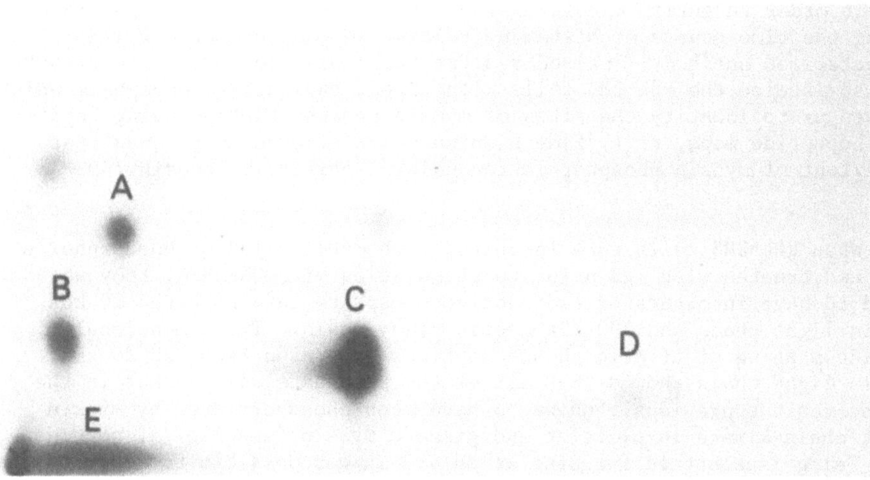

Fig. 4. Autoradiogram of a tryptic peptide map of the phosphorylated platelet 20 kDa myosin light chain following treatment of ^{32}P-labeled platelets with TPA. Peptides B and C are phosphoserine-containing peptides corresponding to peptides containing serine-1 or -2 in the gizzard myosin light chain[4,5]. Peptide E is thought to be a diphosphorylated peptide containing phosphorylated serine-1 and -2. A is a phosphoserine-containing peptide corresponding to a peptide containing serine-19, the site phosphorylated by myosin light chain kinase. This figure is reproduced from reference 6.

In situ phosphorylation of the myosin light chain following treatment with TPA results in phosphorylation of serine-1 or -2 (Fig. 4, peptides B and C), and serine-1 and -2 (peptide E, see ref. 6 for discussion). However, there is very little, if any, phosphorylation of the peptide containing threonine (the equivalent of threonine-9 in the gizzard light chain, see Fig. 4, peptide D and compare to Fig. 1). This raises the possibility that threonine-9 is not a substrate for protein kinase C under *in vivo* conditions and makes it difficult to evaluate the function of protein kinase C phosphorylation, based on *in vitro* experiments. Fig. 4 also shows that there is a small, but significant increase in phosphate at the site known to be phosphorylated by myosin light chain kinase (Fig. 4, peptide A).

In Situ Phosphorylation of Rat Basophilic Leukemia Cell Myosin (RBL-2H3 Cells). In order to relate protein kinase C phosphorylation of the myosin heavy chain and light chain to a cellular function, we chose to study histamine secretion in the rat cell line RBL-2H3. These cells contain receptors for IgE on their surface and can be made to release the contents of their granules by being treated with agents that bring about aggregation of receptor-bound molecules of IgE. One agent that causes this aggregation is dinitrophenylated bovine serum albumin (DNP-BSA), when cells are primed with DNP-specific IgE. Aggregation of receptor-bound IgE molecules results in hydrolysis of phosphatidyl inositol bisphosphate to inositol trisphosphate and diacyglycerol and a rise in the concentration of intracellular calcium and subsequently the release of histamine[16]. We were interested in investigating whether aggregation of RBL-2H3 cell surface receptors caused by DNP-BSA would also result in the phosphorylation of RBL-2H3 myosin.

In order to purify myosin as rapidly and specifically as possible during the time course of histamine release we made use of a well-characterized antibody[17] raised against human platelet myosin in rabbits. We also labeled the RBL-2H3 cells with either radioactive phosphate which allowed us to identify the sites of myosin phosphorylation using tryptic phosphopeptide maps, or [^{35}S]methionine which allowed us to quantitate the extent of myosin phosphorylation using isoelectric focusing/SDS-PAGE[12].

When RBL-2H3 cells were incubated with ^{32}P-labeled orthophosphoric acid and treated with IgE prior to stimulation with DNP-BSA, they were found to have incorporated radioactive phosphate into both the 20 kDa myosin light chain and 200 kDa myosin heavy chain. Two-dimensional peptide mapping of tryptic phosphopeptides generated from the 20 kDa myosin light chain showed that all of the phosphate was present in the serine residue previously shown to have been phosphorylated by myosin light chain kinase in platelet and gizzard myosin (see Fig. 4, peptide A). Using isoelectric focusing at pH 4-6 it was possible to separate the unphosphorylated and monophosphorylated forms of the [^{35}S]methionine labeled 20 kDa myosin light chain. This allowed us to determine that the stoichiometry of phosphorylation of the myosin light chain was 0.4 mol phosphate/mol of myosin light chain in unstimulated cells.

Following antigenic stimulation the amount of phosphate rose from 0.4 mol phosphate/mol myosin light chain to a value of 1.1 mol phosphate/mol light chain, reaching a peak between 10 and 20 min. This increase in the amount of phosphorylation was apparently due to activation of protein kinase C, since tryptic peptide maps of the myosin light chain carried out during this time period showed the appearance of a new phosphopeptide. This peptide comigrated with the phosphopeptide found following phosphorylation of platelet myosin with protein kinase C (see Fig. 1, peptide C).

When the stoichiometry of phosphorylation was determined for the sites phosphorylated by myosin light chain kinase and protein kinase C (i.e., after antigenic stimulation), it was found that the phosphate content of the former site remained at 0.4 mol phosphate/mol light chain throughout the entire 30 min time course. In contrast, the site phosphorylated by protein kinase C which contained no phosphate prior to stimulation increased to 0.7 mol phosphate/mol light chain.

Two-dimensional mapping of tryptic peptides generated from the 200 kDa myosin heavy chain showed that in unstimulated RBL-2H3 cells all of the phosphate was distributed among three major phosphopeptides. Following antigenic stimulation a single new tryptic phosphopeptide appeared. This new peptide comigrated with the tryptic phosphopeptide generated from human platelet myosin that was phosphorylated _in vitro_ by protein kinase C (see Fig. 2, left).

We calculated the stoichiometry of myosin heavy chain phosphorylation using the specific activity derived for the ^{32}P-labeled myosin light chain. We assumed that none of the phosphate found in the myosin heavy chain prior to stimulation was due to the activity of protein kinase C. This assumption is strengthened by the finding that _in vitro_ phosphorylation of human platelet myosin by protein kinase C does not yield any of the peptides present in the unstimulated RBL-2H3 cells following hydrolysis of the heavy chain by trypsin. The stoichiometry of phosphorylation of the myosin heavy chain at the serine site phosphorylated by protein kinase C was calculated to have increased from 0 mol phosphate/mol myosin heavy chain in unstimulated RBL-2H3 cells to

Table I. Stoichiometry of Phosphorylation of Sites in the Myosin Light Chain and Heavy Chain (mol phosphate/mol myosin subunit)

	Light Chain Site		Heavy Chain Site	
	Kinase C	MLCK[1]	Kinase C	Unk[1]
Platelets (stim[1] with TPA)	1	0.2	0.7	0
RBL-2H3 cells (unstim[1])	0	0.4	0	1
RBL-2H3 cells (stim[1])	0.7	0.4	1.0	1

[1]MLCK; myosin light chain kinase site; Unk, unknown site; unstim, unstimulated; stim, stimulated.

1.0 mol phosphate/mol heavy chain after antigenic stimulation.

Correlation of Protein Kinase C Phosphorylation of Myosin With Histamine Secretion in RBL-2H3 Cells. Table I summarizes the data for the stoichiometry of phosphorylation in each of the sites on the myosin light chain and heavy chain. The data show that following TPA treatment of platelets and antigenic stimulation of RBL-2H3 cells there is an increase in myosin heavy chain and light chain phosphorylation at sites phosphorylated by protein kinase C. When histamine release was assayed in RBL-2H3 cells incubated in the absence of antigen it was found to be very low (about 1-3% of the cell's content). After antigenic stimulation it increased to 44% of the total cellular histamine content. Moreover, the time course of protein kinase C phosphorylation of myosin heavy chain and light chain was found to parallel the time course of histamine release. Although the data can be interpreted as evidence for protein kinase C phosphorylation of myosin playing a role in histamine secretion, further experiments will be necessary to establish this relationship.

DISCUSSION

In this paper we have presented data showing that activation of protein kinase C in human platelets and in the rat basophilic leukemia cell line RBL-2H3 results in phosphorylation of both the heavy chain and the 20 kDa light chain of myosin. Activation of protein kinase C in human platelets was initiated by treating them with TPA. In the case of RBL-2H3 cells activation was initiated by crosslinking the receptors for IgE. In both cases the pattern of phosphorylation following protein kinase C activation was the same: phosphorylation of serine-1 or -2 in the myosin light chain and phosphorylation of a serine site(s) that is confined to a single tryptic peptide in the myosin heavy chain. Table I shows that both TPA and DNP-BSA treatment of platelets and RBL-2H3 cells respectively, results in significant amounts of phosphorylation.

Since human platelet myosin can be readily purified in mg quantities it was possible to compare the peptide maps following *in vitro* phosphorylation of human platelet myosin with protein kinase C to *in situ* phosphorylation of both platelet myosin and RBL-2H3 myosin. A major difference between *in vitro* and *in situ* phosphorylation of the myosin light chain is that, whereas serine-1 or -2 and threonine-9 are phosphorylated under *in vitro* conditions, only serine-1 or -2 is phosphorylated under *in situ* conditions. The reason for this major difference is not clear. It is possible that other *in situ* activators of protein kinase C might cause phosphorylation of threonine-9 in the myosin light chain, but these have yet to be described.

The exact site of phosphorylation in the myosin heavy chain catalyzed by protein kinase C has yet to be determined. At present we only know that the same tryptic phosphopeptide map is obtained when platelet myosin is phosphorylated by protein kinase C, when intact platelets are treated with TPA or when RBL-2H3 cells are activated with the antigen, DNP-BSA. In each case a single phosphorylated peptide is found which contains phosphoserine. Whether there is more than one residue of phosphoserine in this peptide is presently unknown.

Finally, the data from RBL-2H3 cells provide a putative function for protein kinase C phosphorylation. The *de novo* phosphorylation of protein kinase C sites in both the myosin heavy chain and light chain is paralleled by the release of histamine from these cells. It is important to remember, however, that other proteins, in addition to myosin, are phosphorylated following activation of protein kinase C. Thus the exact role that protein kinase C phosphorylation of myosin might play in histamine release will have to await more precise experiments, which can either specifically interfere with, or specifically activate myosin phosphorylation. The recent development of a method for rendering RBL-2H3 cells permeable to external proteins and ligands, while leaving them responsive to antigenic stimulation, should make these experiments possible[11].

ACKNOWLEDGEMENT

The authors gratefully acknowledge the skillful editorial assistance of Catherine S. Magruder.

REFERENCES

1. J. R. Sellers and R. S. Adelstein, Regulation of contractile activity, *in*: "The Enzymes", P. D. Boyer and E. G. Krebs, eds., Academic Press, Inc., Orlando, Vol. 18:381 (1987).
2. S. Citi and J. Kendrick-Jones, Regulation of non-muscle myosin structure and function, *BioEssays* 7:155 (1987).
3. M. Naka, M. Nishikawa, R. S. Adelstein, and H. Hidaka, Phorbol ester-induced activation of human platelets is associated with protein kinase C phosphorylation of myosin light chains, *Nature* 306:490 (1983).
4. A. R. Bengur, E. A. Robinson, E. Appella, and J. R. Sellers, Sequence of the sites phosphorylated by protein kinase C in the smooth muscle myosin light chain, *J. Biol. Chem.* 262:7613 (1987).
5. M. Ikebe, D. J. Hartshorne, and M. Elzinga, Phosphorylation of the 20,000-dalton light chain of smooth muscle myosin by the calcium-activated, phospholipid-dependent protein kinase, *J. Biol. Chem.* 262:9569 (1987).

6. S. Kawamoto, A. R. Bengur, J. R. Sellers, and R. S. Adelstein, In situ phosphorylation of human platelet myosin heavy and light chains by protein kinase C, J. Biol. Chem. in press (1989).
7. R. I. Ludowyke, I. Peleg, M. A. Beaven, and R. S. Adelstein, Analysis of phosphorylation of myosin heavy and light chains during antigen stimulation of basophil leukemic cells, J. Cell Biol. 105:118a (1987).
8. J. R. Sellers, M. S. Soboeiro, K. Faust, A. R. Bengur, and E. V. Harvey, Preparation and characterization of heavy meromyosin and subfragment 1 from vertebrate cytoplasmic myosins, Biochemistry 27:6977 (1988).
9. K.-P. Huang, K.-F. J. Chan, T. J. Singh, H. Nakabayashi, and F. L. Huang, Autophosphorylation of rat brain Ca^{2+}-activated and phospholipid-dependent protein kinase, J. Biol. Chem. 261:12134 (1986).
10. U. K. Laemmli, Cleavage of structural proteins during the assembly of the head of bacteriophage T4, Nature 227:680 (1970).
11. E. WoldeMussie, H. Ali, T. Takaishi, R. Siraganian, and M. A. Beaven, Identification of variants of basophilic leukemia (RBL-2H3) cells that have defective phosphoinositide responses to antigen and stimulants of GTP-regulatory proteins, J. Immunol. 139:2431 (1987).
12. P. H. O'Farrell, High resolution two-dimensional electrophoresis of proteins, J. Biol. Chem. 250:4007 (1975).
13. R. E. Shaff and M. A. Beaven, Increased sensitivity of the enzymatic isotopic assay of histamine: Measurement of histamine in plasma and serum, Anal. Biochem. 94:425 (1979).
14. S. Kawamoto and R. S. Adelstein, The heavy chain of smooth muscle myosin is phosphorylated in aorta cells, J. Biol. Chem. 263:1099 (1988).
15. M. Nishikawa, H. Hidaka, and R. S. Adelstein, Phosphorylation of smooth muscle heavy meromyosin by calcium-activated, phospholipid-dependent protein kinase, J. Biol. Chem. 258:14069 (1983).
16. J. R. Cunha-Melo, N. M. Dean, J. D. Moyer, K. Maeyama, and M. A. Beaven, Kinetics of phosphoinositide hydrolysis in rat basophilic leukemic (RBL-2H3) cells varies with the type of IgE-receptor cross-agent used, J. Biol. Chem. 262:11455 (1987).
17. S. Kawamoto and R. S. Adelstein, Characterization of myosin heavy chains in cultured aorta smooth muscle cells, J. Biol. Chem. 262:7282 (1987).

EFFECT OF MULTIPLE PHOSPHORYLATIONS ON MOVEMENT OF SMOOTH MUSCLE AND CYTOPLASMIC MYOSIN

James R. Sellers and Seiji Umemoto

National Heart, Lung and Blood Institute
National Institutes of Health
Bethesda, MD 20892 U.S.A.

Smooth muscle contraction is initiated by phosphorylation of the 20,000 Da light chain subunit of myosin (Sellers and Adelstein, 1987; Hartshorne, 1987). The enzyme responsible for this phosphorylation, myosin light chain (MLC) kinase, is dependent upon both calcium and calmodulin for activity. In addition to its role in muscle contraction, myosin plays an important contractile role in many nonmuscle cells (Korn and Hammer, 1988). This cytoplasmic myosin is also regulated by light chain phosphorylation. Phosphorylation of smooth muscle and cytoplasmic myosin by MLC kinase results in their activation as judged by a large increase in the actin-activated MgATPase activity.

Recently other assays for myosin function have been developed which measure the rate of movement of myosin and actin relative to each other in an *in vitro* assay system (Sheetz and Spudich, 1983; Higashi-Fujime, 1985; Kron and Spudich, 1986; Harada et al., 1987). One such assay utilizes the oriented actin cables which are present in the long cells of the fresh water alga, *Nitella axillaris* (Sheetz et al., 1986). In this system a 2 - 7 cm long single cell of *Nitella* is dissected and pinned to a silicon-coated petri dish and polymer beads coated with myosin are directly added. Upon settling onto the oriented cables of actin which are found at the surface of *Nitella* the beads move at a constant velocity characteristic of the applied myosin.

It was observed that myosin-coated beads from turkey gizzard smooth muscle did not move in the *Nitella* system unless the 20,000 Da light chain subunit was phosphorylated by MLC kinase (Sellers et al., 1985). This is also true for vertebrate smooth muscle myosin-coated beads from bovine aorta and trachea and for vertebrate cytoplasmic myosin-coated beads from human platelets and turkey intestinal epithelial brush border (Table 1). In all of these cases the serine-19 residue of the 20,000 Da light chain was phosphorylated to the extent of 1 mol P_i/mol by MLC kinase. The phosphorylation dependence of the movement under these conditions was directly demonstrated by the use of a calcium-insensitive fragment of MLC kinase prepared by tryptic digestion. If unphosphorylated smooth muscle myosin or cytoplasmic myosin-coated beads were added to the *Nitella* no movment of the beads was observed. Subsequent addition of the calcium-insensitive MLC kinase fragment to the same solution resulted in movement of all of the beads after a brief lag period which is presumably due to the time required for phosphorylation of the myosin. Furthermore the velocities obtained from this experiment were the same as those of control MLC kinase phosphorylated myosins.

Table 1. Velocities of Beads Coated with Various Myosins

Myosin	Velocity (um/s)*	n
Rabbit Skeletal Muscle	1.920 ± 0.072	6
Phosphorylated Turkey Gizzard	0.198 ± 0.007	10
Phosphorylated Bovine Aorta	0.131 ± 0.021	5
Phosphorylated Bovine Trachea	0.117 ± 0.004	6
Phosphorylated Human Platelet	0.039 ± 0.004	4
Phosphorylated Turkey Brush Border	0.043 ± 0.002	2

Conditions: 10 mM KCl, 10 mM imidazole (pH 7.2), 5 mM $MgCl_2$, 1 mM EGTA, 1 mM ATP, 2 mM dithiothreitol, 10 mM sucrose, 0.01 μM phalloidin, 25°C. *Mean ± S.E. n = number of experiments. Data from Umemoto et al. (1988)

While all these myosins required light chain phosphorylation for bead movement, it was observed that each myosin displayed its own characteristic rate (Table 1). Note that the bead velocities of these myosins were considerably slower than those obtained with rabbit skeletal muscle myosin. In general, smooth muscle myosin-coated beads moved at about 5-10% of the velocity of skeletal muscle myosin-coated beads and about 5 times as fast as the cytoplasmic myosin-coated beads. We also observed that the velocity of skeletal muscle myosin-coated beads was not effected by phosphorylation of the light chains by skeletal muscle MLC kinase (data not shown).

```
                    5                   10
Ac-Ser-Ser-Lys-Arg-Ala-Lys-Ala-Lys-Thr-Thr-

                   15                   20
    Lys-Lys-Arg-Pro-Gln-Arg-Ala-Thr-Ser-Asn-
```

Figure 1. Amino-Terminal Sequence of the 20,000 Da Light Chain of Turkey Gizzard Smooth Muscle Myosin

The sequence of the first 20 amino acids from the amino-terminus of the 20,000 Da light chain of turkey gizzard smooth muscle myosin is shown above (Fig. 1)(Pearson et al., 1984). There are other sites of phosphorylation besides serine-19. MLC kinase can also phosphorylated threonine-18 albeit at a considerably slower rate (Ikebe and Hartshorne, 1985). Phosphorylation at this site in addition to serine-19 has been shown to result in a somewhat higher actin-activated MgATPase activity than observed with serine-19 phosphorylation alone (Table 2)(Ikebe et al., 1986; Tanaka et al., 1985). However no effect of threonine-18 phosphorylation on the movement of myosin-coated beads was seen with any of the smooth muscle myosins (Table 3). Note that for the studies of the effect of phosphorylation on MgATPase activity, we have used the soluble, enzymatically active fragment of myosin, heavy meromyosin (HMM) which is

Table 2. Effect of Threonine-18 Phosphorylation on the Actin-Activated MgATPase Activity of Smooth Muscle HMM

Phosphorylated HMM	Vmax (s^{-1})	K_m (μM)
Ser-19	1.05	28
Ser-19, Thr-18	1.17	18

Conditions: 10 mM imidazole (pH 7.0), 2 mM $MgCl_2$, 1 mM ATP, 0.1 mM EGTA, 0.5 mM dithiothreitol, 25°C.

prepared by chymotryptic digestion of gizzard myosin. The reason is that HMM represents a good model for the myosin molecule and, unlike intact myosin, is soluble at low ionic strength which facilitates Michaelis-Menten kinetic analysis (Sellers, 1985).

Protein kinase C can also phosphorylate the 20,000 Da light chain of smooth muscle and cytoplasmic myosins at several sites, threonine-9 and serine-1 and/or -2 (Bengur et al., 1987; Ikebe et al., 1987). Phosphorylation of smooth muscle myosin by protein kinase C alone has little effect on the actin-activated MgATPase activity, whereas phosphorylation of this myosin by both MLC kinase and protein kinase C results in a somewhat reduced actin-activated MgATPase activity compared to that obtained by phosphorylation with MLC kinase alone (Nishikawa et al., 1984; Ikebe et al., 1987). This decrease in activity was due to an increase in the K_{app} for actin with no difference in the V_{max} between MLC kinase phosphorylated HMM and HMM that was phosphorylated by both MLC kinase and protein kinase C.

It is of interest to determine how protein kinase C phosphorylation affects the movement of smooth muscle and cytoplasmic myosin-coated beads in the _in vitro_ motility assay. Table 4 shows that phosphorylation of these myosins by protein kinase C alone did not result in movement of the myosin-coated beads. Phosphorylation of these myosins by both protein kinase C and MLC kinase resulted in bead velocties that did not differ from that obtained by phosphorylation of the myosins by MLC kinase alone.

These results suggest that the velocity of myosin-coated beads in the _in vitro_ motility assay is not directly coupled to the actin-activated MgATPase activity. This finding is not surprising when one considers that separate kinetic steps probably limit the two phenomena. It is well established that the phosphate release (Sellers, 1985) from an actomyosin·ADP·P_i is the rate limiting step in MgATPase cycle (Scheme 1). The _in vitro_ motility assay, however, is probably an _in vitro_ correlate of the unloaded shortening velocity of muscle fiber contraction. It has been proposed on theoretical and experimental grounds that the kinetic step that limits the unloaded shortening velocity is the release of ADP from an actomyosin·ADP complex (Scheme 1)(Siemankowski et al., 1985; Hill and Eisenberg, 1976). Thus, it is not necessary that the two assays respond identically to the effects of these extra phosphorylations.

(Scheme 1)

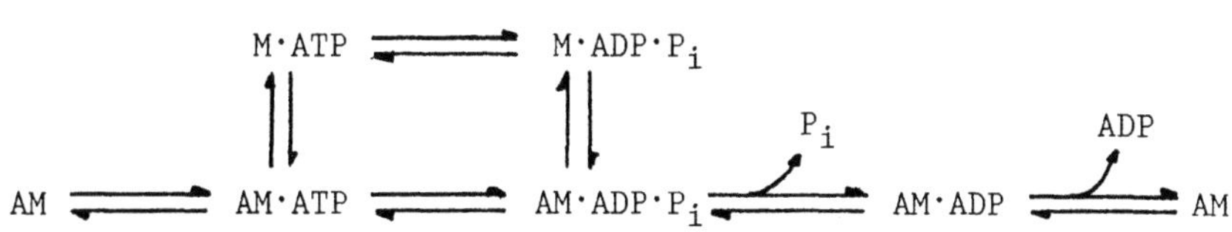

Table 3. Effect of Threonine-18 Phosphorylation on Myosin-coated Bead Velocity

Myosin	Phosphorylation State	Velocity* (μm/s)
Turkey Gizzard	Ser-19	0.188 ± 0.028
	Ser-19, Thr-18	0.191 ± 0.025
Bovine Trachea	Ser-19	0.135 ± 0.006
	Ser-19, Thr-18	0.133 ± 0.013
Bovine Aorta	Ser-19	0.107 ± 0.019
	Ser-19, Thr-18	0.114 ± 0.029

Conditions: Same as Table 1. The data are taken from Umemoto et al., 1988. *Mean ± S.D.

Table 4. Effect of Protein Kinase C Phosphorylation on Myosin-coated Bead Velocity

Myosin	Phosphorylated by	Velocity* (μm/s)
Turkey Gizzard	MLCK alone	0.198 ± 0.023
	MLCK + PKC	0.193 ± 0.028
	PKC alone	none
Bovine Trachea	MLCK alone	0.113 ± 0.022
	MLCK + PKC	0.109 ± 0.039
	PKC alone	none
Bovine Aorta	MLCK alone	0.090 ± 0.013
	MLCK + PKC	0.079 ± 0.025
	PKC alone	none
Human Platelet	MLCK alone	0.035 ± 0.010
	MLCK + PKC	0.034 ± 0.010
	PKC alone	none

Conditions: same as Table 1. *Mean ± S.D.
PKC, protein kinase C
Data are from Umemoto et al. (1988).

Protein kinase C also phosphorylated human platelet myosin on the heavy chain (approximately 0.3 mol P_i/mol)(Bengur et al., 1987a). The site of the phosphorylation was found to be in the LMM portion of the molecule probably near the carboxy-terminus. The results in Table 4 show that there was no obvious effect of this phosphorylation on the bead movement in the in vitro motility assay system.

Multiple phosphorylations of the 20,000 Da light chain do occur in both smooth muscle tissue (Erodi et al., 1987; Kamm et al., 1988) and in nonmuscle cells such as platelets and rat basophil leukemia cells (Naka et al., 1988, Kawamoto et al., 1988; Ludowyke et al., 1987). The results obtained with the in vitro motility assay suggest that these phosphorylations are not likely to effect the relative movement of actin and myosin filaments in these cells. Direct support of these in vitro findings has come from work on skinned smooth muscle fiber strips which shows that the multiple phosphorylation of the 20,000 Da light chain by MLC kinase and protein kinase C has no effect on either unloaded shortening velocity or force production (Haeberle et al., 1988, Sutton and Haeberle, 1988). The effects of phosphorylations of these sites in addition to serine-19 must be more subtle.

REFERENCES

Bengur, A. R., Harvey, E. V., and Sellers, J. R., 1987a, Phosphorylation of platelet myosin by protein kinase C: in vitro and in vivo studies, J. Cell Biol. 105, 656a.

Bengur, A. R., Robinson, E. A., Apella, E., and Sellers, J. R., 1987, Sequence of the sites phosphorylated by protein kinase C in the smooth muscle myosin light chain, J Biol. Chem. 262, 7613.

Colburn, J. C., Michnoff, C. H., Shu, L-C., Slaughter, C. A., Kamm, K. E., and Stull, J. T., 1988, Sites phosphorylated in myosin ligh chain in contracting smooth muscle, J. Biol. Chem., in press.

Erodi, F., Barany, M., and Barany, K., 1987, Myosin light chain isoforms and their phosphorylation in arterial smooth muscle, Cir. Res. 61, 898.

Haeberle, J. R., Sutton, T. A., and Trockman, B. A., 1988, Phosphorylation of two sites on smooth muscle myosin: Effects on contraction of glycerinated vascular smooth muscle, J.Biol.Chem. 263, 4424.

Harada, Y. Noguchi, A., Kishino, A., and Yanagida, T., 1987, Sliding movement of single actin filaments on one-headed myosin filaments, Nature (London) 326, 805.

Hartshorne, D. J., 1987, Biochemistry of the contractile process in smooth muscle, in: "Physiology of the Gastrointestinal Tract", 2nd ed., L. R. Johnson, ed., Raven Press, New York.

Higashi-Fujime, S., 1985, Unidirectional sliding of myosin filaments along the bundle of F-actin filaments spontaneously formed during superprecipitation, J. Cell Biol. 101, 2335.

Eisenberg, E. and Hill, T. L., 1978, A cross-bridge model of muscle contraction, Prog. Biophys. Molec. Biol. 33, 55.

Ikebe, M. and Hartshorne, D. J., 1985, Phosphorylation of smooth muscle myosin at two distinct sites by myosin light chain kinase, J. Biol. Chem. 260, 10027.

Ikebe, M., Hartshorne, D. J., and Elzinga, M., 1986, Identification, phosphorylation and dephosphorylation of a second site for myosin light chain kinase on the 20,000-dalton light chain of smooth muscle myosin, J.Biol. Chem. 261, 36.

Ikebe, M., Hartshorne, D. J., and Elzinga, M., 1987, Phosphorylation of the 20,000 dalton light chain of smooth muscle myosin by the calcium-activated, phospholipid-dependent protein kinase, J. Biol. Chem. 262, 9569.

Kawamoto, S., Bengur, A. R., Sellers, J. R., and Adelstein, R.,S., 1988, Protein kinase C can phosphorylate myosin heavy and light chains in intact human platelets, Biophysical J. 53, 465a.

Korn, E. D. and Hammer J. A., III, 1988, Myosins of nonmuscle cells, Ann. Rev. Biophys. Biophys. Chem. 17, 23.

Kron, S. J. and Spudich, J. A., 1986, Fluorescent actin filaments move on myosin fixed to a glass surface, Proc. Natl. Acad. Sci. (U.S A.) 83, 6272.

Ludowyke, R. I., Peleg, I., Beaven, M. A., and Adelstein, R. S., 1987, Analysis of phosphorylation of myosin heavy and light chains during antigen stimulation of basophil leukemic cells, J. Cell Biol. 105, 118a.

Naka, M. Saitoh, M., Hidaka, H., 1988, Two phosphorylated forms of myosin in thrombin-stimulated platelets, Arch. Bioch. Biophys. 261, 235.

Nishikawa, M., Sellers, J. R., Adelstein, R. S., and Hidaka, H., 1984, Protein kinase C modulates *in vitro* phosphorylation of the smooth muscle heavy meromyosin by myosin light chain kinase, J. Biol. Chem. 259, 8808.

Pearson, R. B., Jakes, R., John, M., Kendrick-Jones, J., and Kemp, B. E., 1984, Phosphorylation site sequence of smooth muscle myosin light chain (M_r = 20,000), FEBS Letters, 168, 108.

Sellers, J. R., 1985, Mechanism of the phosphorylation-dependent regulation of smooth muscle heavy meromyosin, J. Biol. Chem. 260, 15815.

Sellers, J. R. and Adelstein, R. S., 1987, Regulation of contractile processes by phosphorylation, The Enzymes, 18, 381.

Sellers, J. R., Spudich, J. A., and Sheetz, M. P., 1985, Light chain phosphorylation regulates the movement of smooth muscle myosin on actin filaments, J. Cell Biol. 101, 1897.

Siemankowski, R. F., Wiseman, M. O., and White, H. D., 1985, ADP dissociation from actomyosin subfragment 1 is sufficiently slow to limit the unloaded shortening velocity in vertebrate muscle, Proc. Natl. Acad. Sci. (U.S.A.) 82, 658.

Sheetz, M. P. and Spudich, J. A., 1983, Movement of myosin-coated fluorescent beads on actin cables *in vitro*, Nature (London), 303, 31.

Sheetz, M. P., Block, S. M., and Spudich, J. A., 1986, Myosin movement *in vitro*: a quantitative assay using oriented actin cables form *Nitella*, Meth. Enzymol. 134, 531.

Tanaka, T., Sobue, K., Owada, M. K., and Hakura, A., 1986, Linear relationship between diphosphorylation of 20 kDa light chain of gizzard myosin and the actin-activated myosin ATPase activity, Biochem. Biophys. Res. Commun. 131, 987.

Umemoto, S., Bengur, A. R., and Sellers, J. R., 1989, Effect of multiple phosphorylations of smooth muscle and cytoplasmic myosins on movement in an *in vitro* motility assay, J. Biol. Chem., in press.

REGULATION OF EMBRYONIC SMOOTH MUSCLE MYOSIN BY MYOSIN LIGHT CHAIN KINASE AND BY PROTEIN KINASE C

Primal de Lanerolle
Department of Physiology and Biophysics
University of Illinois at Chicago
Chicago, IL 60612

INTRODUCTION

The presence of developmentally regulated myosin isoforms in muscle tissues is well established (1-9). Cardiac myosin has been studied extensively, and three different isoforms have been identified (4,5). These myosin isoforms differ with respect to (a) their mobilities on nondenaturing polyacrylamide gels (3), (b) subunit composition (3,4), and (c) actomyosin ATPase activities (4). Moreover, specific cardiac myosin isoforms are expressed following volume overload of the heart (2) and change depending upon thyroid hormone levels (5). Similarly, fiber-specific myosin isoforms are expressed in skeletal muscle, and the myosin isoform correlates with both the type of innervation and the sequence of electrical stimulation (6).

Fetal smooth muscles also contain a myosin isoform (8,9). Fetal myosin differs from adult smooth muscle myosin in that it contains a 23-kDa light chain (LC_{23})[1]. Studies using monoclonal antibodies (1) and mRNA analyses (10) have demonstrated that embryonic chicken skeletal and cardiac muscles and chicken brains also express LC_{23}. Thus, a specific myosin isoform that contains a 23-kDa light chain appears to be widely distributed among embryonic tissues.

However, little is known about the regulation of fetal[2] smooth muscle myosin. This is in contrast to the extensive characterization of adult smooth muscle myosin which is known to be regulated by phosphorylation of the 20 kD light chain (LC_{20}) by the Ca^{2+}/calmodulin-dependent enzyme myosin light chain kinase (MLC kinase). Therefore, we have used immunological and biochemical approaches to study MLC kinase during embryonic development and characterized the regulation of the actin-activated, Mg^{2+}-dependent ATPase activity of fetal myosin. Our data demonstrate that (a) the expression of MLC kinase is also develop-

[1]The abbreviations used are LC_{20}, the 20-kDa light chain of smooth muscle myosin; MLC kinase, myosin light chain kinase; PK-C, protein kinase C; actomyosin ATPase activity, the actin-activated Mg^{2+}-dependent ATPase activity of myosin; MLC, myosin light chain; LC_{23}, the 23-kDa light chain of embryonic smooth muscle myosin.
[2]The terms "embryonic" and "fetal" are used interchangeably to refer to myosin that contains LC_{23}.

mentally regulated in chicken gizzards and (b) in sharp contrast to adult myosin, fetal myosin is activated by phosphorylation of LC_{20} by protein kinase C (PK-C).

MATERIALS AND METHODS

<u>Protein Purification</u> Fetal myosin was prepared from gizzards removed from 12-day-old chick embryos. The gizzards were homogenized in 0.5 M NaCl, 1 mM dithiothreitol, 1 mM EGTA, 5 mM EDTA, 50 mM Tris-HCl (pH 7.5), 1 mM NaN_3, 0.01 mg/ml each of soybean trypsin inhibitor, N^{α}-p-tosyl-L-lysine chloromethyl ketone, and N-tosyl-L-phenylalanine chloromethyl ketone, 0.005 mg/ml each of leupeptin and pepstatin A, 0.002 mg/ml aprotinin, and 1 mM phenylmethylsulfonyl fluoride and the supernatant was collected by centrifugation. The pellet was re-extracted in homogenization buffer and the supernatant collected and combined with the first extract. The combined extract was made 10 mM MgATP, and the pH was readjusted to 7.5. This combined supernatant was then fractionated by adding ammonium sulfate and the protein that precipitated between 35% and 60% ammonium sulfate was collected. The precipitate was resuspended in homogenization buffer made 10 mM MgATP and applied to a 2.5 x 90 cm Sepharose 4B column equilibrated in homogenization buffer minus the protease inhibitors. The fractions containing myosin (revealed by SDS-polyacrylamide gel electrophoresis) were pooled, concentrated, and re-applied to the Sepharose 4B column to remove minor contaminants. Fetal myosin purified by this method is free of MLC kinase activity. MLC kinase was purified from turkey or chicken gizzards (11) and protein kinase C was purified from human platelets (12) as previously described. Rabbit antibodies to MLC kinase purified from turkey gizzards were raised and affinity purified as previously described (13).

<u>Phosphorylation of Myosins</u> Fetal and adult myosins were phosphorylated with either MLC kinase or PK-C in a basic reaction mixture containing 1 mg/ml myosin, 175 mM NaCl, 5 mM $MgCl_2$, 0.35 mM $CaCl_2$, 2 mM ATP, 10 mM Tris-HCl, pH 7.5. When phosphorylating with MLC kinase, 10^{-7} M calmodulin and 0.02 mg/ml MLC kinase were added to this reaction mixture and incubated for 30 min. When phosphorylating with PK-C, 50 μg/ml phosphatidylserine, 0.8 μg/ml 1,3-diolein, and 0.016 mg/ml PK-C were added to the basic reaction mix and incubated for 60 min at room temperature. The myosins were phosphorylated with both protein kinases by first incubating with MLC kinase for 30 min. PK-C, diolein, and phosphatidylserine were then added to the reaction mixture and incubated for another 60 min.

<u>Quantitation of Stoichiometry of Myosin Phosphorylation</u> Phosphorylated myosins were analyzed by isoelectric focusing polyacrylamide slab gel electrophoresis to separate the phosphorylated from the non-phosphorylated LC_{20} species. The proteins were transferred to nitrocellulose and the nitrocellulose sheet was processed as a Western blot using affinity-purified rabbit antibodies to LC_{20}. The LC_{20} bands were visualized by developing the peroxidase reaction, and the stoichiometry of phosphorylation was determined densitometrically. This procedure is similar to a radioimmunoblotting technique described previously (14).

<u>Quantitation of the Actomyosin ATPase Activity</u> Unphosphorylated and myosins phosphorylated as described above were dialyzed and concentrated in 0.5 M KCl, 1 mM dithiothreitol, 0.1 mM EGTA, 1 mM $MgCl_2$, 1 mM NaN_3, 20 mM Tris-HCl, pH 7.5 using a Centricon 10 Microconcentrator (Amicon Corp., Danvers, MA). The actomyosin ATPase activities were quantitated as described by Sellers et al. (15) at 22°C, in a 1 ml solution that contained 0.25 mg/ml myosin, 0.75 mg/ml actin, 50 mM

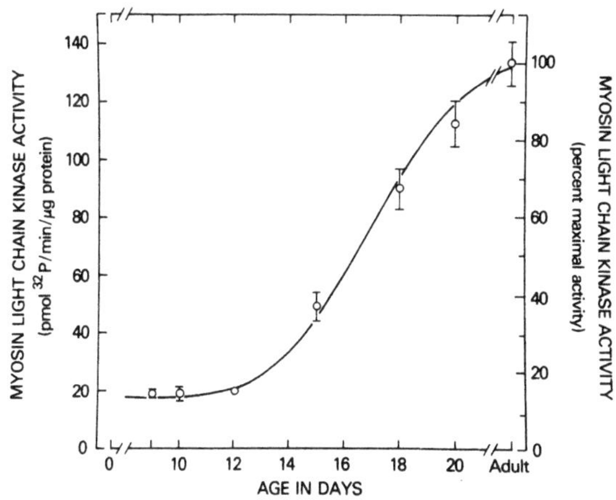

Figure 1. Expression of MLC kinase activity during development of the chicken gizzard: Gizzards removed from fertilized chicken eggs or pieces of adult gizzards were homogenized in ice cold 200 mM NaCl, 2.5 mM DTT, 1 mM EDTA, 50 mM Tris, pH 7.5 and the supernatants were collected by centrifugation. The supernatants were assayed for MLC kinase activity in the presence of 10^{-7} M calmodulin, 0.25 mM Ca^{2+}, 10 mM Mg, and purified turkey gizzard myosin light chains. The reactions were started by adding gamma labeled, ^{32}P-ATP (500 cpm/pmol). The assays were linear with respect to the time and the data were normalized to correct for the amount of protein in each assay. The data represent the mean ± SE and 3 separate fetal gizzards or pieces of adult tissue were assayed at each data point.

KCl, 5 mM Mg^{2+}, 0.1 mM EGTA, 15 mM Tris-HCl, pH 7.5, and 1 mM ATP (250 cpm/pmol). Linear regression analyses were used to calculate the initial rates of ATP hydrolysis. The mean rates for ATPase measurements on myosins from three separate preparations are reported in the text.

Phosphoamino Acid Analyses and Phosphopeptide Maps Myosins were phosphorylated by MLC kinase or PK-C, separately, as described above using γ-labeled [^{32}P]ATP (500 cpm/pmol) as substrate. Phosphopeptide mapping and phosphoamino acid analysis were performed on these myosins as previously described (16).

Other Methods Myosin from adult gizzards was purified as described in Ref. 15. Skeletal muscle actin, purified as described by Spudich and Watt (17), was kindly provided by Jim Sellers (National Institutes of Health) and Ed Taylor (University of Chicago). Protein assays were performed as described by Lowry et al (18). SDS-polyacrylamide gel electrophoresis was conducted as described by Laemmli (19).

RESULTS

Myosin expression is known to be developmentally regulated in the gizzard. That is, the amount of LC_{23} decreases during the later stages of embryonic development and disappears by the 8th day following hatching (1,8,9). Therefore, we began our studies by asking whether MLC kinase activity is also developmentally regulated and, if it is, whether this involves one or more MLC kinase isoforms.

The experiment depicted in Fig. 1 was designed to answer the first of these questions. Gizzards removed from fertilized eggs at different times during embryonic development and pieces of tissue from adult gizzards were homogenized in a standard extraction buffer and the supernatants were collected by centrifugation. These supernatants were assayed for MLC kinase activity using purified turkey gizzard myosin light chains as substrate. As seen in Fig. 1, there is relatively little MLC kinase activity until the last quarter of fetal development.

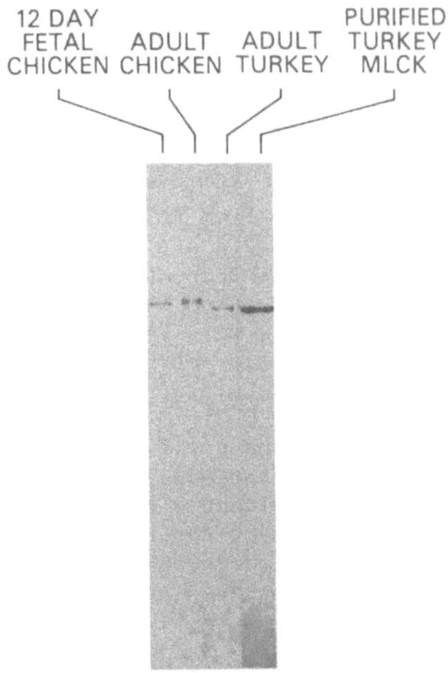

Figure 2. Western blot analysis of fetal and adult MLC kinases: Tissue extracts were prepared from 12 day old fetal chicken gizzards and from adult chicken and turkey gizzards. Along with purified turkey gizzard MLC kinase, they were separated electrophoretically on an SDS, 5-20% gradient polyacrylamide gel. The proteins were transferred to nitrocellulose and processed using 1 μg affinity-purified antibodies to MLC kinase and 25 μg peroxidase-labelled goat anti-rabbit IgG antibodies as previously described (20).

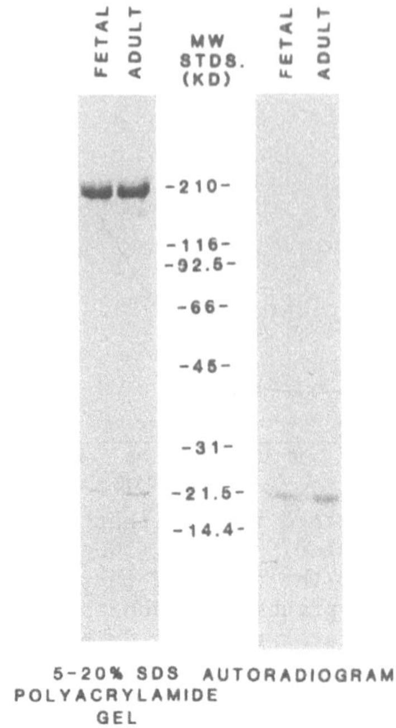

Figure 3. Subunit composition and phosphorylation of fetal and adult myosins: Fetal myosin (purified as described in the text) and adult myosin were phosphorylated with MLC kinase purified from turkey gizzards (11). The phosphorylated myosins were analyzed on a 5-20% polyacrylamide gradient gel in the prescence of SDS (left panel). The gel was stained, destained, and exposed to an x-ray film (right panel). Molecular weight markers are shown in the middle.

We next studied the immunological characteristics of fetal MLC kinase. Studies using affinity-purified antibodies to adult turkey gizzard MLC kinase demonstrated that these antibodies inhibited the low level of MLC kinase activity in extracts prepared from 12 day old fetal gizzards (data not shown). These results are similar to those we have reported on the inhibition of purified adult turkey gizzard MLC kinase (13) by these antibodies.

The immunological characteristics of fetal MLC kinase were further investigated by performing a Western blot analysis (Fig. 2). Interestingly, there appear to be two forms of MLC kinase in adult chicken gizzard and they both have a lower Mr than adult turkey gizzard MLC kinase. Moreover, there is a single immunoreactive species in fetal smooth muscle which has a Mr corresponding to the faster migrating species identified in adult chicken gizzards (Fig. 2).

The presence of low levels of MLC kinase activity early in fetal development, along with the presence of apparently developmentally

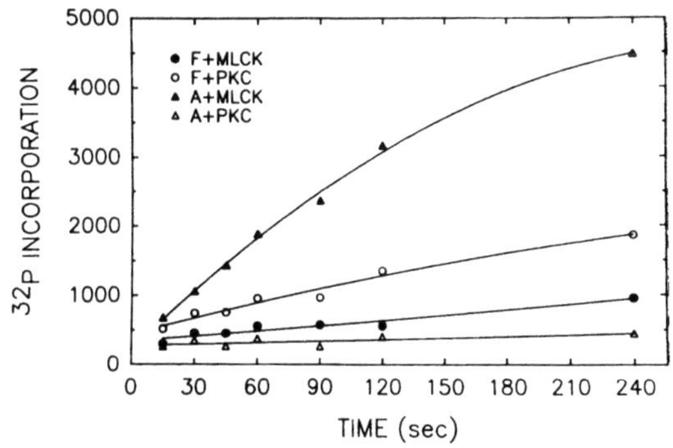

Figure 4. Rates of phosphorylation of fetal myosin by MLC kinase (MLCK) or protein kinase C (PK-C): Fetal or adult myosin was phosphorylated as described in Methods except that ^{32}P-ATP was used as a substrate and aliquots were removed at the times shown, boiled in 1% SDS and analyzed by SDS polyacrylamide gel electrophoresis. The LC_{20} bands were then excised and the radioactivity quantitated using a liquid scintillation counter.

regulated MLC kinase isoforms in fetal gizzards, piqued our curiosity regarding the regulation of fetal myosin. Therefore, we purified fetal myosin from 12 day old chicken gizzards and studied its regulation. As previously reported (1,8,9), myosin purified from 12-day-old chick embryos contains three light chains (Fig. 3). One of the light chains has an Mr = 23,000. A light chain with a similar mobility on an SDS-polyacrylamide gel is not present in adult myosin. Phosphorylation experiments demonstrated that the 20-kDa light chains of fetal and adult myosins are phosphorylated by MLC kinase (Fig. 3). LC_{23} is not phosphorylated under any conditions.

Because of the low level of MLC kinase activity in fetal gizzards, we were curious as to whether other protein kinases were capable of phosphorylating fetal myosin. Both cyclic AMP-dependent protein kinase and the Ca^{2+}/calmodulin-dependent protein kinase II phosphorylate fetal myosin LC_{20} but the stoichiometry of incorporation with either kinase was less than 0.1 mol PO_4/mol LC_{20} at one hour. PK-C, however, incorporated up to 1.5 mol PO_4/mol LC_{20}. Because of this extensive phosphorylation of fetal myosin by PK-C, the rates of fetal and adult myosin phosphorylation by MLC kinase and by PK-C were compared (Fig. 4). Note that the initial rate of phosphorylation of adult myosin by PK-C is much lower than that by MLC kinase (7.29 vs 196.7 pmol/min-μg, respectively). In direct contrast, fetal myosin is phosphorylated at a higher rate by PK-C than by MLC kinase (60.00 vs 21.34 pmol/min-μg, respectively).

The effect of LC_{20} phosphorylation on the actomyosin ATPase activities of adult and fetal myosins was investigated next (Fig. 5). Unphosphorylated adult myosin has a low rate of ATP hydrolysis, and phosphorylation by MLC kinase leads to about a 15-fold increase in this rate (from 0.49 to 7.52 nmol of ^{32}P/min-mg myosin). These rates of ATP hydrolysis are comparable to those previously reported for adult gizzard myosin assayed at 25°C under similar conditions (21). Phosphorylation by PK-C alone has no effect (0.88 nmol of ^{32}P/min-mg myosin) while sequential phosphorylation decreased the MLC kinase-induced increase in the actomyosin ATPase activity of adult myosin (4.5 nmol/min-mg myosin) in agreement with previous reports (16). In contrast, phosphorylation by either MLC kinase or PK-C stimulated the actomyosin ATPase activity of fetal myosin by about 18-fold (from 0.39 to 7.0 nmol of ^{32}P/min-mg myosin). Moreover, sequential phosphorylation of fetal LC_{20} by MLC kinase and PK-C does not lead to a reduction in the MLC kinase-induced increase in actomyosin ATPase activity as with the adult myosin (Fig. 5).

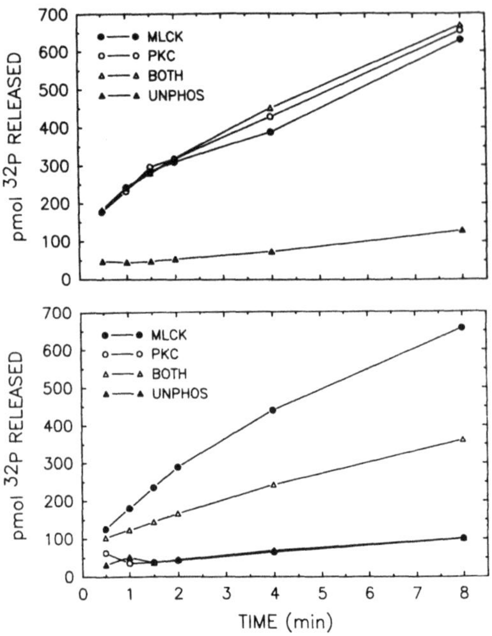

Figure 5. Effect of phosphorylation on the actin-activated Mg^{2+}-ATPase activity of fetal (top) and adult (bottom) myosin: Myosins were phosphorylated with MLC kinase (MLCK), PK-C (PKC), or MLC kinase followed by PK-C (BOTH) as described in the text. MLC kinase and PK-C, individually, and MLC kinase and PK-C, in combination, incorporated 0.8, 1.5, and 2.05 mol PO_4/mol LC_{20} into fetal myosin and 2.03, 0.8, and 1.9 mol PO_4/mol LC_{20} into adult myosin, respectively. The rates of ATP hydrolysis by the phosphorylated and unphosphorylated (UNPHOS) myosins were assayed as described in Methods. This figure depicts the results from a typical experiment. The mean initial rates of ATP hydrolysis from experiments on three separate preparations are given in the text.

In order to understand the differential effects of the two protein kinases, we identified the phosphoamino acids and phosphopeptides phosphorylated by the two protein kinases. MLC kinase phosphorylates serine 19 and, less avidly, threonine 18 on adult LC_{20} when adult myosin is phosphorylated in the presence of a high concentration of MLC kinase (0.1 mg/ml) (22). Using a lower concentration of MLC kinase (0.02 mg/ml) than used in Ref.22, we observed phosphorylation of primarily serine residues with a hint of threonine phosphorylation on adult myosin (23). The phosphorylation of adjacent amino acids may account for the presence of a single phosphopeptide in adult LC_{20} phosphorylated with MLC kinase (Fig. 6, Panel D). PK-C phosphorylates adult LC_{20} at serine and threonine residues on two separate phosphopeptides (Fig. 6, Panel D). PK-C does not phosphorylate the same peptide on adult LC_{20} phosphorylated by MLC kinase (Fig. 6 and Ref. 16).

In contrast, MLC kinase phosphorylates fetal LC_{20} at serine residues (23) on a single phosphopeptide (Fig. 6, Panel A). PK-C phosphorylates both serine and threonine residues at three separates sites on fetal LC_{20} (Fig. 6, Panel B). The autoradiogram of a peptide map of fetal LC_{20} phosphorylated by MLC kinase or PK-C that was run together indicates that both kinases phosphorylate a common peptide (Fig. 6, Panel C).

DISCUSSION

Myosin is widely distributed among eukaryotic cells and its role in muscle contraction is well-established (24). It is also thought to mediate a variety of cellular processes, such as motility, secretion, phagocytosis and cell division (24), that are unrelated to muscle contraction. It is also known that all of these cellular functions are not

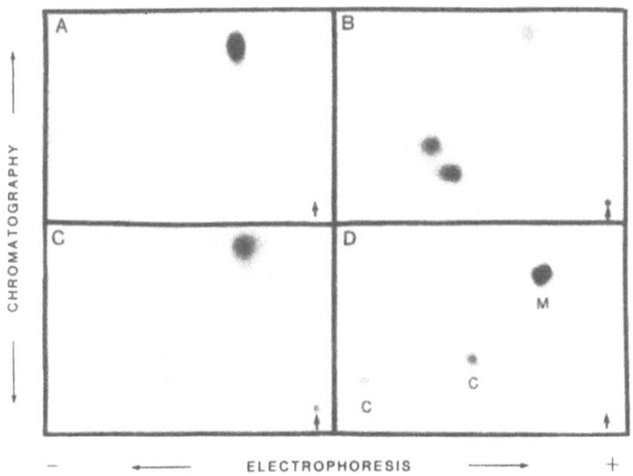

Figure 6. Phosphopeptide maps of fetal and adult LC_{20}: Phosphopeptides from fetal and adult LC_{20} were separated by two-dimensional electrophoresis/ chromatography (16) and then exposed to an x-ray film. Panels A and B are autoradiograms of phosphopeptide maps of fetal myosin phosphorylated with MLC kinase (A) and PK-C (B). Panels C and D show 1:1 mixtures (based on counts/min) of digests of LC_{20} phosphorylated with MLC kinase and PK-C. Panel C shows the fetal LC_{20} mixture, and Panel D shows the adult LC_{20} mixture. M and C designate adult LC_{20} peptides phosphorylated by either MLC kinase (M) or PK-C (C). The arrowhead marks the origin in each plate, and a total of 1500 cpm were applied to each plate.

mediated by a single myosin isoform and myosin structural heterogeneity has been demonstrated at many levels. First, the subunit composition of cardiac, skeletal and smooth muscle myosins and non-muscle myosins are known to be different. Second, antibodies to smooth muscle and non-muscle myosins do not bind to myosins purified from skeletal or cardiac muscles (25). Third, developmentally regulated myosins have been demonstrated in skeletal, cardiac and smooth muscles (1). Fourth, specific myosin isozymes are expressed in certain diseases (12), including human tumors (26). Fifth, myosins are known to be products of multigene families (27).

Regardless of their structural differences, all myosins are characterized by their ability to hydrolyze ATP when they interact with actin (24). In this respect, the data presented in this paper are important for two reasons: They demonstrate for the first time the developmental regulation of smooth muscle MLC kinase (Fig. 1) and that fetal myosin can be activated by phosphorylation by PK-C (Fig. 5).

With respect to the first observation, phosphorylation of serine 19 on adult LC_{20} by MLC kinase is generally considered to be an obligatory event for stimulating ATP hydrolysis by actin and smooth muscle myosin (24). Serine 19 appears to be phosphorylated at a significant rate only by MLC kinase when LC_{20} is associated with the rest of the myosin molecule. In contrast, PK-C phosphorylates intact adult myosin at sites on LC_{20} (Thr 9 and Ser 1 or 2, Ref. 28) that are distinct from the sites phosphorylated by MLC kinase. Most importantly, phosphorylation of LC_{20} by PK-C reduces the MLC kinase-stimulated increase in actomyosin ATPase activity of adult myosin (Fig. 5 and Ref. 16).

On the other hand, phosphorylation of fetal myosin by PK-C, alone or in combination with MLC kinase, stimulates the actomyosin ATPase activity of fetal myosin (Fig. 5). This may be due to the fact that the same site on fetal LC_{20} is phosphorylated by both kinases because (a) MLC kinase phosphorylates a single peptide and PK-C phosphorylates the same peptide and (b) there should be an additive increase in ATPase activity if phosphorylation of two separate sites stimulates ATP hydrolysis by fetal myosin. The absence of an additive effect (Fig. 5) suggests that the two protein kinases phosphorylate a common site. Nevertheless, we cannot exclude the possibility that phosphorylation of one of the sites phosphorylated by PK-C alone is responsible for stimulating the actomyosin ATPase activity. Additional studies, including sequence analysis of the phosphopeptides, are needed to differentiate between these possibilities.

What, then, is the physiological relevance of the observation that fetal myosin is activated by phosphorylation by PK-C? As mentioned above, myosin is thought to be involved in cell division. This is based primarily on the demonstration that myosin is essential for cytokinesis, but not for karyokinesis, in starfish blastomeres (29) and in Dictyostelium discoidium (30,31). By analogy, it is also thought to mediate cell division in higher organisms. Clearly, smooth muscle cells divide during embryogenesis and fetal myosin maybe involved in this process. Moreover, given its wide distribution in embryonic tissues (1), fetal myosin may mediate cytokinesis in other embryonic cells, as well.

Presumably, cytokinesis involves ATP hydrolysis by actin and myosin and this ATPase activity is regulated by phosphorylation of LC_{20}. However, the data in Fig. 1 demonstrate low levels of MLC kinase activity until late in fetal development. This raises the possibility that fetal myosin is activated by another protein kinase. Based on the data in Fig. 5, PK-C is the most likely candidate for this role. Circumstantial evidence for this possibility include the demonstration that PK-C is widely distributed in vertebrate cells (32) and the fact that PK-C is activated by mitogenic compounds such as diacylgycerol and phorbol esters (33).

In conclusion, the data presented above demonstrate (a) a time dependent increase in MLC kinase activity during the embryonic development of the chicken gizzard and (b) that fetal myosin, in contrast to adult myosin, is activated by phosphorylation by PK-C. Thus, these data present a common pathway through which myosin, which is essential for cytokinesis, and PK-C, which has been implicated in a number of growth related processes (32), can affect cell division.

Acknowledgments I thank Janice Gentry for typing the manuscript and Cindy Forgue for her technical assistance. This work was supported, in part, by a grant from the National Institutes of Health (HL 35808). Figures 3,5 and 6 are from Ref. 23 and are printed by permission of the authors.

REFERENCES

1. H. Takano-Ohmuro, T. Obinata, M. Kawashima, T. Masaki, and T. Tanaka, Embryonic Chicken Skeletal, Cardiac, and Smooth Muscles Express a Common Embryo-specific Myosin Light Chain. J. Cell Biol. 100:2025-2030 (1985).
2. A.M. Lompre, K. Schwartz, A. d'Albis, G. Lacombe, N. Van Thiem, and B. Swynghedauw, Myosin Isoenzyme Redistribution in Chronic Heart Overload. Nature 282:105-107 (1979).
3. J.F.Y. Hoh, P.A. McGrath, and R.I. White, Electrophoretic Analysis of Multiple Forms of Myosin in Fast-Twitch and Slow-Twitch Muscles of the Chick. Biochem. J. 157:87-95 (1976).
4. J.J. Schier and R.S. Adelstein, Structural and Enzymatic Comparison of Human Cardiac Muscle Myosins Isolated from Infants, Adults, and Patients with Hypertrophic Cardiomyopathy. J. Clin. Invest. 69:816-825 (1982).

5. I.L. Flink, J.H. Rader, and E. Morkin, Thyroid Hormone Stimulates Synthesis of a Cardiac Myosin Isozyme. J. Biol. Chem. 254:3105-3110 (1979).
6. S. Salmons, and F.A. Sreter, Significance of Impulse Activity in the Transformation of Skeletal Muscle Type. Nature 263:30-34 (1976).
7. D.A. Winkelmann, S. Lowey, and J.L. Press, Monoclonal Antibodies Localize Changes on Myosin Heavy Chain Isozymes during Avian Myogenesis. Cell 34:295-306 (1983).
8. N. Katoh, and S. Kubo, Purfication and Some Properties of Rabbit Stomach Myosin. J. Biochem. 81:1497-1503 (1977).
9. H. Takano-Ohmuro, T. Obinata, T. Mikawa, and T. Masaki, Changes in Myosin Isozymes during Development of Chicken Gizzard Muscle. J. Biochem. 93:903-908 (1983).
10. M. Kawashima, Y. Nabeshima, T. Obinata, and Y. Fujii-Kuriyama, A Common Myosin Light Chain is Expressed in Chicken Embryonic Skeletal, Cardiac, and Smooth Muscles and in Brain Continuously from Embryo to Adult. J. Biol. Chem. 262:14408-14414 (1987).
11. R.S. Adelstein and C.B. Klee, Purification and Characterization of Smooth Muscle Myosin Light Chain Kinase. J. Biol. Chem. 256:7501-7509 (1981).
12. M. Nishikawa, S. Shirakawa, and R.S. Adelstein, Phosphorylation of Smooth Muscle Myosin Light Chain Kinase by Protein Kinase C. J. Biol. Chem. 260:8978-8983 (1985).
13. P. de Lanerolle, R.S. Adelstein, J.R. Feramisco, and K. Burridge, Characterization of Antibodies to Smooth Muscle Myosin Kinase and their use in Localizing Myosin Kinase in Nonmuscle Cells. Proc. Natl. Acas. Sci. USA 78:4738-4742 (1981).
14. D.R. Hathaway and J.R. Haeberle, A Radioimmunoblotting Method for Measuring Myosin Light Chain Phosphorylation Levels in Smooth Muscle. Am. J. Physiol. 249:C345-C351 (1985).
15. J.R. Sellers, M.D. Pato, and R.S. Adelstein, Reversible Phosphorylation of Smooth Muscle Myosin, Heavy Meromyosin, and Platelet Myosin. J. Biol. Chem. 256:13137-13142 (1981).
16. M. Nishikawa, H. Hidaka, and R.S. Adelstein, Phosphorylation of Smooth Muscle Heavy Meromyosin by Calcium-activated, Phospholipid-dependent Protein Kinase. J. Biol. Chem. 258:14069-14072 (1983).
17. J.A. Spudich, and S. Watt, The Regulation of Rabbit Skeletal Muscle Contraction. J. Biol. Chem. 246:1866-1871, (1971).
18. O.H. Lowry, N.J. Rosenbrough, A.L. Farr, and R.J. Randall, Protein Measurement with the Folin Phenol Reagent. J. Biol. Chem. 193:265-275 (1951).
19. U.K. Laemmli, Cleavage of Structural Proteins during the Assembly of the Head of Bacteriophage T4. Nature 227:680-685 (1970).
20. P. de Lanerolle, M. Nishikawa, R. Felsen, and R.S. Adelstein, Immunological Properties of Myosin Light-Chain Kinases. Biochimica et Biophysica Acta 914:74-82 (1987).
21. J.C. Seidel, N. Nath, and S. Nag, Ca^{2+} Dependence of the ATPase Activity of Phosphorylated Smooth Muscle Myosin: Effects of Tropomyosin and Actin. Biochimica et Biophysica Acta 871:93-100 (1986).
22. M. Ikebe, D.J. Hartshorne, and M. Elzinga, Phosphorylation of the 20,000-Dalton Light Chain of Smooth Muscle Myosin by the Calcium-Activated, Phospholipid-Dependent Protein Kinase. J. Biol. Chem. 262:9569-9573 (1987).
23. P. de Lanerolle, and M. Nishikawa, Regulation of Embryonic Smooth Muscle Myosin by Protein Kinase C. J. Biol. Chem. 263:9071-9074 (1988).

24. R.S. Adelstein, Regulation of Contractile Proteins by Phosphorylation. J. Clin. Invest. 72:1863-1866 (1983).
25. P. de Lanerolle, and J.T. Stull, Myosin Phosphorylation during Contraction and Relaxation of Tracheal Smooth Muscle. J. Biol. Chem. 255:9993-10000 (1980).
26. L. Donner, P. de Lanerolle, and J. Costa, Immunoreactivity of Paraffin-embedded Normal Tissues and Mesenchymal Tumors for Smooth Muscle Myosin. Am. J. Clin. Path. 80:677-681 (1983).
27. R.M. Medford, R.M. Wydro, H.T. Nguyen, and B. Nadal-Ginard, Cytoplasmic Processing of Myosin Heavy Chain Messenger RNA: Evidence Provided by using a Recombinant DNA Plasmid. Proc. Natl. Acad. Sci. USA 77:5749-5753 (1980).
28. A.R. Bengur, E.A. Robinson, E. Appella, and J.R. Sellers, Sequence of the Sites Phosphorylated by Protein Kinase C in the Smooth Muscle Myosin Light Chain. J. Biol. Chem. 262:7613-7617 (1987).
29. I. Mabuchi, and M, Okuno, The Effect of Myosin Antibody on the Division of Starfish Blastomeres. J. Cell Biol. 74:251-263 (1977).
30. D.A. Knecht and W.F. Loomis, Antisense RNA Inactivation of Myosin Heavy Chain Gene Expression in Dictyostelium discoideum. Science 236:1081-1086 (1987).
31. A. De Lozanne and J.A. Spudich, Disruption of the Dictyostelium Myosin Heavy Chain Gene by Homologous Recombination. Science 236:1086-1091 (1987).
32. Y. Nishizuka, Studies and Perspectives of Protein Kinase C. Science 233:305-312 (1986).
33. E. Rozengurt, Early Signals in the Mitogenic Response. Science 234:161-166 (1986).

CALCIUM AND POLYPHOSPHOINOSITIDE REGULATION OF ACTIN NETWORK STRUCTURE BY GELSOLIN

Helen L. Yin
Hematology-Oncology Unit
Massachusetts General Hospital
Harvard Medical School
Boston, MA 02114

INTRODUCTION

Receptor-mediated stimulation induces a rapid and transient reorganization of the actin filaments within the cytoskeleton of a variety of cells[1-3]. Although the pathways involved in mediating these changes are not well understood, several modulating proteins are likely to be involved[4,5]. In this paper, I will review the properties of gelsolin, a Ca^{2+} and polyphosphoinositide-modulated actin regulatory protein which has powerful effects on actin filament structure[6]. Because Ca^{2+} and polyphosphoinositide levels change transiently following agonist stimulation, gelsolin may have a pivotal role in modulating cytoskeletal changes.

GELSOLIN-ACTIN INTERACTIONS

Gelsolin is a potent actin-modulating proteins first identified in rabbit lung macrophages[7] and subsequently found in many mammalian cells. The name derives from its ability to mediate the transition of cytosolic extracts from a gel phase to a sol phase, through a reduction in the actin filament length distribution. Cytoplasmic gelsolin consists of a single 80,000 dalton polypeptide. The interactions of gelsolin with actin are multiple and complex[8-10]. It has three main effects on actin. First, gelsolin breaks the noncovalent bond between actin-actin monomers within an actin filament, severing the filament. Second, gelsolin binds to the preferred actin filament end (the "barbed" end with respect to arrowheads formed by binding of myosin subfragments), blocking monomer exchange from that end. This has the effect of raising the critical concentration for actin association into filaments thus causing net actin depolymerization. Third, gelsolin nucleates actin assembly, again promoting formation of short actin filaments.

The immediate effect of actin filament shortening is a decrease in their ability to be crosslinked into a gel network, inhibiting actin gelation and solating preformed gels. In addition, the short capped filaments can serve as a potential source of actin nuclei, provided that gelsolin can be removed from the "barbed" ends of the filaments.

EVIDENCE FOR ROLE OF GELSOLIN IN REGULATING ACTIN STRUCTURES

Several pieces of *in vivo* evidence suggest that gelsolin may indeed be involved in regulating actin assembly-disassembly in cells:

1. Gelsolin is present in motile cells at high concentrations. There is one molecule of gelsolin per 100 actin monomers in macrophages. It is the major Ca^{2+}-dependent actin severing protein in macrophage extracts, and also accounts for most of the actin nucleating and filament end capping activity[11].

2. Most of the gelsolin in resting cells is not bound to actin. When cells are stimulated, gelsolin is complexed to actin reversibly. *In vitro* studies have shown that gelsolin requires uM Ca^{2+} to bind actin to form a 2 actin:1 gelsolin complex[11]. However, subsequent chelation of Ca^{2+} with EGTA releases only one actin[8]. The resultant EGTA-resistant 1:1 complexes (GA_1) do not fragment actin filaments and block actin filaments even at submicromolar Ca^{2+} concentration[8,9]. Therefore, once gelsolin caps the "barbed" end of a filament *in vitro*, it can no longer be dissociated from actin by removal of Ca^{2+}. The polyphosphoinsitides PIP_2 and PIP, can uncap filament ends and dissociate the 1:1 complex *in vitro*[12,13]. Dissociation of GA_1 complex *in vivo* has been demonstrated in several types of cells. Freshly isolated macrophages contain very little GA_1. Ionomycin increases GA_1 complex formation, while the chemotactic peptide formyl-met-leu-phe (fMLP), causes dissociation[14]. Therefore, while gelsolin:actin interaction in cells can be induced by Ca^{2+}, other factors, activated by fMLP, can dissociate GA_1 complexes. Likewise,

thrombin, which promotes actin polymerization in platelets, induces GA_1 complex formation and subsequent dissociation[15], as well as a transient translocation of gelsolin towards platelet plasma membrane[16]. Reversible interaction of gelsolin with the cortical actin filament network has also been reported for erythroblasts[17].

3. When macrophages and fibroblasts are microinjected with gelsolin, there is no detectible change in morphology. This is perhaps to be expected because the PIP2 level is relatively high, and Ca^{2+} level is low. In contrast, microinjection of the Ca^{2+}-independent constitutively active actin severing fragment of gelsolin causes extensive distruption and eventually cell death, suggesting that gelsolin can alter actin filament structures in cells[18].

Taken together, these results suggest that cells possess a mechanism, not directly involving Ca^{2+}, for dissociating actin/gelsolin EGTA-resistant complexes subsequent to agonist stimulation. Since PIP2 and PIP, can uncap actin filaments and dissociate GA_1 complexes _in vitro_[12,13], they may be _in vivo_ regulators of gelsolin function, imposing an additional level of control to that of Ca^{2+} regulation. The impact of PPI regulation of gelsolin is further enhanced by concurrent regulation of profilin[19]. Profilin is an actin monomer sequestering protein which prevent actin incorporation into filaments. Profilin can therefore decrease the steady-state F-actin concentration in the cell cytoplasm and may explain why close to 50% of the actin in the cell extract of unactivated cells is not polymerized in spite of conditions favoring polymerization.

As a working hypothesis, we propose that rapid polymerization of actin filaments can occur through the effects of PPI on gelsolin ,and profilin. In cells at rest, gelsolin is soluble and not bound to actin filaments[14,18]. When cells are activated, the concentration of PPI in the plasma membrane falls as it is converted to diacylglycerol and inositol 1,4,5 triphosphate by activated phospholipase C. The latter product, in turn, mobilizes calcium from internal stores and begins a "gelsolin activation cascade". Gelsolin-Ca^{2+} severs the cortical actin filament network, and rapidly produces numerous short gelsolin-capped actin oligomers which can diffuse freely within the cytoplasm. Some of the oligomers move towards the plasma membrane and contact PIP_2, as its concentration is restored following an initial fall. This leads to dissociation of gelsolin from actin oligomers, exposing nuclei with free "barbed" ends. Simultaneously, PIP_2 releases actin monomers from profilin:actin complexes. The increased availability of polymerization-competent actin monomers, as well as the number of nuclei with free "barbed" ends result in rapid polymerization. Ca^{2+} potentiates the response by allowing gelsolin to sever actin prior to a rise in PIP2 concentration, creating a large number of nuclei, and a large pool of actin monomers.

THE DOMAIN STRUCTURE OF GELSOLIN

The multifunctionality of gelsolin, as well as its differential regulation by Ca^{2+} and PPI, suggest that there are distinct functional and regulatory domains in gelsolin. The domain structure of gelsolin has been analyzed by limited proteolysis[20-24] and the picture that emerges is that gelsolin contains at least three actin binding sites with distinct properties. The actin severing portion of gelsolin is localized to the NH_2-terminal half 45 kDa peptide of gelsolin (residues 1-406 of human plasma gelsolin)[27]. Severing by this gelsolin fragment is inhibited by PIP2, but Ca^{2+} is no longer required for activity. Within the NH_2-terminal peptide, at least two Ca^{2+}-independent actin binding sites have been localized by further chymotrypsin (CT) cleavage. CT17N, a 17 kDa fragment generated from the extreme NH_2-terminus (amino acid residues 1-150), binds actin with a maximum Kd of 4.5×10^{-12} M[24], inhibits actin polymerization and accelerates the depolymerization of F-actin with kinetics consistent with severing. Its severing activity however is weak, being about 1% as effective as gelsolin on a molar basis[22]. The contiguous peptide, CT28N, encompassing residues 151-406, binds actin protomers within actin filaments to a near stoichiometric ratio[23,24], with an estimated Kd of 2.0×10^{-7}M. The stoichiometry of binding implies that CT28N binds randomly along actin filaments, and binding does not result in actin severing. PIP2 blocks this lateral binding with a dose response similar to that observed for PIP2 inhibition of actin severing by intact gelsolin and CT45N[23]. Therefore, CT28N may be the actin binding domain which allows gelsolin to associate with the side of actin filaments, positioning CT14N to bind and sever actin filaments.

The fact that the NH_2-terminal half of gelsolin as well as its actin-binding subfragments are active at submicromolar Ca^{2+} suggests that the stringent Ca^{2+}-regulation of severing by intact gelsolin is attributable to a different part of the molecule. The COOH-terminal half of gelsolin (residues 407-756) which does not sever actin filaments exhibits Ca^{2+}-dependent, reversible binding to actin[20,24], and undergoes Ca^{2+}-induced conformational change[20,25]. Therefore, Ca^{2+}-regulation of intact gelsolin is likely to be mediated by the COOH-half of gelsolin. The precise mechanism by which this occurs has not been elucidated. Ca^{2+} induces a conformational change in gelsolin, causing it to assume a more elliptical shape[26] and to expose a protease-sensitive site in the middle of the molecule[21,22]. Therefore, a reasonable model is that the COOH-half of gelsolin covers up one or both of the NH_2-terminal severing sites, and this block is relieved by Ca^{2+} induced conformational change in the COOH-half. Gelsolin has two Ca^{2+}-binding sites, with $Ka=10^6$ M^{-1}. The location of these sites has not been determined, and there are no strict EF-hand motifs in gelsolin's primary sequence[27].

Less is known about how PPI regulates gelsolin severing. The observation is that PIP2 and PIP, either as pure micelles or in mixed vesicles with other lipids inhibit gelsolin severing, and to a lesser extent, capping and nucleation[12,13]. In addition, PPI can also dissociate EGTA resistant 1:1 G:A complexes and uncap gelsolin blocked actin filaments. These observations raise the possibility that changes in the level of PPI may explain the dissociation of G:A complexes observed _in vivo_, but not _in vitro_ by reduction of Ca^{2+} concentration alone.

GELSOLIN IS A MEMBER OF A LARGE FAMILY OF ACTIN-BINDING PROTEIN

Besides existing as a cytoplasmic protein, a slightly larger form of gelsolin[28] is found in vertebrate plasma at approximately 0.2 mg/ml. Human plasma gelsolin contains a 25-amino acid extension at its NH_2-terminus compared with human cytoplasmic gelsolin, and is a _bona fide_ secreted protein, being encoded by a mRNA with a functional signal peptide sequence immediately upstream of the 25 amino acid plasma extension sequence[27]. Cytoplasmic gelsolin is encoded by a distinct mRNA, with a unique untranslated sequence but a coding sequence identical to that of plasma gelsolin residues 25-756[29]. Genomic analysis shows that the two message forms are derived from a single gene (located on human chromosome 9[30]) in the haploid genome through the use of alternative transcriptional initiation sites and an elaborate pattern of RNA splicing.

The role of plasma gelsolin in blood and extracellular fluid has not been defined. Since gelsolin shortens actin filaments very effectively, one possibility is that it is involved in the clearance of actin from the circulation[31]. During tissue damage or normal cell turnover, actin, a major cell and muscle protein, may be released into the blood stream, causing a transient and localized increase in actin filament concentration. This hypothesis is reasonable because besides gelsolin, another relatively abundant plasma protein, Vitamin D binding protein, can also shorten actin filaments by sequestering actin monomers[32,33]. The coexistence and complementary functions of gelsolin and Vitamin D binding protein suggest that there is a need for efficient removal of actin filaments from blood. Furthermore, there is evidence that skeletal, smooth and cardiac muscles secrete gelsolin[34,35], and skeletal muscles account for half of the total plasma gelsolin synthesized in an adult animal.

Besides the two gelsolins, a number of other Ca^{2+}-regulated actin severing proteins have been identified. These includes villin, a 95 kDa protein found in microvilli of epithelial cells[36], fragmin[37] and severin[38], 45 kDa proteins purified from Physarum polycephalum and Dictyostelium discoideum, respectively. Analysis of the primary sequence of all four proteins reveal a high degree of sequence homology and a common structural theme. The NH_2- and COOH-terminal halves of gelsolin and villin have 50% identical amino acid sequence with respect to each other, indicating that the twofold repeat may have been generated through a gene duplication event. Fragmin and severin, which are half the size of gelsolin/villin, may represent prototype proteins prior to the twofold duplication. Further analysis show that each half of gelsolin contains a threefold repeat, and a similar repeat is observed in fragmin/severin. Therefore, this has led to the speculation that these severing proteins were derived from a small (approximately 14 kDa) ancestral actin binding protein through a series of gene duplication events, and diverged subsequently to adopt their respective unique functions. In the case of villin, which is encoded by a distinct gene[39], there is further specialization by the attachment of an additional COOH-terminal sequence which confers the ability to bundle actin filaments in the absence of Ca^{2+}[40] and is similar to a sequence found in synapsin I, an actin-bundling protein associated with synaptic vesicles. In support of the cassette actin binding domain hypothesis, there is some sequence homology between Acanthamoeba profilin and the various larger Ca^{2+}-dependent actin severing proteins[41].

REFERENCES

1. Howard, T. H. and Wang, D., 1987, Calcium ionophore, phorbol ester, and chemotactic peptide-induced change in F-actin content, F-actin

distribution and the shape of neutrophils, J. Clin. Invest. 79:1359.
2. Sha'afi, R. I. and Molski, T. F. P., 1987, Signalling for increased cytoskeletal actin in neutrophils, Biophys. Res. Commun., 145:934.
3. Sheterline, P., Rickard, J. E., Boothroyd, B. and Richards, R. C., 1986, Phorbol ester induces rapid actin assembly in neutrophil leukocytes independently of changes in [Ca^{2+}] and pHi_s, J. Muscle Res. Cell Motility, 7:405.
4. Stossel, T. P., et al., 1985, Non muscle actin-binding proteins. Ann. Rev. Cell. Biol., 1:353.
5. Pollard, T. D., and Cooper, J. A., 1986, Actin and actin-binding proteins, A critical evaluation of mechanisms and functions, Ann. Rev. Biochem., 55:987.
6. Yin, H. L., 1988, Gelsolin: calcium- and polyphosphoinositide-regulated actin-modulating protein, BioEssays, 7:176.
7. Yin, H. L. and Stossel, T. P., 1979, Control of cytoplasmic actin gel-sol transformation by gelsolin, a calcium-dependent regulatory protein, Nature (Lond.), 281:583.
8. Bryan, J. and Kurth, M., 1984, Actin-gelsolin interactions: evidence for two actin-binding sites, J. Biol. Chem., 259:7480.
9. Janmey, P. A., Chaponnier C., Lind S. E., Zaner K. S., Stossel T. P., and Yin, H. L., 1985, Interactions of gelsolin and gelsolin actin complexes with actin, Effects of calcium on actin nucleation, filament severing and end blocking, Biochemistry, 24:3714.
10. Weeds, A. G., Harriet, H., Gratzer, W., and Gooch, J., 1986, Interactions of pig plasma gelsolin with G-actin, Eur. J. Biochem., 16:77.
11. Yin, H. L. and Stossel, T. P., 1980, Ca^{2+} control of actin gelation, J. Biol. Chem., 255:9494.
12. Janmey, P. A. and Stossel, T. P., 1987, Modulation of gesolin function by phosphatidylinositol-4,5-bis phosphate, Nature, 325: 362.
13. Janmey, P. A., Iida, K., Yin, H. L., and Stossel, T. P., 1987, Polyphosphoinositide micelles and polyphosphoinositide-containing vesicles dissociate endogenous gelsolin-actin complexes and promote actin assembly from the fast-growing end of actin filaments blocked by gelsolin, J. Biol. Chem., 262:12228.
14. Chaponnier, C., Yin, H. L. and Stossel, T. P., 1987, Reversibility of gelsolin/actin interaction in macrophages, J. Exp. Med. 165:97.
15. Lind, S. E., Janmey, P. A., Chaponnier, C., Herbert, T. J., and Stossel, T. P., 1987, Reversible binding of actin to gelsolin and profilin in human platelet extracts, J. Cell Biol., 105:833.
16. Hartwig, J. H., Chambers, K., and Stossel, T. P., Association of

gelsolin with actin filaments and cell membranes of macrophages and platelets, J. Cell Biol. (submitted).

17. Hinssen, H., Vandekerckhove, J., and Lazarides, E., 1987, Gelsolin is expressed in early erythroid progenitor cells and negatively regulated during erythropoiesis, J. Cell Biol., 105:1425.

18. Cooper, J. A., Loftus, D. J., Frieden, C., Bryan, J., and Elson, E. L., Localization and mobility of gelsolin in cells, J. Cell Biol., 106:1229.

19. Lassing, I., and Lindburg, U., 1985, Specific interaction between phosphatidylinositol 4,5-bisphosphate and profilactin, Nature, 318:472.

20. Kwiatkowski, D. J., Janmey, P. A., Mole, J. E., and Yin, H. L., 1985, Isolation and properties of two actin binding domains in gelsolin, J. Biol. Chem., 259:15232.

21. Bryan, J., and Hwo, S., 1986, Definition of an NH_2-terminal actin binding domain and a COOH-terminal Ca^{2+}-regulatory domain in human brevin, J. Cell Biol., 102:1439.

22. Chaponnier, C., Janmey, P. A., and Yin, H. L., 1986, The actin filament severing domain of plasma gelsolin, J. Cell Biol. 103:1473.

23. Yin, H. L., Iida, K., and Janmey, P. A., 1988, Identification of a polyphosphoinositide-modulated domain in glesolin which binds to the sides of actin filaments, J. Cell Biol., 106:805.

24. Bryan, J., 1988, Gelsolin has three actin-binding sites, J. Cell Biol., 106:1553.

25. Hwo, S., and J. Bryan, 1986, Immuno-identification of Ca^{2+}-induced conformation changes in human gelsolin and brevin, J. Cell Biol. 102:227.

26. Rouayrenc, J. F., Fattoum, A., Mejean, C., and Kassab, R., 1986, Characterization of the Ca^{2+}-induced conformational changes in gelsolin and identification of interaction regions between actin and gelsolin, Biochem., 25:3859.

27. Kwiatkowski, D. J., Stossel, T. P., Orkin, S. H., Mole, J. E., Colten, J. R., and Yin, H. L., 1986, Plasma and cytoplasmic gelsolins are encoded by a single gene and contain a duplicated actin-binding domain, Nature (Lond.), 323:455.

28. Yin, H. L., Kwiatkowski, D. J., Mole, J. E., and Cole, F. S., 1985, Structure and biosynthesis of cytoplasmic and secreted variants of gelsolin, J. Biol. Chem., 29:9294.

29. Kwiatkowski, D. J., Mehl, R., and Yin, H. L., 1988, Genomic organization and biosynthesis of secreted and cytoplasmic forms of gelsolin, J. Cell Biol. 106:375.

30. Kwiatkowski, D. J., Westbrook, C. A., Bruns, A. P., and Morton, C.

C., 1988, Localization of gelsolin proximal to ABL on chromosome 9, Am. J. Hum. Genet., 42:565.

31. Lind, S. E., Smith, D. B., Janmey, P. A., and Stossel, T. P., 1986, The role of plasma gelsolin and the vitamin D-binding protein in clearing actin from the circulation, J. Clin. Invest., 78:736.

32. Haddad, J. G., 1982, Human serum binding protein for vitamin D and its metabolites (DBP): Evidence that actin is the DBP binding component in human skeletal muscle. Arch. Biochem. Biophys. 213:538.

33. Van Baelen, H., Bouillon, R., and De Moor, P., 1980, Vitamin D-binding protein (Gc-globulin) binds actin. J. Biol. Chem., 255:2270.

34. Kwiatkowski, D. J., Mehl, R., Izumo, S., Nadal-Ginard, B. and Yin, H. L., 1988, Muscle is the major source of plasma gelsolin, J. Biol. Chem., 263:8239.

35. Nodes, B. R., Shackelford, J. E., and Lebherz, H. G., 1987, Synthesis and secretion of the actin binding protein gelsolin by chicken gizzard muscle, J. Biol. Chem., 262:5422.

36. Bazari, W. L., Matsudaira, P., Wallek, M., Smeal, T., Jakes, R., and Ahmed, Y., The villin sequence and peptide map identical six homologous domains, Proc. Nat. Acad. Sci., In Press.

37. Ampe, C. and Vandekerckhove, 1987, The F-actin capping proteins of Physarum Polycephalum: cap42(a) is very similar, if not identical, to fragmin and is structurally and functionally very homologous to gelsolin: cap42(b) is Physarum actin, EMBO J., 6:4149.

38. Andre, E., Lottspeich, F., Schleicher, M., and Noegel, A, 1988, Severing, gelsolin and villin share a homologous sequence in regions presumed to contain F-actin severing domains, E., J. Biol. Chem., 263:722.

39. Pringault, E., Arpin, M., Garcia, A., Finidori, J., and Louvard, 1986, A human villin cDNa clone to investigate the differentiation of intestinal and kidney cells in vivo and in culture, The EMBO J., 5:3119.

40. Glenney, J. R., Geislen, N., Kaulfus, P., and Weber, K., 1981, Demonstration of at least two different actin binding ssites in villin, a calcium regulated modulator of F-actin organization, J. Biol. Chem., 256:8156.

41. Ampe, C., Mareky, F., Lindberg, U., and Vandekerckhove, 1988, The primary structure of human platelet profilin; reinvestigation of the calf spleen profilin sequence, EMBO J., 6:4149.

SIGNIFICANCES OF TWO DIFFERENT Mr CALDESMONS

Kenji Sobue and Yasushi Fujio

Department of Neurochemistry and Neuropharmacology
Biomedical Research Center, Osaka University Medical School
Osaka, Japan

INTRODUCTION

Regarding the Ca^{2+}-dependent regulatory mechanism of the smooth muscle and nonmuscle actin-myosin interaction, the myosin-linked regulation based on phosphorylation of myosin by Ca^{2+}- and calmodulin-dependent myosin light chain kinase and dephosphorylation by myosin phosphatase has been well documented (1-3). Despite the accumulation of several findings (4-7), however, the precise molecular mechanism of the actin-linked regulation is as yet unclear.

During a search for the targets of calmodulin, we purified a major calmodulin-binding protein with a Mr of 150/147 kDa doublet from chicken gizzard smooth muscles (8). We named this protein "caldesmon", derived from a combination of calmodulin and the Greek *desmos* (binding). Interestingly, caldesmon also binds to F-actin and the binding fashion of this protein to either calmodulin or F-actin is reversibly regulated by the concentration of Ca^{2+}. At less than 1 μM of free Ca^{2+} concentration, caldesmon binds to F-actin. Whereas at a higher concentration of Ca^{2+} (1 μM), calmodulin activated by Ca^{2+} forms a complex with caldesmon and this complex is freed from F-actin (flip-flop binding) (8,9).

Considering a high concentration of caldesmon in smooth muscles (the content of caldesmon in gizzard smooth muscle is equivalent to more than 70 % of the total calmodulin) and its Ca^{2+}-dependent flip-flop binding ability to either calmodulin or F-actin, we first proposed the functional aspect of caldesmon as an actin-linked regulatory protein in the smooth muscle actomyosin system (10). Subsequently, we demonstrated, by using a complete reconstitution system, the molecular mechanism of the caldesmon-involved regulation in the smooth and nonmuscle actin-myosin interaction (7,11). In 1984, Bretscher found the heat-stability of this protein and developed a new method for rapid purification (12). After his stimulating report, several researchers have focused on the function of caldesmon and have proposed some discrepant hypotheses. Because of limitations of space in this brief article, we cannot discuss all of these hypotheses. In this paper, we will demonstrate the relationship between two different Mr forms of caldesmon.

TWO SPECIES OF CALDESMON

We surveyed the distribution of caldesmon in smooth and nonmuscle tissues, and found two different Mr polypeptides which crossreacted with the anti-gizzard caldesmon antibody (Table I) (13-15). The high-Mr (140-150 kDa) protein is exclusively predominant in smooth muscles. On the other hand, in nonmuscle tissues and cells the content of the low-Mr (70-80 kDa) protein is much higher than that of the high-Mr protein. Actually, the low-Mr protein purified from bovine adrenal medulla has an alternative binding ability to either calmodulin or F-actin in a Ca^{2+}-dependent flip-flop manner (14). In addition, the low-Mr protein inhibits the interaction between actin and myosin prepared from bovine

Table I. Distribution of two caldesmons in various tissues

tissues or cells	*h*-caldesmon(μM)	*ℓ*-caldesmon(μM)
chicken		
gizzard	9.2	< 0.06
Bovine		
aorta (media)	3.0	< 0.06
adrenal medulla	0.04	0.82
adrenal cortex	0.05	0.68
anterior pituitary	0.03	0.44
posterior pituitary	0.03	0.64
Rat		
spleen	0.05	0.43
brain	< 0.01	0.25
liver	0.03	0.78
kidney	0.03	0.66
Cultured cells		
balb/c 3T3	0.03	0.62
NRK	< 0.01	0.35
BHK	< 0.01	0.53
MDCK	< 0.01	0.40
CHO-K1	< 0.01	0.48

adrenal medulla (14,16). The similar Mr protein having the flip-flop binding ability was identified in cultured cells (17) and purified from human or porcine platelets (18,19). These results indicate that the low-Mr protein possesses a number of features in common with the gizzard high-Mr protein, such as the Ca^{2+}-dependent flip-flop binding ability to either calmodulin or F-actin and the immunological cross-reactivity with the anti-gizzard caldesmon antibody. We termed the high-Mr protein "*h*-caldesmon" and the low-Mr protein "ℓ-caldesmon", for short.

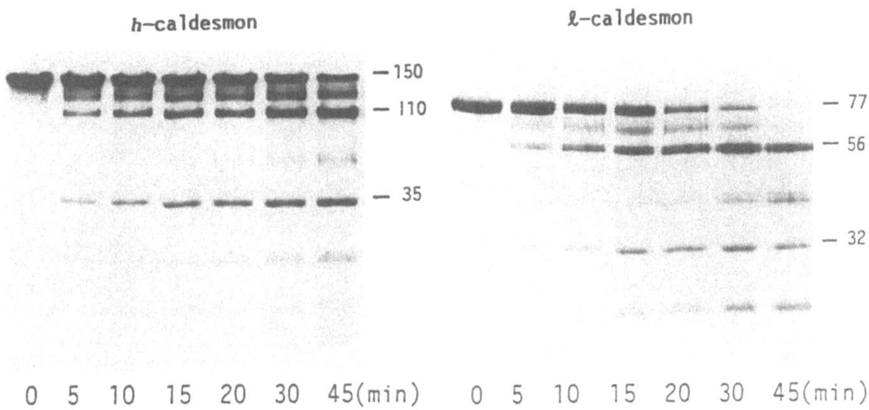

Fig. 1 Time course of the chymotrypstic degestion of *h*- and ℓ-caldesmons. Two caldesmons (0.3mg/ml) was digested with 1/3,000 weight of α-chymotrypsin at 0°C.

COMPARISON OF DOMAIN MAPPING IN TWO CALDESMONS

In order to compare the relationship between *h*- and ℓ-caldesmons, we carried out limited proteolysis with α-chymotrypsin. Limited proteolysis of *h*-caldesmon was originally reported by Szpacenko and Dabrowska (20). Figure 1 shows a typical time course of limited proteolysis as analyzed by SDS-PAGE. It can be seen that with 1/3,000 (by weight) of α-chymotrypsin, limited digestion gave the best result. A comparison of two caldesmons yields similar digestive patterns. *h*-Caldesmon first splits into three major fragments with estimated molecular weights of 120, 110 and 35 kDa and a trace amount of the 25 kDa fragment. Further proteolysis gave rise to other 69, 28 and 22 kDa fragments (21). Using these proteolytic fragments, we attempted to localize the calmodulin- and/or actin-binding domain. It is apparent that the 150 kDa (parent) protein and the 120, 35 and 22 kDa fragments bound to calmodulin and F-actin as determined by calmodulin affinity chromatography and cosedimentation assay. By contrast, the 110, 69 and 28 kDa fragments did not interact with calmodulin or F-actin.

As already emphasized from our experimental evidences (7,10,11), caldesmon is considered to regulate the tropomyosin function through actin filaments. Recently, it has been suggested the direct interaction between *h*-caldesmon and tropomyosin (22-24). We also investigated the tropomyosin-binding domain in the caldesmon molecule. It is noteworthy that the same fragments having the calmodulin- and actin-binding abilities also interact with tropomyosin-coupled affinity column in an ionic strength-dependent manner (Fig. 2). Additionally, amino acid sequence analyzed by automatic amino acid sequencer revealed that the N-terminal of the parent protein and the 110 kDa fragment was blocked. On the other hand, the 120 and 35 kDa fragments could sequence the N-terminal 20-25 amino acids. These results led to the speculation that α-chymotrypsin initially cleavaged caldesmon at two points corresponding to the N-terminal 25 kDa portion and the C-terminal 35 kDa portion. Combining these findings, we propose a tentative domain mapping of *h*-caldesmon (Fig. 3). The calmodulin- and actin-binding abilities in addition to the tropomyosin-interacting capability in the caldesmon molecule is restricted within the 35 kDa fragment at the C-terminal.

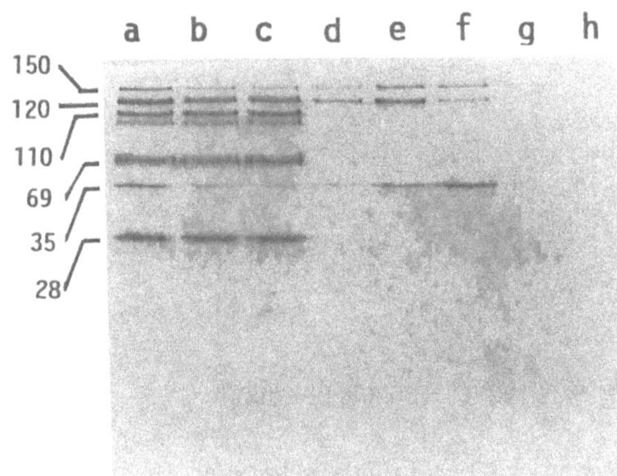

Fig. 2 Interaction of *h*-caldesmon and its proteolytic fragments with tropomyosin. Limited proteolytic products of *h*-caldesmon (lane a) was applied on tropomyosin-coupled Sepharose 4B column. Lane b, flow through fraction; lane c, washing fraction; lane d, 20 mM KCl elution; lane e, 40 mM KCl elution; lane f, 60 mM KCl elution; lane g, 100 mM KCl elution; lane h, 150 mM KCl elution.

In the case of ℓ-caldesmon, it initially degraded into the 56, 32 and 20 kDa fragments. We carried out to localize the functional activity in the ℓ-caldesmon molecule using the same procedures for *h*-caldesmon. The 32 kDa fragment of ℓ-caldesmon also retained the calmodulin-, actin- and tropomyosin-binding abilities (data not shown).

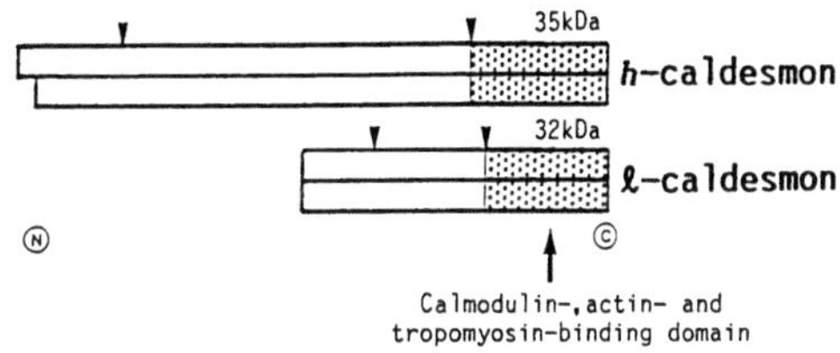

Fig. 3 Domain mapping of caldesmons. ◄ indicates the cleavage sites with α-chymotrypsin.

Table II. Binding of *h*-caldesmon or its 35 kDa fragment to F-actin or F-actin tropomyosin and their inhibitory effects on the actin-activated myosin ATPase activity

Additions (mol/ mol Actin)	*h*-CaD[1)]				35 kDa fragment			
	h-CaD bound (mol/mol Actin)		ATPase activity (%)		35kDa fr. bound (mol/mol Actin)		ATPase activity (%)	
	TM[2)](+)	TM(-)	TM(+)	TM(-)	TM(+)	TM(-)	TM(+)	TM(-)
0	0	0	100	56	0	0	100	56
0.063	0.049	0.035	77	56	0.05	0.03	82	56
0.125	0.072	0.052	62	51	0.10	0.08	64	46
0.250	0.081	0.069	46	44	0.15	0.13	41	41
0.375	0.085	0.075	41	41	0.25	0.22	38	36
0.500	0.086	0.077	41	39	0.28	0.24	36	36

CaD[1)], caldesmon; TM[2)], tropomyosin.

The functional domains purified from *h*- and ℓ-caldesmon also inhibit the tropomyosin-enhanced actin-myosin interaction as determined by myosin ATPase activity and superprecipitation. Table II shows the comparative data for the 35 kDa fragment of *h*-caldesmon and the parent molecule in terms of inhibition of myosin ATPase activity. At lower concentrations, the 35 kDa fragment showed an inhibitory effect on the tropomyosin-enhanced ATPase activity similar to that of the parent molecule only in the presence of tropomyosin. This inhibitory effect of *h*-caldesmon or the 35 kDa fragment was observed to approach the level of ATPase activity in the absence of tropomyosin. On the other hand, they had a less significant effect on ATPase activities in the absence of tropomyosin. However, at higher concentrations, they significantly inhibited ATPase activities regardless of the presence or absence of tropomyosin. This phenomenon may be due to the bundling effect of F-actin induced by *h*-caldesmon and the 35 kDa fragment. Electron microscopic observation with negative staining revealed large actin bundles when the molar ratio of *h*-caldesmon or the 35 kDa fragment to F-actin was high. Like the parent molecule, the 32 kDa fragment purified from ℓ-caldesmon also retained the similar inhibitory effect on the tropomyosin-enhanced ATPase activity. From these observations, we suggest that *h*- and ℓ-caldesmons contain common functional domains with the Mr of 35 kDa and 32 kDa at the C-terminal. As for the functional aspect of two caldesmons, they may be involved in the Ca^{2+}-dependent regulation of the smooth and nonmuscle actomyosin systems as actin-linked regulatory proteins. However, the function of the residual domains in two caldesmons is as yet unknown.

EXPRESSIONAL CHANGE OF TWO CALDESMONS DURING PHENOTYPIC MODULATION OF SMOOTH MUSCLE CELLS

One of the puzzling points is the differential distribution of two Mr forms of caldesmon. Although ℓ-caldesmon is widely distributed in nonmuscle cells, *h*-caldesmon is predominantly expressed in smooth muscles. As a first step to clarify this problem, we investigated the expressional change of two caldesmons in developing chicken gizzards (15). Fig. 4A shows the protein profile of cytoskeletal proteins in developing chicken gizzards as analyzed by SDS-PAGE. The expressional change of two caldesmons at the protein level was determined by means of immunoblotting with the anti-gizzard caldesmon antibody which also recognizes ℓ-caldesmon (Fig. 4B). Patterns of protein accumulation can be classified into three categories. The first category consists of myosin, β-tropomyosin and *h*-caldesmon, which become detectable in 10-day-old embryo gizzards following abrupt increases. The second category comprises actin and desmin, which already appear in early embryo (>7-day old) and gradually increase during development. The third one consists of ℓ-caldesmon which decreases in its amount in association with development of gizzards. Interestingly, the expressional change of ℓ-caldesmon in developing gizzards shows a good contrast to that of *h*-caldesmon. The amount of *h*-caldesmon increases with a concomitant decrease in that of ℓ-caldesmon. This expressional change of two caldesmons is in good agreement with the morphological differentiation of gizzard smoooth muscles. Benett and Cobb clearly demonstrated by their morphological observations that during embryogenesis the massive proliferation of mesenchymal cells causes an enlargement in the bulk of chicken gizzards following differentiation of mesenchymal cells into smooth muscle cells (25). Taken together with their observations, the critical period of cytodifferentiation of smooth muscle cells from their germinal mesenchymal cells is in good agreement with the dramatic increase in the expression of contractile elements such as *h*-caldesmon, myosin and tropomyosin. Our observations lead to the assumption that the expressional change of caldesmon from the *ℓ*- to *h*-form is closely related to the cytodifferentiation of smooth muscle cells.

It is well known that vascular smooth muscle cells in primary culture show the characteristic phenotypic modulation from the contractile to synthetic phase (26). To confirm our forementioned assumption, we examined the differential expression of two caldesmons in a primary culture system (15). Figure 5A shows the growth curve expressed as cell numbers and [^{3}H] thymidine uptake. The cells in the 1 to 2-day culture and the 3 to 7-day culture correspond to the contractile and the synthetic phases, respectively. In fact, the cells in the 1 or 2-day culture showed a contractile response against angiotensin II and a high expression of Ach receptor (Figure 5B and C). In contrast, the cells in the 3 to 7-day culture had no contractile capability and a low level of Ach receptor. The expressional change of two caldesmons during phenotypic modulation of vascular smooth muscle cells was examined by immunoblotting (Fig. 5D). In the contractile phase, the major form of caldesmon detected was the *h*-form. On the

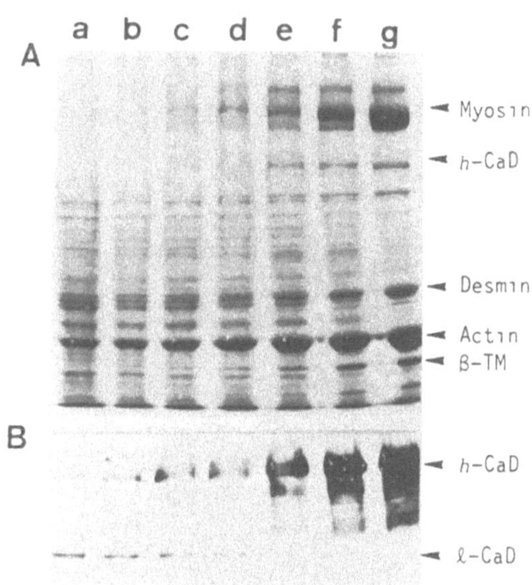

Fig. 4 Expressional changes of cytoskeletal proteins in developing chicken gizzards. (A) Electrophoregram of the whole homogenates of developing gizzards. (B) Immunoreplica of caldesmons corresponding to panel A. CaD, caldesmon; TM, tropomyosin.

other hand, a dramatic decrease in the amount of *h*-caldesmon with a concomitant increase in that of ℓ-caldesmon was observed in the synthetic phase.

These expressional changes of caldesmon during phenotypic modulation of smooth muscle cells are at the protein level. Similar changes were observed at the RNA level (15). Poly(A)$^+$ RNAs from 12-day-old embryonic and 3-day-old posthatched gizzards and from 2- and 7-day cultures of aortic smooth muscle cells were translated in cell-free mRNA-dependent reticulocyte lysates. Figure 6 shows autoradiographs of

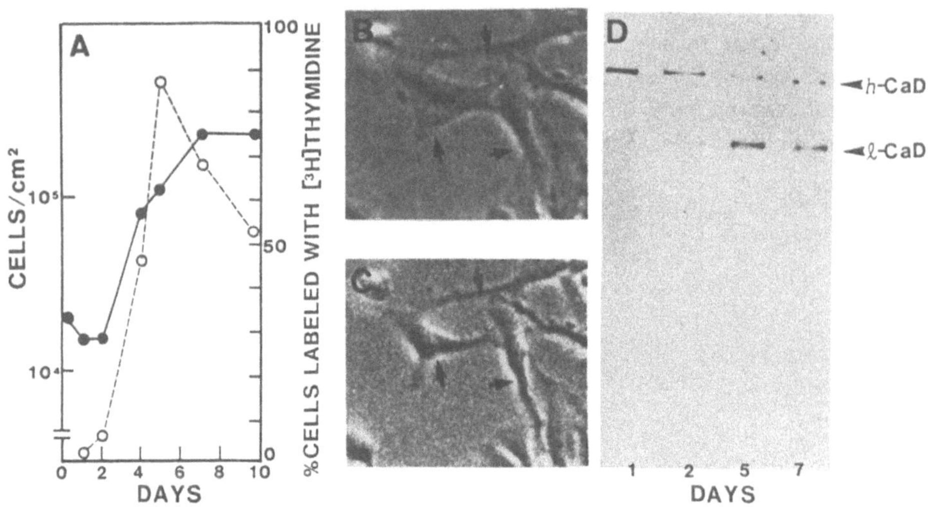

Fig. 5 Expressional change of *h*- and ℓ-caldesmons during phenotypic modulation (dedifferentiation) of rat aortic smooth muscle cells in primary culture. (A) Growth curve (●) and [^{3}H] thymidine uptake (○). (B and C) Phase-contrast micrographs of contractile response in the 2-day cultured cells before (B) and after (C) addition of angiotensin II. (D) Conversion of *h*-caldesmon to ℓ-caldesmon during dedifferentiation of aortic smooth muscle cells in primary culture.

the translational products immunoprecipitated with the caldesmon antibody. Although poly(A)$^+$ RNAs from 12-day-old embyonic gizzard contained a detectable amount of ℓ-CaD mRNA, poly(A)$^+$ RNA-directed synthesis of *h*-caldesmon was the sole RNA in 3-day-old posthatched gizzards. Similary, the RNA from the cultured cells at the contractile phase synthesized *h*-caldesmon in addition to a small amount of ℓ-caldesmon. However, the RNA from the synthetic phase programmed mainly the synthesis of ℓ-caldesmon in contrast to a decreasing synthesis of *h*-caldesmon. Additionally, in our preliminary experiment we have identified two species of mRNA with different nucleotide lengths corresponding to *h*- and ℓ-caldesmons as determined by the method of

Northern blotting using caldesmon cDNA probe. These results indicate that the expressional change of two caldesmons during phenotypic modulation of smooth muscle cells may be regulated at the level of mRNA.

The similar expressional change in two caldesmons was also observed in the BC_3H1 mouse cell line, which possesses many properties characteristic of smooth muscle cells (15). This cell line shows typical phenotypic modulation from the undifferentiated growing cell to the fully differentiated smooth muscle-like cell after transfer of growing cells in a high serum-containing medium to a low-serum medium. In the growing cells, ℓ-caldesmon was mainly detected together with a small amount of *h*-caldesmon. In the differentiated smooth muscle-like cells induced by serum depletion, the amount of *h*-caldesmon increased 3- to 4-fold with a concomitant decrease of ℓ-caldesmon.

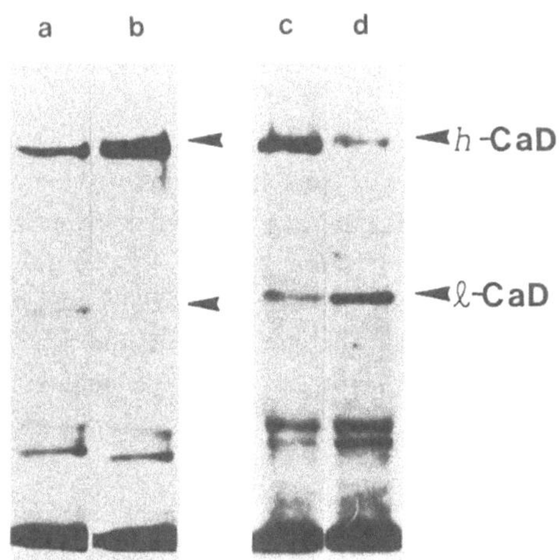

Fig. 6 *In vitro* translation of poly(A)$^+$ RNAs from 12-day-old embryonic gizzard (a), 3-day-old posthatched gizzard (b), 2-day cultured aortic smooth muscle cells (c) and 7-day cultured aortic smooth muscle cells (d).

CONCLUDING REMARKS

There are three possibilities regarding two different Mr caldesmons. i)ℓ-caldesmon is a proteolytic fragment of *h*-caldesmon. ii)Both caldesmons are derived from the different genes. iii)Both caldesmons are derived from a single gene by an alternative splicing. As forementioned in this article, the expressional levels of two caldesmons are distinctively different in the tissue distribution and in phenotypic modulation of smooth muscle cells. And, we have found that there are two mRNAs with different nucleotide lengths corresponding to *h*- and ℓ-caldesmons. From these findings, we exclude the possibility that ℓ-caldesmon is a proteolytic fragment of *h*-caldesmon. Since the 35

Table III. Molecular events during phenotypic modulation of smooth muscle cells

	Undifferentiated or Dedifferentiated Cell	Differentiated Smooth Muscle Cell
Contractile response (angiotensin II)	(-)	(+)
Ach Receptor	(±~+)	(+++)
Contractile elements (myosin, actin, TM)	(±~+)	(+++)
Caldesmon (and its mRNA) predominance	ℓ-type	*h*-type

kDa fragment from *h*-caldesmon and the 32 kDa fragment from ℓ-caldesmon have similar functional properties, the third possibility is the most likelihood. Most recently, Ball and Kovala have suggested by their mapping data that the sequences of ℓ-caldesmon locate in both ends of *h*-caldesmon, and ℓ-caldesmon misses the central area of *h*-caldesmon (27). However, it is necessary to await for the conclusion of later two possibilities until the determination of complete sequencing data as for two caldesmons.

Finally, we summarize the molecular events in smooth muscle cells during phenotypic modulation (Table III). In the undifferentiated or dedifferentiated cells, they have low levels of Ach receptor and contractile elements including myosin and tropomyosin. These cells don't show a contractile response against angiotensin II. In addition to these characteristics, the expression of caldesmon is predominantly the ℓ-form at the levels of protein and mRNA. By contrast, in the differentiated cells, the contents of Ach receptors and contractile elements dramatically increase. Anigiotensin II also induces a contractile responce to these cells. The expression of *h*-caldesmon is predominant at the levels of protein and mRNA. From these observations, it is suggested that the expressional change of caldesmons is a promising marker for studies on smooth muscle differentiation and dedifferentiation.

ACKNOWLEDGEMENTS

This work was supported by grants from the Scientific Research Fund of Ministry of Education, Science and Culture, Japan; National Cardiovascular Center; the Inamori Foundation.

REFERENCES

1. Small, J. V., Sobieszek, A., Int. Rev. Cytol., 64: 241-306 (1980)
2. Adelstein, R. S., Eisenberg, E., Annu. Rev. Biochem., 49: 921-956 (1980)
3. Hartshorne, D. J., Siemenkowski, R. F., Annu, Rev. Physiol., 43: 519-530 (1980)
4. Mikawa, T., Toyo-oka, T., Nonomura, Y., Ebashi, S., J. Biochem., 81: 273-275 (1977)
5. Sparrow, M. P., Maxwell, L. C., Ruegg, J. C., Bohr, D. F., Am. J. Physiol, 219: 1366-1372 (1970)
6. Marston, S. B., Trevett, R. M., Walters, M., Biochem. J., 185: 355-365 (1980)
7. Sobue, K., Kanda, K., Tanaka, T., and Ueki, N., J. Cell. Biochem., 37: 317-325 (1988)
8. Sobue, K., Muramoto, Y., Fujita, M, Kakiuchi, S.: Proc. Natl. Acad. Sci. USA, 78: 5652-5655 (1981)
9. Kakiuchi, S., Sobue, K, Trends Biochem. Sci, 8: 59-62 (1983)
10. Sobue, K., Morimoto, K., Inui, M., Kanda, K., Kakiuchi, S., Biomed. Res., 3: 188-196 (1982)
11. Sobue, K., Takahashi, K., Wakabayashi, I., Biochem. Biophys. Res. Commun., 132: 645-651 (1985)
12. Bretscher, A., J. Biol. Chem., 259: 12873-12880 (1984)
13. Owada, M. K., Hakura, A., Iida, K., Yahara, I., Sobue, K., Kakiuchi, S., Proc. Natl. Acad. Sci. USA, 81: 3133-3137 (1984)
14. Sobue, K, Tanaka, T., Kanda, K., Ashino, K., Kakiuchi, S., Proc. Natl. Acad. Sci. USA, 82: 5025-5029 (1985)
15. Ueki, N., Sobue, K., Kanda, K., Hada, T., Higashino, K., Proc. Natl. Acad. Sci. USA 84: 9049-9053 (1987)
16. Sobue, K., Tanaka, T., Kanda, K., Takahashi, K., Ito, K., Kakiuchi, S., Biomed. Res., 6: 93-102 (1985)
17. Bretscher, A., Lynch, W. J., J. Cell Biol., 100: 1656-1663 (1985)
18. Dingus, J., Hwo, S., Bryan, J., J. Cell Biol., 102: 1748-1757 (1986)
19. Onji, T., Takagi, M, Shibata, N., Biochem, Biophys, Res, Commun., 143 (1987)
20. Szpacenko, A., Dabrowska, R., FEBS Lett., 202: 182-186 (1986)
21. Yazawa, M., Yagi, K., Sobue, K., J. Biochem., 102: 1065-1073 (1987)
22. Smith, C. W. J., Pritchard, K., Marston, S. B., J. Biol. Chem., 262: 116-122 (1987)
23. Graceffa, P., FEBS Lett., 218: 139-142 (1987)
24. Fujii, T., Imai, M., Rosenfeld, G. C., and Bryan, J., J. Biol. Chem., 262: 2757-2763 (1987)
25. Benett, T. and Cobb, J. L. S., Z. Zelforsch., 98: 559-561 (1969)
26. Chamley-Campbell, J., Compbell, G. R., and Ross, R.,Physiol. Rev., 59: 1-61 (1979)
27. Ball, E. H. and Kovala, T., Biochemistry, 27: 6093-6098 (1987)

A MODEL FOR CALDESMON IN LATCH-BRIDGE FORMATION IN SMOOTH MUSCLE

Michael P. Walsh and Cindy Sutherland

Department of Medical Biochemistry
University of Calgary
Calgary, Alberta, Canada T2N 4N1

INTRODUCTION

The primary mechanism of regulation of smooth muscle contraction involves the Ca^{2+}-dependent and reversible phosphorylation of myosin[1]. Stimulation of the smooth muscle cell leads to an elevation of cytosolic free Ca^{2+} concentration whereupon Ca^{2+} binds to calmodulin. Formation of the Ca^{2+}-calmodulin complex is accompanied by a conformational change in the Ca^{2+}-binding protein which is then capable of interaction with a target enzyme, in this case myosin light chain kinase (MLCK). The binding of calmodulin converts the kinase from the inactive apoenzyme form found in the resting cell to the active Ca^{2+}-calmodulin-MLCK ternary complex. The active kinase catalyzes transfer of the terminal phosphoryl group of ATP to serine-19 on each of the two 20 kDa light chains of myosin. In the phosphorylated state, the MgATPase activity of myosin is substantially enhanced by actin. This biochemical parameter is thought to reflect a cross-bridge cycling mechanism whereby actin and myosin filaments slide relative to one another, the required energy being provided by the hydrolysis of ATP. At the level of the intact cell this process would result in force development.

A substantial body of evidence derived from biochemical and physiological experimental approaches supports this phosphorylation theory. Thus stimulation of intact smooth muscle cells by, e.g., carbachol or K^+ depolarization, results in a transient rise in cytosolic free $[Ca^{2+}]$ as shown by the use of Ca^{2+} indicators such as aequorin[2] and quin 2[3]. This in turn leads to the phosphorylation of myosin and subsequent force development[4].

However, it is not possible to explain all the biochemical and physiological observations on the basis of myosin phosphorylation-dephosphorylation. For example, application of the myosin competition test to vertebrate smooth muscle indicates the existence of a Ca^{2+}-dependent thin filament-linked regulatory mechanism[5]. Furthermore, Murphy and coworkers[6] have described the maintenance of force at low levels of myosin phosphorylation during prolonged stimulation of vascular smooth muscle. This so-called "latch state" is Ca^{2+}-dependent and is characterized by a substantially reduced actomyosin ATPase rate and non- or slowly-cycling cross-bridges.

We have been interested for some time in the biochemical basis of latch-bridge formation and initiated our investigations by searching the literature for proteins which could potentially be involved in regulating actin-myosin interactions in a Ca^{2+}-dependent manner. It should be noted, however, that Murphy's group[7] has recently suggested that latch-bridge formation could be explained solely on the basis of myosin phosphorylation so that it is not necessary to propose a second mechanism. However, this does not mean that a secondary mechanism does not exist.

CALDESMON

Caldesmon was first described in 1981 by Kakiuchi and coworkers[8] as a major protein component of smooth muscle which interacts with calmodulin in a Ca^{2+}-dependent manner and with F-actin in a Ca^{2+}-independent manner. It is now known that caldesmon is located on the thin filament _in situ_[9,10]. It therefore has the properties of a potential Ca^{2+}-dependent regulator of actin-myosin interaction.

Under reducing conditions caldesmon exists as a monomer of ~140 kDa and is highly asymmetric (146 nm x 3 nm)[11]. Limited proteolysis studies indicate that caldesmon is anchored to the actin filament near its carboxy-terminus: the isolated protein can be cleaved by chymotrypsin into a 115 kDa polypeptide which does not bind to actin and a 24 kDa polypeptide which retains the actin-binding ability[12-14]. The calmodulin-binding site is also retained in this C-terminal peptide. A comparison of the peptide patterns observed when free caldesmon and actin-bound caldesmon are treated with chymotrypsin indicates that identical peptides are generated in each case, although the rate of digestion is moderately reduced when the caldesmon is bound to actin[14]. Such results would be expected if caldesmon were anchored to actin at one end of the molecule, the bulk of the polypeptide chain extending away from the axis of the thin filament, rather than being bound to actin along its considerable length in a manner analogous to tropomyosin.

FUNCTIONAL PROPERTIES OF CALDESMON

Sobue et al[15] demonstrated that isolated caldesmon inhibits superprecipitation of desensitized smooth muscle actomyosin. This was the first demonstration that caldesmon could regulate actin-myosin interaction and, furthermore, they found that this effect could be reversed by Ca^{2+}-calmodulin. This led to the proposal of the "flip-flop" theory whereby caldesmon in the resting cell at low [Ca^{2+}] would be bound to actin, preventing actin-myosin interaction. Stimulation of the cell leading to an elevation of cytosolic free [Ca^{2+}] and formation of the Ca^{2+}-calmodulin complex would result in dissociation of caldesmon from actin as the Ca^{2+}-calmodulin-caldesmon complex is formed. Consequently, actin-myosin interaction could occur leading to force development. While such a mechanism was supported by in vitro experiments using purified actin, caldesmon and calmodulin, it does not appear to operate in the intact system for the following reasons: (i) maximal force can be developed in a skinned (demembranated) smooth muscle fibre system in the absence of Ca^{2+} upon the addition of a Ca^{2+}/calmodulin-independent MLCK generated by partial proteolysis of the native enzyme[16]; (ii) the ratio of Ca^{2+}-calmodulin to actin required to dissociate caldesmon from the thin filament is much higher than the known physiological ratio (~78 mol actin:1 mol caldesmon:3 mol calmodulin)[17]; (iii) native thin filaments isolated in the presence and absence of Ca^{2+} both retain the full complement of caldesmon[10]; and (iv) treatment of native thin filament preparations with a physiologically relevant concentration of Ca^{2+}-calmodulin results in no dissociation of caldesmon (M.S. Lim and M.P. Walsh, unpublished observation). These observations therefore suggest that caldesmon remains attached to actin in the intact smooth muscle cell and some alternative mechanism of Ca^{2+} regulation of actin-myosin interaction by caldesmon must be envisaged.

We observed that when purified caldesmon was added to a reconstituted system composed of purified smooth muscle actin, myosin, tropomyosin, calmodulin and MLCK, substantial inhibition of the actin-activated myosin MgATPase occurred without any reduction in myosin phosphorylation[9]. We also observed that caldesmon can be phosphorylated by a Ca^{2+}/calmodulin-dependent protein kinase[9,18], to a maximal level of 4 mol P_i/mol caldesmon, and that the phosphorylated protein, unlike unphosphorylated caldesmon, does not inhibit the actomyosin ATPase[19]. The presence of caldesmon phosphatase activity in smooth muscle was also demonstrated indicating the presence of all the enzymatic machinery necessary for the Ca^{2+}-dependent and reversible phosphorylation of caldesmon[18]. Recent evidence[20] suggests that the Ca^{2+}/calmodulin-dependent kinase activity may be a property of caldesmon itself, although it shows some resemblance, e.g., substrate specificity, to calmodulin-dependent kinase II.

CALDESMON-MYOSIN INTERACTION

We reported that caldesmon can inhibit the MgATPase activity of purified skeletal muscle myosin in the absence of actin suggesting a direct interaction between caldesmon and myosin[21]. Interaction between caldesmon and smooth muscle myosin was demonstrated directly by Ikebe and Riordan[22] by affinity chromatography. We have confirmed this observation and shown that phosphorylated caldesmon does not interact with smooth muscle myosin (Fig. 1). Caldesmon, either unphosphorylated (in the presence of EGTA and calmodulin) or phosphorylated (in the presence of Ca^{2+} and calmodulin), was applied to an affinity column of myosin-Sepharose in the presence of EGTA at approximately physiological ionic strength. Bound protein was eluted with 0.4M NaCl. In the case of unphosphorylated caldesmon, the protein was bound to the column and eluted at high ionic strength. On the other hand, phosphorylated caldesmon did not interact with immobilized myosin and was recovered in the flow-through fractions. Identical results were obtained using phosphorylated myosin coupled to Sepharose.

LOCALIZATION OF THE MYOSIN-BINDING SITE IN CALDESMON

Advantage was taken of the fact that caldesmon contains only 2 cysteine residues[11] in order to localize the myosin-binding site. Specific cleavage at cysteine residues with 2-nitro-5-thiocyanobenzoic acid (NTCB) generated three limit peptides of M_r 85, 23.5 and 21.2 kDa (Fig. 2, lane 1). From analysis of the time-course of NTCB digestion of caldesmon, initial cleavage occurred preferentially at one of the cysteine sites to generate 108 and 21.2 kDa fragments; the 108 kDa peptide was subsequently cleaved at the second cysteine residue to generate the 85 and 23.5 kDa peptides. When the NTCB digest of caldesmon was subjected to myosin-Sepharose affinity chromatography, only the 23.5 kDa peptide was bound to the column

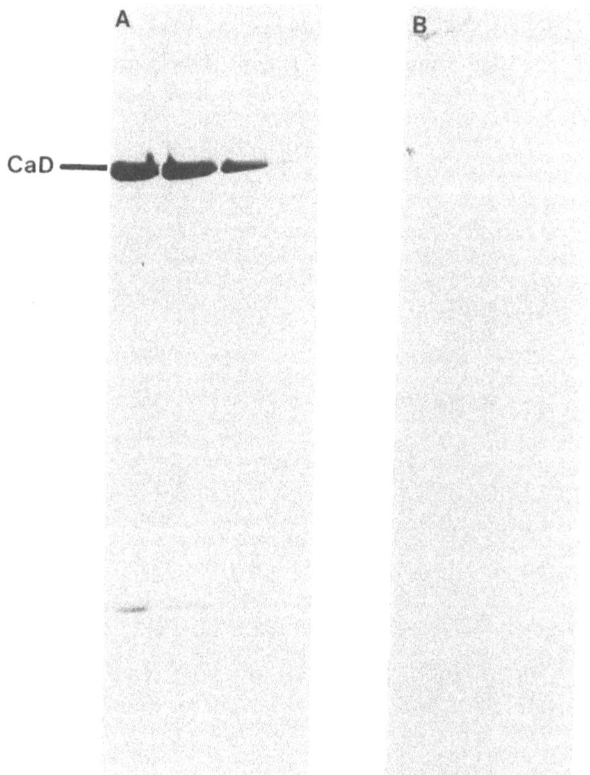

Fig. 1. Effect of phosphorylation of caldesmon on its interaction with myosin. Caldesmon was phosphorylated in the presence of Ca^{2+} and calmodulin; a control in which EGTA replaced Ca^{2+} so that no phosphorylation occurred was run in parallel. Each sample was applied separately to a column of myosin-Sepharose at approximately physiological ionic strength. After washing off unbound proteins, elution of electrostatically bound proteins was achieved with 0.4M NaCl. A shows the SDS-PAGE pattern of unphosphorylated caldesmon, B of phosphorylated caldesmon. Only the fractions of proteins eluted by 0.4M NaCl are shown. In the case of phosphorylated caldesmon, all the protein was recovered in the low ionic strength flow-through fractions.

and could be eluted with 0.4M NaCl (Fig. 2, lanes 3 and 4). This peptide, therefore, contains the myosin-binding site.

The order of the three NTCB peptides was established by subjecting well-defined chymotryptic fragments of caldesmon to digestion by NTCB. When caldesmon is treated with chymotrypsin, an initial major cleavage generates 95 and 42 kDa peptides[12-14]. The 95 kDa peptide is the N-terminal

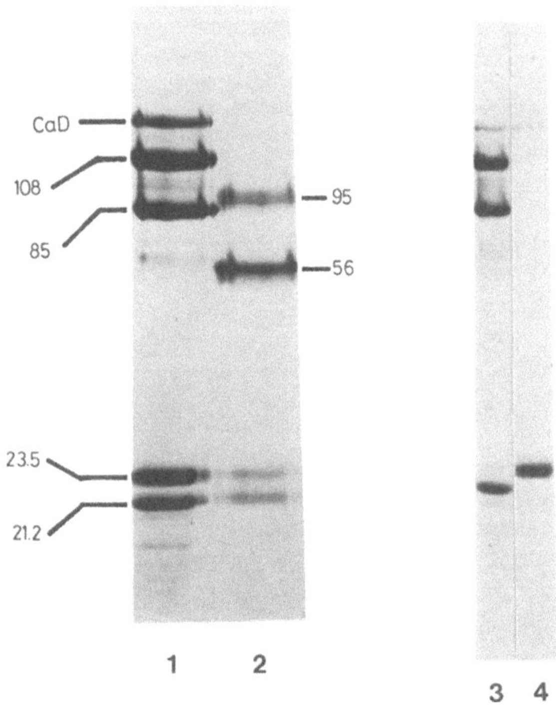

Fig. 2. NTCB digestion of caldesmon (SDS-PAGE analysis). Key to lanes: 1 = partial NTCB digest of caldesmon; 2 = NTCB digest of 95 kDa chymotryptic peptide of caldesmon; 3 = NTCB peptides of caldesmon not bound to myosin-Sepharose; 4 = NTCB peptide of caldesmon bound to myosin-Sepharose.

peptide and the 42 kDa peptide represents the C-terminus which retains the actin- and calmodulin-binding sites. We prepared and isolated these two chymotryptic peptides by DEAE-Sephacel ion-exchange chromatography and subjected them to digestion by NTCB. The 42 kDa C-terminal chymotryptic peptide was degraded by NTCB to 56, 23.5 and 21.2 kDa peptides. We already know from the kinetics of NTCB digestion of intact caldesmon that the 21.2

kDa peptide is a terminal peptide (see above) so that these observations place it at the N-terminus. The 23.5 kDa peptide must be adjacent to this N-terminal peptide leaving the 85 kDa NTCB peptide at the C-terminus.

Analysis of phosphorylated caldesmon by NTCB digestion, SDS-PAGE and autoradiography indicated that three of the four phosphorylation sites are located in the myosin-binding 23.5 kDa NTCB peptide and the fourth site is located in the N-terminal 21.2 kDa NTCB peptide. No $[^{32}P]P_i$ was recovered in the C-terminal 85 kDa NTCB peptide.

We can, therefore, deduce the following organization within the caldesmon molecule:

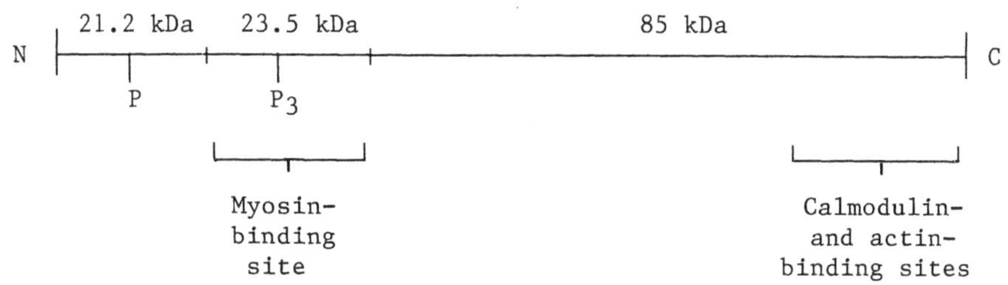

CONCLUSION

Caldesmon appears to be anchored to the actin filament at its C-terminal end, the rest of the very elongated molecule extending away from the thin filament. Caldesmon can also interact with myosin and the site of interaction is located much closer to the N-terminus of caldesmon. It appears that caldesmon is bound to actin under all physiological conditions but the interaction of caldesmon with myosin is blocked by Ca^{2+}/calmodulin-dependent phosphorylation of caldesmon. Consistent with this observation, the four phosphorylation sites are located at or near the myosin-binding site. From a number of in vitro binding studies using purified actin, myosin and caldesmon we can identify four possible states of complex formation between these proteins (Fig. 3). Studies of the Ca^{2+}-dependence of caldesmon phosphorylation[14] indicate that the maximal rate of phosphorylation occurs at low $[Ca^{2+}]$, ~150 nM, and the rate of phosphorylation falls dramatically at higher $[Ca^{2+}]$. It is possible, therefore, that caldesmon is phosphorylated in the resting smooth muscle cell. In this condition, it will not interact with myosin but will remain attached to the thin filament.

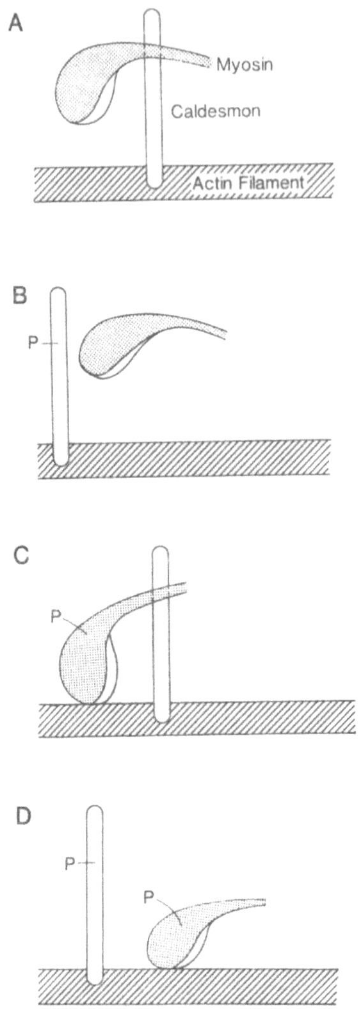

Fig. 3. Possible states of actin-myosin interaction in smooth muscle in the presence of phosphorylated and unphosphorylated caldesmon. The arrangement of the NTCB fragments of caldesmon is shown to the right of A. Phosphorylation is indicated by "P" with no consideration of stoichiometry.

This is represented by Fig. 3B. In an actively contracting muscle, myosin will be phosphorylated and, so long as caldesmon remains phosphorylated, cross-bridge cycling and force development will continue (Fig. 3D). When caldesmon is dephosphorylated it can bind to myosin (specifically the subfragment 2 region[22]) and cross-link actin and myosin. This could occur whether or not myosin is phosphorylated. Such arrangements, which are represented in Fig. 3A and C, may correspond to two alternative states of a latch-bridge. It will be very important to quantify the levels of caldesmon phosphorylation at rest, during force development and during latch in intact and skinned fiber preparations in order to evaluate the physiological significance of these observations.

ACKNOWLEDGEMENTS

This work was supported by a grant from the Medical Research Council of Canada. We are very grateful to Marilyn Devlin for word-processing.

REFERENCES

1. D. J. Hartshorne, Biochemistry of the contractile process in smooth muscle, in: "Physiology of the Gastrointestinal Tract", L. R. Johnson, ed., pp. 423-482, Raven Press, New York (1987).
2. T. T. DeFeo and K. G. Morgan, Calcium-force relationships as detected with aequorin in two different vascular smooth muscles of the ferret, J. Physiol. 369:269 (1985).
3. D. A. Williams and F. S. Fay, Calcium transients and resting levels in isolated smooth muscle cells as monitored with quin 2, Am. J. Physiol. 250:C779 (1986).
4. K. E. Kamm and J. T. Stull, The function of myosin and myosin light chain kinase phosphorylation in smooth muscle, Annu. Rev. Pharmacol. Toxicol. 25:593 (1985).
5. P. K. Ngai, G. C. Scott-Woo, M. S. Lim, C. Sutherland, and M. P. Walsh, Activation of smooth muscle myosin Mg^{2+}-ATPase by native thin filaments and actin/tropomyosin, J. Biol. Chem. 262:5352 (1987).
6. M. O. Aksoy, S. Mras, K. E. Kamm, and R. A. Murphy, Ca^{2+}, cAMP and changes in myosin phosphorylation during contraction of smooth muscle, Am. J. Physiol. 245:C255 (1983).
7. C-M. Hai and R. A. Murphy, Cross-bridge phosphorylation and regulation of latch state in smooth muscle, Am. J. Physiol. 254:C99 (1988).
8. K. Sobue, Y. Muramoto, M. Fujita, and S. Kakiuchi, Purification of a calmodulin-binding protein from chicken gizzard that interacts with F-actin, Proc. Natl. Acad. Sci. U.S.A. 78:5652 (1981).
9. P. K. Ngai and M. P. Walsh, Inhibition of smooth muscle actin-

activated myosin Mg^{2+}-ATPase activity by caldesmon, J. Biol. Chem. 259:13656 (1984).

10. W. Lehman, Caldesmon association with smooth muscle thin filaments isolated in the presence and absence of calcium, Biochim. Biophys. Acta. 885:88 (1986).

11. W. P. Lynch, V. M. Riseman, and A. Bretscher, Smooth muscle caldesmon is an extended flexible monomeric protein in solution that can readily undergo reversible intra- and intermolecular sulfhydryl cross-linking, J. Biol. Chem. 262:7429 (1987).

12. A. Szpacenko and R. Dabrowska, Functional domains of caldesmon, FEBS Lett. 202:182 (1986).

13. T. Fujii, M. Imai, G. C. Rosenfeld, and J. Bryan, Domain mapping of chicken gizzard caldesmon, J. Biol. Chem. 262:2757 (1987).

14. G. C. Scott-Woo and M. P. Walsh, Characterization of the autophosphorylation of chicken gizzard caldesmon, Biochem. J. 255, in press (1988).

15. K. Sobue, K. Morimoto, M. Inui, K. Kanda, and S. Kakiuchi, Control of actin-myosin interaction of gizzard smooth muscle by calmodulin- and caldesmon-linked flip-flop mechanism, Biomed. Res. 3:188 (1982).

16. M. P. Walsh, R. Bridenbaugh, D. J. Hartshorne, and W. G. L. Kerrick, Phosphorylation-dependent activated tension in skinned gizzard muscle fibers in the absence of Ca^{2+}, J. Biol. Chem. 257:5987 (1982).

17. M. P. Walsh, Caldesmon, a major actin- and calmodulin-binding protein of smooth muscle, Prog. Clin. Biol. Res. 245:119 (1987).

18. P. K. Ngai and M. P. Walsh, Properties of caldesmon isolated from chicken gizzard, Biochem. J. 230:695 (1985).

19. P. K. Ngai and M. P. Walsh, The effects of phosphorylation of smooth-muscle caldesmon, Biochem. J. 244:417 (1987).

20. G. C. Scott-Woo and M. P. Walsh, Autophosphorylation of smooth-muscle caldesmon, Biochem. J. 252:463 (1988).

21. M. S. Lim and M. P. Walsh, The effects of caldesmon on the ATPase activities of rabbit skeletal-muscle myosin, Biochem. J. 238:523 (1986).

22. M. Ikebe and S. Riordan, Binding of caldesmon to smooth muscle myosin, J. Biol. Chem. 263:3055 (1988).

CHARACTERIZATION OF THE CALMODULIN-BINDING DOMAIN OF CALCINEURIN DEDUCED FROM A COMPLEMENTARY DNA CLONE

Randall L. Kincaid and Brian M. Martin*

Section on Immunology, Laboratory of Physiologic and Pharmacologic Studies, National Institute on Alcohol Abuse and Alcoholism; * Molecular Neurogenetics Section Clinical Neurosciences Branch, National Institute of Mental Health; Alcohol, Drug Abuse, and Mental Health Administration, Bethesda, MD 20892

INTRODUCTION

An essential role for Ca^{2+}- and calmodulin (CaM)-dependent events in neuronal transduction is suggested by the high amounts of this Ca^{2+}-binding protein in the central nervous system. Further, several enzymes that it modulates (adenylate cyclase, cyclic nucleotide phosphodiesterase, protein kinase and the phosphoprotein phosphatase, calcineurin) have their highest concentrations in the brain, implicating Ca^{2+}- regulated cyclic nucleotide and phosphoprotein metabolism in intracellular signaling. Importantly, the multifunctional CaM-dependent protein kinase is known to be the major protein in post-synaptic densities[1,2], consistent with a critical function for phosphoproteins in neurotransmission.

The predominant CaM-binding protein in brain, first characterized as an inhibitor protein[3-5], has been identified as a protein phosphatase[6-8]. A number of its enzymatic and structural properties suggest that its regulation may be complex: 1) it is a heterodimer made up of a catalytic subunit (60 kDa) and an intrinsic Ca^{2+}-binding[9] subunit (18 kDa); 2) the interaction with CaM that stimulates phosphatase activity also leads to a kinetic deactivation of the enzyme that can be restored by incubation with certain divalent metals[10,11]; 3) in the absence of CaM, micromolar concentrations of Mn^{2+}, Ni^{2+}, and Co^{2+} greatly enhance basal activity[10-12]; 4) the 60 kDa subunit is a metalloprotein[10] containing tightly bound Fe^{2+} and Zn^{2+}; and 5) the protein is activated by phospholipids in a substrate-specific manner[13]. It appears that distinct isoforms of this phosphatase may be present in other tissues[14,15] and immunocytochemical studies have indicated that the protein can be localized in both cytosolic and particulate compartments[16,17].

In order to examine the structural features responsible for the unique regulatory properties of this enzyme and to investigate the molecular basis of its expression, we have undertaken studies to clone the complementary DNA for the catalytic subunit. The present paper describes the properties of the calmodulin-binding domain of this enzyme, based on analysis of a partial cDNA clone isolated from a lambda phage expression library.

METHODS

A murine brain lambda GT-11 expression library was obtained from the American Type Culture Collection (Rockville, MD), and screened for immunopositive plaques essentially as described by Young and Davis[18]. The primary antibody was a goat IgG fraction that was affinity-purified on calcineurin-Sepharose[19] and adsorbed with immobilized E.coli lysate (Promega Biotec, Madison, WI) prior to use. Detecting antibody was an alkaline phosphatase-conjugated rabbit anti-goat IgG (Kirkegaard-Perry Laboratories, Gaithersburg, MD); both antibodies were diluted into blocking solution (5% non-fat dry milk in Tris buffered saline plus 10 mM sodium azide and 0.01% thimerosal). After washing with buffer containing 0.05% Tween-20, immunopositive signals were detected using the chromogens nitroblue tetrazolium (NBT) and bromochloroindolyl phosphate (BCIP) and phage were then plaque-purified.

To assess recombinant fusion proteins, positive phage were introduced into the lysogenic host BNN-103 as follows: an overnight culture of bacteria grown in NZCYM medium lacking ampicillin, was diluted 200-fold (10^7 cells/ ml) and 0.2 ml was infected with 0.1 ml of amplified phage stock (30 min, 30° C). After diluting 5000-fold, 0.1 ml was spread onto fresh plates and grown at 30° C. Individual colonies were transferred to duplicate plates at 30 and 43° C and samples that did not grow at 43° C (lysogens) were selected. For analytical preparations, 5 ml of NZCYM medium were inoculated with 50 µl of an overnight culture and grown at 30° C to an OD_{660} of 0.5 (2-2.5 hr). After addition of 50 µl of 1 M IPTG, cultures were "heat-shocked" for 20 min at 43 °C and removed to 37 °C for an additional 1-2 hr. Bacterial pellets were freeze-thawed and sonicated on ice in 6 volumes of TE medium plus 10µg/ ml soybean trypsin inhibitor, and 20µg/ ml each of leupeptin and pepstatin A. For larger preparations, procedures were identical except that volumes were increased 100 times.

Recombinant phage (100,000 p.f.u.) were amplified on a 150 mm plate to give total lysis of the bacterial lawn and the plate extracted with 10ml of sterile SM containing a few drops of chloroform. The extract was treated with 10 drops of chloroform, clarified by centrifugation and 200 µl of immobilized anti-phage antibody (LambdaSorb, Promega Biotec) was added. After 30 min on ice, the adsorbent was collected by centrifugation, washed and eluted as recommended by the manufacturer. A portion of the purified phage DNA (5-6 µg) was restricted with 100 units of Eco R1, extracted with phenol-chloroform and precipitated with an equal volume of 15% polyethylene glycol in 1.5 M NaCl. Portions of this restriction digest were ligated, without further purification, into dephosphorylated pUC 19 (Bethesda Research Laboratories) with T-4 ligase and DNA transfected into competent cells. DNA sequencing was done using plasmids isolated by the alkaline lysis method[20] except that an additional digestion with RNAse followed by polyethylene glycol/ NaCl precipitation was often carried out. Double-stranded DNA was denatured with 0.2 M NaOH[21] prior to dideoxy sequencing[22] using reagent kits from either New England BioLabs (Waverly, MA) or Promega Biotec. Specific oligonucleotide primers were synthesized with a Beckman System-1 Plus DNA synthesizer. Synthetic peptides were made using a BioSearch Model 9600 peptide synthesizer, purified by high performance liquid chromatography, and the sequences confirmed by gas-phase sequencing.

Fragments of calcineurin were prepared by digesting isolated catalytic subunit with trypsin (1:100 W/W, trypsin: subunit) after solubilization in buffer containing 0.1% sodium dodecyl sulfate (SDS). These were then separated by C-18 reverse phase chromatography and sequenced using an Applied Biosystems 470A gas-phase sequenator (Foster City, CA) with a Model 120 phenylthiohydantoin amino acid analyzer. The concentration of peptide used in assays was based on data obtained from amino acid hydrolysis.

RESULTS AND DISCUSSION

To obtain partial peptide sequence data needed for oligonucleotide probes and for confirmation of deduced amino acid sequence, the 60kDa (α) and 18kDa (β) subunits of bovine brain calcineurin were first isolated using a mild reverse-phase chromatography procedure (Fig. 1). This approach allowed the preparation of pure peptides, although a peak representing unresolved αβ dimer was always seen even after prior incubation of protein with 6 M guanidine HCl (Fig. 1, panel B). Since the latter treatment is known to dissociate the subunits, it appears that their reassociation can occur during reverse-phase chromatography.

The isolated 60kDa subunit was partially digested with trypsin and fragments were separated by chromatography on a C-18 reverse-phase column (data not shown). Individuals peaks were lyophylized and analyzed by gas-phase microsequencing, yielding ten unique peptide sequences. Analysis of their codon degeneracy suggested that they would be not be suitable for design of high specificity oligonucleotide probes, however seven of these peptides were later found in the deduced amino acid sequence from a cDNA cloned by expression vector immunoscreening (see below).

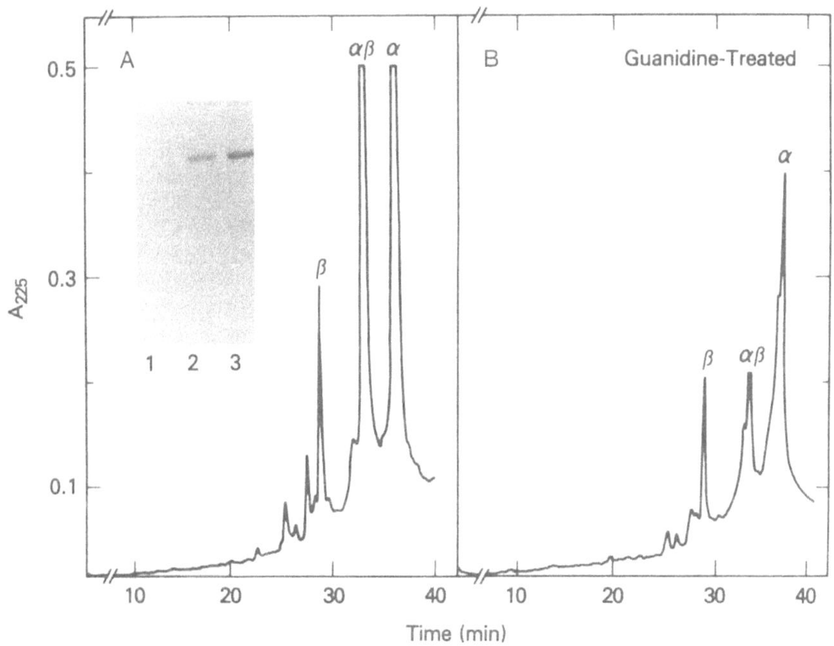

Fig.1 Separation of subunits of calcineurin by reverse-phase high performance liquid chromatography Homogeneous calcineurin (2 mg/ml) purified as described[23] was diluted with 10 vol of 30% acetonitrile in aqueous 0.1% trifluoroacetic acid and 50 μl injected onto a C-4 reverse-phase column (Vydac, 25 x 0.46 cm). Peptides were eluted (1 ml/min) with a gradient of 30-60% acetonitrile. Panel A- Elution profile showing separation of subunits; photographic inset shows an SDS gel of β- lane 1, αβ- lane 2 and α- lane 3 peaks. Panel B- Profile after prior teatment of enzyme with 6M guanidine HCl; the small shoulders on the αβ- and α- peaks resulted from apparent proteolysis of the α subunit based on analysis of SDS gels (data not included).

To facilitate the cloning of calcineurin, expression vector screening of a lambda GT-11 murine brain library was carried out with affinity-purified antibody against the bovine brain enzyme. To reduce the background in such experiments, it was essential to adsorb the antibody with E.coli lysate prior to incubation with plaque-containing filters. Several immunopositive clones were isolated and the phage introduced into the lysogen BNN-103 for controlled expression of the ß-galactosidase fusion protein. After preparation of extracts from induced lysogens followed by Western blotting, one clone showed a peptide exhibiting immunoreactivity toward both calcineurin and ß-galactosidase as well as showing Ca^{2+}-dependent interaction with a biotinylated derivative of CaM[24]. Large cultures of the lysogen were subsequently prepared in an attempt to purify the fusion protein. Affinity chromatography employing immobilized monoclonal antibody against ß-galactosidase (ProtoSorb, Promega Biotec) was unsuccessful under a variety of loading conditions, suggesting that the native conformation of the fusion protein prevented recognition of the specific epitope. However, when Ca^{2+}-dependent chromatography on CaM-Sepharose was carried out, approximately 5.5 mg of protein was eluted with EGTA with a peptide of M_r 155,000 being the predominant species.

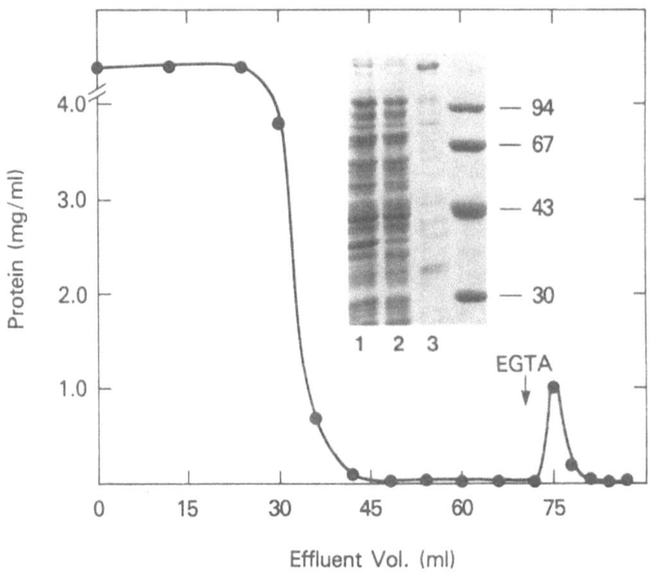

Fig. 2 <u>Calmodulin-Sepharose chromatogaphy of extract from lysogen BNN-CNα-1</u> A 500 ml culture was induced with IPTG, harvested and extracted as described under Methods. The 20,000 x g supernatant (25 ml, 15 mg/ml) was adjusted to 2 mM Ca^{2+} and 150mM NaCl and incubated with 5 ml calmodulin-Sepharose. The suspension was poured into a column, and washed with Ca^{2+}-containing buffer until the effluent protein reached baseline. After further washing with 30 ml buffer, elution buffer containing 5 mM EGTA and 20% glycerol[25] was added (arrow). Photographic inset: SDS gel of starting material, 15 µg (lane 1); fraction not retained, 15 µg (lane 2); and the EGTA eluate, 1 µg (lane 3). Molecular weight standards (in kDa) are shown on the right.

This peptide corresponded to a Coomassie-stained band in the starting material that was reduced substantially in the unretained fraction (Fig. 2, lanes 1, 2). Indeed, when the "pass-through" fraction was chromatographed a second time, virtually all of this band was removed and a second eluate containing about 2 mg protein was obtained. Thus, 5-7 mg of the 155 kDa-enriched protein fraction was recovered starting with 500 ml of bacterial culture. The CaM-Sepharose eluate was electroblotted and shown to be immunoreactive with calcineurin and ß-galactosidase antibodies, consistent with its being the phosphatase/ galactosidase fusion protein. Furthermore, when partially digested with Staphylococcal V-8 protease and blotted, the fusion protein showed several immunoreactive peptides that were identical to fragments produced from the bovine brain enzyme[26]. Such "immunofingerprinting" evidence strongly supported the notion that the fusion protein contained a bona fide fragment of calcineurin.

To verify that the cDNA encoding the fusion protein corresponded to that of calcineurin, the Eco R1 restriction fragment from BNN-CNα-1 was subcloned into the plasmid pUC 19 and sequenced using conventional dideoxy termination procedures with specific oligonucleotide primers. This 1111 base pair (bp) insert contained an open reading frame coding for a polypeptide of 307 amino acids (≈34 kDa) followed by a 3' untranslated region of 190 bp (Fig. 3). Seven tryptic peptides (Table 1) were located in the deduced amino acid sequence, spanning nearly the entire coding region, providing unambiguous proof of the identity of the cloned cDNA.

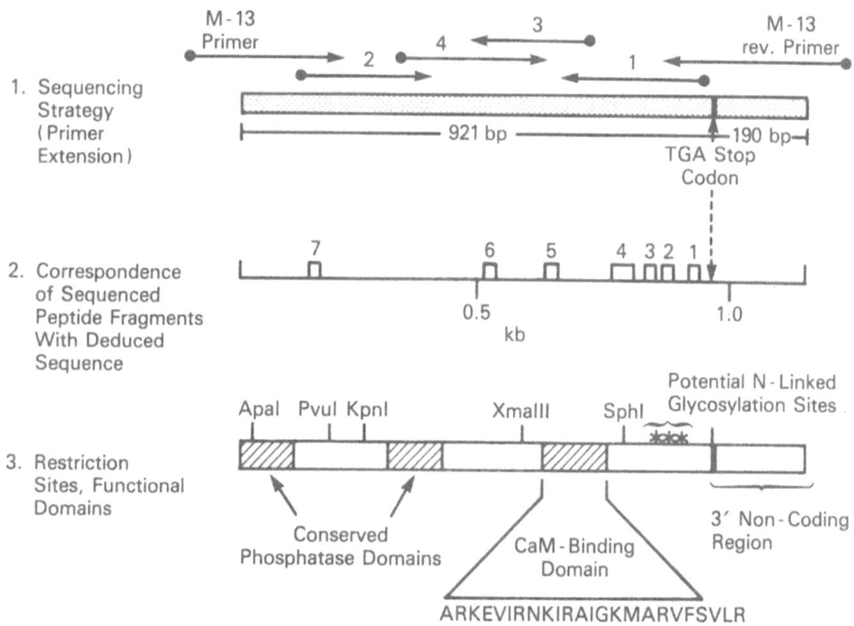

Fig. 3 Analysis of a partial cDNA sequence (CNα-1) for mouse brain calcineurin The figure shows a schematic diagram of (1) the sequencing of the Eco R1 insert using dideoxy procedures; the positions of arrows correspond to the regions sequenced with specific oligonucleotide primers (2) the positions of sequenced tryptic peptides in the deduced amino acid sequence (see Table 1) and (3) the positions of unique restriction enzyme sites, regions corresponding to conserved phosphatase sequences and the putative calmodulin-binding domain.

Table 1. Partial sequences of bovine brain calcineurin (α-subunit) obtained by gas-phase microsequencing of tryptic digests

Peptide #	Amino acid sequence[a]	Tryptic cleavage sites[b]
1	D A M P S D A N L N S I N R	R/D; K/A
2	I T S F E E A K	K/I; K/A
3	Q T L Q S A T V E A I E A D E A I	K/Q; ---
4	G L T P T G M L P S G V L -	K/G; ---
5	V F S V L R E E S E S	R/V; ---
6	- - P N Y L D V Y N N K	---; K/A
7	T N E H F T H N P V	K/T; ---

[a]Amino acid sequences of isolated tryptic fragments were determined as described under Methods. Amino acids are represented by the standard single letter code; ambiguous residues are indicated by spaces.
[b]The tryptic cleavage sites are those deduced from the nucleic acid sequence of cloned cDNA and correspond to the NH_2-terminal and COOH-terminal boundaries of the fragment.

The cloned cDNA contained two tandem polyadenylation signals (AATAAA) approximately 30 nucleotides from the 3' end of the sequence but no stretch corresponding to the poly A tail was observed; this may suggest that the cDNA was primed off an internal adenine-rich region. Interestingly, the coding region just preceding the TGA stop codon contains three consensus sites for N-linked glycosylation (NxS or NxT); since to date there is no evidence that the enzyme is glycosylated, this may be purely coincidental. Near the 5' end of the cDNA, the coding region contains two limited regions of identity with protein phosphatases 1 and 2A[27,28], indicating that calcineurin is ancestrally related to these other enzymes. However, the exact number of amino acids separating these conserved domains is somewhat greater for calcineurin than in the other two enzymes, and appears to reflect the insertion of 13 additional amino acids in the sequence of calcineurin[26]. Thus, although the preservation of catalytic subdomains appears important, there may be subtle ways of arranging them to provide properties that are unique to a specific enzyme.

A region approximately 110 amino acids from the carboxyl terminus of the protein possessed several features that suggested it may be the CaM-binding domain. Perhaps most importantly, this was the only region having clusters of positively-charged amino acids in an otherwise hydrophobic environment; this general property has been observed for other CaM-binding domains[29-31]. However, it should be pointed out that within this putative domain there was a single glutamic acid residue as well, this being the first reported example of an anionic side chain in a binding site. In addition to the overall character of this region, a remarkable similarity was noted between the deduced amino acid sequence I R A I G K M, seen in calcineurin, and that for myosin light chain kinase V R A I G R L (see Table 2). When the nucleotide sequences over this stretch were compared, there were only 4 differences over the 21 nucleotides, or 81% identity. Such an agreement suggests that there may have been a common regulatory domain that was incorporated by both enzymes, despite their clearly opposed enzymatic roles. Also of note is the presence of a consensus site for protein phosphorylation (RxxSxxR) near the C-terminal end of the domain that could serve as a substrate for several protein kinases. In the cases of other CaM-dependent enzymes, phosphorylation of the binding domain is known to occur, exerting an effect on the CaM regulation of enzyme activity[32]; whether this will be true for calcineurin remains to be seen.

Table 2. Comparison of the CaM-binding domain of several enzymes

Enzyme	Sequence[a]
Calcineurin	A R K E V I R N K I R A I G K M A R V F S V L R
MLCK (Sm.musc)	A R R K W Q K T G H A V R A I G R L S S M A M I
CaM-kinase II	A R R K L K G A I L T T M L A T R N F S G G K G
Ca^{2+}- ATPse	L R R G Q I L W F R G K N R I Q T Q I K V V N A

[a]Amino acid sequence data for the enzymes other than calcineurin are taken from Guerriero, et. al.[29], Hanley, et. al.[30] and James, et. al.[31], respectively. Underlined sequences show potential phosphorylation sites for serine/threonine kinases.

The presence of a glutamic acid residue in the deduced sequence of the CaM-binding region prompted us to confirm this region using alternate methods for DNA sequencing. When several different clones from the murine library were sequenced on both strands using automated fluorescent dideoxy nucleotide procedures (DuPont Genesis 2000 system), all gave the identical sequence (D. Amorese, R. Kincaid, unpublished data). Thus, this assignment apparently did not reflect a simple sequencing artifact or a peculiarity of one specific clone. The possibility still exists, however, that an error in the construction of the original library (presumably in the reverse transcribing of mRNA) may have given rise to such cDNAs. However, additional evidence supporting the authenticity of the CaM-binding sequence was obtained by synthesizing the peptide and testing its effect on protein-protein interactions (Fig. 4). When purified bovine brain calcineurin was electroblotted onto nitrocellulose and probed for binding to biotinylated CaM, there was a dose-dependent inhibition of functional interaction with the immobilized α-subunit using nanomolar concentrations of the peptide.

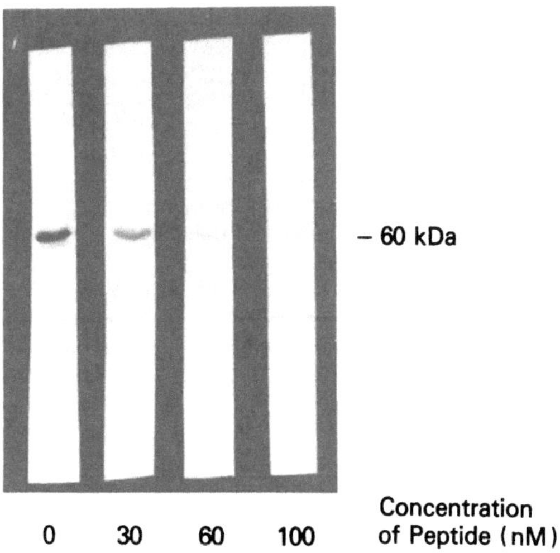

Fig. 4 Inhibition of CaM-calcineurin binding by synthetic peptide
Strips (100 ng calcineurin) were reacted with 40 nM biotinylated CaM with the indicated amounts of peptide as described[24].

The amount of synthetic CaM-binding peptide necessary for complete inhibition of binding was two to three times the amount of biotinylated CaM under the specified conditions. This may reflect a lower association constant of CaM for the peptide relative to the renatured α-subunit although this is difficult to determine since free-solution considerations are not applicable. Also, differences between species (bovine brain subunit vs. mouse brain peptide), multiple (less effective) solution conformations of the peptide, or depletion of synthetic peptide due to non-specific interactions with immobilized proteins may contribute to the lack of a strict stoichiometric inhibition. Although such experiments clearly show competition, the precise determination of binding peptide inhibitory efficacy must await enzymatic studies.

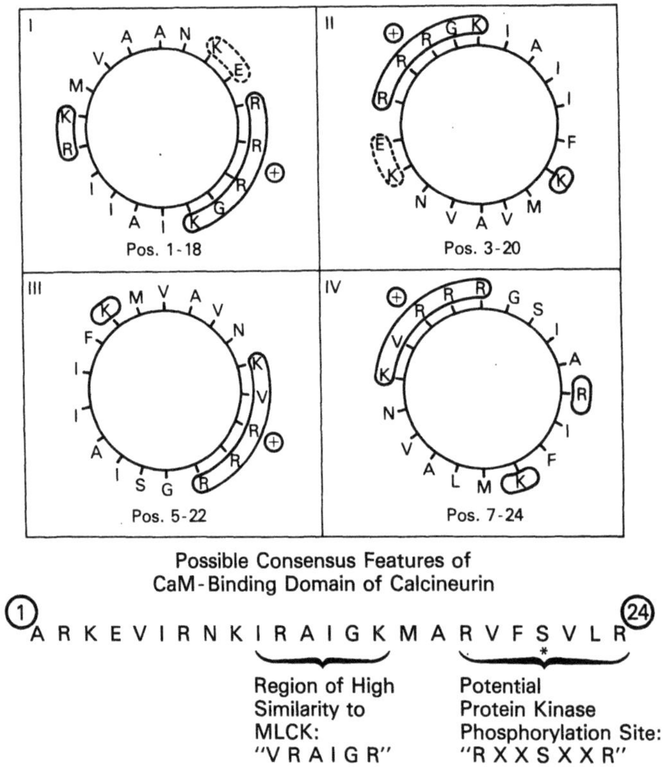

Fig. 5 Helical wheel representations of secondary structure in successive regions of the CaM-binding domain
The 24 amino acid sequence of the putative CaM-binding domain (see lower part of the figure) was divided into 18 residue segments and helical wheels constructed; positions 1-18, 3-20, 5-22 and 7-24 in panels I- IV. Regions of positively charged side chains are outlined with solid lines while adjacent anionic and cationic side chains are indicated with broken lines. Consensus features of the domain are shown at the bottom.

It has been proposed that interaction of CaM with its binding domain induces α-helix in the target region, in agreement with the predicted secondary structure in a number of model amphipathic peptides and protein domains known to interact with CaM[33]. Helical wheel representations based on helices having 3.6 residues per turn (i.e, 18 positions per wheel) showed a general agreement of the calcineurin domain with the properties suggested for CaM-binding sites, having a dominant positively-charged face apposed to a region of hydrophobic character (Fig. 5). The exact picture varied, of course, depending on the residue at which the "helix" was arbitrarily initiated and in some of these representations, the single glutamic acid was aligned next to a lysine in the configuration. When searches for helix initiation and helix breaking residues were carried out with data bases (Gene-Master, Bio-Rad) using the general model of Chou and Fasman[34], it was observed that the positions between residues 5 and 22 had a high probability of α-helix and were flanked by regions that interrupt or "break" helical structure. It therefore seems probable that this region corresponds to the consensus domain for interaction.

It appears that features of the binding domain present in other enzyme sites, such as the presence of aromatic residues in the first half of the region, may not be absolutely required. Indeed, the marked diversity in exact amino acid composition may reflect subtle requirements relating to the function of the particular enzyme in question or to other properties of enzyme microenvironment. In this regard it will be of interest to examine the effect of different synthetic calcineurin-like peptides on inhibition of interaction as well as the use of genetic engineering of the CaM-binding domain of this enzyme using site-directed mutagenesis. Since calcineurin is thought to have a dissociation constant for CaM in the low to sub-nanomolar range, it may be possible to generate interactions that are extremely high affinity wherein CaM becomes essentially an intrinsic modulator, as in the case of phosphorylase kinase[35]. Among CaM-dependent enzymes with the same function (e.g., cyclic nucleotide phosphodiesterases) the activation constants are known to vary substantially[36,37], with one isoform reportedly having CaM bound as an intrinsic subunit[38]. Whether a similar spectrum of isoenzyme forms exists in the case of calcineurin is unknown at present.

It has been suggested recently that one mechanism involved in CaM regulation involves the participation of vicinal internal- or pseudo-substrate sites that affect expression of enzyme activity[39,40]. To date, there are no appropriate models with which to assess this possibility in protein phosphatases although the presence of a potential phosphorylation site in the CaM-binding region might suggest a plausible locus for self-inhibition. It may be of interest that the regions immediately before and after the CaM-binding domain in calcineurin are extremely high in acidic amino acid residues and these may perhaps interact with exposed cationic side chains of the regulatory domain. However, it is not possible to estimate their actual proximity in the native conformation of the protein and their adjacent location in the linear sequence may be fortuitous.

One final unanswered question relates to the site of association of the catalytic (α) subunit with the intrinsic Ca^{2+}-binding (ß) subunit that appears essential for catalytic competence[41]. Thus far we have no definitive evidence demonstrating complex formation with fusion proteins, although negative results may reflect an inability to satisfy some unknown structural requirements (e.g. proper incorporation of tightly bound metal into the recombinant α-subunit). Since the detection of such a complex requires a sensitive and reliable assay, it may be that a more appropriate test is required. Ongoing studies on the expression of recombinant forms of the catalytic subunit are focusing on this central issue in hopes of providing insights into the complex regulation of phosphatase activity.

REFERENCES

1. M. B. Kennedy, M. K. Bennet, and N. E. Erondu, Biochemical and immunochemical evidence that the major postsynaptic density protein is a subunit of a calmodulin-dependent protein kinase, Proc. Nat. Acad. Sci. U.S.A. 80:7357 (1983).
2. P. T. Kelly, T. L. McGuiness, and P. Greengard, Evidence that the major postsynaptic density protein is a component of a Ca^{2+}/calmodulin-dependent protein kinase, Proc. Nat. Acad. Sci. U.S.A. 81:945 (1984).
3. J. H. Wang and R. Desai, Modulator binding protein: Bovine brain protein exhibiting the Ca^{2+}-dependent association with the protein modulator of cyclic nucleotide phosphodiesterase, J. Biol. Chem. 252:4175 (1977).
4. C. B. Klee and M. H. Krinks, Purification of the cyclic 3',5'-nucleotide phosphodiesterase inhibitory protein by affinity chromatography on activator protein coupled to Sepharose, Biochemistry 17:120 (1978).
5. R. W. Wallace, T. J. Lynch, E. A. Tallant, and W. Y. Cheung, Purification and characterization of an inhibitor protein of brain adenylate cyclase and cyclic nucleotide phosphodiesterase, J. Biol. Chem. 254:377 (1979).
6. A. A. Stewart, T. S. Ingebritsen, A. Manalan, C. B. Klee, and P. Cohen, Discovery of a Ca^{2+}- and calmodulin-dependent protein phosphatase. Probable identity with calcineurin (CaM-BP_{80}), FEBS Lett. 137:80 (1982).
7. S. -D. Yang, E. A. Tallant, and W. Y. Cheung, Calcineurin is a calmodulin-dependent protein phosphatase, Biochem. Biophys. Res. Commun. 106:1419 (1982).
8. M. M. King, C. Y. Huang, P. B. Chock, A. C. Nairn, H. C. Hemmings, Jr., K. -F. J. Chan, and P. Greengard, Mammalian brain phosphoproteins as substrates for calcineurin, J. Biol. Chem. 259:8080 (1984).
9. C. B. Klee, T. H. Crouch, and M. H. Krinks, Calcineurin: A calcium- and calmodulin-binding protein of the nervous system, Proc. Nat. Acad. Sci. U.S.A. 79:6270 (1979).
10. M. M. King and C. Y. Huang, The calmodulin-dependent activation and deactivation of the phosphoprotein phosphatase, calcineurin, and the effects of nucleotides, pyrophosphate and divalent metal ions, J. Biol. Chem. 259:8847 (1984).
11. C. J. Pallen and J. H. Wang, Regulation of calcineurin by metal ions. Mechanism of activation by Ni^{2+} and an enhanced response to Ca^{2+}/calmodulin, J. Biol. Chem. 259:6134 (1984).
12. M. M. King and C. Y. Huang, Activation of calcineurin by nickel ions, Biochem. Biophys. Res. Commun. 114:956 (1983).
13. M. Politino and M. M. King, Calcium- and calmodulin-sensitive interactions of calcineurin with phospholipids, J. Biol. Chem. 262:10109 (1987).
14. A. A. Stewart, T. S. Ingebritsen, and P. Cohen, The protein phosphatases involved in cellular regulation. 5. Purification and properties of a Ca^{2+}/calmodulin-dependent protein phosphatase (2B) from rabbit skeletal muscle, Eur. J. Biochem. 132:289 (1983).
15. R. L. Kincaid, H. Takayama, M. L. Billingsley, and M. V. Sitkovsky, Differential expression of calmodulin-binding proteins in B-, T-lymphocytes and thymocytes, Nature 330:176 (1987).
16. J. G. Wood, R. W. Wallace, J. N. Whitaker, and W. Y. Cheung, Immunocytochemical localization of calmodulin and a heat-labile calmodulin-binding protein ($CaMBP_{80}$) in basal ganglia of mouse brain, J. Cell Biol. 84:66 (1980).

17. R. L. Kincaid, C. D. Balaban, and M. L. Billingsley, Differential localization of calmodulin-dependent enzymes in brain: Evidence for selective expression of cyclic nucleotide phosphodiesterase in specific neurons, Proc. Nat. Acad. Sci. U.S.A. 84:1118 (1987).
18. R. A. Young and R. W. Davis, Efficient isolation of genes by using antibody probes, Proc. Nat. Acad. Sci. U.S.A. 80:1194 (1983).
19. R. L. Kincaid, Preparation, characterization and properties of affinity-purified antibodies to calmodulin-dependent cyclic nucleotide phosphodiesterase and the protein phosphatase calcineurin, Methods in Enzymol. (Corbin and Johnson, eds.) 159:627 (1988).
20. T. Maniatis, E. F. Fritsch, and J. Sambrook in: "Molecular cloning; a laboratory manual" Cold Spring Harbor, NY (1982).
21. E. Y. Chen and P. H. Seeburg, Supercoil sequencing: a fast and simple method for sequencing plasmid DNA, DNA 4:165 (1985).
22. F. Sanger, S. Nicklen, and A. R. Coulson, DNA sequencing with chain-terminating inhibitors, Proc. Nat. Acad. Sci. U.S.A. 74:5463 (1977).
23. R. L. Kincaid, V. C. Manganiello, C. E. Odya, J. C. Osborne, Jr., I. E. Stith-Coleman, M. A. Danello, and M. Vaughan, Purification and properties of calmodulin-stimulated phosphodiesterase from mammalian brain, J. Biol. Chem. 259:5158 (1984).
24. M. L. Billingsley, K. R. Pennypacker, C. G. Hoover, D. J. Brigati, and R. L. Kincaid, A rapid and sensitive method for detection and quantification of calcineurin and calmodulin-binding proteins using biotinylated calmodulin, Proc. Nat. Acad. Sci. U.S.A. 82:7585 (1985).
25. R. L. Kincaid and M. Vaughan, Purification and properties of calmodulin-activated cyclic nucleotide phosphodiesterase from brain, Methods in Enzymol. (Corbin and Johnson, eds.) 159:557 (1988).
26. R. L. Kincaid, M. S. Nightingale, and B. M. Martin, Characterization of a cDNA clone encoding the calmodulin-binding domain of mouse brain calcineurin, Proc. Nat. Acad. Sci. U.S.A., in press (1988).
27. N. Berndt, D. G. Campbell, F. B. Caudwell, P. Cohen, E. F. da Cruz e Silva, O. B. da Cruz e Silva, and P. T. W. Cohen, Isolation and sequence analysis of a cDNA clone encoding a type-1 protein phosphatase catalytic subunit: homology with protein phosphatase 2A, FEBS Lett. 223:340 (1987).
28. O. B. Da Cruz e Silva, S. Alemany, D. G. Campbell, and P. T. W. Cohen, Isolation and sequence analysis of a cDNA encoding the entire catalytic subunit of a type-2A protein phosphatase, FEBS Lett. 221:415 (1987).
29. V. Guerriero, Jr., M. A. Russo, N. J. Olson, J. A. Putkey, and A. R. Means, Domain organization of chicken gizzard myosin light chain kinase deduced from a cloned cDNA, Biochemistry 25:8372 (1986).
30. R. M. Hanley, A. R. Means, T. Ono, B. E. Kemp, K. E. Burgin, N. Waxham, and P. T. Kelly, Functional analysis of a complementary DNA for the 50 kDa subunit of calmodulin kinase II, Science 237:293 (1987).
31. P. James, M. Maeda, R. Fischer, A. K. Verma, J. Krebs, J. T. Penniston, and E. Carafoli, Identification and primary structure of a calmodulin binding domain of the Ca^{2+} pump of human erythrocytes, J. Biol. Chem. 263:2905 (1988).
32. M. A. Conti and R. S. Adelstein, The relationship between calmodulin binding and phosphorylation of smooth muscle myosin kinase by the catalytic subunit of 3':5' cAMP-dependent protein kinase, J. Biol. Chem. 256:3178 (1981).
33. J. A. Cox, M. Comte, J. E. Fitton, and F. Degrado, The interaction of calmodulin with amphiphilic peptides, J. Biol. Chem. 260:2527 (1985).
34. P. Y. Chou and G. D. Fasman, Prediction of the secondary structure of proteins from their amino acid sequence, Adv. in Enzymol. 47:45 (1978).

35. P. Cohen, A. Burchell, J. G. Foulkes, P. T. W. Cohen, T. C. Vanaman, and A. L. Nairn, Identification of the Ca^{2+}-dependent modulator protein as the fourth subunit of rabbit skeletal muscle phosphorylase kinase, FEBS Lett. 92:287 (1978).
36. D. C. LaPorte, W. A. Toscano, Jr., and D. A. Storm, Cross-linking of iodine-125-labelled, calcium-dependent regulatory protein to the Ca^{2+}-sensitive phosphodiesterase purified from bovine heart, Biochemistry, 18: 2820 (1979).
37. R. L. Kincaid and M. Vaughan, Molecular and regulatory properties of calmodulin-dependent phosphodiesterase from brain, in: "Calcium and Cell Function", vol.VI, W. Y. Cheung, ed. Academic Press, Orlando (1986).
38. R. K. Sharma and J. H. Wang, Purification and characterization of bovine lung calmodulin-dependent cyclic nucleotide phosphodiesterase. An enzyme containing calmodulin as a subunit, J. Biol. Chem. 261: 14160 (1986).
39. A. P. Kwiatkowski, D. J. Shell, M. M. King, The role of autophosphorylation in activation of the type II calmodulin-dependent protein kinase, J. Biol. Chem. 263:6484 (1988).
40. B. Kemp. R. Pearson, V. Guerriero, Jr., I. Bagchi, and A. R. Means, The calmodulin binding domain of chicken smooth muscle myosin light chain kinase contains a pseudosubstrate sequence, J. Biol. Chem. 262:2542 (1987).
41. D. L. Merat, Z. Y. Hu, T. E. Carter, and W. Y. Cheung, Bovine brain calmodulin-dependent protein phosphatase: regulation of subunit A activity by calmodulin and subunit B, J. Biol. Chem. 260:11053 (1985).

THE ROLE OF BRAIN PROTEIN PHOSPHATASES 1 AND 2A IN THE REGULATION OF MICROTUBULE ASSEMBLY

Hideyuki Yamamoto, Yoshiki Saitoh, Kohji Fukunaga, and Eishichi Miyamoto

Department of Pharmacology, Kumamoto University Medical School, Kumamoto, Japan

Protein phosphatase C was purified from bovine brain and was considered to consist of 10 and 90 %, respectively, of the catalytic subunits of protein phosphatases 1 and 2A on the basis of the effects of ATP and inhibitor-2. We investigated the dephosphorylation of microtubule proteins and the autophosphorylated Ca^{2+}/calmodulin-dependent protein kinase II (CaM kinase II) by protein phosphatase C.

INTRODUCTION

There is now a great deal of evidence that protein phosphorylation is an important mechanism by which second messengers act to regulate a variety of cellular processes (Nairn et al., 1985). CaM kinase II is a major class of protein kinase in the CNS as well as cyclic AMP-dependent protein kinase (A-kinase), protein kinase C and cyclic GMP-dependent protein kinase (Nairn et al., 1985; Miyamoto 1986). In previous work, we reported that CaM kinase II purified from brain had the ability to phosphorylate microtubule-associated protein 2 (MAP2), tau factor, and tubulin and that the phosphorylation of each protein resulted in the inhibition of microtubule assembly (Yamamoto et al., 1985).

CaM kinase II was autophosphorylated in a Ca^{2+}/calmodulin-dependent manner (Fukunaga et al, 1982; Bennett et al., 1983; Goldenring et al., 1983). The effects of autophosphorylation on enzyme activity are still controversial.

In this communication, we purified protein phosphatase C from bovine brain and examined the dephosphorylation of microtubule proteins and its effect on microtubule assembly. Furthermore, we investigated the dephosphorylation of CaM kinase II by protein phosphatase C. The autophosphorylation and dephosphorylation resulted in the inactivation and reactivation of the enzyme, respectively.

MATERIALS AND METHODS

Preparation of proteins CaM kinase II was purified from rat brain (Fukunaga et al., 1982). The specific activity of the enzyme was 320-760 nmol/mg/min with chicken gizzard myosin light chain as substrate. The catalytic subunit of A-kinase was purified by the method previously described by Miyamoto et al. (1973). The specific activity of the enzyme was 190-220 nmol/mg/min when whole histone was used as substrate. MAP2, tau factor and tubulin were purified from porcine brain (Yamamoto et al., 1985). Inhibitor-2 was purified from skeletal muscle (Yang et al., 1981). Protein phosphatase 2A was partially purified from porcine

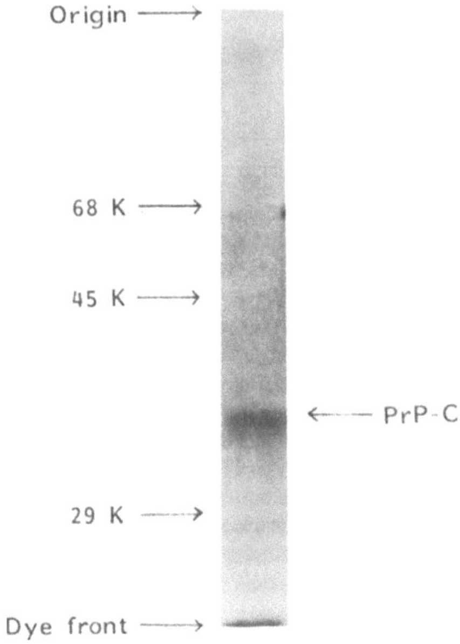

Fig. 1. Protein staining of protein phosphatase C with Coomassie Blue. Purified protein phosphatase C (0.5 µg) was subjected to SDS-PAGE in 13.5 % acrylamide. Standard proteins used include bovine serum albumin (68 K), ovalbumin (45 K), and carbonic anhydrase (29 K). PrP-C, protein phosphatase C.

brain by the method of Tung et al. (1985) at the final step of hydrophobic chromatography on aminohexyl-Sepharose 4B. The specific activity of the enzyme was 220 nmol/mg/min when estimated with histone H1 phosphorylated by the catalytic subunit of A-kinase as substrate. Phosphoprotein substrates such as MAP2, tau factor, and tubulin were prepared by the method previously described (Yamamoto et al., 1988).

Assay for dephosphorylation of protein The standard assay system for protein phosphatase activity was described elsewhere (Goto et al., 1985; Saitoh et al., 1986, 1987; Yamamoto et al., 1988). The procedures for collection of TCA-insoluble protein and for count of the precipitate were as described previously (Miyamoto et al., 1973, 1978).

Assay for microtubule assembly The determination of microtubule assembly was carried out in the course of time by the turbidity measurement at 350 nm in a thermostatically regulated Hitachi 124 spectrophotometer (37°C) with a chart recorder, according to the method of Gaskin et al. (1974).

Purification of protein phosphatase C from bovine brain Protein phosphatase C was purified from bovine brain by a modification of the method originally used for rabbit skeletal muscle (Resink et al., 1983). The specific activity of the enzyme was 720-1,100 nmol Pi/mg/min with histone H1 as substrate under standard conditions.

Other procedures Molecular weight standards include chymotrypsinogen A (25,000), carbonic unhydrase (29,000), ovalbumin (45,000), and bovine serum albumin (68,000). Autoradiography of the gel was performed as described previously (Miyamoto et al., 1981). [γ-^{32}P] ATP was prepared by the method of Post and Sen (1967). Protein was determined by the method of Bradford (1976) with bovine serum albumin as standard.

RESULTS

Purification of protein phosphatase C from bovine brain We purified 140-fold protein phosphatase C from bovine brain with 8 % yield

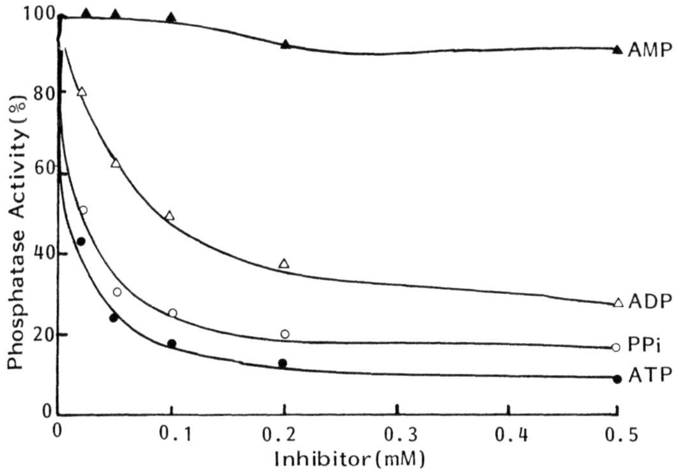

Fig. 2. Effects of various compounds on the activity of protein phosphatase C. Protein phosphatase C (1.9 μg) was preincubated at 30°C for 10 min with ATP (○), ADP (△), AMP (▲), and pyrophosphate (●). Protein phosphatase activity was measured with 22 μg of tubulin phosphorylated by CaM kinase II under standard assay conditions. The activity of protein phosphatase C determined in the absence of any test compound was taken as 100 %, and from this value, each activity was calculated.

with histone H1 as substrate. The specific activity of the enzyme was 720 nmol/mg/min. Analysis of the purified protein phosphatase C by SDS-PAGE demonstrated the presence of a dominent polypeptide of Mr 36,000 (Fig. 1). The enzyme was eluted as a single peak of activity at a position of Mr 35,000 on Sephacryl S-200 column.

Effects of phosphate analogues and NaF on the activity of the purified protein phosphatase C Protein phosphatase C dephosphorylated MAP2, tau factor, and tubulin phosphorylated by CaM kinase II and the catalytic subunit of A-kinase. Protein phosphatase C also dephosphorylated CaM kinase II. The activity of protein phosphatase C toward tubulin phosphorylated by CaM kinase II was inhibited strongly by ATP, ADP, and PPi (Fig. 2). The concentrations of ATP, ADP, PPi and NaF required to inhibit 50 % of the activity were 16, 90, 19 μM, and 6 mM, respectively. Similar results were obtained using CaM kinase II and histone H1 phosphorylated by the catalytic subunit of A-kinase. These inhibitory effects were also reported using the catalytic subunit of protein phosphatase 2A from turkey gizzard (Pato and Adelstein, 1983), rabbit liver (Ingebritsen et al., 1980), and bovine cardiac muscle (Hsiao et al., 1978; Noiman and Chock, 1983). These results suggest that brain protein phosphatase C mainly contains the catalytic subunit of protein phosphatase 2A.

Effect of inhibitor-2 The effect of inhibitor-2 on brain protein phosphatase C activity was investigated using histone H1, tubulin and CaM kinase II as substrates with 1 μg of protein phosphatase C. The maximal inhibition was only 10 % with any of substrates in the presence of about 500 nM inhibitor-2.

Dephosphorylation of MAP2, tau factor, and tubulin by protein phosphatase C   Dephosphorylation of microtubule proteins by protein phosphatase C reached a plateau in 30-min incubation. Table 1 shows the dephosphorylation of microtubule proteins phosphorylated by the two kinases. Protein phosphatase C could dephosphorylate MAP2, tau factor, and tubulin phosphorylated by both protein kinases, at the same level. Tau factor phosphorylated by CaM kinase II was more dephosphorylated than that phosphorylated by the catalytic subunit of A-kinase. The dephosphorylation was confirmed by autoradiography of the gel obtained on SDS-PAGE (data not shown).

Comparison of kinetic properties of protein phosphatase C for dephosphorylation of microtubule proteins   We determined the Vmax and Km values of protein phosphatase C for each microtubule protein (Table 2). The Vmax value for tau factor was greater than those for MAP2 and tubulin. The Km values for MAP2 and tau factor were similar and 12 times lower than that for tubulin. Prominent difference was not observed for the Vmax and Km values for each microtubule protein between the substrates by CaM kinase II and the catalytic subunit of A-kinase.

Dephosphorylation of microtubule proteins by the holoenzyme of protein phosphatase 2A   From the investigation of the effects of ATP and inhibitor-2, dephosphorylation of microtubule proteins by protein phosphatase C was found to be cunducted mainly by the catalytic subunit of protein phosphatase 2A. It was reported that the substrate specificity of protein phosphatase 2A was different between the high-molecular weight form (holoenzyme) and the catalytic subunit (Pato and Adelstein, 1983; Imaoka et al., 1983). Therefore, we investigated whether or not the holoenzyme of brain protein phosphatase 2A could dephosphorylate microtubule proteins (Table 3). The holoenzyme of brain protein phosphatase 2A dephosphorylated microtubule proteins phosphorylated by the two kinases (Table 3).

Effect of dephosphorylation of MAP2, tau factor, and tubulin on microtubule assembly   We reported that the phosphorylation of MAP2, tau factor, and tubulin by CaM kinase II separately induced the inhibition of microtubule assembly (Yamamoto et al., 1985). Therefore, we investigated whether or not the dephosphorylation of MAP2, tau factor, and tubulin by protein phosphatase C decreased the inhibitory effect by CaM kinase II. Table 4 shows that the phosphorylation of MAP2 and tau factor by CaM kinase II induced the inhibition of microtubule assembly, and the dephosphorylation by protein phosphatase C removed its inhibitory effect. The effects of phosphorylation and dephosphorylation of MAP2 and tau factor are more prominent on the initial rate than on the final extent of microtubule assembly (Table 4). These results suggest that the nucleation reaction is more dependent on phosphorylation than the elongation reaction. When tubulin was phosphorylated (0.8 mol/mol tubulin) by CaM kinase II, the initial rate and the final extent were 70 and 55 % of the control, respectively, using tau factor as an inducer for microtubule assembly. When tubulin was dephosphorylated (0.5 mol/mol tubulin) by protein phosphatase C, its inhibitory effect was not removed.

Dephosphorylation of CaM kinase II by protein phosphatase C   Fig. 3 shows the time course of autophosphorylation and dephosphorylation of CaM kinase II. Both 49K and 60K subunits of CaM kinase II were autophosphorylated with changes in apparent molecular weights to 53K and 64K on SDS-PAGE. Protein phosphatase C dephosphorylated CaM kinase II with concomitant restoration of original molecular weights on SDS-PAGE (Fig. 3). The amounts of phosphate incorporated were calculated to be 6.5 and 15.7 mol/mol for the 49K and 60K subunits, respectively, on 5 min-incubation. The phosphate amounts were decreased to 3.9 and 6.2

Table 1. Dephosphorylation by protein phosphatase C of microtubule proteins phosphorylated by CaM kinase II and the catalytic subunit of A kinase.

	CaM kinase II		Catalytic subunit	
Substrate	Phospho-rylation (mol/mol)	Dephospho-rylation (%)	Phospho-rylation (mol/mol)	Dephospho-rylation (%)
MAP2	15.2	50.8	12.1	51.8
Tau factor	8.2	83.9	8.5	64.4
Tubulin	0.6	42.0	1.1	47.0

MAP2, tau factor, and tubulin were phosphorylated by CaM kinase II or by the catalytic subunit of A-kinase, as indicated. Since CaM kinase II phosphorylates each microtubule protein more than the catalytic of A-kinase, we adjusted the amount of phosphate incorporated into each protein to a similar range. MAP2 (7.4 μg) or tau factor (7.5 μg) was dephosphorylated for 60 min by protein phosphatase C (0.9 μg) under standard assay conditions. Tubulin (22 μg) was dephosphorylated for 60 min by protein phosphatase C (4.4 μg). Under these conditions, the dephosphorylation reached the maximal level for each protein. Values have been corrected for those determined in the absence of protein phosphatase C and represents the mean of duplicate determinations. Molecular weights were 270,000, 60,000, and 55,000 for MAP2, tau factor, and tubulin, respectively.

Table 2. Vmax and Km values of protein phosphatase C for microtubule proteins phosphorylated by CaM kinase II and the catalytic subunit of A-kinase.

	CaM kinase II		Catalytic subunit	
Substrate	Vmax (nmol/mg/min)	Km (μM)	Vmax (nmol/mg/min)	Km (μM)
MAP2	463	2.7	217.8	1.9
Tau factor	1,185.1	2.1	1,059.2	1.6
Tubulin	101.2	25	84.1	25

MAP2, tau factor, and tubulin were phosphorylated by CaM kinase II and the catalytic subunit of A-kinase, as indicated. Incubation was carried out for 3 min under standard conditions. The amount of protein phosphatase C used was 0.9 μg for MAP2 and tau factor, and 2.2 μg for tubulin. The Vmax and Km values were determined from the double-reciprocal plot. Values represent the means of duplicate determinations.

Table 3. Dephosphorylation by the holoenzyme of protein phosphatase 2A of microtubule proteins phosphorylated by CaM kinase II (A) and the catalytic subunit of A-kinase (B).

Substrate	Phosphorylation (mol/mol)	Dephosphorylation Enzyme amount (μg)	Dephosphorylation 30 min (%)	Dephosphorylation 60 min (%)
A. CaM kinase II				
MAP2	10.8	1.5	23	20
		3.8	32	26
Tau factor	12.2	1.5	25	27
		3.8	40	39
Tubulin	0.5	4.4	37	50
B. A-kinase				
MAP2	10.0	1.5	19	24
		3.8	31	28
Tau factor	13.0	1.5	39	53
		3.8	53	58
Tubulin	0.6	4.4	47	59

MAP2, tau factor, and tubulin were phosphorylated by CaM kinase II (A) or the catalytic subunit of A-kinase (B), as indicated. MAP2 (10 μg), tau factor (8.6 μg), and tubulin (16 μg) were dephosphorylated for 30 or 60 min by the holonzyme of protein phosphatase 2A (1.5-4.4 μg), under standard assay conditions. Values have been corrected for those determined in the absence of protein phosphatase 2A, and represent the means of duplicate determinations.

Table 4. Effect of dephosphorylation of MAP2 or tau factor on microtubule assembly.

CaM kinase II	PrP-C	MAP2: MAP2 phosphorylation (mol/mol)	MAP2: MT assembly Rate (%)	MAP2: MT assembly Extent (%)	Tau factor: Tau factor phosphorylation (mol/mol)	Tau factor: MT assembly Rate (%)	Tau factor: MT assembly Extent (%)
-	+	0	112.5	96.4	0	89.0	123.0
-	-	0	100.0	100.0	0	100.0	100.0
+	-	10.9	62.5	72.7	14.1	37.0	61.7
+	+	3.4	93.4	94.5	2.5	82.0	107.0

MAP2 (123 μg) and tau factor (30 μg) were incubated at 30°C for 30 min with (+) or without (with boiled enzyme)(-) 2 μg of CaM kinase II, under standard conditions. Phosphorylation was stopped by the addition of 30 μl of 25 mM EGTA. Dephosphorylation was carried out at 30°C for 15 min with 7.8 μg of protein phosphatase C (+) or boiled protein phosphatase C (-). After dephosphorylation, incubation for microtubule (MT) assembly was started with 1 mg/ml of tubulin, under the conditions described in MATERIALS AND METHODS. Rate and extent indicate the initial rate and the final extent of microtubule assembly compared with the control value, respectively. Values of the phosphate amount incorporated into MAP2 and tau factor have been corrected for those determined without each substrate and represent the means of duplicate determinations.

mol/mol for the 49K and 60K subunits, respectively, by a 15-min incubation with protein phosphatase C. Using tau factor and chicken gizzard light chain as substrates, the autophosphorylation resulted not only in the inactivation, but also in the loss of dependence of enzyme activity on Ca^{2+} and calmodulin. Dephosphorylation of CaM kinase II by protein phosphatase C could reactivate the enzyme activity. The holoenzyme of brain protein phosphatase 2A dephosphorylated CaM kinse II (Fig. 4).

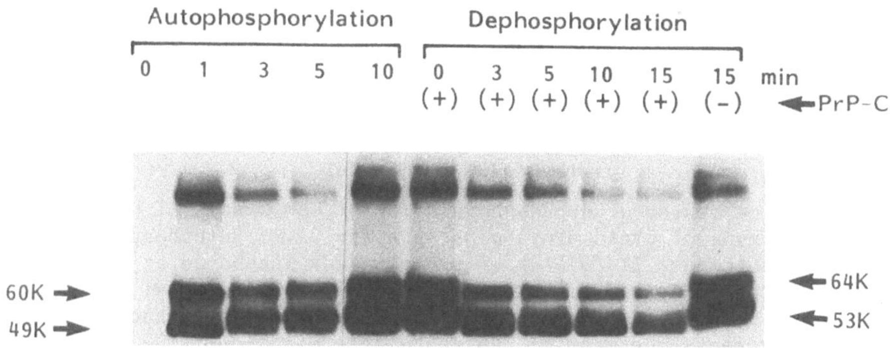

Fig. 3. Autophosphorylation and dephosphorylation of CaM kinase II with time. CaM kinase II (2.4 μg) was autophosphorylated at 30°C for the indicated times under standard conditions. After autophosphorylation for 10 min, the dephosphorylation of the enzyme was started with (+) or without (-) protein phosphatase C (PrP-C)(1.0 μg) under standard conditions. The reaction of autophosphorylation or dephosphorylation was terminated at the indicated times by the addition of the samlpe buffer (Laemmli, 1970). The aliquots were subjected to SDS-PAGE, followed by autoradiography. The arrows of 49K and 60K represent the positions of the 49K and 60K subunits of CaM kinase II. The respective subunits migrated to the positions of 53 K and 64 K after autophosphorylation.

DISCUSSION

Ca^{2+}, calmodulin-dependent phosphorylation of MAP2 (Fukunaga et al., 1982; Yamauchi and Fujisawa, 1982; Yamamoto et al., 1983, 1985), tau factor (Yamamoto et al., 1983, 1985), and tubulin (Burke and DeLorenzo, 1981; Fukunaga et al., 1982; Yamauchi and Fujisawa, 1984; Yamamoto et al., 1985) were reported by several laboratories. The phosphorylation of proteins induced the inhibition of microtubule assembly (Jameson et al, 1988; Jameson and Caplow, 1981; Nishida et al., 1981; Yamamoto et al., 1983, 1985). In the present study, we attempted to elucidate the switch-off mechanism of the inhibition for microtubule assembly.

Protein phosphatase C purified from bovine brain dephosphorylated microtubule proteins phosphorylated by CaM-kinase II and the catalytic subunit of A-kinase. No difference in kinetic properties of dephosphorylation was observed between the substrates phosphorylated by the two kinases. We reported that MAP2 and tau factor were good substrates for calcineurin (Goto et al., 1985). The kinetic properties of dephospho-

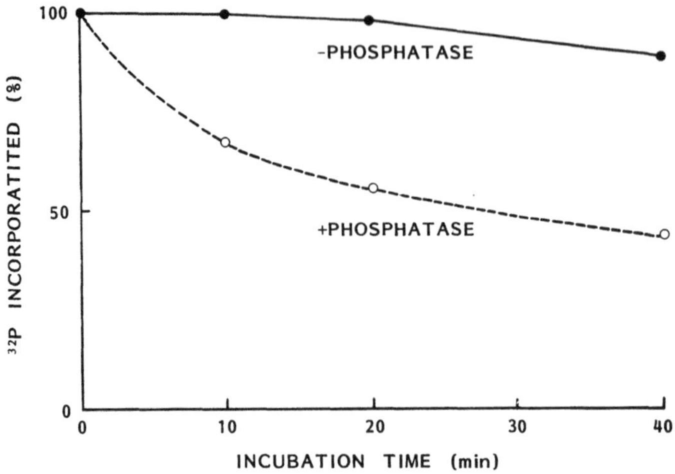

Fig. 4. Time course of dephosphorylation of CaM kinase II by the holoenzyme of brain protein phosphatase 2A. Autophosphorylated CaM kinase II (5.5 μg of protein) was incubated at 30°C for the indicated times with (o---o) or without (•—•) 4.4 μg of the holoenzyme of brain protein phosphatase 2A. Each value represents the mean of duplicate determinations.

rylation of MAP2 and tau factor by calcineurin were also similar between the substrates phosphorylated by CaM kinase II and A-kinase. Thus the switch-off mechanism may be the same between Ca^{2+}/ calmodulin and cyclic AMP systems.

The dephosphorylation of MAP2 and tau factor each resulted in the removal of inhibition of the microtubule assembly (Table 4). Murthy and Flavin reported that protein phosphatase C from sheep brain reversed the effects of MAP2, phosphorylated by A-kinase, on microtubule assembly (1983). Thus, the phosphorylation-dephosphorylation regulation by protein kinases and protein phosphatases may be an important mechanism for the control of microtubule assembly-disassembly cycle.

We found that autophosphorylation of CaM kinase II changed the two subunits to the higher molecular weight forms on SDS-PAGE. Protein phosphatase C dephosphorylated CaM kinase II with concomitant restoration of original molecular weights. CaM kinase II was inactivated and reactivated by autophosphorylation and dephosphorylation, respectively. These data may suggest that the conformational changes of CaM kinase II by autophosphorylation and dephosphorylation are partly related to the enzyme activities.

Brain protein phosphatase C consisted of 10 and 90 %, respectively, of the catalytic subunits of protein phosphatases 1 and 2A on the basis of the effects of ATP and inhibitor-2. The substrate specificity of protein phosphatase 2A was reported to differ between the catalytic subunit and the holoenzyme (Pato and Adelstein, 1983; Imaoka et al., 1983). The holoenzyme of brain protein phosphatase 2A could also dephosphorylate microtubule proteins and CaM kinase II (Table 3 and Fig. 4). These data suggest that protein phosphatases 1 and 2A, mainly protein phosphatase 2A, are involved in the switch-off mechanism of Ca^{2+}/ calmodulin-dependent regulation of microtubule formation via dephosphorylation of microtubule proteins and CaM kinase II. This supports the idea that the phosphorylation and dephosphorylation of microtubule proteins regulate microtubule assembly-disassembly *in vivo*.

ACKNOWLEDGMENT

This work was supported in part by a Grant-in-Aid for Scientific Research from the Ministry of Education, Science and Culture, Japan.

REFERENCES

Bennett M.K., Erondu N.E., and Kennedy M.B., 1983, Purification and characterization of a calmodulin-dependent protein kinase that is highly concentrated in brain. J. Biol. Chem., 258:12735.

Bradford M.M., 1976, A rapid and sensitive method for the quantitation of microgram quantities of protein utilizing the principle of protein-dye binding. Anal. Biochem., 72:248.

Burke B.E. and DeLorenzo R.J., 1981, Ca^{2+}- and calmodulin-stimulated endogenous phosphorylation of neurotubulin. Proc. Natl. Acad. Sci. USA, 78:991.

Fukunaga K., Yamamoto H., Matsui K., Higashi K., and Miyamoto E., 1982, Purification and characterization of a Ca^{2+}- and calmodulin-dependent protein kinase from rat brain. J. Neurochem., 39:1607.

Gaskin F., Cantor C.R., and Shelanski M.L., 1974, Turbidimetric studies of the in vitro assembly and disassembly of porcine neurotubules. J. Mol. Biol., 89:737.

Goldenring J.R., Gonzalez B., McGuire J.S.Jr., and DeLorenzo R.J., 1983, Purification and characterization of a calmodulin-dependent kinase from rat brain cytosol able to phosphorylate tubulin and microtubule associated proteins. J. Biol. Chem., 258:12632.

Goto S., Yamamoto H., Fukunaga K., Iwasa T., Matsukado Y., and Miyamoto E., 1985, Dephosphorylation of microtubule-associated protein 2, τ factor, and tubulin by calcineurin. J. Neurochem., 45:276.

Hsiao K.J., Sandberg A.R., and Li H.C., 1978, The role of ATP and divalent cations in the regulation of a cardiac phosphorylase phosphatase (phosphoprotein phosphatase) of Mr=35,000. J. Biol. Chem., 253:6901.

Imaoka T., Imazu M., Usui H., Kinohara N., and Takeda M., 1983, Resolution and reassociation of three distinct components from pig heart phosphoprotein phosphatase. J. Biol. Chem., 258:1526.

Ingebritsen T.S., Foulkes J.G., and Cohen P., 1980, The broad specificity protein phosphatase from mammalian liver: separation of the Mr 35,000 catalytic subunit into two distinct enzymes. FEBS Lett., 119:9.

Jameson L. and Caplow M., 1981, Modification of microtubule steady-state dynamics by phosphorylation of the microtubule-associated proteins. Proc. Natl. Acad. Sci. USA, 78:3413.

Jameson L., Frey T., Zeeberg B., Dalldorf F., and Caplow M., 1980, Inhibition of microtubule assembly by phosphorylation of microtubule-associated proteins. Biochemistry, 19:2472.

Miyamoto E., Petzold G.L., Kuo J.F., and Greengard P., 1973, Dissociation and activation of adenosine 3',5'-monophosphate-dependent and guanosine 3',5'-monophosphate-dependent protein kinases by cyclic nucleotides and by substrate proteins. J. Biol. Chem., 248:179.

Miyamoto E., Miyazaki K., Hirose R., and Kashiba A., 1978, Multiple forms of protein kinases in myelin and microsomal fractions of bovine brain. J. Neurochem. 31:269.

Miyamoto E., Fukunaga K., Matsui K., and Iwasa Y., and Miyamoto E., 1981, Occurrence of two types of Ca^{2+}-dependent protein kinases in the cytosol fraction of the brain. J. Neurochem. 37:1324.

Miyamoto E., 1986, Characterization of a multifunctional Ca^{2+}-calmodulin-dependent protein kinase in the brain, in Neuromethods, Vol. 5, Boulton A.A., Baker G.B., and Yu P.H., eds, pp. 519, The Humana Press, New Jersey.

Murthy A.S.N. and Flavin M., 1983, Microtubule assembly using the microtubule-associated protein MAP-2 prepared in defined states of phosphorylation with protein kinase and phosphatase. Eur. J. Biochem. 137:37.
Nairn A.C., Hemmings B.A.Jr., and Greengard P., 1985, Protein kinases in the brain. Annu. Rev. Biochem. 54:931.
Nishida E., Kuwaki T., and Sakai H., 1981, Phosphorylation of microtubule-associated proteins (MAPs) and pH of the medium control interaction between MAPs and actin filaments. J. Biochem. 90:575.
Noiman E.S. and Chock P.B., 1983, Properties of a Mr = 38,000 phosphoprotein phosphatase: modulation by divalent cations, ATP, and fluoride. J. Biol. Chem. 258:4214.
Pato M.D. and Adelstein R.S., 1983, Purification and characterization of a multisubunit phosphatase from turkey gizzard smooth muscle. J. Biol. Chem. 258:7047.
Post R.L. and Sen A.K. 1967, ^{32}P-labeling of a (Na^+ + K^+)-ATPase intermediate. Methods. Enzymol. 10:773.
Resink T.J., Hemmings B.A., Tung H.Y.L., and Cohen P., 1983, Characterization of a reconstituted Mg-ATP-dependent protein phosphatase. Eur. J. Biochem. 133:455.
Saitoh Y., Yamamoto H., Fukunaga K., Matsukado Y., and Miyamoto E., 1986, Autophosphorylation and dephosphorylation of multifunctional Ca^{2+}-calmodulin-dependent protein kinase from rat brain. Biomed. Res. 7:399.
Saitoh Y., Yamamoto H., Fukunaga K., Matsukado Y., and Miyamoto E., 1987, Inactivation and reactivation of the multifunctional calmodulin-dependent protein kinase from brain by autophosphorylation and dephosphorylation: involvement of protein phosphatases from brain. J. Neurochem. 49:1286.
Tung H.Y.L., Alemany S., and Cohen P., 1985, The protein phosphatases involved in cellular regulation. 2. Purification, subunit structure and properties of protein phosphatases-$2A_0$, $2A_1$, and $2A_2$ from rabbit skeletal muscle. Eur. J. Biochem. 148:253.
Yamauchi T. and Fujisawa H., 1982, Phosphorylation of microtubule-associated protein 2 by calmodulin-dependent protein kinase (kinase II) which occurs only in the brain tissues. Biochem. Biophys. Res. Commun. 109:975.
Yamauchi T. and Fujisawa H., 1984, Calmodulin-dependent protein kinase (kinase II) which is involved in the activation of tryptophan 5-monooxygenase catalyzes phosphorylation of tubulin. Arch. Biochem. Biophys. 234:89.
Yamamoto H., Fukunaga K., Tanaka E., and Miyamoto E., 1983, Ca^{2+}- and calmodulin-dependent phosphorylation of microtubule-associated protein 2, τ factor, and inhibition of microtubule assembly. J. Neurochem. 41:1119.
Yamamoto H., Fukunaga K., Goto S., Tanaka E., and Miyamoto E., 1985, Ca^{2+}, calmodulin-dependent regulation of microtubule formation via phosphorylation of microtubule-associated protein 2, τ factor, and tubulin, and comparison with the cyclic AMP-dependent phosphorylation. J. Neurochem. 44:759.
Yamamoto H., Saitoh Y., Fukunaga K., Nishimura H., and Miyamoto E., 1988, Dephosphorylation of microtubule proteins by brain protein phosphatases 1 and 2A, and its effect on microtubule assembly. J. Neurochem. 50:1614.
Yang S.D., Vandenheede J.R., and Merlevede W., 1981, A simplified procedure for the purification of the protein phosphatase modulator (inhibitor-2) from rabbit skeletal muscle. FEBS Lett. 132:293.

DEMONSTRATION OF DIFFERENT REGIONAL DISTRIBUTIONS OF CALCINEURIN SUBUNITS USING MONOCLONAL ANTIBODIES

Hideki Matsui, Toshifumi Itano, Seiji Etoh, Masaaki Tokuda, Jerry H. Wang*, and Osamu Hatase

Dept. of Physiology Kagawa Med. Sch,. Kagawa 761-07 Japan
*Dept. of Medical Biochemistry, Univ. of Calgary Calgary, Alberta, T3B, 4W7, Canada

SUMMARY

Immunohistochemical localizations of calcineurin subunits A (60KDa) and B (20KDa) were examined in a rat brain using subunit specific monoclonal antibodies. The immunoreactivity of the subunit A was abundant in the hippocampus, in the striatum, and in the thalamus, but weak in the neocortex. On the contrary, the immunoreactivity of the subunit B was more abundant in the cortex, and it was more ubiquitous than subunit A.

The distribution of subunit A in the rat brain very well agreed with that of zinc, which is an intrinsic metal ion and a potent inhibitor of calcineurin phosphatase.

INTRODUCTION

Calcineurin is the major calmodulin-binding protein in the brain, and possesses metal ion and calmodulin-stimulated phosphatase activity (4) (9) (8). This enzyme is composed of two subunits: Subunit A (60KDa) has the catalytic domain and interacts with calmodulin, while Subunit B (20KDa) has the Ca binding domain which shows E-F band structure like calmodulin. This means that the activity of calcineurin is controlled by two distinct calcium binding proteins: calmodulin and B subunit. Calcineurin is a metalloenzyme containing tightly bound zinc and iron (6) (14). The physiological significance of these metal ions in calcineurin activity is not well established. One mol of calcineurin tightly and stoichiometrically binds one mole of both zinc and iron; furthermore, the addition of zinc ions strongly inhibits the activity of calcineurin phosphatase.

There is a controversy over the subunit structure of the enzyme. Wang et al. suggested a one A and two B structure (9). On the other hand, Klee et al. reported a one to one structure from the result of their cross-linking experiment (4). In the present report, we studied the regional distribution of subunits A and B using the monoclonal antibodies specific for the subunits A and B, independently.

MATERIALS AND METHODS

Calcineurin and rat brain calmodulin-binding proteins were purified, as previously described (10). The monoclonal antibodies were prepared and purified from mouse ascites fluid using a DEAE-Affigel blue column. Slab gel electrophoresis, in the presence of 0.1% sodium dodecyl sulfate (SDS), was carried out according to the procedure of Laemmli using a 12% acrylamide concentration (5). Essentially, immunoblotting was done as described by Towbin et al (12). The electrophoretic transfer of proteins was carried out in the presence of 0.1% SDS to increase the transfer efficiency. The detection of the cross-reactive protein bands was performed using a protein A-gold immunoblot assay kit (Bio Rad) or alkaline-phosphatase conjugated second antibody.

For light microscopy adult rats were anesthetized by a intraperitoneal injection of pentobarbital (50 mg/kg), and perfused with a 0.1 M Na-phosphate buffer (pH 7.4) containing 10% formaldehyde from the heart for 10 min . After the perfusion, the cerebral hemispheres were incised and fixed for 24 h at 4 °C in a fixation solution. The samples were sectioned in slices 10 μm thick after paraffin embedding and set on gelatin-coated slide glass. The samples were deparaffinized and stained by the PAP (peroxidase antiperoxidase) method using a DAKO monoclonal Kit (DAKO, U.S.A.). The purified A subunit specific monoclonal antibody VD_3 or B subunit specific VA_1 was used as the first antibody at a concentration of 200 μg/ml. Non-specific mouse IgG at the same concentration was used as a control. First antibody incubation was continued for 16 h at 4 °C. The colorization was done with 0.2 mg/ml of diaminobenzidine (DAB) and 0.009% hydrogenperoxide in a 0.1 M Tris-HCl buffer (pH 7.6). Osmification (0.1% OsO_4) was done in a 0.05 M cacodylate buffer (pH 7.4) (15). Counter-staining was done using Cresyl violet.

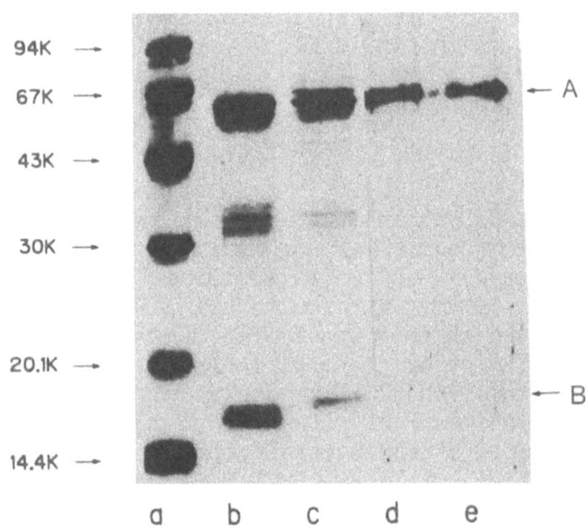

Fig. 1. Immunoreactivity of VD_3 monoclonal antibody. d, e: immunoblots of rat brain calmodulin-binding proteins (lane e, 4.2 μg) and bovine brain calcineurin (lane d, 7.1 μg). Proteins were subjected to SDS-PAGE and electrophoretically transblotted onto a nitrocellulose membrane. The resulting electroblot was incubated with VD_3 (10 μg/ml) overnight, and bound antibody was visualized by the protein A-Gold method at room temperature. b, c: protein-staining with Coomassie brilliant blue on SDS-PAGE. Rat brain calmodulin binding proteins (lane c, 4.2 μg) and bovine brain calcineurin (lane b, 7.1 μg). a: molecular weight markers.

RESULTS AND DISCUSSION

Specificity of the monoclonal antibodies

Immunoblotting analysis of VD_3 reactivity to bovine brain calcineurin and rat brain calmodulin binding proteins is shown in Fig. 1. In both cases, only a polypeptide with a molecular weight identical to the A subunit of calcineurin strongly reacted with the antibody, while the B subunit did not react at all. This showed that rat brain calcineurin and bovine brain calcineurin had common antigenic sites which bound to VD_3 and also confirmed that VD_3 was specific for the A subunit of calcineurin.

The specificity of the monoclonal antibody VA_1 is shown in Fig. 2. Bovine brain calcineurin and rat brain crude extract were immunoblotted with the monoclonal antibody VA_1. A polypeptide with a molecular weight identical to the B subunit strongly reacted with the antibody, while the A subunit did not. A protein band with a molecular weight slightly lower than B also reacted with VA_1. This indicates that VA_1 is specific for B but is also reactive with a minor component of the brain, probably the degradative product of the B subunit. Moreover, the B subunit of the bovine brain and the rat brain share the same antigenic site.

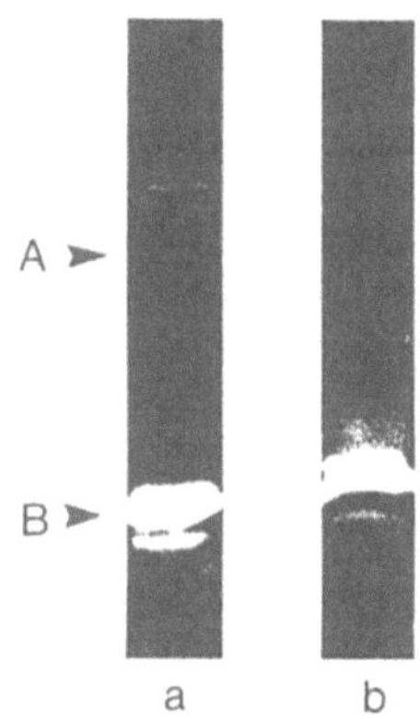

Figs 2. Immunoblotting analysis of the monoclonal antibody VA_1. Purified bovine brain calcineurin (1 μg) and rat brain crude extract (105,000 x g supernatant, 350 μg) were electrophoresed on SDS-PAGE. Proteins were electrophoretically transblotted onto nitrocellulose filters. The nitrocellulose blots were incubated with 27 μg/ml monoclonal antibody VA_1 in Tris saline with 3% BSA for 2h. The blot was then washed with Tris saline and incubated with alkaline phosphatase-conjugated rabbit anti-mouse immunoglobulin G (Sigma). As a control, the same samples were processed without incubation with the monoclonal antibody VA_1 solution. In this case no protein bands could be seen. Arrows show the position of each subunit (A and B) of calcineurin on the nitrocellulose. a, purified bovine brain calcineurin; b, rat brain crude extract.

Immunohistochemical localization of calcineurin A

Immunohistochemical localization of calcineurin A in the rat brain frontal section is shown in Figs. 3A, 4, 5. Heavy deposits of PAP reaction products were localized in two areas; one was at the hippocampal formation, and the other was at the basal ganglia. In contrast, besides thses two areas, other parts of the brain, such as the neocortex and the thalamus, were very lightly stained (Fig. 4). In the hippocampus, the

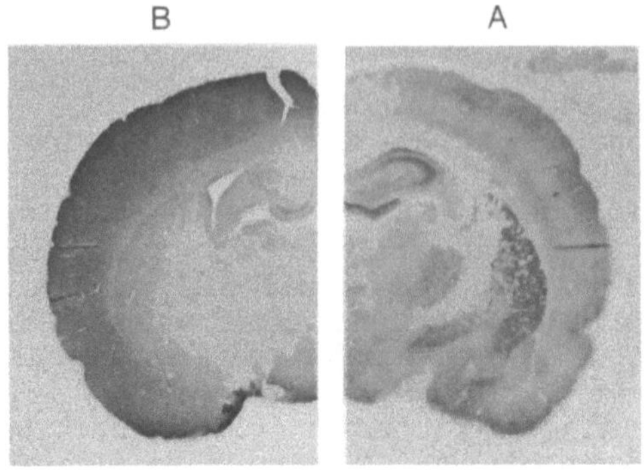

Fig. 3. The comparison of the immunohistochemical localization of calcineurin A and B in the rat brain frontal section. The rat brain was sectioned and stained with VD_3 (A) or VA_1 (B) and visualized by the PAP method. Counterstaining was done with cresylviolet.

heaviest deposit was in the stratum lucidum in CA_3 (Fig. 5). Stratum moleculare of the dentate gyrus and stratum orience were lightly stained, while the control experiment was not stained (Figs. 4, 5). Deposits in the stratum lucidum were confined within the CA_3 region, and the stratum radiatum and the stratum lacunosum were not stained. Higher magnification of this region showed specific localization, and the perikaryon of the pyramidal cells was almost devoid of the deposits (Fig. 6). In the dentate gyrus, the apical dendrites of granule cells were stained in the stratum moleculare whereas the granule cells themselves were not stained (Fig. 4).

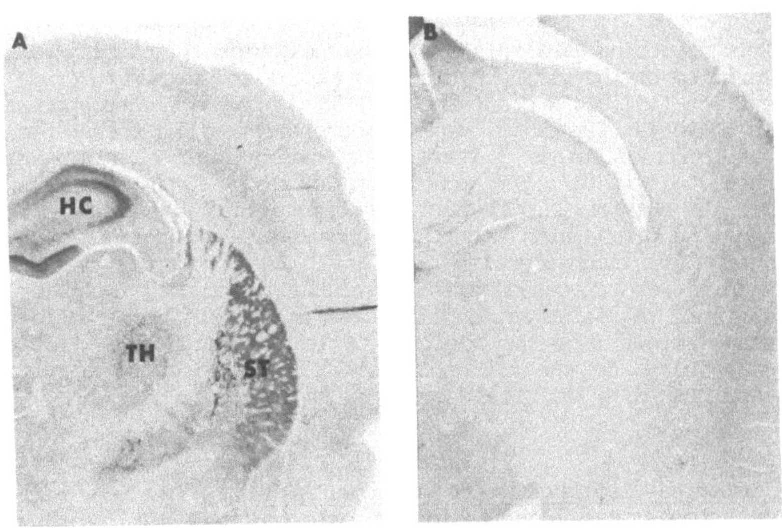

Fig. 4. The distribution of calcineurin A in the rat brain frontal staining. The rat brain was sectioned and stained with VD_3 antibody (A) and non-specific mouse IgG (B). HC Hippocampus; ST striatum; TH thalamus. Counter staining was done with cresylviolet.

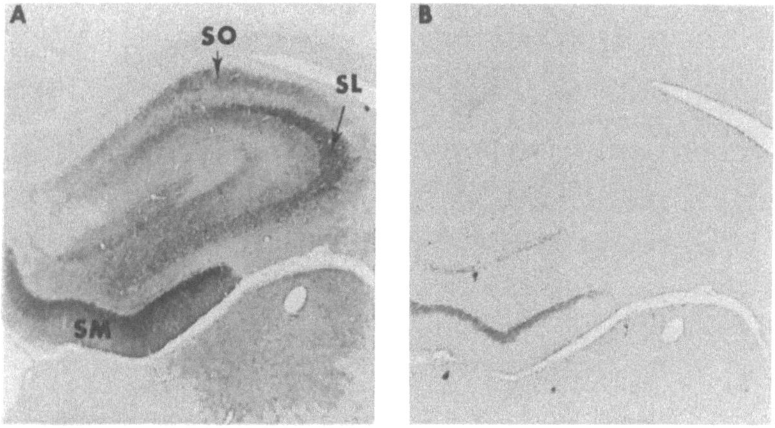

Fig. 5. The distribution of calcineurin A in the rat brain hippocampus. The staining procedure is the same as in Fig. 4. VD_3 antibody staining (A), non-specific mouse IgG (B). The stratum lucidum (SL) was most heavily stained. The stratum moleculare (SM) and the stratum orience were positively stained but lighter than the stratum lucidum. Counterstaining was cone with cresylviolet.

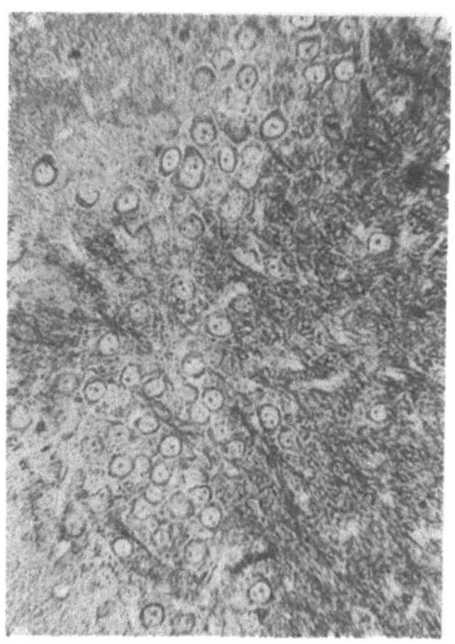

Fig. 6. A higher magnification of the CA_3 region stained with VD_3. The stratum lucidum (SL) was strongly stained, but the cytoplasm of granular cells (G) was negative. The stratum radiatum (SR) was positive but was very lightly stained compared to the stratum lucidum. Counterstaining was done with cresylviolet.

Immunohistochemical localization of B subunit

Fig. 3B shows the frontal section of a rat brain immunostained with the monoclonal antibody VA_1, showing the B subunit distribution. There was a striking difference in the immunohistochemical distribution between the A subunit (Fig. 3A) and B subunit (Fig. 3B). Although the striatum and the hippocampus were not negative, the stainability of these regions was very low. In contrast, the cortex showed more positive deposits than either the striatum or the hippocampus.

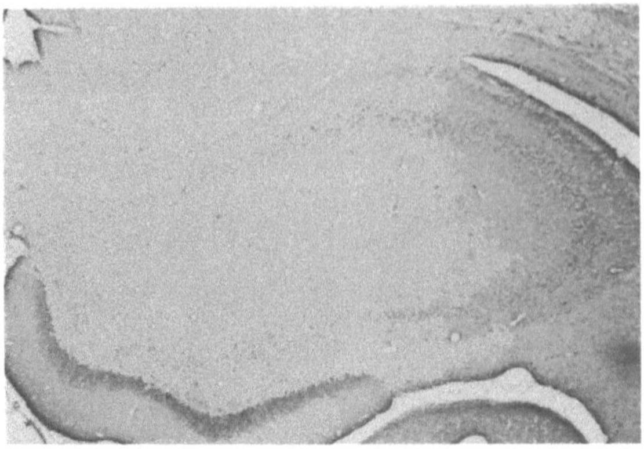

Fig. 7. The immunostaining of calcineurin B by VA_1 in the rat brain hippocampus. Counterstaining was done with cresylviolet.

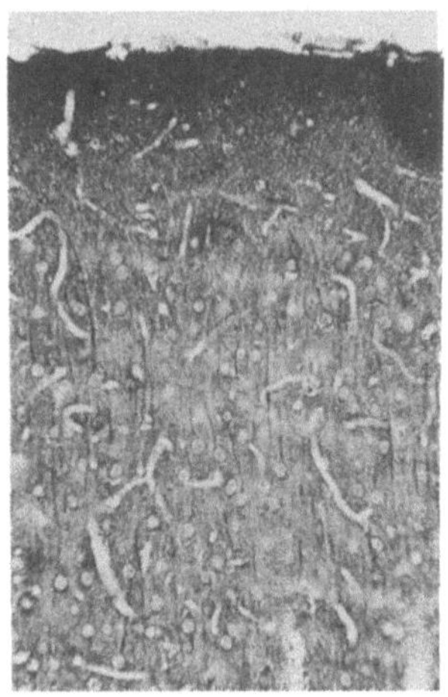

Fig. 8. The distribution of calcineurin B by VA_1 in the rat brain cortex. The apical dendrite and the perikaryon of the pyramidal cells in the third layer showed strong deposits.

At a higher magnification of the hippocampus, the stratum lucidum, the stratum moleculare, and the stratum orience were positive, but there was no striking difference between these areas and other areas of the brain as could be seen with A specific VD_3 (Fig. 7). On the contrary, the cortex was well stained (Fig. 8). The apical dendrite and the perikaryon of the pyramidal cells in the third layer of the cortex especially showed strong deposits, and this indicates that the B subunit is rich in these regions of the cortex (Fig. 8). However, we detected only a small amount of the A subunit in the cortex (Fig. 4). This suggests either of following possibilities: (1) The antigenicity of the A subunit in the cortex is masked, or (2) the cortex contains more B subunits free from A subunits, suggesting that the subunit structure in the cortex may be different from that in the other parts of the brain.

Miyamoto et al. reported the immunohistochemical distribution of calcineurin in the rat brain. Since their work was done with polyclonal antibodies which recognized both A and B subunits, they detected a wider distribution of calcineurin. The summation of the results of our findings using monoclonal antibodies VA_1 and VD_3 is consistent with the data using polyclonal antibodies previously reported by Miyamoto et al. (2) (16).

Calcineurin A was abundant in the stratum lucidum where mossy fiber terminals form giant synaptic boutons with pyramidal cell dendrites, but there was very little in the stratum radiatum and the stratum lacunosum where mossy fibers run with very few synapses. This suggests that calcineurin A associates with giant synaptic formation (15).

Otsuka et al. and Stengaard-Pedersen et al. showed that zinc was also rich in the hippocampal mossy fiber boutons in CA_3 (3) (13). King et al. and Pallen et al. reported that calcineurin was a metalloenzyme containing tightly bound zinc, which was a good inhibitor of calcineurin activity (6) (8). The coexistence of calcineurin and zinc in the hippocampus may suggest that some of the physiological roles of zinc in the hippocampus are likely to be associated with calcineurin and probably control calcineurin activity in the synaptic function.

Reference

1. Cooper, N.G.F., McLauglin, B.J., Tallant, E.A. and Cheung, W.Y., J. Cell. Biol., 101 (1985) 1212-1218.
2. Goto, S., Matsukado, Y., Mihara, Y., Inoue, N. and Miyamoto, E., Brain Research, 371 (1986) 237-243.
3. Ibata, Y. and Staelelin, T., J. Histochem. Cytochem., 17 (1969) 171-175.
4. Klee, C.B., Crough, T.H. and Krinks, M.H., Proc. Natl. Acad, Sci. USA, 76 (1979) 6270-6273.
5. Laemmli, U.K., Nature (London), 227 (1970) 680-685.
6. King, M.M. and Huang, C.Y., J. Biol. Chem., 259 (1984) 8847-8856.
7. Matsui H., Pallen, C.J., Adachi, A.-M., Wang J.H. and Lam, Peter H.-Y., J. Biol. Chem., 260 (1985) 4174-4179.
8. Pallen, C.J., Sharma, R.K., Matsui, H. and Wang J.H., Adv. Prot. Phosphatases, Vol. I, (1985) 147-162.
9. Sharma, R.K., Desai, R., Waisman, D.M. and Wang, J.H., J. Biol. Chem., 254 (1979) 4276-4282.
10. Sharma, R.K., Taylor, W.A. and Wang J.H., Meth. Enzymol., 102 (1983) 210-219.
11. Stengaard-Pedersen, K., Redens. K. and Larsson, L.-L., Brain Research, 212 (1981) 230-233.
12. Towbin, H., Staehelin, T. and Gordon, J., Proc. Natl. Acad. Sci. U.S.A., 76 (1979) 4350-4354.
13. Wood, J.G., Wallace, R.W., Whitaker, J.N. and Cheung, W.Y., J. Cell. Biol., 84 (1980) 66-76.
14. Rao, J. and Wang, J.H., J. Biol. Chem., (1988) in press.
15. Matsui, H., Doi, A., Itano, T., Shimada, M., Wang, J.H. and Hatase, O., Brain Research, 402 (1987) 193-196.
16. Miyamoto, E., Goto, S., Yamamoto, H., Fukunaga, K. and Matsukado, Y., Adv. Prot. Phosphatases, Vol. III (1986) 187-200.

STIMULATION BY INOSITOL TRISPHOSPHATE AND TETRAKISPHOSPHATE OF A PROTEIN PHOSPHATASE

Jean Zwiller, Stuart S. Nakamoto and Alton L. Boynton*

Cancer Research Center
1236 Lauhala Street
University of Hawaii
Honolulu, Hawaii 96813

INTRODUCTION

Hormone-receptor coupling often results in the activation of phospholipase C which in turn hydrolyzes phosphatidylinositol 4,5-bisphosphate into two second messengers, 1,2-diacylglycerol (DAG) and inositol 1,4,5-trisphosphate [Ins(1,4,5)]P_3. It is well known that DAG remains in the membrane domain and activates protein kinase C while Ins(1,4,5)P_3 is soluble and is known to increase the intracellular level of Ca^{2+}.[1,2] Ins(1,4,5)P_3 may subsequently be converted to Ins(1,3,4,5)P_4 by a soluble inositol trisphosphate 3-phosphokinase.[3] Both Ins(1,4,5)P_3 and Ins(1,3,4,5)P_4 are substrates of a membrane-bound 5-phosphomonoesterase being degraded to Ins(1,4)P_2 and Ins(1,3,4)P_3 respectively.[4,5] Ins(1,4,5)P_3 and Ins(1,3,4)P_3 both mobilize Ca^{2+},[2,6] while the physiological function of Ins(1,3,4,5)P_4 is unknown although indirect evidence suggests that it may affect plasma membrane Ca^{2+} channels.[7,8]

In this chapter we will review evidence which suggests that Ins(1,4,5)P_3 and Ins(1,3,4,5)P_4 may mediate their physiological effects by activating a phosphoprotein phosphatase that is present mainly in the particulate fraction of rat brain.

MATERIALS AND METHODS

Reagents. Ins(1,3,4,5)P_4, Ins(1,4,5)P_3, Ins(1,3,4)P_3, Ins(2,4,5)P_3, and Ins(1,4)P_2 were kindly supplied by Dr. R.F. Irvine (Cambridge, U.K.). Ins(1,4,5)P_3 was also from Sigma and Amersham and Ins(1,3,4,5)P_4 was also from Boehringer. The other inositol phosphates and chemicals were from Sigma. Ammonium molybdate was obtained from Kodak. Superose 12, monoQ and FPLC equipment were from Pharmacia. [γ-^{32}P]ATP was obtained from New England Nuclear.

---

*To whom correspondence should be addressed.

Enzymatic Preparation. Sprague-Dawley rats were sacrificed and the brains rapidly removed and homogenized with a Potter-Elvehjem homogenizer in 6 volumes of 20 mM Tris-Cl buffer, pH 7.5, containing 0.3 M sucrose, 2 mM EDTA, 10 mM EGTA, 2 mM DTT, 2 mM phenylmethylsulfonyl fluoride, 10 mg/l leupeptin, 10 mg/l trypsin inhibitor, and 10% glycerol. The homogenate was filtered through four layers of cheesecloth, then centrifuged for 10 min at 1,000 x g. The pellet (P1) was suspended in 20 mM Tris-Cl buffer, pH 7.5, containing 1 mM DTT, 0.1 mM EDTA, and 10% glycerol (buffer A), and the supernatant fluid was sedimented at 8,000 x g for 10 min. The pellet (P2) was resuspended in buffer A, and the supernatant fluid was centrifuged at 107,000 x g for 20 min. The final pellet (P3) was also resuspended in buffer A using a hand-held homogenizer.

For further characterization and purification of the protein phosphatase, the enzyme was extracted from P1 by resuspending it in buffer A with 0.2% (w/v) CHAPS detergent, followed by incubation for 30 min at 0°C. The mixture was then centrifuged at 107,000 x g for 20 min and the supernatant used as the enzyme source.

Partial Purification of the Protein Phosphatase. The protein phosphatase was partially purified from the P1-CHAPS extract by FPLC-gel filtration. 0.2 ml of the CHAPS extract was loaded on a Superose 12 column (1 x 30 cm) and eluted with buffer A containing 0.01% CHAPS at a flow rate of 9 ml/hr, 0.6 ml fractions being collected.

In a separate experiment, the protein phosphatase was precipitated by ethanol according to Brandt et al.[9] prior to the gel filtration step. Briefly, 4 volumes of ethanol at room temperature were added to an aliquot of the P1-CHAPS extract. The denatured proteins were pelleted at 10,000 x g for 10 min and extracted with buffer A. The mixture was centrifuged at 107,000 x g for 20 min and the resulting supernatant applied to the Superose column as described before.

Preparation of ^{32}P-Histone. Histone type II-A was phosphorylated with rabbit muscle cAMP-dependent protein kinase (type-I). Incubation mixtures (4 ml) contained 20 mg histone, 1 mg cAMP-dependent protein kinase, 20 mM Tris-Cl, pH 7.5, 5 mM $MgCl_2$, 0.5 mM DTT, 0.7 μM cAMP, and 0.15 mM [γ-^{32}P] ATP (350 Ci/mol). The reaction mixture was incubated at 30°C for 3 hr and terminated by the addition of trichloroacetic acid (2% final concentration). The phosphorylated histone was recovered as reported by Meisler and Langan.[10]

Preparation of ^{32}P-Phosphorylase Kinase. Phosphorylase kinase (from Sigma) was purified by monoQ ion exchange on FPLC. The fractions containing phosphorylase kinase activity, as measured by Killilea et al.[11], pooled and phosphorylated by cAMP-dependent protein kinase according to Antoniw and Cohen.[12] The incubation contained, in addition to the two proteins, 10 mM sodium glycerol 2-phosphate, 0.2 mM EGTA, 0.4 mM EDTA, 1 μM cAMP, 2 mM $MgCl_2$ and 0.16 mM [γ-^{32}P] ATP (65 Ci/mol). After 30 min at 30°C, the reaction was terminated by the addition of a solution containing 80% ammonium sulfate, 8 mM EDTA and 40 mM NaF, pH 7.0 and the phosphorylase kinase recovered as in (12).

Assay of Protein Phosphatase Activity. Histone phosphatase activity was assayed by measuring the release of $^{32}P_i$. The assay mixtures (80 μl) contained 50 mM Tris-Cl, pH 7.5, 2 mM $MnCl_2$, 0.5 mM DTT, 1 mM EDTA, 2 μM [^{32}P] histone (based on incorporated $^{32}P_i$) plus enzyme. After 4 min of incubation at 30°C, the reactions were stopped by the addition of 0.1 ml of 1 N H_2SO_4 containing 1 mM potassium phosphate. The $^{32}P_i$ released was extracted as the phosphomolybdate complex and measured according to

Killilea et al.[11] The dephosphorylation was kept to a limit of 20% conversion from the phosphorylated form. Proteins were assayed using the Bio-Rad protein assay with bovine γ-globulin as a standard.

Dephosphorylation of Phosphorylase Kinase. ^{32}P-labeled phosphorylase kinase, prepared as described above, was dephosphorylated in the incubation mixture used for the phosphatase activity determination, except that the Mn^{2+} concentration was 1 mM. The protein phosphatase source used was the active fraction of the FPLC-gel filtration step after ethanol precipitation, prepared as described above. The reaction (320 μl final volume) was initiated by addition of the enzyme at 30°C. At the times indicated, 70 μl aliquots were removed for the determination of phosphorylase kinase β-subunit phosphorylation. The reaction was terminated with $NaDodSO_4$-sample buffer and the proteins separated by $NaDodSO_4$ 7.5% polyacrylamide gel electrophoresis. Dephosphorylation was determined by densitometric scanning of the autoradiogram. Results were expressed as the percentage phosphorylation of phosphorylase kinase at the indicated time, compared to the initial phosphorylation.

RESULTS

The 1000 x g rat brain pellet (P1) was extracted with CHAPS and contained protein phosphatase activity when incubated in a standard phosphatase medium containing [^{32}P]-histone (Fig. 1). The release of $^{32}P_i$ was linear with respect to the protein concentration up to 50 μg (Fig. 1A) and with respect to time up to about 8 min. (Fig. 1B). All the subsequent assays were carried out for 4 min with less than 45 μg protein.

Figure 2 shows the effects of several inositol phosphates on this protein phosphatase activity. All the $InsP_3$ tested stimulated the histone phosphatase activity with a similar pattern. Ins(1,4,5)P_3 and Ins(1,3,4,5)P_4 were the most potent inositol phosphates, stimulating

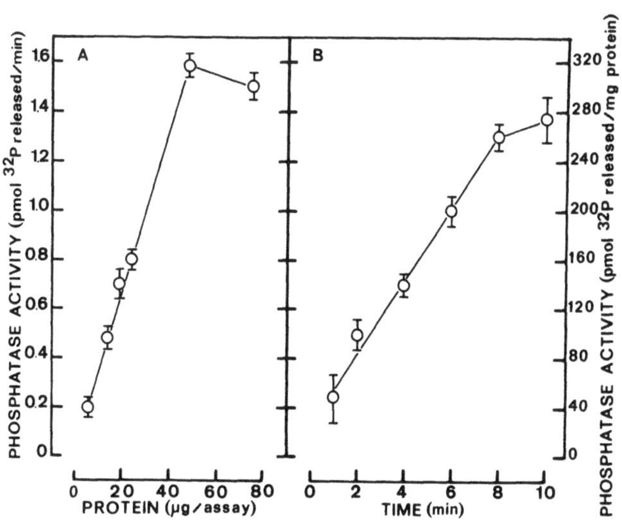

Figure 1. Characterization of the protein phosphatase activity. The P1 pellet was extracted with CHAPS detergent and its histone phosphatase activity assayed as described in Materials and Methods. (A) Effect of protein concentration. The incubation time was 4 min. (B) Time course. The protein content was 19 μg/assay. Each point is the average ± SD of triplicate values.

phosphatase activity at 100 μM, followed by $Ins(2,4,5)P_3$, a nonphysiological occurring isomer, and $Ins(1,3,4)P_3$ (Fig. 2). No activation occurred with inositol phosphates tested containing less than 3 phosphate groups. Inositol monophosphates showed no effect on the enzyme activity even when added to the incubation mixture at concentrations up to 1 mM. Furthermore, this high concentration of monophosphate derivative had no effect on the $Ins(1,4,5)P_3$ stimulation (data not shown), suggesting that the observed $Ins(1,4,5)P_3$ effect is not substrate-directed through ionic interaction.

The inositol phosphate effect on the protein phosphatase, usually assayed in the presence of 2 mM Mn^{2+}, could be observed in the absence of any added metal ion. On the contrary, the effect was totally abolished by a Mn^{2+} concentration greater than 10 mM and by a Mg^{2+} concentration greater than 1 mM (date not shown). This observation can be tentatively explained by the presence in the particulate fraction of 5-phosphomonoesterase which degrades $Ins(1,4,5)P_3$ and whose activity was described as strictly dependent on Mg^{2+}, or to a lesser extent on Mn^{2+}.[6]

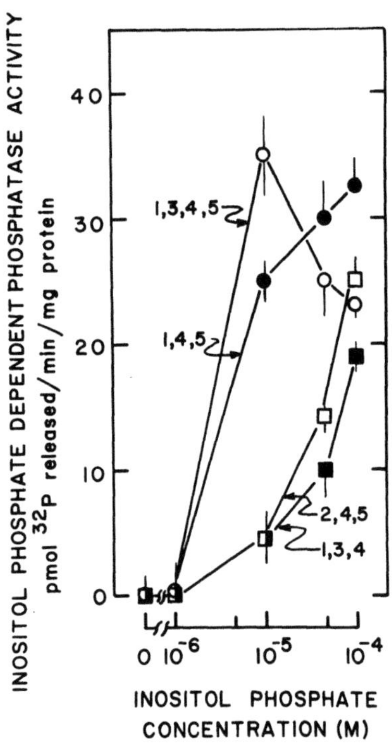

Figure 2. Effect of several inositol phosphates on the histone phosphatase activity in the P1 pellet. CHAPS extracted material (12 μg protein) was assayed in the presence of the indicated inositol phosphates. $Ins(1,4)P_2$, cyclic inositol 1,2-monophosphate, inositol 1-monophosphate and inositol 2-monophosphate had no effect. Each point is the average ± SD of triplicate values of 3 separate experiments. Histone phosphatase activity was assayed by measuring the releasse of $^{32}P_i$. The assay mixtures (80 μl) contained 50 mM Tris-Cl, pH 7.5, 2 mM $MnCl_2$, 0.5 mM dithiothreitol (DTT), 1 mM EDTA, 2 μM [^{32}P] histone (based on incorporated $^{32}P_i$). After 4 min of incubation at 30°C, the reactions were stopped with H_2SO_4 and the $^{32}P_i$ released measured according to Killilea et al.(11). Histone type II-AS (Sigma) was phosphorylated with cAMP-dependent protein kinase and recovered as described by Meisler and Langan (10).

Heparin has been reported as an inhibitor of type-1 protein phosphatase activity.[12] We found heparin to inhibit the phosphatase activity in the presence or absence of Ins(1,4,5)P_3 (Fig. 3). The activity was abolished when heparin was added to the assay mixture at a concentration of 50 μg/ml, with a half-maximal effect at about 15 μg/ml. Dose response curves were similar to those reported for muscle and liver phosphatase type-1.[12].

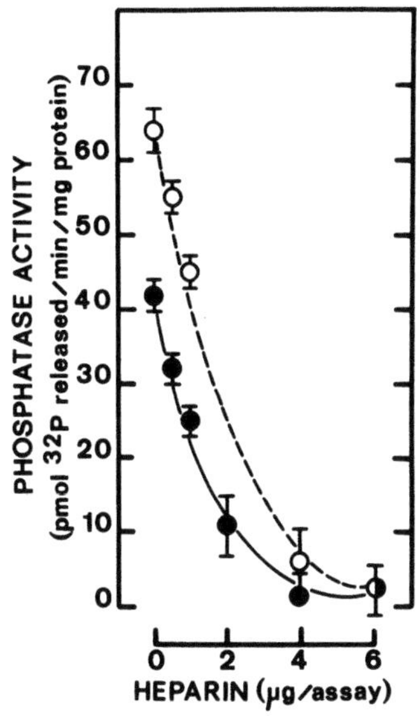

Figure 3. Effect of heparin on Ins(1,4,5)P_3 activated protein phosphatase. The P1 pellet-CHAPS extracted material (19 μg protein) was assayed for protein phosphatase activity in the absence (●) or presence (○) of 50 μM Ins(1,4,5)P_3 with the indicated amount of heparin. Each point is the average ± SD of 4 separate experiments. From Zwiller, et al. (20) and reproduced with permission from Academic Press.

The subcellular distribution of the protein phosphatase in the rat brain is summarized in Table 1. About one-third of the total activity of the protein phosphatase is recovered in the cytosolic fraction, P1 being the richest enzyme source. The low recovery of the enzyme activity in P3 (microsomal fraction) parallels the low protein content (3%). The specific activity is highest in the cytosol followed by the P_3 fraction. Ins(1,4,5)P_3 produced a comparable stimulation (about 90%) in all the subcellular fractions.

Table 1. Subcellular distribution of rat brain protein phosphatase activated by Ins(1,4,5)P_3.

	Activity (pmol ^{32}P/min)	Yield %	Specific Activity (pmol ^{32}P/min/mg prot.) -InsP_3	+InsP_3
P1	3,093	52	18.8	36.1 (92%)*
P2	493	8.3	21.8	41.4 (90%)
P3	172	2.9	26.1	50.4 (93%)
Supernatant	2,160	36	36.1	66.1 (83%)

P1, P2, P3 and supernatant were obtained as described under Materials and Methods. Histone phosphatase activities were measured at initial rates using 15 to 45 μg/assay. Yield is percent of the total activity measured in the homogenate.
*Indicates percent of increase upon 100 μM Ins(1,4,5)P_3 addition.

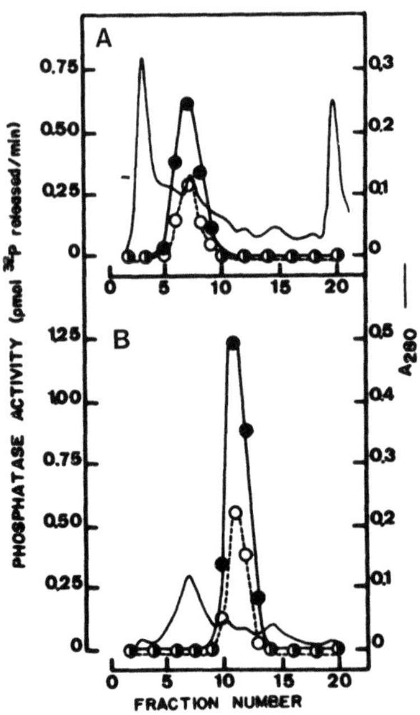

Figure 4. FPLC gel filtration of Ins(1,4,5)P_3 activated protein phosphatase. (A) An aliquot of the CHAPS extracted material from the P1 pellet was submitted to FPLC-Superose 12 (Pharmacia) as described in Materials and Methods. (B) An aliquot of the same extract was precipitated by ethanol before loading on the Suprose 12 column. The protein phosphatase activity was measured in each fraction in the absence (○) or the presence (●) of 100 μM Ins(1,4,5)P_3. The void volume (6.5 ml) is not shown. From Zwiller, et al. (20) and reproduced with permission from Academic Press.

The P1-CHAPS extracted material was submitted to FPLC-gel filtration and a single peak of histone phosphatase activity was observed (Fig. 4A). Ins(1,4,5)P_3 stimulated phosphatase activity of this fraction to a similar extent to that observed in the crude preparation. It has been reported that denaturation of the native enzyme by ethanol precipitation causes the release of the catalytic subunit of the phosphatase from several tissues which is characterized by a reduction in size.[9,13]. When the P1-CHAPS material was precipitated by ethanol before loading on the gel filtration column, such a shift in the phosphatase activity towards lower molecular weight was observed (Fig. 4B). Ins(1,4,5)P_3 still stimulated the catalytic subunit to a level comparable to that of the holoenzyme.

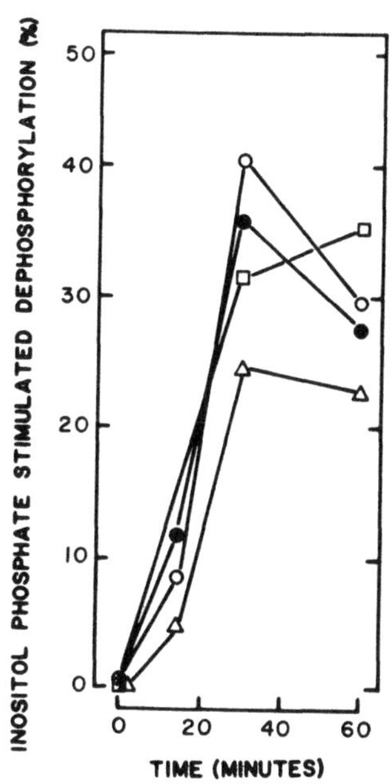

Figure 5. Effect of inositol phosphates on phosphorylase kinase dephosphorylation. ^{32}P labeled phosphorylase kinase was incubated in the presence or absence of inositol phosphates with protein phosphatase partially purified by chromatography on FPLC-Superose 12 after ethanol precipitation, as described in Materials and Methods. At the indicated time, aliquots were removed and subjected to $NaDodSO_4$ polyacrylamide gel electrophoresis. The autoradiogram of the gel was scanned and the radioactivity in the phosphorylase kinase β-subunit expressed as the percentage of the initial phosphorylation. The results represent the dephosphorylation obtained in presence of the inositol phosphate subtracted from the dephosphorylation in its absence. These results are representative of three separate experiments. (●), 1 μM Ins(1,4,5)P_3; (○), 10 μM Ins(1,4,5)P_3; (△), 100 μM Ins(1,4,5)P_3; (□) 10 μM Ins(1,3,4,5)P_4. From Zwiller, et al. (20) and reproduced with permission from Academic Press.

Figure 5 shows that $Ins(1,4,5)P_3$ and $Ins(1,3,4,5)P_4$ also stimulate the protein phosphatase when phosphorylase kinase was used as the substrate instead of histone. Dephosphorylation of the phosphorylated B-subunit of phosphorylase kinase by the catalytic subunit of the phosphatase was optimum in the presence of 1 μM-10 μM $Ins(1,4,5)P_3$. $Ins(1,3,4,5)P_4$ showed a similar stimulation of the dephosphorylation of phosphorylase kinase.

DISCUSSION

We present evidence for a rat brain phosphoprotein phosphatase that can be activated by $Ins(1,4,5)P_3$ and $Ins(1,3,4,5)P_4$. No effect was observed with $Ins(1,4)P_2$ and various inositol monophosphate. The activated phosphatase may be type-1 according to the classification of Ingebritsen and Cohen[14] since it meets the following criteria: a) phosphohistone and phosphorylase kinase are substrates; b) the phosphatase activity and its activation by the inositol phosphates could be observed in the absence of any added metal ion (i.e., in the presence of EDTA); c) the activity is mainly found in the membranes, as has been described for type-1 enzymes[15], while protein phosphatases -2A, -2B and -2C are located almost exclusively in the cytosol[15]; d) the $Ins(1,4,5)P_3$ stimulated protein phosphatase was inhibited by heparin a concentration at which phosphatase type-1 is specifically inhibited.[12,16]

The denaturation of the protein phosphatase by ethanol causes a release of the catalytic subunit which has a lower molecular weight than the holoenzyme complex.[9,13] The observation that $Ins(1,4,5)P_3$ activates both the native enzyme and the catalytic subunit of lower size demonstrates that the very nature of the $Ins(1,4,5)P_3$ target is the catalytic subunit of the protein phosphatase.

To date, the only physiological effect described for $InsP_3$ is its ability to release Ca^{2+} from the endoplasmic reticulum. As we have no evidence at this point that the $InsP_3$ effect on the phosphatase is involved in the process of calcium release, we can only speculate about such a role. Although the concentration of $Ins(1,4,5)P_3$ producing a dephosphorylation of phosphorylase kinase is somewhat in agreement with that needed for Ca^{2+} mobilization, higher concentrations of $Ins(1,4,5)P_3$, but not of $Ins(1,3,4,5)P_4$, are necessary to observe a stimulation of the phosphohistone phosphatase activity. No clear explanation can be presented for this discrepancy, although some non-specific binding of the inositol phosphates to the rather basic protein histone could be evoked. We should point out that some cofactor may be necessary for the full physiological activity of this phosphatase. Interestingly enough, one could observe a parallelism in the ability of the several $InsP_3$ in activating the phosphatase and in releasing calcium from permeabilized 3T3 cells.[6,17] Furthermore, heparin, which blocks $Ins(1,4,5)P_3$ activation of the phosphatase, also inhibits $Ins(1,4,5)P_3$ induced release of Ca^{2+} in permeabilized liver cells and pancreatic B-cells.[18,19] The subcellular distribution of the inositol phosphate-activated protein phosphatase presents another argument that the enzyme might be involved in the mechanism of calcium release. It appears that its specific activity was highest in the cytosol and in the microsomal fraction. It should be noted that Ingebritsen et al.[15] observed that 20% of protein phosphatase type-1 (which may be the phosphatase described here) in liver was associated with the microsomal fraction. Furthermore, the solubility of an enzyme does not preclude the possibility of its substrate being membrane bound. This is the case for the soluble enzyme phospholipase C, whose substrate phosphatidylinositol 4,5-bisphosphate is an obvious plasma membrane component and likewise for protein kinase C and its effector 1,2-diacylglycerol. Clearly, the nature and the localization of the physiological substrates of the inositol phosphate activated protein phosphatase are critical for the elucidation of its possible implication in the mechanism of calcium mobilization. Nevertheless, the effect we describe here may concern a cellular function for which no implication of inositol phosphatase has yet been described.

ACKNOWLEDGMENTS

This work was supported by research grant CA-39745 from the National Cancer Institute.

REFERENCES

1. Nishizuka, Y., 1984, The role of protein kinase C in cell surface signal transduction and tumor promotion. Nature, 308:693.
2. Streb, H., Irvine, R.F., Berridge, M.J. and Schulz, I., 1983, Release of Ca^{2+} from a nonmitochrondrial intracellular store in pancreatic acinar cells by inositol-1,4,5-trisphosphate. Biochem. J. 232:211.
3. Batty, I.R., Nahorski, S.R. and Irvine, R.F., 1985, Rapid formation of inositol 1,3,4,5-tetrakisphosphate following muscarinic receptor stimulation of rat cerebral cortical slices. Biochem. J., 233:211.
4. Downes, C.P., Mussat, M.C. and Michell, R.H., 1982, The inositol trisphosphate phosphomonesterase of the human erythrocyte membrane. Biochem. J., 203:169.
5. Majerus, P.W., Connolly, T.M., Deckim, H., Ross, T.S., Boss, T.E., Ishii, H., Bansal, V.S. and Wilson, D.B., 1986, The metabolism of phosphoinositide-derived messenger molecules. Science, 234:1519.
6. Irvine, R.F., Letcher, A.J., Lander, D.J. and Berridge, M.J., 1986, Specificity of inositol phosphate-stimulated Ca^{2+} mobilization from Swiss-mouse 3T3 cells. Biochem. J., 240:301.
7. Irvine, R.F. and Moor, R.N., 1986, Micro-injection of inositol 1,3,4,5-tetrakisphosphate activates sea urchin eggs by a mechanism dependent on external Ca^{2+}. Biochem. J., 240:917.
8. Morris, A.P., Gallacher, D.V., Irvine, R.F. and Petersen, O.H., 1987, Synergism of inositol trisphosphate and tetrakisphosphate in activating Ca^{2+}-dependent K^+ channels. 330:653.
9. Brandt, H., Killilea, S.D. and Lee, E.Y.C., 1974, Activation of phosphorylase phosphatase by a novel procedure: Evidence for a regulatory mechanism involving the release of a catalytic subunit from enzyme-inhibitor complex(es) of higher molecular weight. Biochem. Biophys. Res. Commun., 61:598.
10. Meisler, M.H. and Langan, T.A., 1969, Characterization of a phosphatase specific for phosphorylated histones and protamine. J. Biol. Chem., 244:4961.
11. Killilea, S.D., Aylward, J.H., Mellgren, R.L. and Lee, E.Y.C., 1978, Purification and properties of bovine myocardial phosphorylase phosphatase (protein phosphatase C). Arch. Biochem. Biophys., 191:638.
12. Gergely, P., Erdodi, F. and Bot, G., 1984, Heparin inhibits the activity of protein phosphatase-1. FEBS Lett. 169:45.
13. Mumby, M.C., Green, D.D. and Russell, K.L., 1985, Structural characterization of cardiac protein phosphatase with a monoclonal antibody. J. Biol. Chem., 260:13763.
14. Ingebritsen, T.S. and Cohen, P., 1983, The protein phosphatases involved in cellular regulation. 1. Classification and substrate specificity. Eur. J. Biochem., 132:255.

15. Ingebritsen, T.S., Stewart, A.A. and Cohen, P., 1983, The protein phosphatases involved in cellular regulation. 6. Measurement of type-1 and type-2 protein phosphatases in extracts of mammalian tissue: An assessment of their physiological roles. Eur. J. Biochem., 132:297.
16. Brautigan, D.L., Gruppuso, P.A. and Mumby, M., 1986, Protein phosphatase type-1 and type-2 catalytic subunits both bind inhibitor-2 and monoclonal immunoglobulins. J. Biol. Chem., 261:14924.
17. Irvine, R.F., Brown, K.D. and Berridge, M.J., 1984, Specificity of inositol trisphosphate-induced calcium release from permeabilized Swiss-mouse 3T3 cells. Biochem. J., 222:269.
18. Hill, T.H., Berggren, P.O. and Boynton, A.L., 1987, Heparin inhibits inositol trisphosphate-induced calcium release from permeabilized rat liver cells. Biochem. Biophys. Res. Commun., 149:897.
19. Nilsson, T., Zwiller, J., Boynton, A.L. and Berggren, P.O., 1988, Heparin inhibits IP_3-induced Ca^{2+} release in permeabilized pancreatic β Cells. FEBS Letts, 229:221-214.
20. Zwiller, J., Ogasawara, E.M., Nakamoto, S.S. and Boynton, A.L. 1988, Stimulation by inositol trisphosphate and tetrakisphosphate of a protein phosphatase. Biochem. Biophys. Res. Commun., 155:767.

THE DOMAIN STRUCTURE OF THE CALMODULIN-DEPENDENT PHOSPHODIESTERASE ISOZYMES

Jeffrey P. Novack[1], Harry Charbonneau[2], Donald K. Blumenthal[3], Kenneth A. Walsh[2], and Joseph A. Beavo[1]

Departments of Pharmacology [1] and Biochemistry [2]
University of Washington, Seattle, WA
Dept. of Biochemistry[3], Univ. of Texas Health Center, Tyler, Texas

I. INTRODUCTION

The calmodulin-dependent cyclic nucleotide phophodiesterases (CaM-PDEs) hydrolyze both cAMP and cGMP and are stimulated 5-15 fold by the binding of calcium and calmodulin (CaM)[1,2]. These isozymes provide a mechanism whereby calcium, acting as a second messenger, can alter the intracellular levels of cAMP and cGMP. The CaM-PDEs are members of a large family of PDE isozymes that differ in molecular weight, substrate specificity, kinetic parameters and most significantly in mode of regulatory control. Protein and cDNA sequence studies show that PDEs from diverse species (Drosophila, Yeast, and Bovine) contain a 270 amino acid segment displaying significant sequence similarities. We have postulated[3,4] that this homologous domain comprises the catalytic site of these isozymes, while the non-homologous regions have domains that are involved in regulatory function (see Figure 1). Evidence that the conserved domain is associated with the catalytic site is derived from limited proteolysis of the 61K CaM-PDE which produces a 36K fragment that is fully active, yet CaM-independent. This 36K fragment, which is produced by trypsin treatment in a method similar to Kincaid[5], can be immunoblotted by anti-peptide antisera directed against sequence from the conserved domain[4].

There is substantial evidence for the existence of multiple isozymes within the CaM-PDE family (see Table 1). Four bovine isozymes of CaM-PDE have been purified and characterized. Two isozymes (designated 63K and 61K) have been purified from bovine brain. These two isozymes are distinguished primarily by molecular weight and antibody reactivity[7,8]. A third CaM-PDE isozyme isolated from bovine heart (designated 59K) differs not only in molecular weight, but also has a 10-20 fold higher affinity for calmodulin[9]. Hansen and Beavo[7] have characterized a monoclonal antibody

(ACAP-1) that binds both the 61K brain and the 59K heart isozymes. ACAP-1 has a 100-fold higher affinity for the Ca^{2+}/calmodulin-PDE complex than for the free PDE; a

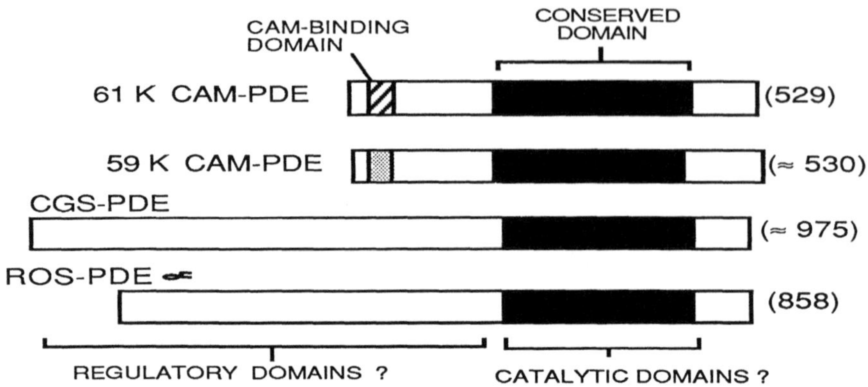

Figure 1.
Schematic representation of mammalian (bovine) phosphodiesterase sequences including conserved homologous domains (black) and regions of difference (white). The number of residues, where known, is indicated on the right in parentheses. CGS-PDE refers to the cyclic GMP-stimulated PDE from bovine heart and ROS-PDE α refers to the rod outer segment PDE from bovine retina[6].

feature that has been exploited for immuno-affinity chromatography. The 63K brain isozyme is not recognized by the ACAP-1 antibody (see Table 1). Recently, Sharma and Wang[10] have purified a 58K isozyme from bovine lung using a monoclonal antibody that recognizes the 61K isozyme. The lung 58K isozyme has a high affinity for calmodulin, which may actually be tightly bound as a subunit in a manner similar to that observed for phosphorylase kinase[11].

Although the CaM-PDE isozymes are widely distributed in mammalian tissues, few specific functional roles have been defined. In cultured astrocytoma cells cyclic AMP levels are decreased by cholinergic agonists in what appears to be a calcium mediated activation of a PDE[12]. Saitoh et. al.[13] have found that more CaM is associated with the PDE in contracted pig atrial strips than when the atria are relaxed. This implies that calcium/CaM can associate in vivo with the PDE in response to increased calcium levels. It is not known, however, which CaM-PDE isozymes are present in these tissues.

We are currently determining the sequence of the brain 61K and the heart 59K CaM-PDE isozymes in order to understand the structural differences between the isozymes, to explore the mechanism of calmodulin activation, and to gain some insight into specific isozyme functions. The relationship between the 61K brain and the 59K heart proteins is particularly intriguing because of the marked difference in calmodulin affinity between these isozymes. It has been suggested that the heart isozyme could be a proteolytic form of the 61K brain isozyme[14]. This is supported by the observation of similar peptide maps for the brain and heart isozymes[14].

II. ANTIBODY REACTIVITY OF THE LUNG 58K CAM-PDE

As previously described, the ACAP-1 monoclonal antibody recognizes both the 61K brain and the 59K heart isozymes but not the 63K brain form. It was of interest to also examine the antibody reactivity of the 58K lung isozyme purified by Sharma and Wang[10]. In lung homogenates the ACAP-1 antibody immunoprecipitated about 80-90% of the calmodulin-dependent activity. This is similar to the percentage immunoprecipitated from a heart extract homogenized under identical conditions. The ACC monoclonal antibody, which recognizes Ca^{2+}/CaM with a much higher affinity when it is complexed with the PDE, also immunoprecipitated about 80% of the CaM-dependent activity from both lung and heart (see Table 1). The 58K, 61K, and 59K CaM-PDE isozymes all appear to share at least one common antigenic determinant. Thus, it is likely that the lung 58K isozyme shares some structural similarities with the heart and brain isozymes. However, the exact structural relationship between these isozymes awaits further sequencing studies. The CaM-binding domain of the lung isozyme will be of particular interest.

Table 1.

A comparison of the subunit molecular weights and antibody reactivity of the purified bovine isozymes of the calmodulin-stimulated phosphodiesterases.

TISSUE	MOLECULAR WEIGHT	ANTIBODY REACTIVITY	
		ACAP	ACC
Bovine Brain	63K	-	+
	61K	+	+
Bovine Heart	59K	+	+
Bovine Lung	58K [a]	+	+

a Data from Sharma and Wang (1986) [10]

III. PRIMARY STRUCTURE COMPARISON OF THE BRAIN 61K AND THE HEART 59K ISOZYMES

The complete protein sequence of the brain 61K CaM-PDE has been determined and greater than 90% of the heart 59K CaM-PDE residues have been identified by methods previously described[3]. The sequence data for the 59K CaM-PDE were obtained from the Edman degradation of peptides produced by cleavage at lysine and methionine residues. The available peptide sequences have been aligned and overlapped to generate two larger segments of sequence of about 150 residues. The remainder of the peptide sequences have been overlapped to produce 7 segments of 20-40 residue in length. All available segments of sequence can be aligned with the 61K CaM-PDE as shown in figure 2. This alignment leaves 7 gaps where there is no corresponding 59K sequence. With the exception of a single gap of 12 residues in the conserved domain, all other gaps are less than three residues in length. About 15 residues at the N-terminus have not yet been sequenced and the C-terminus has not been unambiguously identified. However, sequence data and amino acid composition from the whole protein as well as from the putative N- and C- terminal peptides (data not shown) suggest that the 59K and 61K isozymes are nearly the same length.

Every amino acid residue of the heart 59K isozyme that has been observed is identical to the brain 61K except for 15 of 18 residues in a contiguous stretch near the N-terminal of both proteins (Figure 2). The difference region corresponds to a portion of the 61K isozyme that has been preliminarily identified as a CaM-binding domain. By analogy, it is likely that this difference region also comprises the CaM-binding site of the 59K CaM-PDE. Also, the sequence differences observed for the 59K isozyme could explain the increased affinity of the heart 59K CaM-PDE for calmodulin. In order to address this

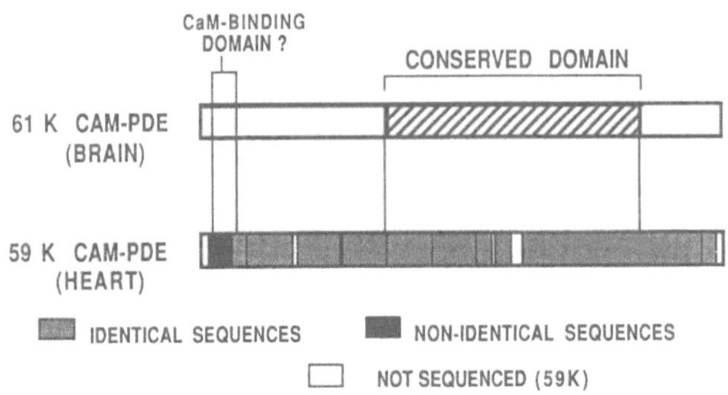

Figure 2.

Schematic representation of the sequence identities between the brain 61K and heart 59K CaM-PDEs. The hatched area in the 61K isozyme represents the 270 amino acid conserved domain[3].

question, we prepared a synthetic peptide corresponding to this region of the heart 59K isozyme and assessed the ability of this peptide to bind CaM. These experiments are described in section IV.

The existence of a single defined region of non-identity strongly implies an alternative RNA splicing mechanism for the generation of the heart 59K and brain 61K isozymes. It is unlikely that two separate genes could encode these isozymes since no divergence of the amino acid sequence appears to have occurred in the identical regions. Tissue-specific alternative RNA splicing of exons encoding these CaM-binding domains that have differential affinity for CaM would generate a mechanism for the differential response of these isozymes to calcium/CaM.

IV. CALMODULIN-BINDING DOMAINS OF THE 61K BRAIN AND 59K HEART ISOZYMES

The M-13 peptide from rabbit skeletal muscle myosin light chain kinase is probably the best characterized CaM-binding domain[15]. Like most peptides found to bind calmodulin in a calcium-dependent manner, the MLCK peptide has a large number of basic and hydrophobic residues which have the potential to form an amphipathic helix[16]. In an attempt to identify regions of the brain 61K CaM-PDE that could comprise a CaM-binding site, the entire sequence was searched for segments of sequence having the ability to form a helix with basic and amphipathic character. A twenty residue region beginning at residue 22 and lying outside the conserved domain has these properties, making it a good candidate for the CaM-binding domain (see figure 3).

The preliminary identification of the N-terminal regions from the 61K brain and the 59K heart as the CaM-binding domains is consistent with results from limited proteolysis[4]. Limited proteolysis of the 61K brain isozyme with trypsin produces an irreversible calmodulin-independent activation that yields a stable 36K fragment. This 36K CaM-independent fragment also was immunoblotted by an anti-peptide antibody to the conserved domain. Allowing for the inaccuracy of molecular weight determination by SDS-PAGE, this evidence places the region responsible for CaM-stimulation in the N-terminal third of the protein.

Regions from the heart 59K isozymes were also scanned for amphipathic character, helix forming ability, and a net number of basic residues. The N-terminal difference region in the heart isozyme has the properties of a CaM-binding domain. A comparison of proposed CaM-binding domains from the heart and 61K brain CaM-PDE isozyme with that of rabbit skeletal muscle myosin light chain kinase (MLCK) is shown in figure 3. As has been seen with CaM-binding domains from other proteins, little sequence identity is found and there is no evidence for a common evolutionary relationship.

ENZYME	RES.# START		END
61K CaM-PDE	22	QTE**K**MWQ**R**L**K**GIL**R**CLV**K**QLE**K**	44
59K CaM-PDE	23	I**RRK**HLQ**R**PIF**R**L**R**CLV**K**QLE**K**	45
Rabbit SkM-MLCK	575	LM**KRR**W**KK**NFIAVSAAN**R**F**KK**I	597

Figure 3.

A comparison of the calmodulin-binding domains from the 61K brain and 59K heart CaM-PDEs with the CaM-binding domain from myosin light chain kinase[15]. Basic residues have been highlighted in bold print. Residue numbers for the 59K CaM-PDE are approximations based on amino acid composition from a blocked N-terminal peptide.

We have prepared synthetic peptides with sequences corresponding to the proposed heart and brain CaM-binding domains (see figure 5). These peptides have been tested for their ability to inhibit the activation of the heart 59K CaM-PDE by calmodulin (see figure 4). The synthetic peptide analogue for the 61K CaM-PDE does not correspond exactly to the putative CaM-binding region in the native protein . A methionine residue has been replaced with the isosteric analogue norleucine and an N-terminal arginine replaces a glutamate (figure 4, bold type-face). The heart 59K synthetic peptide is one residue shorter and corresponds exactly to the native protein sequence. As is shown in figure 4, nanomolar concentrations of these two peptides can inhibit the CaM-activation of purified heart 59K CaM-PDE. The 61K synthetic peptide also inhibits the purified brain 61K-CaM-PDE isozyme at similar concentrations (data not shown). This inhibition can be overcome by the addition of excess calmodulin. Interestingly, the inhibitory effects of the synthetic peptides reflects the affinities of the native isozymes for calmodulin (e.g.- the 59K heart analogue is a more potent inhibitor than the brain 61K synthetic peptide). This preliminary synthetic peptide inhibition data also supports the proposed location of the CaM-binding site.

The differences in inhibition by the heart and brain synthetic peptides is consistent with the idea that the observed sequence changes result in a higher CaM-affinity for the intact isozymes. A closer examination of the differences between the CaM-PDE isozymes reveals that the dissimilar region encompasses about two thirds of the proposed calmodulin binding region (see Figure 5). However, it is not clear which sequence changes confer a higher affinity for calmodulin. The heart CaM-binding domain has a greater number of basic residues than the analogous 61K brain region and the distribution pattern of hydrophobic and basic residues is altered (see figure 5).

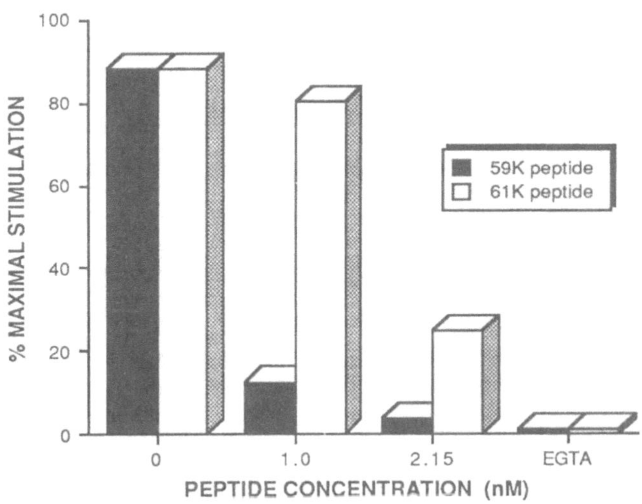

Figure 4.

Inhibition of the heart CaM-PDE by synthetic peptide analogues of the heart 59K and brain 61K CaM-PDEs. The peptide analogue sequences: brain 61K: ***RKN***$_{L}$*WQRLKGILRCLVKQLEKG-NH*$_2$
heart 59K: RKHLQRPIFRLRCLVKQLEKG-NH$_2$

Peptide concentrations were determined by amino acid analysis.

Interestingly, the heart domain also has a proline residue in the middle of the proposed CaM-binding domain. It is also not clear how the presence of a proline residue in the middle of the proposed CaM-binding domain affects CaM affinity. In order to address some of these questions additional synthetic peptide analogs of different lengths and with specific deletions are being made in order to determine which residues or combinations of residues are most important for CaM-affinity.

BRAIN - Q T E **K** M W Q **R** L **K** G I L **R** C L V **K** Q L E **K**

HEART - I **R** **R** **K** H L Q **R** P I F **R** L **R** C L V **K** Q L E **K**

Figure 5.

A comparison of the CaM-binding domains of the 59K heart and 61K brain CaM-PDEs. The underlined residues are identical in both isozymes and the basic residues are highlighted in bold print.

V. CONCLUSIONS

We have clearly demonstrated that the heart 59K is a distinct protein from the brain 61K CaM-PDE and is not a proteolytic product of a 61K precursor. The primary sequence difference between these isozymes is localized to a small N-terminal region that has been tentatively identified as part of the CaM-binding domain in both isozymes. The structural differences in the CaM-binding domains may explain the higher CaM-binding affinity of the heart 59K isozyme. We propose that the 59K heart and 61K brain isozymes are produced from the same gene by a tissue-specific alternative RNA splicing mechanism. The rat pyruvate kinase gene is another example of tissue-specific alternative splicing of N-terminal domains that leads to functional differences[17].

We suggest that the proposed tissue-specific alternative RNA splicing mechanism provides the means for generating isozymes with identical kinetic parameters (e.g.- V_{max}, K_m for substrate) that differ only in their responsiveness to calcium and/or CaM. This allows for the production of specific PDE isozymes in individual tissues. For example, alternative splicing could produce CaM-PDE isozymes that are capable of responding to different ranges calcium, since an increased affinity for CaM implies a corresponding increase in calcium sensitivity[18]. This feature would be important if the maximal concentration of calcium produced in response to stimuli were different in the brain and heart tissues. Alternatively, the increased affinity of the 59K heart isozyme for CaM might offset the effects of having lower available CaM in the heart versus the brain[2]. Bovine cardiac troponin T isoforms have been shown to exhibit differential calcium-sensitive binding to tropomyosin due to alternative RNA splicing[19].

Further work using type-specific anti-peptide antibodies is in progress to determine if other CaM-PDE isozymes are similar to the heart, the 61K brain, or are distinct forms.

REFERENCES

1. Klee, C.B., Crouch, T.H., Krinks, M.H., Subunit structure and catalytic properties of bovine brain Ca^{2+}-dependent cyclic nucleotide phosphodiesterase, Biochemistry 18:722 · (1979)
2. Klee, C.B., Crouch, T.H., Richman, P.G., Calmodulin, in "Ann. Rev.of Biochemistry", 49:489 (1980)
3. Charbonneau, H., Beier, N., Walsh, K.A., and Beavo, J.A., Identification of a conserved domain among cyclic nucleotide phosphodiesterases, Proc. Natl. Acad. Sci. USA 83:9308 (1986).
4. Charbonneau, H., Novack,J.P., MacFarland, R.T., Walsh, K.A., and Beavo, J.A., Structure and function of calmodulin-dependent phosphodiesterase isozymes, in: "Calcium-Binding Proteins in Health and Disease", Norman, A.W., Vanaman, T.C., Means, A.R., eds., Academic Press, San Diego USA 505 (1987).
5. Kincaid, R.L., Stith-Coleman, I.E., and Vaughan, M., Proteolytic activation of calmodulin-dependent cyclic nucleotide phosphodiesterase, J. Biol. Chem., 260:9009 (1984).
6. Ovchinnikov, Y.A., Lipkin, V.M., Kumarev,V.P., Gubanov, V.V., Khramstov, N.V., Akhmedov, N.B., Zagranichny, V.E., Muradov, K.G., Cyclic GMP phosphodiesterase from cattle retina: Amino acid sequence of the γ-subunit and nucleotide sequence of the corresponding cDNA, Fed. of Eur. Biochemical Soc., 204:288 (1986).
7. Hansen, R.S., and Beavo, J.A., Purification of two calcium/calmodulin-depndent forms of the cyclic nucleotide phosphodiesterase by using conformation-specific monoclonal antibody chromatography, Proc. Natl. Acad. Sci. USA 79:2788 (1982).
8. Sharma, R.K., Adachi, A.M., Adachi, K., Wang, J.H., Demonstration of bovine brain calmodulin-dependent cyclic nucleotide phosphodiesterase isozymes by monoclonal antibodies, J. Biol. Chem., 259:9248 (1984).
9. Hansen, R.S., and Beavo, J.A., Differential recognition of calmodulin-enzyme complexes by a conformation-specific anti-calmodulin monoclonal antibody, J. Biol. Chem., 261:14636 (1986).
10. Sharma, R.K., and Wang, J.H., Purification and characterization of bovine lung calmodulin-dependent cyclic nucleotide phosphodiesterase J.Biol.Chem., 261:14160 (1986).
11. Cohen, P., Burchell, A., Foulkes, J.G., Cohen, P.T.W., Vanaman, T.C., Nairn, A.C., Identification of the Ca^{2+}-dependent modulator protein as the fourth subunit of Rabbit skeletal muscle phosphorylase kinase, Fed.of Eur. Biology Soc.Lett., 92:287 (1978).
12. Meeker, R.B., Harden, T.K., Muscarinic cholinernergic receptor-mediated activation of phosphodiesterase, Mol. Pharm., 22:310 (1982)
13. Saitoh, Y.,, Hardman, J.G., Wells, J.N., Differences in the association of calmodulin with cyclic nucleotide phosphodiesterase in relaxed and contracted atrial strips, Biochemistry, 24:1613 (1985).
14. Krinks, M.H., Haiech, J., Rhoads, A., Klee, C.B., Reversible and irreversible activation of cyclic

nucleotide phosphodiesterase: Separation of the regulatory and catalytic domains by limited proteolysis, in: Adv.in Cyclic Nucleotide and Protein Phosphorylation Res., eds., Strada, S.J., and Thompson, W.J., 16:31 (1984).
15. Blumenthal, D.K., Takio, K., Edelman, A.M., Charbonneau, H., Titani, K., Walsh, K.A., Krebs, E.G., Identification of the calmodulin-binding domain of skeletal muscle myosin light chain kinase, Proc. Natl. Acad. Sci., 82:3187 (1985).
16. DeGrado, W.F., Prendergast, F.G., Wolfe, H.R., Cox, J.A., The design, synthesis, and characterization of tight-binding inhibitors of calmodulin, J. Cell. Biochem., 29:83 (1985).
17. Noguchi, T., Yamada, K., Inoue, H., Matsuda, T., Tanaka, T., The L- and R-type isozymes of rat pyruvate kinase are produced from a single gene by the use of different promoters, J. Biol. Chem., 262:14366 (1987)
18. Olwin, B.B., Storm, D.R., Calcium binding to complexes of calmodulin and calmodulin binding proteins, Biochemistry, 24:8081 (1985)
19. Tobacman, L.S., Lee, R., Isolation and function comparison of bovine cardiac troponin T isoforms, J. Biol. Chem., 262:4059 (1987).

REGULATION OF THE 63-KDA SUBUNIT-CONTAINING CALMODULIN-DEPENDENT CYCLIC NUCLEOTIDE PHOSPHODIESTERASE ISOZYME

Rajendra K. Sharma, Guang Yi Zhang, Marilyn J. Mooibroek and Jerry H. Wang

Cell Regulation Group
Department of Medical Biochemistry
The University of Calgary
3330 Hospital Drive N.W.
Calgary, Alberta, T2N 4N1, Canada

INTRODUCTION

Calmodulin-stimulated cyclic nucleotide phosphodiesterase, originally thought to be a single enzyme species in all tissues has been shown to exist in isozymic forms [for review, 1-5]. Homogeneous preparations of calmodulin-dependent phosphodiesterases showing distinct kinetic, regulatory, molecular and immunological properties have been obtained [for review, see (1)]. The well-characterized mammalian isozymes possess low basal (Ca^{2+}-independent) activity and exhibit relatively low affinity towards cyclic AMP. Thus, it appears that these enzymes function most efficiently when cells are stimulated to increase the concentration of both Ca^{2+} and cAMP and, as such, they are expected to play important roles in integrating the regulatory actions of Ca^{2+} and cAMP.

There are two major calmodulin-dependent cyclic nucleotide phosphodiesterase isozymes in bovine brain, designated 60-kDa and 63-kDa phosphodiesterase isozymes according to their subunit molecular weight[1]. Although the two isozymes show similar kinetic properties, they are regulated differentially by second messenger-dependent phosphorylation reactions. The main difference is that while the 60-kDa isozyme is a substrate of the cAMP-dependent protein kinase[1], the 63-kDa isozyme is phosphorylated in Ca^{2+}/calmodulin-dependent reactions[1]. Both isozymes are under complex regulation involving multiple second messenger-dependent reactions. To achieve meaningful regulation, the multiple regulatory activities of each of the isozymes have to intereact under certain defined conditions. Working hypotheses have been proposed to integrate the multiple second messenger-dependent regulatory activities directed towards the individual isozymes during the surge in cell Ca^{2+} and cAMP[6].

In this present communication, the regulation of the 63-kDa brain calmodulin-dependent phosphodiesterase isozyme by Ca^{2+} and cAMP is used to illustrate the role of the enzyme in signal integration.

MULTIPLE CALMODULIN ACTIONS ON 63-KDA ISOZYME

Table 1 summarizes the multiple second messenger-dependent regulatory reactions towards the two bovine brain cyclic nucleotide phosphodiesterase isozymes. One unique feature of the multiple regulatory actions towards 63-kDa phosphodiesterase isozyme is that all three actions are dependent on Ca^{2+} and calmodulin. Furthermore, these reactions can produce opposite effects on phosphodiesterase activity. The direct action of calmodulin on the 63-kDa is opposed by the kinase-mediated calmodulin action while the kinase-mediated and phosphatase-mediated calmodulin regulatory reactions are oppositely directed. Superimposed on such antagonistic calmodulin actions toward the phosphodiesterase isozyme is the stimulation of adenylate cyclase by calmodulin, which counteracts the calmodulin stimulation of the phosphodiesterase. The apparent contradictions between the various calmodulin actions on cAMP metabolism may be removed if these reactions are temporally separated. Such a temporal separation may be proposed to exist if the different regulatory actions of calmodulin are turned on by different concentrations of Ca^{2+}. Based on this consideration, a working hypothesis may be proposed which suggests that adenylate cyclase and the protein kinase can be turned on at lower Ca^{2+} concentrations, whereas, the activities of the phosphodiesterase isozyme and phosphatase require higher concentrations of Ca^{2+} (Fig. 1)[16]. The first two reactions, which act concertedly to increase cAMP concentration, will predominate at the early stage of a surge in cytosolic Ca^{2+}, while the other two reactions will prevail to reduce cAMP concentrations at the height of the Ca^{2+} flux. One possible mechanism whereby the multiple calmodulin actions may be temporally separated is that brain adenylate cyclase and the calmodulin-dependent protein kinase be activated by lower concentrations of Ca^{2+} than the phosphatase and phosphodiesterase isozyme. Indeed, it has been shown that brain adenylate cyclase is activated by a much lower concentration of Ca^{2+} than brain phosphodiesterase[7].

Table 1. Regulation of PDE Isozymes by Ca^{2+} and cAMP

	60-kDa PDE Isozyme	63-kDa PDE Isozyme
Regulatory Properties	1. Activation by Ca^{2+} and calmodulin.	1. Activation by Ca^{2+} and calmodulin.
	2. Phosphorylation by cAMP-dependent protein kinase with reduction of PDE affinity towards calmodulin.	2. Phosphorylation by calmodulin-dependent protein kinase with consequent decrease in affinity towards calmodulin.
	3. Blockage of PDE phosphorylation by Ca^{2+} and calmodulin.	3. Reversal of PDE-phosphorylation by calmodulin-dependent phosphatase.
	4. Reversal of PDE phosphorylation by Ca^{2+} and calmodulin-dependent phosphatase.	

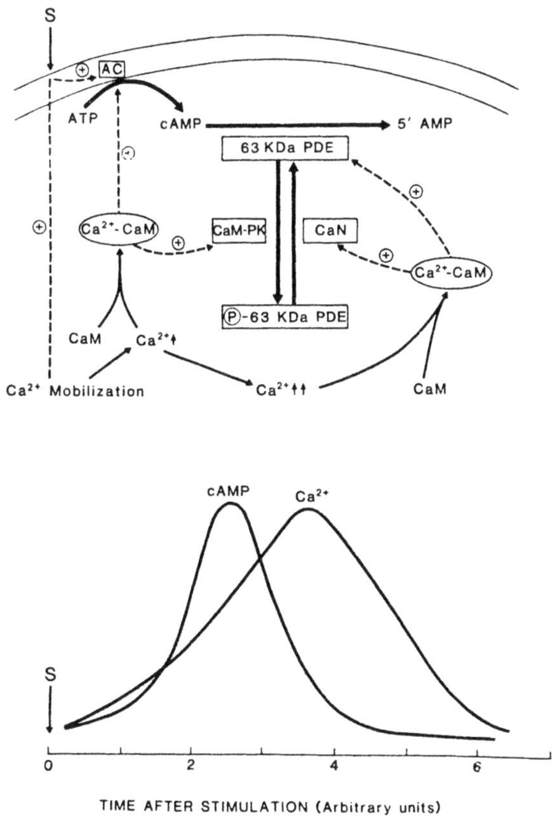

Fig. 1. Schematic representation of the relationship between the Ca^{2+} and cAMP mediated regulation of the 63-kDa subunit-containing phosphodieterase isozyme and the coupling between Ca^{2+} and cAMP fluxes. AC, adenylate cyclase, PDE, phosphodiesterase; CaN, calcineurin; CaM-PK, calmodulin-dependent protein kinase; $Ca^{2+}\uparrow$, small increase in Ca^{2+} concentration; $Ca^{2+}\uparrow\uparrow$, large increase in Ca^{2+} concentration.

To test further the validity of the working hypothesis depicted in Fig. 1, we have purified and characterized the 63-kDa phosphodiesterase kinase from bovine brain. Bovine brain contains two calmodulin-dependent phosphodiesterase kinases which can be separated on a Sephacryl S-300 column (Fig. 2). The higher molecular weight, 500,000 dalton phosphodiesterase kinase has been purified to close to homogeneity, showing essentially a single polypeptide on SDS-PAGE with a molecular weight about 50,000 (Fig. 2). The purified phosphodiesterase kinase was found to phosphorylate histones, casein, MAPS and tubulin in addition to the 63-kDa isozyme. The phosphorylation of 63-kDa phosphodiesterase isozyme by the purified kinase resulted in the maximal incorporation of 1 mol phosphate per mol phosphodiesterase subunit. While it was shown previously[6] that phosphorylation of the 63-kDa phosphodiesterase isozymes by the total calmodulin binding protein fraction resulted in a maximal incorporation of 2 mol phosphate per mol phosphodiesterase subunit, preliminary results indicated that the phosphodiesterase kinase represented by the second activity peak on the S-300 gel filtration column phosphorylates the phosphodiesterase isozyme at a distinct site.

Based on molecular weight, subunit structure and substrate specificity, it may be suggested that the purified phosphodiesterase kinase is a member of the calmodulin-dependent protein kinase II group. Members of this group have been purified from various tissues and shown to exist in isozyme forms [for review, see (8)]. The identification of the phosphodiesterase kinase as calmodulin-dependent protein kinase II is in agreement with the earlier observation of Fukunaga et al[9] who have shown that a 640-kDa calmodulin-dependent protein kinase from rat brain catalyzes the phosphorylation of a rat brain calmodulin-dependent cyclic nucletoide phosphodiesterase.

DEPENDENCE OF PHOSPHODIESTERASE PHOSPHORYLATION ON CA^{2+} AND CALMODULIN

The phosphorylation of 63-kDa phosphodiesterase isozyme by the purifield kinase shows an almost total dependence on Ca^{2+}. The dose-dependence of Ca^{2+}-activation of the purified phosphodiesterase kinase was characterized using both casein and 63-kDa phosphodiesterase as substrates (Fig. 3). The dose-dependence of Ca^{2+}-activation of both phosphorylations were determined at 20 μM of calmodulin. Activation of the phosphodiesterase phosphorylation required higher Ca^{2+} concentration than that of casein

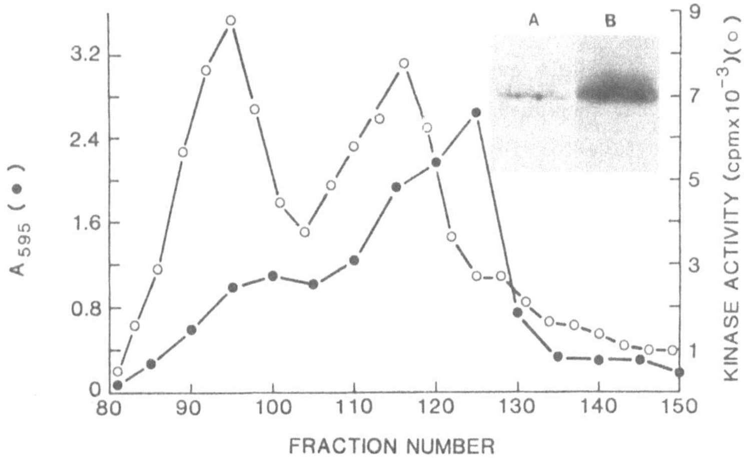

Fig. 2. Gel filtration chromatography of bovine brain calmodulin-dependent PDE protein kinases. Total calmodulin binding proteins were applied to the Sephacryl S-300 column. Calmodulin-dependent protein kinase activity was monitored by using the 63-kDa cyclic nucleotide phosphodiesterase isozyme as substrate in the presence of Ca^{2+} and calmodulin (o---o) and protein was determined by dye-binding method (●---●). Sodium dodecyl sulfate gel electrophoresis of purified calmodulin-dependent protein kinase II. Lane A, purified calmodulin-dependent protein kinase II; lane B, autoradiograph of purified calmodulin-dependent protein kinase II.

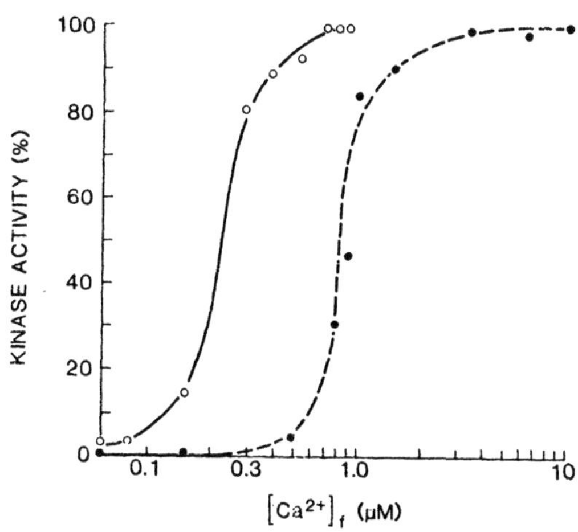

Fig. 3. Effect of Ca^{2+} on calmodulin-dependent protein kinase II activity. Casein (o---o) or 63-kDa phosphodiesterase isozyme (●---●) were phosphorylated for 10 min by calmodulin-dependent protein kinase II in the presence of 20 μM calmodulin and various concentrations of Ca^{2+}.

phosphorylation. A likely explanation is that the effective concentration of calmodulin in the phosphodiesterase phosphorylation reaction was reduced due to the competitive binding of calmodulin to the phosphodiesterase isozyme, the substrate protein. Contrary to the expectation[6], the dose dependence of Ca^{2+} activation of phosphodiesterase phosphorylation is not lower than that required for phosphodiesterase activity[10] at comparable calmodulin concentrations.

AUTOPHOSPHORYLATION OF THE PROTEIN KINASE

The purified bovine brain calmodulin-dependent protein kinase II undergoes a rapid autophosphorylation reaction which is complete with a maximal incorporation of about 1 mol phosphate per mol protein subunit in a few minutes. The reaction is essentially completely dependent on the presence of Ca^{2+} and calmodulin; however, once initiated, it converts to a Ca^{2+}-independent reaction. In one experiment, autophosphorylation was initiated by [γ-^{32}P]ATP addition, and immediately following mixing, excess EGTA was added. This EGTA addition was found to have little effect on the autophosphorylation reaction. An attempt was then made to slow down the reaction by lowering the reaction temperature to 10°C. Fig. 4 shows that even at 10°C, the reaction was rendered essentially Ca^{2+}-independent within 15 seconds at which time the level of phosphate incorporation was about 1 mol phosphate per mol enzyme subunit.

This is in agreement with work by other investigators[11-14] who have shown that calmodulin-dependent protein kinase II's from different sources are converted to Ca^{2+}-independent forms upon autophosphorylation and that autophosphorylation itself becomes Ca^{2+} independent when only 2 of 12 subunits of the enzyme have been phosphorylated. The phosphodiesterase phosphorylating activity of bovine brain calmodulin-dependent protein kinase II was also rendered essentially Ca^{2+}-independent upon autophosphorylation (Fig. 5). Maximal phosphate incorporation into the phosphodiesterase isozyme, catalyzed by the autophosphorylated protein kinase, was about 1 mol per mol of 63-kDa subunit, irrespective of the presence of Ca^{2+} or EGTA. Thus, both the rate and extent of phosphorylation of the 63-kDa phosphodiesterase by the autophosphorylated protein kinase is independent of Ca^{2+}.

KINASE AUTOPHOSPHORYLATION AND THE REGULATION OF 63-KDA PHOSPHODIESTERASE

The working hypothesis depicting the regulation of 63-kDa phosphodiesterase (Fig. 1) requires a temporal separation of the multiple calmodulin actions, with the phosphorylation of the phosphodiesterase isozyme preceding

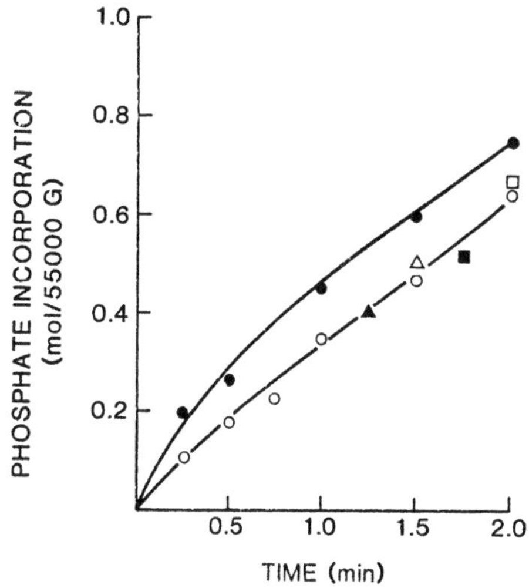

Fig. 4. Effect of EGTA on the autophosphorylation of calmodulin-dependent protein kinase II. Autophosphorylation reactions were carried out at 30°C (●---●) or 10°C (o---o) in the presence of Ca^{2+} and calmodulin. In separate autophosphorylation reactions, performed at 10°C, EGTA was added at 15 (▲), 30 (△), 45 (■), or 60 sec (□) and phosphate incorporation was determined after 1 min.

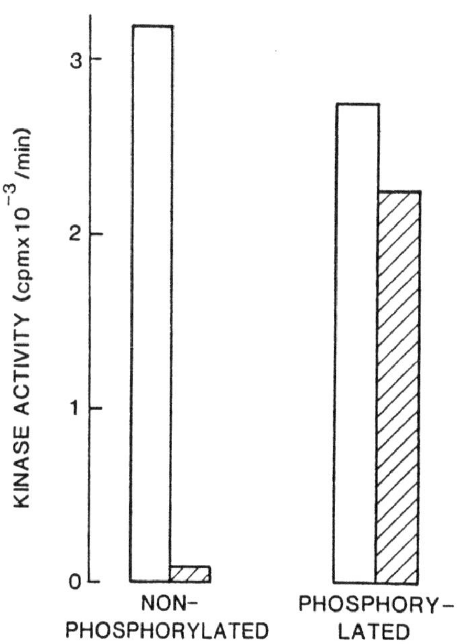

Fig. 5. Conversion of calmodulin-dependent protein kinase II to a calmodulin-independent form by autophosphorylation. The 63-kDa isozyme phosphorylation activity was determined in the presence of calmodulin and Ca^{2+} □ or EGTA ▨.

direct activation of the enzyme by calmodulin. The suggestion that such temporal separation of calmodulin actions may arise from a differential Ca^{2+} sensitivity of the protein kinase and the phosphodiesterase isozyme is untenable since the protein kinase is not activated by lower concentrations of Ca^{2+} compared with the calmodulin-dependent phosphodiesterase (see above). However, the observation that the protein kinase becomes Ca^{2+}-independent and fully activated upon autophosphorylation suggests that, as an alternative mechanism, the protein kinase may use the autophosphorylation reaction to override its requirement for high concentrations of Ca^{2+} (Fig. 6).

While initiation of autophosphorylation of the kinase still requires high Ca^{2+} concentrations for the phosphorylation of PDE, recent studies of Ca^{2+} transients in single cells suggest that a Ca^{2+} surge may actually be composed of a series of rapid Ca^{2+} transients. As a result, the protein kinase could be autophosphorylated, which we have shown to be a very efficient reaction and converted to a Ca^{2+}-independent state at the initial Ca^{2+} transient. Its activity would be maintained independently of subsequent oscillations in Ca^{2+} concentration. Another possibility is that the protein kinase may be localized proximal to Ca^{2+} entry sites and briefly exposed to a high concentration of Ca^{2+} at the onset of a cell Ca^{2+} surge. Consequently, the enzyme would become autohposphorylated and Ca^{2+}-independently activated.

In addition to providing a mechanism for activating protein kinase prior to the activation of the phosphodiesterase and calcineurin phosphatase, autophosphorylation of the protein kinase as a regulatory device also increases the flexibility of the regulation of 63-kDa phosphodiesterase. The monocyclic cascade becomes a bicyclic cascade[17], which introduces a number of additional regulatory possibilities. For example, the reversal of calmodulin-dependent protein kinase II autophosphorylation has been suggested to be catalyzed by protein phosphatase I which is subject to the action of a number of protein inhibitors. Some of these inhibitors, e.g., protein inhibitor I and DARPP 32, can be converted from inactive to active phosphatase inhibitors upon phosphorylation by cAMP-dependent protein kinase. As a result, cAMP may exert an inhibitory effect on 63-kDa phosphodiesterase through a regulatory cascade involving phosphatase inhibitor I, phosphatase I and calmodulin-dependent protein kinase II. In addition, this cAMP effect can be reversed by calcineurin, since the phosphorylated phosphatase inhibitor 1 is one of the best substrates known for calcineurin[18]. This complex regulatory interaction, which is schematically shown in Fig. 7, shows good agreement with the previously suggested role for 63-

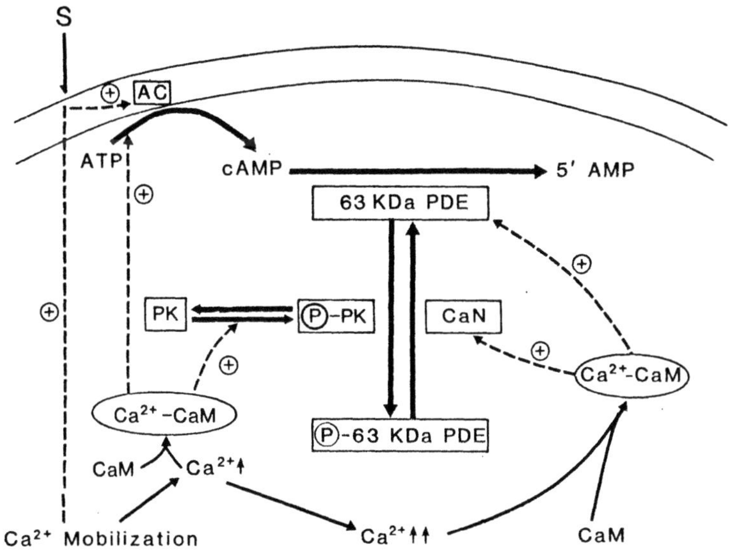

Fig. 6. The role of autophosphorylation of calmodulin-dependent protein kinase II on 63-kDa phosphodiesterase isozyme regulation. PK, calmodulin-dependent protein kinase; P-PK, phosphorylated calmodulin-dependent protein kinase. Others as in Fig. 2.

kDa phosphodiesterase in the dynamic coupling of cAMP and Ca^{2+} fluxes in the cell[6]. In the early stage of cell stimulation, the initial increase in cAMP and Ca^{2+} causes a temporary suppression of phosphodiesterase activity to maintain the rise in cAMP concentration. As the Ca^{2+} concentration in the cell is further elevated, calcineurin is activated to reverse the phosphorylation of 63-kDa phosphodiesterase and relieve the enzyme inhibition. Since calcineurin also dephosphorylates phosphatase inhibitor 1 to cause the reactivation of phosphatase I, autophosphorylation of calmodulin-dependent protein kinase II is also reversed. Thus rephosphorylation of 63-kDa phosphodiesterase will no longer occur as Ca^{2+} concentrations subside in the cell.

ACKNOWLEDGEMENTS

G.Y.Z. is on leave from Xuzhou Medical University. M.J.M. is a recipient of an Alberta Heritage Foundation for Medical Research (AHFMR) Studentship. J.H.W. is an AHFMR Medical Scientist. This work is supported by operating grants from the Medical Research Council of Canada and the Canadian Heart and Stroke Foundation.

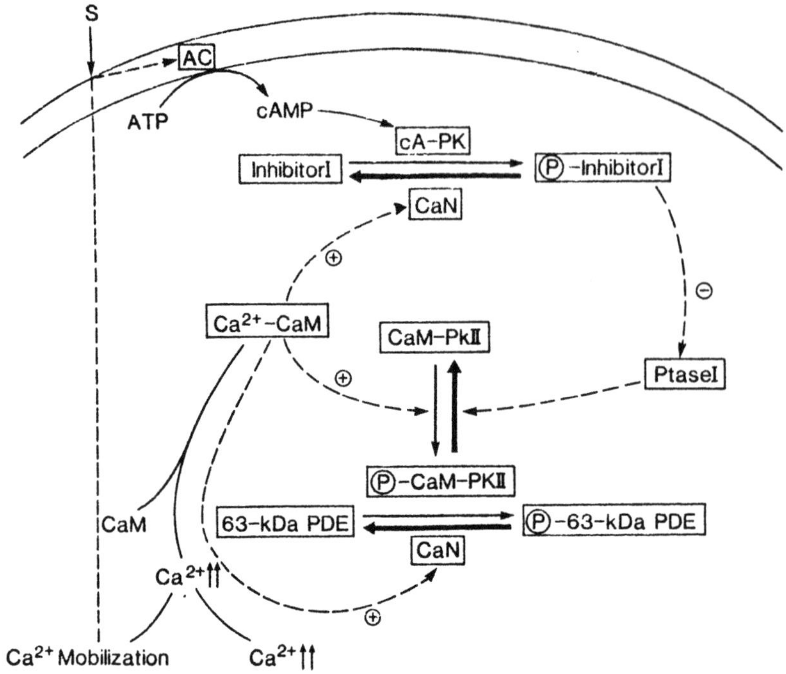

Fig. 7. Schematic diagram of the regulation of 63-kDa phosphodiesterase isozyme by Ca^{2+} and cAMP as mediated by the autophosphorylation reaction of CaM-dependent protein kinase II. The scheme depicts the interactions of cAMP-dependent protein kinase, phosphatase I, phosphatase inhibitor I and calcineurin on the bicyclic regulatory cascads. Light arrow, early events and heavy arrow, late events.

REFERENCES

1. R. K. Sharma, M. Mooibroek, and J. H. Wang, Calmodulin-stimulated cyclic nucleotide phosphodiesterase isozymes, Mol. Asp. Cell. Regulation (Calmodulin) 5:265 (1988).
2. A. Vandermeers, M-C. Vandermeers-Piret, J. Rathe, and J. Christophe, Purification and kinetic properties of two soluble forms of calmodulin-dependent cyclic nucleotide phosphodiesterase from rat pancreas, Biochem. J. 211:341 (1983).
3. K. Purvis, A. Olsen, and V. Hansson, Calmodulin-dependent cyclic nucleotide phosphodiesterase in the immature rat testies, J. Biol. Chem. 256:11434 (1981).
4. H. Hidaka, T. Yamaki, and M. Yamabe, Two forms of Ca^{2+}-dependent cycl 3':5'-nucleotide phosphodiesterase from human aorta and effect of free fatty acids, Arch. Biochem. Biophys. 187:315 (1978).

5. R. S. Hansen and J. A. Beavo, Purification of two calcium/calmodulin-dependent forms of cyclic nucleotide phosphodiesterase by using conformation-specific monoclonal antibody chromatography, Proc. Natl. Acad. Sci. USA 79:2788 (1982).

6. R. K. Sharma and J. H. Wang, Regulation of cAMP concentration by calmodulin-dependent cyclic nucleotide phosphodiesterase, Biochem. Cell Biol. 64:1072 (1986).

7. M. T. Piascik, P. L. Wisler, C. L. Johnson, and J. D. Potter, Ca^{2+}-dependent regulation of guinea pig brain adenylate cyclase, J. Biol. Chem. 255:4176 (1980).

8. A. C. Nairn, H. C. Hemming, and P. Greengard, Protein kinases in the brain, Annu. Rev. Biochem. 54:931 (1985).

9. K. Fukunaga, M. Yamamoto, E. Tanaka, T. Iwasa, and E. Miyamoto, Phosphorylation and activation of calmodulin-sensitive cyclic nucleotide phosphodiesterase by a brain Ca^{2+}, calmodulin-dependent protein kinase, Life Sci. 35:493 (1984).

10. C. Y. Huang, V. Chan, P. B. Chock, J. H. Wang, and R. K. Sharma, Mechanism of activation of cyclic nucleotide phosphodiesterase: requirement of the binding of four Ca^{2+} to calmodulin for activation, Proc. Natl. Acad. Sci. USA 78:871 (1981).

11. J. Kuret and H. Schulman, Mechanism of autophosphorylation of the multifunctional Ca^{2+}/calmodulin dependent protein kinase, J. Biol. Chem. 266:6427 (1985).

12. T. Yamouchi and H. Fujisawa, Self-regulation of calmodulin-dependent protein kinase II and glycogen synthase by autophosphorylation, Biochem. Biophys. Res. Commun. 129:213 (1985).

13. Y. Hashimoto, C. M. Schworer, R. J. Colbran, and T. R. Soderling, Autophosphorylation of Ca^{2+}/calmodulin-dependent protein kinase II, J. Biol. Chem. 262:8051 (1987).

14. C. M. Schworer, R. J. Colbran, and T. R. Soderling, Reversible generation of a Ca^{2+}-independent form of Ca^{2+} (calmodulin)-dependent protein kinase II by an autophosphorylation mechanism, J. Biol. Chem. 261:8581 (1986).

15. N. M. Woods, K. S. R. Cuthbertson, and P. H. Cobbold, Repetitive transient rises in cytoplasmic free calcium in hormone-stimulated hepatocytes, Nature 319:600 (1986).

16. J. R. Monck, E. E. Reynolds, A. P. Thomas, and J. R. William, Novel kinetics of single cell Ca^{2+} transients in stimulated hepatocytes and A10 cells measured using fura-2 and fluorescent videomicroscopy, J. Biol. Chem. 263:4569 (1988).

17. P. B. Chock, S. G. Rhee, and E. R. Stadtman, Interconvertible enzyme cascades in cellular regulation, Ann. Rev. Biochem. 49:813 (1980).

18. A. A. Stewart, T. S. Ingebritsen, A. Manalan, C. B. Klee, and P. Cohen, Discovery of a Ca^{2+}- and calmodulin-dependent protein phosphatase, FEBS Lett. 137:80 (1982).

MOLECULAR PROPERTIES OF CYCLIC NUCLEOTIDE PHOSPHODIESTERASE ISOZYMES

Samuel J. Strada, Philip A. Kithas,
Michael E. Whalin and W. Joseph Thompson

Department of Pharmacology
University of South Alabama
College of Medicine
Mobile, Alabama 36688, U.S.A.

ABSTRACT

Mammalian cells contain multiple molecular forms of cyclic nucleotide phosphodiesterase that differ in substrate specificity and kinetic and regulatory properties. Calcium/calmodulin and cyclic GMP are important regulators of the hydrolysis of cyclic AMP by either stimulating or inhibiting the activity of distinct forms of phosphodiesterase. Several isozymes of cyclic nucleotide phosphodiesterase have been purified to apparent homogeneity. Although some sequence homology is observed the isozymes appear genetically distinct by immunological criteria. Cyclic AMP- and calmodulin-dependent protein kinases can phosphorylate these enzymes and alter their kinetic and regulatory properties. Both tissue specificity and pharmacological selectivity of isozymes have been demonstrated for several drugs. In certain cases, e.g. cardiac muscle, the selective inhibition of a high affinity cAMP phosphodiesterase activity in a specific subcellular fraction correlates with pharmacologic responses. The results from molecular and pharmacologic studies of cyclic nucleotide phosphodiesterases have indeed expanded the role this system of isoenzymes exerts in the regulation of cellular function.

INTRODUCTION

Multiple molecular forms of cyclic nucleotide phosphodiesterase (CN PDE) represent the only established degradative pathway catalyzing the breakdown of cyclic 3',5'-adenosine monophosphate (cAMP) and/or cyclic 3',5'-guanosine monophosphate (cGMP) to their corresponding 5'-nucleotide. The enzyme system consists of a a family of isozymes that differ in substrate affinities and kinetic characteristics, physicochemical properties, responsiveness to various effectors, including hormones, sensitivity to a host of pharmacological agents, and regulation by both general and specific mechanisms. PDE isozymes are present in most cell types. They are found in association with membranes as well as in the cytoplasm. Their relative amounts and proportions, however, vary in different cells. Considerably less is known about the membrane-associated enzymes than those present in the cytoplasm of cells.

Purification and characterization of these isozymes has been difficult and tedious but diligent efforts have yielded significant information about four major classes of enzyme forms in mammalian cells (see Table 1). Table I shows a subdivision of the isozymes on the basis of their substrate preferences and allosteric regulatory properties (1). Type I PDE is activated by calmodulin plus Ca^{2+} and has a slightly higher affinity for cGMP than for cAMP. Type II PDE also hydrolyzes cGMP and cAMP, but activity toward the latter substrate is markedly stimulated by low (micromolar) concentrations of cGMP. Type III PDE's are GMP-specific enzymes. One example is from retinal rod outer segments. Its activation by light involves rhodopsin and the guanine nucleotide binding protein transducin. An enzyme with high specificity for cGMP both in catalysis and allosteric binding has been isolated from lung and platelets. Type IV PDE has a higher affinity for cAMP than do the other isozymes and appears to be responsive to certain hormones, eg. insulin. Within this class there appears to exist two different isozymes which demonstrate high affinity for cAMP but differ in their ability to utilize cGMP as an alternate substrate, allosteric regulation by low concentrations of cGMP, and selectivity towards drugs.

Immunological and sequence analyses have provided some basis for structural homologies and relationships; the data so far are consistent with the possibility that divergent forms evolved from an ancestral enzyme. Although considerable progress has been made towards the purification and characterization of these four classes of enzymes, and in certain cases their amino acid and/or relevant nucleotide sequences are known, our knowledge is limited, especially with respect to the membrane-associated forms. Our most recent studies have involved the investigation of Type IV enzymes that are localized to particulate subcellular fractions since these forms appear to occupy crucial sites for modulation by hormones and by pharmacologic agents. In addition to describing the data from our own laboratory, critical aspects of the biochemistry, regulation and pharmacology of the individual isozymic forms listed in Table I will be discussed. Studies of the cyclic nucleotide degradative pathway can be used to probe fundamental biological processes since it, like the synthetic pathway, is regulated by hormones and other biological effectors. For example, knowledge of the biochemistry of PDE isozymes has shed light on the mechanism of action of insulin (2,3), contributed to the discovery of calmodulin (4), provided a more complete understanding of visual transduction systems (5), and led to the development of novel therapeutic agents (6-8).

Table 1. Distinct Forms of Cyclic Nucleotide Phosphodiesterase

Enzyme Form	Characteristic
Type I	Calmodulin-Stimulated Cyclic Nucleotide PDE
Type II	cGMP-Stimulated Cyclic Nucleotide PDE
Type III	cGMP-Specific PDE
	a. Rhodopsin-Stimulated cGMP PDE
	b. cGMP-Binding Protein PDE
Type IV	cAMP-Specific PDE
	a. cGMP-Inhibited cAMP PDE
	b. cGMP-Insensitive cAMP PDE

Characteristics of Individual Cyclic Nucleotide Phosphodiesterase Isozymes

A. Type I Ca^{2+}/Calmodulin Stimulated CN-PDEs

The enzyme is present in most tissues, is principally cytosolic and has its highest specific activities in brain. Type I PDE has been purified from brain (9,10), heart (11,12) and lung (13). The isozyme may consists of several isoforms that are comprised of different subunits (14). Most studies agree that the native Mr of the enzyme is near 140,000 and is composed of 2 catalytic subunits and 2 calmodulins. The catalytic subunit has been reported to vary in different tissues from 59 to 74 kD. Insight into the potential regulation of this enzyme by Ca^{2+} and calmodulin has come from studies by Sharma and Wang (14). They showed that the different subunits (60 vs 63 kD) were differentially phosphorylated by cAMP and calmodulin-dependent protein kinases. The effect of phosphorylation by either kinase is to reduce the affinity of the enzyme for calmodulin, rendering it less sensitive to changes in cytosolic calcium. A degree of pharmacological selectivity e.g. with phenothazines (15), and W-7 and related compounds (16), has been demonstrated for this form of the enzyme. CG 9343B (17) is a novel selective calmodulin antagonist that lacks the protein kinase C inhibitory and dopamine blocking properties of the phenothiazines and other calmodulin antagonists.

B. Type II cGMP Stimulated CN PDEs

This isozyme is found in many tissues but its distribution between cytosolic and membrane-bound fractions appears to vary amongst cells. The cytosolic enzyme has been purified from adrenal (18,19), heart (20), liver (21) and lung (22); particulate enzyme forms from liver (23) and brain (24). The average Km reported is near 30 µM for cAMP and 10 µM for cGMP; positive homotropic cooperativity is observed for both substrates. The V_{max} is similar for both substrates (120 µM/min/mg). The effect of cGMP on cAMP hydrolysis is to linearize the kinetic behavior of the enzyme with cAMP as substrate (ie. alter the degree of cooperativity). Evidently, cGMP promotes a conformational state showing higher affinities for cAMP; this state is also sensitive to effects of low temperature, high pH and sulfhydryl reagents (25). Since the Kact for cGMP on cAMP hydrolysis is submicromolar, tissue levels of cyclic nucleotide are consistent with the idea that these kinetic interactions may have important physiological consequences. The nature and loci of the cGMP regulating site are not yet known; studies by Erneaux and colleagues (26) have shown a close correlation between the displacement of cGMP binding and cGMP activation constants for a series of CN analogs. The physiological role of this enzyme is also unknown; however, recent studies by Hartzell and Fischmeister (27) have suggested its involvement in controlling the modulation of a slow inward calcium current in cardiac tissue. Similarily, Levitan et al. have implicated a role in CN regulation of neuronal channels in aplysia (28). Papaverine, dipyridamole and IBMX, though not selective, are fairly potent inhibitors of cAMP and cGMP hydrolysis by this enzyme (29). However, in general, the Type II enzyme shows poor pharmacological selectivity with most currently available drugs (30).

The purified cytosolic forms are reported not to be substrates for protein kinases (20). We recently found that a purified particulate enzyme of brain could be phosphorylated by cAMP-dependent protein kinase (24). Interestingly, whilst the subunit size of this enzyme appears similar to the 105 kD Mr reported for the cytosolic enzyme, the native Mr of the particulate enzyme is near 400,000 daltons; in contrast, a native size of 240,000 daltons is observed for the cytosolic form (Table 2). The hydrodynamic properties of the cytosolic liver enzyme are consistent with a dimeric structure (Mr=210,000) having a fractional ratio of 1.7 and an elongated geometry (31). In contrast, gel filtration data on the particulate enzyme show a Mr=380,000, which is consistent with a tetrametric structure. These results raise important questions about the natural localization and composition of Type II PDE, its membrane association mechanism, the nature of the native structure and the significance of protein phosphorylation of the isozyme.

C. Type III cGMP Specific PDEs

The Km of the Rhodopsin-stimulated enzyme of rod outer segments membranes is near 4 mM for cAMP and 150 µM for cGMP. This enzyme is reportedly among the most active enzymes known (32). It is a major extrinsic protein of disk membranes and is easily solubilized by hypotonic lysis (33). The average Mr of this Type III enzyme is 170,000 with heteromeric subunits of 88,84,10 kD. The enzyme catalytical subunits are the larger Mr α and β, whereas the smaller γ-subunit is an internal inhibitor of the protein enzymatic activity. A significant homology is seen between the α- and β-subunits of cGMP PDE. Comparison of the α-subunit amino acid sequence from bovine retina with the known partial amino acid sequences of CN PDE's from other sources (34) revealed close homology in peptide chains adjoining the C terminus and analogous segments of other PDE believed to be associated with the catalytical center. The catalytical center in the α-subunit of ROS-cGMP-specific PDE is presumed, therefore, to be located in the C terminus of the molecule, while regulatory regions interacting with the γ-subunit of the enzyme and transducin are likely to be in the N-terminal part (35). The heterocomplex is activated by transducin, a quanine nucleotide binding protein with homologies to other coupling proteins, Ns, Ni, No (36,37). The activation mechanism is believed to involve subunit dissociation of at least one subunit. Bleached rhodopsin increases GTP binding by transducin which, inturn, activates

Table 2. Properties of Type II cGMP-Stimulated Cytosolic and Particulate Phosphodiesterases

Properties	Cytosolic[a]	Particulate[b]
Monomer NW	105,000	105,000
Native MW	240,000	380,000
Subunit Structure	Dimer	Teramer?
K_m cAMP[1]	30 µM	28 µM
K_m cGMP[2]	10 µM	16 µM
V_{max} cAMP	105 U/mg	160 U/mg
V_{max} cGMP	105 U/mg	120 U/mg
Modulator	cGMP (Stimulation)	cGMP (Stimulation)
Phosphorylation[3]	No	Yes
Kinetics	Positive Cooperativity	Positive Cooperativity

[a]From Harrison et al. (20).
[b]From Whalin et al. (24).
[1]U = µmol cAMP hydrolyzed/min at 30°.
[2]U = µmol cGMP hydrolyzed/min at 30o.
[3]In vitro phosphorylation by cAMP-dependent protein kinase.

cGMP hydrolysis by dissociation of the inhibitory (gamma) subunit of the enzyme complex. Rhodopsin, transducin, and Type III PDE have been successfully reconstituted as a functional, light-sensitive cGMP cascade in liposomes and preformed lipid bilayers (38). The presence of transducin binding domains on the holoenzyme isolated from retinal rod outer segment membranes suggests the possibility that analogous domains could be present in other types of membrane-bound PDEs.

A second form of Type III PDE has been demonstrated in lung and platelets and may exist in other tissues (39). In addition to its specifity for cGMP as substrate, it has the distinct feature of stimulation of cGMP binding by IBMX (hence, it has been termed a cGMP binding protein PDE). The Ka for cGMP binding is ≈0.5 μM whereas the Km for cGMP hydrolysis is near 4 μM. The enzyme has recently been purified to apparent homogeneity and some of its physicochemical properties have been described (40). This PDE may be a substrate for cGMP-dependent protein kinase; phosphorylation of the enzyme reportedly influences the cGMP binding properties (40). There are no selective pharmacological inhibitors of any of the Type III enzymes.

D. Type IV cAMP Specific PDEs

These isozymes are present in most cells, and have a lower apparent Km, or higher affinity, for cAMP than do the other forms of the enzyme. Recent data have indicated the presence of at least two distinct types of Type IV isoenzymes. These high affinity forms differ in: (a) their ability to utilize cGMP as an alternate substrate in the hydrolytic reaction; (b) regulation by cGMP inhibition; and (c) sensitivity to certain pharmacological agents. Another important characteristic of Type IV PDE activities is their regulation by hormones and/or protein phosphorylation. The properties of the two isozymes are compared in Table 3.

Potent pharmacologic inhibitors can distinguish between these two forms of Type IV activity. For example, the pyrazolopyridine derivative, SQ65442 is 355 times more potent (Table 4) than is the antithrombotic drug Cilostamide in inhibiting the cGMP insensitive

Table 3. Properties of Type IV cAMP-Specific Phosphodiesterases

Properties	cGMP-Inhibited Phosphodiesterase[a]	Non-cGMP Inhibited Phosphodiesterase[b]
Monomer MW	110,000	81,000
Subunit Structure	Dimer?	Monomer?
K_m cAMP	0.15 μM	1.6 μM
K_m cGMP[1]	0.06 μM	312
V_{max} cAMP[2]	6.0 U/mg	8-20 U/mg
V_{max} cGMP[3]	0.6 U/mg	5-10 U/mg
Modulator	cGMP (inhibition)	Unknown
Phosphorylation[4]	Yes	No

[a]From Harrison et al. (20).
[b]From Thompson et al. (42).
[1]Ki of cGMP as a competitive inhibitor of cAMP hydrolysis.
[2]U = μmol cAMP hydrolyzed/min at 30°.
[3]U = μmol cGMP hydrolyzed/min at 30°.
[4]In vitro phosphorylation by cAMP-dependent Protein Kinase.

isozyme. Conversely, the IC_{50} value for inhibition of rabbit liver cGMP Type IV PDE for cilostamide is 13,500 times lower than for the SQ compound (Table 5).

In an earlier study, we reported the characteristics of a cytosolic dog kidney enzyme, which appeared homogenous by classical criteria, and demonstrated specific activities much lower than those of other PDE forms. cGMP is a poor substrate for this enzyme and will inhibit cAMP hydrolysis only at very high concentrations (Ki=312μM). Recent modifications of earlier purification procedures (41) have enabled the separation of higher affinity activity from a 60 KD protein previously thought to be the enzyme. This protein, initially detected by immunological techniques, binds the enzyme avidly by both hydrophobic and ionic interactions. The kinetic and pharmacological data are identical to those published previously except for a 200 fold higher maximum velocity at ~5 μmol/min/mg. The purified enzyme shows an apparent native size of 71-74 kD by gel filtration whereas SDS-PAGE indicates one 82 kD protein with >90% purity. Physiological regulation of this enzyme by covalent modification or other mechanisms has yet to be established. With the exception of the kidney cytosolic enzyme very little is known regarding the distribution of this isozyme; however, brain tissue does contain an enzyme with similar properties, e.g. it shows marked sensitivity to inhibition by RO 20-1724. The drug, Rolipram (43) is even more potent as an inhibitor in brain, but does not inhibit activity as well in other tissues.

The other major Type IV isoenzyme is the cGMP inhibitable form. Although an enzyme activity demonstrating similar characteristics is present in the cytoplasm of cells, the natural localization of this isozyme is in association with membranes in heart and liver, and possibly other tissues. In the liver the enzyme is enriched in light microsomal fractions and in the heart it is present in the sarcoplasmic reticulum (SR). In either case, the activities recovered in these subcellular fractions represent only a small percentage of the total cAMP hydrolytic activity measured in cell homogenates. For example only 1-2% of the total cardic Type IV PDE activity of rabbit heart resides in the SR fraction. This activity is almost totally inhibitable by 5 μM cGMP and is much more sensitive to inhibition by selective cardiotonic agents than is cytosolic Type IV activity (44).

Table 4. Inhibition Constants (IC_{50}) for Drugs Affecting Purified Dog Kidney Type IV Phosphodiesterase Activity

Agent	IC_{50} (μM;	% error)
Cilostamide	35.50	(3.0%)
Milrinone	50.50	(6.3%)
CI 914	102.00	(4.0%)
RO 20-1724	4.00	(5.7%)
SQ 65442	0.10	(3.7%)

The activity of purified dog kidney Type IV PDE was measured with 0.25 μM cAMP. IC_{50} values were calculated from inhibition curves using 12 to 15 concentrations from a four parameter logistic function using a 2 + 2 linear regression approach developed by P. Munson, D. Rodbard and M.L. Jaffe and adapted for an IBM PC by M.L. Jaffe. Percent error is that of the 95% confidence limits at the IC_{50}.

Table 5. Inhibition Constants (IC_{50}) for Drugs Affecting Rabbit Liver Microsomal Type IV Phosphodiesterase Activity

Agent	IC_{50} (μM; % error)	
Cilostamide	0.0440	(19.7%)
Milrinone	0.6450	(21.6%)
CI 914	2.4900	(10.5%)
RO 20-1724	233.0000	(23.3%)
SQ 65442	596.0000	(9.1%)

1. Selective Effects of Cardiotonic Drugs on Isozymes of CN PDE

Different tissues vary in the content of the two isozymes of Type IV PDE as well as the amounts of Types I and II. We have compared the soluble Type IV PDE in liver, kidney and heart using similar homogenization conditions to minimize proteolysis, centrifugation to clarify membranous elements and DEAE-cellulose to fractionate the various forms (Figure 1). The high affinity activities all require higher concentrations of sodium acetate for release from DEAE (Peak III) and thus elute after Types I and II PDE forms. The content of cGMP inhibitable versus cGMP insensitive activity in Peak III varies from little in kidney to readily distinguishable activities in liver (fractions IIIA and IIIB) and intermediate values in cardiac tissue.

Figures 2 and 3 show the relative cGMP sensitivities and the SQ-65442 sensitivity of Peak III activities. Thus, the major subtypes of higher affinity PDE appear to be different isozymes and their variation in amounts and subcellular locations suggests the possibility of different regulatory functions in these tissues.

A series of studies were conducted to test the hypothesis that inhibition of this specific form of high affinity Type IV cAMP PDE can account for the inotropic effects of several novel cardiotonic drugs that have recently been developed (44-48). These drugs are more potent inhibitors of the relatively specific Peak III cAMP PDE activity (see Figure 1 below) resolved from Peaks I and II from the cytosolic fraction of ventricular myocardium. Correlation of the effects of these drugs on contractility, protein kinase activation, and Type IV inhibition support and extend the hypothesis that specific inhibitors of cardiac Type IV PDE in adult rabbits exert their contractile effects through changes in the metabolism of cAMP (49). The correlation was much better for inhibition of the particulate Type IV PDE in the SR than the cytosolic Type IV PDE (Table 6).

2. Regulation of Type IV cN PDE's by Insulin

Although cellular levels of Type IV PDE occurs in limited amounts, many hormones (e.g. insulin) and cellular processes such as differentiation (50) are known modifiers of this enzyme system. Only recently has it been clarified that the insulin activated enzyme is probably a cGMP inhibitable Type IV PDE, associated with the cell membrane. Even though it seems unlikely that regulation of a sole pathway can account for all of the physiological actions of insulin (51,52), high affinity cAMP Type IV PDE represents one of the few enzymes that has clearly been established as an integral part of the action of this peptide hormone especially with respect to the antagonism of insulin action by glucagon,

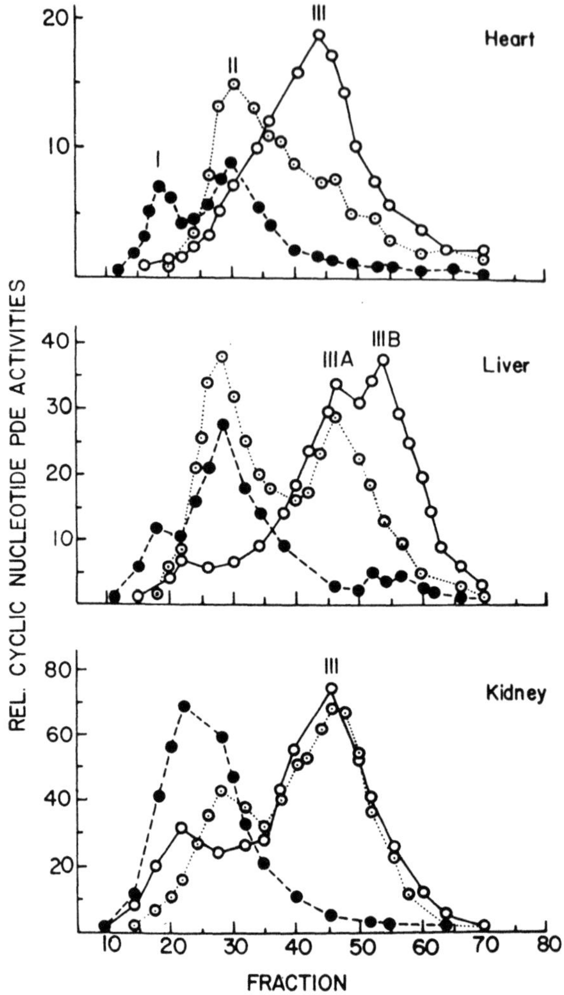

Fig. 1. Rat tissues were homogenized and fractionated under identical conditions. One gm tissue was homogenized in 4 vol of buffer containing 4mM EDTA, 0.25M sucrose, 15mM 2-βSH, 1mM benzamidine, 0.1mM PMSF, 0.02mM TLCK, pH 7.0. Homogenates were prepared with a Dounce homogenizer and sonicated for 2x 30 sec bursts (50% setting, Biosonic IV, microtip). Sonicates were centrifuged for 30 min at 78,000 x g, pellets rehomogenized in original buffer and recentrifuged. the supernatants were combined and gel filtered on PD-10 columns. The gel filtered extracts were applied to a DEAE cellulose column previously equilibrated with 70mM sodium acetate buffer, pH 6.5, 30% ethylene glycol, 5mM 2-βSH, benzamidine, PMSF and TLCK. The column was washed with 4 bed vol of buffer and frctionated with a linear acetate gradient (70-1000 mM) with a flow rate of 0.5 ml/min. Eighty one-half ml fractions were collected and assayed for cyclic nucleotide phosphodiesterase activities as follows: Recovery of Type IV homogenate cAMP phosphodiesterase activity at 0.25 μM cAMP in the supernatant fraction was 95% for kidney, 84% for liver, and 65% for heart, respectively.

0.25 μM cAMP o--o; 0.25 μM cAMP + 5 μM cGMP ⊙--⊙; 1.0 μM cGMP ●--●

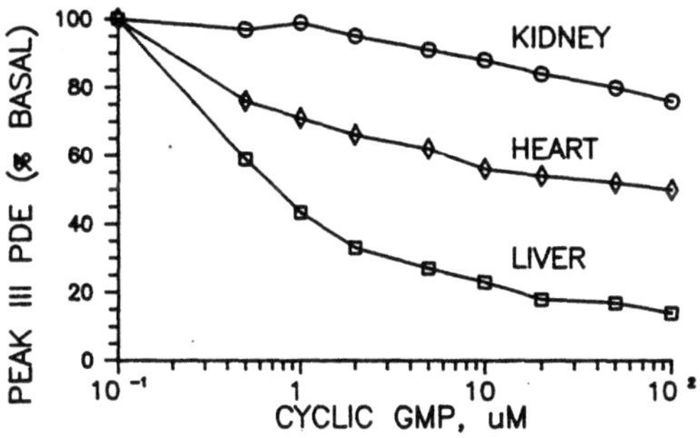

FIGURE 2

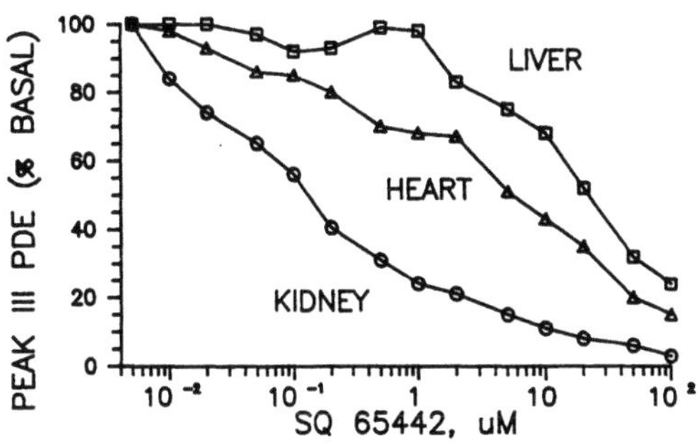

FIGURE 3

epinephrine and ACTH. Studies by Corbin and colleagues found a clear correlation between the IC_{50} values for PDE inhibition by cAMP analogs and the blockade of the analog responses in adipocytes and hepatocytes (53-54) by insulin. Weber et al. have reported that activation of Type IV PDE by insulin is independent of G-protein coupling (55) since the response is not sensitive to pertussis toxin.

Type IV PDE has been reported to be modulated by protein phosphorylation in platelets (56) hepatocytes (57) and adipocytes (58). The requirement of an intact cell system for Type IV PDE activation by insulin has proven a significant impediment towards elucidating the membrane components and cellular mechanisms involved in insulin receptor activation of the enzyme. In this regard our finding that vanadate complexed with glutathione activates particulate adipocyte and liver microsomal, cGMP inhibitable Type IV PDE *in vitro* has provided a tool to study insulin activation of the enzyme (59).

Table 6. Cardiotonic Drug Selectivity for Particulate (SR-Associated) Phosphodiesterase Activity

	K_i (µM)*	
Drug	PDE III	SR-PDE
Milrinone	11	0.18
Amrinone	280	8.5
Sulmazole	155	...
Isomazole	...	24
Imazodan	18	1.5
Piroximone	15	6

*K_i values determined by Dixon analysis at four different substrate concentrations and six inhibitor concentrations.

3. Purification of Hormone-Sensitive Type IV Isozyme

The cGMP inhibitable Type IV PDE constitutes a small proportion of total cAMP hydrolysis measurable in tissues. The particulate nature of the enzyme may have important pharmacolgical and physiological relevance as discussed previously; however, this also makes it difficult to purify and to study its physico-chemical characteristics. Harrison et al. (60) highly purified a soluble, bovine heart enzyme that contained a major 110 kD protein, which they reported was a cGMP inhibitable Type IV isozyme. Degerman et al. (61) obtained 17 µg of enzyme from 432 rats after a 65,000 fold purification using a derivative of cilostamide as an affinity ligand. SDS-page of the purified protein identified a 64 kD subunit. This value is in close agreement with the 61 kD value reported by Grant and Coleman (62) for a platelet enzyme after a 2800 fold purification. The reported specific activities of the 110, 64 and 61 kD enzymes vary from 1-8 µmol/min/mg; these values are comparable to those obtained for the dog kidney cGMP-insensitive Type IV enzyme described above. The relationship of these cytosolic activities to each other or to the particulate activities, which share similar characteristics are not yet clear.

4. Pathologic Alterations in Type IV cN PDE Isozyme

Diabetic animals, including chemically-induced or genetic experimental models of diabetes, have consistently shown changes in the activity of Type IV enzymes (63,64). Preliminary studies in the insulin-resistant Zucker rat model have shown diminished vandyl/ glutathione activatable liver microsomal Type IV PDE. Whether these biochemical alterations reflect secondary manifestations of other metabolic pertubations or represent critical determinants in the etiology of diabetes is still unresolved. Alteration in Type IV PDE enzyme activities have been noted in atopic dermatitis (65), asthma (66,67) and leukemias (68,69).

CONCLUSIONS

This brief overview has tried to illustrate how advances from research on CN-PDE isozymes have led to fundamental information of biological and chemical significance. Extensive progress has been made on the characteristics of these individual forms with respect to their molecular and regulatory properties. However, many important questions remain unanswered, and this is especially true for those activities associated with membranes. The role of reversible phosphorylation-dephosphorylation reactions or other covalent mechanisms can be studied more completely as more stable and purer preparations of the different isozymes become widely available. This enzyme system has undergone a renewed interest by the pharmaceutical industry. The development of novel drugs with tissue specificity and enzyme selectivity for the different isozymes comprising the CN PDE system continues to reveal important therapeutic modalities for a variety of diseases.

ACKNOWLEDGMENTS

The experimental work was performed in the University of South Alabama Laboratory for Cyclic Nucleotide Research and was supported by USPHS Grant GM33538, and a contract #49620-87-C-0071 from the Air Force Office of Scientific Research.

REFERENCES

1. Strada, S. J. and Thompson, W. J., "Advances cyclic Nucleotide Phosphorylation Research 16, Nomenclature Recommendation," p. vi (1984).
2. Appleman, M. M., Allan, E. H., Ariano, M.A., Ong, K. K., Tusang, C. A., Weber, H. W., and Whitson, R. H., Insulin Control of Cyclic AMP Phosphodiesterase, Adv. Cyc. Nuc. Prot. Phos. Res., 16:149-158 (1984).
3. Weber, H. W. and Appleman, M. M., Insulin-Dependent and Insulin-Independent Low K_m Cyclic AMP Phosphodiesterase from Rat Adipose Tissue, J. Biol. Chem., 257:5339-5341 (1982).
4. Wang, J. H. and Waisman, D. M., Calmodulin and Its Role in the Second Messenger System, Curr. Topics in Cell Res., 15:47-107 (1979).
5. Hurley, J. B. and Stryer, L., Purification and Characterization of the Gamma Regulatory Subunit of the Cyclic GMP Phosphodiesterase from Retinal Rod Outer Segments, J. Biol. Chem., 257:11094-11099 (1982).
6. Alvarez, R., Banerjee, G. L., Bruno, J. J., Jones, G. L., Liittschwager, K., Strosberg, A. M., and Venuti, M. C., A Potent and Selective Inhibitor of Cyclic AMP Phosphodiesterase with Potential Cardiotonic and Antithrombic Properties, Molec. Pharm., 29:554-560 (1986).
7. Weishaar, R. E., Cain, M. H., and Bristol, J.A., A new Generation of Phosphodiesterase Inhibitors: Multiple Molecular Forms of Phosphodiesterase and the Potential for Drug Selectivity, J. Med. Chem., 28:537-545 (1985).
8. Pang, D. C., Cyclic AMP and Cyclic GMP Phosphodiesterases: Target for Drug Development, Drug Devel. Res., 12:85-92 (1988).
9. Sharma, R. K., Wang, T. H., Wirch, E., and Wang, J. H., Purification and Properties of Bovein Brain Calmodulin-Dependent Cyclic Nucleotide Phosphodiesterase, J. Biol. Chem., 255:5916-5923 (1980).
10. Shenolikar, S., Thompson, W. J., and Strada, S. J., Characterization of a Novel Form of Ca^{2+}-Calmodulin Stimulated Cyclic GMP Phosphodiesterase from Bovine Brain, Biochemistry, 24:672-678 (1985).
11. Ho, H. C., Wirch, E., Stevens, F. C., and Wang, J. H., Purification of a Ca^{++}-Activatable Cyclic Nucleotide Phosphodiesterase from Bovine Heart by Specific Interaction with its Calcium Dependent Modulator Protein, J. Biol. Chem., 252:43-50 (1977).

12. Hanson, R.S., Charbonneau, H., and Beavo, J.A., Purification of Calmodulin-Stimulated Cyclic Nucleotide Phosphodiesterase by Monoclonal Antibody Affinity Chromatography, Meth. Enzymol., 159:543-557 (1988).
13. Sharma, R. K. and Wang, J. H., Purification and Characterization of Bovine Lung Calmodulin-Dependent cyclic Nucleotide Phosphodiesterase, J. Biol. Chem., 261:14160-14166 (1986).
14. Sharma, R. K. and Wang, J. H., Calmodulin and Ca^{++}-Dependent Phosphorylation and Dephosphorylation of 63-kDa Subunit-Containing Bovine Brain Calmodulin-Stimulated cyclic Nucleotide Phosphodiesterase Isozyme, J. Biol. Chem., 261:1322-1328 (1986).
15. Saitoh, Y., Hardman, J. G., and Wells, J. N., Differences in the Association of Calmodulin with Cyclic Nucleotide Phosphodiesterase in Relaxed and Contracted Arterial Strips, Biochemistry, 24:1613-1618 (1985).
16. Hidaka, H., Inagaki, M., Nishikawa, M. and Tanaka, T., Selective Inhibitors of Calmodulin-Dependent Phosphodiesterase and Other Enzymes, Meth. Enzymol., 159:652-660 (1988).
17. Norman, J. A., Ansell, J., Stone, G. A., Wennogle, L. P., and Wasley, J. W. F., CGS 9343B, a Novel, Potent, and Selective Inhibitor of Calmodulin Activity, Mol. Pharmacol., 31:535-540 (1987).
18. Martins, T. J., Mumby, M. D., and Beavo, J. A., Purification and Characterization of a Cyclic GMP-Stimulated Cyclic Nucleotide Phosphodiesterase from Bovine Tissues, J. Biol. Chem., 257:1973-1979 (1982).
19. Miot, F., Van Haastert, J. M., and Erneaux, C., Specificity of cGMP Binding to a Purified cGMP-Stimulated Phosphodiesterase from Bovine Adrenal Tissue, Eur. J. Biochem., 149:59-65 (1985).
20. Harrison, S. A., Beier, N., Martins, T. J., and Beavo, J. A., Isolation and Comparison of Bovine Heart cGMP-Inhibited and cGMP-Stimulated Phosphodiesterases, Meth. Enzymol., 159:685-702 (1988).
21. Yamamoto, T., Manganiello, V. C., and Vaughan, M., Purification and Characterization of Cyclic GMP-Stimulated Cyclic Nucleotide Phosphodieterase From Cat Liver, J. Biol. Chem., 258:12526-12533 (1983).
22. Mumby, M. C., Martins, T. J., Chang, M. L., and Beavo, J. A., Identification of cGMP-Stimulated Cyclic Nucleotide Phosphodiesterase in Lung Tissue with Monoclonal Antibodies, J. Biol. Chem., 257:13283-13290 (1982).
23. Pyne, N. J., Cooper, M. E., and Houslay, M. D., Identification and Characterization of Both the Cytosolic and Particulate Forms of Cyclic GMP-Stimulated Cyclic AMP Phosphodiesterase from Rat Liver, Biochem. J., 234:325-334 (1986).
24. Whalin, M. E., Strada, S. J., and Thompson, W. J., Purification and Characterization of Membrane Bound, Rabbit Brain cGMP Activatable (Type II) Cyclic Nucleotide Phosphodiesterase, Biochim. Biophys. Acta, In Press (1988).
25. Wada, H., Osborne, J. C., Jr., and Manganiello, V. C., Effects of Temperature on Allosteric and Catalytic Properties of the cGMP-Stimulated Cyclic Nucleotide Phosphodiesterase From Calf Liver, J. Biol. Chem., 262:5139-5144 (1987).
26. Erneaux, C., Miot, F., Van Haastert, P. J. M., and Jastoff, B., The Binding of Cyclic Nucleotide Analgos to a Purified Cyclic GMP-Stimulated Phosphodiesterase from Bovine Adrenal Tissue, J. Cyclic Nuc. Res., 10:463-472 (1985).
27. Hartzell, H. C. and Fischmeister, R., 1986, Opposite Effects of Cyclic GMP and Cyclic AMP on Ca^{2+} Current in Single Heart Cells, Nature, 323:273-275 (1986).
28. Levitan, I. B., Modulation of Ion Channels in Neurons and Other

Cells, Ann. Rev. Neurosci., 11:119-136 (1988).
29. Yamamoto, T., Yamamoto, S., Osborne, J. C., and Manganiello, V. C., Complex Effects of Inhibitors on Cyclic GMP-Stimulated Cyclic Nucleotide Phosphodiesterase, J. Biol. Chem., 258:14173-14177 (1983).
30. Erneaux, C., Couchie, D., Dumont, J. E., Baraniak, J., Stec, W.J., Abbad, E. G., Petridis, G., and Jastorff, B., Specificity of Cyclic GMP Activation of a Multi-Substrate Cyclic Nucleotide Phosphodiesterase from Rat Liver, Eur. J. Biochem., 115:503-510 (1981).
31. Yamamoto, T., Yamamoto, S., Manganiello, V. C., and Vaughan, M., Effects of Fatty Acids on Activity of cGMP-Stimulated Cyclic Nucleotide Phosphodiesterase From Calf Liver, Arch. Biochem. Biophys., 229:81-89 (1984).
32. Stryer, L. The Cyclic GMP Cascade of Vision, Ann. Rev. Neurosci. 9:87-119 (1986).
33. Baehr, W., Morita, E. A., Swanson, R. J., and Applebury, M. L., Characterization of Bovine Rod Outer Segment G-protein, J. Biol. Chem., 257:6452-6460 (1982).
34. Charbonneau, H., Beir, N., Walsh, K., and Beavo, J.A., Identification of a Conserved Domain Among Cyclic Nucleotide Phosphodiesterases from Diverse Species, Proc. Nat. Acad. Sci., 83:9308-9312 (1986).
35. Ovchinnikov, Yu A., Gubanov, V. V., Khramtsou, N. V., Ischenko, K. A., Zagranichny, V. E., Muradov, K. G., Shuvaeva, T. M., Lipkin, V. M., Cyclic GMP Phosphodiesterase from Bovine Retina-Amino Acid Sequence of the α-Subunit and Nucleotide Sequence of the Corresponding cDNA, FEBS Lett., 223:169-173 (1987).
36. Ovchinnikov, YuA., Lipkin, V. M., Kumarev, V. P., Gubanov, V. V., Khramtsov, N. V., Akhmedov, N. B., Zagranichny, V. E. and Muradov, K. G., Cyclic GMP Phosphodiesterase from Cattle Retina-Amino Acid Sequence of the γ-Subunit and Nucleotide Sequence of the Corresponding cDNA, FEBS Lett., 204:288-292 (1986).
38. Tyminski, P. N., Latimer, L. H., and O'Brien, D. F., Reconstitution of Rhodopsin and the cGMP Cascade in Polymerized Bilayer Membranes, Biochem., 27:2696-2705 (1988).
39. Francis, S. H., Effectors of Rat Lung cGMP Binding Protein-Phosphodiesterase, Curr. Top. Cell. Reg., 26:247-262 (1985).
40. Francis, S. H. and Corbin, J., 1988, purification of cGMP-Binding Protein Phosphodiesterase From Rat Lung, Meth. Enzymol., 159:722-729 (1988).
41. Epstein, P. M., Strada, S. J., Sarada, K., and Thompson, W. J., Catalytic and Kinetic Properties of Purified High-Affinity Cyclic AMP Phosphodiesterase from Dog Kidney, Arch. Bioc. Biops., 218:119-133 (1982).
42. Thompson, W. J., Shen, C-C., and Strada, S. J., Preparation of Dog Kidney High-Affinity cAMP Phosphodiesterase, Meth. Enzymol., 159:760-766 (1988).
43. Schultz, J. E. and Folkers, G., Unusual Stereospecificity of the Potential Antidepressant Rolipram on the cyclic AMP Generating System from Rat Brain Cortex, Pharmacopsychiat., 21:83-86 (1988).
44. Kithas, P. A., Artman, M., Thompson, W. J., and Strada, S. J., Subcellular Distribution of High-Affinity Type IV Cyclic AMP Phosphodiesterase Activities in Rabbit Ventricular Myocardium: Relations to the Effects of Cardiotonic Drugs, Circ. Res., 62:782-789 (1988).
45. Robertson, D. W., Jones, N. D., Krushinski, J. H., Pollock, G. D., Swartzendruber, J. K., and Hayes, J. S., Molecular Structure of the Dihydropyridazinone Cardiotonic 1,3-Dihydro-3,3-Dimethyl-5-(1,4,5,6-tetrahydro-6-oxo-3-pyridazinyl)-2H-indol-2-one, a Potent Inhibitor of Cyclic AMP Phosphodiesterase, J. Med. Chem., 30:623-627 (1987).

46. Moos, W. H., Humblet, C. C., Sircar, I., Rithner, C., Weishaar, R. E., Bristol, J. A., and McPhail, A. T., Cardiotonic Agents 8. Selective Inhibitors of Adenosine 3',5'-Cyclic Phosphate Phosphodiesterase III. Elaboration of a Five-Point Model for Positive Inotropic Activity, J. Med. Chem., 30:1963-1972 (1987).
47. Earhart, P. W., Hagedorn, III, A. A., and Sabio, M., Cardiotonic Agents 3. A Topographical Model of the Cardiac cAMP Phosphodiesterase Receptor, Mol. Pharmacol., 33:1-13 (1988).
48. Kauffman, R. F., Crowe, G. V., Utterback, B. G., and Robertson, D. W., LY195115: A Potent, Selective Inhibitor of Cyclic Nucleotide Phosphodiesterase Located in the Sarcoplasmic Reticulum, Mol. Pharmacol., 30:609-616 (1987).
49. Kithas, P. A., Artman, M., Thompson, W. J., and Strada, S. J., Subcellular Distribution of High-Affinity Type IV Cyclic AMP Phosphodiesterase Activities in Rabbit Ventricular Myocardium: Relations to Postnatal Maturation, J. Molec. Cell. Cardiol., In Press (1988).
50. Elks, M. L., Manganiello, V. C., and Vaughan, M., Effect of Dexamethasone on Adenosine 3',5'-Monophosphate Content Phosphodiesterase Activities in 3T3-L1 Adipocytes, Endocrinol., 115:1350-1356 (1984).
51. Gabbay, R. A. and Lardy, H. A., Site of Insulin Inhibition of cAMP-Stimulated Glycogenolysis, J. Biol. Chem., 259:6052-6055 (1984).
52. Larner, J., Insulin Mediator--Fact or Fancy?, J. Cyclic Nucleo., 8:289-296 (1982).
53. Gettys, T. W., Blackmore, P. F., and Corbin, J. D., An Assessment of Phosphodiesterase Activity in situ After Treatment of Hepatocyte with Hormones, Am. J. Physiol., 254:(Endocrinol. Metab. 17):E449-E453 (1988).
54. Gettys, T. W., Blackmore, P. F., Redmon, J. B., Beebe, S. J. and Corbin, J. D., Short-Term Regulation of cAMP by Accelerated Degradation in Rat Tissues, J. Biol. Chem., 262:333-339 (1987).
55. Weber, H. W., Chung, F-Z, Day, K., and Appleman, M. M., Insulin Stimulation of Cyclic AMP Phosphodiesterase is Independent from the G-Protein Pathways Involved in Adenylate Cyclase Regulation, J. Cyc. Nuc. Prot. Phos. Res., 11:345-354 (1987).
56. Macphee, C. H., Reifsnyder, D. H., Moore, T. A., and Beavo, J. M., Intact Cell and Cell-Free Phosphorylation and Concomitant Activation of a Low K_m, cAMP Phosphodiesterase Found in Human Platelets, J. Cyclic Nucleo. Prot. Phos. Res., 11:487-496 (1987).
57. Houslay, M. D., Wallace, A. V., Marchmont, R. J., Martin, B. R., and Heyworth, C. M., Insulin Controls Intracellular Cyclic AMP Concentrations in Hepatocytes by Activating Specific Cyclic AMP Phosphodiesterases: Phosphorylation of the Peripheral Plasma Membrane Enzyme, Adv. Cyclic Nucleo., 16:159-176 (1984).
58. Cordle, S.R. and Corbin, J.D., Activation of cGMP-Insensitive Low Km Phosphodiesterase by cAMP-Dependent Protein Kinase, FASEB J., 2:1739 (1988).
59. Souness, J. E., Thompson, W. J., and Strada, S.J., Adipocyte Cyclic Nucleotide Phosphodiesterase Activation by Vanadate, J. Cyclic Nuc. Prot. Phos. Res., 10:383-396 (1985).
60. Harrison, S. A., Reifsnyder, D. H., Gallis, B., Cadd, G. G., and Beavo, J. A., Isolation and Characterization of Bovine Cardiac Muscle cGMP-Inhibited Phosphodiesterase: A Receptor for New Cardiotonic Drugs, Mol. Pharm., 29:506-514 (1986).
61. Degerman, E., Belfrage, P., Newman, A. H., Rice, K. C., and Manganiello, V. C., Purification of the Putative Hormone-Sensitive Cyclic AMP Phosphodiesterase from Rat Adipose Tissue Using a Derivative of Cilostamide as a Novel Affinity Ligand, J. Biol. Chem., 262:5797-5807 (1987).

62. Grant, P. G. and Coleman, R. W., Purification and Characterization of Human Platelet Cyclic Nucleotide Phosphodiesterase, Biochem., 23:1801-1807 (1984).
63. Benelli, C., Lopez, S., Desbuquois, B., and Achagiotis, C., Changes in Low-K_m cAMP Phosphodiesterase Activity in Liver Golgi ractions from Hyper- and Hypoinsulinemic Rats, Diabetes, 37:171-722 (1988).
64. Fatemi, S. H., Insulin-Dependent Cyclic AMP Turnover in Isolated Rat Adipocytes, Cell. Mol. Biol., 31:153-161 (1985).
65. Hanifin, J. M., Atopic Dermatitis, J. All. Clin. Immun., 73:211-222 (1984).
66. Holden, C. A., Chan, S. C., Norris, S., and Hanifin, J. M., Histamine Induced Elevation of Cyclic AMP Phosphodiesterase Activity in Human Monocytes, Agents and Action, 22:36-42 (1987).
67. Chan, S. C., Hanifin, J. M., Holden, C., Peters, J. E., Thompson, W. J., and Hirshman, C. A., Elevated Leukocyte Phosphodiesterase as a Basis for Depressed Cyclic AMP Responses in the Basenjigreyhound Dog Model of Asthma, J. Allergy and Clin. Immunol., 76:148-158 (1985).
68. Onali, P., Strada, S. J., Epstein, P. M., and Thompson, Purification and Characterization of High affinity Cyclic Nucleotide Phosphodiesterase from Human Myelogenous Leukemic Cells, Cancer Res., 45:1384-1391 (1985).
69. Epstein, P. M. and Hachisu, R., Cyclic Nucleotide Phosphodiesterase in Normal and Leukemic Human Lymphocytes and Lymphoblasts, Adv. Cyc. Nuc. Prot. Phos. Res., 16:303-324 (1984).

ROLES OF CALCIUM IN THE REGULATION OF TYROSINE HYDROXYLASE*

T. Nagatsu[a], K. Kiuchi[b], H. Hidaka[c], and K. Suzuki[d]

[a]Department of Biochemistry, [b]Radioisotope Center Medical Division, [c]Department of Pharmacology, Nagoya University School of Medicine, Nagoya 466, Japan, and [d]Tokyo Metropolitan Institute of Medical Science, Tokyo 113, Japan

INTRODUCTION

The biosynthesis of catecholamines in catecholaminergic neurons and adrenal medullary cells is mainly regulated by the pterin-dependent monooxygenase, tyrosine hydroxylase (TH) (Nagatsu *et al.*, 1964).

TH requires (6*R*)-L-*erythro*-tetrahydrobiopterin as the natural cofactor. TH is tetramer composed of four homologous subunits. In any species, Mr of TH subunit is approximately 60 kDa. The primary structures of TH subunit were determined by cDNA cloning in the rat, human, and bovine. Only in human TH, four mRNA species were identified, and they were shown to be formed by alternative mRNA splicing from a single gene (Grima *et al.*, 1987; Kaneda *et al.*, 1987; O'Malley *et al.*, 1987; Kobayashi *et al.*, 1988).

TH is known to be activated by stimulation of catecholaminergic neurons concomitantly with catecholamine release from and Ca^{2+} influx into the neurons at nerve endings. It is speculated, therefore, that increased intracellular Ca^{2+} may activated TH to increase the biosynthesis of catecholamines. There are three possible mechanisms of activation of TH *in vivo* by increased Ca^{2+} concentrations, which can be suggested by *in vitro* experiments. The first possibility is activation as a result of phosphorylation of TH by Ca^{2+}/calmodulin-dependent protein kinase II (Ca/CaM kinase) (Yamauchi and Fujisawa, 1981; Atkinson *et al.*, 1987, Griffith and Schulman, 1988). The second possibility is activation as a result of phosphorylation of TH by Ca^{2+}-stimulated, phospholipid-dependent protein kinase, i.e., protein kinase C (C-kinase) (Chowdhury and Fillenz, 1988).

We previously reported that TH may be activated *in situ* by Ca/CaM kinase and cyclic AMP-dependent protein kinase (A-kinase) and that the two processes may act in different fashions. The Ca^{2+}-calmodulin system may result in interconversion from a less active form to an active form in a basal state, and the cyclic AMP system may maximally activate TH (Nagatsu *et al.*, 1985; Hirata and Nagatsu, 1985). On the other had, activation of TH by C-kinase *in situ* remains to be further clarified.

The third possible mechanism is irreversible activation after limited proteolysis by Ca^{2+}-activated neutral protease (CANP, calpain) (Togari *et al.*, 1985). TH contains the inhibitory regulatory domain in the N-terminal

---

[a]To whom correspondence should be addressed. *This work is supported by Grant-in-Aid for Science Research on Priority Areas, Ministry of Education, Science and Culture, Japan.

region and the catalytic domain in the C-terminal region, and the limited proteolysis of the N-terminal regulatory domain is considered to result in irreversible activation of TH. In our previous study, we proved that TH, calpain, and the endogenous inhibitor (calpastatin) co-localized in the cytosolic fraction of bovine adrenal medulla. Both calpain I (μCANP) and calpain II (mCANP) with low and high Ca^{2+} requirements activated TH following limited proteolysis, and the activation was inhibited by leupeptin and calpastatin (Togari et al., 1985). Limited proteolysis of TH molecule by calpain in vitro may activate TH by removing the N-terminal regulatory domain. In this experiment, however, purified TH was incubated with calpain II partially purified from bovine adrenal medulla. It is possible, therefore, that TH is further hydrolyzed not only by calpain, but also by other proteases. In the present experiment, we have examined proteolysis of purified TH with purified calpain.

In this study we have attempted to elucidate further the mechanism of regulation of TH activity by Ca^{2+}.

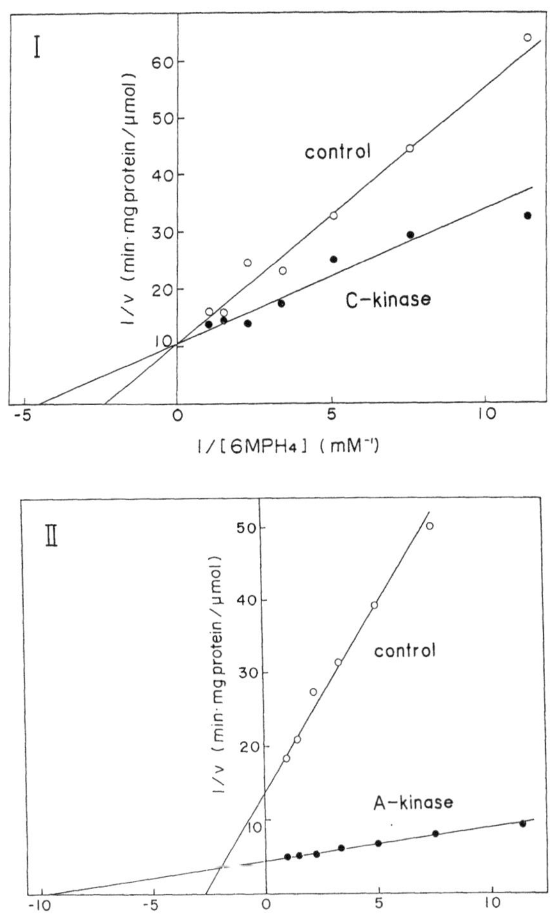

Fig. 1. Kinetic comparison of enzymatic activity of TH after incubation under the phosphorylating conditions without (○) or with (●) protein kinase: (I) C-kinase; (II) A-kinase. TH activity was assayed in 0.1 M MES-NaOH buffer (pH 6.8) containing 0.2 mM L-tyrosine and $6MPH_4$ from 87.8 μM to 1 mM.

REGULATION OF TH ACTIVITY BY Ca^{2+}-DEPENDENT PHOSPHORYLATION

Activation of TH by protein kinase C *in vitro*

TH purified from bovine adrenal medulla was phosphorylated *in vitro* by C-kinase and A-kinase. Approximately 0.33 mol and 0.67 mol of ^{32}P were incorporated per 60 kDa subunit of TH by C-kinase and A-kinase. Changes in kinetic properties toward a pterin cofactor, (6RS)-methyl-5,6,7,8-tetrahydropterin ($6MPH_4$) without (control) and with phosphorylation of TH were different between C-kinase (Fig. 1, I) and A-kinase (Fig. 1, II), respcctively As shown in Fig. 1 (I and II), A-kinase increased Vmax and decreased Km value toward $6MPH_4$, whereas C-kinase only decreased Km value without any change in Vmax. The results indicate that C-kinase activates TH by increasing the affinity toward the pterin cofactor, at least *in vitro*. Atkinson *et al*. (1987) reported that purified rat striatal TH is activated by an activator protein after phosphorylation of TH by Ca/CaM kinase with increased Vmax and without affecting the Km value for the pterin cofactor. Therefore, the two Ca^{2+}-dependent protein kinases, Ca/CaM kinase and C-kinase, may activate purified TH in different fashions, at least *in vitro*.

Effects of specific protein kinase inhibitors on *in situ* TH activity

To determine if both Ca/CaM kinase and C-kinase may regulate TH activity *in situ*, rat striatal tissue slices that contain all of the components of TH, cAMP and Ca/CaM kinase and C-kinase were subjected to experimental manipulations using specific inhibitors for Ca/CaM kinase and C-kinase.

ML-9 [1-(5-chloronaphthalene-1-sulfonyl)-1H-hexahydro-1,4-diazepine] is a specific inhibitor of Ca/CaM kinase, having Ki values of 5 µM for Ca/CaM kinase and 54 µM for C-kinase (Saitoh *et al*., 1987). In contrast, H-7 [1-(5-isosquinolinesulfonyl)-2-methylpiperazine] is a specific inhibitor for C-kinase; Ki values are 6 µM for C-kinase and 97 µM for Ca/CaM kinase (Hidaka *et al*., 1984).

As shown in Fig. 2, ML-9 at 100 µM inhibited TH system by 50%, whereas H-7 at 100 µM did not affect TH system. Fig. 3 indicates a dose-dependent inhibition of TH system by ML-9 from 10 µM to 1 mM. ML-9 from 1 µM to 100 µM did not affect *in vitro* activity of purified bovine adrenal TH without or with maximal phosphorylation by A-kinase.

Kinetic analysis revealed that the enzyme in homogenates of control tissue slices had two different Km values for $6MPH_4$, indicating the presence of two forms of TH in striatal tissue slices: a less active form with a relatively low affinity for the pterin cofactor and a more active form with a relatively high affinity. ML-9 produced an increase of the high Km form and a decrease of the low Km form, respectively. These results are similar to our previous results (Hirata and Nagatsu, 1985) using another Ca/CaM kinase inhibitor, W-7 [N-(6-aminohexyl)-5-chloro-1-naphthalenesulfonamide] (Hidaka *et al*., 1978), or trifluoroperazine.

The present results suggest that, although both Ca/CaM kinase and C-kinase phosphorylates and activates TH *in vitro*, only Ca/CaM kinase may activate TH *in vivo*. This conclusion is consistent with a recent report on rat pheochromocytoma PC12 cells that Ca/CaM kinase mediates phosphorylation of TH by hormonal and electrical stimuli which elevate intracellular Ca^{2+} in PC12 cells (Griffith and Schulman, 1988).

EFFECT OF CALPAIN ON TH

We previously reported that the two Ca^{2+}-activated neutral protease, calpain, with high (calpain I or µCANP) or low (calpain II or mCANP) Ca^{2+} affinity activated pure bovine adrenal TH by limited proteolysis (Togari *et al*., 1985). After full activation of TH due to limited proteolysis by

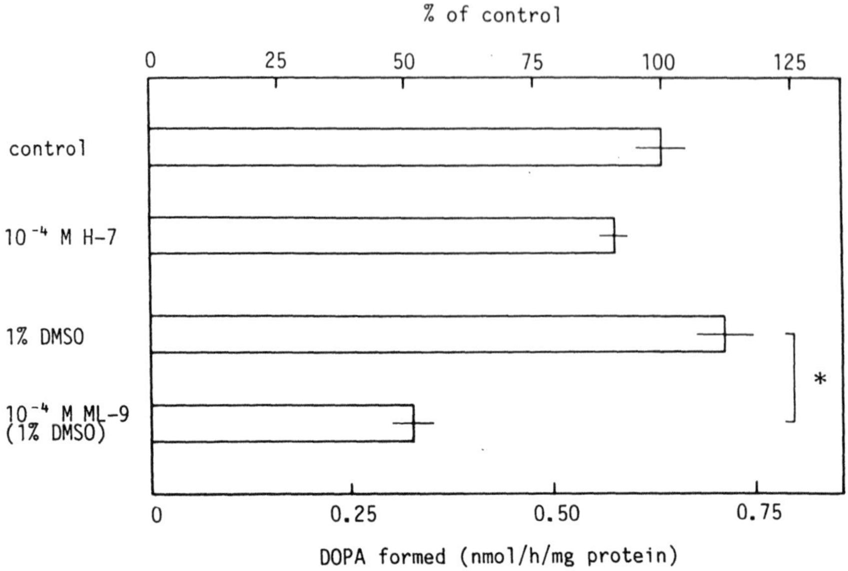

Fig. 2. Effect of H-7 or ML-9 on DOPA formation in rat striatal slices. The slices were incubated at 37°C for 1 h in a Krebs-Ringer medium containing NSD-1055 and drugs. Each value is the mean ± S.E.M. of 4 experiment. * $p < 0.01$ for difference from control (1% DMSO).

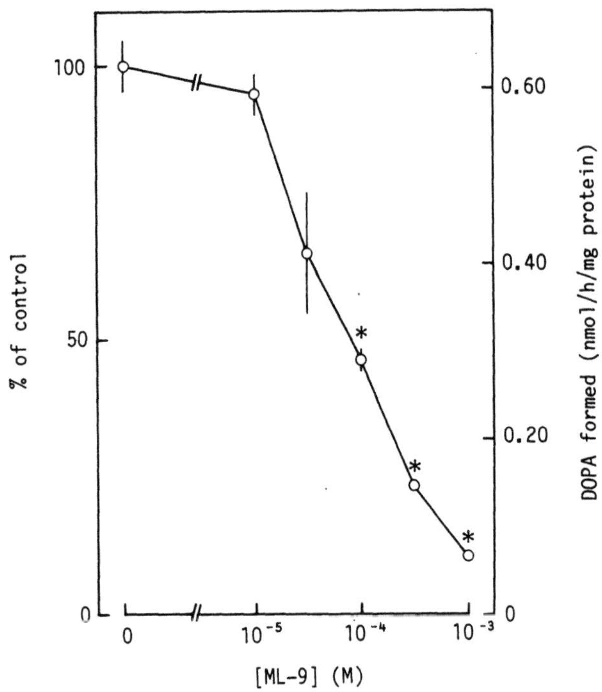

Fig. 3. Effect of ML-9 concentration on DOPA formation in rat striatal slices. The slices were incubated at 37°C for 1 h in a Krebs-Ringer medium containing NSD-1055 and various concentrations of ML-9. Each value is the mean ± S.E.M. of 3 or 4 experiments. * $p < 0.01$ for difference from control.

partially purified calpain II, Mr of TH decreased from 240 kDa to 120 kDa as measured by HPLC gel filtration, indicating an intensive degradation of TH molecule. However, we could not find any 120 kDa species of TH in the brain (striatum) or adrenal medulla of the rat, bovine, or human. This indicates that an extensive proteolysis and the resultant activation of TH may occur only in vitro, but not in vivo.

We have tried to examine the effect of minimum proteolysis of TH by pure calpain on the properties of TH. Calpain II was completely purified from chiken skeletal muscle (Ishiura et al., 1978), and the minimum amount was used for limited proteolysis of TH. TH was completely purified from bovine adrenal medulla, as described by Nagatsu and Oka (1986).

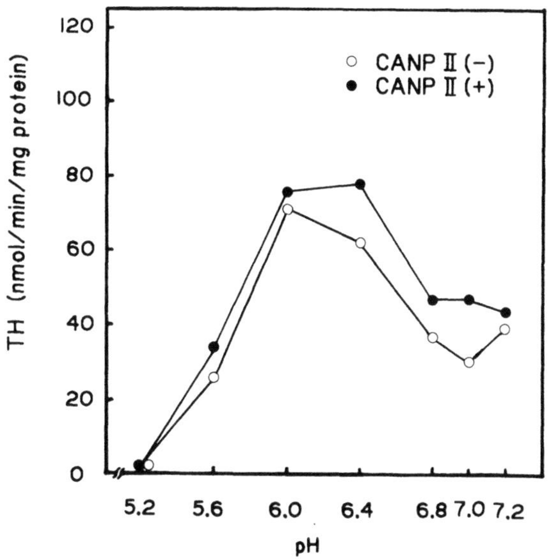

Fig. 4. The pH-profile of TH activity after incubation for 20 min at 25°C in the raction mixture for proteolysis without (○) or with (●) calpain II. TH activity was assayed in 0.1 M MES-NaOH buffer (pH 7.0) containing 0.2 mM L-tyrosine and 1 mM $6MPH_4$.

Preincubation of TH with calpain II was terminated by leupeptin, and TH activity and Mr were measured by HPLC-electrochemistry (Nagatsu and Oka, 1986) and sodium dodecylsulfate polyacrylamide gel electrophoresis (SDS-PAGE) respectively.

A slight but significant activation of TH activity at pH 7.0 was observed after incubation at 25°C for 20 min (Fig. 4). SDS-PAGE (Fig. 5) showed formation of a major band, which Mr was estimated to be approximately 56 kDa (Fig. 6). Fig. 7 shows kinetic studies of calpain-treated TH for the pterin cofactor ($6MPH_4$). As shown in Fig. 7, the activation of TH was not clearly observed. A dramatic change was the insensitivity to dopamine inhibition. Dopamine inhibited native TH in competition with $6MPH_4$, as we had reported previously (Nagatsu et al., 1964).

The present results indicate that pure calpain cleaves only a small domain with Mr of 4 kDa which regulates feed-back inhibition of TH by the end product, catecholamine.

However, the 56 kDa species of TH besides the main 60 kDa species cannot be detected yet in the adrenal medulla and brain of the rat, bovine, and human.

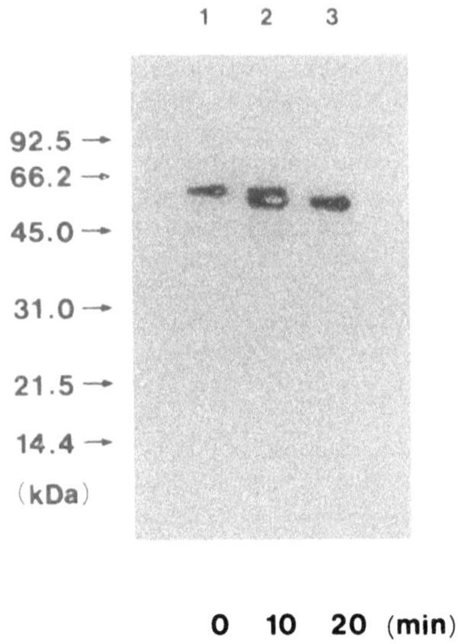

Fig. 5. Autoradiogram of Western blotting analysis of TH treated with calpain II for (1) 0 min, (2) 10 min and (3) 20 min. TH was detected using the anti-rat adrenal TH IgG and [^{125}I] anti-rabbit Ig antibody.

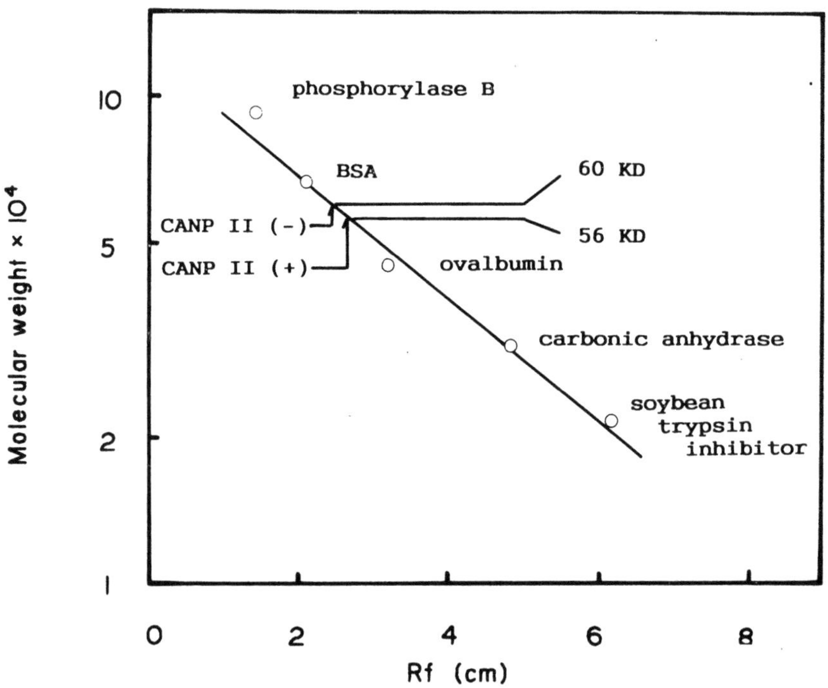

Fig. 6. Determination of molecular weight of TH treated with calpain II by SDS-PAGE (10% gel). Molecular weights of standards are as follows: phosphorylase B (92,500); BSA (66,200); ovalbumin (45,000); carbonic anhydrase (31,000); soybean trypsin inhibitor (21,500).

CONCLUSION

Among three Ca^{2+}-related mechanisms of activation of TH, i.e., Ca/CaM kinase, C-kinase, and calpain, only Ca/CaM kinase seems to be operated *in vivo*. Fig. 8 shows a schematic presentation of the mechanism of TH regulation in the nerve endings of dopaminergic neurons in the brain. Dopamine and ATP are co-released from the nerve terminal by nerve impulse.

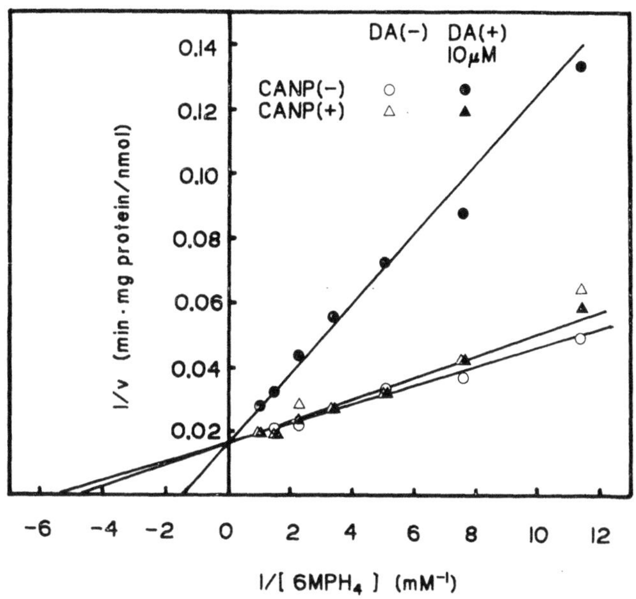

Fig. 7. Kinetic comparison of the enzymatic activity of TH incubated in the reaction mixture for proteolysis without (○ , ●) or with (△ , ▲) calpain II. TH activity was measured in the absence (○ , △) or presence (● , ▲) of 10 μM dopamine.

Dopamine may stimulate a D_2-type autoreceptor located in presynaptic nerve terminals to inhibit Ca^{2+} influx. As the consequence, dopamine release is inhibited, and TH activity is inhibited owing to the decreased activity of Ca/CaM kinase which phosphorylates and activates TH. Adenosine formed from released ATP may also stimulate presynaptic adenosine receptor to activate TH due to phosphorylation by A-kinase. Calpain that is activated by increased Ca^{2+} influx into the nerve endings could also modulate TH molecule, but it remains to be elucidated whether or not calpain regulates TH *in vivo*.

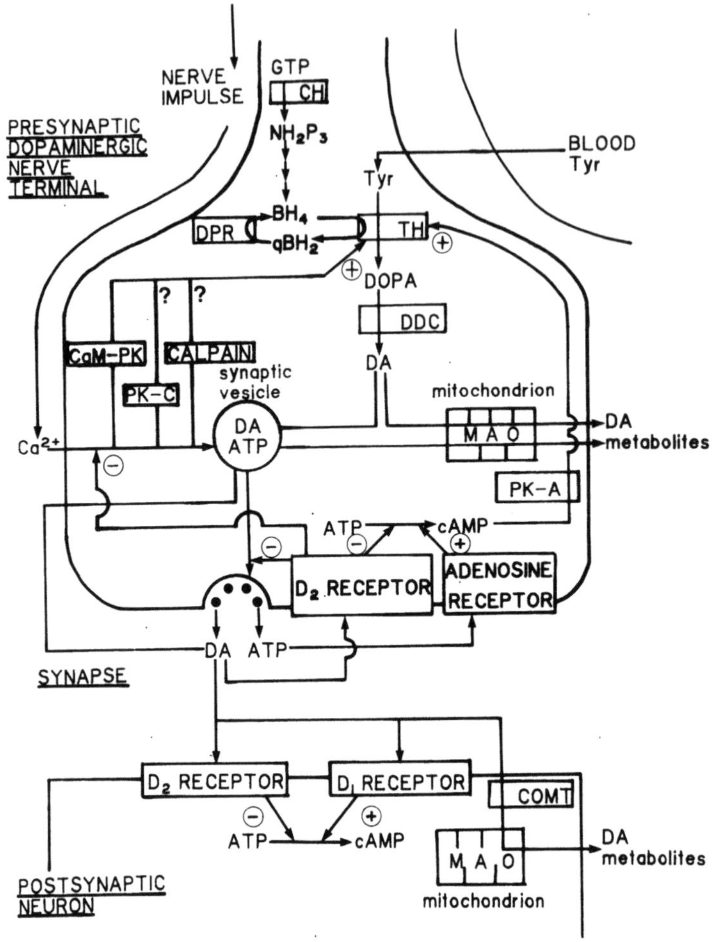

Fig. 8. Scheme of the mechanism of Ca^{2+}-related TH regulation in the presynaptic dopaminergic neurons.

REFERENCES

Atkinson, J., Richtand, N., Schworer, C., Kuczenski, R., and Soderling, T. 1987, Phosphorylation of purified rat striatal tyrosine hydroxylase by Ca^{2+}/calmodoulin-dependent protein kinase II: effect of an activator protien, J. Neurochem., 49:1241.

Chowdhury, M. and Fillenz, M., 1988, K^+-Dependent stimulation of dopamine synthesis in striatal synaptosomes is mediated by protein kinase C, J. Neurochem., 50:624.

Grima, B., Lamouroux, A., Boni, C., Julien, J.-F., Javoy-Agid, F., and Mallet, J., 1987, A single human gene encoding multiple tyrosine hydroxylases with different predicted functional characteristics, Nature, 326:707.

Griffith, L.C. and Schulman, H., 1988, The multifunctional Ca^{2+}/calmodulin-dependent protein kinase mediates Ca^{2+}-dependent phosphorylation of tyrosine hydroxylase, J. Biol. Chem., 263:9542.

Hidaka, H., Asano, M., Iwadare, S., Matsumoto, I., Totsuka, T., and Aoki, N., 1978, A novel vascular relaxing agent, N-(6-aminohexyl)-5-chloro-1-naphthalenesulfonamide which affects vascular smooth muscle actomyosin J. Pharmac. Exp. Ther., 207:8.

Hidaka, H., Inagaki, M., Kawamoto, S., and Sasaki, Y., 1984, Isoquinolinesulfonamides, novel and potent inhibitors of cyclic nucleotide dependent protein kinase and protein kinase C, Biochemistry, 23:5036.
Hirata, Y. and Nagatsu, T., 1985, Evidence for the involvement of Ca^{2+}-calmodulin and cyclic AMP in the regulation of the tyrosine hydroyxlase system in rat striatl tissue slices, Biochem. Pharmacol., 34:2637.
Ishiura, S., Murofushi, H., Suzuki, K., and Imahori, K., 1978, Studies of a calcium-activated neutral protease from chicken skeletal muscle, J. Biochem., 84:225.
Kaneda, N., Kobayashi, K., Ichinose, H., Kishi, F., Nakazawa, A., Kurosawa, Y., Fujita, K., and Nagatsu, T., 1987, Isolation of a novel cDNA clone for human tyrosine hydroxylase: alternative RNA splicing produces four kinds of mRNA from a single gene, Biochem. Biophys. Res. Commun., 146:971.
Kobayashi, K., Kaneda, N., Ichinose, H., Kishi, F., Nakazawa, A., Kurosawa, Y., Fujita, K., and Nagatsu, T., 1988, Structure of the human tyrosine hydroxylase gene: alternative splicing from a single gene accounts for generation of four mRNA types, J. Biochem., 103:907.
Nagatsu, T., Levitt, M., and Udenfriend, S., 1964, Tyrosine hydroxylase, the initial step in norepinephrine biosynthesis. J. Biol. Chem., 239:2910.
Nagatsu, T., Hirata, Y., Sawada, M., and Hidaka, H., 1985, Elucidation of regulatory mechanism of tyrosine hydroxylase and tryptophan hydroxylase by calmodulin antagonists, in: "Calmodulin Antagonists and Cellular Physiology", H. Hidaka and D. J. Hartshorne, eds., Academic Press, New York, p. 423.
Nagatsu, T. and Oka, K., 1986, Tyrosine 3-monooxygenase from bovine adrenal medulla in: Meth. Enzymol. vol. 142, S. Kaufman, ed., Academic Press, New York, p. 56.
O'Malley, K.L., Anhalt, M.J., Martin, B.M., Kelsoe, J.R., Winfield, S.L., and Ginns, E.I., 1987, Isolation and characterization of the human tyrosine hydroxylase gene: identification of 5' alternative splice sites responsible for multiple mRNAs, Biochemistry, 26:6910.
Saitoh, M., Ishikawa, T., Matsushima, S., Naka, M., and Hidaka, H., 1987, Selective inhibition of catalytic activity of smooth muscle myosin light chain kinase, J. Biol. Chem., 262:7796.
Togari, A., Ichikawa, S., and Nagatsu, T., 1986, Activation of tyrosine hydroxylase by Ca^{2+}-dependent neutral protease, calpain, Biochem. Biophys. Res. Commun., 134:749.
Yamauchi, T. and Fujisawa, H., 1981, Tyrosine 3-monooxygenase is phosphorylated by Ca^{2+}-, calmodulin-dependent protein kinase, followed by activation by activator protein, Biochem. Biophys. Res. Commun., 100:807

CALBINDIN-D28 IN MAMMALIAN BRAIN, RETINA, AND ENDOCRINE PANCREAS : IMMUNOHISTOCHEMICAL COMPARISON WITH CALRETININ

R. Pochet[1], F. Blachier[1], W. Malaisse[2], M. Parmentier[3], B. Pasteels[1], V. Pohl[1], A. Résibois[1], J. Rogers[4] and A. Roman[1]

[1] Lab of Histology, Fac. of Medicine, Université Libre de Bruxelles (ULB), Belgium
[2] Lab of Experimental Medicine, ULB
[3] I.R.I.B.H.N., ULB
[4] Physiological Lab, University of Cambridge, U.K.

ABSTRACT

Calbindin 28K and calretinin are very similar calcium binding proteins which are both present in the central nervous system (CNS). They respectively bind 4 and 5 Ca^{++} ions. We have compared by immunohistochemistry and in situ hybridization their localisation in the brain and the retina. The two proteins are generally expressed in different neurons with a few neurons containing both calcium binding proteins. Calbindin 28K is also present in the endocrine system. We have examined the cellular distribution of calbindin in the pancreatic endocrine cells of chick, rat and human and found variable distribution among the different endocrine cell types. We also describe the presence of calbindin in RINm5F cells, an insulin-producing tumor cell line derived from a radiation-induced rat insulinoma.

INTRODUCTION

Calbindin 28K is a cytosolic calcium binding protein of 261 amino-acids whose exact MW is 30,300 daltons in human (Parmentier et al, 1987a) and 30,420 daltons in chick (Wilson et al, 1985). It is present in all classes of vertebrates (Parmentier et al, 1987b) and in a wide range of tissues (Norman et al, 1982). Calbindin belongs to the EF hand containing proteins family and binds four atoms of calcium per mole. Its role within the cell is unknown.

Calretinin is another EF hand containing calcium binding protein localized in the nervous system and having 58% homology with calbindin (Rogers, 1987). Its

MW is 31,760 daltons in human with five Ca^{++} binding sites (Parmentier et al, 1988). Calretinin and Calbindin are generally expressed in different neurons in chick brain (Rogers, 1987). We have compared the localization of both proteins in the rat brain and have examined its localization within the endocrine pancreas of chick, rat and human as well as in transformed cells from a rat insulinoma (Gazdar et al, 1980).

Antisera against calbindin stain a wide variety of neurons in rat brain, as surveyed by Jande et al (1981), Baimbridge and Miller (1982), Feldman and Christakos (1983) and Garcia-Segura et al (1984). Some of these surveys may also have detected cross-reaction with calretinin, as in the chick (Rogers, 1987).

MATERIAL AND METHODS

Antibodies

Primary antibodies were used as follows : Rabbit antiserum against chick intestinal calbindin given by DEM Lawson (Spencer et al, 1978) at 1:6000 dilution and rabbit antiserum against a ß-galactosidase-calretinin fusion protein at 1:2000 dilution (Rogers, manuscript submitted; see Rogers, 1987 for description of the fusion protein from plasmid pURXR.3).

Cross reactivity

Preincubation of anti-calbindin with purified calbindin, or of anti-calretinin with the calretinin fusion protein, completely abolished the labelling in all neurons except in Purkinje cells where a faint labelling for calbindin is still present (cellular body and processes). On the contrary preincubation of anti-calbindin with the calretinin fusion protein, or anti-calretinin with calbindin, does not modify the labelling.

Immunohistochemistry

Adult rats were perfused with either 4% paraformaldehyde in 0.1 M phosphate buffer pH 7.4 (PAF) or with "Helly" fixative (Langeron, 1949) and then embedded in paraffin. Eyes from monkey (cervopithèque barbades) were dissected and fixed for 6 hours in Helly fixative. The immunocytochemical reactions were performed as described earlier (Roman et al, 1988). Briefly, the 5 μm sections were rehydrated and incubated 48 h with the primary antibody, which was followed by secondary antibody and then by peroxidase/-diaminobenzidine. For pancreas, a small part of the organ was immersed in Bouin Hollande fixative (BHS) for three days (Pochet et al, 1987).

Western Blotting

The detailed procedure is published elsewhere (Pochet et al, 1985). Briefly the pelleted cells were homogenized by boiling for 5 min a denaturing medium, then electrophoresed on a discontinuous slab gel, before electroblotting onto nitrocellulose sheets.

In situ hybridization

Wistar rat brains were used fresh, frozen in Tissue-Tek and cut into serial 15 μm sections, fixed in 4% paraformaldehyde in 0.1 M sodium phosphate buffer pH 7.4 (PBS) + 5 mM $MgCl_2$ + 15 mM Vanadyl sulfate, rinsed in PBS, dehydrated, then kept frozen (-20°C) and dessicated.

Before hydridization, sections have undergone several treatment : saponin (0.01% in PBS), pronase (50 μg/ml in 50 mM Tris pH 7.5, 5 mM EDTA), acetic anhydride (0.25% V/V), dehydrated and air dried.

Sense (negative) and antisense (hybridizing) probes were obtained following in vitro transcription of plasmids containing either human calbindin 27 K or 29 K cDNA ($\pm$250 base pairs) using T_3 and T_7 RNA polymerases respectively (Stratagene) in the presence of 4 μM ^{35}S uridine triphosphate nucleotide. Hybridization mixture (50% formamide, 2 x SSC, 1 x Denhardt, ribonuclease inhibitor, 5 mM dithiotreitol, 10% dextran sulfate, 100 μg/ml E. Coli tRNA, 15 μg/ml sonicated DNA from salmon testes, 10^{-5} M uridine 5´triphosphate) was applied (5.10^5 cpm/25 μl/section) and sections were overlaid for protection with Thermanox coverslips left at 55°C-60°C overnight. After ribonuclease treatment, they were washed in formamide 50%, 0,25 SSC (2 x at 45°C, 4 x at 20°C) and rinsed in 0,5 SSC without formamide. Slides were exposed for 2-3 weeks at 4°C to Ilford G5 emulsion and stained with 5% Giemsa after fixation.

Cell culture

Tumoral islet cells of the RINm5F line were cultured in RPMI 1640 medium supplemented with 2 mM glutamine, antibiotics (penicillin 100 U/ml, streptomycin 100 μg/ml) and decomplemented fetal calf serum (10%, V/V) as described earlier (Giroix et al, 1985).

RESULTS AND DISCUSSION

Rat brain : immunohistochemistry

The antisera against chick calbindin and calretinin give clearly different staining patterns in rat brain, demonstrating that calbindin and calretinin are detected with only weak cross-hybridization. Both antisera label a large number of neurons and their fibres. They often co-exist in a given nucleus but usually not in the same cell. The Purkinje cells and dentate gyrus granule cells are clearly labelled with anti-calbindin and not with anti-calretinin; the reverse is true of the reticular nucleus of the thalamus.

There is neither calbindin-like immunoreactivity (-LIR) nor calretinin-LIR in the *corticospinal tract* and in *motoneurons* (nuclei of nerves III, IV, VI, V-motor, and XII).

On the other hand, *sensory nerves* (VIII and sensory root of V) are labelled by both antibodies, as are most of the nuclei in which they terminate (superior colliculus, dorsal part of the pretectal nuclei, cochlear nuclei, sensory nuclei of the trigeminal nerve, medial vestibular nucleus). All these nuclei contain many calretinin-positive cells or many calbindin-positive cells, or both; few are double-positive.

Many *hypothalamic nuclei* are strongly labelled by both antisera, especially anti-calbindin, and some contain many double-positive neurons. In the *thalamus*, the nuclei of the posterior complex are mostly negative for both, but calbindin-LIR is present in numerous cells of the more anterior and medial nuclei, while calretinin-LIR is very conspicuous in bands of cells that seem to form a partial shell around the thalamus (including the reticular nucleus). Finally, in the *reticular formation* of the brainstem, few neurons are calbindin-positive while many are calretinin-positive.

Rat brain : in situ hybridization

Human probes hybridized with rat RNA under stringent conditions (see

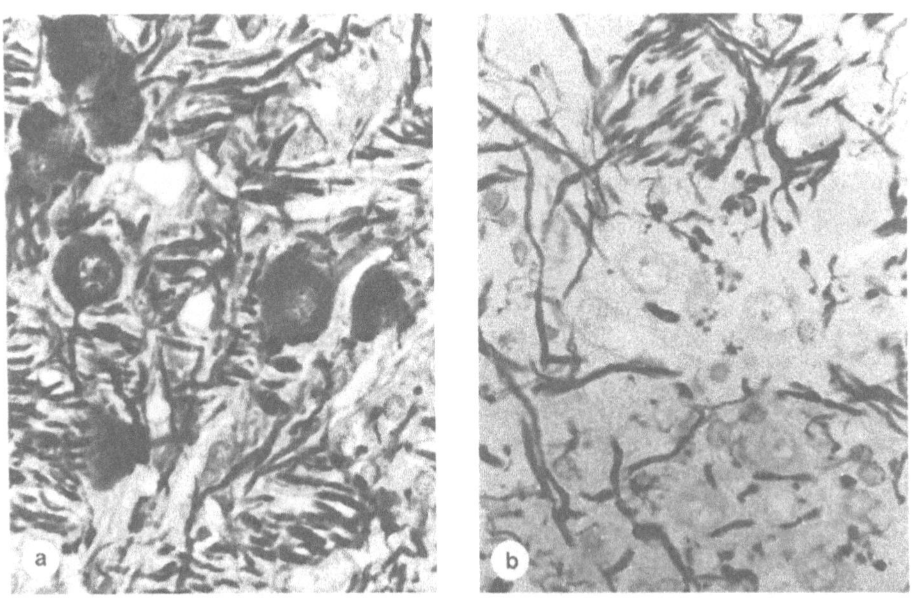

Fig. 1 Immunohistochemistry with antisera to chick proteins of the middle one third of the ventral cochlear nucleus. Calretinin-LIR and calbindin-LIR are both present in nerve VIII. At this level of the nucleus, most of the neurons are calretinin-LIR positive (a) and calbindin negative (b).

method) which completely eliminate any cross reaction. For instance in cerebellum Purkinje cells are negative for the calretinin probe. Fig. 2 illustrates ISH for both probes in a rat thalamus section.

There is a large consensus between ISH and immunohistochemistry. A complete survey of ISH mapping of both calbindin 28 and calretinin will be published elsewhere.

Monkey retina

In the photoreceptor layer, cones are labelled with both antibodies. At the level of the fovea (fig. 3), cones are calbindin-negative but still calretinin-positive. This staining difference may represent a functional difference in cones of the fovea. It may be comparable to the absence of calbindin from the cones of the red field of the pigeon retina, which is analogous to the fovea (Pasteels et al, 1987; Gallifret, 1968). Rods are calbindin-negative, but calretinin-LIR can be detected in the inner segment.

In the outer plexiform layer and the inner nuclear layer, horizontal cells contain both calretinin-LIR and calbindin-LIR, the latter being more intense. In amacrine cells, the concentration of calretinin and calbindin label seem rather similar.

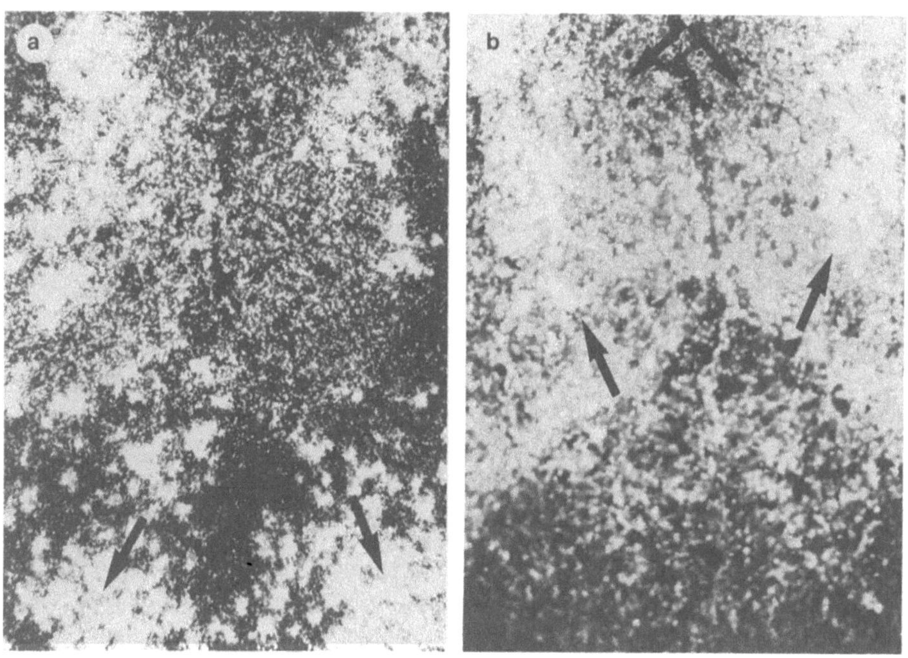

Fig. 2 In situ hybridization of rat thalamus with probes from human genes.
Fig.2a Part of the habenula and the posterior nucleus (arrows) of the thalamus are calretinin positive.
Fig.2b Only the habenula nucleus is calbindin positive (arrows).

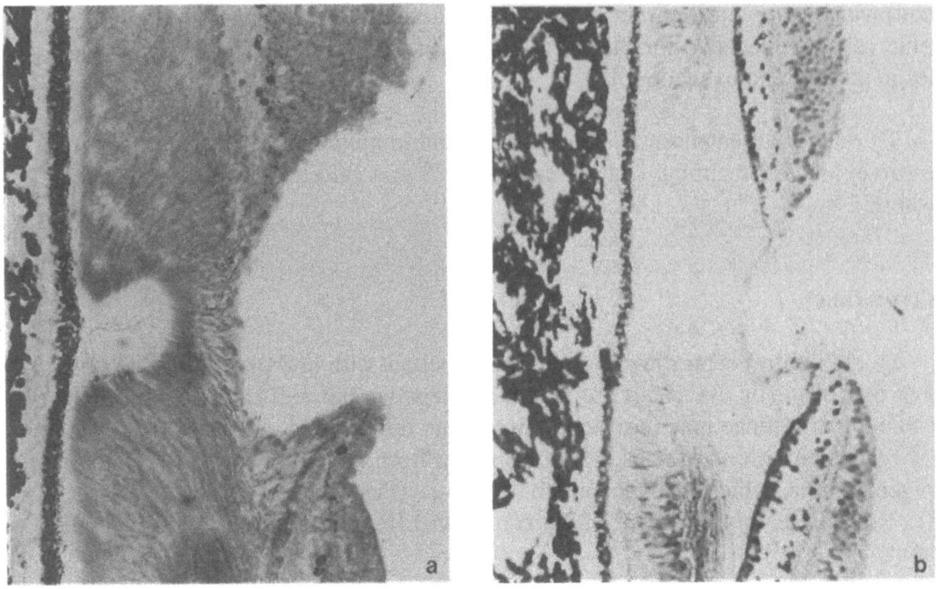

Fig. 3 Immunohistochemistry of the monkey retina. Sections of the fovea labelled with anti-calretinin (a) and with anti-calbindin (b).

Endocrine pancreas

We already described the distribution of calbindin in the endocrine pancreas of chick and rat (Pochet et al, 1987) where it appears that both insulin secreting cells and non insulin secreting cells contain calbindin. In human pancreas, again calbindin is present (fig.4) in islets. The distribution is not homogenous, some cells are rich in calbindin, some contain a low amount and others are calbindin negative. Double labelling using anti-glucagon or anti-insulin or anti-somatostatin serum (not shown) suggest that calbindin is not associated preferentially with a specific secreting cell population.

Transformed cells

Western blot experiments using RINm5F cells and anti-chick calbindin demonstrate the presence of a band comigrating with calbindin (Blachier et al, 1988). This suggests expression of the calbindin gene in this transformed cell lines. The band is still recognized after heating the 100,000 g supernatant at 65°C indicating that, as calbindin, the protein is heat stable. HPLC experiments (not shown) comparing the profile of elution of calbindin of rat brain homogenate with RINm5F cell homogenate show an identical time of elution for both proteins recognized by anticalbindin. The detection of calbindin in culture cells will allow further experiments to search for an eventual regulation of the gene expression by Vitamin D and to investigate whether insulin secretion can be modulated by calbindin.

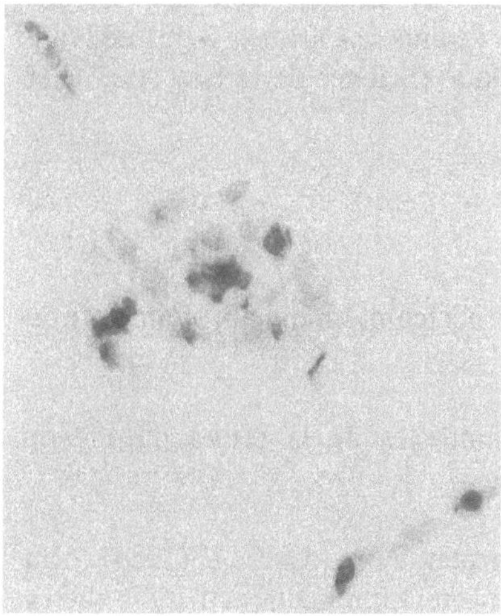

Fig. 4 Immunohistochemistry of a human pancreas section labelled with anti-calbindin.

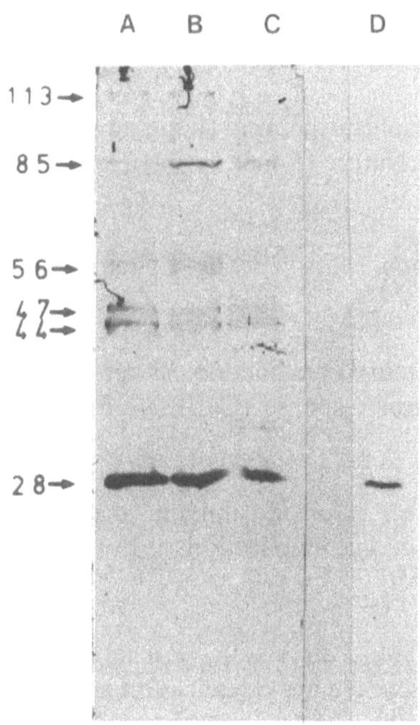

Fig. 5 Western blot of RINm5F cell extrat using anti-calbindin 3 x 10^6 cells in lane A, B, C.

A. 100,000 g (60 min) heated (65°C) supernatant.
B. 100,000 g (60 min) supernatant.
C. Total homogenate
D. 15 isolated rat islet cells.

ACKNOWLEDGMENTS

We thank P. Miroir for typing the manuscript. F.B. is an EEC fellow, M.P. is Chargé de Recherche (FNRS). This work was supported by grants from the Belgian Foundation for Scientific Medical Research (grants 3.4501.84, 3.4511.88 and 3.4531.88), the Ministère de la Politique Scientifique (Science de la vie), the EEC (Stimulation grant) and the British Council.

REFERENCES

Baimbridge KG, Miller JJ and Parkes CO (1982). Calcium-binding protein distribution in the rat brain. Brain Res 239: 519-525.

Gallifret Y (1968). Les diverses aires fonctionnelles de la rétine du Pigeon. Zeitschrift für Zellforschung 86: 535-545.

Garcia-Segura LM, Baetens D, Roth J, Norman AW, Orci L (1984). Immunohistochemical mapping of calcium-binding protein immunoreactivity in the rat central nervous system. Brain Res 296: 75-86.

Gazdar AF, Chick WL, Oie HK, Sims HL, King DL, Weir GC and Lauris V (1980). Continuous, clonal, insulin- and somatostatin-secreting cell lines established from a transplantable rat islet cell tumor. Proc Natl Acad Sci, USA, 77: 3519-3523.

Giroix MH, Malaisse-Lagae F, Sener A and Malaisse WJ (1985). Hescose metabolism in pancreatic islet. Galactose transport, Phosphorylation and oxidation. Mol Cell Biochem, 66: 61-64.

Jande SS, Maler L and Lawson DEM (1981). Immunohistochemical mapping of vitamin D-dependent calcium-binding protein in brain. Nature (London) 294: 765-767.

Langeron M (1949). Précis de microscopie. 7e edition, Masson et Cie, Editeurs, Paris.

Norman AW, Roth J and Orci L (1982). The vitamin D endocrine system : steroid metabolism, hormone receptors and biological response (CaBPs). Endocrinol Rev, 3: 331-366.

Parmentier M, Lawson DEM, and Vassart G (1987a). Human calbindin complementary DNA sequence. Evolutionary and functional implications. Eur J Biochem, 170: 207-215.

Parmentier M, Ghysens M, Rypens F, Lawson DEM, Pasteels JL, Pochet R (1987b). Calbindin in vertebrate classes. Immunohistochemical localization and Western blot analysis. Gen Comp Endocrinol 65: 399-407.

Parmentier M, Pohl V and Vassart G (1988). A calbindin 29 KDA : a new neuronal protien. Abstract p. 181, Sixth International Symposium on Calcium-binding proteins in Health and Disease, Nagoya, Japan, July 1988.

Pasteels B, Parmentier M, Lawson DEM, Verstappen A and Pochet R (1987). Calcium binding protein immunoreactivity in pigeon retina. Investigative Ophthal. Visual Sci., 28: 658-664.

Pochet R, Parmentier M, Lawson DEM, Pasteels JL (1985). Rat brain synthesizes two vitamin D-dependent calcium-binding proteins. Brain Res 345: 251-256.

Pochet R, Pipeleers DG and Malaisse WJ (1987). Calbindin D 27 kDa : preferential localization in non-B islet cells of the rat pancreas. Biol. of the Cell, 61: 155-161.

Résibois A, Rypens F and Pochet R (1988). Epithelial and neuronal calbindin in avian intestine : an immunohistochemical study. Cell and Tissue Res 251: 611-620.

Rogers JH (1987). Calretinin : a gene for a novel calcium-binding protein expressed principally in neurons. The J Cell Biol, 105: 1343-1353.

Roman A, Brisson P, Pasteels B, Demol S, Pochet R and Collin JP (1988). Pineal-retinal molecular relationship; immunocytochemical evidence of calbindin 27-kDa in pineal transducers. Brain Res., 442: 33-42.

Spencer R, Charman M, Wilson PW, Lawson DEM (1978). The relationship between vitamin D stimulated calcium transport and intestinal CaBP in the chicken. Biochem J 170:93-101.

Wilson PW, Harding M, Lawson DEM (1985). Putative amino acid sequence of chick calcium-binding protein deduced from a complementary DNA sequence. Nucleic Acid Res. 13: 8867-8881.

THE CALPAIN-CALPASTATIN SYSTEM IN HEMATOPOIETIC CELLS*

Takashi Murachi, Toshio Murakami,[1] Michiko Ueda, Ichiro Fukui, Takao Hamakubo, Yoshifumi Adachi,[2] and Masakazu Hatanaka[2]

Department of Clinical Science and Laboratory Medicine, Kyoto University Faculty of Medicine, Kyoto 606, [2]Department of Home Economics, Kyoto Bunkyo Junior College, Uji, Kyoto 611, and [2]Department of Hematology and Immunology, Institute for Virus Research, Kyoto University, Kyoto 606, Japan

INTRODUCTION

Calpain is a Ca^{2+}-dependent cysteine endopeptidase [EC 3.4.22. 17] and calpastatin is a calpain-specific endogenous inhibitor protein (Murachi et al., 1981b). At least two molecular species of calpain are known to exist. Calpain I requires low or μM concentration of Ca^{2+} for activation, and calpain II requires high or mM Ca^{2+}. Both calpains I and II are heterodimers, each consisting of one heavy (approximately 80 kDa) and one light (approximately 30 kDa) subunits. The heavy subunits of calpains I and II are different genetic products, while the light subunits of both calpains are identical (Murachi, 1983; Suzuki, 1987). Calpastatin is an extremely heat-stable protein, which can be separated from calpains by appropriate chromatographies in the presence of Ca^{2+}-chelating agent such as EGTA (ethylene glycol bis(β-aminoethyl ether)-*N*,*N*,*N*',*N*'-tetraactic acid)(Nishiura et al., 1978; Waxman and Krebs, 1978). Calpastatin can bind to and thus inhibit calpain, either I or II, only when the enzyme is activated by Ca^{2+} (Murachi et al., 1981b). Isolation and nucleotide sequencing of cDNAs for several molecular species of calpains and calpastatins have recently been completed, establishing the primary structures of these molecules (Ohno et al., 1984; Sakihama et al., 1985; Takano et al., 1986; Emori et al., 1987; Takano et al., 1988).

Calpains I and II and calpastatin are very widely, but diversely, distributed in animal cells. A large number of cellular component proteins have been reported to be susceptible to calpains which include several kinases, cytoskeletal and membrane proteins, and hormone receptor proteins (Murachi 1984; Murachi et al., 1987). Because of such wide and diverse distribution with so many different candidates for the natural substrate, the true biological function of calpain still remains obscure.

*This work was supported in part by a grant for Scientific Research from the Ministry of Education, Science and Culture, Japan and a grant for Basic Research from the Muscular Dystrophy Association, U.S.A. (to T.M.)

Accumulating evidence indicates that the autolysis of a calpain molecule results in significant lowering of the level of Ca^{2+} concentration that is needed for the activity of calpain (Inomata et al., 1986; Melloni et al., 1985). The concentration of Ca^{2+} required for autolysis is markedly decreased in the presence of phospholipid (Imajoh et al., 1986) and cell membrane preparation (Melloni et al., 1985). The molecular mechanisms of activation has been extensively studied (Suzuki et al., 1988).

CALPAINS IN NUCLEATED ERYTHROCYTES

In the course of distribution studies of calpain and calpastatin in various tissues and cells, we found that erythrocytes in rat peripheral blood contained calpain I and calpastatin, lacking calpain II (Murachi et al., 1981a). This was rather unique, since all other tissues of rats tested at the same time contained both calpains I and II, although relative abundance was variable from one tissue to the other. That only calpain I is present in erythrocytes was found to be true also with human, monkey and guinea pig (Murachi et. al., 1981b). The absence of calpain II in human erythrocytes was further verified using specific antibody directed against calpain II heavy subunit (Hatanaka et al., 1984). Since chicken erythrocytes are known to be nucleated, in contrast to denucleated erythrocytes in mammalian blood, we became interested in investigating whether these nucleated erythrocytes contain both calpains I and II. Another reason for our being particularly interested in chickens was related to a previous report by Suzuki et al. (1985). These authors have concluded from their study on messenger RNA for calpain that only a single molecular species of calpain heavy subunit exist in chickens.

Fresh chicken blood was collected from a white Leghorn, 3 to 4 months old. Leukocytes-and-platelets-free erythrocytes were hemolyzed, and the hemolysate was centrifuged at 105000 x g for 60 min. The soluble fraction thus obtained was chromatographed on a DEAE-cellulose column at pH 7.5 with a linear gradient of NaCl from 0 to 0.4 M. When the elution fractions were assayed for their caseinolytic activity in the presence of a fixed amount of human erythrocyte calpain I previously added to each test tube, one negative and two positive peaks were recorded, as shown in the inset of Fig. 1 (Murakami et al., 1988). The first negative peak should

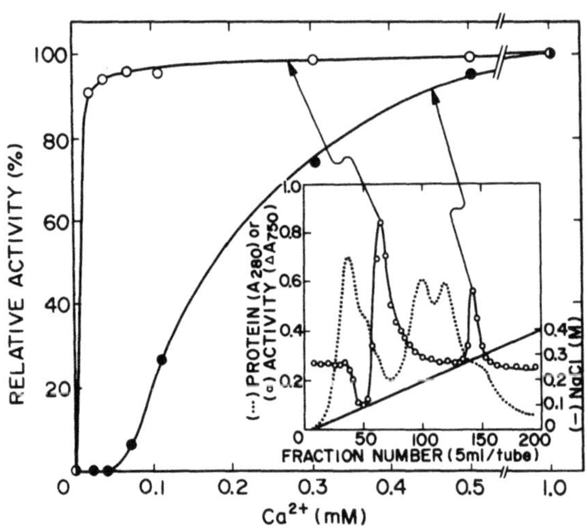

Fig. 1. Identification of both calpains I and II in nucleated chicken erythrocytes (Murakami et al., 1988).

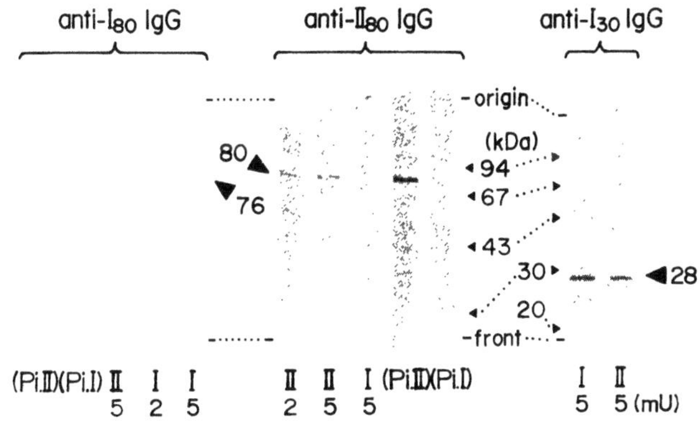

Fig. 2. Detection of chicken erythrocyte calpains by Western blot analysis (Murakami et al., 1988).

correspond to that of calpastatin. The Ca^{2+} requirements of the enzymes that appeared as two positive peaks clearly differed from one another (Fig. 1). For half-maximal activity, the former enzyme required 10 μM Ca^{2+}, while the latter required 180 μM Ca^{2+}. These results led us to conclude that the two enzymes are to be respectively classified as calpain I and calpain II.

Figure 2 are the results of Western blot analysis of the calpain proteins collected from each one of the two peaks of Fig. 1, using antibodies directed respectively against pig (Pi) calpain I-heavy (I80), calpain II-heavy (II80), and calpain I-light (I30) subunits. The staining patterns shown in the figure indicate mutually exclusive and dose-dependent cross-reactivity with specific antibodies that should discriminate between calpains I and II. It was thus concluded that chicken erythrocytes do contain both calpain I and calpain II, hence that chicken do have at least two individual genes for calpain I and for calpain II as other higher animals have (Murakami et al., 1988).

Denucleated erythrocytes contain only calpain I, whereas nucleated erythrocytes contain both calpains I and II, with calpain I predominating in amount. These findings in different erythrocytes might suggest that the expression of calpain II protein has been weakened or even missed during the course of maturation process of erythrocytes (Murakami et al., 1988). The regualtion of gene expression as related to the differentiation of erythroid lineage cells must be one of the target of future investigations.

CALPAINS IN POLYMORPHONUCLEAR LEUKOCYTES

There was some controversy in the early literature as to whether or not calpain is a heterodimer. However, the results of the later studies are mostly supportive to the heterodimeric nature. The parallel purification of calpains I and II from rat kidney yielded heterodimeric products of both, each being composed of 80 kDa and 25 kDa subunits (Yoshimura et al., 1983). The existence of heterodimeric calpain molecules in solution was demonstrated using bifunctional cross-linking reagents such as dimethyl suberimidate (Mellgren et al., 1982) and disuccimidyl suberate (Hatanaka et al., 1985). The experiments on dissociation of heterodimeric calpain into its heavy and light subunits, followed by reassociation of the latters disclosed the concerted action of both subunits in enhancing the efficiency of restoring the catalytic activity (Tsuji and Imahori, 1981; Kikuchi et al., 1984). The light subunit was found to be readily proteolyzed, either by autolysis or the action of other proteases, so that it can be lost during the purification procedures (Mellgren et al., 1982; Inoamta et al., 1986). The heterodimeric nature of calpain has finally been confirmed by the determination of primary structures of the heavy and light subunits, which must be regarded as the products sharing common genetic information (Suzuki, 1987).

It was reported, on the other hand, that a monomeric calpain of 85 kDa was isolated from human polymorphonuclear (PMN) cells (Pontremoli et al., 1985; Pontremoli and Melloni, 1986). These authors claimed that this 85 kDa monomer was the only molecular species of calpain found in human PMN cells. This does not seem to be in accord with the above concept that the intact form of calpain must be a heterodimer. It also differs from the general finding that all animal tissues, except mammalian erythrocytes, contain calpains I and II.

We isolated PMN cells from pig peripheral blood as intact as possible, so as to avoid any hazard of artificial breakdown of cells as well as their constituents during prepartion (Fukui et al., 1988). The average cell populations were: PMN cells, 95.5%; eosinophils, 3.5%; monocytes, 0.3%; others, 0.7%. The cells were disrupted either by pressure treatment, called nitrogen cavitation (N-CV or -CAV) method or by homogenization (homogenzn). Nitrogen cavitation at 335 pounds per square inch (psi) was chosen as the condition under which the destruction of membranes and organelles were minimal.

Figure 3 shows the results of Western blot analysis on the cytosol preparations obtained by different methods. By nitrogen cavitation at 335 psi, anti-calpain I heavy (I80) subunit antibody stained one single band of 70 kDa, while anti-calpain II heavy subunit (II80) antibody stained one single band of 80 kDa. With the supernatant of the homogenate, 83, 80, and 30 kDa bands were found. The lysate obtained by nitrogen cavitation at 475 psi gave a somewhat intermediate picture. These results indicate that the original form of the heavy subunit of calpain I in PMN cells must have been of 83 kDa and that it can undergo breakdown to varying degrees into a 70 kDa protein depending upon the conditions used for disruption of the cells. It is also important that both calpains I and II were found to be present (Fukui et al., 1988).

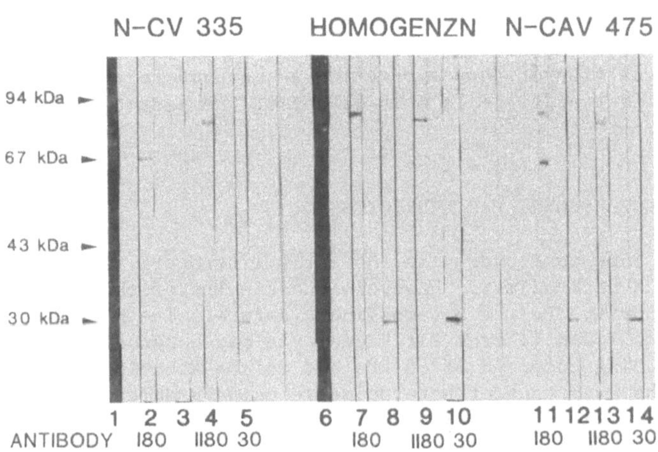

Fig. 3. Western blot analysis of PMN cell cytosol preparations (Fukui et al., 1988).

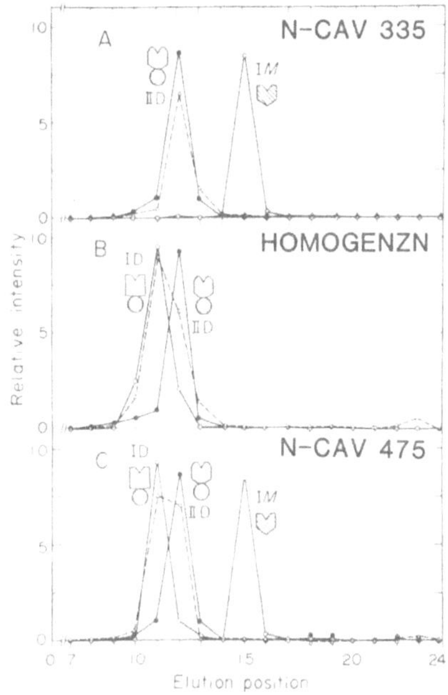

Fig. 4. Location of calpain antigenicities associated with the HPLC elution fractions of PMN cytosol preparations (Fukui et al., 1988).

The existence of both calpains I and II in ordinary heterodimeric forms (ID and IID) in PMN cells were further demonstrated by gel permeation HPLC analysis as is illustrated in Fig. 4. Since the HPLC was carried out in the absence of any detergent, one can expect concomitant elution of a heavy and a light subunits from the column, if the stuff under test should have been a heterodimer. The graph shows the intensities of the dot stainings with specific antibodies which were quantified by a reflectometer. Concomitant elution of the heavy (solid line) and light (broken line) subunits can be seen for both calpains I and II in the case of homogenization, clearly rejecting the concept that PMN cells contain only one species of calpain that is a monomer of 85 kDa (Pontremoli and Melloni, 1986). The 70 kDa calpain I (IM) does not accompany the light subunit (Fig. 4, panels A and C), resembling a monomeric 85 kDa species reported by Pontremoli et al. (1985). These monomers must be artificial products formed during the course of preparation. Thus, we have been able to settle the issue, concluding that PMN cells are not exceptional (Fukui et al., 1988).

CALPAINS AND CALPASTATIN IN T CELL LINE CELLS+

In the course of the screening of human hematopoietic system cells with respect to their expressiveness of calpains I and II using respectively specific antibodies, Adachi et al. (1985) found that all the cells tested contained calpain I, whereas the expression of calpain II was not

+A part of this work was also presented at the 14th International Congress of Biochemistry, Prague, July 10-15, 1988 (Chester Beatty Lecture; Murachi, 1989) and at the 627th Meeting of the Biochemical Society, Nottingham, July 20-22, 1988 (Symposium on the Gene Expression: Regulation at RNA and Protein Levels; Murachi et al., 1989).

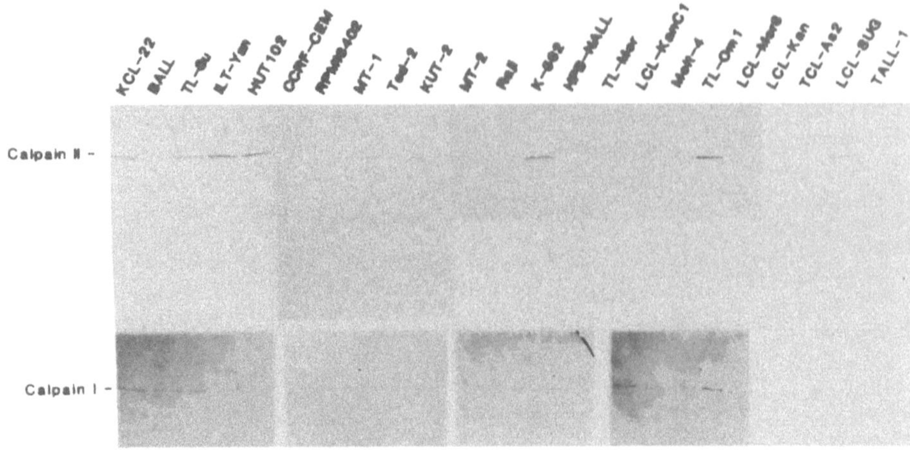

Fig. 5. Expression of calpains I and II in human hematopoietic system cells (Adachi et al., 1985; Adachi, Y., Murachi, T., and Hatanaka, M., manuscript in preparation).

uniform. Figure 5 is the results of Western blot analysis. Actually, there were nine cell lines that showed marked expression of calpain II. Furthermore, all these nine cell lines were found to be T-cell line cells which had been infected by HTLV-I, a retrovirus responsible for human adult T-cell leukemia. T-cell line cells, which had not been infected by HTLV-I, did not express appreciable amount of calpain II.

We also performed chromatographic analysis on HTLV-I-infected and non-infected T-cell lineages. Two examples of the results are shown in Fig. 6. MT-2 cells are known to carry HTLV-I gene and therefore express interleukin (IL) 2 receptor protein. The production of calpain II and calpastatin by MT-2 can be seen. By contrast, CCRF-CEM cells, which are not HTLV-I-infected and therefore do not express IL-2 receptor protein, produced calpain I, but neither calpain II nor calpastatin.

We then further carried out a DNA transfection experiment by the procedures outlined in Fig. 7. Starting from an HTLV-I-uninfected, natu-

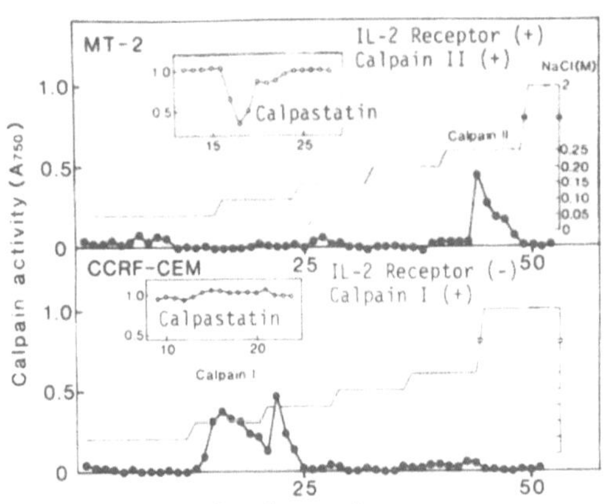

Fig. 6. Chromatographic analysis on calpain and calpastatin activities (Adachi et al., 1985; Adachi, Y., Murachi, T., and Hatanaka, M., manuscript in preparation).

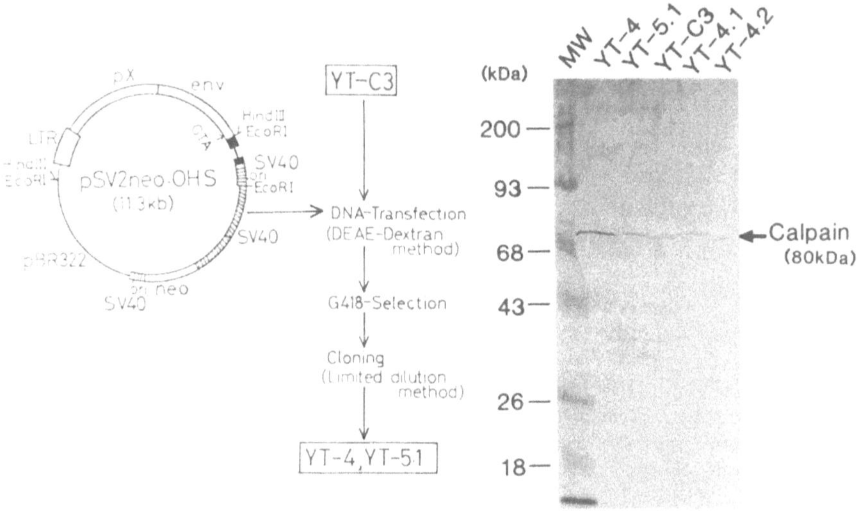

Fig. 7. Enhanced expression of calpain II after transfection of HTLV-I DNA to YT-C3 cell (Adachi, Y., Murachi, T., and Hatanaka, M., manuscript in preparation).

ral killer-like T-cell line cell, YT-C3, stable and transformed cell lines, YT-4 and YT-5.1 were established. As shown in the figure, these transformed cells now produced fairly large amounts of calpain II, suggesting the same mechanism operative on the expression of calpain II as on that of IL-2 receptor protein.

Figure 8 shows the Western blot data on calpastatin with various hematopoietic system cells, demonstrating stronger expression of calpastatin concomitant with that of IL-2 receptor protein (Adachi et al., 1988b). These results were in full accord with what we had found by chromatographic analysis (see above, Fig. 6).

From these and other observations, we have come to propose a hypothetical model as is shown in Fig. 9. In the event of HTLV-I infection, it has been known that a product protein from pX region of the viral genome, called <u>tat</u>, activates the transcription of IL-2 receptor protein. This is

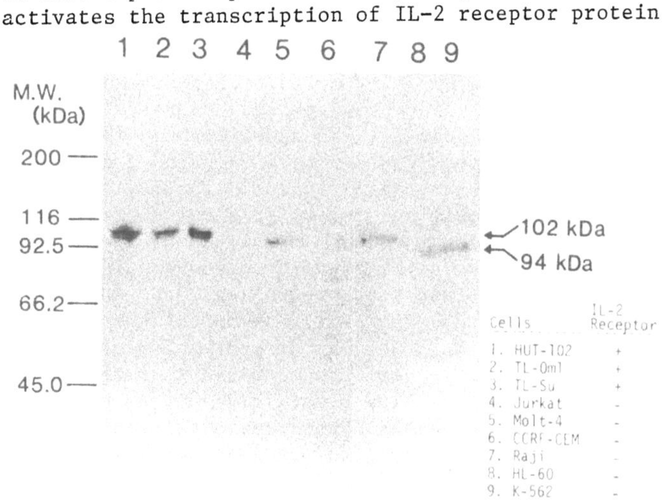

Fig. 8. Expression of calpastatin in human hematopoietic system cells (Adachi et al., 1988b).

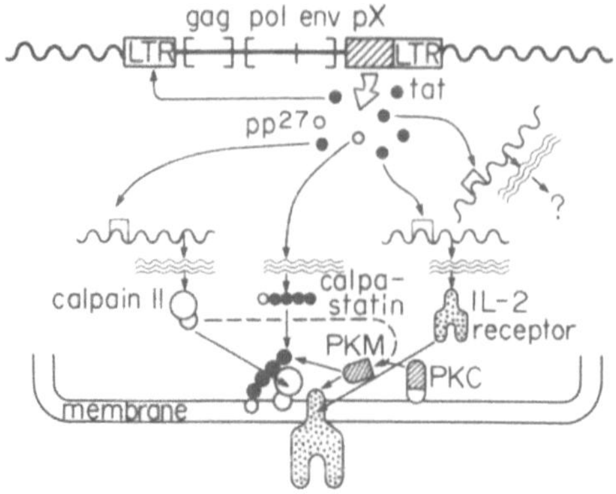

Fig. 9. Hypothetical model for the expression and interaction of calpain II, calpastatin, IL-2 receptor, and protein kinase C (Adachi, Y., Hatanaka, M., and Murachi, T., manuscript in preparation).

known as transacting activation. We presume that the same transacting activation mechanism may function for the enhanced expression of calpain II. On the other hand, we have some reasons for speculating that the enhanced expression of calpastatin protein by HTLV-I infection is brought about by another pX product, known at pp^{27}. No matter how different these mechanisms were, it is the solid observation that the viral infection gives rise to enhanced expression of these three proteins concomitantly. IL-2 receptor protein must be translocated to the membrane so as to be properly exposed toward the outside of the cell membrane. It has also been known that calpain is to be translocated to the cell membrane (Pontremoli and Melloni, 1986; Suzuki et al., 1987), and so is calpastatin, particularly upon its being phosphorylated (Adachi et al., 1988a; Murachi et al., 1989). It is also known that protein kinase C (PKC) can be transformed into its active form, called protein kinase M (PKM) by the action of calpain (Inoue et al., 1977). The model in Fig. 9 was designed to include all these pieces of evidence into one, which must be, of course, a subject to future debate.

SUMMARY

Calpain I requires low Ca^{2+} for activation and calpain II requires high Ca^{2+}. It was generally accepted that erythrocytes contain calpain I and calpastatin, but no calpain II. We have recently found, however, that nucleated chicken erythrocytes contain both calpains I and II in addition to calpastatin. The finding is significant in rectifying the previous view that the chicken has only one molecular species of calpain, whereas mammals have two. Another erroneous view which prevailed previously was that polymorphonuclear (PMN) cells contain only one calpain species. We could also recently demonstrate that pig PMN cells do contain both calpains I and II. The cloning of cDNAs for calpastatin enabled us to utilize them as the probes in studying the expression of calpastatin in various hematopoietic cell-line cells. We found that several T cells infected with human retrovirus HTLV-I markedly increased the production of calpastatin, which could be measured both by calpain-inhibition assay and by Western blot analysis, but the level of mRNA for calpastatin did not significantly change when compared with noninfected T cells. The increase in calpastatin protein always parallels with the expression of interleukin 2 receptor protein by the HTLV-I-infected T cells, although the biological implication of such phenomena is almost entirely unknown yet.

REFERENCES

Adachi, Y., Kobayashi, N., Hatanaka, M., and Murachi, T., 1985, Expression of calpain II in human T-cell lineage after a human retrovirus (HTLV) infection, Abstr., 13th Int. Congr. Biochem., Amsterdam, August, p. 816.

Adachi, Y., Hatanaka, M., Takano, E., and Murachi, T., 1988a, Phosphorylation of calpastatin in human hematopoietic cells, Abstr., 14th Int. Congr. Biochem., Prague, July, TH:0 76.

Adachi, Y., Takano, E., Murachi, T., and Hatanaka, M., 1988b, Distribution and expression of calpastatin in human hematopoietic system cells, Biol. Chem. Hoppe-Seyler, 369 (Suppl.):223.

Emori, Y., Kawasaki, H., Imajoh, S., Imahori, K., and Suzuki, K., 1987, Endogenous inhibitor for calcium-dependent cysteine protease contains four internal repeats that could be responsible for its multiple reactive sites, Proc. Natl. Acad. Sci., U.S.A., 84:3590.

Fukui, I., Toyohara, H., Ito, K., Hamakubo, T., and Murachi, T., Molecular and catalytic characterization of intact heterodimeric and derived monomeric calpains isolated under different conditions from pig polymorphonuclear leukocytes, Biochemistry, 27:3260.

Hatanaka, M., Yoshimura, N., Murakami, T., Kannagi, R., and Murachi, T., 1984, Evidence for membrane-associated calpain I in human erythrocytes. Detection by an immunoloelectrophoretic blotting method using monospecific antibodies, Biochemistry, 23:3272.

Hatanaka, M., Yumoto, N., Sakihama, T., and Murachi, T., 1985, Evidence for heterodimeric nature of calpain molecules as studied by cross-linking modification, Biochem. Int., 10:187.

Imajoh, S., Kawasaki, H., and Suzuki, K., 1986, The amino terminal hydrophobic region of the small subunit of calcium-activated neutral protease (CANP) is essential for its activation by phosphatidylinositol, J. Biochem. (Tokyo), 99:1281.

Inomata, M., Imahori, K., and Kawashima, S., 1986, Autolytic activation of calcium-activated neutral protease, Biochem. Biophys. Res. Commun., 138:638.

Inoue, M., Kishimoto, A., Takai, Y., and Nishizuka, Y., 1977, Studies on a cyclic nucleotide-independent protein kinase and its proenzyme in mammalian tissues. II. Proenzyme and its activation by calcium-dependent protease from rat brain, J. Biol. Chem., 252:7610.

Kikuchi, T., Yumoto, N., Sasaki, T., and Murachi, T., 1984, Reconstritution of calpain I and calpain II from their subunits: Interchangeability of the light subunits, Arch. Biochem. Biophys., 234:639.

Mellgren, R. L., Repetti, A., Muck, T. C., and Easly, J., 1982, Rabbit skeletal muscle calcium-dependent protease requiring millimolar Ca^{2+}. Purification, subunit structure, and a Ca^{2+}-dependent autoproteolysis, J. Biol. Chem., 257:7203.

Melloni, E., Pontremoli, S., Michetti, M., Sacco, O., Sparatore, B., Salmino, F., and Horecker, B. L., 1985, Binding of protein kinase C to neutrophil membranes in the presence of Ca^{2+} and its activation by a Ca^{2+}-requiring proteinase, Proc. Natl. Acad. Sci., U.S.A., 82:6435.

Murachi, T., Hatanaka, M., Yasumoto, Y., Nakayama, N., and Tanaka, K., 1981a, A quantitative distribution study on calpain and calpastatin in rat tissues and cells, Biochem. Int., 2:651.

Murachi, T., Tanaka, K., Hatanaka, M., and Murakami, T., 1981b, Intracellular Ca^{2+}-dependent protease (calpain) and its high-molecular weight endogenous inhibitor (calpastatin), Adv. Enzyme Regul., 19:407.

Murachi, T., 1983, Calpain and calpastatin, Trends Biochem. Sci., 8:167.

Murachi, T., 1984, Calcium-dependent proteinases and specific inhibitors: Calpain and calpastatin, Biochem. Soc. Symp., 49:149.

Murachi, T., Hatanaka, M., and Hamakubo, T., 1987, Calpains and neuropeptide metabolism, in: "Neuropeptides and Their Peptidases," A. J. Turner, ed., Ellis Horwood, Chichester, p. 202.

Murachi, T., 1989, Intracellular regulatory system involving calpain and calpastatin, Biochem. Int., in press.

Murachi, T., Takano, E., Maki, M., Adachi, Y., and Hatanaka, M., 1989, Cloning and expression of the genes for calpains and calpastatin, Biochem. Soc. Symp., in press.

Murakami, T., Ueda, M., Hamakubo, T., and Murahci, T., 1988, Identification of both calpains I and II in nucleated chicken erythrocytes, J. Biochem. (Tokyo), 103:168.

Nishiura, I., Tanaka, K., Yamato, S., and Murachi, T., 1978, The occurrence of an inhibitor of Ca^{2+}-dependent neutral protease in rat liver, J. Biochem. (Tokyo), 84:1657.

Ohno, S., Emori, Y., Imajoh, S., Kawasaki, H., Kisaragi, M., and Suzuki, K., 1984, Evolutionary origin of a calcium-dependent protease by fusion of genes for a thiol protease and a calcium-binding protein?, Nature (London), 312:566.

Pontremoli, S., Sparatore, B., Salmino, F., Michetti, M., Sacco, O., and Melloni, E., 1985, Reversible activation of human neutrophil calpain promoted by interation with plasma membrane, Biochem. Int., 11:35.

Pontremoli, S., and Melloni, E., 1986, Extralysosomal protein degradation, Annu. Rev. Biochem., 55:455.

Sakihama, T., Kakidani, H., Zenita, K., Yumoto N., Kikuchi, T., Sasaki, T., Kannagi, R., Nakanishi, S., Ohmori, M., Takio, K., Titani, K., and Murachi, T., 1985, A new putative Ca^{2+}-binding protein: Structure of the light subunit of porcine calpain elucidated by molecular cloning and protein sequence analysis, Proc. Natl. Acad. Sci., U.S.A., 82:6075.

Suzuki, K., Ohno, S., Imajoh, S., Emori, Y., and Kawasaki, H., 1985, Identification and distribution of mRNA for calcium-activated neutral protease (CANP), Biomed. Res., 6:323.

Suzuki, K., 1987, Calcium activated neutral protease: Domains structure and activation regulation, Trends Biochem. Sci., 12:103.

Suzuki, K., Imajoh, S., Emori, Y., Kawasaki, H., Minami, Y., and Ohno, S., 1987, Calcium-activated neutral protease and its endogenous inhibitor. Activation at the cell membrane and biological function, FEBS Lett., 220:271.

Suzuki, K., Imajoh, S., Emori, Y., Kawasaki, H., Minami, Y., and Ohno, S., 1988, Regulation of activity of calcium activated neutral protease, Adv. Enzyme Regul., 27:153.

Takano, E., Maki, M., Hatanaka, M., Mori, H., Zenita, K., Sakihama, T., Kannagi, R., Marti, T., Titani, K., and Murachi, T., 1986, Evidence for the repetitive domain structure of pig calpastatin as demonstrated by cloning of complementary DNA, FEBS Lett., 208:199.

Takano, E., Maki, M., Mori, H., Hatanaka, M., Marti, T., Titani, K., Kannagi, R., Ooi, T., and Murachi, T., 1988, Pig heart calpastatin: Identification of repetitive domain structure and anomalous behavior in polyacrylamide gel electrophoresis, Biochemistry, 27:1964.

Tsuji, S., and Imahori, K., 1981, Studies on the Ca^{2+}-activated neutral protease of rabbit skeletal muscle. I. The characterization of the 80K and the 30K subunits, J. Biochem. (Tokyo), 90:233.

Waxman, L., and Krebs, E. G., 1978, Identification of two protease inhibitors from bovine cardiac muscle, J. Biol. Chem., 253:5888.

Yoshimura, N., Kikuchi, T., Sasaki, T., Kitahara, A., Hatanaka, M., and Murachi, T., 1983. Two distinct Ca^{2+} proteases (calpain I and calpain II) purified concurrently by the same method from rat kidney, J. Biol. Chem., 258:8883.

ROLE OF ANIONS IN PLATELET INTRA CELLULAR CALCIUM ION RISE AND ARACHIDONIC ACID RELEASE INDUCED BY THROMBIN

Yukio Ozaki and Shoji Kume

Dept. Clinical Central Lab.
Yamanashi Medical College
Yamanashi Pref., Japan

INTRODUCTION

Calcium mobilization following agonist-receptor interactions has been shown to play a pivotal role in signal transduction in a variety of cells. Recently, it has been suggested that anion transport is closely associated with intracellular calcium rises; depletion of chloride ions from the extracellular medium suppressed calcium uptake and subsequent maturation of sea urchins, and more recently an anion-sensitive Ca^{++} channel has been found in mouse pancreatic beta cells[1].

In human platelets, the importance of anion transport has already been noted in the release of intracellular granule contents[2,3], while its association with calcium mobilization has not been evaluated thus far. In the present report, we investigated the role of anion transport in thrombin-induced intracellular calcium ion ($[Ca^{++}]i$) rises and arachidonic acid release which is known to be dependent upon the intracellular calcium ion level.

METHODS

1. Platelet preparation.

Venous blood was obtained from healthy donors who had not taken any drugs for 2 weeks. Platelets were purified with centrifugation, and unless otherwise stated, suspended at a concentration of 2 x 10^8 cells/ml in a Hepes buffer containing 140 mM NaCl, 4 mM KCl, 1 mM $MgCl_2$ (pH 7.2).

2. Measurement of intracellular calcium ion concentrations. Platelets were loaded either with fura 2 or aequorin for the determination of intracellular calcium ion concentrations. Since the two anion channel blockers used in the present study possessed strong autofluorescence, the effects of anion channel blockers were evaluated with aequorin-loaded platelets. Aequorin loading was performed with the EGTA-ATP method as described by Johnson et al.[4], and the emitted light was measured with a platelet ionized calcium aggregometer (Baxter-Travenol). Fura 2-loaded cells were prepared by incubating platelet-rich plasma with 3 uM fura 2 A/M for 1 hour at 37C. The cells were washed twice and the intracellular calcium ion concentra-

tion were determined by the ratio of fluorescence at 340 nm excitation wavelength to that of 380 nm, measured with a Hitachi F-4010 fluorescence spectrophotometer.

3. Measurement of arachidonic acid metabolites.

The release of arachidonic acid was estimated with the production of two arachidonic acid metabolites, 12-hydroxy-5,8,10-heptadecatrienoic acid (HHT), and 12-hydroxy-5,8,10,14-eicosatetraenoic acid (12-HETE). The amount of HHT represents the arachidonic acid metabolite production via the cyclooxygenase pathway, and 12-HETE, via the 12-lipoxygenase pathway. Unless otherwise stated, the platelet suspension was incubated for 10 min at 37C with 0.1 U/ml thrombin, and the reaction was stopped by the addition of ethylacetate. The lipid fraction was extracted, and applied to the reverse phase high performance liquid chromatography. The production of HHT and 12-HETE were standardized with the known amounts of synthetic HHT and 12-HETE.

RESULTS AND DISCUSSION

Two stilbene analogues, 4,4-diisothiocyanatostilbene-2,2'-disulfonic acid (DIDS) and 4-acetamido-4'-isothiocyanatostilbene-2,2'-disulfonic acid (SITS) have been found to block anion channels in a variety of cell types[5]. These two anion channel blockers in a dose-dependent manner inhibited thrombin-induced $[Ca^{++}]i$, DIDS being more potent than SITS (Figure 1). The ID_{50} value for DIDS was approximately 65 uM.

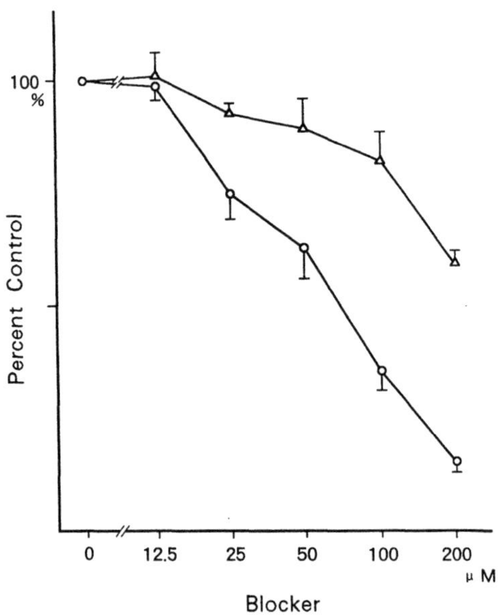

Figure 1. Effect of anion channel blockers on $[Ca^{++}]i$ rise induce by 0.1 U/ml thrombin. Aequorin-loaded platelets were preincubated with an appropriate amount of DIDS or SITS for 5 min, and then 0.1 U/ml thrombin was added to the platelet suspension to stimulate aequorin luminescence. Open circle, DIDS; open triangle, SITS. (means±SD of 4 experiments).

The inhibitory effects of these two anion channel blockers were more remarkable with the production of arachidonic acid metabolites. DIDS inhibited the production of HHT and 12-HETE almost equally with an ID_{50} value of 25 uM, and SITS, 50-100 uM (Figure 2). Since the products of the cyclooxygenase pathway and 12-lipoxygenase pathway were equally reduced, it is suggested that DIDS and SITS exert their inhibitory effects at a site proximal to phospholipase A_2 or at phospholipase A_2 itself. However, at the concentrations which effectively lowered the production of arachidonic acid metabolites, both anion channel blockers showed only minimal inhibitory effects on intra-

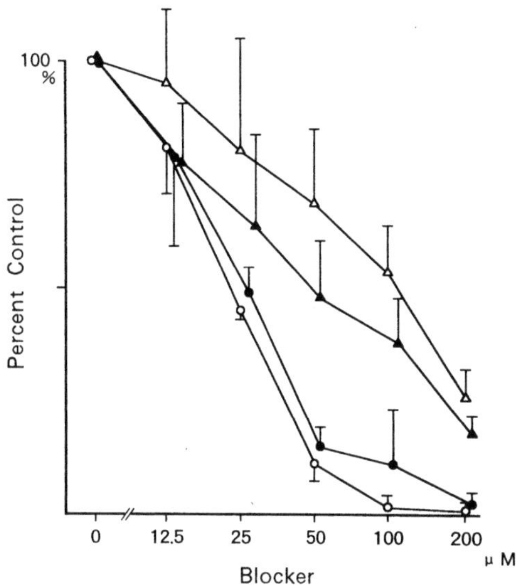

Figure 2. Effects of anion channel blockers on HHT and 12-HETE production induced by thrombin. Platelets preincubated with DIDS or SITS were activated with 0.1 U/ml thrombin, and after 10 min, lipids were extracted for HPLC analysis of HHT and 12-HETE. Open circle, DIDS on HHT; closed circle, DIDS on 12-HETE; open triangle, SITS on HHT; close triangle, SITS on 12-HETE. The data are means SD±of 4 experiments.

cellular calcium ion rises, as shown in Figure 1. These findings combined together suggests that these anion channel blockers acted on a certain site distal to the mechanism regulating the intracellular calcium ion levels and proximal to the arachidonic acid releasing mechanism, although a part of the inhibitory effect may be proximal or at the site of calcium mobilization. Thus, phospholipase A_2 activation induced by thrombin in human platelets appears to be regulated by some anion channel-dependent mechanism, but less so on Ca^{++} mobilization.

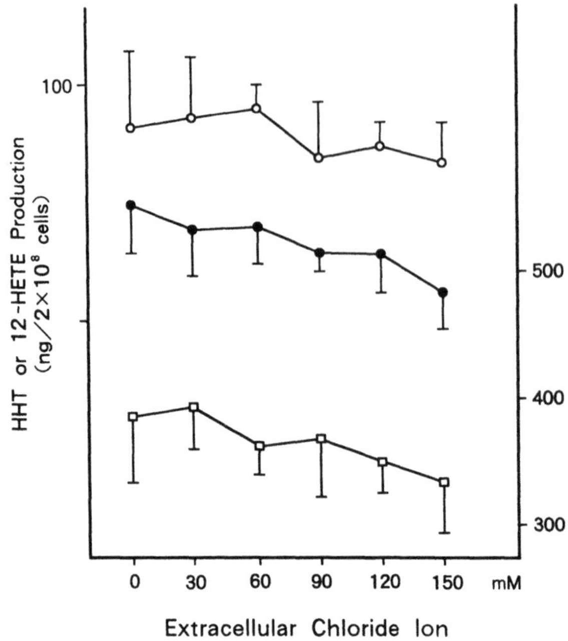

Figure 3. Effects of chloride ion depletion on thrombin-induced $[Ca^{++}]i$ rise and HHT, 12-HETE production.

Various ratios of chloride ions in extracellular fluid were replaced with impermeant isethionate. Platelets were then activated with 0.1 U/ml thrombin, and $[Ca^{++}]i$ rises, HHT and 12-HETE production was measured. Open circle, HHT; closed circle, 12-HETE; open box, $[Ca^{++}]i$. The data presented as means±SD of 4 experiments.

Since the major anion in the extracellular fluid is chloride ion, we then evaluated the effect of chloride ion depletion from the extracellular fluid on thrombin-induced responses. Various ratios of chloride ions were replaced with membrane-impermeable isethionate, and the thrombin-induced intracellular calcium ion rise and HHT, 12-HETE production were measured. As shown in Figure 3, depletion of chloride ions had no significant effect on these cellular responses, suggesting that DIDS and SITS suppress the release of arachidonic acids by blocking the transport of other anions. Since hydroxyl ions and bicarbonate ions are two other anions present in the extracellular fluid. we then evaluated the effects of these anions on cellular responses.

The addition of bicarbonate ions to the extracellular fluid significantly increased the production of arachidonic acid metabolites (43 21% increase in HHT production with 10 mM bicarbonate), while the intracellular calcium ion rises were not altered (data not shown). The increase in pH of the extracellular fluid, which is equivalent to the increased amount of hydroxyl ions, also potentiated the production of arachidonic acid metabolites induced by thrombin, while the intracellular calcium ion rises induced by thrombin were not significantly changed by the extracellular pH levels (Figure 5).

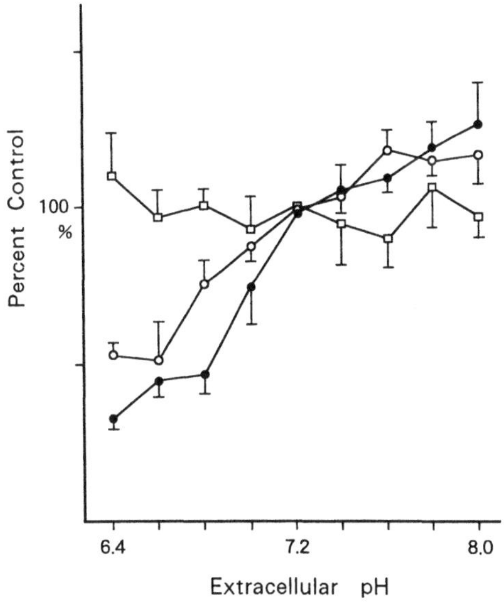

Figure 4. Effect of extracellular pH on $[Ca^{++}]i$ rise and HHT, 12-HETE production.

Platelets were suspended in the buffer set at various pH. After addition of 0.1 U/ml thrombin, $[Ca^{++}]i$ rise and HHT, 12-HETE production was measured. Open circle, HHT; closed circle, 12-HETE; open box, $[Ca^{++}]i$. The data are means SD±of 4 experiments.

These findings, taken together, suggest that bicarbonate or hydroxyl ions, transported via anion channels, regulate the production of arachidonic acid metabolites induced by thrombin. HOwever, mobilization of anions appears to have little, if not at all, effect on Ca^{++} mobilization induced by thrombin. Since the changes in the HHT production were essentially the same as that of 12-HETE, it is likely that the activity of phospholipase A_2, not that of cyclooxygenase or 12-lipoxygenase, was affected by the anion mobilization. Anion channels are known to participate in regulation of intracellular pH, and in some case, anion mobilization effectively elevates intracellular pH[6,7]. Since the optimal pH for platelet phospholipase A_2 is shifted toward alkali pH (around 9.5)[8], intracellular alkalinization would offer a favorable environment for phospholipase A_2 activity. Thus, it is possible that anion mobilization activated with thrombin elevates intracellular pH, and facilitate arachidonic acid release by phospholipase A_2. Although we have some evidence suggesting that the intracellular pH is increased by the mobilization of these two anions, intracellular alkalinization is also regulated by the powerful Na^+/H^+ exchanger in a Na^+-containing buffer[9]. Whether the change in intracellular pH plays a major role in anion channel-mediated potentiation of arachidonic acid release awaits further evaluations.

REFERENCES

1. S. Janove, and H. P. Meissner, Effects of Cl^- deficiency on the membrane potential in mouse pancreatic beta cells, Biochim. Biophys. Acta 937:309 (1988)

2. H. B. Pollard, K. Tack-Goldman, C. J. Pazoles, C. E. Creutz, and N. R. Shulman, Evidence for control of serotonin secretion from human platelets by hydroxyl ion transport and osmotic lysis, Proc. Natl. Acad. Sci. USA 74:5295 (1977)
3. W. F. Bennet, J. S. Belville, and G. Lynch, The relationship between extracellular pH and control of blood platelet serotonin secretion, Mol. Pharmacol. 17:356 (1980)
4. P. C. Johnson, J. A. Ware, P. B. Cliveden, M. Smith, A. M. Dvorak,and E. W. Salzman, Measurement of ionized calcium in blood platelets with the photoprotein aequorin. Comparison with quin 2, J. Biol. Chem. 260:2069 (1985)
5. Z. I. Cabantchik, P. A. Knauf, and A. Rothstein, The anion transport system of the red blood cell. The role of membrane protein evaluated by the use of probes, Biochim. Biophys. Acta, 515:239 (1978)
6. A. Ladoux, I. Krawice, E. J. Cragoe, Jr., J. P. Abita, and C. Frelin, Properties of the Na^+-dependent Cl^-/HCO_3 exchange systen in U937 human leukemic cells, Eur. J. Biochem. 170:43 (1987)
7. I. H. Madshus, and S. Olsnes. Selective inhibition of sodium-linked and sodium-independent bicarbonate/chloride antiport in Vero cells, J. Biol. Chem. 262:7486 (1987)
8. R. F. Irvine, How is the level of free arachidonic acid controlled in mammalian cells?, Biochem. J. 204:3 (1982)
9. A. Livne, S. Grinstein, and A. Rothstein, Characterization of Na^+/H^+ exchange in platelets, Thromb. Haemostas. 58:971 (1987)

FREQUENCY-DEPENDENT INHIBITION OF THE INTRACELLULAR CALCIUM TRANSIENTS BY CALMODULIN ANTAGONISTS IN THE AEQUORIN-INJECTED RABBIT PAPILLARY MUSCLE

Masao Endoh

Department of Pharmacology
Yamagata University School of Medicine
990-23 Yamagata, Japan

INTRODUCTION

Calmodulin, a ubiquitous functional protein in eukaryotic cells, has been found to play an essential role in regulation of enzymes and cellular processes in various tissues including smooth muscle cells (Cheung, 1980; Klee and Newton, 1985). Although it has been shown that calmodulin exists in myocardial cells and suggested that it may be involved in regulation of myocardial functions, based on the effects of calmodulin on enzyme activities and Ca^{2+} uptake, its regulatory role in intact myocardium remains unclear. In order to get insight into the role of calmodulin in regulation of myocardial contractility, the effects of calmodulin antagonists, W-7 and trifluoperazine (TFP), on the intracellular calcium transients and isometric contractions were assessed simultaneously in the isolated rabbit papillary muscle, superficial cells of which had been microinjected with the Ca^{2+} sensitive bioluminescent protein aequorin.

There are a variety of manipulations to modify the intracellular Ca^{2+} mobilization including cardiotonic agents, and changes in extracellular Ca^{2+} concentration ($[Ca^{2+}]_o$), experimental temperature, or frequency of stimulation. Among these interventions, changes in frequency are chosen to examine the involvement of calmodulin in regulation of the calcium transients and strength of isometric contractions by use of calmodulin antagonists. In mammalian ventricular myocardium of most species, elevation of frequency results in an increase in force of contraction, which has long been known as frequency-force relationship. When the interval between contractions is prolonged, the strength of contraction reaches a low level of rested state contraction; by abbreviating the interval, the strength of contraction becomes higher essentially due to increased Ca^{2+} release from sarcoplasmic reticulum (SR) resulting from increased Ca^{2+} influx through voltage-dependent Ca^{2+} channels, and/or by intracellular accumulation of Na^+ (Koch-Weser and Blinks, 1963; Sheu and Blaustein, 1986). On the other hand, it has been shown in isolated SR vesicles that calmodulin catalyzes phosphorylation of phosphoramban and thereby increases the rate of Ca^{2+} uptake into the SR vesicles (Katz and Remtulla, 1978; Le Peuch et al., 1979; Tada et al., 1983). Therefore, provided that calmodulin antagonists modulate the contractility by interfering with calmodulin function in contracting myocardium, it appears to be reasonable to expect that the interaction may be assessed by investigating the influence of these compounds on frequency-force relationship, which is ascribed to the graded contribution of SR function and voltage-dependent Ca^{2+} channel activity to regulation of myocardial contractility. It is known that most calmodulin antagonists possess nonspecific effects that belong to individual compounds (Stoclet et al., 1985). Nevertheless, if different calmodulin antagonists affect the calcium transients and isometric contractions in an identical manner, it may be taken as indirect evidence suggesting potential common mechanism of action of these compounds involving calmodulin as the functional regulator.

METHODS

Albino rabbits of either sex (1.8 and 3.8 kg) were used in these experiments. Thin papillary muscles were excised from the right ventricle and mounted horizontally in an organ bath especially constructed for aequorin injection. Aequorin was microinjected into multiple (50-100) superficial cells of papillary muscles by the application of gas pressure (Blinks et al., 1978; Blinks, 1984) at 32°C. The aequorin was extracted, purified, and prepared for injection according to techniques that have been described elsewhere in detail (Blinks et al., 1978). It was dissolved at a concentration of about 2mg/ml in solution containing 150 mM KCl and 5 mM HEPES (N-2-hydroxyethylpiperazine-N'-2-ethanesulfonic acid) buffer, pH 7.5. The aequorin solution was loaded into fine-tiped (35-50 megohms in 150 mM KCl) micropipetes through which potential was monitored to confirm the penetration. After aequorin injection, when a satisfactory light signal was obtained on electrical stimulation at 0.5 Hz, the muscle was transferred to an apparatus with 50 ml organ bath designed to record light signals with high efficacy and to minimize motion artifacts in the aequorin signals (Blinks, 1984; Blinks and Endoh, 1986). The apparatus contained a vertically mounted ellipsoidal reflector in which the aequorin-injected muscle was positioned at the upper focal point and the photocathode of the photomultiplier (EMI 9635A, Thorn-EMI, Plainview, New York) at the lower. The muscle was suspended inside a glass tube that extended axially into the top of the ellipsoidal reflector from a water-jacketed organ bath. The lower end of the muscle was clamped to a miniature version of an electrode assembly described previously (Blinks, 1966), and the top of the muscle was connected with 9-0 Tevdek thread to the arm of a Servo-operated electromagnetic muscle lever (Brutsaert and Claes, 1974; Brutsaert et al., 1978) operated in the isometric mode. The experiments were performed at 37.5°C. The pH of the solution was 7.4 when bubbled with 95% O_2 and 5% CO_2. The concentrations of the various constituents in Krebs-Henseleit solution were as follows (mmol/l): Na^+ 142.9, K^+ 5.9, Mg^{2+} 1.2, Ca^{2+} 2.5, $H_2PO_4^-$ 1.2, HCO_3^- 24.9, SO_4^{2-} 1.2, Cl^- 127.8, and glucose 11.1. The muscle was electrically stimulated by square wave pulses of 5 ms duration and a voltage about 20% above threshold, at a frequency of 1 Hz in basal conditions. During an equilibration period of 30 min the muscle was stretched initially by a tension of 5 mN and the length was adjusted to give 90% of the maximal developed tension. In some muscles, (±)-bupranolol (0.3 or 1.0 µM) was present in the organ bath throughout the experiments to confirm that norepinephrine released by electrical stimulation did not modify the responses induced by changes in frequency of stimulation. Signal averaging of between 32 and 512 contractions was carried out to obtain a satisfactory signal-to-noise ratio in the aequorin signals. Averaged signals were recorded in digital form on floppy disks.

In determination of frequency-force relationship, the interval between each contraction was prolonged from 1 s in the basal condition to the longest interval (32 or 16 s). When the strength of isometric contractions reached a steady level at each interval, the interval was abbreviated stepwise up to 0.25 s (4 Hz). After determination of control frequency-force relationship (FFR), the interval was returned to 1 s. When the force developed reached a steady level at 1 Hz, a calmodulin antagonist was allowed to act for at least 30 min before the determination of second FFR in the presence of the calmodulin antagonist. Concentration-dependent effects of calmodulin antagonists on FFR were determined after addition of the compounds in increasing concentrations.

Drugs used in these experiments were obtained from the following sources: W-7 [N-(6-aminohexyl)-5-chloro-1-naphthalenesulfonamide] (gift from Banyu Pharmaceutical Co., Tokyo); trifluoperazine [2-trifluoromethyl-10-[3-(1-methylpiperazinyl-4)propyl]-phenothiazine dihydrochloride, TFP] (Smith, Kline and French Labs., Philadelphia); (±)-bupranolol hydrochloride (gift from Sanol, Monheim, FRG); (-)-isoproterenol bitartrate, (-)-epinephrine bitartrate (gifts from Sterling-Winthrop, Rensselaer, New York).

RESULTS

Influence of frequency of stimulation. In the isolated rabbit papillary muscle, the peak calcium transients and strength of isometric contractions were modified prominently by changing the frequency of electrical stimulation (stimulus interval). When the stimulus interval was prolonged to 16 s and longer both calcium transients and isometric contractions were scarcely affected by changing intervals in most muscles (Figs. 1 and 2). As shown in figures 1 and 2, in the interval of between 16 and 2 s, the strength of isometric contractions was increased definitely by raising frequency of stimulation, while the peak calcium transients being little affected. In this range of intervals, the duration of calcium transients was prolonged markedly, and this seems to be responsible for the increase in force of contraction. When the interval was abbreviated further to 1 s and shorter up to 0.35 s, both the peak calcium transients and isometric contractions were increased substantially in parallel. When the interval was further shortened, the peak calcium transients increased further, while the strength of isometric contractions become steady or were reduced slightly after having reached the maximum at 0.4 or 0.35 s.

Influence of Frequency of Stimulation

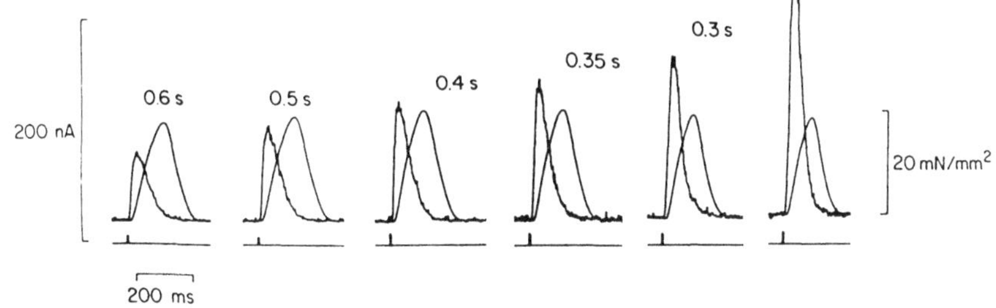

Rabbit papillary muscle, 37.5°C, numbers given are stimulus interval.

Fig. 1. Influence of frequency of stimulation on calcium transients and isometric contractions in the aequorin-injected rabbit papillary muscle. For aequorin (noisy) light signals, 32 (0.35-0.25 s), 64 (0.5 and 0.4 s), 128 (16, 8, 0.8 and 0.6 s) and 256 (4, 2 and 1 s) contractions were signal averaged. Muscle length: 7.5 mm; cross-sectional area: 0.85 mm^2.

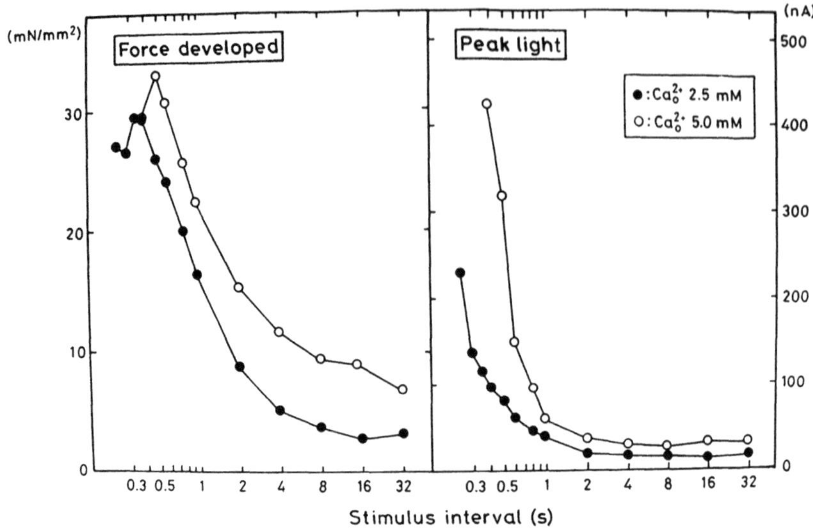

Fig. 2. Influence of elevation of $[Ca^{2+}]_o$ to 5.0 mM from 2.5 mM in the normal Krebs-Henseleit solution on the frequency-induced changes in force developed and peak calcium transients. Control data in 2.5 mM $[Ca^{2+}]_o$ were calculated from actual tracings presented in figure 1. Aequorin signals in 5.0 mM $[Ca^{2+}]_o$: signal-averaged data of 32 to 128 contractions.

Figure 2 shows the influence of elevation of $[Ca^{2+}]_o$ to 5.0 mM from 2.5 mM in the normal Krebs-Henseleit solution on the frequency-induced changes in force developed and peak calcium transients. The frequency-force relationship was shifted upward substantially in parallel by raising $[Ca^{2+}]_o$, while the increase in peak calcium transients induced by raising frequency was potentiated more markedly at shorter intervals.

Influence of calmodulin antagonists. One of classical calmodulin antagonists, TFP (0.1-0.3 mM), produced a frequency-dependent inhibitory effect on calcium transients and isometric contractions in a concentration-dependent manner. Figure 3 shows the effect of TFP (0.3 mM) on the frequency-force relationship. At stimulus intervals of 4 s and longer, TFP scarcely affected the calcium transients and isometric contractions. Both parameters were suppressed more when the interval was abbreviated to 1 s and shorter.

Elevation of $[Ca^{2+}]_o$ from control (2.5 mM) to 5.0 and 7.5 mM in the presence of 0.3 mM TFP, increased markedly the peak calcium transients and strength of isometric contractions at longer stimulus intervals. Dissociation of isometric contractions from the peak calcium transients was observed when the interval was shortened: in the presence of 0.3 mM TFP and 5.0 mM $[Ca^{2+}]_o$ the peak calcium transients increased up to the intervals 0.3 and 0.35 s, respectively, while the force reached the peak at 0.4 and 0.5 s, and then decreased. In the presence of 7.5 mM $[Ca^{2+}]_o$ the muscle became inexcitable at 0.5 s (Fig. 3). In the presence of TFP, the force of contraction was readily increased by elevation of $[Ca^{2+}]_o$, which is in strong contrast to the case with W-7 as will be shown in the next.

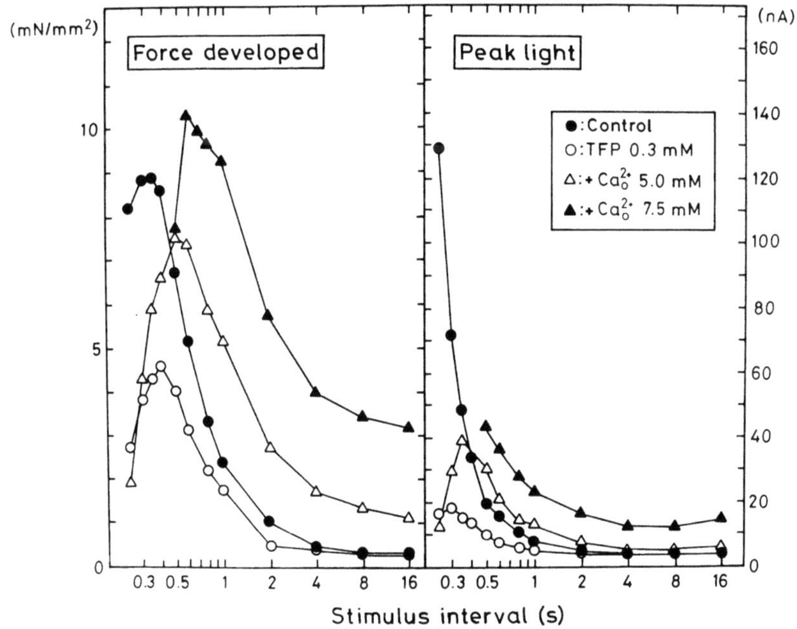

Fig. 3. Reversal by elevation of $[Ca^{2+}]_o$ of frequency-dependent inhibitory effects of TFP (0.3 mM) on the isometric contractions and peak calcium transients in the aequorin-injected rabbit papillary muscle. Peak aequorin signals were calculated from signal-averaged data of 64 to 128 contractions. Muscle length: 6.5 mm; cross-sectional area: 0.68 mm^2.

W-7 (0.05-0.3 mM) decreased the peak calcium transients and isometric contractions also in a frequency-dependent manner. Figure 4 shows the influence of 0.1 mM W-7 on the frequency-force relationship in the aequorin-injected rabbit papillary muscle. At intervals of 4 s and longer, the peak calcium transients were not affected by W-7, while the strength of isometric contractions were slightly decreased. When the interval of stimulation was abbreviated to 2 s and shorter, both force of contraction and peak calcium transients were not increased appreciably. Thus, the frequency-force relationship was mostly abolished in the presence of 0.1 mM W-7. Elevation of $[Ca^{2+}]_o$ to 7.5 mM readily recovered the peak calcium transients depressed by W-7 especially in the range of lower frequencies, while the force of contraction did not recover even when the peak calcium transients were raised to a level higher than the control (Fig. 4). Thus, the dissociation of the force of contraction from the peak calcium transients was more pronounced in the presence of W-7 than with TFP. The muscle became inexcitable at an interval of 0.5 s in the presence of 0.1 mM W-7 and 7.5 mM $[Ca^{2+}]_o$.

Isoproterenol (3 x 10^{-7} M) enhanced the frequency-induced increases in both peak signals substantially to a similar extent at intervals of 16 to 1 s (Fig. 5). In this condition W-7 (0.1 mM) shifted the relation downwards in parallel. It was examined whether $[Ca^{2+}]_o$ could antagonize the depressant effect of W-7. $[Ca^{2+}]_o$ was elevated stepwise by 2.5 mM up to 15 mM. While the peak calcium transients at lower frequencies (interval: 16-4 s) was increased even to a level higher than that with isoproterenol alone, the force of contraction scarcely recovered. Thus, the force of contraction was completely dissociated from the peak calcium transients by W-7 in the presence of isoproterenol and the elevated $[Ca^{2+}]_o$.

Influence of W-7 on the duration of aequorin signals and isometric contractions were examined in the presence of catecholamines (3 x 10^{-7} M isoproterenol and 10^{-5} M epinephrine). Changes caused by W-7 were qualitatively similar either in the absence or presence of the catecholamines.

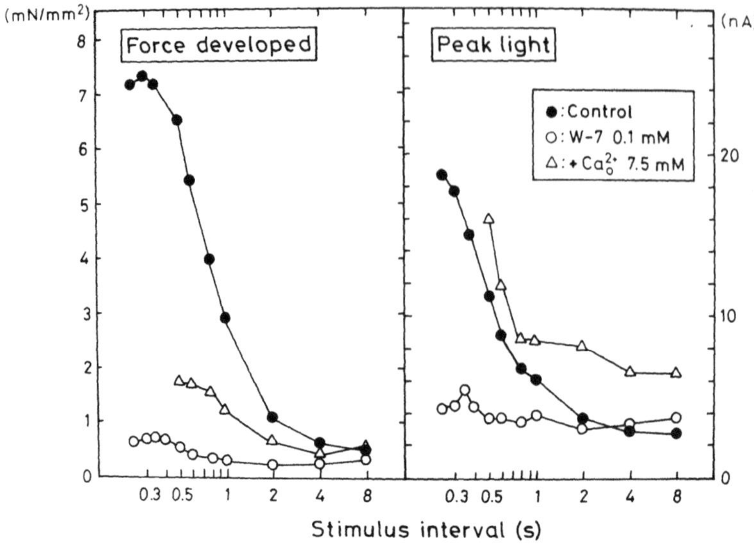

Fig. 4. Influence of elevation of $[Ca^{2+}]_o$ on frequency-dependent inhibitory effects of W-7 (0.1 mM) on the isometric contractions and peak calcium transients in the aequorin-injected rabbit papillary muscle. Peak aequorin signals were calculated from signal-averaged data of 128 contractions. Muscle length: 6.5 mm; cross-sectional area: 0.85 mm^2.

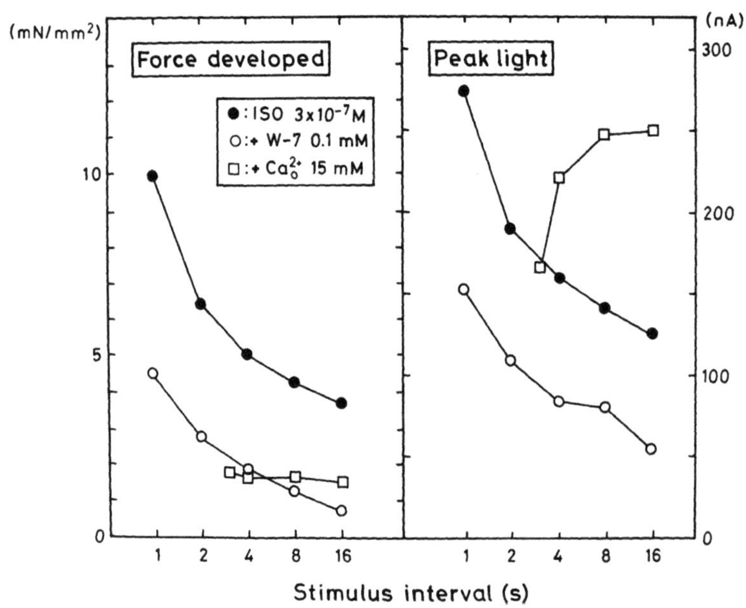

Fig. 5. Influence of $[Ca^{2+}]_o$ on the inhibitory effects of W-7 (0.1 mM) on the frequency-induced changes in isometric contractions and peak calcium transients in the presence of 3×10^{-7} M isoproterenol in the aequorin-injected rabbit papillary muscle. Peak aequorin signals were calculated from signal-averaged data of 32 to 128 contractions. Muscle length: 5.5 mm; cross-sectional area: 0.56 mm^2.

W-7 at 0.1 and 0.3 mM abbreviated both the signals in a concentration-dependent manner at a stimulus interval of 1 s. At an interval of 16 s, the peak aequorin signals were reduced without abbreviation of aequorin signals, while the isometric contractions being abbreviated, but to a lesser extent than at 1 s (data not shown).

DISCUSSION

Mechanisms involved in the frequency-induced regulation of myocardial contractility have long been fields of widespread interest and activity since Bowditch (1871) described first the phenomenon. It has been postulated that the frequency-induced increase in myocardial contractility is associated with an increase in the amount of Ca^{2+} entering the myocardial cells with each beats, since the excitation-contraction coupling in cardiac muscle depends essentially on Ca^{2+} influx during the phase of electrical activity (Koch-Weser and Blinks, 1963). The mechanism of increase in the intracellular Ca^{2+} concentration ($[Ca^{2+}]_i$) has been ascribed to accumulation of $[Na^+]_i$: the increased Na^+ influx that accompanies the increase in frequency leads to an increase in $[Na^+]_i$, and in turn this may inhibit Ca^{2+} efflux and promote Ca^{2+} influx via the Na^+/Ca^{2+} exchange mechanism, so that $[Ca^{2+}]_i$ will rise (Sheu and Blaustein, 1986).

The influence of frequency on the peak and duration of calcium transients appears to be different depending on the range of frequency. At intervals of 2 s and shorter, the increase in force of contraction was associated with the frequency-induced increase in peak calcium transients, while at intervals of 4 to 2 s, prolongation of calcium transients being related more closely to the increase in force. Thus, different factors may be involved in the regulation of frequency-induced changes depending on the range of frequency. The increased $[Ca^{2+}]_i$ by raising frequency may have to be lowered rapidly during relaxation to maintain the diastolic $[Ca^{2+}]_i$ level sufficiently low. In relatively low frequency range (interval: 2-4 s) there seems to be conditions in which the $[Ca^{2+}]_i$ during contraction was increased without acceleration of rate of decline of calcium transients, while the rate of decline was accelerated at the interval of 1 s and shorter. It appears likely that calmodulin is involved in this frequency-induced acceleration of rate of decline in calcium transients and of relaxation, since it is considered that the frequency-induced changes are exerted essentially by elevation of $[Ca^{2+}]_i$ due to physical maneuver and therefore may trigger the calmodulin-mediated regulation of SR function that has been demonstrated in the in vitro experiments. The results by use of calmodulin antagonists in the present study failed to obtain evidence to support this postulate as will be discussed later.

Calmodulin antagonists, W-7 and TFP in concentrations that scarcely affected the peak calcium transients and force of isometric contractions at long stimulus intervals (4 to 32 s) suppressed them markedly at higher frequencies (stimulus intervals of 2 s and shorter). This implies that calmodulin antagonists have exerted essentially their inhibitory effects on the Ca^{2+} mobilization induced by raising the frequency resulting in abolition of frequency-force relationship. Concerning sites of action at which calmodulin antagonists, W-7 and TFP, had effectively modified the process involved in the frequency-force relationship, it is noteworthy that inhibition caused by these compounds resembles closely to the mode of inhibition caused by Ca^{2+} antagonists, most of which have been shown to decrease the Ca^{2+} influx via voltage-dependent Ca^{2+} channels in a use-dependent manner (Bayer et al., 1975; Fleckenstein, 1977). These similarities imply that the essential site of action of W-7 and TFP may be voltage-dependent Ca^{2+} channels. This possibility is supported by the previous findings that TFP displaced $[H^3]$nitrendipine binding in smooth and cardiac muscle with high affinity (Luchowski et al., 1984), and inhibited slow action potentials of cultured chick heart cells (Bkaily et al., 1984). Greenberg and co-workers

(1987) have investigated that calmodulin antagonists including W-7 and TFP inhibited $^{45}Ca^{2+}$ influx through voltage-dependent Ca^{2+} channels in PC12, a clonal neural cell line, and emphasized that the Ca^{2+} channel antagonist properties should be considered in interpreting the effects of calmodulin antagonists on Ca^{2+}-dependent cellular events. However, there is no evidence at present that calmodulin is involved in the functional modulation of voltage-dependent Ca^{2+} channels. Luchowski and co-workers (1984) postulated that calmodulin antagonists bind to a protein associated with the dihydropyridine binding site that has a hydrophobic domain similar to that exposed on calmodulin by Ca^{2+}, but not to calmodulin itself.

In addition to the potential inhibitory effect on Ca^{2+} influx via sarcolemmal Ca^{2+} channels, W-7 (and TFP to a lesser extent) appears to have exerted intracellular effects which cannot be accounted for by the inhibition of $[Ca^{2+}]_i$ mobilization. Especially in the presence of W-7, elevation of $[Ca^{2+}]_o$ at longer intervals readily increased the peak calcium transients without or with relatively less recovery of strength of associated contractions. In this respect, the effects of calmodulin antagonists in rabbit papillary muscles are different from those of Ca^{2+} antagonists such as D600, the inhibitory effects of which on both signals were reversed by elevation of $[Ca^{2+}]_o$ toward the control substantially in parallel. [This finding with D600 in the rabbit papillary muscle is not consistent with the previous observations in the cat papillary muscle (Morgan et al., 1983), the reason of which is unknown.] Thus, calmodulin antagonists have caused dissociation of force developed from peak calcium transients, indicating to have produced an effect at a step distal from Ca^{2+} mobilization. Similar findings have been reported in other tissues such as cultured chromaffin cells (Kenigsberg et al., 1982) and smooth muscle cells (Karaki and co-workers, 1982). In smooth muscle, existing experimental evidence supports the view that calmodulin antagonists may inhibit smooth muscle contraction by interfering with calmodulin-dependent myosin light chain phosphorylation that plays a central role for regulation of smooth muscle contraction (Cheung, 1980; Klee and Newton, 1985). In cardiac muscle, it is also shown that TFP inhibits the myosin light chain phosphorylation in cultured rat myocytes (Klein, 1983). However, the role of myosin light chain in regulation of myocardial contractility is not clear. Since most calmodulin antagonists bind also to troponin C (that plays a key role in regulation of myocardial contractility) with slightly lower affinity than to calmodulin (Tkachuk et al., 1987), dissociation of force from peak calcium transients observed in cardiac muscle may be interpreted by the interaction of these compounds with troponin C.

It has been shown in the in vitro experiments that calmodulin is involved in facilitation of Ca^{2+} transport by increasing Ca^{2+} sensitivity of sarcolemmal Ca^{2+} pump ATPase (Caroni and Carafoli, 1981) or by phosphorylation of phospholamban in SR vesicles (Katz and Remtulla, 1978; Le Peuch et al., 1979; Tada et al., 1983). Aass and co-workers (1983) have suggested that TFP (10^{-5} M) may have enhanced the effect of isoproterenol probably through inhibition of calmodulin-induced activation of these enzymes in the intact rat papillary muscle. There is little evidence, however, that the inhibition of calmodulin-induced modulation of these enzymes is reflected to contraction-relaxation cycle in intact myocardial cells. Provided that calmodulin modulates the SR function in a similar manner to cyclic AMP (Le Peuch et al., 1979; Tada et al., 1983), inhibition of this process may result in reduction of force associated with prolongation of calcium transients and isometric contractions. On the contrary, present findings that the decrease in force developed by both W-7 and TFP was associated with abbreviation of both signals indicate that the primary site of action of these compounds other than the SR may be more critical for the changes observed. Calmodulin antagonists appear not to be so specific as β-adrenoceptor antagonists that are capable of inhibiting both the cyclic AMP-mediated abbreviation and increase of calcium transients.

It is noteworthy that W-7 decreased the peak calcium transients and isometric contractions also at long intervals (8 and 16 s) when the increase in peak calcium transients and force induced by raising frequency had been potentiated by β-adrenoceptor activation. It will be of interest to examine in more detail whether or not these enhanced inhibitory effects of W-7 reflect the involvement of calmodulin in the regulation due to cyclic AMP-induced increase in $[Ca^{2+}]_i$.

In summary, calmodulin antagonists, W-7 and TFP, modified the frequency-force relationship and associated frequency-induced changes in calcium transients in a frequency-dependent manner in intact myocardial cell preparations. These findings are in accordance with the view that W-7 and TFP exert more pronounced inhibitory effects when $[Ca^{2+}]_i$ is elevated by raising frequency. It is not clear, however, these effects are exerted through the specific hindrance of calmodulin-mediated regulation of myocardial contraction. In the presence of W-7, elevation of $[Ca^{2+}]_o$ or β-adrenoceptor activation was capable of increasing the peak calcium transients, but with less increase in strength of isometric contractions, indicating the additional intracellular site of action of W-7 to result in dissociation of force development from increase in peak calcium transients.

Acknowledgements The present study was supported partly by Grant-in Aid for Scientific Research on Priority Area No. 62624004 from the Ministry of Education, Sciences, and Culture, Japan, and National Institute of Health grant HL-12186. The author is grateful to Professor J. R. Blinks for continuous encouragements and advice during the course of carrying out the study.

REFERENCES

Aass, H., Skomedal, T., and Osnes, J.-B., 1983, Effects of trifluoperazine on β-adrenergic responses of rat papillary muscle: related to calmodulin? Acta pharmacol. et toxicol., 53:265.

Bayer, R., Hennekes, R., Kaufmann, R., and Mannhold, R., 1975, Inotropic and electrophysiological actions of verapamil and D600 in mammalian myocardium: I. Pattern of inotropic effects of the racemic compounds, Naunyn-Schmiedeberg's Arch. Pharmacol., 290:49.

Bkaily, G., Sperelakis, N., Eldefrawi, M., 1984, Effects of the calmodulin inhibitor, trifluoperazine, on membrane potentials and slow action potentials of cultured heart cells, Eur. J. Pharmacol., 105:23.

Blinks,J. R., 1984, Methods for monitoring Ca^{++} concentrations with photoproteins in living cardiac cells, in: Methods in Studying Heart Membranes, vol. II, N. S. Dhalla, ed., Boca Raton, Fla., CRC Press, p. 237.

Blinks, J. R., 1966, Field stimulation as a means of effecting the graded release of autonomic transmitters in isolated heart muscle, J. Pharmacol. Exp. Ther., 151:221.

Blinks, J. R., Mattingly, P. H., Jewell, B. R., van Leeuwen, M., Harrer, G. C., and Allen, D. G., 1978, Practical aspects of the use of aequorin as a calcium indicator: assay, preparation microinjection, and interpretation of signals, Meth. Enzymol., 57:292.

Blinks, J. R., and Endoh, M., 1986, Modification of myofibrillar responsiveness to Ca^{++} as an inotropic mechanism, Circulation, 73(suppl. III):III-85.

Bowditch, H. P., 1871, Uber die Eigenthumlichkeiten der Reizbarkeit welche die Muskelfasern des Herzens zeigen, Ber. Sach. Wiss., 23:652.

Brutsaert, D. L., and Claes, V. A., 1974, Onset of mechanical activation of mammalian heart muscle in calcium and strontium-containing solutions, Circ. Res., 35:345.

Brutsaert, D. L., De Clerck, N. M., Goethals, M. A., and Housmans, P. R., Relaxation of ventricular cardiac muscle, J. Physiol. (Lond.)., 283:469.

Caroni, P., and Carafoli, E., 1981, The Ca^{2+}-pumping ATPase of heart sarcolemma: characterization, calmodulin dependence, and partial purification, J. Biol. Chem., 256:3263.

Cheung, W. Y., 1980, Calmodulin plays a pivotal role in cellular regulation, Science, 207:19.

Fleckenstein, A., 1977, Specific pharmacology of calcium in myocardium, cardiac pacemakers, and vascular smooth muscle, Ann. Rev. Pharmacol. Toxicol., 17:149.

Greenberg, D. A., Carpenter, C. L., and Messing, R. O., 1987, Interaction of calmodulin inhibitors and protein kinase C inhibitors with voltage-dependent calcium channels, Brain Res., 404:401.

Karaki, H., Murakami, K., Nakagawa, H., Ozaki, H., and Urakawa, N., 1982, Effects of calmodulin antagonists on tension and cellular calcium content in depolarized vascular and intestinal smooth muscles, Brit. J. Pharmacol., 77:661.

Katz, S. and Remtulla, M. A., 1978, Phosphodiesterase protein activator stimulates calcium transport in cardiac microsomal preparations enriched in sarcoplasmic reticulum, Biochem. Biophys. Res. Commun., 83:1373.

Kenigsberg, R. L., Côté, A., Trifaró, J. M., 1982, Trifluoperazine, a calmodulin inhibitor, blocks secretion in cultured chromaffin cells at a step distal from calcium entry, Neuroscience, 7:2277.

Klee, C. B., and Newton, D. L., 1985, Calmodulin: an overview, in: Control and Manipulation of Calcium Movement, Parratt, J. R., Raven Press, New York, p. 131.

Klein, I., 1983, Trifluoperazine inhibits the contraction of cultured rat cardiac cells and the phosphorylation of myosin light chain, J. Clin. Invest., 71:518.

Koch-Weser, J., and Blinks, J. R., 1963, The influence of the interval between beats on myocardial contractility, Pharmacol. Rev., 15:601.

Le Peuch, C. J., Haiech, J., Demaille, J. G., 1979, Concerted regulation of cardiac sarcoplasmic reticulum calcium transport by cyclic adenosine monophosphate dependent and calcium-calmodulin-dependent phosphorylations, Biochemistry, 18:5150.

Luchowski, E. M., Yousif, F., Triggle, D. J., Maurer, S. C., Sarmiento, J. G., and Janis, R. A., 1984, Effects of metal cations and calmodulin antagonists on [^{3}H] nitrendipine binding in smooth and cardiac muscle, J. Pharmacol. Exp. Ther., 230:607.

Morgan, J. P., Wier, W. G., Hess, P., and Blinks, J. R., 1983, Influence of Ca^{++}-channel blocking agents on calcium transients and tension development in isolated mammalian heart muscle, Circ. Res., 52(suppl. I):47.

Sheu, S.-S., and Blaustein, M. P., 1986, Sodium/calcium exchange and regulation of cell calcium and contractility in cardiac muscle, with a note about vascular smooth muscle, in: The Heart and Cardiovascular System, Fozzard, H. A., Haber, E., Jennings, R. B., Katz, A. M., Morgan, and H. E., ed., Raven Press, New York, p. 509.

Stoclet, J. C., Lugnier, C., Follenius, A., Scheftel, J. M., and Gérard, D., 1985, Calmodulin and calcium regulation: effect of antagonists, in: Calcium Entry Blockers and Tissue Protection, Godfraind, T., Vanhoutte, P. M., Govoni, S., and Paoletti, R., ed., Raven Press, New York, p. 31.

Tada, M., Inui, M., Yamada, M., Kadoma, M., Kuzuya, T., Abe, H., and Kakiuchi, S., 1983, Effects of phospholamban phosphorylation catalyzed by adenosine 3':5'-monophosphate- and calmodulin-dependent protein kinases on calcium transport ATPase of cardiac sarcoplasmic reticulum, J. Mol. Cell. Cardiol., 15:335.

Tkachuk, V. A., Baldenkov, G. N., Feoktistov, I. A., Men'shikov, M. Y., Quast, U., and Herzig, J. W., 1987, Metofenazate as a more selective calmodulin inhibitor than trifluoperazine, Arzneim.-Forsch/Drug Res., 37:1013.

CALCIUM IN MITOSIS: ROLE OF 51-KD PROTEIN IN THE CENTROSOME OF SEA URCHIN EGG IN ASTER FORMATION

Hikoichi Sakai, Kunihiro Ohta, Masaru Toriyama, and Sachiko Endo

Department of Biophysics and Biochemistry
Faculty of Science, University of Tokyo
Hongo, Tokyo 113

INTRODUCTION

It has been proposed that cytoplasmic free calcium is a regulator of many cellular events during cell cycle. Signal transduction in the trigger of cell multiplication includes transient increase in free calcium through the inosital turnover (Berridge & Irvine, 1984; Rozengurt, 1986) similar to a burst of free calcium observed at fertilization of eggs (Whitaker & Steinhardt, 1985). This is probably followed by a variety of chains of reactions towards replication of genomes and construction of mitotic machineries. Since microtubules which are susceptible to calcium have been shown to be main skeletons of the mitotic apparatus, calcium has been proposed as a regulator in mitosis.

Since Mazia and Dan (1952) carried out the pioneering work on the isolated mitotic apparatus, many studies have been focused on the properties of the mitotic apparatus (Sakai, 1978) and dynamic characteristics of the spindle (Salmon et al., 1984; Saxton et al., 1984; Hamaguchi et al., 1985; Soltys & Borisy, 1985). The mitotic apparatus contains the centrosomes that organize astral and spindle microtubules. Isolated centrosomes also radially nucleate microtubules (Weisenberg & Rosenfeld, 1975; McGill & Brinkley, 1975; Snyder & McIntosh, 1975; Pepper & Brinkley, 1979; Gould & Borisy, 1977). The nucleating site is not the centriole but the pericentriolar material both *in vitro* (Gould & Borisy, 1977) and *in vivo* (Berns et al., 1977; Berns & Richardson, 1977) and the nucleation ability is in a cell cycle-dependent manner (Telzer & Rosenbaum, 1979; Kuriyama & Borisy, 1981). The pericentriolar materials in cultured cells are shown to be amorphous assembly (Robbins et al., 1968). In contrast, those in sea urchin eggs are composed mainly of clusters of granules (Endo, 1979) called microtubule-organizing granules (MTOGs) (Endo et al., 1985). Electron microscopy shows that the MTOGs play a role in nucleating and organizing microtubules from prophase through metaphase (Endo, 1980; Paweletz et al., 1984; Mazia, 1984). This brief review describes possible role of calcium in mitosis and the 51-kD protein in the MTOGs which are susceptible to micromolar level of free calcium.

POSSIBLE REGULATORY ROLE FOR INTRACELLULAR CALCIUM IN MITOSIS

Several lines of evidence suggest that cytoplasmic free calcium is important in the regulation of mitosis. Although microtubules are quickly depolymerized when calcium is microinjected into the mitotic spindles of echinoderm and mammalian cells (Kiehart, 1981; Izant, 1983), experimental manipulations that lower intracellular free calcium or inhibit calcium transport across the plasma membrane delay the transition from metaphase to anaphase (Izant, 1983; Hepler, 1985). In contrast, microinjection of calcium during mataphase stimulates the transition to anaphase (Izant, 1983), and elevation of intracellular free calcium during anaphase accelerates chromosome-to-pole movement (Cande, 1981; Hauser & Beier, 1980). Recently, a calcium-binding 15-kD protein which is different from calmodulin was purified from sea urchin eggs (Hosoya et al., 1986). The 15-kD protein was shown to localize in the mitotic apparatus as visualized by indirect immunofluorescence staining using antibodies against the 15-kD protein. The antibody totally blocked cell division of sea urchin eggs when it was microinjected (Hosoya et al., personal communication).

The mitotic apparatus contains membrane elements that are capable of sequestering and probably locally releasing calcium. In spermatogenous cells of the water fern, *Marsilea vestita*, and stomatal complex cells of barley, *Hordeum vulgare*, localization of calium is seen along the tubules and cisternae of the extensive smooth endoplasmic reticulum that surrounds and invades the mitotic spindle (Wick & Hepler, 1980). In early anaphase of a subsidiary cell of barley, membrane accumulation in the spindle pole is especially pronounced (Hepler, 1980). It was suggested that the membrane system may function to control the concentration of free calcium in the mitotic apparatus (Hepler, 1980). In fact, calcium sequestration was demonstrated with ATP-dependent accumulation of ^{45}Ca by isolated mitotic apparatus whose vesicles were left intact (Silver et al., 1980). Furthermore, the localization of calcium-sequestering activity (Hepler & Wolniak, 1984) or the calcium transport system (Petzelt & Hafner, 1986) was found in the mitotic apparatus. The calcium-sequestering activity was shown to change during cell cycle of sea urchin embryos (Suprynowicz & Mazia, 1985). It was shown that antibodies against the calcium transport enzyme of smooth muscle inhibit mitosis when microinjected into dividing sand dollar embryos (Silver, 1986). In addition, inositol phosphate-mediated release of calcium during mitosis was reported (Sillers & Forer, 1985). For the change in intracellular distribution of free calcium analyzed using fluorescent calcium indicators such as Quin-2 or fura-2 (Keith et al., 1985 ; Ratan et al., 1986; Poenie et al., 1986), we will discuss it later.

MICROTUBULES AND CALMODULIN IN THE MITOTIC APPARATUS

In a series of work on the regulation of microtubule assembly (Sakai, 1980, a review), Nishida and Kumagai (1980) found that microtubules in the sea urchin egg are very susceptible to free calcium. Half maximal inhibition of tubulin polymerization occurs at a free calcium concentration of 10^{-7}M in a solution of a low ionic strength. When the ionic strength increases, a high calcium sensitivity is still observed with a half maximal inhibition of polymerization at 2-3 μM calcium. Calmodulin does not increase the calcium sensitivity of polymerization of egg tubulin different from brain tubulin (Nishida et al., 1979). This was later confirmed by Keller et al. (1982). Instability of microtubules in the isolated mitotic apparatus to micromolar levels of calcium was shown by Salmon and Segall (1980). High calcium sensitivity of microtubules in the sea urchin egg was further reported by Keller et al. (1982) who demonstrated that spindle microtubules are sensitive to free calcium at micromolar concentrations such that 3-20 μM causes complete depolymerization at physiological temperature for some species of the sea urchin. Microinjection of calcium causes depolymerization of the microtubules of the mitotic apparatus in a localized region surrounding the injection site (Kiehart, 1981), and microinjection of calcium buffers at intracellular concentration of 4 μM causes birefringence of the mitotic apparatus to decrease and the spindle becomes shortened, followed by recovery a few min after injection (Hamaguchi, 1982).

Calmodulin has been shown to localize in the mitotic spindle region (Welsh et al., 1978; Zavortink et al., 1983; Willingham et al., 1983; Vantard et al., 1985) and a 52-kD calmodulin-binding protein was shown to be associated with the mitotic spindle in CHO cells (Brady et al., 1986). That fluorescently labeled calmodulin injected into cells associates with the mitotic spindle was first demonstrated by Hamaguchi (1982). Microinjection of DACM[N-(7-dimethylamino-4-methylcoumarinyl)-maleimide]-labeled calmodulin into live sea urchin eggs reveals that the mitotic apparatus has many binding sites for calmodulin. At metaphase, the fluorescence of the spindle poles and the area around the poles become the most intense. Such distribution of calmodulin in the centrosome persists until telophase. Similar distribution was reported by Zavortink et al.(1983). It should be stressed that calmodulin in sea urchin eggs does not seem to stimulate microtubule disassembly in a calcium-dependent manner, although microinjection of Ca/calmodulin into gerbil fibroma cells causes depolymerization of microtubules around the site of injection (Keith et al., 1983). Instead, the possibility remains to be solved that calmodulin-dependent protein kinase will phosphorylate MAPs as shown by Yamamoto et al.(1983, 1985), thereby suppressing the ability of MAPs to promote microtubule assembly (Nishida et al., 1982) or accelerating microtubule disassembly (Yamamoto et al., 1983, 1985). Activation of the type II kinase is shown to be in a calcium/calmodulin-dependent cooperative manner, enhancing the sensitivity of the enzyme to small changes at the intracellular calcium level (Le Vine et al., 1986). On the other hand, there is a report showing that cold-stable microtubules are sensitive in the presence of calcium/calmodulin (Job et al., 1981). However, the half-maximal effective calcium concentration is 100 μM in the presence of calmodulin, which seems too high to imply physiological significance.

Ca/calmodulin-dependent depolymerization of microtubules *in vitro* was first demonstrated using calmodulin-dependent kinase II by Yamauchi & Fujisawa (1983). Yamamoto et al.(1985) presented evidence that phosphorylation of tubulin by calcium/calmodulin-dependent kinase induces inhibition of assembly. This was followed by Wandosell et al.(1986) who have shown that phosphorylation of α- and β-tubulin at their caroboxyl-terminal region by brain Ca/calmodulin-dependent kinase causes decrease in the rate and the extent of microtubule assembly. Phosphorylation results in a decrease in the binding of MAP2 to microtubules. That phosphorylation of microtubule-associated proteins, MAP2 and tau, causes decrease in the ability of MAPs to promote microtubule initiation has been demonstrated by Nishida et al. (1982), which was later confirmed by Burns et al.(1984). Recently, calcium and calmodulin-dependent phosphorylation of a 62 kD protein was shown to correlate with depolymerization of microtubules in the isolated mitotic apparatus of sea urchin eggs (Dinsmore & Slobada, 1988). They assume that phosphorylation of the 62-kD protein destabilizes microtubules in the mitotic apparatus, although the mechanism by which the 62-kD protein acts as a depolymerizing factor is not known.

ROLE OF THE CENTROSOME IN MITOSIS

The centrosome is a well known organelle, *the real division organ of the cell* by Boveri (1901), as the center of mitosis [extensively reviewed by D. Mazia (1987)]. It is a center for organization of microtubule cytoskeletons in cells. The centrosome cycle was extensively studied on sea urchin eggs (Paweletz et al., 1984) and CHO cells (Kuriyama & Borisy, 1981). By comparing rates of elongation of centrosomal and flagellar-seeded microtubules, Bergen et al.(1980) demonstrated that centrosomal microtubules grow at a rate corresponding to that at the plus end of flagellar axonemes.

In vitro experiments show that microtubule assembly off the centrosome is kinetically preferred (Mitchison & Kirschner, 1984).

With an elaborate procedure, Sluder and Rieder (1985) demonstrated that chromosomes cannot organize a spindle in the absence of centrosomes and only the sperm provides the centrosomes in the development of the sea urchin zygote. Microinjection of isolated human centrosomes into mature starfish oocytes was shown to induce release from arrest at interphase (Picard et al., 1987). Furthermore, epidermal growth factor (Sherline & Mascardo, 1982) and the tumor promoter 12-*O*-tetradecanoylphorbol-13-acetate (Euteneuer & Schliwa, 1985) induce centrosomal separation in HeLa cells and polymorphonuclear leukocytes, respectively. The centrosome separation in the former case was shown to be blocked by trifluoperazine, chlorpromazine and W-7 as calmodulin inhibitors, cytochalsins B and D as actin severing agents, and taxol as microtubule stabilizing agent.

The centrosome has so far been known to contain several proteins besides calmodulin. A monoclonal antibody against MAP1 stained interphase nucleus of HeLa S3 cells and PtK2 cells in a punctate pattern. At mitotic stage of these cells, the centrosome was intensely stained, indicating that some MAP-like protein is localized in the centrosome (Sato et al., 1983). Human autoimmune sera specific for the nuclear mitotic apparatus protein (NuMA protein, a phosphoprotein of 250,000 MW) (Lydersen & Pettijohn, 1980; Price & Pettijohn, 1986) detected that at the time of nuclear breakdown the NuMA protein redistributes from the nucleus to the spindle poles. It was shown that once the NuMA protein is associated with the spindle polar region, colcemid does not cause it to dissociate. In contrast colcemid prevented accumulation of the NuMA protein around the centrosomes at the onset of mitosis. This was likely due to depolymerization of astral microtubules, because colcemid totally blocks movement of beads toward the center of the mitotic aster and UV-irradition recovers it (Hamaguchi et al., 1986). It was shown that the staining of interphase PtK1 cells with a monoclonal antibody which is reactive with phosphoproteins of HeLa mitotic cells was restricted to discrete patches or spots within the nucleus (Vandre et al., 1984) similar to the staining pattern obtained with the same PtK1 cells with anti-NuMA protein antibody (Price & Pettijohn, 1986) or with PtK2 cells stained with anti-MAP1 antibody (Sato et al., 1983). However, at mitosis, the phosphoprotein was detected at the kinetochores and midbodies, in addition to the centrosome. Different from animal cells, plant cells have no centrosome. However, human scleroderma serum was found to stain the spindle pole region of meristematic higher plants from preprophase to late anaphase (Clayton et al., 1985).

Identification of centrosomal proteins has been carried out using enriched preparations of centrosomes from human T-lymphoblasts and rabbit anti-centrosome serum (Gosti-Testu et al., 1986). The pericentriolar material appears as granules of ~80 nm loosely associated with each other, as in the sea urchin centrosome (Endo, 1979). A family of high molecular weight protein ranging from 180,000 to 250,000, a 130,000 MW protein and a 60,000-65,000 MW doublet were detected. These proteins localize in the pericentriolar material and the centrosomal staining changed significantly during cell cycle, showing the maximum accumulation at the pole at metaphase. Although these antigens seem to be involved in the centrosomal function, it is unknown at present which protein is actually associated with tubulin to initiate microtubules. Another approach to identify the centrosomal proteins has been performed using monoclonal antibodies against mitotic spindles or taxol asters isolated from sea urchin eggs (Kuriyama & Borisy, 1985). The antibodies preferentially stained the centrosphere, recognizing proteins of 190-kD, 180-kD and 20-kD, or 50-kD protein the chain weight of which is close to the 51-kD protein reported by Toriyama et al.(1988). Using the serum of a patient with the CREST(calcinosis,

Raynaud's phenomenon, esophageal dysmotility, schlerodacytyly, and telangiectasias) syndrome, Sager et al. (1986) characterized the serum which was shown to label at the spindle poles of meta-anaphase HeLa cells and showed reactions with several proteins of 115, 110, 50, 36, and 28-kD chain weights. Of these, only 115/110-kD protein was shown to localize at the spindle pole and prophase aster after affinity-purification of the antibodies. In a different approach, a monoclonal antibody which was produced against *Drosophila* intermediate filament protein was found to cross-react specifically with the centrosome in the sea urchin egg (Schatten et al., 1987). The antigen recognized by the antibody was a protein of 68,000 molecular weight, which seems to be a component of sea urchin centrosomes from its persistent localization within the centrosome at all stages of the centrosome cycle. All of these works mentioned above, however, have been unable to identify protein components responsible for initiation of microtubule assembly.

Recently, a centrosomal protein (51-kD protein) was identified in MTOGs from the mitotic apparatus. Evidences were presented that the 51-kD protein is involved in aster formation. The centrosome of the sea urchin egg was first shown to be composed of clusters of electron dense granules (Endo, 1979). These granules were shown to begin to assemble at the poles from prophase. The maximum accumulation of the granules were shown at metaphase and then changed to a flat configuration after anaphase. These changes in the shape of the centrosomes were confirmed by Schatten et al. (1986) for the centrosomes in mouse oocytes and in sea urchin eggs identified with autoimmune antisera from scleroderma patients. Isolated mitotic apparatuses provide a large number of functional centrosomes. Homogenization of the mitotic apparatus causes fragmentation of the centrosomes, resulting in the formation of numerous small asters when incubated with tubulin (Toriyama et al., 1984). An indirect immunofluorescence image of the aster is shown in Fig. 1. Measurement of the growth rate of the microtubules reveals that the plus end is distal to the center (Toriyama et al., 1988) in the same manner as in the growth of microtubules from the centrosome in live cells (Eutenerer et al., 1982; Soltys & Borisy, 1985). Electron microscopy discloses that the center of these asters consists of small clusters with 10 to 20 granules having diameters ranging from 40 to 140 nm (Toriyama et al., 1988), corresponding to those of MTOGs in the centrosome (Endo, 1979).

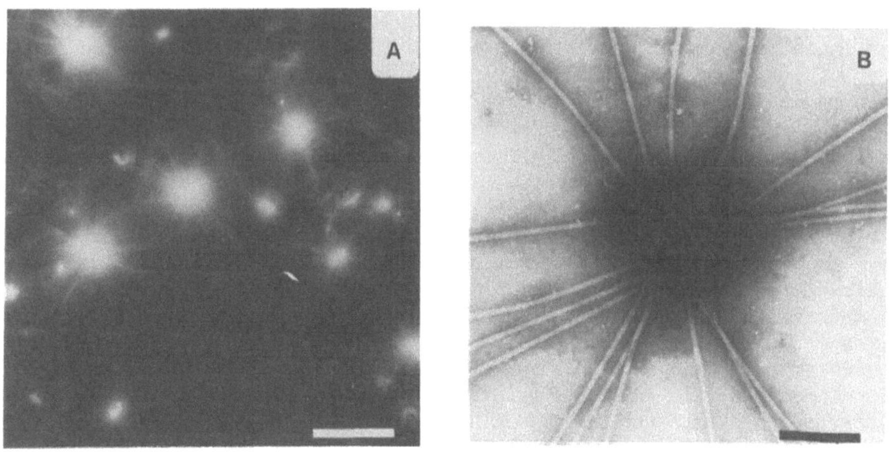

Fig. 1. A. Dark-field contrast micrograph of reconstructed asters. Bar, 20 μm.
B. Electron micrograph of a reconstructed aster negatively stained. Bar, 500 nm.

Aster forming activity in the isolated mitotic apparatuses can be solubilized in a solution of higher ionic strength in the presence of glycerol. Dialysis of a 0.5 M KCl extract of mitotic apparatuses causes formation of granular aggregates which are capable of initiating astral microtubules when incubated with tubulin (Toriyama et al., 1988). A protein fraction containing a major protein component of 51,000 Mr.Wt. can be prepared from solubilized MTOG fractions by phosphocellulose column chromatography, which forms granules by dialysis against a solution of low ionic strength and is capable of initiating astral microtubules when incubated with tubulin (MTOG activity). Polarity of the astral microtubules is the same as that of microtubules initiated from the MTOGs of the centrosome (Toriyama et al., 1988) as demonstrated by a pulse-label using biotynylated tubulin (Ohta et al., 1988a). Electron microscopy shows that the granular aggregate in the astral center resembles MTOGs in the centrosome with diameters ranging from 100 to 300 nm slightly larger than those of the centrosomal MTOGs. The 51-kD protein as a major component of MTOGs is a basic protein with a pI of 9.8. Amino acid analysis of the 51-kD protein reveals that lysine is the most prominent amino acid residue (Toriyama et al., 1988).

Polyclonal antibodies raised in rabbits against the 51-kD protein and affinity purified stains only the center of asters reconstructed from tubulin and MTOGs, indicating that the 51-kD protein is a major component of MTOGs in the mitotic apparatus. Antibody blocking experiments show that the affinity-purified antibody almost totally inhibits the aster forming ability of MTOGs from the isolated mitotic apparatus (Ohta et al., 1988c). Therefore, the 51-kD protein seems to be involved in the initiation of microtubules from the MTOGs *in vitro*.

CHANGES IN FREE CALCIUM DURING MITOSIS

Calcium seems to be sequestered at the onset of the formation of the mitotic apparatus. However, it has been well established that spindle microtubules depolymerize after anaphase. The mechanism by which microtubules depolymerize has been ascribed to the effect of calcium which will be released after anaphase. Several reports describe changes in intracellular free calcium during cell cycle. Yoshimoto et al.(1985) treated sea urchin eggs with a calcium-ionophore (A23187) and calcium efflux into the ambient solution was measured by the luminescence of aequorin. The calcium efflux was shown to change cyclically during early cleavage cycles, consistently decreasing at the time of furrowing of every cleavage. Using a fluorescent calcium chelator, fura-2, Poenie et al.(1985) succeeded in continuous measurement of intracellular free calcium from fertilization through the first cleavage of individual eggs of the sea urchin. Calcium fluctuations (0.1 to 0.4 μM) were observed during the cell cycle with multiple peaks at fertilization, pronuclear migation, streak stage, muclear membrane breakdown, metaphase/anaphase transition, and cleavage. On the other hand, Keith et al.(1985) found that intracellular free calcium concentrations in PtK2 epithelial cells loaded with Quin-2 drop by 50 % at meta-anaphase (from 53 to 28 nM). In contrast, abrupt and transient increase in the intracellular free calcium ion concentration was found in PtK1 cells at the metaphase-anaphase transition (Poenie et al., 1986). The mitotic cells loaded with fura-2 showed a rapid elevation of calcium from the basal level of 130 nM to 500-800 nM for 20 seconds. From the fura-2 fluorescence ratio imaging, they concluded that the elevated calcium appeared uniformly distributed over the entire spindle and central region of the cell. In stamen hair cells of *Tradescantia*, injection of arsenazo III and measurements of changes in differential absorbance revealed that free calcium concentration begins to increase at the onset of anaphase and continues to increase for 10-15 min while the chromosomes move to the poles (Hepler & Callaham, 1987). They estimated that the absorbance changes during late anaphase correspond to a calcium concentration of 1-3 μM.

It seems to be important to localize the changes in the intracellular free calcium ion concentration to elucidate possible sites of its action. To this end, Keith et al.(1985) used mitotic endosperm cells of *Haemanthus* loaded with acetoxymethylester of Quin-2. In anaphase, there was a marked departure from the homogeneous distribution of free calcium seen in the previous stages. They observed regions of higher free calcium near the mitotic spindle poles. The gradient lasted from early anaphase through late anaphase. Furthermore, Ratan et al.(1986) found that free calcium in PtK2 epithelial cells at metaphase is significantly lower than that at interphase and the distribution of calcium through most of the metaphase cell is uniform. The metaphase-anaphase transition was accompanied by a rise in the calcium concentration in the spindle pole region with a peak of 183 nM, aporoximately 2- to 3-fold higher than that within the midzone.

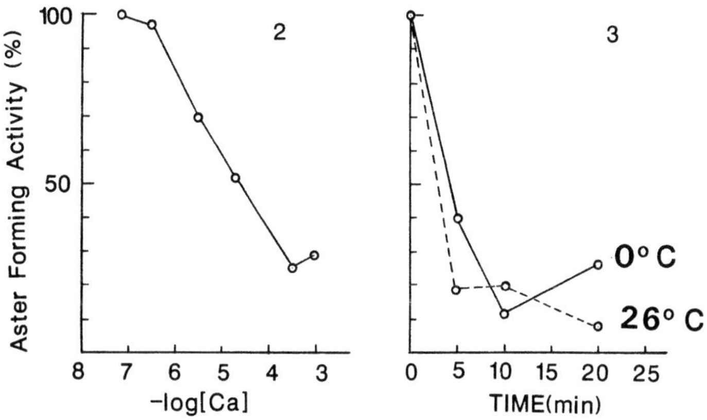

Fig. 2. Dependence of inhibition of the MTOG activity on free calcium concentration.

Fig. 3. Time course of inhibition of the MTOG activity by free calcium.

INHIBITION BY CALCIUM OF ASTER FORMING ABILITY OF MTOGs AND POSSIBLE ROLE OF 51-KD PROTEIN IN MITOSIS

The above results suggest that the centrosome is exposed to increasing free calcium concentration after anaphase. This stimulated to examine the effect of free calcium on aster forming ability using MTOGs prepared from isolated mitotic apparatus (Ohta et al.,1988a). The MTOGs was first incubated in a solution containing various concentrations of free calcium. This was followed by the addition of a tubulin solution containing EGTA to measure microtubule initiation. Figure 2 shows that concentrations of calcium as low as 1 µM begin to block the aster forming ability of MTOGs. This calcium sensitivity might become much more pronounced with some other factors yet unidentified. The time course of inhibition shown in Fig. 3 indicates that a preincubation of ~5 min with 20 µM calcium is critical to the MTOGs loss of ability. A longer incubation of the calcium-treated MTOGs with an excess amount of EGTA prior to the addition of tubulin caused the MTOGs to restore the ability to initiate astral microtubules. Although MTOGs from the isolated mitotic apparatus show high calcium sensitivity, MTOGs reconstructed from the 0.5 M KCl extract decrease the calcium sensitivity. Since the ^{45}Ca overlay method (Maruyama et al., 1984) failed to detect Ca-binding to the 51-kD protein, we can assume that a calcium-binding protein which has higher affinity to MTOGs in the presence of calcium than does tubulin would be released during fractionation of the 51-kD protein fraction.

Using monoclonal antibodies and affinity-purified polyclonal antibodies against the 51-kD protein, evidences have been presented that the 51-kD protein is localized in the centrosomal regions of the MAs from prophase through anaphase (Ohta et al., 1988bc). At metaphase, the 51-kD protein becomes localized in the spindle as well as in the centrosome and the proximal region of the asters. Although the function of the 51-kD protein in the spindle is unknown at present, it may have a role in stabilizing spindle microtubules. It was also shown that the 51-kD protein is localized in the center of the cytasters induced in unfertilized eggs by taxol (Ohta et al., 1988b). Using monoclonal antibodies which do not inhibit the aster forming activity of the MTOGs, it was shown that the antibodies suppressed the formation of the MA when injected before prophase (Ohta et al., 1988a). Therefore, cleavage did not occur. When the antibody was injected at prometaphase, growth of the astral microtubules was not inhibited, but the formation of the spindle was suppressed, causing total suppression of the karyokinesis. It can be assumed that the antibody blocks the assembly of MTOGs into the centrosome, thereby blocking the formation of the mitotic apparatus.

The 51-kD protein was shown to be a component of MTOGs in the centrosome. This protein is most likely an initiator of microtubule assembly in the formation of asters and the spindle when it is incorporated into MTOGs which are highly sensitive to free calcium ions. The Ca-sensitivity of MTOGs may indicate an interesting implication that MTOGs would decrease its ability to initiate microtubules after anaphase. Local increase in free calcium at the polar region of cells after anaphase (Keith et al., 1985; Ratan et al., 1986) seems to be consistent with the above possibility.

ACKNOWLEDGMENTS

This work was supported by a Grant-in-Aid for scientific research from the Ministry of Education, Science and Culture of Japan (60065005).

REFERENCES

Bergen, L.G., Kuriyama, R. and Borisy, G.G., 1980, J.Cell Biol., 84:151.
Berns, M.W., Rattner, J.B., Brenner, S. and Meredith, S., 1977, J.Cell Biol. 72:351.
Berns, M.W. and Richardson, S.M., 1977, J.Cell Biol., 75:977.
Berridge, M.J. and Irvine, R.F., 1984, Nature, 312:315.
Boveri, T. "Zellen-studien IV. Über die Natur der Centrosomen". Fischer, Jena (1901).
Brady, R.C., Cabral, F. and Dedman, J.R., 1986, J.Cell Biol., 103:1855.
Burns, R.G., Islam, K. and Chapman, R., 1984, Eur. J. Biochem., 141:609.
Cande, W.Z., 1981, *in* "International Cell Biology", H.G. Schweiger, ed., Springer, Berlin, pp. 382-391.
Clayton, L., Black, C.M. and Lloyd, C.W., 1985, J.Cell Biol., 101:319.
Dinsmore, J.H. and Sloboda, R.D., 1988, Cell, 53:769.
Endo, S., 1979, Cell Struct. Funct., 4:71.
Endo, S., 1980, Devel., Growth & Differ., 22:509.
Endo, S., Toriyama, M. and Sakai, H., 1985, *in* "Cell Motility: Mechanism and Regulation", H. Ishikawa, S. Hatano and H. Sato, ed., Univ. Tokyo Press, Tokyo, pp. 403-414.

Euteneuer, U. and Schliwa, M., 1985, J.Cell Biol., 101:96.
Euteneuer, U., Jackson, W.T. and McIntosh, J.R., 1982, J.Cell Biol., 94:644.
Gosti-Testu, F., Marty, M.-C., Berges, J., Maunoury, R. and Bornens, M., 1986, EMBO J., 5:2545.
Gould, R.R. and Borisy, G.G., 1977, J.Cell Biol., 73:601.
Hamaguchi, Y. and Iwasa, F., 1982, *in* "Biological Functions of Microtubules and Related Structures", H. Sakai, H. Mohri and G.G. Borisy, ed., Academic Press, Tokyo, pp. 199-210.
Hamaguchi, Y., Toriyama, M., Sakai, H. and Hiramoto, Y., 1985, J.Cell Biol., 100:1262.
Hamaguchi, M.S., Hamaguchi, Y. and Hiramoto, Y., 1986, Devel., Growth & Differ., 28:461.
Hauser, M. and Beier, A.M., 1980, Eur. J. Cell Biol., 22:313.
Hepler, P.K., 1980, J.Cell Biol., 86:490.
Hepler, P.K. and Wolniak, S.M., 1984, Internatl. Rev. Cytol., 90:169.
Hepler, P.K., 1985, J.Cell Biol., 100:1363.
Hepler, P.K. and Callaham, D.A., 1987, J.Cell Biol., 105:2137.
Hosoya, H., Iwasa, F., Ohnuma, M., Mabuchi, I., Mohri, H., Sakai, H. and Hiramoto, Y., 1986, FEBS Lett., 205:121.
Izant, J.G., 1983, Chromosoma, 88:1.
Job, D., Fischer, E.H. and Margolis, R.L., 1981, Proc. Natl. Acad. Sci., USA, 78:4679.
Keith, C., DiPaola, M., Maxfield, F.R. and Shelanski, M.L., 1983, J.Cell Biol., 97:1918.
Keith, C.H., Ratan, R., Maxfield, F.R., Bajer, A. and Shelanski, M.L., 1985, Nature, 316:848.
Keith, C.H., Maxfield, F.R. and Shelanski, M.L., 1985, Proc. Natl. Acad. Sci., USA, 82:800.
Keller, T.C.S., Jemiolo, D.K., Burgess, W.H. and Rebhun, L.I., 1982, J.Cell Biol., 93:797.
Kiehart, D.P., 1981, J.Cell Biol., 88:604.
Kuriyama, R. and Borisy, G.G., 1981, J.Cell Biol., 91:822.
Kuriyama, R. and Borisy, G.G., 1985, J.Cell Biol., 101:524.
Le Vine, H., III, Sahyoun, N.E. and Cuatrecasas, P., 1986, Proc. Natl. Acad. Sci., USA, 83:2253.
Lydersen, B.K. and Pettijohn, D.E., 1980, Cell, 22:489.
Maruyama, K., Mikawa, T. and Ebashi, S., 1984, J. Biochem. (Tokyo), 95:511.
Mazia, D. and Dan, K., 1952, Proc. Natl. Acad. Sci., USA, 38:826.
Mazia, D., 1984, Exp. Cell Res., 153:1.
Mazia, D., 1987, Internatl. Rev. Cytol., 100:49.
McGill, M. and Brinkley, B.R., 1975, J.Cell Biol., 67:189.
Mitchison, T. and Kirschner, M., 1984, Nature, 312:232.
Nishida, E., Kumagai, H., Ohtsuki, I. and Sakai, H., 1979, J. Biochem. (Tokyo), 85:1257.
Nishida, E. and Kumagai, H., 1980, J. Biochem. (Tokyo), 87:143.
Nishida, E., Kotani, S., Kuwaki, T. and Sakai, H., 1982, *in* "Biological Functions of Microtubules and Related Structures", H. Sakai, H. Mohri and G.G. Borisy, ed., Academic Press, Tokyo, pp. 285-295.
Ohta, K., Toriyama, M., Endo, S. and Sakai, H., 1988a, Zool. Sci., 5:613.
Ohta, K., Toriyama, M., Endo, S. and Sakai, H., 1988b, Cell Motil. Cytoskel., in press.
Ohta, K., Toriyama, M., Endo, S. and Sakai, H., 1988c, Protoplasma, in press.
Paweletz, N., Mazia, D. and Fonze, E.M., 1984, Exp. Cell Res., 152:47.
Pepper, D.A. and Brinkley, B.R., 1979, J.Cell Biol., 82:585.
Petzelt, C. and Hafner, M., 1986, Proc. Natl. Acad. Sci., USA, 83:1719.
Picard, A., Karsenti, C., Dabauvalle, M.C. and Doree, M., 1987, Nature, 327:170.
Poenie, M., Alderton, J., Tsien, R.Y. and Steinhardt, R.A., 1985, Nature, 315:147.
Poenie, M., Alderton, J., Steinhardt, R. and Tsien, R., 1986, Science, 233:886.

Price, C.M. and Pettijohn, D.E., 1986, Exp. Cell Res., 166:295.
Ratan, R.R., Shelanski, M.L. and Maxfield, F.R., 1986, Proc. Natl. Acad. Sci., USA, 83:5136.
Robbins, E., Jentzsch, G. and Micali, A., 1968, J.Cell Biol., 36:329.
Rozengurt, E., 1986, Science, 234:161.
Sager, P.R., Rothfield, N.L., Oliver, J.M. and Berlin, R.D., 1986, J.Cell Biol., 103:1863.
Sakai, H., 1978, Internatl. Rev. Cytol., 55:23.
Sakai, H., 1980, Biomed. Res., 1:359.
Salmon, E.D. and Segall, R.R., 1980, J.Cell Biol., 86:355.
Salmon, E.D., Leslie, R.J., Saxton, W.M., Karow, M.L. and McIntosh, J.R., 1984, J.Cell Biol., 36:329.
Sato, C., Nishizawa, K., Nakamura, H., Komagoe, Y., Shimada, K., Ueda, R. and Suzuki, S., 1983, Cell Struct. Funct., 8:245.
Saxton, W.M., Stemple, D.L., Leslie, R.J., Salmon, E.D., Zavortink, M. and McIntosh, J.R., 1984, J.Cell Biol., 99:2175.
Schatten, H., Schatten, G., Mazia, D., Balczon, R. and Simerly, C., 1986, Proc. Natl. Acad. Sci., USA, 83:105.
Schatten, H., Walter, M., Mazia, D., Biessmann, H., Paweletz, N., Coffe, G. and Schatten, G., 1987, Proc. Natl. Acad. Sci., USA, 84:8488.
Sherline, P. and Mascardo, R., 1982, J.Cell Biol., 95:316.
Sillers, P. and Forer, A., 1985, Cell Biol. Int. Rep., 9:275.
Silver, R.B., Cole, R.D. and Cande, W.Z., 1980, Cell, 19:505.
Silver, R.B., 1986, Proc. Natl. Acad. Sci., USA, 83:4302.
Sluder, G. and Rieder, C.L., 1985, J.Cell Biol., 100:897.
Snyder, J.A. and McIntosh, J.R., 1975, J.Cell Biol., 67:744.
Soltys, B.J. and Borisy, G.G., 1985, J.Cell Biol., 100:1682.
Suprynowicz, F.A. and Mazia, D., 1985, Proc. Natl. Acad. Sci., USA, 82:2389.
Telzer, B.R. and Rosenbaum, J.L., 1979, J.Cell Biol., 81:484.
Toriyama, M., Endo, S. and Sakai, H., 1984, Cell Struct. Funct., 9:213.
Toriyama, M., Ohta, K., endo, S. and Sakai, H., 1988, Cell Motil. Cytoskel., 9:117.
Vandre, D.D., Davis, F.M., Rao, P.N. and Borisy, G.G., 1984, Proc. Natl. Acad. Sci., USA, 81:4439.
Vantard, M., Lambert, A.-M., De Mey, J., Picquot, P. and Van Eldik, L.J., 1985, J.Cell Biol., 101:488.
Wandosell, F., Serrano, L., Hernandez, M.A. and Avila, J., 1986, J. Biol. Chem., 262:10332.
Weisenberg, R.C. and Rosenfeld, A.C., 1975, J.Cell Biol., 64:146.
Welsh, M.J., Dedman, J.R., Brinkley, B.R. and Means, A.R., 1978, J.Cell Biol., 81:624.
Whitaker, M.J. and Steinhardt, R.A., 1985. *in* "Biology of Fertilization", vol. 3, C.B. Metz and A. Monroy, ed., Academic Press, Orland, Florida, pp. 167-221.
Wick, S.M. and Hepler, P.K., 1980, J.Cell Biol., 86:500.
Willingham, M.C., Wehland, J., Klee, C.B., Richert, N.D., Rutherford, A.V. and Pastan, I.H., 1983, J. Histochem. Cytochem., 31:445.
Yamamoto, H., Fukunaga, K., Tanaka, E. and Miyamoto, E., 1983, J. Neurochem., 41:1119.
Yamamoto, H., Fukunaga, K., Goto, S., Tanaka, E. and Miyamoto, E., 1985, J. Neurochem., 44:759.
Yamauchi, T. and Fujisawa, H., 1983, Biochem. Biophys. Res. Commun., 110:287.
Yoshimoto, Y., Iwamatsu, T. and Hiramoto, Y., 1985, Biomed. Res., 6:387.
Zavortink, M., Wellsh, M.J. and McIntosh, J.R., 1983, Exp. Cell Res., 149: 375.

CONDITIONAL-LETHAL MUTANT OF CALMODULIN IN YEAST

Yoshikazu Ohya and Yasuhiro Anraku

Department of Biology, Faculty of Science
University of Tokyo
Hongo, Bunkyo-ku, Tokyo, 113, Japan

INTRODUCTION

Molecular biological and genetic analyses have proved useful in studies on the roles of calmodulin in cell proliferation. Calmodulin genes or its cDNA were isolated from a number of eukaryotes such as electric eel, chicken, human, rat, *Xenopus*, *Dictyostelium*, *C. elegance* and yeasts. Among them, yeasts are the most suitable organisms for genetic analysis. Like *E. coli*, any gene in yeasts can be replaced with its mutational allele that is constructed *in vitro*, if its gene has been cloned. Essential gene can also be analyzed by using combination of the gene-replacement in a diploid strain and dissecting its tetrad.

Calmodulin was purified to complete homogeneity from the yeast *Saccharomyces cerevisiae* (Ohya et al., 1987; Luan et al., 1987), and the genes were isolated from *S. cerevisiae* (Davis et al., 1986) and *Schizosaccharomyces pombe* (Takeda and Yamamoto, 1987). In these two organisms, haploid cells containing a disrupted calmodulin gene were found not to grow from spores. This observation, at first, clearly indicated that calmodulin is an essential protein. On microscopic analysis, growth-arrested cells from spores containing disrupted calmodulin gene were observed on spore-dissection plate (Davis et al., 1986; Takeda and Yamamoto, 1987). But the function of calmodulin in cell proliferation is hard to analyzed by microscopic examination of spores, because large numbers of arrested cells cannot be examined by this procedure.

To examine the step in cell proliferation at which yeast calmodulin is specifically involved, we constructed a conditional-lethal mutant of calmodulin in the yeast *S. cerevisiae*. In our mutant, the expression of calmodulin can be controlled by *GAL1* promoter: the mutant grew normally in galactose medium, but in glucose medium, in which the promoter was repressed, its growth stopped. The growth arrest was associated with decrease in intracellular calmodulin. We, then, analyzed the defects of these calmodulin-deficient cells on comparing landmarks of the cell cycle (Ohya and Anraku, manuscript submitted). In this paper we will describe the specific defect of this calmodulin mutant on the progression of nuclear division. Furthermore, we show that the null mutation of yeast calmodulin is suppressed by expression of chicken calmodulin, although these primary structures were different from each other (Ohya and Anraku, manuscript in preparation).

MATERIALS AND METHODS

Media

YPD or glucose medium contained 1% Bacto-yeast extract (Difco, Detroit), 2% polypeptone (Daigo Chemicals, Saitama), and 2% glucose (Wako Chemicals, Tokyo). YPGal or galactose medium contained 1% Bacto-yeast extract, 2% polypeptone, and 2% galactose (Wako Chemicals, Tokyo). YPD or YPGal supplemented with 0.4 mg of G418 (GIBCO Laboratories, Grand Island, N.Y.) per ml was used for testing antibiotic G418 sensitivity of yeast strains. Presporulation medium containing 0.8% Bacto-yeast extract, 0.3% polypeptone and 10% glucose, and sporulation medium containing 1% potassium acetate (Wako Chemicals) were used for tetrad analysis.

Construction of pGCAM104 and pGCAM201 plasmid

The yeast calmodulin gene (CMD1) was cloned from an X2180-1A cosmid bank using oligonucleotide probes corresponding to the partial amino acid sequence of purified yeast calmodulin (Ohya et al., 1987). The DNA sequence of the cloned gene was identical to that determined previously (Davis et al., 1986).

For construction of the pGCAM104 plasmid (Fig. 1a), the CMD1 gene without nucleotides corresponding to 12 amino acid residues was fused to the GAL1 promoter (from Dr. T. Oshima) at the EcoRI site located in the N-terminal CMD1 coding region. Then, 40mer synthetic DNA (made by Genet AII, Nippon Zeon), corresponding to the N-terminal 12 amino acid residues, was inserted into the the EcoRI site. The sequence of the 40mer DNA was modified at two positions from that of the native calmodulin gene so that the synthetic DNA possessed an Nsp(7524)V digestion site, making it easy to select a proper insert by digestion with restriction endonuclease. This modification did not cause any change in the amino acid sequence, and had little effect on codon usage in the CMD1 gene. The pGCAM104 plasmid was obtained by insertion of the resultant fusion gene into centromere vector YCpG11 (Ohya et al., 1986).

In the pGCAM201 plasmid (Fig. 1b), a chicken calmodulin cDNA was fused to GAL1 promoter. For construction of pGCAM201 plasmid, TaqI-ScaI fragment of chicken calmodulin cDNA in the pCB12 plasmid (Putkey et al., 1983) was inserted into Nsp(7524)V-PvuII gap in the pRS1 plasmid that contained GAL1 promoter, G418^r marker and replicative origin of 2 um plasmid.

Construction of YOG18

Plasmid pGCAM104 was used to transform a heterozygous cmd1 diploid, YMMT131 (+/cmd1::URA3). A G418^r transformant was selected and after sporulation, tetrad analysis was performed. Spores showing the Ura$^+$, G418^r phenotype possessed the cmd1 allele and the plasmid pGCAM104: they grew on YPGal but did not grow on YPD. Two such haploids of different mating types (YOG18-2D and YOG18-3C) were crossed, and the diploid strain YOG18 was obtained.

Observation of microtubule and nuclear morphology

Microtubules were stained using monoclonal antibody against tubulin as described in Adams and Pringle (1984). Cells were fixed with 3.7% formaldehyde and stained with 1 µg/ml DAPI (4'6-diamidino-2-phenylindole) for examination of nuclear morphology. They were examined using an Olympus BHS-RFK fluorescent microscope equipped with a high-pressure mercury vapor lamp (HBO, 100W) and UVFL 100/1.30 objective.

RESULTS

Construction of a strain with calmodulin regulated by galactose

For analysis of the key step requiring calmodulin, we constructed a conditional-lethal mutant in which the expression of calmodulin was controlled by a galactose-inducible GAL1 promoter (Johnston and Davis, 1984; West et al. 1984). The GAL1 promoter turns on in galactose medium and turns off in glucose medium, its level of expression in galactose medium being about 1000 times that in glucose medium. Since the expression of an essential calmodulin gene could easily be controlled by this fusion gene, we could analyze the consequences of loss of calmodulin function in a large population of cells.

For construction of a strain in which calmodulin was expressed only by the GAL1P-CMD1 fusion gene, the chromosomal CMD1 gene had to be disrupted. Then, pGCAM104 (Fig. 1a) containing the GAL1P-CMD1 fusion gene was used to transform a heterozygous CMD1/cmd1::URA3 diploid (YMMT131). A G418^r transformant (YMMT104-8) was selected and after sporulation, tetrad analysis was performed. When heterozygous CMD1/cmd1 diploid cells (YMMT131) and those with YCpG11 vector (YMMT132) were sporulated, there were two viable spores and two nonviable spores in many tetrads (Table 1). All viable spores showed the Ura$^-$ phenotype because cells containing cmd1::URA3 allele (Ura$^+$) could not grow from spores (Davis et al., 1986). On the other hands, of the G418^r transformant with the pGCAM104 plasmid (YMMT104-8), three or four spores in many tetrads were able to grow on galactose plate (Table 1). Some spores showed the Ura$^+$ phenotype, indicating the existence of a disrupted chromosomal cmd1 gene. All Ura$^+$ spores showed G418^r and did not grow on glucose plate, indicating that these Ura$^+$ cells contained the pGCAM104 plasmid and that the expression of calmodulin in these strains was regulated by galactose.

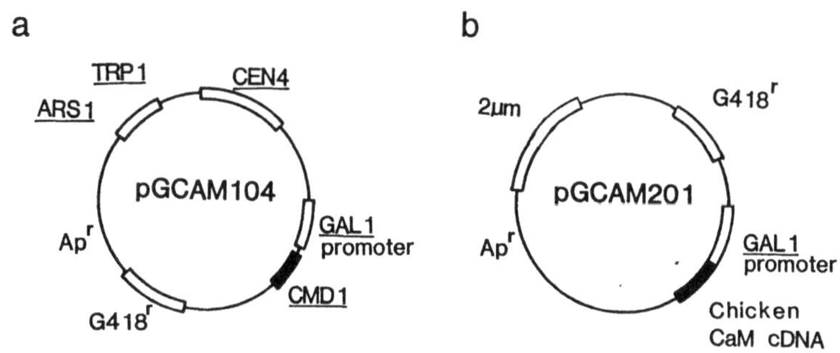

Figure 1
Plasmid containing fusion gene capable of controlling calmodulin synthesis by GAL1 promoter.

a, Plasmid pGCAM104 containing the fusion gene in which the coding region of yeast calmodulin (CMD1) is fused to the GAL1 promoter. pGCAM104 is a derivative of the centromere vector YCpG11 (Ohya et al., 1986). b, Plasmid pGCAM201 containing the fusion gene in which the cDNA of the chicken calmodulin is fused to the GAL1 promoter.

Table 1

Tetrad analysis of +/cmd1 heterozygous diploids containing fusion plasmids[a]

	Spore viability				
Diploid	4+:0-	3+:1-	2+:2-	1+:3-	0+:4-
YMMT131 (+/cmd1)	0	0	45	4	0
YMMT132 (+/cmd1 [YCpG11])	0	0	21	3	0
YMMT104-8 (+/cmd1 [pGCAM104])	32	9	2	0	0
YMMT201 (+/cmd1 [pGCAM201])	9	16	14	2	0

a Spores were grown on YPGal plate for 3 days after tetrad dissection.

Growth properties of the conditional lethal mutant of calmodulin

To examine the role of calmodulin in cell proliferation, we used the conditional-lethal diploid strain YOG18, as diploid cells could easily be analyzed on comparing landmarks of the cell cycle. The YOG18 strain had both a disrupted chromosomal calmodulin gene (cmd1::URA3) and the pGCAM104 plasmid. This YOG18 strain grew normally with a generation time of about 120 min in galactose medium, in which the GAL1 promoter is active. But in glucose medium, in which the GAL1 promoter is repressed, cell growth ceased: cell growth, measured as optical density, reached a plateau after 14-16 h (Fig. 2). Growth arrest in glucose medium was associated with decrease in the intracellular calmodulin level. In galactose medium, YOG18 cells contained 2.3 µg calmodulin/mg protein. After incubation in glucose medium for 8 h, their calmodulin content decreased to about 0.1 µg/mg protein, which was the same as that of normally growing wild-type cells. And after 12 h, no calmodulin (<0.01 µg/mg protein) was detectable by Western blotting analysis.

Terminal phenotype of the mutant

On microscopic examination of cells 14 h after calmodulin expression had stopped, two types of cells were observed. The main type (65 ± 3%) had daughter cells of the same size as the mother cells. The other, minor type (27 ± 3%) had small buds of less than one third the size of the mother cells. The rest of the cells (8%) were unbudded or half-budded. The volumes of the both arrested cells were about 5 times that of normally budding cells. In control condition (galactose medium), these cells were not observed. When DNA was stained with DAPI (4'6-diamidino-2-phenyl indole), one condensed nucleus was seen, located near the neck of the bud in the major type of cells, but not always near the neck in the minor type of cells (Fig. 3, DNA).

The DNA content of these nuclei, stained with propidium iodide, was determined with a fluorescence microscope photometer. Cells of the diploid strain YOG18 growing in galactose medium contained 2c and 4c DNA per nucleus, like growing wild-type cells, but in glucose medium all the cells contained 4c DNA . Both types of arrested cells also contained 4c DNA, which corresponded to those in nuclei at G2 or M, indicating that in both types of cells, arrest occurred after the G2 phase. We next examined the intracellular distribution of tubulin with indirect immunofluorescence antibody (Fig. 3, MT), and found that the mitotic spindle did not elongate, indicating that the arrested cells could not undergo nuclear division. Another feature of these cells was the appearance of several long cytoplasmic microtubules extending from the spindle pole body.

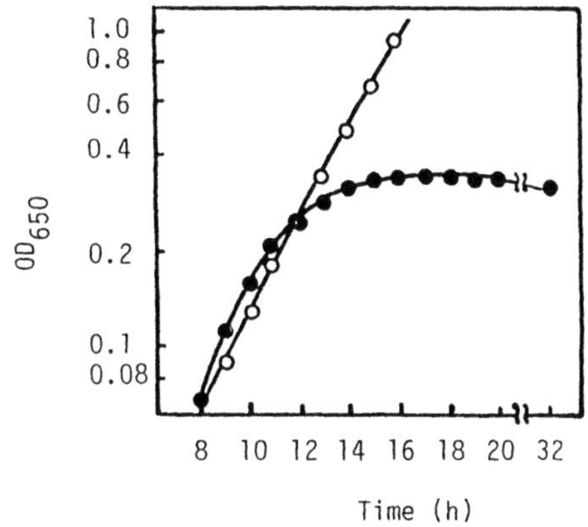

Figure 2

Growth of YOG18 strain cells with calmodulin controlled by galactose

Cells of the YOG18 strain (cmd1/cmd1 [pGCAM104]) were grown in galactose medium at 30 °C. Exponentially growing cells were collected and transferred to galactose (○) or glucose (●) medium at time 0 h. Cultures were incubated at 30 °C with shaking, and the OD_{650} was measured at the indicated times.

Morphological change of the mutant after calmodulin expression stopped

Analysis of the terminal phenotype of the mutant showed that all cells contained G2/M nuclei, there being two types of buds. At arrest, most cells had a large bud, but a few had a small one. In order to know why the minor population of arrested cells existed in the culture and whether the appearance of the terminal phenotype was associated with loss of calmodulin, we examined the morphological transition of the cell population by time-lapse photomicrography (Hartwell, 1978). As shown in Fig. 2, YOG18 cells which had been incubated for 9 h in glucose medium, underwent one more cell division, at which time, calmodulin had not been completely deprived, although its level was less than 0.1 µg/mg protein. Then, we spread the cells on glucose plate and observed their morphological transition and cell division by photomicrography at the time of plating (9 h), after incubation on the plate for 5 h (corresponding to 14 h in glucose medium) and 23 h, respectively. Result showed that 107 of the 128 major type of cells with large buds and 63 of the 68 minor type of cells with small buds underwent a last cell division after 9 to 14 h: this time corresponded exactly to that when calmodulin was completely lost from the cells. Therefore, the terminal phenotype described above seemed to be caused primarily by deprivation of calmodulin, and calmodulin may be involved in regulation of bud enlargement.

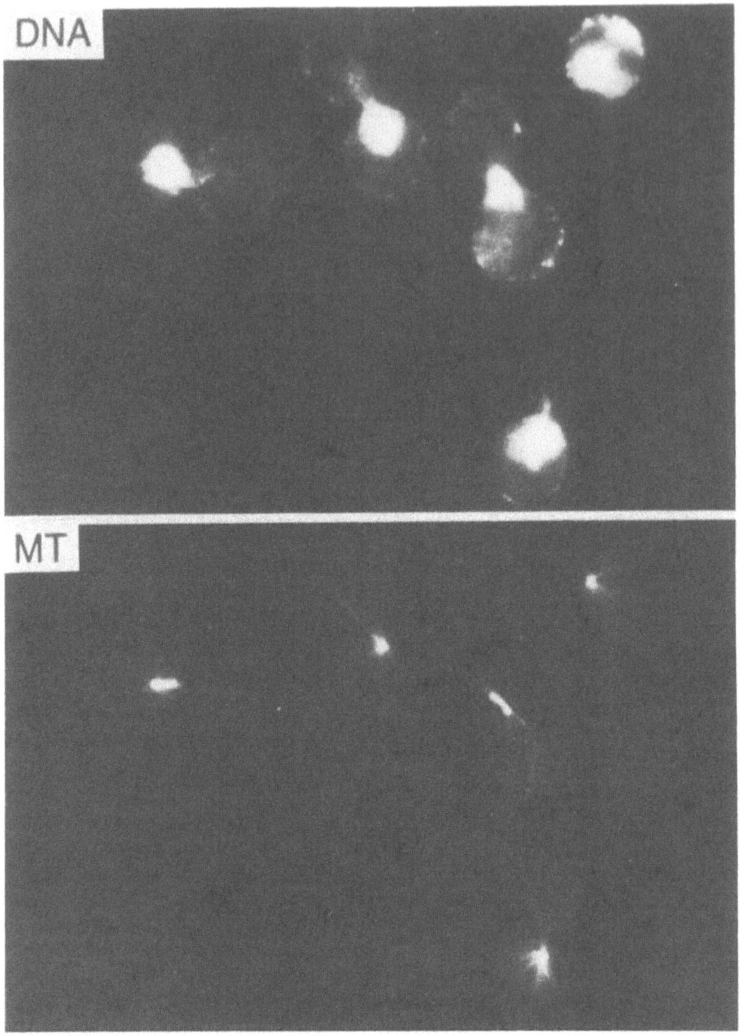

Figure 3

Terminal morphology of cells of the YOG18 strain.

Cells of the YOG18 strain grown in galactose medium were transferred to fresh glucose medium. After 14 h-incubation at 30°C, the morphologies of microtubules revealed by indirect immunofluorescent photomicrograph (MT), and nuclei stained with DAPI (DNA) are shown.

Chromosome loss in calmodulin-deficient cells

The fidelity of mitotic chromosome transmission in S. cerevisiae is known to be high: the spontaneous rate of loss of chromosome V is 8.3×10^{-6} events/cell division (Hartwell and Smith, 1985). This fidelity is due to precise chromosome segregation as well as complete chromosome replication and condensation. Understandably, several yeast cdc mutants deficient in nuclear division showed increased rates of chromosome loss both at the permissive temperature (Hartwell and Smith, 1985) and after prolonged incubation at their restrictive temperature. Thus, since decrease of calmodulin resulted in cell arrest at nuclear division, cells deficient in calmodulin may lose chromosomes more readily. When we analyzed the chromosomal stability by counting the appearance of recessive marker phenotype from heterozygous diploid, we found that the rate of loss of chromosome III, IV and V in glucose medium was much higher than that in galactose medium (0.1 - 2.4 % loss). This result indicated that calmodulin-deficient cells showed increased rates of chromosome loss as the result of a defect in precise progression of chromosomal segregation.

Complementation of yeast calmodulin by chicken calmodulin

To know whether yeast calmodulin is functionally homologous to vertebrate calmodulin, we examined the complementation of yeast calmodulin mutant by chicken calmodulin. We, therefore, constructed a pGCAM201 plasmid in which a chicken calmodulin was controlled by the GAL1 promoter (Fig. 1b). So, it would be expected that cells containing the pGCAM201 plasmid expressed chicken calmodulin in galactose medium.

After transformation of a heterozygous CMD1/cmd1::URA3 diploid (YMMT131) with pGCAM201 plasmid, the resultant transformant (YMMT201) was sporulated and subjected to tetrad analysis. As shown in Table 1, the transformant (YMMT201) produced many tetrads in which three or four spores were able to grow on galactose plate, similar to the transformant with pGCAM104. Some spores showed the Ura^+ phenotype, and the all Ura^+ spores showed $G418^r$, indicating the disrupted chromosomal cmd1 gene was complemented by the pGCAM201 plasmid.

We, then checked whether complementation by pGCAM201 was really achieved by expression of chicken calmodulin. If so, cells with disrupted chromosomal cmd1 gene and the pGCAM201 plasmid should not grow in glucose plate, since chicken calmodulin was controlled by GAL1 promoter. As shown in Fig. 4-1d, these cells could grow in galactose plate, but could not grow to form colonies on glucose plate (Fig. 4-2d). The growth rate in galactose plate was similar to that of wild-type strain (Fig. 4-1a), calmodulin-disrupted strain complemented by CMD1 plasmid (Fig. 4-1b) and calmodulin disrupted strain complemented by GAL1p-CMD1 plasmid (Fig. 4-1c). This experiment indicated that the null mutation of yeast calmodulin is suppressed by expression of chicken calmodulin, meaning that yeast calmodulin is functionally homologous to vertebrate calmodulin.

DISCUSSION

To determine the step at which calmodulin is essential for yeast cell growth, we constructed a GAL1P-CMD1 fusion gene in which the galactose-inducible GAL1 promoter regulates the expression of calmodulin. In glucose medium, in which the GAL1 promoter is repressed, cells harboring the GAL1P-CMD1 fusion gene could not grow because of lack of calmodulin. On comparing landmarks of the yeast cell cycle and monitoring morphological transition of the cell population, we concluded that cells deficient in expression of calmodulin showed bud emergence and DNA synthesis, but did not accomplish nuclear division. Spindle did not elongate in these cells, clearly indicating that their growth stopped in the middle stage of nuclear division. Recently, Eilam and Chernichovsky (1988) reported that trifluoperazine inhibited yeast cell growth at two stages of cell cycle, G1 and nuclear division. Taking our results into account, it seems likely that trifluoperazine inhibits growth by an antagonistic effect on calmodulin as well as on other cellular Ca^{2+}-binding proteins.

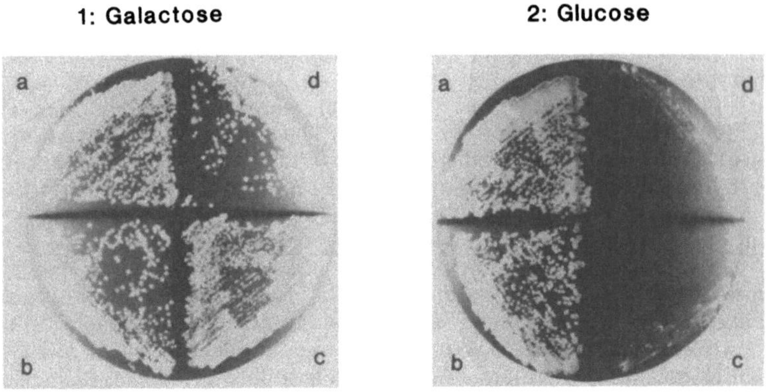

Figure 4

Complementation of yeast calmodulin mutant by chicken calmodulin

Cells of each strain were streaked on galactose (1) or glucose (2) plate, and their growth was analyzed after 2 days at 30 °C. a, Wild-type strain (YMMT100); b, cmd1::URA3 strain with pCAM124 plasmid containing wild-type CMD1 gene (YMMT124-2A); c, cmd1::URA3 strain with pGCAM104 plasmid containing GAL1 promoter-CMD1 fusion gene (YOG18-3C); d, cmd1::URA3 strain with pCAM201 plasmid containing, GAL1 promoter-chicken calmodulin fusion gene (YOG21-1A).

The finding that terminal phenotype of the conditional-lethal mutant of calmodulin was unique was unexpected, as calmodulin is thought to regulate a variety of essential enzyme systems and protein components within cells. We think that the present results with the GAL1P-CMD1 gene may not indicate that calmodulin is required only for nuclear division during cell cycle. When expression of calmodulin was repressed in glucose medium, the amount of calmodulin within the cells decreased gradually (Figure 3). Therefore, cells continued to grow until the intracellular level of calmodulin decreased to a lower limit for growth, and growth was arrested at a stage where further progression was not possible at the calmodulin level remaining in the cells. Therefore, the cells were probably not arrested at a stage where less calmodulin was needed for progression than for nuclear division. In other words, nuclear division may be the step in the yeast cell cycle requiring most calmodulin for its progression. In this sense, it is interesting that overexpression of calmodulin lead to shorten the G1 period in mouse cells (Rasmussen and Means, 1987), because it suggests that calmodulin has a regulatory effect on the rate of cell cycle progression in mammalian cells.

Hartwell and his colleagues isolated temperature-sensitive cell cycle mutants of over 50 complementation groups (Pringle and Hartwell,

1981). At the restrictive temperature, growth of each mutant stopped at a specific stage of the cell cycle. Some of these mutants, such as cdc13, cdc16, cdc17, cdc20 and cdc23 can not undergo nuclear division at the restrictive temperature, like cells deficient in calmodulin. Another type of mutant, a Ca^{2+}-dependent mutant that has defects in nuclear division, has also been isolated (Ohya et al., 1984). This mutant (cal1) could not grow in Ca^{2+}-poor medium, in which it stopped growing at the middle stage of nuclear division with a small bud. Trifluoperazine-dependent pseudorevertants were obtained from the cal1 mutants. This pseudorevertant was shown to have suppressor mutation tfr1 which brings Tfp-dependence to the cal1 strain and simultaneously causes Tfp-resistance itself (Ohya et al., 1984). These cdc and cal1 cells are thought to have defects in function of the mitotic apparatus, including the kinetochore, pole-to-pole spindle, kinetochore spindle, nuclear membrane, mitotic ER and some unknown cytoplasmic factor. Therefore, the target of calmodulin in the mitotic apparatus may be a product of the cdc or cal1 gene.

In higher eukaryotic cells, Ca^{2+} and calmodulin are thought to be the key elements in local depolymerization of kinetochore microtubules during anaphase. Microtubule disassembly was promoted by calmodulin both in vitro (Marcum et al., 1978) and in vivo (Keith et al., 1983). In the mitotic apparatus of mammalian (Welsh et al., 1979) and plant (Vantard et al., 1985) cells, calmodulin is localized at the spindle, especially the kinetochore spindle, as shown by indirect immunofluorescent antibody staining. Localized increase in the intracellular concentration of Ca^{2+} has been observed during mitosis (Poenie et al., 1986; Ratan et al., 1986; Hepler and Callaham, 1987). Furthermore, microinjection of a specific antibody against the Ca^{2+}-transport system in the membranes of Ca^{2+}-sequestering organelles inhibited mitosis (Silver, 1986; Hafner and Petzelt, 1987). These results all suggested that Ca^{2+} and calmodulin are essential for regulation of microtubule assembly during nuclear division. In addition, calmodulin is thought to regulate other targets in mitotic cells. Calmodulin and a number of calmodulin-binding proteins are located in liver nuclei (Bachs and Carafoli, 1987). By analogy with other mechano-chemical contractile systems, such as the myosin-actin system in smooth muscle (Sobue et al., 1981), the increase in intracellular concentration of Ca^{2+} during anaphase may cause mitotic movement via calmodulin. In this study, we established that calmodulin is essential for nuclear division and chromosomal stability. Judging from our observations on nuclear and spindle morphology, yeast calmodulin seems to play important roles in early mitosis. Although cell cycle control in yeast cannot be directly comparable to that in higher eukaryotic cells, our genetic results are consistent with the idea that calmodulin plays essential roles in the mitotic apparatus. Further genetical analyses with yeast cells should indicate the target(s) of calmodulin in nuclear division.

The structure and function of calmodulin have been conserved well in eukaryotic cells. Amino acid sequences of mammalian calmodulins, indeed, show 91% homology with that of spinach calmodulin, and 92% homology with that of calmodulin from Tetrahymena. However, calmodulin from S. cerevisiae is shown to be different from other calmodulin: amino acid sequence of yeast calmodulin shows only 60% homology with other calmodulins. Because of difference in its primary structure, the biological properties of yeast calmodulin is also different from those of mammalian calmodulin. Actually, yeast calmodulin activates bovine phosphodiesterase, pea NAD kinase and myosin light chain kinase at a similar Ca^{2+} concentration, but level of activation and/or dose of calmodulin is quite different. Moreover, yeast calmodulin binds three Ca^{2+} per calmodulin protein, while vertebrate calmodulins bind four (Luan et al., 1987). Despite of these biochemical and structural differences, yeast calmodulin mutant was shown to be complemented by chicken calmodulin in this study. This implies that essential function of calmodulin for cell proliferation is conserved throughout evolutional development from yeast to vertebrate. And we now convince that we could prepare the novel system that is used to identify functional domain of vertebrate calmodulin essential for nuclear division.

ACKNOWLEDGMENTS

We thank Drs. Taro Nakamura and Shin-ichi Kawamoto at Hokkaido University for supplying the plasmid pCB12. This work was supported by a Grant-in-Aid for Scientific Research from the Ministry of Education, Science, and Culture of Japan. Part of this work was carried out under the NIBB Cooperative Research Program (88-108).

REFERENCES

Adams, A. E. M. and Pringle, J. R., 1984, J. Cell Biol., 98:934.
Bachs, O. and Carafoli, E., 1987, J. Biol. Chem., 262:10786.
Davis, T. N., Urdea, M. S., Masiarz, F. R. and Thorner, J., 1986, Cell, 47:423.
Eilam, Y. and Chernichovsky, D., 1988, J. Gen. Microbiol., 134:1063.
Hafner, M. and Petzelt, C., 1987, Nature, 330:264.
Hartwell, L. H., 1978, J. Cell Biol., 77:627.
Hartwell, L. H. and Smith, D., 1985, Genetics, 110:381.
Hepler, P. K. and Callaham, D. A., 1987, J. Cell Biol., 105:2137.
Johnston, M. and Davis, R. W., 1984, Mol Cell Biol., 4:1440.
Keith, C., Dipaola, M., Maxfield, F. R. and Shelanski, M. L., 1983, J. Cell Biol., 97:1918.
Luan, Y., Matsuura I., Yazawa, M., Nakamura, T., and Yagi, K., 1987, J. Biochem., 102:1531.
Marcum, J. M., Dedman, J. R., Brinkley, B. R. and Means, A. R., 1978, Proc. Natl. Acad. Sci. U.S.A., 75:3771.
Ohya, Y., Ohsumi, Y., and Anraku Y., 1984, Mol. Gen. Genet., 193:389.
Ohya, Y., Miyamoto, S., Ohsumi, Y., and Anraku, Y., 1986, J. Bacteriol., 165:28.
Ohya, Y., Uno, I., T. Ishikawa, T., and Anraku, Y., 1987, Eur. J. Biochem., 163: 13.
Poenie, M., Alderton, J., Steinhardt, R. and Tsien, R., 1986, Science, 233:886.
Pringle, J. R. and Hartwell, L. H., 1981 In Strathern, J. N. et al. (ed.), The Molecular Biology of the Yeast Saccharomyces. Cold Spring Harbor Laboratory, New York, Vol I, pp97.
Putkey, J. A., Ts'ui K. F., Tanaka, T., Lagace, L., Stein, J. P., Lai, E. C., Means, A. R., 1983, J. Biol. Chem., 258:11864.
Rasmussen, C. D. and Means, A. R., 1987, EMBO J, 6:3961.
Ratan, R. R., Shelanski, M. L. and Maxfield, F. R., 1986, Proc. Natl. Acad. Sci. U.S.A., 83:5136.
Sasaki, Y. and Hidaka, H., 1982, Biochem. Biophys. Res. Commun., 104:451.
Silver, R. B., 1986, Proc. Natl. Acad. Sci. U.S.A., 83:4302.
Sobue, K., Muramoto, Y., Fujita, M. and Kakiuchi, S., 1981, Proc. Natl. Acad. Sci. U.S.A., 78:5652.
Takeda, T. and Yamamoto, M., 1987, Proc. Natl. Acad. Sci. U.S.A., 84:3580.
Vantard, M., Lambert, A. -M., Mey, J. D., Picquot, P. and Van Eldik, L. J., 1985, J. Cell. Biol., 101:488.
Welsh, M. J., Dedman, J. R., Brinkley, B. R. and Means, A. R., 1979, J. Cell Biol., 81:624-634.
West, R. W., Yocum, R. R., and Ptashne, M. (1984) Mol. Cell Biol., 4:2467.

HUMAN VITAMIN D RECEPTOR MUTATIONS: IDENTIFICATION OF MOLECULAR DEFECTS IN HYPOCALCEMIC VITAMIN D RESISTANT RICKETS

Mark Hughes, Peter Malloy, Dirk Kieback, Donald McDonnell, David Feldman, J. Wesley Pike, and Bert O'Malley

Departments of Cell Biology, Genetics, and Pediatrics
Baylor College of Medicine
Houston, Texas
and
Department of Medicine
Stanford University Medical Center
Stanford, California

INTRODUCTION

Our laboratories have been interested in the mechanism by which steroid receptors in general, and the 1,25-dihydroxyvitamin D (1,25-$(OH)_2D$) receptor specifically, interact with sequences in the target cell genome and regulate the transcription of specific gene products. Elucidation of these structure-function relationships of the 1,25-$(OH)_2D_3$ receptor macromolecule have been hampered primarily because of its extremely low intracellular concentration (0.001%), even in primary target tissues such as intestine. This concentration is less than one-tenth the cellular concentration of other steroid receptor molecules. Nonetheless, Pike and Haussler succeeded in purifying the chicken intestinal vitamin D receptor (1) and raised monoclonal antibodies which were reactive to both the chicken and mammalian isoforms (2). The monoclonals were used to recover vitamin D receptor (VDR) complementary DNA (cDNA) from a chicken intestine expression library (3). Subsequently, the chicken clones were used as probes to screen several human cDNA libraries and isolate the mammalian full length cDNA encoding the VDR (4). The nucleotide sequence of the 4605 base pair human cDNA was shown to include 1281 bp of open reading frame, 115 bp of noncoding leader sequence, and 3209 bp of 3'- noncoding sequence. The cloned sequence was subsequently transfected into COS-1 receptor negative monkey kidney cells and a single VDR species was produced that was indistinguishable from the native receptor (4).

It became clear from sequence comparisons that the VDR belonged to a superfamily of steroid receptor genes including receptors for estrogen, progesterone, glucocorticoids and the thyroid hormone receptor (its closest neighbor; 5). Because a common structural motif exists between VDR and the other hormone receptors, there is a strong argument that the mechanism of action of 1,25-$(OH)_2D$ parallels that of the other steroid hormone systems. It has now become possible to dissect the D-receptor macromolecule and to define its regulatory domains. Briefly, this chapter will discuss mutations which have been generated synthetically in VDR by *in vitro* molecular techniques, and then describe naturally occurring human gene mutations which are present in two families with hereditary tissue resistance to 1,25-$(OH)_2$D.

HUMAN VITAMIN D RECEPTOR DELETION ANALYSIS

In order to define the boundaries of the functional domains of VDR, a series of 5-prime and 3-prime deletion mutants were generated by cDNA manipulation, cloned into an expression vector (4), and transfected into COS-1 cells for analysis. 5'-deletions in the wild type cDNA were constructed by placing an in-frame artificial start site on the 5'-end of truncated cDNAs such that increasing amounts of the amino terminal region would be missing from the COS-1 generated receptors. Deletions from the 3'-end were produced using the natural start codon and engineered stop codons at various 3-prime locations along the cDNA. The proteins produced from the deleted recombinant cDNAs were tested for hormone binding activity using hydroxylapatite assay and DNA binding by chromatography over columns containing calf thymus DNA-cellulose. A summary of the generated mutations and binding results is shown in Figure 1. The only recombinant protein which was able to effectively bind 1,25-$(OH)_2[^3H]$D was the wild type receptor. In contrast, when these constructions were tested for DNA-cellulose activity, all of the engineered plasmids with 3'-deletions were capable of interacting with DNA. These receptor subfragments which contained the amino terminal portion of the molecule bound well to calf thymus DNA and were shown to elute at 0.22M KCl, similar to wild type protein.This was not the case for mutants with 5'-deletions (Δ1-104, Δ1-114, Δ1-166, Δ1-282). None of the receptor subfragments generated from these transfections displayed DNA-binding activity.

These data are consistent with the overall structural domains that have been identified for other steroid receptors (5). The data demonstrate that the 5'-portion of the cDNA encodes the amino acids most critical for receptor-DNA interaction. The hormone binding domain is not

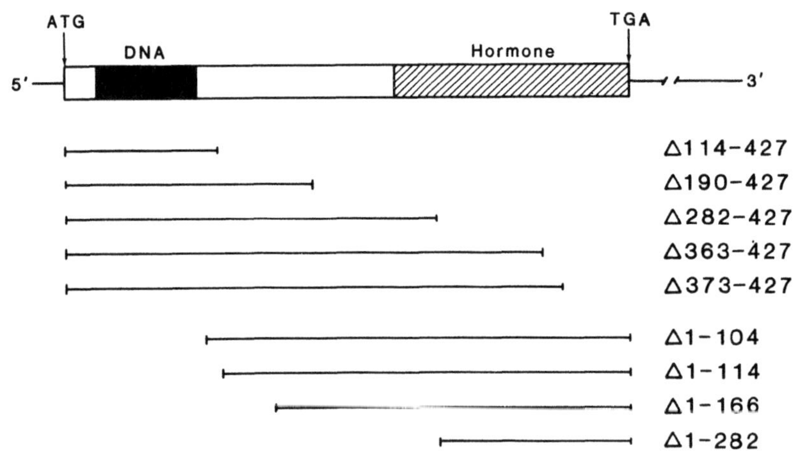

Figure 1. *In vitro* generated deletion mutations of the 1,25-$(OH)_2$D receptor. A series of 5' and 3' mutations were used to define the DNA and hormone binding domains. The numbers at the right correspond to the removed amino acids from the resulting protein.

precisely delineated by these experiments, but if VDR is similar to the thyroid receptor and other steroid receptors in this regard, the amino acids most important for hormone binding reside toward the carboxy terminus of the molecule. The similarity between VDR and the other receptors studied to date is most striking when the amino acid sequences of these proteins are compared. There are two areas of high sequence homology in this family of proteins. They have been designated as evolutionarily conserved regions one (C1) and two (C2), and are depicted at the top of Figure 1. The most highly conserved region is C1, a 70 amino acid cysteine rich area near the amino terminus of VDR, but more centrally located in the glucocorticoid, mineralcorticoid, progesterone and estrogen receptors. Using this sequence as the basis for comparison with other receptors, there is a 39-49% homology across a region of ~ 70 amino acids comprising C1. The sequence contains a clustering of basic residues likely to interact with DNA, and there is a remarkable positional conservation of nine cysteine residues which are thought to form fingerlike motifs that coordinately bind zinc and intercalate into the major groove of DNA (6). This hypothesis is supported by *in vitro* generated mutants of this region in the other receptors (see Ref. 5 for review) which, while continuing to bind hormone, have lost their ability to bind DNA.

The second region of homology (C2) lies toward the carboxy terminus of the protein, and while less invariant in sequence across the receptor family of proteins, it reveals significant similarity (19-27%) over these evolutionary divergent molecules. Based on mutational analysis of estrogen and glucocorticoid receptors (5), this domain contains essential residues involved in steroid hormone binding. The more graded degree of homology in the receptor C2 region may parallel the sructural similarities of the hormones themselves.

The above *in vitro* mutagenesis experiments, combined with sequence comparisons of the deduced VDR amino acids, clearly demonstrate that this protein belongs to the superfamily of *trans*-acting transcriptional regulatory factors defined by the thyroid and steroid hormone receptors. While the DNA binding domain is localized to the C1 region, this approach currently leaves unanswered the question of which amino acids in this area are critical for interaction with nucleic acid. Point mutagenesis of these residues is ongoing in many laboratories in an attempt to better identify those which are the most significant receptor function. The interpretation of such studies has been difficult, however, because of the limitations in the available assay systems. For example, following point mutagenesis, the receptors are cotransfected into heterologous cells along with a reporter gene containing a steroid response element linked to a gene which codes for an enzyme marker. These cotransfected genes are not in a native chromatin structure. Furthermore, only a fraction of cells (<5%) actually take up the DNA, and because the cDNA is in an overexpression vector these cells produce the mutagenized receptors in vast excess to what occurs in cells naturally. While this type of assay can indicate only whether the synthesized mutant protein maintains crude activity, more subtle assessments regarding functionality cannot be made with this methodology.

An alternate approach to this problem is to identify naturally occurring receptor mutations which lack some functional activity intrinsic to steroid receptor action. This may be difficult, however, since the production of nonfunctional receptors may be lethal. Lethality might be expected in most congenital defects of the glucorticoid or thyroid receptor systems, especially if the mutation was in a crucial gene domain. Naturally occurring mutations of the progesterone receptor undoubtedly exist, but may be difficult to identify since the primary manifestation is infertility, a diagnosis with many causes and in which genetic transmission of the trait would, by definition, be rare. The vitamin D receptor provides an ideal research opportunity in this regard since well described vitamin D receptor mutations have been identified and grouped according to their biochemical defects. It is one of the few steroid hormone systems which has naturally occurring, easily identifiable, genetically inherited, non-lethal human mutations. It provides a model system not only for analysis of critical gene loci and functional domains in the D-receptor but, presumably, for the other steroid hormone receptor systems as well. The study of nature's mutational experiments have been invaluable in understanding many human disease conditions (sickle cell, alpha-1-antitrypsin disease, beta-thalassemia, phenylketonuria, hemophilia, etc.). We anticipate that the analysis of the human mutations of VDR will improve our knowledge of the function of all steroid hormone receptors. The remainder of this chapter will outline our attempts at describing the molecular basis for tissue vitamin D-resistance in two kindreds with defective receptor-DNA interaction. We will briefly review this human syndrome, describe the phenotype and nomenclature, and then outline our approach in identifying the precise amino acids responsible for this disorder of bone and mineral metabolism.

HEREDITARY HYPOCALCEMIC VITAMIN D RESISTANT RICKETS

In 1937 Albright first suggested that rickets could arise not only from nutritional deficiency of vitamin D (as classically taught), but also from the inadequate action of vitamin D (7). Fraser (8) is generally credited with the first report in 1958 of a patient with a unique form of vitamin resistance which has come to be variably known as pseudovitamin D-deficiency rickets, vitamin D-dependency rickets, hypocalcemic vitamin D-resistant rickets and vitamin D-dependent rickets (VDDR). The nomenclature has been confusing owing to a poor understanding of the pathophysiology behind the disease condition. However, a genetic lesion in the pathway of vitamin D action has long been suspected given the autosomal recessive inheritance pattern seen in many families. Two distinct subtypes of this disorder are known. One results from hereditary defects in synthesis of 1,25-$(OH)_2$D (VDDR-type I), and the other from hereditary resistance to the action of normal or elevated blood levels of the vitamin D hormone (VDDR-type II). The clinical features of hereditary vitamin D dependent rickets type I are usually not apparent at birth but begin developing during the first year of life with the onset of hypocalcemia leading to muscular weakness, hypotonia, poor growth, enamel hypoplasia, and motor retardation. Tetany can result in seizures, laryngospasm and death. Alternatively, children present with pathologic bone fractures mistakenly diagnosed as x-linked hypophosphatemia, osteogenesis imperfecta or "battered child" syndrome. Biochemically, the features include hypocalcemia, secondary hyperpara-thyroidism, hypophosphatemia, elevated alkaline phosphatase, appropriate 25(OH)D and undetectable plasma 1,25-$(OH)_2$D. The pathogenesis of vitamin D dependent rickets-type I is consistent with an inborn error of metabolism involving a defective renal 25-hydroxyvitamin-D-1-alpha-hydroxylase. Patients respond to therapeutic administration of vitamin D_2 (0.5-1.5 mg/day) or 25-(OH)D (0.1-0.5 mg/day) despite the fact that serum 1,25-$(OH)_2$D levels remain low suggesting an enzymatic block in hormone production (9). This is presumably because pharmacological levels of 25-(OH)D are able to overwhelm the receptor binding site for 1,25-$(OH)_2$D and produce some amount of receptor activity. Ideal treatment is with 1,25-$(OH)_2$D (0.2-2.0 mg/day) and these physiological doses result in rapid normalization of hypocalcemia and bone healing after just several days of therapy.

This satisfactory therapeutic approach is not available to patients suffering from the alternate form of this disorder: Target tissue resistance to 1,25-$(OH)_2D$ or vitamin D-dependent rickets-type II. While the literature generally refers to this disease subtype as vitamin D-dependent rickets-type II, we prefer to use the term "hypocalcemic vitamin D-*resistant* rickets" since these patients are clearly resistant to (and not dependent on) 1,25-$(OH)_2D$. Since 1978 there have been numerous reports (10-31) describing a clinically heterogeneous group of patients with apparent target tissue resistance to this hormone (10-31). While the phenotype is variable, many of the findings are similar to those described above for VDDR-type I disease. This includes an autosomal recessive mode of genetic transmission, secondary hyperparathyroidism and hypocalcemia at some point during the course of the disease. In contrast, however, 1,25-$(OH)_2D$ levels are markedly elevated in hypocalcemic vitamin D-resistant rickets (HVDRR). Other observations which are frequently reported include normal plasma levels of 25-(OH)D, hypophosphatemia and complete alopecia which becomes evident during the first year of life. While all families with HVDRR do not display alopecia, if one affected member exhibits hair loss, this finding is present in all family members with the disease.

Progress in unraveling the molecular pathogenesis of HVDRR was significantly accelerated when it was found that fibroblasts cultured from normal human skin contain the 1,25-$(OH)_2D$ receptor (32-34) and show stimulation of 24-hydroxylase activity by hormone administration (20,21). This provided a suitable source for "target" tissue in which to analyze the vitamin D receptors from this patient population. Skin fibroblasts have been evaluated from affected members of at least 11 kindreds with resistance to 1,25-$(OH)_2D$ (10-31). Analysis has included cellular uptake measurements of 1,25-$(OH)_2D$, induction of the 24-hydroxylase by vitamin D hormone, assay of high affinity receptors by radiolabeled hormone binding and specific monoclonal antibody association, and receptor-DNA interaction. A spectrum of molecular defects in the vitamin D receptor from these patients has emerged. Some have a quantitative deficiency of receptors while other kindreds display abnormal hormone binding, depressed DNA interaction all in concert with an inability to activate the 24-hydroxylase enzyme in their fibroblast cells. As with other steroid hormones, a multistep cellular process must be intact for successful hormone-receptor regulation of appropriate gene activity. On the protein level, these patients have been shown to lack strategic receptor functions which result in this severe disorder of bone and mineral metabolism. Our goal has been to uncover the precise genetic mutations accounting for these abnormalities in receptor function.

KINDREDS WITH TISSUE RESISTANCE TO VITAMIN D

Two families have been studied recently at both the protein and genomic levels for clues to their D-receptor defects. The first family, identified as the G-kindred (Figure 2), are Arabs living in the Middle East. The parents, denoted as G3 and G4, are first cousins and are phenotypically normal with no calcium or bone abnormalities. Of the six children (4 male, 2 female), 2 males appear to have classical type II disease with alopecia and early onset rickets. G1, born in 1981, was treated with oral calcium, phosphate and 1-alpha-hydroxyvitamin D, a synthetic analog. He exhibited clinical improvement in both his serum chemistries and rickets at 6-7 years of age. G2, born in 1985, has been treated with oral calcium which has improved the serum calcium but not the rickets. Serum 1,25-$(OH)_2D$ levels were 300 pg/ml and 500 pg/ml for G1 and G2, respectively; this is a significant elevation over the normal range of 20-80 pg/ml. The second kindred, a family from Boston denoted the D-kindred, has been previously described in detail (30), and consists of five members. The parents are phenotypically normal, as is one unaffected daughter. Two affected daughters displayed severe rickets and

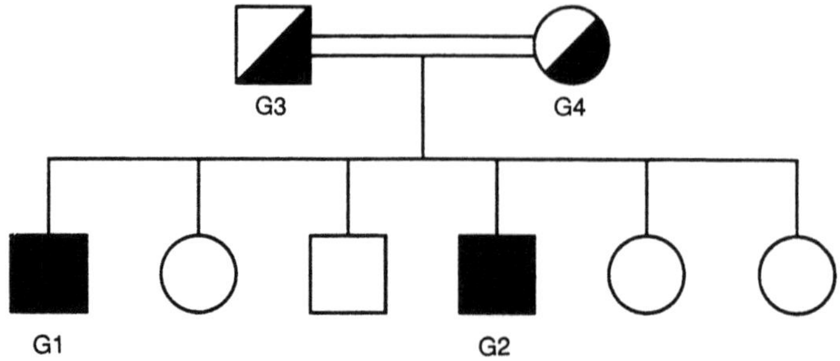

Figure 2. Pedigree of the G-kindred. Squares are males and circles are females with the fully-shaded symbols (G1, G2) displaying the disease phenotype and the half-filled symbols (G3, G4) being normal clinically but carrying the disease allele. The double line between the parents signifies consanguinity in the mating.

the classical phenotype of HVDRR. Receptor analysis of this family has revealed decreased affinity of the hormone-receptor complex for DNA (30).

ANALYSIS OF THE RECEPTOR PHENOTYPE FOR THE G-KINDRED

Fibroblasts from skin biopsies were obtained and grown in culture for measurement of 1,25-$(OH)_2[^3H]D$ binding (27,30), 24-hydroxylase activity (20,27,36) DNA-cellulose affinity (29), and western blot analysis using anti-VDR monoclonal antibody (2,27,37). The cells from the affected children and obligate heterozygous parents displayed normal 1,25-$(OH)_2D$ binding that was saturable and of high affinity (27; data not shown). However, when bioresponsiveness (20) was tested by the ability of the hormone to induce 24-hydroxylase activity, the cells from G1 and G2 failed to exhibit 24-hydroxylase activity even when treated with concentrations as high as 100 nM 1,25-$(OH)_2D$. The cells from the parents exhibited a normal dose-dependent 24-hydroxylase enzyme response activity.

Given the findings of normal hormone binding but defective bioresponsiveness, we next explored the possibility that the VDR in these cells may have altered hydrodynamic or physical properties. A western blot of hypertonic cell extracts from G1, G3 and a normal control indicated that the VDR from each cell line migrated at the expected molecular weight of about 48,000. (27, data not shown) consistent with the value determined from the human VDR cDNA clone and endogenous human receptor (4). Sucrose gradient analysis of receptors from both kindreds displayed a sedimentation coefficient (3.3S) identical to the normal human VDR (27,30). We conclude that the size and hormone binding of these receptors appeared intact despite an inability of the cells to respond to 1,25-$(OH)_2D$ as measured by 24-hydroxylase.

DNA binding was measured in the G-kindred (Figure 3) and D-kindred (30) by analyzing the ability of VDR to associate with DNA-cellulose. Normal VDR elutes from the column at about 0.2 M KCl (Figure 4, bottom panel, and 1,2,30,32,38) but receptor from these patients elutes prematurely at 0.1 M KCl (Figure 3, top panel). Interestingly, receptors from the parents (G3, middle panel) display two distinct peaks, one at the normal 0.2 M KCl and the other at 0.1 M KCl. These data demonstrated that the phenotypic abnormality in these kindreds is likely to result from a defective VDR-DNA interaction in the affected children. It further suggests that VDR is expressed from both normal and affected alleles in the obligate heterozygous parent.

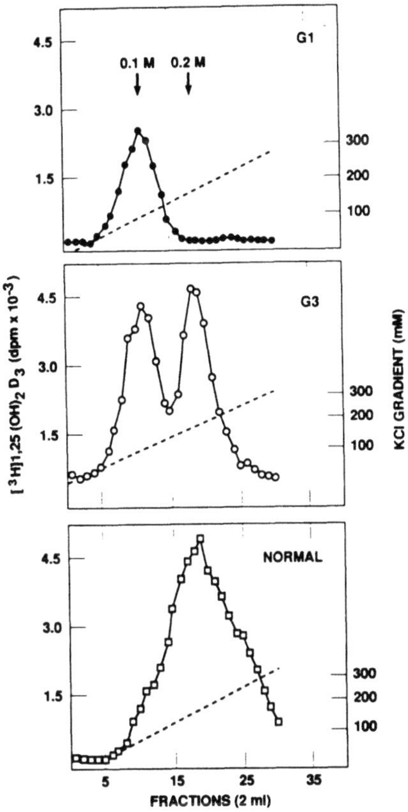

Figure 3. DNA-cellulose chromatography with KCl gradient elution of the 1,25-$(OH)_2[^3H]$-labeled cell extracts from the affected child (G1, top) and the normal parent (G3, bottom). A normal unrelated control is shown for comparison (bottom panel).

MOLECULAR DETERMINATION OF THE DEFECTIVE RECEPTOR GENOTYPE

We reasoned from results of the *in vitro* mutagenesis experiments on this receptor as well as mutations created in other receptors (5), that a reasonable place to initiate the search for the molecular gene defect would be in the conserved region one (C1), the region required for DNA binding. Work by Pike and coworkers (in preparation) had defined the intron-exon boundaries of the VDR gene and determined the nucleotide sequence of these borders. Furthermore, these results show that the C1 domain is coded by two separate exons in the genome. One approach to analysis of the patient's genome in this region involves the synthesis of genomic libraries from each family member followed by laborious screening, amplification, mapping and sequencing of each clone. A more rapid approach involves the use of the new technique termed polymerase chain reaction (PCR) (39), now in wide use for genetic diagnosis of disease. This novel technique is illustrated in Figure 4. Briefly, genomic DNA (500 nanograms) is heat denatured and then allowed to hybridize to a vast excess of oligonucleotide probe synthesized complementary to the ends of the DNA of interest. The large (Klenow) fragment of DNA-polymerase is added which uses the oligonucleotides as a primer, synthesizing the genomic DNA 5-prime to 3- prime. The oligos are made complementary to

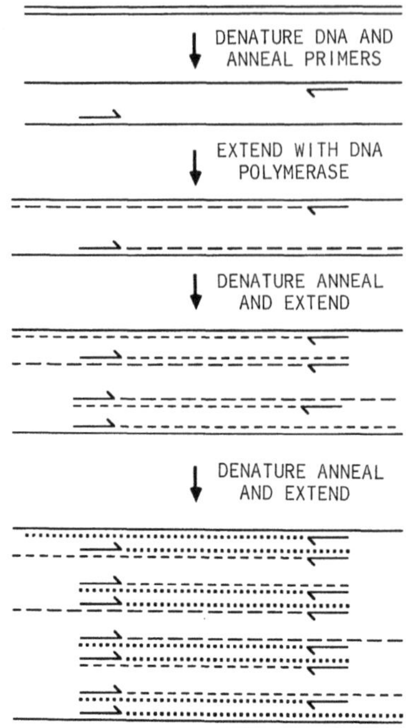

Figure 4. The polymerase chain reaction technique. See text for description.

different strands of intron sequence flanking exons to be examined and at relative positions along the sequence such that the polymerase extension product of the one, when denatured, can serve as a template for the other, and vice versa. The polymerase is added and the reaction is allowed to occur for 2 minutes. The reaction products are heat denatured for 1 min, allowed to reanneal, and repeated by the addition of fresh enzyme. Repeated cycles (3-6 min each) result in the amplification of the genomic DNA of known sequence between the oligonucleotide primers. The only requirement is knowledge of the intron DNA sequence at the ends of the exon sequences of interest to be amplified. New modifications include the use of *TaqI* polymerase, a heat stable enzyme that is not denatured at 94° and which obviates the requirement for adding fresh

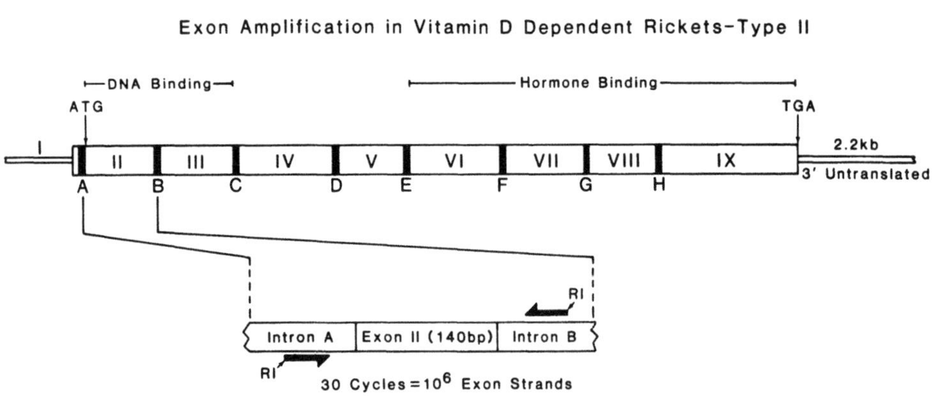

Figure 5. Strategy for amplification of patient genomic VDR sequences using polymerase chain reaction.

enzyme at each cycle. Literally millions of DNA fragments can be amplified in several hours and made ready for subcloning or even direct sequencing.

Figure 5 illustrates the strategy used on our patients. Four intron-annealing oligonucleotides were synthesized to the borders of VDR exons II and III, corresponding to the known DNA-binding domain and potentially the site of the molecular defect in the affected patients. An *EcoRI* restriction recognition sequence was artificially included at the extreme 5'-ends of the oligonucleotides for ease of subcloning after the PCR cycling was complete.

Thirty cycles of amplification were performed on 500 ng of patient genomic DNA by annealling the oligonucleotides for two minutes at 55°, followed by elongation with *TaqI* polymerase for three minutes at 72°, and then one minute at 94° to denature all products and restart the cycle. It should be noted from Figure 6 that the oligonucleotides were selected to hybridize well within the flanking intron sequence. This strategy required the polymerase to generate some of the remaining intron sequence, the intron-exon junction, and the entire exon coding region of the gene. Any junctional splice-site mutations in the patient DNA would consequently be identified.

The samples carried through this amplification were subsequently digested with restriction enzyme *EcoRI* to produce easily clonable single-stranded overhangs on both ends of the DNA of interest. These were ligated into a plasmid vector at the *EcoRI* site and subcloned. Resulting colonies were screened using either an end-labeled [^{32}P]-oligonucleotide complementary to the middle of the amplified DNA, or internally labeled (nick-translated) authentic VDR cDNA. Positive hybridizing clones were picked and sequenced to analyze for the potential mutation(s) in the patient.

A reproducible single base mutation (GGC to GAC) was identified in the triplet codon for glycine in exon II of both patients G1 and G2. The

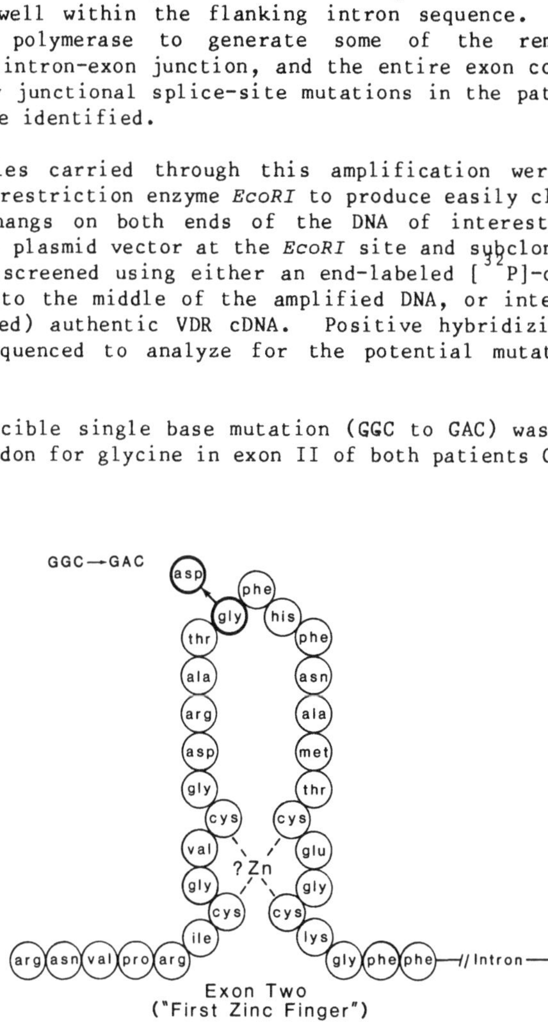

Figure 6. Molecular defect in the G-kindred. A single base substitution results in the conversion of glycine to aspartic acid at the tip of the first VDR zinc finger of exon II.

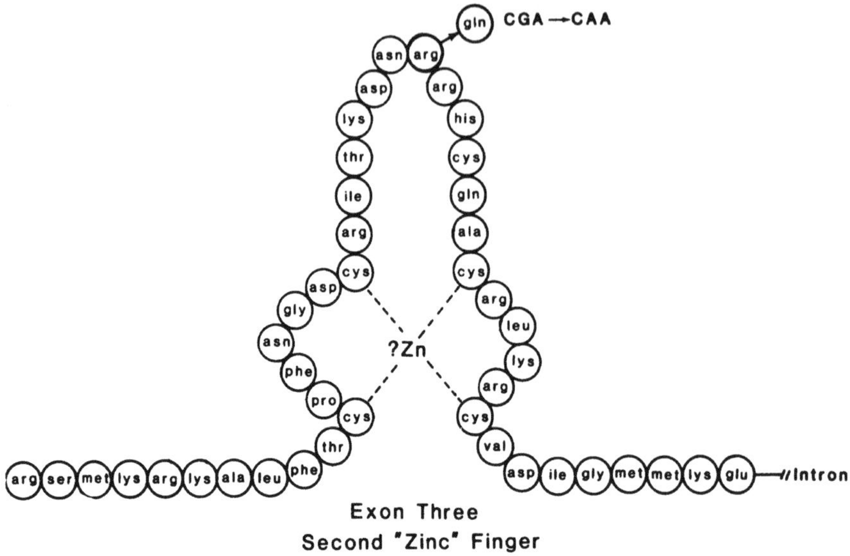

Figure 7. Mutation identified in the D-Kindred. The amino acid substitution of glutamine for arginine in exon III abolishes receptor-DNA interaction resulting in tissue resistance to the hormone.

base change alters the amino acid at the tip (middle) of the first DNA-binding "zinc finger" (Figure 6) from a neutrally charged glycine to a negatively charged aspartic acid. The glycine at this position is conserved in evolution throughout all of the steroid hormone receptors sequenced to date including progesterone, estrogen, glucocorticoid, the vitamin D receptor from chicken, and the thyroid hormone receptor. While only the mutant codon was identified in patients G1 and G2, their parents provided genomic DNA which generated both the mutant and normal sequence; a random selection of PCR clones revealed 58% normal and 42% abnormal codons, consistent with the expected equal presence of both alleles in the heterozygote parents.

Similar experiments were conducted on members of the D-kindred. While VDR exon II was entirely normal in this family, a point mutation at the tip of the second zinc finger (exon III) was identified in these affected children (Figure 7); the triplet codon change of CGA to CAA results in an amino acid substitution of arginine to glutamine. This is also a conserved positional amino acid in all the steroid and thyroid receptors, and strongly suggests it imparts a crucial role in protein-DNA interaction. As before, the parents display both the normal and disease alleles. An unaffected child in this family received both normal alleles from her parents and is not a carrier of the disorder. It is unclear at present how these amino acid substitutions actually produce defective protein-DNA interaction but the effect is profound physiologically and identified readily by analysis of binding to nonspecific DNA such as calf thymus DNA-cellulose. In the G-kindred the mutation changes a neutrally charged glycine to a negative aspartic acid at a point where the protein is hypothesized to interact with negatively charged DNA (5,6). The mutation in the D-family replaces the very basic arginine (pK=12) for an uncharged glutamine. Either mutation appears equally destructive to receptor function, *in vivo*.

The vitamin D receptor mutations described above result in human disease and represent the first naturally occurring mutants identified for any of the members of the steroid/thyroid receptor superfamily of genes. As members of a class of gene *trans*-acting factors, these mutations identify critical amino acids required for interaction with DNA. Further experiments designed to delineate actual protein-DNA contact points will provide insight into the precise nature of the defective biochemical interaction. Studies evaluating mutant receptor association with upstream regulatory elements of target vitamin D genes may lead to a better understanding of how the protein recognizes DNA and leads to gene activation. We are currently evaluating patients with the HVDRR phenotype who display normal receptor binding to DNA but absent or depressed 1,25-$(OH)_2D$ binding. Mutations in this area of the receptor gene may provide clues to the configuration of the steroid binding pocket on the receptor thereby providing insight into the molecular mechanism by which hormonal ligands lead to receptor activation.

Lessons in clinical medicine clearly show that a detailed study of nature's mistakes often provides critical inroads into our understanding of normal cellular processes. Hypocalcemic vitamin resistant rickets may prove to be that important key in steroid receptorology.

REFERENCES

1. J.W. Pike, S. Marion, C.A. Donaldson, M.R. Haussler, Serum and Monoclonal Antibodies against the chick intestinal receptor for 1,25-dihydroxyvitamin D_3, J Biol Chem 258:1289 (1983).
2. J.W. Pike, Monoclonal antibodies to chick intestinal receptors for 1,25-dihydroxyvitamin D, J Biol Chem 259:1167 (1984).
3. D. McDonnell, D. Mangelsdorf, J. Pike, M. Haussler, B.W. O'Malley, Molecular cloning of complementary DNA encoding the avian receptor for vitamin D, Science 235:1214 (1987).
4. A. Baker, D. McDonnell, M. Hughes, T. Crisp, D. Mangelsdorf, M. Haussler, J. Pike, J. Shine, B. O'Malley, Cloning and expression of full length cDNA encoding human vitamin D receptor, Proc Natl Acad Sci, USA 85:3294 (1988).
5. R. Evans, The steroid and thyroid receptor superfamily, Science 240:889 (1988).
6. J. Miller, A. McLachlan, A. Klug, EMBO Journal 4:1609 (1985).
7. F. Albright, A.M. Butler, E. Bloomberg, Rickets resistant to vitamin D therapy, Am J Dis Child 54:531 (1937).
8. D. Fraser, R.B. Salter, The diagnosis and management of the various types of rickets, Pediatr Clin N Am 5:417 (1958).
9. T.M. Reade, C.R. Scriver, F.H. Glorieus, B. Nogrady, E.E. Delvin, M.F. Holick, H.F. DeLuca, Response to crystalline 1α-hydroxyvitamin D_3 in vitamin D dependency, Pediatr Res 9:593 (1975).
10. S.J. Marx, A.M. Speigel, E.M. Brown, D.G. Gardner, R.W. Pouns, Jr., M. Affie, A.J. Hamstra, H.F. DeLuca, A familial syndrome of decrease in sensitivity to 1,25-dihydroxyvitamin D, J Clin Endocrinol Metab 47:1303 (1978).
11. J.S. Adams, T.O. Wahl, W.V.Moore, W.A. Horton, B.P. Lukert, Familiar vitamin D-dependent rickets: further evidence for end organ resistance to active vitamin D metabolites. Program and Abstracts, 61st Annual Meeting of the Endocrine Society, 1979 767 (abstract).
12. M.H. Brooks, N.H. Bell, L. Love, P.H. Stern, E. Orfei, S.F. Queener, A.J. Hamstra, H.F. DeLuca, Vitamin-D-dependent rickets type II, resistance of target organs to 1,25-dihydroxyvitamin D, N Engl J Med 298:996 (1978).
13. J.E. Zerwekh, K. Glass, J.Jowsey, C.Y.C. Pak, An unique form of osteomalacia associated with end organ refractoriness to 1,25-

dihydroxyvitamin D and apparent defective synthesis of 25-hydroxyvitamin D. J. Clin Endocrinol. Metab 49:171 (1979).
14. T. Fujita, M.Nomura, S. Okajima, H. Furuya, Adult onset vitamin D-resistant osteomalacia with the unresponsiveness to parathyroid hormone, J Clin Endocrinol Metab 50:927 (1980).
15. T. Kudoh, T. Kumagai, N. Uetsuji, S. Tsugawa, K. Oyanagi, Y. Chiba, T. Minami, T. Nakao, Vitamin D dependent rickets: decreased sensitivity to 1,25-dihydroxyvitamin D, Eur J Pediatr 137:307 (1981).
16. S. Yoshikawa, T. Nakamura, Y. Nishii, Vitamin D dependent rickets with limited response to 1αOHD$_3$ and high serum 1,25-(OH)$_2$D levels--long term follow-up, in: "Vitamin D: Chemical, Biochemical and Clinical Endocrinology of Calcium Metabolism", A.W. Norman K. Schaefer, D.V. Herrath, H-G Grigoleit (eds.). Walter de Gruyter, New York (1982).
17. J.F. Rosen, A.R. Fleischman, L. Finberg, A. Hamstra, H.F. DeLuca, Rickets with alopecia: an inborn error of vitamin D. metabolism, J Pediatr 94:729 (1979).
18. U.A. Liberman, R. Samuel, A. Halabe, R. Kauli, S. Edelstein, Y. Weisman, S. Papapoulos, T.L. Clemens, L.J. Fraher, J.L.H. O'Riordan, End-organ resistance to 1,25-dihydroxycholecalciferol, Lancet I:504-(1980).
19. Y. Tsuchiya, N. Matsuo, H. Cho, M. Kumagai, M. Yasaka, T. Suda, H. Orimo, M. Shiraki, An unusual form of vitamin D-dependent rickets in a child: alopecia and marked end-organ hyposensitivity to biolgically active Vitamin D. J Clin Endocrinol Metab 51:685 (1980).
20. D. Feldman, T. Chen, C. Cone, M. Hirst, S. Shani, A. Benderli, Z. Hochberg, Vitamin D resistant rickets with alopecia: cultured skin fibroblasts exhibit defective cytoplasmic receptors and unresponsiveness to 1,25-(OH)$_2$D$_3$, J Clin Endocrinol 55:1020 (1982).
21. J.J. Sockalosky, R.A. Ulstrom, H.F. DeLuca, D.M. Brown, Vitamin D-resistant rickets: end organ unresponsiveness to 1,25-(OH)$_2$D$_3$, J Pediatr 96:701 (1980).
22. S. Beer, M. Tieder, D. Kohelet, V.A. Liberman, E. Vure, G. Bar-Joseph, D. Gabizon, V. Borochowitz, M. Varon, D. Modai, Vitamin-D resistant rickets with alopecia: a form of end organ resistance to 1,25 dihydroxy vitamin D, Clin Endocrinol 14:395 (1981).
23. J.E. Griffin, J.S. Chandler, M.R. Haussler, J.E. Zerwekh, Receptor-positive resistance to 1,25 dihydroxyvitamin D$_3$: a new cause of osteomalacia associated with impaired induction of 24-hydroxylase in fibroblasts, Clin Res 30:524A (1982).
24. T. Chen, Z. Hochberg, A. Benderli, S. Stanford, C. Cone, D. Feldman, Vitamin D-resistant rickets with alopecia: defective cytoplasmic receptors for 1,25-(OH)$_2$D$_3$ in cultured skin fibroblasts, Abstracts at the Fourth Annual Meeting of the American Society for Bone and Mineral Research, S-49 (1982).
25. U.A. Lieberman, S. Balsan, S.J. Marx, True resistance to 1,25 dihydroxyvitamin D--cellular basis and implication of a new congenital syndrome, Abstracts of the 64th Meeting of the Endocrine Society (1982).
26. C. Eil, U.A. Lieberman, J.F. Rosen, S.J. Marx, A cellular defect in hereditary vitamin D-dependent rickets type II: defective nuclear uptake if 1,25 dihydroxyvitamin D in cultured skin fibroblasts, N Engl J Med 304:1588 (1981).
27. P.J. Malloy, Z. Hochberg, J.W. Pike, D. Feldman, Vitamin D dependent rickets, type II: vitamin D receptor analysis in cultured fibroblasts from patients and parents exhibiting normal steroid binding but decreased DNA binding, Clin Endocrinol and Metab (submitted).
28. T.L. Chen, M.A. Hirst, C.M. Cone, Z. Hochberg, H-U Tietze, D. Feldman, 1,25 dihydroxyvitamin D resistance, rickets, and alopecia:

analysis of receptors and bioresponse in cultured fibroblasts from patients and parents, J Clin Endocrinol Metab 59:383 (1984).

29. Z. Hochberg, A. Benderlı, J.Levy, P. Vardi, Y Weisman, T. Chen, D. Feldman, 1,25-dihydroxyvitamin D resistance, rickets and alopecia, Am J Med 77:805 (1984).

30. M.A. Hirst, H.I. Hochman, D. Feldman, Vitamin D resistance and alopecia: a kindred with normal 1,25 dihydroxyvitamin D binding, but decreased receptor affinity for deoxyribonucleac acid, J Clin Endocrinol Metab 60:490 (1985).

31. T.L. Clements, J.S. Adams, N. Horiuchi, B.A. Cilchrest, H. Cho, Y. Tsuchiya, N. Matsuo, T. Suda, M.F. Holick, Interaction of 1,25 dihydroxyvitamin D_3 with keratinocytes and fibroblasts from skin of normal subjects and a subject with vitamin D-dependent rickets, type II: a model for study of the mode of action of 1,25-dihydroxyvitamin D_3, J. Clin Encodrinol Metab 56:824 (1983).

32. W.E. Stumpf, M. Sar, F.A. Reid, Y. Tanaka, H.F. DeLuca, Target cells for 1,25-dihydroxyvitamin D in intestinal tract, kidney, skin, pituitary and parathyroid, Science 206: 1188 (1979).

33. K. Colston, M. Hirst, D. Feldman, Organ distribution of the cytoplasmic 1,25-dihydroxyvitamin D receptor in various mouse tissues, Endocrinology 107:1916 (1980).

34. D. Feldman, T. Chen, M. Hirst, K. Colston, M. Karasek, C. Cone, Demonstration of 1,25-dihydroxyvitamin D receptors in human skin biopsies. J. Clin. Endocrinol. Metab. 51:1463 (1980).

35. D. Feldman, T.A. McCain, M.A. Hirst, T.L. Chen, K.W. Colston, Characterization of a cytoplasmic receptor-like binder for 1,25 dihydroxycholecalciferol in rat intestinal mucosa, J Biol Chem 254:10378 (1979).

36. J.S. Chandler, S.K. Chandler, J.W. Pike, M.R. Haussler, 1,25 dihydroxyvitamin D_3 induces 25-dihydroxyvitamin D_3-24-hydroxylation in a cultured monkey kidney cell line (LLC-MK_2) apparently deficient in the high affinity receptor for the hormone, J. Biol Chem 259:2214.

37. D.J. Mangelsdorf, J.W. Pike, M.R.Haussler, Avian and mammalian receptors for 1,25-dihydroxyvitamin D: In vitro translation to characterize size and hormone-dependent regulation. Proc Natl Acad Sci USA 84: 354 (1987).

38. U.A. Lieberman, C. Eil, S.J. Marx, Receptor-positive hereditary resistance to 1,25-dihydroxyvitamin D: chromatography of hormone receptor complexes on deoxyribonucleic acid-cellulose shows two classes of mutation. J Clin Endocrinol Metab (1986).

39. R. Saiki, S. Scharf, F. Faloona, K. Mullis, G.T. Hoorn, N. Arnheim, Polymerase chain reaction, Science 230:1350-1354 (1985).

DISREGULATION OF CELL CALCIUM AND CALCIUM-BINDING PROTEINS IN EXPERIMENTAL HYPERTENSION

Ramachandra M. Rao, Eric W. Young, and David A. McCarron

Division of Nephrology and Hypertension
Institute for Nutrition and Cardiovascular Research
Oregon Health Sciences University
Portland, Oregon USA

INTRODUCTION

Hypertension in experimental animal models is often associated with several distinct abnormalities in calcium metabolism from the level of the cell to the whole organism. The physiological and biochemical defects in the regulation of calcium include: low serum ionized calcium, elevated serum PTH, hypercalciuria, decreased intestinal calcium absorption, altered vitamin D metabolism, decreased calcium reabsorption, altered membrane- binding, and decreased binding to intracellular calcium-binding proteins. These derangements of systemic and intracellular calcium regulation lead to the overall calcium imbalance best reflected by reduced bone density and bone mineralization.

Most of the evidence collected to date on experimental hypertension and altered Ca^{2+} metabolism has been gathered in the Aoki Okamato strain of the spontaneously hypertensive rat (SHR). Important additional information also exists in other types of hypertensive animal models including the salt-sensitive strain of SHR, the Milan strain of SHR, the Dahl sodium chloride-dependent genetic rat, the DOCA-saline rat (mineralocorticoid/NaCl-dependent), and the renal vascular two-kidney, one-clip rat. The information accumulated on calcium disturbances in experimental hypertension is highlighted by surprising consistency in most areas, but with apparent discrepancies in others.

Like any other physiological phenomenon, the regulation of blood pressure and cellular calcium handling alters as the hypertensive animal matures. The variations in results between studies are due to the differences in the age, gender, hormonal status of the animals studied. Thus in discussing the findings, the numerous biological variables must be considered in order to provide a comprehensive view of disturbances of calcium homeostasis in hypertension.

The current view accepts the fact that the homeostasis of calcium under normal conditions is determined by the translocation of calcium among three components of the body: intestine, bone, and kidney. The intestinal absorption of calcium is regulated by $1,25(OH)_2$ vitamin D_3 although a large portion of calcium absorption is either vitamin D-independent or passive uptake. The primary stimuli of $1,25(OH)_2D_3$ synthesis and release from the kidney are PTH, phosphate depletion, and calcium depletion. Bone calcium deposition is largely a mass action process while bone calcium resorption is controlled by PTH levels. Renal calcium excretion is a function of filtered load (determined by glomerular filtration rate (GFR) and serum ultrafilterable calcium) and tubular reabsorption which will be activated by PTH. A defect in subcellular Ca^{2+} metabolism that would affect Ca^{2+} transport in each of the cell or organ systems would account for the relationship of increased arterial pressure to disturbed Ca^{2+} homeostasis. Ultimately this body of research has important implications for clinical medicine as hypertension in humans has comparable defects in Ca^{2+} metabolism associated with it. The established observation that provision of additional dietary calcium will lower blood pressure in humans provides compelling evidence that the organism's ability to properly regulate cellular Ca^{2+} handling may be a critical determinant of whether systemic blood pressure is normally controlled.

SYSTEMIC CALCIUM BALANCE

PTH and Ca^{2+}

Several lines of evidence demonstrate that a number of defects in biochemical and metabolic processes are contributing factors to the induction of a major disturbance in the whole animal calcium homeostasis, leading to hypertension. The circulating levels of PTH are elevated in the male SHR as early as 6 weeks of age[1] measured by carboxy-terminal PTH assay while all WKY of the same age had undetectable carboxy-terminal PTH. In the same study, serum calcium was significantly lower in 10-week old male SHR than in WKY. Similar observations reported by McCarron et al.[2] showed that amino-terminal PTH was significantly elevated in male SHR at 18, 24, and 29 weeks of age. A c-terminal assay also revealed higher serum PTH in the SHR at 29 weeks. Again, the serum ionized calcium was significantly lower in SHR at these ages. A mid-molecule PTH assay failed to find a difference between SHR and WKY, possibly due to the lesser sensitivity of this assay[3]. Recently, the SHR has been shown to have an increase in the parathyroid gland mass at 11 weeks of age[4] consistent with its enhanced PTH production. A reduction in the clearance of immunoreactive PTH fragments is unlikely mostly because the GFR is consistently normal in SHR[5,6]. Serum PTH levels are elevated in the SHR, in spite of a decrease in serum ionized calcium and a reduction in serum phosphorus with normal serum total calcium levels[2].

$1,25(OH)_2$ Vitamin D_3

Inappropriately low circulating levels of $1,25(OH)_2D_3$ have recently been reported in the adolescent SHR in the setting of elevated PTH and low serum ionized calcium and phosphorus concentrations. This constellation of findings suggests a fundamental defect in the regulation of $1,25(OH)_2D_3$ metabolism in the SHR[7,9]. In contrast, in young male SHR the serum levels of $1,25(OH)_2D_3$ are increased[10,11] or equal to that of the WKY[1,9,12,13]. Serum 25(OH) vitamin D_3 tends to be elevated in the SHR, indicating that precursor availability is not compromised, but that metabolic conversion is impaired[12-14]. Additional evidence for the altered vitamin D metabolism in SHR is demonstrated by the subnormal increments in serum 1,25(OH) D_3 in response to PTH infusion[15], cyclic AMP[9], and phosphate depletion[8]. The low basal and PTH-stimulated serum $1,25(OH)_2D_3$ in SHR is explainable entirely by reduced production rate because metabolic clearance of

$1,25(OH)_2D_3$ is not different between SHR and WKY[3]. The defective mechanism for abnormal $1,25(OH)_2D_3$ production is uncertain, but intrinsic enzyme activity from isolated mitochondria appears to be normal or even enhanced[9]. However, an increased intracellular or mitochondrial concentration of calcium, inorganic phosphate, or hydrogen ion concentrations could possibly suppress the activity of renal α hydroxylase. There is some evidence for systemic acidosis[16] and phosphate retention[10,17] in the SHR, although intracellular ionic status has not been determined apart from preliminary reports indicating unchanged or low level status in renal proximal tubules[18,19]. In addition, the fact that the PTH infusion could induce only subnormal increments of $1,25(OH)_2D_3$ from the kidney strongly indicates a defect in end-organ responsiveness to that calcium-regulating hormone in the SHR.

Renal Calcium Excretion

McCarron et al.[2] reported significantly increased urinary calcium excretion in 17-, 22-, 28-, and 43-week old male SHR with simultaneous decreases in serum ionized calcium and elevated serum PTH. Hypercalciuria under these conditions was interpreted to reflect a renal calcium leak. Data in young SHR is conflicting: 24-hour urinary calcium has been reported as increased[5,20,21], decreased[10,11], and not different[2,5,20,22]. This indicates that the kidneys of adolescent and mature SHR are unable to normally reabsorb calcium in spite of lower filtered calcium load and higher PTH, which stimulates calcium reabsorption. Impaired renal tubular calcium reabsorption may arise from the suppressed cyclic AMP generation[23], although alternative mechanistic explanations must be considered. In the female SHR, higher 24-hour calcium excretion was observed at 25 weeks and 1 year of age[5]. Parathyroidectomy augmented hypercalciuria in the 25-week old female SHR[5]. After fasting, urinary calcium excretion was decreased in 23-week old females and in 8- to 14-week old males[5,6]. Based on the results on low, fasting calcium excretion and normal urinary cyclic AMP (PTH was not measured), the hypercalciuria was interpreted as a result of increased intestinal calcium absorption[5], not to diminished renal tubular reabsorption. This distinction between absorptive and renal hypercalciuria in the SHR remains unclear because of sex and age differences in the animals used in the various studies.

Hypercalciuria also develops in the uninephrectomized, DOC-treated rat model in response to sodium chloride loading but not sodium bicarbonate[24]. In this model, hypercalciuria precedes the onset of hypertension. Also, calcium excretion is increased in the Milan rat strain of spontaneous hypertension[25]. This is evidence that hypercalciuria is common to several types of experimental hypertension. Which appears to be based on the renal abnormality. The hypercalciuria may be secondary to a suppressed renal adenylate cyclase response to PTH as demonstrated in the Dahl-sensitive and DOC-NaCl hypertensive rats[26]. The Milan hypertensive rat strain possess a decreased Ca-ATPase activity[27]. Each of these subcellular abnormalities could contribute to the development of a renal calcium leak.

Intestinal Ca^{2+} Transport

The absorption of calcium by the intestine has been a subject of substantial controversy. Various techniques that have been utilized to study calcium handling by the duodenum have yielded conflicting results. Using *in vitro* everted duodenal sac preparations different investigators have reported varied findings, including decreased absorption in 5- and 12-week old male SHR[12], increased absorption in 12-week old male SHR[28], and no difference in absorption in 5- and 10-week old male SHR[1,28]. Studies utilizing only male SHR and intact duodenal tissue under controlled electrical conditions (Ussing chamber) have consistently produced results indicating a significant reduction in active intestine calcium transport (net) in 12-[7,29] and 24-week old rats fed a normal calcium diet[29].

Isolated duodenal enterocytes provide still another technique for the analysis of intestinal calcium transport. Influx of radiolabeled Ca^{2+} was decreased in enterocytes isolated from 12- to 14- and 28- to 32-week old male SHR as compared with aged matched WKY. Also, $^{45}Ca^{2+}$ efflux was decreased in enterocytes from 12- to 14-week old SHR[30]. These results indicate that transcellular intestinal Ca2+ transport is abnormal in adolescent as well as mature male SHR.

In vitro and segmental in vivo techniques cannot fully explain the differences in the transcellular and segmental intestinal calcium absorption, and measurements of whole animal calcium absorption have therefore been attempted. Balance studies have shown an increase in absorption in 4-week old SHR[11,31], but by 12 weeks of age calcium absorption is reduced in SHR compared to WKY[31]. As a whole, the data from several labs using different diets indicate that male SHR may hyperabsorb calcium when they were young, but malabsorb calcium as they mature and develop hypertension.

Bone Mineral Metabolism

Bone mineralization provides an accurate index of whole animal calcium balance. In adult male SHR, bone calcium content is significantly decreased at 23[7] and 54[32] weeks of age. At earlier ages SHR show no difference compared with WKY[6,10]. Similarly, 26 weeks old male SHR had decreased cortical thickness, ash weight/volume and ash weight/dry weight of the femoral bone[33]. In contrast, a study of hydralazine-treated female SHR showed increased fractional bone mass and calcium density at one year of age [5]. These results in male SHR support a disregulation of whole animal calcium balance in growing and mature animals.

Inadequate calcium retention at the level of the whole animal or within organ systems of several animal models must ultimately be the result of one or more alterations in subcellular calcium metabolism of the tissues involved. How such a putative defect(s) relates to an increase in arterial pressure is an area of active research investigation. The focus of that effort has been on the vascular smooth muscle because it is the cell system most directly involved in the pathogenesis of hypertension. The involvement of vascular smooth muscle may be secondary to calcium depletion or another primary manifestation of a subcellular defect in calcium metabolism.

SUBCELLULAR CA^{2+} METABOLISM

Intracellular free calcium levels contribute to the regulation of a number of biochemical processes in the vascular smooth muscle cells involved in both contraction and relaxation. Based upon the theoretical construct of Blaustein[34], it had been anticipated that with the advent of new techniques to measure intracellular free calcium in living cells that cytoplasmic free Ca^{2+} would be shown to be elevated. Such a finding would have been consistent with measurements of intracellular calcium in other non-vascular cells isolated from both hypertensive humans and animals.

Intracellular Free Calcium Levels

Several investigators have reported an increase in the intracellular free Ca^{2+} levels in platelets of SHR and hypertensive patients compared to controls[35-39]. Erne et al.[35], and Le Quan Sang and Devynck[38] observed a positive relationship between blood pressure and platelet intracellular free calcium concentration. Furthermore, Erne et al.[35] observed a fall in intracellular calcium levels upon the lowering of blood pressure with antihypertensive therapy. Baba et al.[40] also observed an increase in the free calcium levels in platelets of 4-, 11-, and 28-week old SHR, but no significant difference was observed in other types of experimental animals models such as the DOCA salt-sensitive and two-kidney one-clip rat models. Studies on cytoplasmic calcium levels in renal tubules demonstrated either a decrease or no difference in the SHR compared to the WKY[18,19]. Cytoplasmic free calcium levels have been reported to be elevated in red blood cells[41] and lymphocytes[42] isolated from the SHR.

Examination of vascular smooth muscle cells have not revealed the anticipated increase in cytoplasmic free Ca^{2+} levels. The studies by Bukoski et al.[43] of vascular smooth muscle cells isolated from the SHR failed to identify any differences in the basal intracellular free calcium levels when compared to the WKY. In isolated mesenteric resistance vessels of the SHR and the WKY, calcium transients were lower in response to agonists in the former (Bukoski RD, unpublished observations). These findings are precisely the opposite of the previously hypothesized increase in the intracellular free calcium levels in vascular tissue proposed by Blaustein et al.[34]. The reports from the other investigators showed an increase in free calcium levels in both primary and passaged vascular smooth muscle cells of 12-week old SHR[44] and no differences in monolayers of vascular smooth muscle cells obtained from SHR and WKY[45].

The observed increases in the cytoplasmic calcium concentrations in some types of cells and tissues derived from hypertensive animals and humans must reflect one or a combination of the following: 1) increased influx or diminished efflux of calcium, 2) altered binding of Ca^{2+} to calcium-binding proteins, or 3) changes in sequestration of calcium by intracellular organelles. Several of the reported abnormalities in calcium-binding proteins and plasma membrane-bound enzymes of tissue derived from hypertensive animals might be contributing to the altered calcium fluxes; deranging several physiological and biochemical events in the cells of hypertensive animals. To the extent that these abnormalities of calcium binding may produce a defect in active Ca^{2+} transport in the intestine and/or kidney, the ability to retain calcium would be impaired, accounting for the evidence of Ca^{2+} depletion in both human and experimental hypertension.

Calcium Fluxes

Calcium fluxes and membrane binding have been studied in vascular smooth muscle cells, enterocytes and red blood cells of SHR. Studies of vascular smooth muscle isolated from SHR which have examined $^{45}Ca^{2+}$ influx into the whole tissue have shown an increase[46], while no differences and even a decrease in influx have been reported by other investigators[47,48]. Studies done on conduit arteries revealed that unidirectional influx of $^{45}Ca^{2+}$ into mesenteric resistance vessels of SHR is enhanced[49,50]. These results support the finding of Mulvany and Nyborg[51] that resistance arteries of mesenteric have an increased sensitivity to extracellular calcium and also similar to the results of Kozniewska[52] reporting enhanced reactivity to flunazinine in the cerebro-vascular bed of the SHR.

In contrast, calcium fluxes in the duodenum seems quite opposite to the trend observed in the vascular smooth muscle. Isolated enterocytes exhibited not only lower influx rate, but diminished intracellular calcium pools (both bound and free). Calcium efflux rate constants are also decreased in SHR enterocytes, perhaps as a functional response to the decrease in the calcium influx rate[53]. This pattern of altered calcium fluxes in different tissues suggests a functional and/or structural change in altered membrane-bound calcium translocating enzymes and/or calcium-binding proteins in the SHR.

Studies of Ca2+ transport by isolated membrane fractions of vascular smooth muscle have been consistent in demonstrating a depressed ability of smooth muscle cell membrane to actively transport calcium. When ATP-dependent calcium uptake was measured in microsomal fractions of aorta isolated from SHR[46,54,55] or cell membrane enriched fractions of mesenteric arteries of the SHR[56,57], uptake was attenuated compared with fractions isolated from the normotensive controls. It is possible that this decrease in uptake is a reflection of an abnormal cell membrane Ca-ATPase system. Earlier studies demonstrated that the Ca Mg,ATPase activity and calcium-binding ability of subcellular membranes of arterial smooth muscle was decreased in the SHR[54]. Based on the fact that the calcium uptake by subcellular vesicles isolated from vascular smooth muscle is decreased in the SHR, it is reasonable to assume that this reflects an abnormality of calcium pump.

Calcium Binding Proteins

Preliminary studies from our laboratory suggest that two proteins with molecular weights of 110 Kd and 17 Kd were expressed significantly less in the membrane isolate of vascular smooth muscle of the SHR compared to WKY (Rau, unpublished observations). Based on their molecular weights, the 110 Kd protein might reflect a reduction in the Ca-Mg ATPase while the 17 Kd may reflect a reduction in calmodulin in the arterial smooth muscle of SHR. Higaki et al.[58] also reported that calmodulin levels were decreased in aorta, kidney, heart, and brain tissues of SHR. But when they studied, DOCA-salt sensitive rat, only brain calmodulin levels were decreased without any changes in heart, aorta and kidney levels when compared to Wistar-Kyoto normals.

Baba et al.[40] reported decreased calmodulin levels in the mesenteric arteries of SHR in comparison to WKY. In the DOCA-salt sensitive hypertensive rat there was no observed difference in the mesenteric artery calmodulin levels. Pokudin et al.[59] isolated calmodulin from the brain of SHR and WKY and were unable to find any differences in the quantity of this calcium-binding protein or its ability to bind calcium. But Huang et al.[60] reported that a soluble activator of calmodulin activity is present in the heart, kidney and smooth muscle of the SHR.

These defects in the calcium-binding proteins are not limited to vascular tissue. Other cells and organs critical to calcium homeostasis of hypertensive animals have been reported to be involved. In the duodenal cells isolated from the SHR, the vitamin D-dependent calcium-binding protein content is decreased[31] with a simultaneous drop in the renal calcium-binding protein. Kowarski et al.[62] reported that the integral membrane calcium-binding protein (IMCAL) was significantly decreased in the intestinal mucosal cells isolated from the SHR besides in heart, kidney, testis compared to WKY.

Sokabe and co-workers[63] found that in the quasi-calmodulin gene there was a deletion of six nucleotides inside the coding region which corresponds to ASN-97 and GLY-98 in the calcium-binding sub-domain III of normal calmodulin. Northern blot analysis showed that the quasi-calmodulin gene is expressed in the SHR. Alteration of one of the functional calcium-binding sub-domains that would be produced by deletion of two amino acids could result in abnormal calcium binding and sequestration in the brain of the SHR or model with a well established central neural component to its increased arterial pressure.

CONCLUSION

The evidence for abnormal cellular calcium metabolism, including transport, membrane binding, and sequestration in a variety of tissues suggests that a fundamental defect in the biochemical and functional properties of some intracellular and/or membrane-bound calcium-binding proteins exists in hypertensive animals. These abnormalities of cell calcium-binding protein metabolism would provide a pathophysiologic basis for disregulation of normal calcium homeostasis at the systemic level in experimental hypertension. Whether these abnormalities are responsible for altered arterial pressure regulation remains to be determined; if they are responsible for the increase in vascular resistance that underlies hypertension, the precise cellular mechanisms are as yet unknown. Future research elucidating the alterations - whether at primary, secondary, or tertiary levels of protein configuration - will be critical to our understanding the etiology of altered cellular and organ calcium metabolism and how it affects the control of arterial pressure.

REFERENCES

1. Stern N, Lee DBN, Silis V., et al. Effects of high calcium intake on blood pressure and calcium metabolism in young SHR. Hypertension 1984;6:639-646.
2. McCarron D, Yung NN, Ugoretz BA, Krutzik S. Disturbances of calcium metabolism in the spontaneously hypertensive rat: attenuation of hypertension by calcium supplementation. Hypertension 1981;3(Suppl I:I162-I167.
3. Young EW, Hsu CH, Patel S, et al. Metabolic degradation and synthesis of calcitriol in the spontaneously hypertensive rat. Am J Physiol 1987;252:E778-E782.
4. Merke J, Slotkowski A, Mann H, et al. Abnormal $1,25(OH)_2D_3$ receptor status in genetically hypertensive rats. Kidney Int 1987;31:303.
5. Lau K, Chen S, Eby B. Evidence for an intestinal mechanism in hypercalciuria of the spontaneously hypertensive rat. Am J Physiol 1984;247:E625-E633.
6. Hsu CH, Chen PS, Smith DE, Yang CS. Pathogenesis of hypercalciuria in spontaneously hypertensive rats. Miner Electrolyte Metab 1986;12:130-141.
7. Lucas PA, Brown RC, Drueke T, Lacour B, Metz JA, McCarron DA. Abnormal vitamin D metabolism, intestinal calcium transport, and bone status in the spontaneously hypertensive rat compared with

its genetic control. J Clin Invest 1986;78:221-227.
8. Kurtz TW, Portale AA, Morris RC. Evidence for a difference in vitamin D metabolism between spontaneously hypertensive and Wistar-Kyoto rats. Hypertension 1986;8:1015-1020.
9. Kawashima H. Altered vitamin D metabolism in the kidney of the spontaneously hypertensive rat. Biochem J 1986;237:893-897.
10. Bindels RJM, van den Brock LAM, Jongen MJM, et al. Increased plasma calcitonin levels in young spontaneously hypertensive rats: role in disturbed phosphate homeostasis. Pflugers Arch 1987;408:395-400.
11. Lau K, Langman CB, Gafter U, et al. Increased calcium absorption in prehypertensive spontaneously hypertensive rats. J Clin Invest 1986;78:1083-1090.
12. Schedl HP, Miller DL, Paper JM, et al. Calcium and sodium transport and vitamin D metabolism in the spontaneously hypertensive rat. J Clin Invest 1984;73:980-986.
13. Schedl HP, Miller DL, Horst RL, et al. Intestinal calcium transport in the spontaneously hypertensive rat: response to calcium depletion. Am J Physiol 1986;250:G412-G419.
14. Hsu CH, Yang CS, Patel SR, Stevens MG. Calcium and vitamin D metabolism in spontaneously hypertensive rats. Am J Physiol 1987;253:F712-F718.
15. Young EW, Patel SR, Hsu CH. Plasma $1,25(OH)_2$ vitamin D_3 response to parathyroid hormone, cyclic AMP, and phosphorus depletion in the spontaneously hypertensive rat. J Lab Clin Med 1986;6:562-566.
16. Lucas PA, Lacour B, McCarron DA, Drueke T. Disturbances of acid-base balance in the young spontaneously hypertensive rat. Clin Sci 1987;73:211-215.
17. Eby B, Salvi D, Lau K. Pathophysiology and consequence of reduced PO_4 excretion in the spontaneously hypertensive rat. Proc First Annu Mtg Soc Hypertens 1986;29A.
18. Jacobs WR, Brazy PC, Mandel LJ. Fura-2 measurements of intracellular free calcium (Ca_f) in renal cortical tubules from SHR and WKY rats. Kidney Int 1987;31:350.
19. Llibre J, LaPointe M, Batlle DC. Fura-2 measurements at cytosolic cell Ca^{2+} in renal proximal tubules and circulating lymphocytes of rats with genetic hypertension. Kidney Int 1987;31:302.
20. Ayachi S. Increased dietary calcium lowers blood pressure in the spontaneously hypertensive rat. Metabolism 1979;28:1234-1238.
21. Hsu CH, Chen PS, Caldwell RM. Renal phosphate excretion in spontaneously hypertensive and Wistar Kyoto rats. Kidney Int 1984;25:789-795.
22. Hsu CH, Patel S, Young EW. Calcemic response to parathyroid hormone in spontaneously hypertensive rats: role of calcitriol. J Lab Clin Med 1987;110:682-689.
23. McCarron DA. Impaired nephrogenous cAMP response in the spontaneously hypertensive rat. Kidney Int 1983;23:106.
24. Kurtz TW, Morris RC. Dietary chloride as a determinant of disordered calcium metabolism in salt-dependent hypertension. Life Sci 1985;36: 921-929.
25. Cirillo M, Galletti F, Corrado MF, Strazzulo P. Disturbances of renal and erythrocyte calcium handling in the Milan hypertensive strain. J Hypertens 1986;4:443-449.
26. Umemura S, Smythe DD, Pettinger WA. Renal adenylate cyclase in Dahl and DOC-Na hypertensive rats: defective response to parathyroid hormone with calcium leak. J Hypertens 1986;4:S291-S293.
27. Bianchi G, Ferrari P, Salvati P, et al. A renal abnormality in the Milan hypertensive of rats and humans predisposed to essential hypertension. J Hypertens 1986;4:533-536.

28. Toraason MA, Wright GL. Transport of calcium by duodenum of spontaneously hypertensive rat. Am J Physiol 1981;241:G344-G347.
29. McCarron DA, Lucas PA, Shneidman RS, Drueke T. Blood pressure development of the spontaneously hypertensive rat following concurrent manipulation of the dietary Ca^{2+} and Na^{+}: relation to intestinal Ca^{2+} fluxes. J Clin Invest 1985;76:1147-1154.
30. McCarron DA, Lucas P, Lacour B, Drueke T. Ca^{2+} efflux rate constant in isolated SHR enterocytes. Kidney Int 1986;29:252.
31. Drueke T, Lucas PA, Bourgouin P, et al. Changes in calcitriol status and related parameters in the young hypertensive rat (SHR). Kidney Int 1988;33:294.
32. Metz JA, Karanja N, McCarron DA. Characterization of bone calcium, magnesium, and density in the spontaneously hypertensive rat: differential effects of dietary calcium and sodium. (submitted).
33. Izawa Y, Sagara K, Kadota T, Makita T. Bone disorders in spontaneously hypertensive rat. Calcif Tissue Int 1985;37:605-607.
34. Blaustein MP. Sodium ions, calcium ions, blood pressure regulation and hypertension: a reassessment and a hypothesis. Am J Physiol 1977;232: C165-C173.
35. Erne P, Bolli P, Burgisser E, Buhler FR. Correction of platelet calcium with blood pressure: effect of antihypertensive therapy. N Engl J Med 1984;310, 1084-1088.
36. Brushi G, Brushi ME, Caroppo M, Orlandini G, Spaggiari M, Cavatorta A. Cytoplasmic free $[Ca^{2+}]$ is increased in the platelets of spontaneously hypertensive rats and essential hypertensive patients. Clin Sci 1985; 68:179-184.
37. Le Quan Sang KH, Montenay-Garestier T, Devynck MA. Platelet cytosolic free calcium concentration in essential hypertension. Nouv Rev Fr Hematol 1985;27:279-283.
38. Le Quan Sang KH, Devynck MA. Increased platelet cytosolic free calcium concentration in essential hypertension. J Hypertens 1986;4:567-574.
39. Lechi A, Lechi C, Bonadonna G, et al. Increased basal and thrombin-induced free calcium in platelets of essential hypertensive patients. Hypertension 1987;9:230-235.
40. Baba A, Fukuda K, Kuchii M, et al. Intracellular free calcium concentration, Ca^{++} channel and calmodulin level in experimental hypertension in rats. Jpn Circ J 1987;51:1216-1222.
41. Larsen FL, Katz S, Roufogallis BD, Brooks DE. Physiologic shear stresses enhance the Ca^{2+} permeability of human erythrocytes. Nature (London) 1981;294:667-668.
42. Brushi G, Brushi ME, Caroppo M, Orlandini G, Pavarani C, Cavatorta A. Intracellular free $[Ca^{2+}]$ in circulating lymphocytes of spontaneously hypertensive rats. Life Sci 1984;35:535-542.
43. Bukoski RD, Pressley MS, McCarron DA. Intracellular Ca^{2+} ($[Ca]_i$) measured in single aortic myocytes from spontaneously hypertensive (SH) and normotensive Wistar Kyoto (WK) rats using fura-2. Am J Hypertens (in press).
44. Sugiyama T, Yoshizumi M, Takaku F, et al. The elevation of the cytoplasmic calcium ions in vascular smooth muscle cells in SHR. Measurement of the free calcium ions in single living cells by laser microfluorospectrometry. Biochem Biophys Res Comm 1986;141:340-345.
45. Nabika T, Velletri PA, Beaven MA, Endo J, Lovenberg W. Vasopressin-induced calcium increases in smooth muscle cells from spontaneously hypertensive rats. Life Sci 1985;37:579-584.
46. Bhalla RC, Webb RC, Ashley T, Brock T. Calcium fluxes, calcium binding and adenosine 3',5'-monophosphate dependent protein kinase activity in aorta of spontaneously hypertensive and Wistar

normotensive rats. Mol Pharmacol 1978;14:468-477.
47. Shibata S, Kochii M, Taniguchi T. Calcium fluxes and binding in the aortic smooth muscle from the spontaneously hypertensive rat. Blood Vess 1975;12:279-289.
48. Zsoter TT, Wetchinsky C, Henein NF, Ho LC. Calcium kinetics of the aorta of spontaneously hypertensive rat. Cardiovasc Res 1977;11:353-357.
49. Cauvin C, van Breemen C. Altered ^{45}Ca fluxes in isolated mesenteric resistance vessels from SHR. Fed Proc 1985;44:1008.
50. Cauvin C, Hwang BS, Yamamoto M, van Breemen C. Effects of dihydropyuridines on tension and calcium-45 influx in isolated mesenteric resistance vessels from spontaneously hypertensive and normotensive rats. Am J Cardiol 1987;59:116B-122B.
51. Mulvany MJ, Nyborg N. An increased calcium sensitivity of mesenteric resistance vessels in young and adult spontaneously hypertensive rats. Br J Pharmacol 1980;71:585-596.
52. Kozniewska E. Enhanced reactivity towards flunazinine in cerebro-vascular bed of spontaneously hypertensive rats. Experientia 1988;44:221-222.
53. Lacour B, Roullet CM, Lucas PA, McCarron DA, Drueke T. Impaired calcium efflux in enterocytes of spontaneously hypertensive rat (SHR). Kidney Int 1988;33:300.
54. Aoki K, Yamashita Y, Tomita N, Tazumi K, Hotta K. ATPase activity and Ca^{2+} binding ability of subcellular membrane of arterial smooth muscle in spontaneously hypertensive rat. Jpn Heart J 1974;15:180-181.
55. Webb RC, Bhalla RC. Altered calcium sequestration by subcellular fractions of vascular smooth muscle from spontaneously hypertensive rats. J Mol Cell Cardiol 1976;8:651-661.
56. Kwan CY, Belbeck L, Daniel EE. Abnormal biochemistry of vascular smooth muscle plasma membrane isolated from hypertensive rats. Mol Pharmacol 1980;77:137-140.
57. Kwan CY, Daniel EE. Arterial muscle abnormalities of hydralazine treated spontaneously hypertensive rats. Eur J Pharmacol 1982;82:1878-1890.
58. Higaki J, Ogihara T, Kumahara Y, Bravo EL. Calmodulin levels in hypertensive rats. Clin Sci 1985;68:407-410.
59. Pokudin NI, Orlov SN, Ryashsky GG, Menshikov NY, Tkachuk VA, Postnov YV. Isolation and characteristics of calmodulin from the brain of rats with spontaneous genetic hypertension. Kardiologiya 1985;25:72-77.
60. Huang SL, Wen YI, Kripranycz DB, et al. Abnormality of calmodulin activity in hypertension. Evidence of the presence of an activator. J Clin Invest (in press).
61. Kowarski S, Cowen LA, Schachter D. Decreased content of integral membrane calcium-binding protein (IMCAL) in tissues of the spontaneously hypertensive rat. Proc Natl Acad Sci USA 1986;83:1097-1100.
62. Nojima H, Kishi K, Sokabe H. Organization of calmodulin genes in the spontaneously hypertensive rat. J Hypertens 1986;4(Suppl 3):S275-S277.

A POSSIBLE MECHANISM FOR INCREMENTS OF CALCIUM AND PROTEIN KINASE C ACTIVITY IN KIDNEY OF EXPERIMENTAL ACUTE FLUOROSIS

Yasunobu Suketa and Yoshio Sakai

Department of Environmental Biochemistry and Toxicology
University of Shizuoka School of Pharmaceutical Sciences
Shizuoka, Shizuoka 422, Japan

INTRODUCTION

Chronic fluoride toxicity has studied in livestock and dental fluorosis. Experimental-induced chronic toxicity in rodents is also associated with nephrotoxicity. In human being, acute toxicity usually occurs as a result of accidental or suicidal ingestion of fluoride, and it results in gastrointestinal effects, severe hypercalcemia, hyperglycemia, nephrotoxicity, and shock (Shupe et al., 1984; Taylor et al., 1961). On the other hand, we found that a single large dose of fluoride (NaF 35 mg/kg, ip) to rats markedly increased the calcium content of the kidney (Suketa et al., 1977, 1986; Suketa and Kanamoto, 1983). Moreover, we demonstrated fluoride-induced hyperglycemia to be due to elevation of cAMP levels in serum, liver and kidney by the fluoride dose (Suketa and Sato, 1980; Suketa et al., 1980, 1982, 1985, 1986).

It was recently reported that sodium fluoride stimulates calcium mobilization in human neutrophils (Strnard and Wong, 1985). Guillon et al. (1986) was found that phosphatidyl inositol-4,5-bisphosphate (PIP_2) phospholipase C activity in WRK-1 cell membranes is activated by fluoride. The GTP-binding protein is known to be involved in the functional coupling of several receptors to PIP_2-phospholipase C (Okajima and Ui, 1984). Sternwise and Gilman (1982) was described that G_s is activated by fluoride ion (in vitro) and that very small amount of aluminum is necessary for the activation. To understand the possible mechanism for the changes in renal calcium mobilization induced by the fluoride dose (NaF 35 mg/kg, ip) is essential to seek for more appropriate medical care for the acute fluorosis. Thus, changes in renal protein kinase C activity and calcium mobilization by fluoride dose were examined concerning with the related regulatory system.

MATERIALS AND METHODS

Experimental animals were male Wistar albino rats, weighing 95-105 g and housed in an air-conditioned room at 22°C. They had access to basal diet (MF: Oriental Yeast Co., Tokyo, Japan) and water ad libitum for 1 week. All animals were fasted for 24 hr before the experiments to minimize the effects of electrolyte absorption from the bowels and stabilize the urinary excretion of electrolyte. The rats were killed at various times after a single ip injection of NaF (35 mg/kg) or NaCl (48.7 mg/kg).

Membrane Preparations. The plasma membrane fraction was prepared according to the methods of Forte et al., (1973) and Guillon et al., (1986a). Kidney slices were homogenized at 4°C in a medium composed of 10 mM LiCl, 0.5 mM EDTA and 10 mM Tris-HCl by 35-40 strokes of Dounce tight-pstle homogenizer. After a 5-min centrifugation (0°C) at 200 x g, the supernatant was subgected to a second centrifugation (600 x g at 0°C for 12 min). The pellet was washed with the homogenization buffer and was subjected to a centrifugation (38000 x g at 0°C, for 15 min). The final pellet was resuspended in the homogenization buffer and used as plasma membrane fraction. The renal cortical brush border membranes and basolateral membranes were prepared by magnesium precipitation method (Molitoris and Simon, 1985).

Preparations of Microsomes and Cytosol. The kidney slices were immediately homogenized at 4°C in a Potter-Elvehjem Teflon-glass homogenizer with 5 ml of ice-cold 0.25 M sucrose-0.03 M histidine buffer (pH 7.5) according to the method of Jørgensen (1968). The cytosol fraction was obtained by centrifugation (105000 x g for 1 hr) of the homogenate at 0°C.

PIP_2-Phospholipase C Activity Assay. The PIP_2-phospholipase C activity was assayed by incubation of plasma membranes in a medium composed of 10 mM LiCl, 0.1 mM ATP, 10 mM Tris-HCl (pH 8) and [inositol-2-^3H(N)]-phosphatidyl inositol-4,5-bisphosphate. The reaction was stopped by the addition of 500 µl TCA (15%) and 100 µl bovine serum albumin (20 mg/ml) and cooling the tube to 0°G. The supernatant was collected by centrifugation (2000 x g for 10 min). TCA in the supernatant was eliminated by extraction of diethylether. The water layer was adjusted to pH 7.0 by adding 1N NaOH. The extract was diluted 10 times with water. Labeled IP, IP_2 and IP_3 containing in the diluted extract were separated by chromatography on Dowex 1 x 10 (100-200 mesh, formate form) columns (1 x 2 cm) as described by Berridge et al., (1983).

Assays of Protein Kinase C Activity and Cyclic AMP. Protein kinase C activity (Mazurek et al., 1987) was carried out in a final volume of 250 µl containing 50 µg of histone (type V-S), 20 mM Tris-HCl at pH 7.4, 10 mM $Mg(NO_3)_2$, 100 µg of L-α-phosphatidyl-L-serine, 10 µg of diolein, 24 µM ATP (0.5 Ci of $\gamma^{32}P$-ATP), 300 µM $CaCl_2$ and 25 µl of cytosol fraction (protein 1 mg/ml). The enzyme was incubated at 37°C for 15 min. Cyclic AMP concentration was assayed in kidney and serum according to the radioimmunoassay method of Honma et al., (1977).

GTPγS-Binding Assay. Binding of [γ-^{35}S]GTP to the plasma membranes and cytosol was assessed with a rapid filtration technique using membrane filter (pore size, 0.45 µm) by the method of Northup et al., (1982). Samples (1 mg protein/ml) of 100 µl in solution containing 0.1% lubrol 12A9 were diluted with 2 ml of 20 mM Tris-HCl (pH 8), 100 mM NaCl, and 25 mM $MgCl_2$ (buffer A) at 4°C. Samples were then applied to a 25 mm filter which had been rinsed with 2 ml of buffer A, and the filter bound sample was rapidly washed 3 times with 2-ml volumes of buffer A.

Determinations of Calcium and Protein. Calcium estimation was carried out by the method of Willis (1960) using a Hitachi model 518 digital atomic absorption spectrophotometer. Protein was determined by the method of Lowry et al., (1951).

Materials. ^{35}S-GTPγS and ^3H-phosphatidylinositol-4,5-bisphosphate were purchased from New England Nuclear (Boston, M.A). $\gamma^{32}P$-ATP was obtained from Hungarian Academy of Sciences (Budapest, Hungary). L-α-Phosphatidyl-L-serine and phosphatidylinositol-4,5-bisphosphate were from Sigma Chemical Co., (St. Louis, M.O.).

Statistical significance of the data (mean standard error) was analyzed by Student's t-test.

RESULTS

Effect of Fluoride Administration on Calcium Mobilization in Kidney

The response of renal calcium content was examined 24 hr after administration of various doses of fluoride (NaF 10-35 mg/kg, ip) to rats (Fig. 1A). The elevation of renal calcium content with time after the injection of a single dose of fluoride (NaF 35 mg/kg, ip) was then examined (Fig. 1B). Renal calcium content reached maximum 24 hr after the administration of fluoride. At that time, changes in subcellular calcium contents in kidney by fluoride administration were presented in Table 1. The elevations of

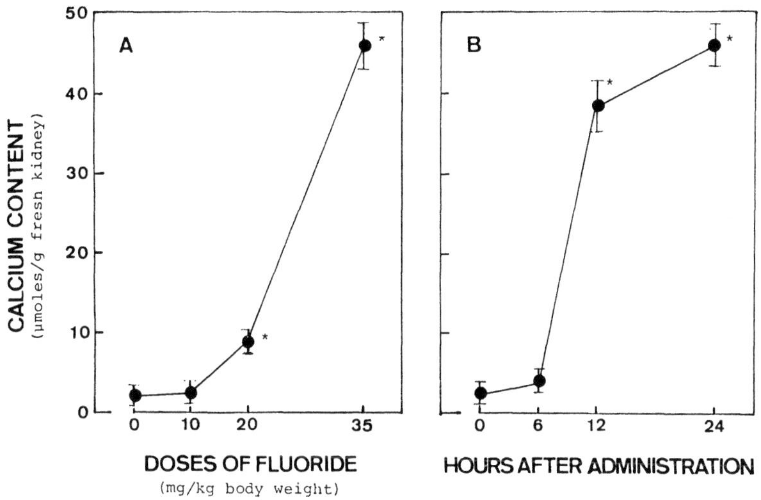

Fig. 1. Response of renal calcium content to fluoride dose (A) and changes in the calcium content after the fluoride dose (B). The rats were killed 24 hr after various fluoride doses as shown in Fig. 1A. Changes in the calcium content were determined at various times after a single fluoride injection (NaF 35 mg/kg, ip). Values are averages obtained from four rats. Standard errors are indicated vy vertical lines. Asterisks indicate significant difference from control ($p<0.02$).

Table 1. Changes in Subcellular Calcium Contents in Kidney by Fluoride Administration

subcellular fraction	Control (C)	Fluoride-administered (F)	F/C
		μmoles/g fresh weight	
Whole cells	0.97±0.06	36.26±2.53	37.4[a]
Nuclei and cell debris	0.27±0.02	23.99±0.25	88.9[a]
Mitochondria	0.19±0.01	4.37±0.45	23.0[a]
Microsomes	0.27±0.03	2.75±0.27	10.2[a]
Cytosol	0.24±0.01	5.17±2.42	21.5[a]

The rats were killed 24 hr after the dose of NaF (35 mg/kg, ip) or NaCl (48.7 mg/kg, ip as control). Values are means obtained from five rats±SE. Significant difference from controls ([a]$p<0.01$).

calcium in renal cytosol and microsomes were 21.5 and 10.2 times, respectively.

Changes in Renal Protein Kinase C Activity after Fluoride Administration

Protein kinase C is known to be stimulated by micromolar calcium. Thus, we have investigated the possibility that renal protein kinase C activity may be enhanced by slight increment of calcium with 6 hr after the fluoride dose. The renal protein kinase C activity was prompted and gave a maximal increment 3 hr after the dose (Fig. 2). At 6 hr after the dose, the renal protein kinase C activity was recovered to the control level despite the midway to further augment of renal calcium content. In this experiment, L-α-phosphatidyl-L-serine was added as the membrane factor to the incubation media.

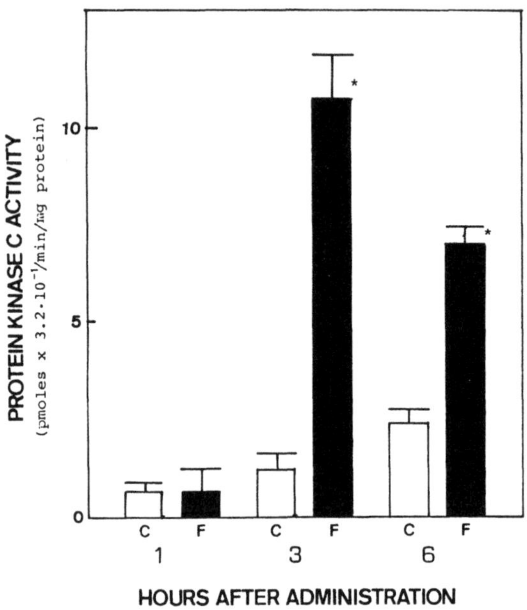

Fig. 2. Effect of fluoride administration on renal protein kinase C activity. Changes in the protein kinase C activity of kidney were determined at various times after a single dose of fluoride (NaF 35 mg/kg, ip) or control (NaCl 48.7 mg/kg, ip). Values are averages obtained from four rats. Standard errors are indicated by vertical bars. Asterisks indicate significant difference from control ($p<0.02$).

Effect of Fluoride Administration on Renal Phospholipase C Activity

The possibility that enhancement of renal phospholipase C activity by fluoride might precede to the elevation of the protein kinase C activity, was deduced from the above observations (Fig. 1 and 2). Inositol-1,4,5-triphosphate (IP_3) was reported to release calcium from endoplasmic reticulum into cytosol (Streb et al., 1983). Diacylglycerol (DG) acts as a signal molecule by enhancing the Ca^{++} sensitive of protein kinase C (Streb et al., 1983). Thus, changes in renal phospholipase C activity were examined after the fluoride dose. The formations of IP, IP_2 and IP_3 from phosphatidyl inositol-bisphosphate were slightly increased 30 min after the fluoride dose and then decreased beyond expectation as shown in Fig. 3.

The GTP-binding protein is described to play an essential role in mediating interactions between hormone receptors and the catalyst (Ross and Gilman, 1980). Moreover, it has been postulated that the GTP-binding protein is a GTPase and that hydrolysis of nucleotide terminates the normal activation of adenylate cyclase by GTP.

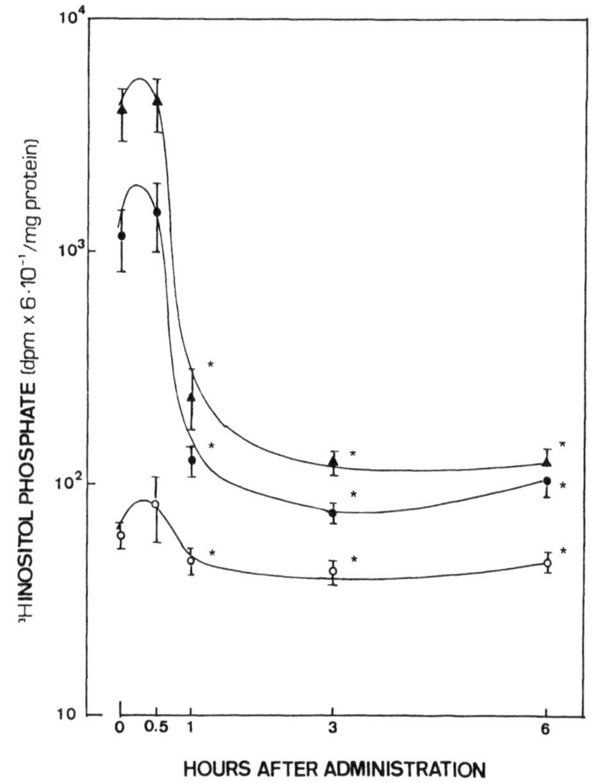

Fig. 3. Effect of fluoride administration on renal PIP_2-phospholipase C activity. Changes in the enzyme activity of kidney were determined as the productions of IP (○), IP_2 (●) and IP_3 (▲) at various times after a single large dose of fluoride (NaF 35 mg/kg, ip). Values are averages obtained from four rats. Standard errors are indicated by vertical bars. Asterisks indicate significant difference from control ($p<0.02$).

On the other hand, translocation of cytosolic protein kinase C to membranes is thought to reflect the intracellular activation of the enzyme (Anderson et al., 1985). Thus, to understand a translocation of membrane-bound GTP-binding protein to cytosol, the GTP-binding protein activity was examined with plasma membrane and cytosol after the fluoride dose. The GTP-binding activity in the plasma membranes was a maximum i hr after the fluoride dose, but that in the cytosol was contimuously elevated to 6 hr following the fluoride dose. The total GTP-binding protein activity was reached to a plateau at 1 hr after the fluoride dose as shown in Fig. 4.

Effect of Fluoride Administration on Cyclic AMP Contents in Kidney and Serum

The GTP-binding protein is known to regulate adenylate cyclase activity (Ross and Gilman, 1980). Thus, changes in cyclic AMP contents in kidney and serum were examined following the fluoride dose. The contents of cAMP in kidney and serum were maximum increases 1 hr after the fluoride dose as shown in Fig. 5. The maximum increase of cAMP in kidney and serum was associated with the time when the total GTP-binding protein activity was reached to the plateau as hown in Fig. 4.

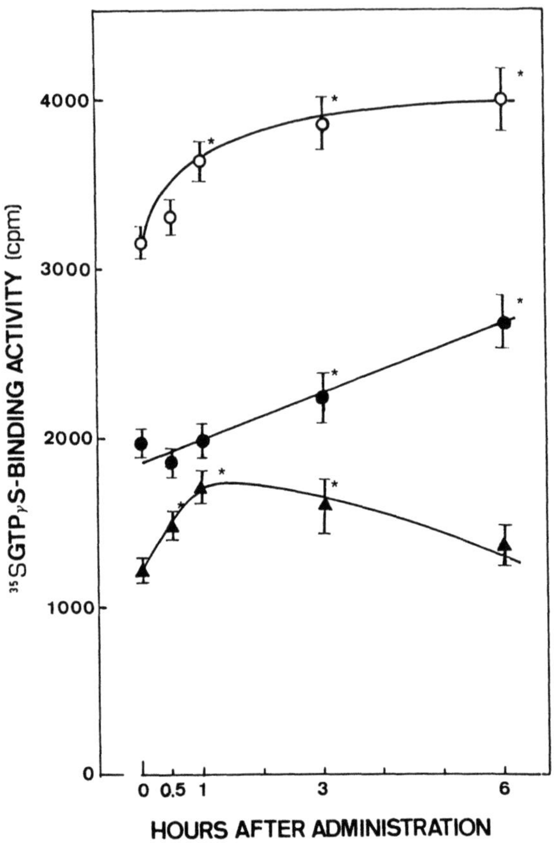

Fig. 4. Effect of fluoride administration on GTP-binding protein in plasma membranes and cytosol of kidney. Changes in the protein in plasma membranes (▲) and cytosol (●) of kidney were determined at various times after a single large dose of fluoride (NaF 35 mg/kg, ip). Values are averages obtained from four rats±SE. Asterisks indicate significant difference from control ($p<0.02$).

○ , Total GTP-binding protein.

Suppression of Fluoride-Induced Renal Calcium Increment by Nifedipine and Thyroid-Parathyroidectomy

Elevation of renal calcium observed immediately after the fluoride dose was suggested to be due to Ca^{++}-channel gate opening by a direct reaction of fluoride or increment of DG and IP_3 induced by the fluoride dose.

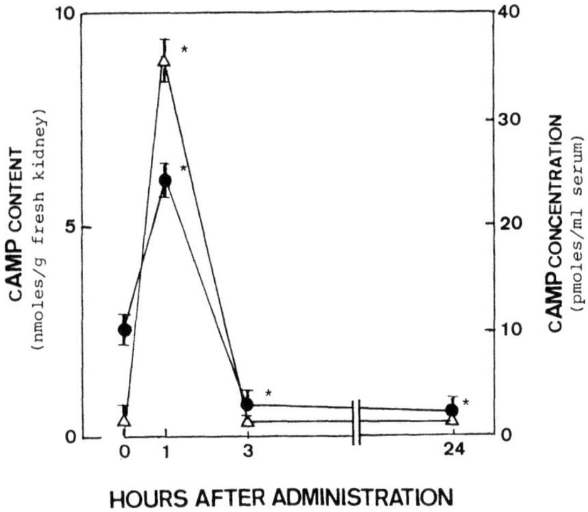

Fig.5. Changes in cyclic AMP in kidney and serum after fluoride administration. The rats were killed at various times after single large dose of fluoride (NaF 35 mg/kg, ip). Values are averages obtained from four rats±SE. Asterisks indicate significant difference from control ($p<0.02$). ●, Kidney; △, serum.

Thus, suppressive effect of fluoride-induced renal calcium increment by nifedipine was examined as shown in Table 2. The increment of calcium in renal brush border membranes (BBM) was significantly suppressed by nifedipine. Moreover, the remarkable increment of renal calcium induced by the fluoride dose was presumed due to enhancement of parathyroid hormone (PTH) secretion caused by the fluoride dose. The increment of renal calcium was significantly suppressed by thyroid-parathyroidectomy as shown in Table 3.

Table 2. Effect of Nifedipine-Treatment on Elevation of Calcium Contents in Renal Basal Membranes and Brush Border Membranes by Fluoride Administration Compared to the Corresponding Magnesium Contents

Treatment	Cation contents of membrane fractions			
	BBM-Ca	BLM-Ca	BBM-Mg	BLM-Mg
		nmoles/mg protein		
Control	23.2±4.59	25.5±2.94	16.0±1.92	40.3±1.84
Fluoride (F)	46.7±1.70	17.0±3.20	24.4±1.66	53.8±2.29
F + nifedipine	13.6±5.56	21.7±1.07	15.0±1.15	41.1±2.00

BBM: brush border membranes, BLM: basal lateral membranes.
Nifedipine (1.5 mg/kg) was injected ip immediately after the administration of NaF (35 mg/kg, ip) or NaCl (48.7 mg/kg, ip as control). The rats were killed 4 hr after the fluoride dose.
Values are averages obtained from five rats±SE.

Table 3. Suppressive Effect of Thyroid-Parathyroidectomy on Renal Microsomal and Cytosolic Calcium Contents of Fluoride-Administered Rats

Subcellular fraction	Control			Fluoride-administered		
	Intact	TPTX	Sup. by TPTX	Intact	TPTX	Sup. by TPTX
	μmoles/g		%	μmoles/g		%
Whole cells	1.26±0.04	1.26±0.04	-	60.21±3.97	23.07±3.98	63.0[a]
Microsomes	0.32±0.03	0.31±0.02	3.1	2.81±0.24	1.80±0.15	40.2[a]
Cytosol	0.41±0.03	0.43±0.04	-	8.17±4.45	4.82±0.29	43.3[a]

The rats were killed 24 hr after the administration of NaF (35 mg/kg, ip) or NaCl (48.7 mg/kg, ip as control). Values are averages obtained rom five rats±SE. Significant suppression by thyroid-parathyroidectomy ([a]$p<0.01$).

DISCUSSION

Renal protein kinase C activity was found to be markedly elevated by a single large dose of fluoride (NaF 35 mg/kg, ip). The elevation was demonstrated to be due to augment of renal cytosolic calcium induced by the fluoride dose. Protein kinase C is known to play an important role in hormonal regulation and to be activated by diacylglycerol (DG) and inositol triphosphate (IP_3) (Ross and Gilman, 1980). In this study, elevation of renal PIP_2 phospholipase C activity ($IP + IP_2 + IP_3$) was reached to maximum within 30 min after the fluoride dose, but the maximal enzyme activity was potentiated less than our expectation. One of the reasons might be due to dilution of fluoride during membrane preparation. On the contrary, the strong suppression of PIP_2 phospholipase C activity was emerged after the maximal activation of the enzyme. The suppression of the PIP_2 phospholipase C activity can be accounted for feedback inhibition due to a large increment of cytosolic calcium and a consecutive elevation of cytosolic GTP-binding protein as shown in Fig. 4.

The finding about the translocation of membrane-bound GTP-binding protein to cytosolic GTP-binding protein is very interested in intracellular regulatory system. GTP-binding protein with a GTPase activity is thought to be involved in the functional coupling of several receptors to PIP_2 phospholipase C (Guillon et al., 1986). The increment of the renal membrane-bound GTP-binding protein was associated with the enhancement of the corresponding renal cAMP content immediately after the fluoride dose. Moreover, the elevation of renal cytosolic calcium immediately after the fluoride dose was concluded to be due to opening of Ca^{++}-channel gate from the finding that the calcium elevation was suppressed by nifedipine. Actually, Comperts (1983) speculated in the Ca^{++}-channel of mast cell to be regulated by guanine nucleotide-binding protein.

On the other hand, a retarded large increment of renal calcium 24 hr after the fluoride dose could be suppressed by thyroid-parathyroidectomy. From the results, the retarded large increment was demonstrated to be mainly due to stimulation of parathyroid function as a secondary action of the fluoride dose. In this experiment, increment of PIP_2 phospholipase C activity by the fluoride dose was found to be small beyond expectation. Thus, another factors (Hidaka and Hartshorne, 1985) may be contributed to a large increment of renal protein kinase C activity. We examined changes in renal cytosolic protein kinase C activity after the fluoride dose, and also the corresponding membrane-bound protein kinase C activity should be measured in order to understand the renal intracellular regulation.

REFERENCES

Anderson, W.B., Estival, A., Tapiovaara, H., and Gopalekrishna, R., 1985, Altered subcellular distribution of protein kinase C (a phorbol ester receptor). Possible role in tumor promotion and the regulation of cell growth relationship to changes in adenylate cyclase activity, Adv. Cyclic Nucleotide Protein Phosphorylation Res., 19: 287.

Berridge, N.J., Dawson, R.M.C., Downes, C.P., Heslop, J.P., and Irvine, R.F., 1983, Changes in the levels of inositol phosphates after agonist-dependent hydrolysis of membrane phosphoinositides, Biochem. J., 212: 473.

comparts, B.D., 1983, Involvement of guanine nucleotide-binding protein in the gating of Ca^{++} by receptors, Nature, 306: 64.

Forte, J.G., Forte, T.M., and Heinz, E., 1973, Isolation of plasma membranes from Ehrlich ascites tumor cells. Influence of amino acids on (Na^+ + K^+)-ATPase and K^+-stimulated phosphatase, Biochim. Biophys. Acta, 298: 827.

Guillon, G., Mouillac, B., and Balestre, M.N., 1986a, Activation of polyphosphoinositide phospholipase C by fluoride in WRK-1 cell membranes, FEBS Lett., 204: 183.

Guillon, G., Balestre, M.N., Mouillac, B., and Devilliers, G., 1986b, activation of membrane phospholipase C by vasopressin. A requirement for guanyl nucleotides, FEBS Lett., 196: 155.

Hidaka, H., and Hartshorne, J.D., 1985, "Calmodulin Antagonists and Cellular Physiology," Academic Press, New York.

Honma, M., Satoh, T., Takesawa, J., and Ui, M., 1977, An ultrasensitive method for the simultaneous determination of cyclic AMP and cyclic GMP in small-volume samples from blood and tissue, Biochem. Med., 18: 257.

Jørgensen, P.L., 1968, Regulation of the (Na^+ + K^+)-activated ATP hydrolyzing enzyme system in rat kidney. 1. The effect of adrenalectomy and supply of sodium on the enzyme system, Biochim. Biophys. Acta, 151: 212.

Lowry, O.H., Rosebrough, N.J., Farr, A.F., and Randall, R.J., 1951, Protein measurement with the Folin phenol reagent, J. Biol. Chem., 193: 265.

Mazurek, N., Regazzi, R., Borner, C., Wyss, R., Conscience, J.F., Erne, P., Eppenberger, U., and Fabbrs, D., 1987, Altered protein kinase C in mast cell variant defective in exocytosis, Proc. Natl. Acad. Sci., 84: 1277.

Molitoris, B.A., and Simon, F.R., 1985, Renal cortical brush-border and basolateral membrane: Cholesterol and phospholipid composition and relative turnover, J. Membrane Biol., 83: 207.

Northup, J.K., Smigel, M.D., and Gilman, A.G., 1982, The guanine nucleotide activating site of the regulatory component of adenylate cyclase. Identification by ligand binding, J. Biol. Chem., 257: 11416.

Okajima, F., and Ui, M., 1984, ADP-ribosyllation of the specific membrane protein by islet activating protein, pertussis toxin, associated with inhibition of a chemotactic peptide-induced arachidonate release in neutrophils, J. Biol. Chem., 259: 13863.

Ross, E.M., and Gilman, A.G., 1980, Biochemical properties of hormone-sensitive adenylate cyclase, Ann. Rev. Biochem., 49: 533.

Shupe, J.L., Leone, N.E., and Peterson, H.B., 1984, "Fluorides - Effects on vegetation, animals and human," Peragon Press, Salt Lake.

Sternwise, P.C., and Gilman, A.G., 1982, Aluminum: A requirement for activation of the regulatory component of adenylate cyclase by fluoride, Proc. Natl. Acad. Sci., 79: 4888.

Streb, H., Irvine, R.F., Berridge, M.J., and Schulz, I., 1983, Release of Ca^{++} from a non-mitochondrial intracellular store in pancreatic acinar cells by inositol-1,4,5-triphosphate, Nature, 306: 67.

Strnard, C.F., and Wong, K., 1985, Calcium mobilization in fluoride activated human neutrophils, Biochim. Biophys. Res. Comm., 133: 161.

Suketa, Y., Mikami, E., and Hayashi, M., 1977, Changes in calcium and magnesium in the kidneys of rats intoxicated with a single large dose of fluoride, Toxicol. Appl. Pharmacol., 39: 313.

Suketa, Y., and Sato, M., 1980, Changes in glucose-6-phosphatase activity in

liver and kidney of rats treated with a single large dose of fluoride, Toxicol. Appl. Pharmacol., 52: 386.
Suketa, Y., Sato, M., and Kura, M., 1980, Effect of fluoride administration on renal glucose-6-phosphatase activity in rats, Experientia, 36: 438.
Suketa, Y., Yamada, M., Hasegawa, J., and Asao, Y., 1982, A possible mechanism for elevation of glucose-6-phosphatase activity in kidney and liver of fluoride-treated rats, Mol. Pharmacol., 22: 116.
Suketa, Y., and Kanamoto, Y., 1983, A role of thyroid-parathyroid function in elevation of calcium content in kidney of rats after a single large dose of fluoride, Toxicology, 26: 335.
Suketa, Y., Asao, Y., Kanamoto, Y., Sakashita, T., and Okada, S., 1985, Changes in adrenal function as a possible mechanism for elevation of serum glucose by a single large dose of fluoride, Toxicol. Appl. Pharmacol., 80: 199.
Suketa, Y., Kanamoto, Y., Satoh, Y., and Okada, S., 1986, Changes in the metabolism of glucose and calcium following a single large dose of fluoride to rats, in: "Studies in Environmental Science 27 - Fluoride Research 1985," H, Tsunoda, and M.-H. Yu, ed., Elsevier, Amsterdam.
Taylor, J.M., Scott, J.K., Maynard, E.A., Smith, F.A., and Hodge, H.C., 1961, Toxicic effects of fluoride on the rat kidney. 1. Acute injurry from single large dose, Toxicol. Appl. Pharmacol., 3: 278.
Willis, J.B., 1960, The determination of metals in blood serum by atomic absorption spectroscopy, 1. Calcium, Spectrochim. Acta, 16: 259.

AUTHOR INDEX

SUBJECT INDEX

MIX
Papier aus verantwortungsvollen Quellen
Paper from responsible sources
FSC® C105338

If you have any concerns about our products,
you can contact us on
ProductSafety@springernature.com

In case Publisher is established outside the EU,
the EU authorized representative is:
Springer Nature Customer Service Center GmbH
Europaplatz 3, 69115 Heidelberg, Germany

Printed by Libri Plureos GmbH
in Hamburg, Germany